Asymmetric Synthesis with Chemical and Biological Methods

Edited by
Dieter Enders and Karl-Erich Jaeger

1807–2007 Knowledge for Generations

Each generation has its unique needs and aspirations. When Charles Wiley first opened his small printing shop in lower Manhattan in 1807, it was a generation of boundless potential searching for an identity. And we were there, helping to define a new American literary tradition. Over half a century later, in the midst of the Second Industrial Revolution, it was a generation focused on building the future. Once again, we were there, supplying the critical scientific, technical, and engineering knowledge that helped frame the world. Throughout the 20th Century, and into the new millennium, nations began to reach out beyond their own borders and a new international community was born. Wiley was there, expanding its operations around the world to enable a global exchange of ideas, opinions, and know-how.

For 200 years, Wiley has been an integral part of each generation's journey, enabling the flow of information and understanding necessary to meet their needs and fulfill their aspirations. Today, bold new technologies are changing the way we live and learn. Wiley will be there, providing you the must-have knowledge you need to imagine new worlds, new possibilities, and new opportunities.

Generations come and go, but you can always count on Wiley to provide you the knowledge you need, when and where you need it!

William J. Pesce
President and Chief Executive Officer

Peter Booth Wiley
Chairman of the Board

Asymmetric Synthesis with Chemical and Biological Methods

Edited by
Dieter Enders and Karl-Erich Jaeger

WILEY-VCH Verlag GmbH & Co. KGaA

The Editors

Prof. Dr. Dieter Enders
Institut für Organische Chemie
RWTH Aachen
Landoltweg 1
52074 Aachen
Germany

Prof. Dr. Karl-Erich Jaeger
Institut für Molekulare Enzymtechnologie
der HHU Düsseldorf im FZ Jülich
52426 Jülich
Germany

All books published by Wiley-VCH are carefully produced. Nevertheless, authors, editors, and publisher do not warrant the information contained in these books, including this book, to be free of errors. Readers are advised to keep in mind that statements, data, illustrations, procedural details or other items may inadvertently be inaccurate.

Library of Congress Card No.:
applied for

British Library Cataloguing-in-Publication Data
A catalogue record for this book is available from the British Library.

Bibliographic information published by the Deutsche Nationalbibliothek
Die Deutsche Nationalbibliothek lists this publication in the Deutsche Nationalbibliografie; detailed bibliographic data are available in the Internet at <http://dnb.d-nb.de>.

© 2007 WILEY-VCH Verlag GmbH & Co. KGaA, Weinheim

All rights reserved (including those of translation into other languages). No part of this book may be reproduced in any form – by photoprinting, microfilm, or any other means – nor transmitted or translated into a machine language without written permission from the publishers. Registered names, trademarks, etc. used in this book, even when not specifically marked as such, are not to be considered unprotected by law.

Typesetting SNP Best-set Typesetter Ltd., Hong Kong
Printing betz-druck GmbH, Darmstadt
Binding Litges & Dopf GmbH, Heppenheim
Cover Design Adam Design, Weinheim
Wiley Bicentennial Logo Richard J. Pacifico
Printed in the Federal Republic of Germany

Printed on acid-free paper

ISBN: 978-3-527-31473-7

Foreword

Stereochemistry has been an important topic for more than a hundred years. Nevertheless, as far as the chemist's everyday life was concerned, it was mainly of interest to natural product chemists for most of the time. This changed in the 1950s when synthetic chemists, following the example of R. B. Woodward, G. Stork and others, began to boldly address complex natural product targets. At this time the racemic compound was targeted, i.e. it was diastereoselectivity that counted. In the 1970s it became increasingly clear that biological activities of enantiomers could differ to the extent that one member of a pair is toxic or generally harmful. In this respect, the Contergan disaster was a signal. Pharmacologic testing of both individual enantiomers rather than the racemic agent became a common practice.

Demand created interest in the development of new methods for syntheses of *enantiomerically pure compounds*, termed *EPC*-syntheses by Dieter Seebach. First auxiliary controlled, stoichiometric asymmetric synthesis began to flourish in the second half of the 1970s. Dieter Enders, the spiritus rector of this book, with his SAMP/RAMP method, was one of the pioneers of this field. At about the same time the potential of enzyme-catalyzed enantioselective reactions became more and more visible, not least through pioneering early work of the M. R. Kula/ C. Wandrey team in Düsseldorf/Jülich and of Hans-Joachim Gais in Darmstadt, later Aachen. In the 1980s, very few people dared to address transition metal catalyzed asymmetric synthesis. This changed in the 1990s after work of Kagan, Knowles, Sharpless, Noyori and others had shown that results useful for organic synthesis can be obtained.

Thus, in the early 1990s the stage was set for an Aachen/Jülich group of chemists to launch a collaborative program in the field of EPC synthesis that led to a prestigious Collaborative Research Center (Sonderforschungsbereich, SFB) "Asymmetric Syntheses with Chemical and Biological Methods", which was to become operative for 12 years (1994–2005). In this book the main results, obtained by ca. 20 research groups, are reviewed. A very large and colorful landscape of methods and applications is presented.

The first part of the book is devoted to auxiliary controlled reactions using the SAMP/ RAMP method (D. Enders) and metallated allylsulfoximines (H.-J. Gais).

Asymmetric Synthesis with Chemical and Biological Methods. Edited by Dieter Enders and Karl-Erich Jaeger
Copyright © 2007 WILEY-VCH Verlag GmbH & Co. KGaA, Weinheim
ISBN 978-3-527-31473-7

Syntheses of an impressive array of natural products, including medicinally interesting alkaloids, underline the usefulness of these methods. The following part deals with enantioselective reactions catalyzed by transition metal complexes. Chiral ligands with a modular make-up are of crucial importance here and many new classes are described: phosphines containing an arenechromiumtricarbonyl moiety ("Daniphos" ligands, A. Salzer), phosphaferrocenes (C. Ganter), sulfoximine-based N,N- and P,N-ligands (C. Bolm), P,C- and N,O-ligands containing a [2,2]paracyclophane skeleton (C. Bolm, S. Bräse) and phosphines based on dihydroquinolines ("Quinaphos" ligands, W. Leitner). Catalyst immobilization on or in a zeolite matrix was much debated in the SFB; finally, W. F. Hoelderich's group has been able to obtain highly active, reusable hydrogenation as well as Jacobsen type epoxidation catalysts.

The next part of the book deals with enzyme catalysis and bioorganic synthesis. An important aim of this research has been the preparation of enantiomerically pure small molecules that are useful in general organic synthesis and as intermediates in drug process synthesis. It is apparent that there has been fruitful and remarkably successful collaboration between ca. 10 groups, led by established as well as junior group leaders. The first three articles, with authors from the groups of K.-E. Jaeger, M.-R. Kula, M. Pohl, M. Müller and G. A. Sprenger, deal with applications of techniques of enzyme biochemistry, for example site-directed mutagenesis and directed evolution based on recombinant DNA technology. The following articles describe asymmetric syntheses of a large variety of chiral alcohols using R-specific alcohol dehydrogenases (W. Hummel), aldolases and related types of C-C bond forming enzymes (W.-D. Fessner) as well as sucrose synthase I (L. Elling). An article naming 17 authors on asymmetric synthesis of 1,3-diols and propargylic alcohols concludes the section.

An asset of the Aachen/Jülich bioorganic synthesis approach is technology transfer, which is testified by no less than five start-up companies. Scale-up requires stable and highly efficient enzymes as well as appropriate reaction technology. The development of membrane reactors has been a key to success. Reaction technology is outlined by C. Wandrey and co-workers in the final article.

Reading this book is worthwhile for anybody seeking an impression of the state of the art of the entire field of asymmetric synthesis. A lot of interesting material is offered to the expert from academia or industry as well as to the student looking for an interesting field of graduate research.

Günter Helmchen

Contents

Foreword V

Preface XVII

List of Contributors XIX

1 **Stoichiometric Asymmetric Synthesis** 1

1.1 Development of Novel Enantioselective Synthetic Methods 1
Dieter Enders and Wolfgang Bettray

1.1.1 Introduction 1
1.1.2 α-Silyl Ketone-Controlled Asymmetric Syntheses 1
1.1.2.1 Regio- and Enantioselective α-Fluorination of Ketones 2
1.1.2.2 α-Silyl Controlled Asymmetric Mannich Reactions 3
1.1.3 Asymmetric Hetero-Michael Additions 5
1.1.3.1 Asymmetric Aza-Michael Additions 5
1.1.3.2 Asymmetric Oxa-Michael Additions 10
1.1.3.3 Asymmetric Phospha-Michael-Additions 11
1.1.4 Asymmetric Syntheses with Lithiated α-Aminonitriles 14
1.1.4.1 Asymmetric Nucleophilic α-Aminoacylation 14
1.1.4.2 Asymmetric Nucleophilic Alkenoylation of Aldehydes 16
1.1.5 Asymmetric Electrophilic α-Substitution of Lactones and Lactams 18
1.1.6 Asymmetric Synthesis of α-Phosphino Ketones and 2-Phosphino Alcohols 22
1.1.7 Asymmetric Synthesis of 1,3-Diols and *anti*-1,3-Polyols 24
1.1.8 Asymmetric Synthesis of α-Substituted Sulfonamides and Sulfonates 26

1.2 Asymmetric Synthesis of Natural Products Employing the SAMP/RAMP Hydrazone Methodology 38
Dieter Enders and Wolfgang Bettray

1.2.1 Introduction 38
1.2.2 Stigmatellin A 38

Asymmetric Synthesis with Chemical and Biological Methods. Edited by Dieter Enders and Karl-Erich Jaeger
Copyright © 2007 WILEY-VCH Verlag GmbH & Co. KGaA, Weinheim
ISBN 978-3-527-31473-7

1.2.3   Callistatin A   41
1.2.4   Dehydroiridodiol(dial) and Neonepetalactone   51
1.2.5   First Enantioselective Synthesis of Dendrobatid Alkaloids Indolizidine 209I and 223J   53
1.2.6   Efficient Synthesis of (2S,12′R)-2-(12′-Aminotridecyl)pyrrolidine, a Defense Alkaloid of the Mexican Bean Beetle   57
1.2.7   2-epi-Deoxoprosopinine   58
1.2.8   Attenol A and B   62
1.2.9   Asymmetric Synthesis of (+)- and (−)-Streptenol A   64
1.2.10 Sordidin   66
1.2.11 Prelactone B and V   69

1.3   Asymmetric Synthesis Based on Sulfonimidoyl-Substituted Allyltitanium Complexes   75
Hans-Joachim Gais

1.3.1   Introduction   75
1.3.2   Hydroxyalkylation of Sulfonimidoyl-Substituted Allylltitanium Complexes   80
1.3.2.1 Sulfonimidoyl-Substituted Bis(allyl)titanium Complexes   80
1.3.2.2 Sulfonimidoyl-Substituted Mono(allyl)tris(diethylamino)titanium Complexes   82
1.3.3   Aminoalkylation of Sulfonimidoyl-Substituted Allyltitanium Complexes   85
1.3.3.1 Sulfonimidoyl-Substituted Bis(allyl)titanium Complexes   85
1.3.3.2 Sulfonimidoyl-Substituted Mono(allyl)tris(diethylamino)titanium Complexes   86
1.3.4   Structure and Reactivity of Sulfonimidoyl-Substituted Allyltitanium Complexes   88
1.3.4.1 Sulfonimidoyl-Substituted Bis(allyl)titanium Complexes   88
1.3.4.2 Sulfonimidoyl-Substituted Mono(allyl)titanium Complexes   91
1.3.5   Asymmetric Synthesis of Homopropargyl Alcohols   95
1.3.6   Asymmetric Synthesis of 2,3-Dihydrofurans   96
1.3.7   Synthesis of Bicyclic Unsaturated Tetrahydrofurans   98
1.3.8   Asymmetric Synthesis of Alkenyloxiranes   100
1.3.9   Asymmetric Synthesis of Unsaturated Mono- and Bicyclic Prolines   102
1.3.10 Asymmetric Synthesis of Bicyclic Amino Acids   105
1.3.11 Asymmetric Synthesis of β-Amino Acids   108
1.3.12 Conclusion   111

1.4   The "Daniphos" Ligands: Synthesis and Catalytic Applications   115
Albrecht Salzer and Wolfgang Braun

1.4.1   Introduction   115
1.4.2   General Synthesis   116

1.4.3	Applications in Stereoselective Catalysis	120
1.4.3.1	Enantioselective Hydrogenations	120
1.4.3.2	Diastereoselective Hydrogenation of Folic Acid Ester	122
1.4.3.3	Enantioselective Isomerization of Geranylamine to Citronellal	124
1.4.3.4	Nucleophilic Asymmetric Ring-Opening of Oxabenzonorbornadiene	124
1.4.3.5	Enantioselective Suzuki Coupling	126
1.4.3.6	Asymmetric Hydrovinylation	126
1.4.3.7	Allylic Sulfonation	128
1.4.4	Conclusion	129
1.5	New Chiral Ligands Based on Substituted Heterometallocenes	130
	Christian Ganter	
1.5.1	Introduction	130
1.5.2	General Properties of Phosphaferrocenes	131
1.5.3	Synthesis of Phosphaferrocenes	132
1.5.4	Preparation of Bidentate P,P and P,N Ligands	133
1.5.5	Modification of the Backbone Structure	136
1.5.6	Cp–Phosphaferrocene Hybrid Systems	139
1.5.7	Catalytic Applications	145
1.5.8	Conclusion	146
2	**Catalytic Asymmetric Synthesis**	**149**
2.1	Chemical Methods	149
2.1.1	Sulfoximines as Ligands in Asymmetric Metal Catalysis	149
	Carsten Bolm	
2.1.1.1	Introduction	149
2.1.1.2	Development of Methods for Sulfoximine Modification	150
2.1.1.3	Sulfoximines as Ligands in Asymmetric Metal Catalysis	162
2.1.1.4	Conclusions	170
2.1.2	Catalyzed Asymmetric Aryl Transfer Reactions	176
	Carsten Bolm	
2.1.2.1	Introduction	176
2.1.2.2	Catalyst Design	177
2.1.2.3	Catalyzed Aryl Transfer Reactions	180
2.1.3	Substituted [2.2]Paracyclophane Derivatives as Efficient Ligands for Asymmetric 1,2- and 1,4-Addition Reactions	196
	Stefan Bräse	

2.1.3.1   [2.2]Paracyclophanes as Chiral Ligands   196
2.1.3.2   Synthesis of [2.2]Paracyclophane Ligands   199
2.1.3.2.1 Preparation of FHPC-, AHPC-, and BHPC-Based Imines   199
2.1.3.2   Structural Information on AHPC-Based Imines   199
2.1.3.3   Asymmetric 1,2-Addition Reactions to Aryl Aldehydes   200
2.1.3.3.1 Initial Considerations   200
2.1.3.3.2 Asymmetric Addition Reactions to Aromatic Aldehydes: Scope of Substrates   203
2.1.3.4   Asymmetric Addition Reactions to Aliphatic Aldehydes   205
2.1.3.5   Addition of Alkenylzinc Reagents to Aldehydes   206
2.1.3.6   Asymmetric Conjugate Addition Reactions   208
2.1.3.7   Asymmetric Addition Reactions to Imines   208
2.1.3.8   Asymmetric Addition Reactions on Solid Supports   212
2.1.3.8.1 Applications   213
2.1.3.9   Conclusions and Future Perspective   213

2.1.4   Palladium-Catalyzed Allylic Alkylation of Sulfur and Oxygen Nucleophiles – Asymmetric Synthesis, Kinetic Resolution and Dynamic Kinetic Resolution   215
Hans-Joachim Gais

2.1.4.1   Introduction   215
2.1.4.2   Asymmetric Synthesis of Allylic Sulfones and Allylic Sulfides and Kinetic Resolution of Allylic Esters   216
2.1.4.2.1 Kinetic Resolution   216
2.1.4.2.2 Selectivity   220
2.1.4.2.3 Asymmetric Synthesis   220
2.1.4.2.4 Synthesis of Enantiopure Allylic Alcohols   224
2.1.4.3   Asymmetric Rearrangment and Kinetic Resolution of Allylic Sulfinates   225
2.1.4.3.1 Introduction   225
2.1.4.3.2 Synthesis of Racemic Allylic Sulfinates   225
2.1.4.3.3 Pd-Catalyzed Rearrangement   226
2.1.4.3.4 Kinetic Resolution   227
2.1.4.3.5 Mechanistic Considerations   228
2.1.4.4   Asymmetric Rearrangment of Allylic Thiocarbamates   229
2.1.4.4.1 Introduction   229
2.1.4.4.2 Synthesis of Racemic O-Allylic Thiocarbamates   229
2.1.4.4.3 Acyclic Carbamates   229
2.1.4.4.4 Cyclic Carbamates   231
2.1.4.4.5 Mechanistic Considerations   232
2.1.4.4.6 Synthesis of Allylic Sulfides   232
2.1.4.5   Asymmetric Synthesis of Allylic Thioesters and Kinetic Resolution of Allylic Esters   233
2.1.4.5.1 Introduction   233

2.1.4.5.2 Asymmetric Synthesis of Allylic Thioesters 234
2.1.4.5.3 Kinetic Resolution of Allylic Esters 235
2.1.4.5.4 Memory Effect and Dynamic Kinetic Resolution of the Five-Membered Cyclic Acetate 238
2.1.4.5.5 Asymmetric Synthesis of Cyclopentenyl Thioacetate 242
2.1.4.6 Kinetic and Dynamic Kinetic Resolution of Allylic Alcohols 242
2.1.4.6.1 Introduction 242
2.1.4.6.2 Asymmetric Synthesis of Symmetrical Allylic Alcohols 242
2.1.4.6.3 Asymmetric Synthesis of Unsymmetrical Allylic Alcohols 244
2.1.4.6.4 Asymmetric Synthesis of a Prostaglandin Building Block 245
2.1.4.6.5 Investigation of an Unsaturated Analogue of BPA 245
2.1.4.7 Conclusions 246

2.1.5 The QUINAPHOS Ligand Family and its Application in Asymmetric Catalysis 250
Giancarlo Franciò, Felice Faraone, and Walter Leitner

2.1.5.1 Introduction 250
2.1.5.2 Synthetic Strategy 252
2.1.5.3 Stereochemistry and Coordination Properties 254
2.1.5.3.1 Free Ligands 254
2.1.5.3.2 Complexes 256
2.1.5.4 Catalytic Applications 261
2.1.5.4.1 Rhodium-Catalyzed Asymmetric Hydroformylation of Styrene 261
2.1.5.4.2 Rhodium-Catalyzed Asymmetric Hydrogenation of Functionalized Alkenes 263
2.1.5.4.3 Ruthenium-Catalyzed Asymmetric Hydrogenation of Aromatic Ketones 265
2.1.5.4.4 Copper-Catalyzed Enantioselective Conjugate Addition of Diethylzinc to Enones 267
2.1.5.4.5 Nickel-Catalyzed Asymmetric Hydrovinylation 268
2.1.5.4.6 Nickel-Catalyzed Cycloisomerization of 1,6-Dienes 270
2.1.5.5 Conclusions 273

2.1.6 Immobilization of Transition Metal Complexes and Their Application to Enantioselective Catalysis 277
Adrian Crosman, Carmen Schuster, Hans-Hermann Wagner, Melinda Batorfi, Jairo Cubillos, and Wolfgang Hölderich

2.1.6.1 Introduction 277
2.1.6.2 Immobilized Rh Diphosphino Complexes as Catalysts for Asymmetric Hydrogenation 278
2.1.6.2.1 Preparation and Characterization of the Immobilized Rh–Diphosphine Complexes 279

2.1.6.2.2 Enantioselective Hydrogenation over Immobilized Rhodium Diphosphine Complexes 282
2.1.6.3 Heterogeneous Asymmetric Epoxidation of Olefins over Jacobsen's Catalyst Immobilized in Inorganic Porous Materials 284
2.1.6.3.1 Preparation and Characterization of Immobilized Jacobsen's Catalysts 285
2.1.6.3.2 Epoxidation of Olefins over Immobilized Jacobsen Catalysts 287
2.1.6.4 Novel Heterogenized Catalysts for Asymmetric Ring-Opening Reactions of Epoxides 291
2.1.6.4.1 Synthesis and Characterization of the Heterogenized Catalysts 291
2.1.6.4.2 Asymmetric Ring Opening of Epoxides over New Heterogenized Catalysts 293
2.1.6.5 Conclusions 295

2.2 Biological Methods 298

2.2.1 Directed Evolution to Increase the Substrate Range of Benzoylformate Decarboxylase from *Pseudomonas putida* 298
Marion Wendorff, Thorsten Eggert, Martina Pohl, Carola Dresen, Michael Müller, and Karl-Erich Jaeger

2.2.1.1 Introduction 298
2.2.1.2 Materials and Methods 300
2.2.1.2.1 Reagents 300
2.2.1.2.2 Construction of Strains for Heterologous Expression of BFD and BAL 300
2.2.1.2.3 Polymerase Chain Reactions 301
2.2.1.2.4 Generation of a BFD Variant Library by Random Mutagenesis 302
2.2.1.2.5 High-Throughput Screening for Carboligation Activity with the Substrates Benzaldehyde and Dimethoxyacetaldehyde 303
2.2.1.2.6 Expression and Purification of BFD Variants 303
2.2.1.2.7 Protein Analysis Methods 304
2.2.1.2.8 Enzyme Activity Assays 304
2.2.1.3 Results and Discussion 304
2.2.1.3.1 Overexpression of BFD in *Escherichia coli* 304
2.2.1.3.2 Random Mutagenesis of BFD Variant L476Q 305
2.2.1.3.3 Development of a High-Throughput Screening Assay for Carboligase Activity 305
2.2.1.3.4 Identification of a BFD Variant with an Optimized Acceptor Aldehyde Spectrum 306
2.2.1.3.5 Biochemical Characterization of the BFD Variants 308
2.2.1.3.6 Decreased Benzoyl Formate Decarboxylation Activity of Variant 55E4 308

2.2.1.3.7 Formation of 2-Hydroxy-3,3-dimethoxypropiophenone and Benzoin *308*
2.2.1.3.8 Enantioselectivity of the Carboligation Reaction *310*
2.2.1.4 Conclusions *311*

2.2.2 C–C-Bonding Microbial Enzymes: Thiamine Diphosphate-Dependent Enzymes and Class I Aldolases *312*
Georg A. Sprenger, Melanie Schürmann, Martin Schürmann, Sandra Johnen, Gerda Sprenger, Hermann Sahm, Tomoyuki Inoue, and Ulrich Schörken

2.2.2.1 Introduction *312*
2.2.2.2 Thiamine Diphosphate (ThDP)-Dependent Enzymes *312*
2.2.2.2.1 Transketolase (TKT) *313*
2.2.2.2.2 1-Deoxy-D-xylulose 5-Phosphate Synthase (DXS) *317*
2.2.2.2.3 Phosphonopyruvate Decarboxylase (PPD) from *Streptomyces viridochromogenes* *318*
2.2.2.3 Class I Aldolases *318*
2.2.2.3.1 Transaldolase (TAL) *320*
2.2.2.3.2 Fructose 6-Phosphate Aldolase (FSA) *321*
2.2.2.4 Summary and Outlook *321*

2.2.3 Enzymes for Carboligation – 2-Ketoacid Decarboxylases and Hydroxynitrile Lyases *327*
Martina Pohl, Holger Breittaupt, Bettina Frölich, Petra Heim, Hans Iding, Bettina Juchem, Petra Siegert, and Maria-Regina Kula

2.2.3.1 Introduction *327*
2.2.3.2 2-Ketoacid Decarboxylases *327*
2.2.3.2.1 Comparative Biochemical Characterization of Wild-Type PDC and BFD *328*
2.2.3.2.2 Identification of Amino Acid Residues Relevant to Substrate Specificity and Enantioselectivity *330*
2.2.3.2.3 Optimization of the Substrate Range of BFD by Site-Directed Mutagenesis *330*
2.2.3.2.4 Optimization of Stability and Substrate Range of BFD by Directed Evolution *330*
2.2.3.3 Hydroxynitrile Lyases *332*
2.2.3.3.1 HNL from *Sorghum bicolor* *333*
2.2.3.3.2 HNL from *Linum usitatissimum* *337*

2.2.4 Preparative Syntheses of Chiral Alcohols using (R)-Specific Alcohol Dehydrogenases from *Lactobacillus* Strains *341*
Andrea Weckbecker, Michael Müller, and Werner Hummel

2.2.4.1 Introduction *341*
2.2.4.2 (R)-Specific Alcohol Dehydrogenase from *Lactobacillus kefir* *341*

2.2.4.3 Comparison of (R)-Specific ADHs from *L. kefir* and *L. brevis* 342
2.2.4.4 Preparative Applications of ADHs from *L. kefir* and *L. brevis* 345
2.2.4.4.1 Synthesis of (R,R)-Diols 346
2.2.4.4.2 Synthesis of Enantiopure 1-Phenylpropane-1,2-diols 346
2.2.4.4.3 Synthesis of Enantiopure Propargylic Alcohols 346
2.2.4.4.4 Regioselective Reduction of *t*-Butyl 6-chloro-3,5-dioxohexanoate to the Corresponding Enantiopure (S)-5-Hydroxy Compound 346
2.2.4.5 Coenzyme Regeneration and the Construction and Use of "Designer Cells" 347
2.2.4.6 Discussion 349

2.2.5 Biocatalytic C–C Bond Formation in Asymmetric Synthesis 351
Wolf-Dieter Fessner

2.2.5.1 Introduction 351
2.2.5.2 Enzyme Mechanisms 352
2.2.5.2.1 Class II Aldolases 352
2.2.5.2.2 Class I Fructose 1,6-Bisphosphate Aldolase 355
2.2.5.2.3 Sialic Acid Synthase 355
2.2.5.2.4 Rhamnose Isomerase 356
2.2.5.3 New Synthetic Strategies 357
2.2.5.3.1 Sugar Phosphonates 357
2.2.5.3.2 Xylulose 5-Phosphate 359
2.2.5.3.3 RhuA Stereoselectivity 359
2.2.5.3.4 Aldolase Screening Assay 361
2.2.5.3.5 Aldose Synthesis 361
2.2.5.3.6 Tandem Chain Extension–Isomerization–Chain Extension 362
2.2.5.3.7 Tandem Bidirectional Chain Extensions 363
2.2.5.3.8 Non-Natural Sialoconjugates 369
2.2.5.4 Summary and Outlook 373

2.2.6 Exploring and Broadening the Biocatalytic Properties of Recombinant Sucrose Synthase 1 for the Synthesis of Sucrose Analogues 376
Lothar Elling

2.2.6.1 Introduction 376
2.2.6.2 Characteristics of Recombinant Sucrose Synthase 1 (SuSy1) Expressed in *Saccharomyces cerevisiae* 377
2.2.6.2.1 Expression and Purification of SuSy1 from Yeast 377
2.2.6.2.2 The Substrate Spectrum of SuSy1 from Yeast 378
2.2.6.3 Characteristics of Recombinant Sucrose Synthase 1 (SuSy1) Expressed in *Escherichia coli* 381
2.2.6.3.1 Expression and Purification of SuSy1 from *E. coli* 381
2.2.6.3.2 The Substrate Spectrum of SuSy1 from *E. coli* 382
2.2.6.4 Sucrose Synthase 1 Mutants Expressed in *S. cerevisiae* and *E. coli* 383
2.2.6.5 Outlook 384

2.2.7   Flexible Asymmetric Redox Reactions and C–C Bond Formation by
          Bioorganic Synthetic Strategies 386
          *Michael Müller, Michael Wolberg, Silke Bode, Ralf Feldmann,
          Petra Geilenkirchen, Thomas Schubert, Lydia Walter, Werner Hummel,
          Thomas Dünnwald, Ayhan S. Demir, Doris Kolter-Jung, Adam Nitsche,
          Pascal Dünkelmann, Annabel Cosp, Martina Pohl, Bettina Lingen, and
          Maria-Regina Kula*

2.2.7.1   Introduction 386
2.2.7.2   Diversity-Oriented Access to 1,3-Diols Through Regio- and
          Enantioselective Reduction of 3,5-Dioxocarboxylates 386
2.2.7.2.1 Regio- and Enantioselective Enzymatic Reduction 387
2.2.7.2.2 Dynamic Kinetic Resolution 388
2.2.7.2.3 Stereoselective Access to 1,3-Diols by Diastereoselective
          Reduction 389
2.2.7.2.4 Nucleophilic Substitution of Chlorine 390
2.2.7.2.5 Application in Natural Product Syntheses 391
2.2.7.2.6 Discussion and Outlook 392
2.2.7.3   Chemo- and Enantioselective Reduction of Propargylic Ketones 395
2.2.7.3.1 Enantioselective Reduction of Aryl Alkynones 395
2.2.7.3.2 Synthesis of Enantiopure 3-Butyn-2-ol 396
2.2.7.3.3 Enzymatic Reduction of α-Halogenated Propargylic Ketones 397
2.2.7.3.4 Modification of α-Halogenated Propargylic Alcohols 398
2.2.7.3.5 Olefinic Substrates 399
2.2.7.3.6 Discussion and Outlook 401
2.2.7.4   Thiamine Diphosphate-Dependent Enzymes: Multi-purpose Catalysts
          in Asymmetric Synthesis 401
2.2.7.4.1 Formation of Chiral 2-Hydroxy Ketones Through BFD-Catalyzed
          Reactions 402
2.2.7.4.2 BAL as a Versatile Catalyst for C–C Bond Formation and Cleavage
          Reactions 405
2.2.7.4.3 Asymmetric Cross-Benzoin Condensation 407
2.2.7.4.4 Discussion and Outlook 408
2.2.7.5   Summary 409

3   Reaction Technology in Asymmetric Synthesis 415

3.1   Reaction Engineering in Asymmetric Synthesis 415
          Stephan Lütz, Udo Kragl, Andreas Liese, and Christian Wandrey

3.1.1   Introduction 415
3.1.2   Membrane Reactors with Chemical Catalysts 418
3.1.3   Membrane Reactors with Biological Catalysts 420
3.1.3.1   Membrane Reactors with Whole Cells 420
3.1.3.2   Membrane Reactors with Isolated Enzymes 421
3.1.4   Two-Phase Systems 422
3.1.5   Conclusions 425

3.2 Biocatalyzed Asymmetric Syntheses Using Gel-Stabilized Aqueous–Organic Two-Phase Systems *427*
Marion B. Ansorge-Schumacher

3.2.1 Gel-Stabilized Two-Phase Systems *428*
3.2.2 Benzoin Condensation with Entrapped Benzaldehyde Lyase *430*
3.2.3 Reduction of Ketones with Entrapped Alcohol Dehydrogenase *432*
3.2.4 Conclusion *433*

Index *435*

Name Index *443*

Preface

After the pioneering work of Louis Pasteur and Emil Fischer in the middle and at the end of the nineteenth century, respectively, it still took more than fifty years before chemists started to discuss transition state models together with polar and steric effects to gain more insight into the phenomenon of asymmetric induction. Even first observations in organic synthesis of enantioselectivities comparable to those of enzymes in the late fifties and sixties of the 20th century did not convince the chemical community and the term "asymmetric synthesis" was regarded a mechanistic curiosity rather than a practical way to synthesize compounds of high enantiomeric purity.

In the mid-seventies, with the development of generally applicable stoichiometric asymmetric syntheses, especially the Meyers oxazoline methodology as the first one, the scientific community began to believe that asymmetric synthesis really worked resulting in an explosive growth of this new field. Later on, and mainly driven by the fact that the biological activity of enantiomers is usually different, dozens of new chemical companies were founded all over the world in a newly created area called "chirotechnology".

Around that time and after intensive discussions several professors of the RWTH Aachen University and the nearby Jülich Research Center decided to apply at the German Research Council for a so-called Collaborative Research Center on the topic of asymmetric synthesis. Looking back, it was truly a seminal event when the Professors D. Enders, W. Keim, M.-R. Kula, H. Sahm and C. Wandrey stopped their cars at the highway station Köln-Frechen and nailed down the proposed research topic as "Asymmetric Synthesis with Chemical and Biological Methods". After Professor E. Winterfeldt, as an advisor, saw this new initiative "under a good star", indeed the new "Sonderforschungsbereich 380" was funded and started in 1994.

From the very beginning of this long term research endeavor, the aim has been to cover *all* aspects of the *entire* field of asymmetric synthesis including stoichiometric and catalytic asymmetric syntheses with chemical and biological methods as well as the development of new reaction technologies. The interdisciplinary cooperation among the areas of classical organic and inorganic chemistry as well as technical chemistry (RWTH Aachen University) and the various fields of enzyme technology and biotechnology (Research Center Jülich, HHU Düsseldorf) resulted in efficient asymmetric syntheses of synthetic building blocks, fine chemicals, natural products and biologically active compounds in general. Mechanistic and theoretical aspects, organic synthesis, organometallic chemistry, homogeneous and heterogeneous transition metal catalysis, microbiology, enzyme- and biotechnology were all employed and used for stereoselective C-H-, C-C-, and C-heteroatom bond formations.

Besides the scientific success of this Collaborative Research Center as measured in publications, patents and foundation of start-up companies, it should be mentioned that a high percentage of the younger scientific members received and accepted calls for full professorships including D. Vogt (Eindhoven), W.-D. Fessner (Darmstadt), U. Kragl (Rostock), A. Liese (Hamburg), S. Bräse (Karlsruhe), G. Sprenger (Stuttgart) and M. Müller (Freiburg) and also associate professorships as C. Ganter (Düsseldorf), L. Elling (Aachen), M. Ansorge-Schumacher (Berlin) and M. Pohl (Privatdozent, Düsseldorf). A highlight during the twelve years of funding was the "Deutsche Zukunftspreis" awarded by the Federal President of Germany to Prof. Kula and Dr. Pohl and presented in a spectacular nationwide television show broadcasted from Berlin in 2002. Professor Maria-Regina Kula, herself being a chemist, was always aware of the necessity to combine biological and chemical catalytic methods. As her 70th birthday coincides with the appearance of this book, the editors would like to express their warm congratulations and best wishes for her future.

We thank the German Research Council ("Deutsche Forschungsgemeinschaft") for the generous financial support of the Collaborative Research Center "Sonderforschungsbereich, SFB 380" over a period of twelve years. In particular, we are thankful to Dr. H. H. Lindner and Dr. A. Pollex-Krüger as well as Dr. W. Rohe, Dr. P. Schmitz-Möller and Dr. H. Schruff for their organizational help during the course of the priority programme. In addition, on behalf of all participants of the Collaborative Research Center, we would like to thank the scientific referees, the Professors M. Ballauff (Bayreuth), J. E. Bäckvall (Stockholm), A. Böck (München), H. Brunner (Regensburg), H. Buchholz (Erlangen-Nürnberg), W. Buckel (Marburg), G. Dziuk (Freiburg), F. Effenberger (Stuttgart), H. Eschrig (Dresden), H. Fischer (Konstanz), W. Francke (Hamburg), G. Gottschalk (Göttingen), H. Griengl (Graz), G. Helmchen (Heidelberg), U. Kazmaier (Saarbrücken), H. Kessler (München), H. Kunz (Mainz), E. P. Kündig (Genf), J. Mulzer (Wien), H.-U. Reißig (Berlin), K. Sandhoff (Bonn), G. Schulz-Eckloff (Bremen), H. Simon (München), W. Spiess (Mainz), J. Thiem (Hamburg), H. Tschesche (Bielefeld), H. Vahrenkamp (Freiburg), and H. Waldmann (Dortmund) for their help, advice and the many fruitful discussions.

We hope that this book will be useful and a source of inspiration for all those interested in the chemical, biological and technical aspects of asymmetric synthesis in general and will stimulate new ideas and research activities among the young scientists in this rapidly growing field.

Aachen / Jülich, December 2006

Dieter Enders
Karl-Erich Jaeger

List of Contributors

**Prof. Dr. Marion-B.
Ansorge-Schumacher**
Technische Universität Berlin
Institut für Chemie /
Enzymtechnologie
Straße des 17. Juni 124
10623 Berlin
Germany

Melinda Batorfi
Institut für Brennstoffchemie
und Physikalisch-chemische
Verfahrenstechnik
RWTH Aachen
Worringerweg 1
52074 Aachen
Germany

Dr. Wolfgang Bettray
Institut für Organische Chemie
RWTH Aachen
Landoltweg 1
52074 Aachen
Germany

Silke Bode
Lehrstuhl für Pharmazeutische und
Medizinische Chemie
Institut für Pharmazeutische
Wissenschaften
Albert-Ludwigs-Universität Freiburg
Albertstr. 25
79104 Freiburg
Germany

Prof. Dr. Carsten Bolm
Institut für Organische Chemie
RWTH Aachen
Landoltweg 1
52074 Aachen
Germany

Prof. Dr. Stefan Bräse
Institut für Organische Chemie
Universität Karlsruhe (TH)
Fritz-Haber-Weg 6
76131 Karlsruhe
Germany

Dr. Wolfgang Braun
Institut für Anorganische Chemie
RWTH Aachen
Landoltweg 1
52074 Aachen
Germany

List of Contributors

Dr. Holger Breithaupt
Institut für Molekulare
Enzymtechnologie der HHU
Düsseldorf im FZ Jülich
52426 Jülich
Germany

Dr. Annabel Cosp
Institut für Biotechnologie II
Forschungszentrum Jülich
52425 Jülich
Germany

Adrian Crosman
Institut für Brennstoffchemie
und Physikalisch-chemische
Verfahrenstechnik
RWTH Aachen
Worringerweg 1
52074 Aachen
Germany

Dr. Jairo Cubillos
Institut für Brennstoffchemie
und Physikalisch-chemische
Verfahrenstechnik
RWTH Aachen
Worringerweg 1
52074 Aachen
Germany

Prof. Dr. Ayhan S. Demir
Department of Chemistry
Middle East Technical
University
06531 Ankara
Türkei

Carola Dresen
Lehrstuhl für Pharmazeutische und
Medizinische Chemie
Institut für Pharmazeutische
Wissenschaften
Albert-Ludwigs-Universität Freiburg
Albertstr. 25
79104 Freiburg
Germany

Dr. Pascal Dünkelmann
Lehrstuhl für Pharmazeutische und
Medizinische Chemie
Institut für Pharmazeutische
Wissenschaften
Albert-Ludwigs-Universität Freiburg
Albertstr. 25
79104 Freiburg
Germany

Dr. Thomas Dünnwald
Institut für Biotechnologie II
Forschungszentrum Jülich
52425 Jülich
Germany

Dr. Thorsten Eggert
Institut für Molekulare
Enzymtechnologie der HHU
Düsseldorf im FZ Jülich
52426 Jülich
Germany

Prof. Dr. Lothar Elling
Lehr- und Forschungsgebiet
Biomaterialien
Institut für Biotechnologie und
Helmholtz-Institut für
Biomedizinische Technik
RWTH Aachen
Worringerweg 1
52074 Aachen
Germany

Prof. Dr. Dieter Enders
Institut für Organische Chemie
RWTH Aachen
Landoltweg 1
52074 Aachen
Germany

Prof. Dr. Felice Faraone
Dipartimento di Chimica
Inorganica, Chimica Analitica e
Chimica Fisica
Università degli studi di
Messina
Salita Sperone 31 (vill. S. Agata)
98166 Messina,
Italy

Ralf Feldmann
Institut für Biotechnologie II
Forschungszentrum Jülich
52425 Jülich
Germany

Prof. Dr. Wolf-Dieter Fessner
Institut für Organische Chemie
und Biochemie
TU Darmstadt
Petersenstr. 22
64287 Darmstadt
Germany

Dr. Giancarlo Franciò
Institut für Technische und
Makromolekulare Chemie
RWTH Aachen
Worringerweg 1
52074 Aachen
Germany

Prof. Dr. Hans-Joachim Gais
Institut für Organische Chemie
RWTH Aachen
Landoltweg 1
52074 Aachen
Germany

Prof. Dr. Christian Ganter
Institut für Anorganische Chemie der
HHU Düsseldorf
Universitätsstr. 1
40225 Düsseldorf
Germany

Petra Geilenkirchen
Institut für Biotechnologie II
Forschungszentrum Jülich
52425 Jülich
Germany

Dr. Petra Heim
Institut für Molekulare
Enzymtechnologie der HHU
Düsseldorf im FZ Jülich
52426 Jülich
Germany

Prof. Dr. Wolfgang Hölderich
Institut für Brennstoffchemie und
physikalisch-chemische
Verfahrenstechnik
RWTH Aachen
Worringerweg 1
52074 Aachen
Germany

Prof. Dr. Werner Hummel
Institut für Molekulare
Enzymtechnologie der HHU
Düsseldorf im FZ Jülich
52426 Jülich
Germany

Dr. Hans Iding
Institut für Molekulare
Enzymtechnologie der HHU
Düsseldorf im FZ Jülich
52426 Jülich
Germany

Dr. Tomoyuki Inoue
Institut für Mikrobiologie
Universität Stuttgart
Allmandring 31
70550 Stuttgart
Germany

Prof. Dr. Karl-Erich Jaeger
Institut für Molekulare
Enzymtechnologie der HHU
Düsseldorf im FZ Jülich
52426 Jülich
Germany

Dr. Sandra Johnen
Institut für Mikrobiologie
Universität Stuttgart
Allmandring 31
70550 Stuttgart
Germany

Bettina Juchem
Institut für Molekulare
Enzymtechnologie der HHU
Düsseldorf im FZ Jülich
52426 Jülich
Germany

Dr. Doris Kolter-Jung
Institut für Biotechnologie II
Forschungszentrum Jülich
52425 Jülich
Germany

Prof. Dr. Udo Kragl
Institut für Chemie
Universität Rostock
Albert-Einstein-Str. 3a
18059 Rostock
Germany

Prof. em. Dr. Maria-Regina Kula
Institut für Molekulare
Enzymtechnologie der HHU
Düsseldorf im FZ Jülich
52426 Jülich
Germany

Prof. Dr. Walter Leitner
Institut für Technische und
Makromolekulare Chemie
RWTH Aachen
Worringerweg 1
52074 Aachen
Germany

Prof. Dr. Andreas Liese
Institut für Technische Biokatalyse
Technische Universität
Hamburg-Harburg
Denickestr. 15
21073 Hamburg
Germany

Dr. Bettina Lingen
Institut für Molekulare
Enzymtechnologie der HHU
Düsseldorf im FZ Jülich
52426 Jülich
Germany

Dr. Stephan Lütz
Institut für Biotechnologie II
Forschungszentrum Jülich
52425 Jülich
Germany

Prof. Dr. Michael Müller
Lehrstuhl für Pharmazeutische und
Medizinische Chemie
Institut für Pharmazeutische
Wissenschaften
Albert-Ludwigs-Universität Freiburg
Albertstr. 25
79104 Freiburg
Germany

Adam Nitsche
Institut für Biotechnologie II
Forschungszentrum Jülich
52425 Jülich
Germany

Priv. Doz. Dr. Martina Pohl
Institut für Molekulare
Enzymtechnologie der HHU
Düsseldorf im FZ Jülich
52426 Jülich
Germany

Prof. Dr. Hermann Sahm
Institut für Biotechnologie I
Forschungszentrum Jülich
52425 Jülich
Germany

Prof. Dr. Albrecht Salzer
Institut für Anorganische
Chemie
RWTH Aachen
Landoltweg 1
52074 Aachen
Germany

Dr. Ulrich Schörken
Institut für Mikrobiologie
Universität Stuttgart
Allmandring 31
70550 Stuttgart
Germany

Dr. Thomas Schubert
Institut für Biotechnologie II
Forschungszentrum Jülich
52425 Jülich
Germany

Dr. Martin Schürmann
Institut für Mikrobiologie
Universität Stuttgart
Allmandring 31
70550 Stuttgart
Germany

Dr. Melanie Schürmann
Institut für Mikrobiologie
Universität Stuttgart
Allmandring 31
70550 Stuttgart
Germany

Dr. Carmen Schuster
Institut für Brennstoffchemie und
Physikalisch-chemische
Verfahrenstechnik
RWTH Aachen
Worringerweg 1
52074 Aachen
Germany

Dr. Petra Siegert
Institut für Molekulare
Enzymtechnologie der HHU
Düsseldorf im FZ Jülich
52426 Jülich
Germany

Prof. Dr. Georg A. Sprenger
Institut für Mikrobiologie
Universität Stuttgart
Allmandring 31
70550 Stuttgart
Germany

Gerda Sprenger
Institut für Mikrobiologie
Universität Stuttgart
Allmandring 31
70550 Stuttgart
Germany

Dr. Hans-Hermann Wagner
Institut für Brennstoffchemie und
Physikalisch-chemische
Verfahrenstechnik
RWTH Aachen
Worringerweg 1
52074 Aachen
Germany

Lydia Walter
Lehrstuhl für Pharmazeutische
und Medizinische Chemie
Institut für Pharmazeutische
Wissenschaften
Albert-Ludwigs-Universität
Freiburg
Albertstr. 25
79104 Freiburg
Germany

Prof. Dr. Christian Wandrey
Institut für Biotechnologie II
Forschungszentrum Jülich
52425 Jülich
Germany

Dr. Andrea Weckbecker
Institut für Molekulare
Enzymtechnologie der HHU
Düsseldorf im FZ Jülich
52426 Jülich
Germany

Dr. Marion Wendorff
Institut für Molekulare
Enzymtechnologie der HHU
Düsseldorf im FZ Jülich
52426 Jülich
Germany

Dr. Michael Wolberg
Institut für Biotechnologie II
Forschungszentrum Jülich
52425 Jülich
Germany

1
Stoichiometric Asymmetric Synthesis

1.1
Development of Novel Enantioselective Synthetic Methods
Dieter Enders and Wolfgang Bettray

1.1.1
Introduction

Since the pioneering times of the mid-1970s, when the first practical and generally applicable methods in asymmetric synthesis [1] were developed, such as the oxazoline method of Meyers [2] and the SAMP/RAMP hydrazone method [3], there has been a tremendous growth in this research field. One major driving force for this rapid development is of course the different biological activities of enantiomers and thus the need for enantiopure compounds. In this chapter we describe the development of some efficient synthetic methods for asymmetric carbon–carbon and carbon–heteroatom bond formation, which have been carried out within the frame of the "Sonderforschungsbereich 380" (1994–2005) and employing the concept of stoichiometric asymmetric synthesis.

1.1.2
α-Silyl Ketone-Controlled Asymmetric Syntheses

Electrophilic substitutions with carbon and hetero electrophiles α to the carbonyl group of aldehydes and ketones are among the most important synthetic operations. Such regio-, diastereo-, and enantioselective substitutions can be carried out efficiently with the SAMP/RAMP hydrazone methodology [3]. For cases where virtually complete asymmetric inductions could not be attained, an alternative approach based on α-silylated ketones **2** was developed [4]. They can be prepared easily from ketones **1** in high enantiomeric purity ($ee > 98\%$) by asymmetric carbon silylation employing the SAMP/RAMP hydrazone method (Fig. 1.1.1). After the introduction of various electrophiles via classical enolate chemistry with excellent asymmetric inductions, the desired product ketones **3**

Asymmetric Synthesis with Chemical and Biological Methods. Edited by Dieter Enders and Karl-Erich Jaeger
Copyright © 2007 WILEY-VCH Verlag GmbH & Co. KGaA, Weinheim
ISBN 978-3-527-31473-7

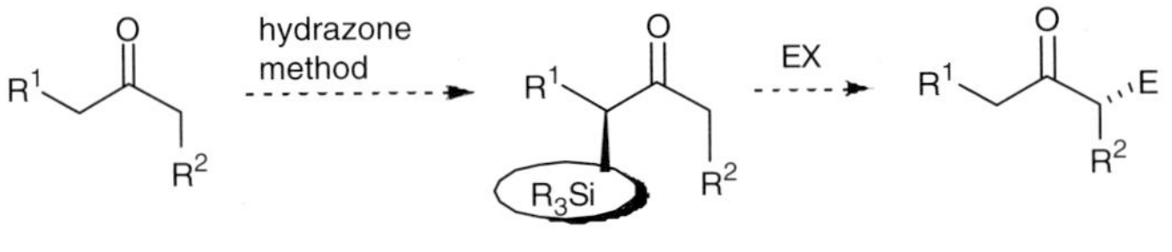

Fig. 1.1.1 α-Silyl-controlled asymmetric synthesis (the "silyl trick").

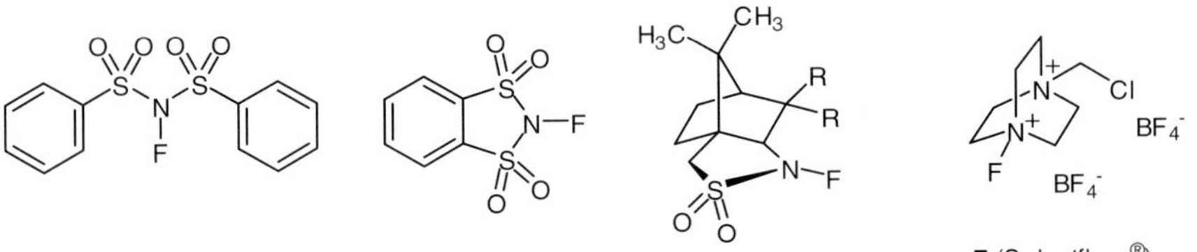

4 (NFSI, Accufluor®) **5** (NFOBS) **6** (Davis, R=H, Cl) **7** (Selectfluor®)

Fig. 1.1.2 NF-reagents for electrophilic fluorination.

are obtained by removal of the "traceless" silyl directing group with various sources of fluoride.

1.1.2.1 Regio- and Enantioselective α-Fluorination of Ketones

Due to the unique properties of organofluorine compounds and their rapidly increasing practical usage in plant protection, medicine, and many other areas, the scientific and economic interest in organofluorine coumpounds has grown immensely over recent decades. With the availability of user-friendly NF reagents such as **4** (NFSI, Accufluor®), **5** (NFOBS), **6** (Davis et al.) and **7** (Selectfluor®) (Fig. 1.1.2), for electrophilic fluorination [5], the efficient synthesis of α-fluorinated ketones, aldehydes, and esters has become possible. However, the asymmetric inductions in enantioselective α-fluorinations of ketones reached no practical values (ee = 10–75%) until the mid-1990s. We were therefore pleased to see that our α-silyl ketone-controlled approach led for the first time to the target α-fluoro ketones in high yields, few steps, and very good enantiomeric excesses [6].

As shown in Scheme 1.1.1, symmetric and unsymmetric ketones (control of regioselectivity) as well as cyclic and acyclic ketones **8** were first converted to the corresponding virtually enantiopure α-silyl ketones **2** (ee > 98%) employing the SAMP/RAMP hydrazone methodology. Metallation with LDA and treatment of the enolates with the N-fluorosulfonamide **4** (NFSI) afforded the α-fluoro-α′-silylated ketones **9** with moderate to excellent diastereomeric excesses. Finally, the racemization-free removal of the sterically demanding silyl directing group was carried out with fluoride sources in almost quantitative yields, leading to the desired α-fluoroketones **10** (ee 55 to >96%). Especially in the case of cyclic ketones almost complete asymmetric inductions could be achieved. As the epimeric

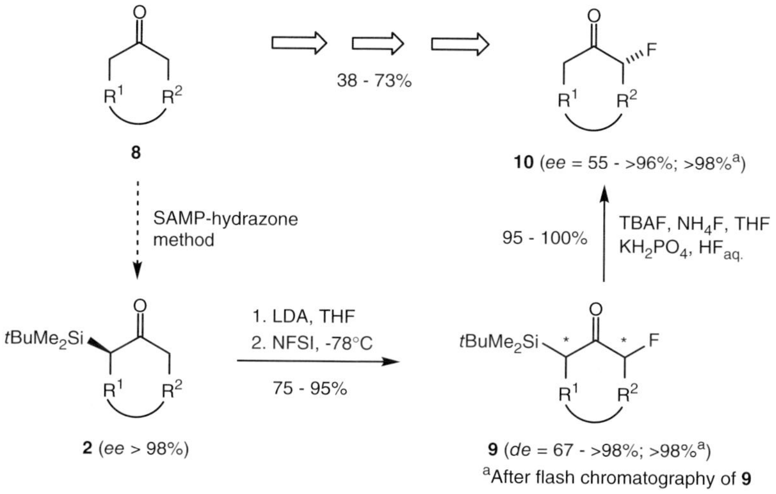

Scheme 1.1.1 Asymmetric synthesis of α-fluoroketones.

fluorinated silyl ketones **9** can be separated easily by flash column chromatography, various enantiopure α-fluoroketones **10** could be obtained in this way.

Although efficient organocatalytic methods for the electrophilic α-fluorination of aldehydes and ketones have recently been developed [7], high enantiomeric excesses can only be reached with aldehydes so far. The asymmetric inductions in the case of ketone fluorinations have remained low ($ee \leq 36\%$) [7a]. Thus, the α-silyl ketone-controlled stoichiometric asymmetric synthesis of α-fluoroketones **10** (Scheme 1.1.1) still constitutes a practical method.

1.1.2.2 α-Silyl Controlled Asymmetric Mannich Reactions

The Mannich reaction, in which an aminomethyl group is introduced in the α-position of the carbonyl function, has been the subject of investigations since the early 20th century [8]. In 1985 our research group, in close cooperation with Steglich and coworkers, developed a first asymmetric Mannich reaction [9]. Some ten years later, with the enantiopure α-silylated ketones **2** in hand, we reported a first practical procedure for the regio- and enantioselective α-aminomethylation of ketones taking advantage of the excellent asymmetric inductions with the help of the "traceless" silyl control group [10].

As depicted in Scheme 1.1.2, the silyl ketones (*S*)-**11** of high enantiomeric purity were converted into the *Z*-configured silyl enol ethers (*S*)-**12**, which were used in the aminomethylation step by treatment with dibenzyl(methoxymethyl)amine in the presence of a Lewis acid. The silylated Mannich bases (*S*,*R*)-**13** were obtained in excellent yields and diastereomeric excesses ($de = 92–96\%$). Finally,

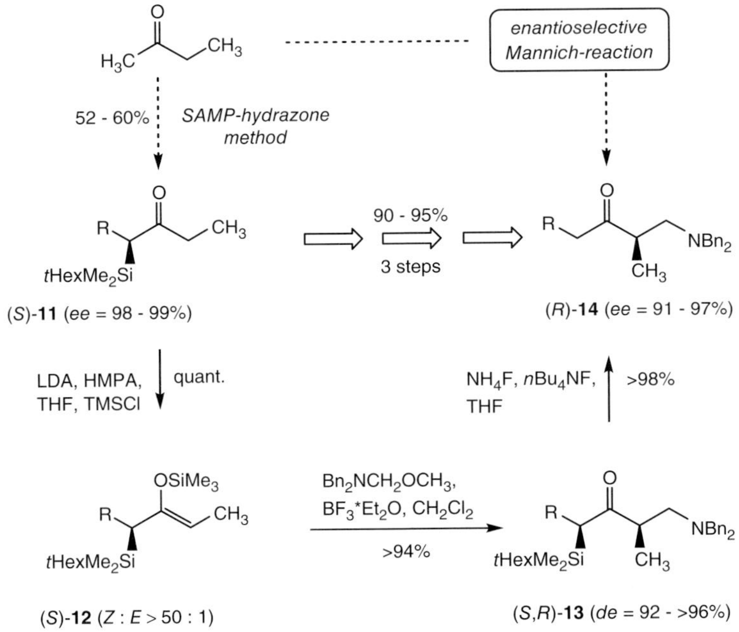

1.1 Development of Novel Enantioselective Synthetic Methods

Scheme 1.1.2 Enantioselective synthesis of β-dibenzylamino ketones.

the silyl directing group was removed tracelessly by employing a fluoride source. In this way, the α-substituted β-amino ketones (R)-**14** were obtained in three steps with superb overall yields of 90–95% and, most importantly, with very high enantiomeric excesses ee of 91–97%. To explain the almost complete diastereofacial selectivity of the Mannich key step, two transition states can be discussed: a closed one along the lines of the Zimmerman–Traxler model, and an open one with the iminium ion formed in situ, explaining in both cases the R configuration at the newly generated stereocenter (Scheme 1.1.3).

After the successful asymmetric synthesis of α-substituted β-amino ketones (R)-**14**, we envisaged the diastereo- and enantioselective synthesis of α,β-disubstituted Mannich bases. As shown in Scheme 1.1.4, we were able to use benzaldehyde-N-phenylimine [11] as well as α-alkoxycarbonylaminosulfones [12] as Mannich electrophiles to synthesize in good overall yields and high de- and ee-values the anti-configured β-amino ketones (R,S)-**15** and (R,S)-**16**, respectively [13].

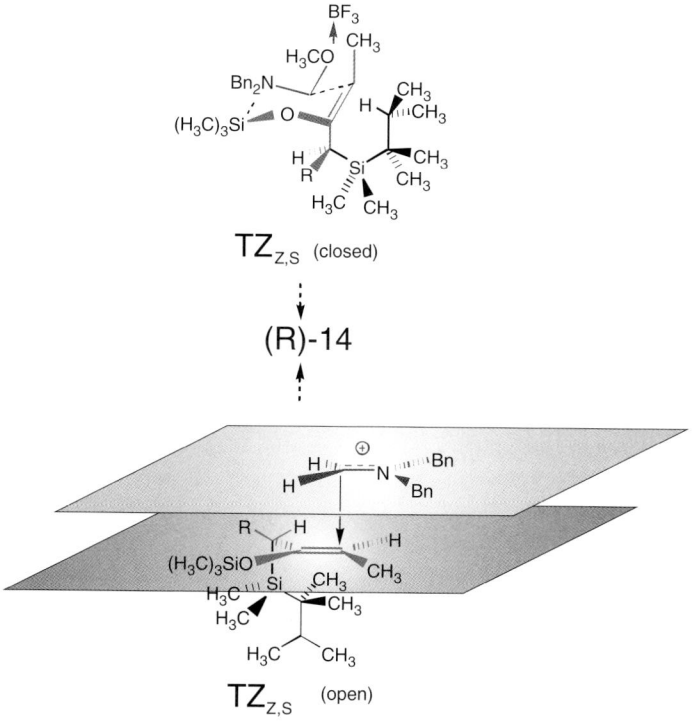

Scheme 1.1.3 Postulated transition states for the asymmetric Mannich reaction.

1.1.3
Asymmetric Hetero-Michael Additions

Besides the successful development of asymmetric syntheses with carbon–carbon bond formation, methods for carbon–heteroatom bond formation were also investigated intensively. In this context we developed several practical protocols for carbon–nitrogen, carbon–oxygen, and carbon–phosphorus bond formation.

1.1.3.1 Asymmetric Aza-Michael Additions

During our investigations on asymmetric C—C bond formation reactions via conjugate addition of SAMP hydrazones to various α,β-unsaturated Michael acceptors, it occurred to us to use the chiral hydrazine auxiliary SAMP as a nitrogen nucleophile and a chiral equivalent of ammonia in aza-Michael additions. Thus, we developed diastereo- and enantioselective 1,4-additions for the synthesis of β-amino acids and β-aminosulfonates [14, 15].

β-Amino acids, even if they are thought to be of less importance than the corresponding α-amino acids, can be found in various biologically active natural

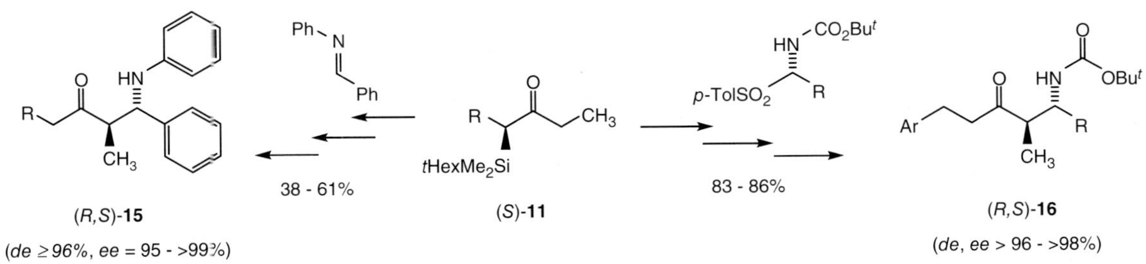

Scheme 1.1.4 Asymmetric synthesis of *anti*-configured Mannich bases.

1.1.3 Asymmetric Hetero-Michael Additions

products and are key building blocks for β-lactam antibiotics. The new world of β-peptides especially is of enormous importance [16]. This is reflected by the increasing research activity in this field and new stereoselective syntheses of β-amino acids [17].

The hydrazine auxiliaries (S)-2-methoxymethyl-1-trimethylsilylaminopyrrolidine (TMS-SAMP) and its enantiomer TMS-RAMP are easily accessible via deprotonation of the parent hydrazines SAMP or RAMP and subsequent silylation with chlorotrimethylsilane. As shown in Scheme 1.1.5, the nitrogen nucleophile was generated by deprotonation of TMS-SAMP and added to the enoate **17**, leading, after desilylation of the 1,4-adduct **18**, to the β-hydrazino esters **19**. The removal of the chiral auxiliary proceeded via N—N bond cleavage with Raney nickel, providing access to the corresponding β-amino acids **20** in good yields and excellent enantioselectivities (ee = 90–98%). Since the protocol can be extended in the sense of a tandem hetero-Michael addition/α-ester-enolate alkylation or aldol reaction, the synthesis of α-substituted β-amino acid derivatives is possible, too. As shown in Scheme 1.1.6, after initial 1,4-addition of the chiral N-nucleophile the α-alkylation/α-aldol reaction using the appropriate halides R^2X or aldehydes R^2CHO leads to the α-substituted β-hydrazino esters **22** with good to excellent stereoselectivity. Based on the results of Yamamoto et al. [18] demonstrating the control of the *syn/anti* stereoselectivity via the enolate geometry [18], we were able to synthesize – after deprotonation of the β-hydrazino ester (R = Ph) with LDA and subsequent trapping with allyl bromide, followed by desilylation – the *syn*-β-hydrazino ester **22** (R^1 = Ph, R^2 = allyl) with high diastereo- and enantiomeric excesses. This example demonstrates that, in principle, both diastereomers of the β-hydrazino esters **22** are accessible.

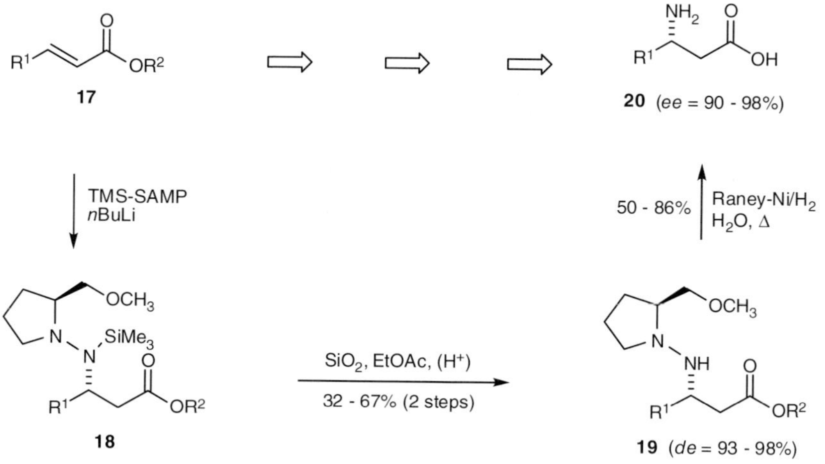

R^1 = Me, Et, nPr, iPr, nBu, iBu, nHept, nUndec, Ph
R^2 = Me, tBu

Scheme 1.1.5 Enantioselective synthesis of β-amino acids.

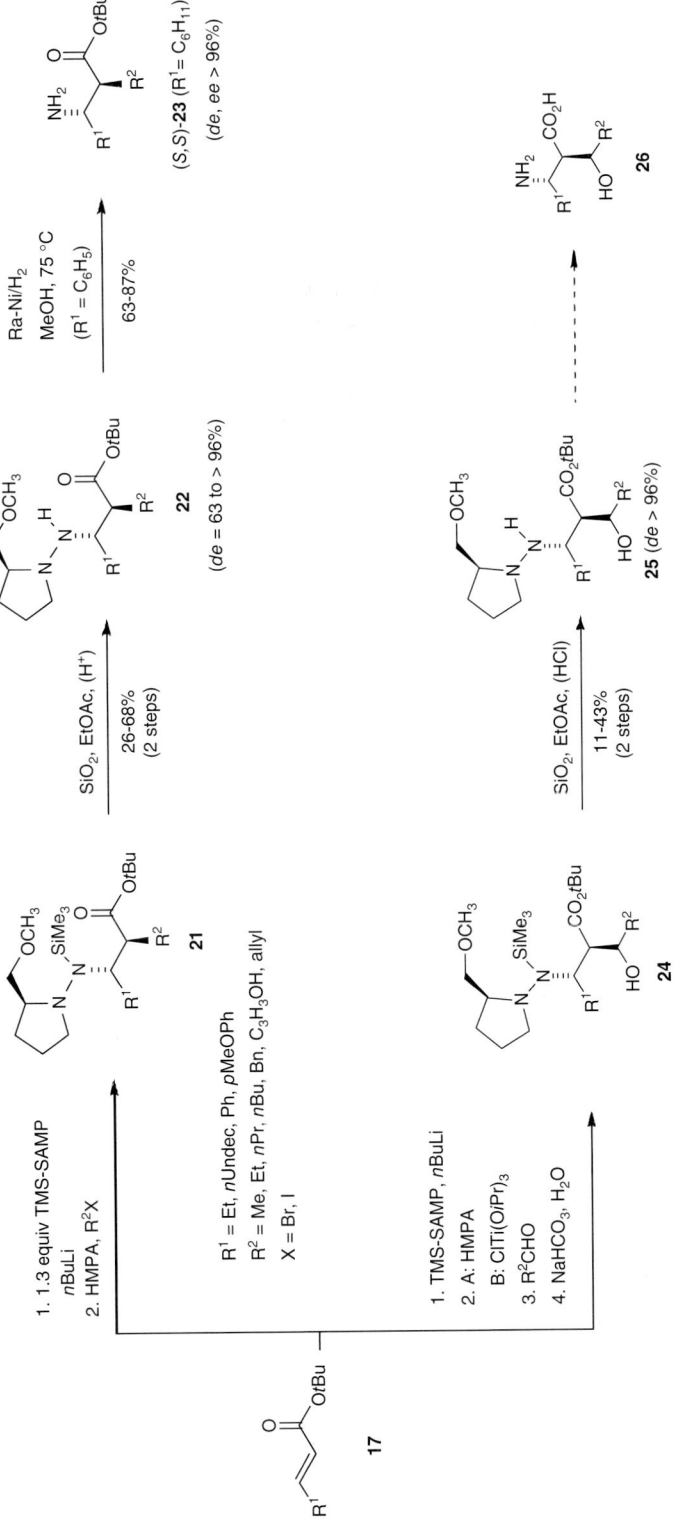

Scheme 1.1.6 Tandem aza-Michael addition/α-ester enolate alkylation or aldol reaction.

1.1.3 Asymmetric Hetero-Michael Additions

In addition, we were able to extend the tandem hetero Michael addition/α-ester-enolate alkylation protocol by an intramolecular variant via a Michael-initiated ring closure (MIRC) reaction leading to diastereo- and enantiomerically pure trans-configured 2-amino-cycloalkanoic acids 30 (Scheme 1.1.7) [14c,d].

Sulfones have become increasingly important in organic synthesis in recent years and α,β-unsaturated sulfones especially are known to be excellent Michael acceptors. Following our concept of using SAMP derivatives as chiral equivalents of ammonia, the enantioselective aza-Michael addition has been investigated in order to provide a new method for the synthesis of β-aminosulfones [15]. As shown in Scheme 1.1.8, the conjugate addition of SAMP to the (E)-alkenylsulfones 31 afforded the Michael adducts 32 in the presence of catalytic amounts of ytterbium(III) triflate with moderate yields and moderate to good diastereoselectivities. Nevertheless, the virtually diastereomerically pure β-hydrazinosulfones 32 could be obtained by chromatographic separation of the epimers. The alternative use of (R,R,R)-2-amino-3-methoxymethylazabicyclo[3.3.1]octane [RAMBO] as a nitrogen nucleophile led to higher selectivities. Thus, after reductive N—N bond cleavage (thereby removing the chiral auxiliary), the β-aminosulfones 33 could be obtained without racemization; this led after direct Boc protection to the N-Boc-protected β-aminosulfones 34 without prior purification of the amines.

In order to extend this method in the sense of a tandem Michael/α-alkylation protocol, in a first step the intermediate β-aminosulfones had to be N,N-dibenzylated, followed by metallation with LDA in the presence of TMEDA and

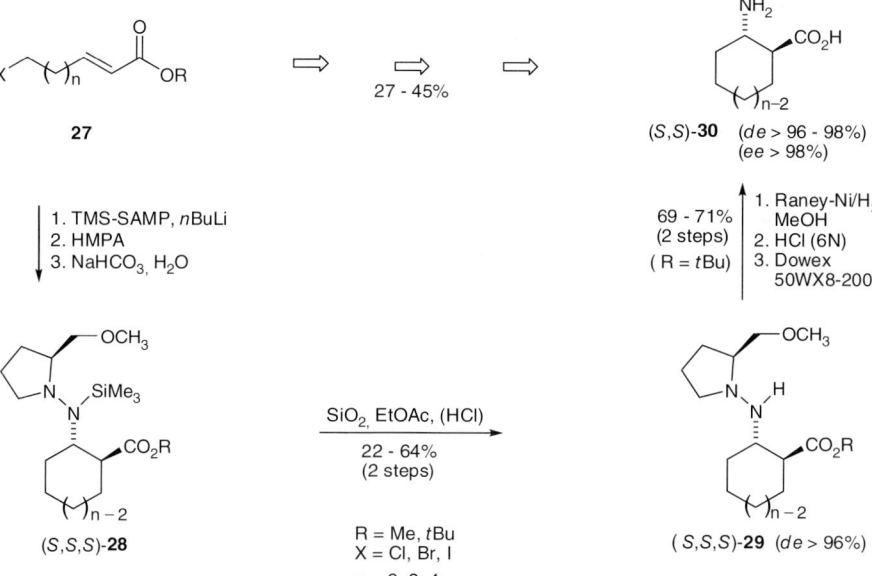

Scheme 1.1.7 Asymmetric aza-MIRC reaction.

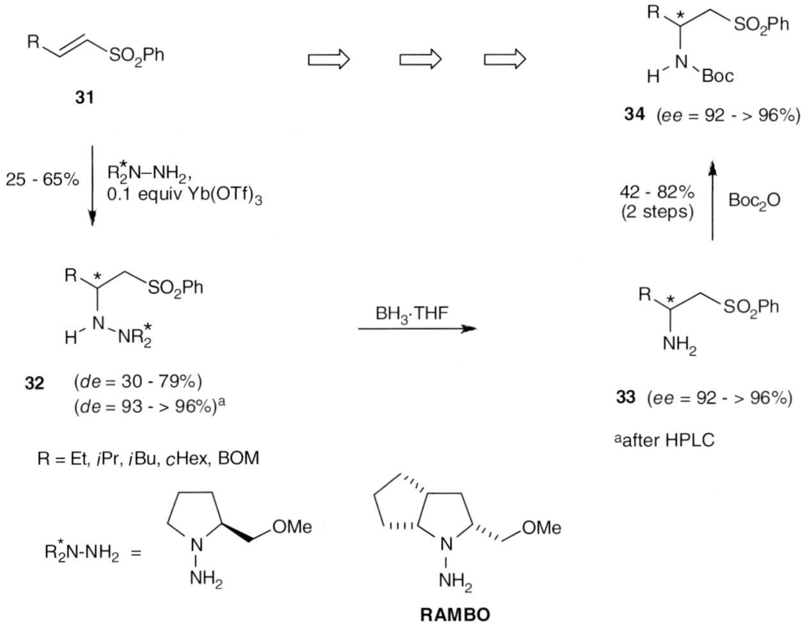

Scheme 1.1.8 Enantioselective synthesis of protected β-aminosulfones.

trapping with electrophiles. The desired *anti* products **38** were obtained in good yields (81–97%) and high stereoselectivities (*de, ee* > 96% after chromatography) (Scheme 1.1.9).

1.1.3.2 Asymmetric Oxa-Michael Additions

It is interesting to note that the oxa-analogous Michael addition was reported for the first time in 1878 by Loydl et al. [19] in their work on the synthesis of artificial malic acid, which was five years ahead of the discovery of the actual Michael reaction described first by Komnenos [20], Claisen [21], and later Michael in 1887 [22] as one of the most important methods for C—C bond formation. In continuation of the early work on the oxa-Michael addition [23], the inter- and intramolecular additions of alkoxides to enantiopure Michael acceptors has been investigated, leading to the diastereo- and enantioselective synthesis of the corresponding Michael adducts [24]. The intramolecular reaction has often been used as a key step in natural product synthesis, for example as by Nicolaou et al. in the synthesis of Brevetoxin B in 1989 [25]. The addition of oxygen nucleophiles to nitroalkenes was described by Barrett et al. [26], Kamimura et al. [27], and Brade and Vasella [28].

At the beginning of our investigations enantioselective intermolecular oxa-Michael additions with removable chirality information in the oxygen-nucleophile had not been reported. Therefore we developed a highly diastereo- and enantiose-

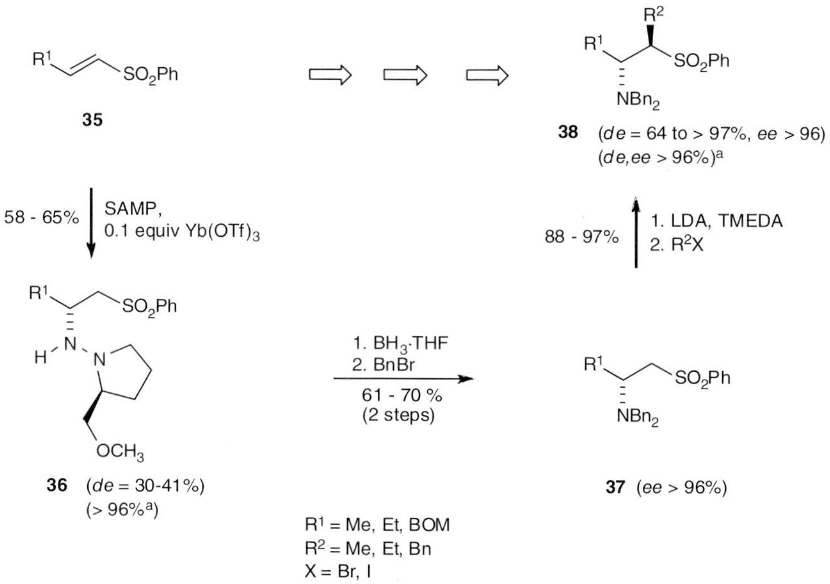

1.1.3 Asymmetric Hetero-Michael Additions

Scheme 1.1.9 Enantioselective synthesis of α,β-disubstituted aminosulfones.

lective conjugate hydroxylation method by 1,4-addition of an enantiopure oxygen nucleophile to nitroalkenes [29]. When the enantiopure sodium alkoxide of (1R,2S)-(−)-N-formylnorephedrine (**40**) was employed as a nucleophile, the conjugate addition to (E)-nitroalkenes led to the corresponding Michael adducts **41** in good yields (56–85%) with high diastereoselectivities (de 94 to >98%), which could be converted to the protected amino alcohols **43** by reduction of the nitro function, protection of the amino group to form **42**, and removal of the chiral auxiliary without racemization. Thus, the vicinal amino alcohols could be synthesized in an overall yield of 34–53% and enantiomeric excesses ee of 94 to >98% (Scheme 1.1.10).

1.1.3.3 Asymmetric Phospha-Michael Additions

The conjugate addition of phosphorus nucleophiles of various oxidation states and in neutral or metallated form constitutes an efficient and well-known method for C—P bond formation [30]. In the case of phosphanes as nucleophiles especially, the corresponding phosphane–borane adducts have been used in 1,4-additions to Michael acceptors. Following the idea to use a chirally modified phosphorus nucleophile in asymmetric Michael additions to aromatic nitroalkenes, we synthesized the new enantiopure phospite **45** starting from TADDOL (**44**) with nearly quantitative yield. Due to the C_2 symmetry, of the

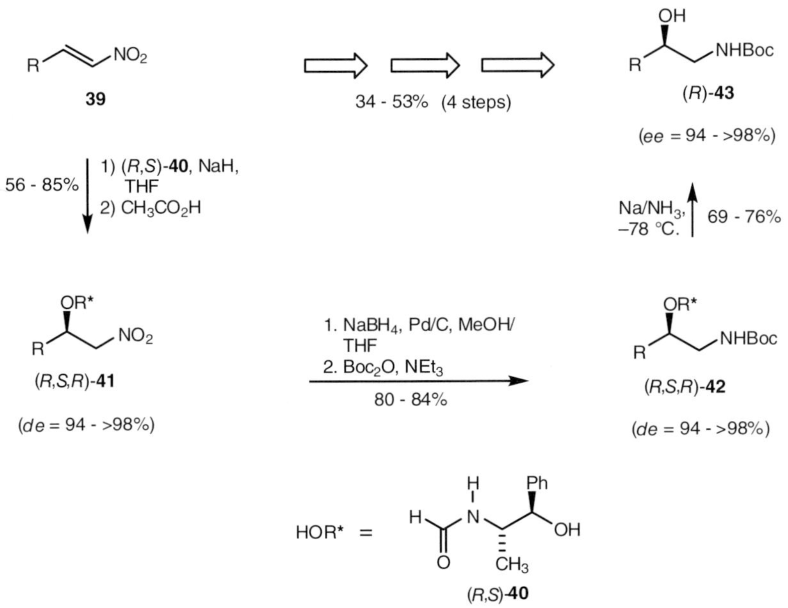

Scheme 1.1.10 Asymmetric oxa-Michael additions.

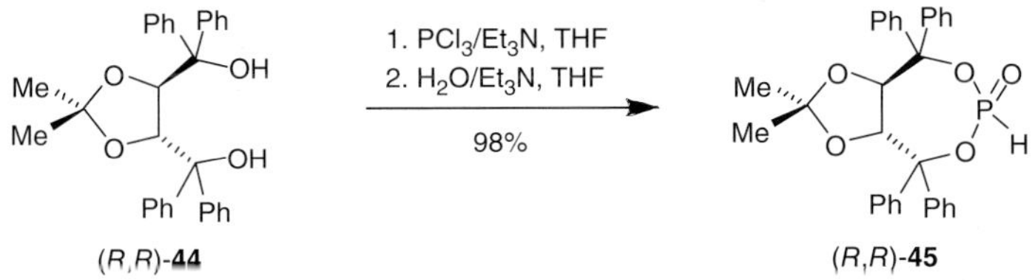

Scheme 1.1.11 Synthesis of the TADDOL-derived enantiopure phosphite.

ligand the formation of a new stereocenter at phosphorus is avoided, which would otherwise have made a separation of the diastereomers necessary (Scheme 1.1.11).

Noltes et al. [31] have already described the reaction of diethylzinc with diphenylphosphane leading to organozinc–phosphorus adducts which have been shown to be very reactive but very insoluble. In our synthesis of the phosphite **45** with the nitroalkenes **46** in the presence of Et_2Zn the precipitation of a solid could be observed as well, which was unreactive at –78 °C but underwent addition to the nitroalkenes at 0 °C. In the presence of TMEDA even at these low temperatures the desired Michael adducts **47** were formed with very good yields and high diastereomeric excesses (de = 84–96%) (Scheme 1.1.12). After removal of the

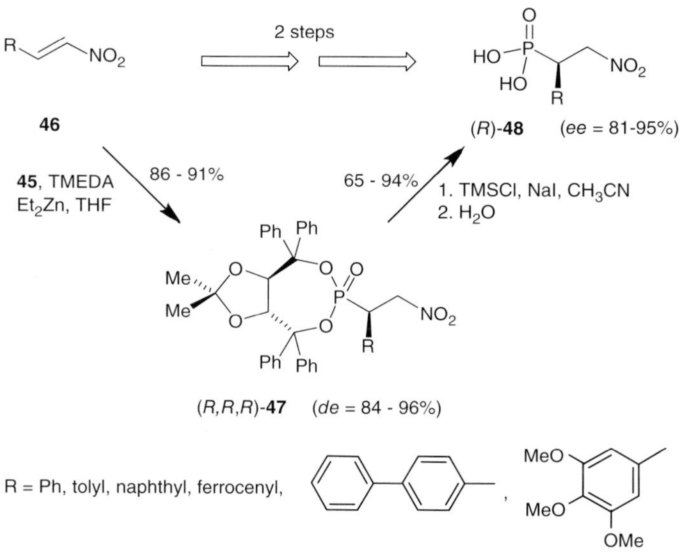

Scheme 1.1.12 Asymmetric phospha-Michael additions.

chiral auxiliary following the modified conditions of Morita et al. [32], the β-nitrophosphonic acids **48** could be obtained with virtually no racemization.

In recent years the addition of phosphorus compounds containing a labile P—H bond to double and triple bonds under heterogeneous conditions using a solid base, as for example Al_2O_3/KOH in the Pudovik reaction, was shown by Koenig et al. [33], and by other groups in the case of C—C bond formation [34]. Based on these findings we decided to develop the first asymmetric P—C bond formation reaction under heterogeneous conditions with the KOH/Fe_2O_3 system as a solid base in the conjugate addition of chiral phosphites to alkylidene malonates leading to optically active β-substituted β-phosphonomalonates **52** (Scheme 1.1.13) [30c]. Due to the metal oxide activation of the P(O)H group the deprotonation of the P—H bond can be performed even with weak bases. In our studies we investigated the effect of different metal oxides as solid bases and in all cases the phosphonate **50a** (R = Ph) could be observed. The best results could be achieved with the Fe_2O_3/KOH system.

Using these optimized conditions, the conjugate addition of **45** to the malonates **49** in the presence of the solid base provided access to the Michael adducts in good yields and high diastereoselectivities. The removal of the chiral TADDOL auxiliary could be accomplished without epimerization or racemization in this case, too. After esterification of the very polar acid intermediates with diazomethane the corresponding dimethyl esters **52** were obtained with yields of 72–94% and enantiomeric excesses ee of 84–94% (Scheme 1.1.13).

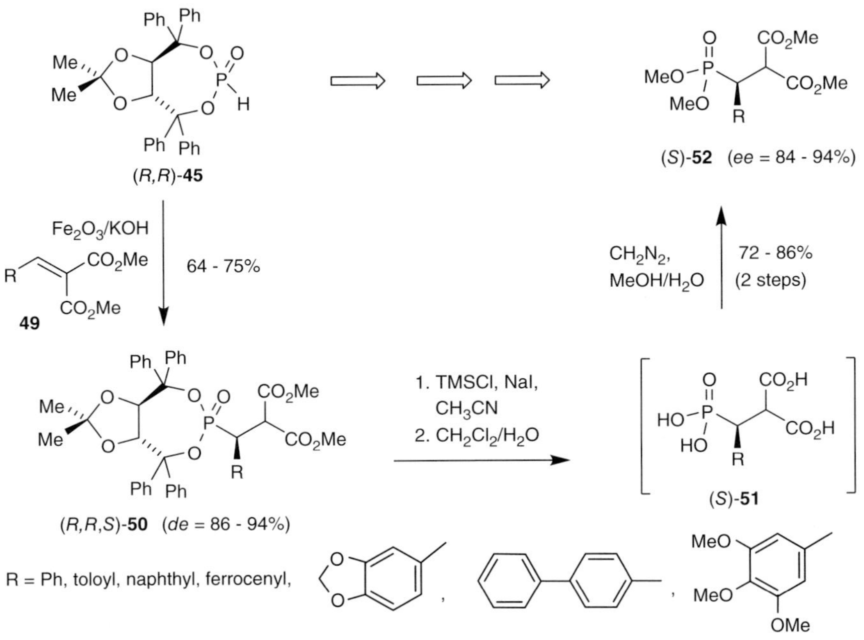

Scheme 1.1.13 Solid-base-mediated phospha-Michael additions.

1.1.4
Asymmetric Syntheses with Lithiated α-Aminonitriles

1.1.4.1 Asymmetric Nucleophilic α-Aminoacylation

α-Aminonitriles have occupied an important position historically as versatile intermediates in organic synthesis [35, 36], especially since the aminonitrile moiety may be hydrolyzed to the corresponding carbonyl compound under extremely mild conditions [37]. After metallation they represent masked acyl carbanion equivalents with nonclassical d^1 reactivity. Thus metallated aminonitriles can be used as nucleophilic acylating agents in carbon–carbon bond forming processes leading to 1,2- [38] and 1,4-difunctionalized substances, with a great deal of attention given to conjugate additions to α,β-unsaturated carbonyl compounds [39].

In continuation of our investigations on asymmetric nucleophilic acylations with lithiated α-aminonitriles [40], we envisaged the asymmetric synthesis of 3-substituted 5-amino-4-oxo esters **A**, bearing both α-amino ketone and δ-amino ester functionalities (Scheme 1.1.14) [41]. Since α-amino ketones are precursors of chiral β-amino alcohols [42, 43] and chiral amines [43], their asymmetric synthesis has the potential to provide valuable intermediates for the synthesis of biologically active compounds, including peptidomimetics [44]. The retrosynthetic analysis of **A** leads to the α-aminoacyl carbanion **B** and β-ester carbocation

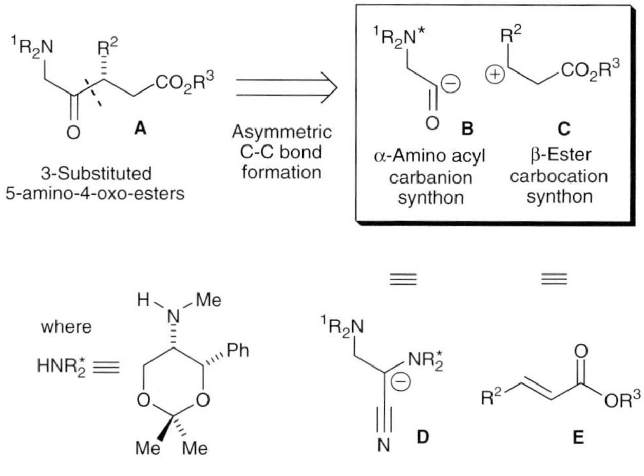

Scheme 1.1.14 Retrosynthetic strategy for the asymmetric nucleophilic α-aminoacylation.

C synthons, synthetic equivalents of which being the amino-acetaldehyde-derived metallated aminonitrile **D** bearing the chiral auxiliary (*S*,*S*)-**53** and an α,β-unsaturated ester **E**, respectively. This should make it possible to open up a pathway to an enantioselective conjugate addition of an α-aminoacyl carbanion equivalent **D** to enoates in order to access the target 3-substituted 5-amino-4-oxo esters.

Thus, the *N*,*N*-dibenzyl-protected aminonitrile **55** was prepared via Swern oxidation of *N*,*N*-dibenzylaminoethanol **54** followed by treatment with the enantiopure amine auxiliary (*S*,*S*)-**53** and HCN, resulting in the formation of a 3:2 epimeric mixture of the aminonitriles **55** in 55% yield, from which the single diastereomers could be isolated by chromatography. After lithiation with LDA, addition to the requisite (*E*)-α,β-unsaturated esters and hydrolysis of the aminonitrile moiety with silver nitrate, the desired α-amino keto esters (*R*)-**56** were obtained with yields of 65–81% and enantiomeric excesses *ee* of 78–98%, which could be improved to *ee* > 98% by a simple recrystallization. Since the amino ketone functionality can be cleaved oxidatively, the 5-amino-4-oxo-esters **56** could be transformed to the corresponding succinic half-esters **57** with hydrogen peroxide in methanol in good to excellent yields (68–90%) (Scheme 1.1.15).

Because 3-substituted 1,2-amino alcohols and even β-alkyl-γ-hydroxy-δ-amino esters are potentially precursors to pharmacologically interesting materials, further investigations have been carried out to extend the methodology in this direction. Thus, the reduction of the ketone moiety of **56** by applying L-selectride or lithium tri-*t*-butoxyaluminum hydride opened access to the *cis*-amino lactones **58** (45–54% yield; *de* = 90–94%) and the *trans*-amino lactones **58** (67–76% yield; *de* > 98%), respectively (Scheme 1.1.16). Monodebenzylation with cerium ammoni-

Scheme 1.1.15 Enantioselective synthesis of 3-substituted 5-amino-4-oxo esters and succinic half-esters.

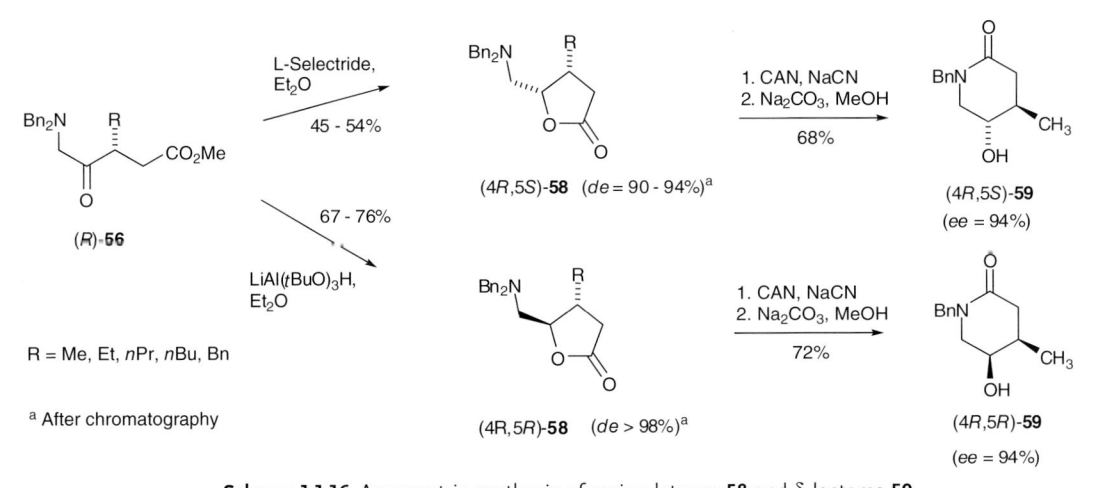

Scheme 1.1.16 Asymmetric synthesis of amino latones **58** and δ-lactams **59**.

um nitrate (CAN) and then treatment with sodium carbonate and methanol allowed the transformation into the respective hydroxy δ-lactam **59**.

1.1.4.2 Asymmetric Nucleophilic Alkenoylation of Aldehydes

α-Hydroxyenones **F** are versatile building blocks, structural subunits of many natural products, and suitable precursors for Nazarov cyclizations. The retrosyn-

1.1.4 Asymmetric Syntheses with Lithiated α-Aminonitriles

thetic analysis shown in Scheme 1.1.17 indicates that the targets **F** can be traced back to the addition of a synthetic equivalent of the α,β-unsaturated acyl anion **G** to aldehydes **H**. Toward this goal, a great deal of attention has been devoted to the development of both achiral and chiral equivalents of **G**, such as heteroatom-stabilized allylic anions [45], organometallic zirconium [46], or vinyl-lithium reagents [47]. With respect to asymmetric syntheses, metallated β,γ-unsaturated α-aminonitriles **I** should be a good choice since the chirality information can easily be introduced via the amino group. Using this methodology based on the pioneering work of Ahlbrecht et al. [48], Fang et al. [49] only observed the γ-addition with aldehydes as electrophiles.

Based on our previous results on the nucleophilic alkenoylation of aldehydes via metallated α,β-unsaturated aminonitriles [50], we now envisaged an enantioselective variant. Thus, the enantiopure α-aminonitriles **60** were metallated with LDA and by reaction with aldehydes the adducts **61** could be obtained. Subsequent cleavage of the aminonitrile function with silver nitrate led to the desired α′-hydroxyenones **62** in overall yields of 29–80% and enantiomeric excesses *ee* of 50–69%. Alternatively, the conjugate addition of the lithiated aminonitrile **63** to *t*-butyl crotonate led to the γ-keto ester **63** in 35% yield and an enantiomeric excess *ee* of >90% (Scheme 1.1.18).

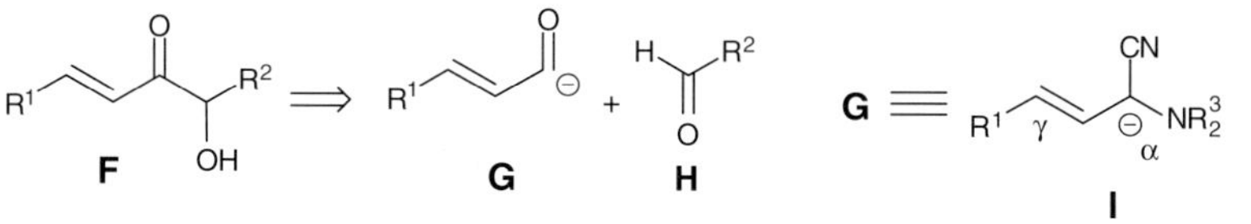

Scheme 1.1.17 Retrosynthetic analysis.

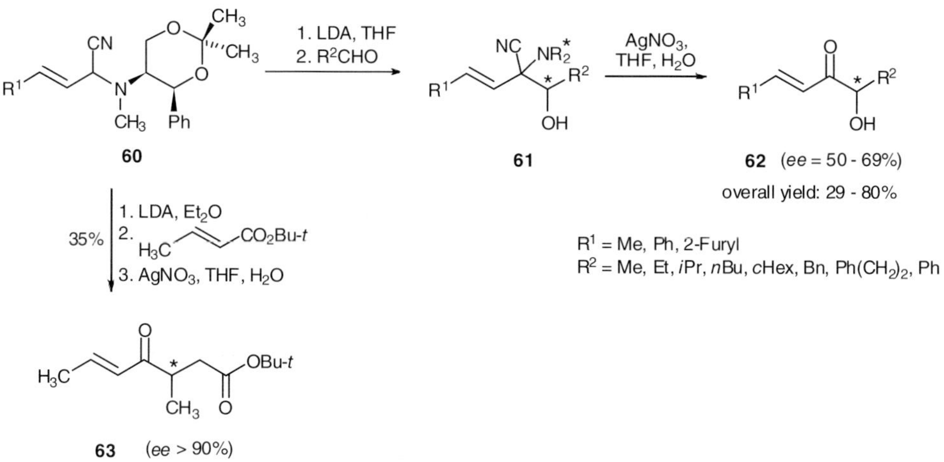

Scheme 1.1.18 Asymmetric nucleopilic alkenoylation of aldehydes and enoates.

1.1.5
Asymmetric Electrophilic α-Substitution of Lactones and Lactams

Lactones [51] and lactams [52] of different ring sizes play an important role as characteristic structural units in natural and biologically active compounds. With respect to lactones, the stereoselective syntheses are based on C—C or C—O bond formation reactions starting from acyclic precursors [53] or from cyclic substrates, which are transformed to the optically active lactones by enantioselective deprotonation/oxidation [54], metal-catalyzed enantioselective Baeyer–Villiger oxidation [55], enantioselective protonation [56], γ-alkylation [57], and hydroformylation [58]. For the asymmetric synthesis of α-substituted γ- and δ-lactams the α-alkylation of lactams derived from (R)-(−)-phenylglycinol has been investigated by Royer, Quirion, Husson and coworkers [59], while Koga, Kobayashi et al. used the presence of tetradentate ligands for the enantioselective alkylation of lactam enolates [60]. Furthermore, the hydrogenation of a 3-alkylidene-2-piperidone was used by Chung et al. [61] and the Meerwein–Eschenmoser rearrangement [62] was reported by Stevenson et al. Additionally, the bicyclic lactam-method of *Meyers* has to be mentioned [63].

With the goal of an efficient and generally applicable asymmetric synthesis of α-substituted lactones, in 1985 we developed a first practical protocol for the synthesis of N,N-dialkyl-lactone hydrazones [64]. In this manner the lactone SAMP hydrazones (S)-**66** were synthesized in a two-step procedure starting from the chloroalkanoyl chlorides **64** via an ambidoselective cyclization of the intermediate ω-chlorohydrazides **65** in the presence of silver tetrafluoroborate (C—O bond formation). These hydrazone lactones undergo α-alkylation reactions with good yields and high diastereoselectivity [65]. Subsequent oxidative cleavage of the hydrazones furnished the α-substituted lactones **67** in enantiomeric excesses *ee* of 90–94% and overall yields of 32–58% (Scheme 1.1.19).

Under slightly different conditions via an ambidoselective cyclization of the chloroalkanoyl hydrazides **68** with sodium hydride, the lactams **69** were obtained with C—N bond formation. Electrophilic substitutions in the α-position with various electrophiles, including alkylation and Michael additions, and subsequent reductive N—N bond cleavage with lithium in liquid ammonia generated the α-substituted lactams **71** [66]. In order to achieve better asymmetric inductions the chiral auxiliary SAMP was changed to SADP and SAEP, which possess a greater steric demand in the pyrrolidine side-chain. When necessary, the undesired diastereomers of **70** could be separated by chromatography. During the treatment with lithium in ammonia, a slight loss of enantiomeric purity was observed for the lactams derived from tosylaziridine. Finally, after optimization of the protocol, overall yields of 37–65% with moderate to high *ee* values were reached.

γ-Butyrolactams, as well as β-aminoethylated lactones, are γ-aminobutyric acid (GABA) derivatives, which are of great importance in the regulation of neurological disorders. We therefore extended our investigations toward a nitroalkene Michael addition/reduction sequence.

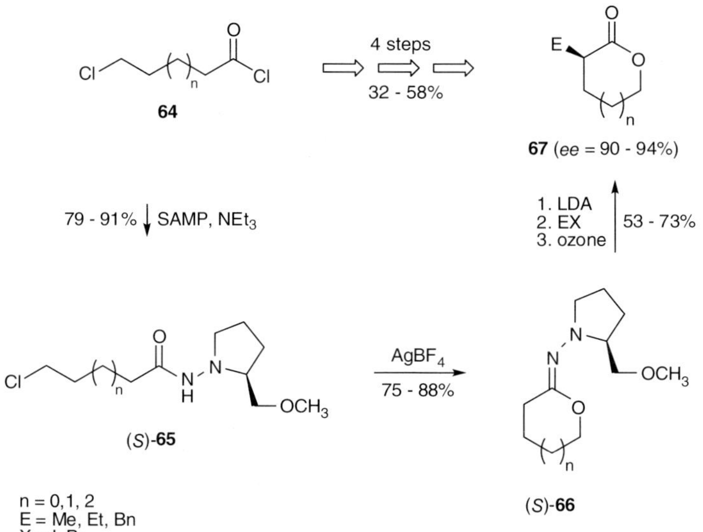

1.1.5 Asymmetric Electrophilic α-Substitution of Lactones and Lactams | 19

Scheme 1.1.19 Asymmetric synthesis of α-alkylated lactones.

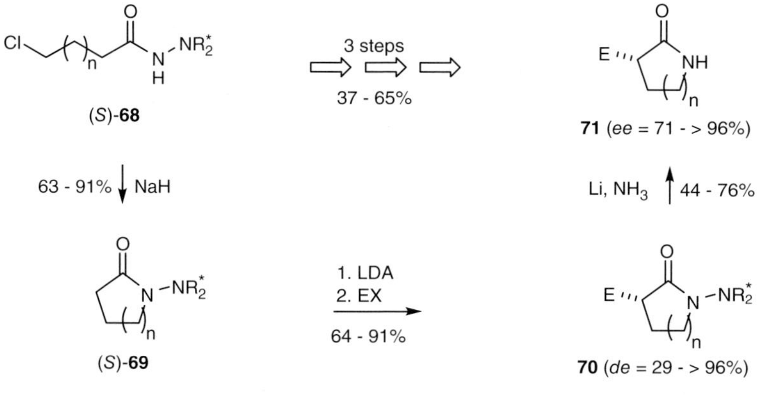

n = 0, 1, 2
E = Me, Et, Bn
X = I, Br

Scheme 1.1.20 Asymmetric synthesis of α-substituted lactams.

n = 1, 2, 3
NR_2^* = SMP, SDP, SEP
E = Me, Et, Pr, Bu, allyl, $SiMe_3$, $(CH_2)_2OSiMe_2tBu$, CH_2CO_2Me,
   Bn, 3,4,5-tri-MeO-Bn, 2,4-di-Cl-Bn, $(CH_2)_2NHTs$, $CH(Ph)CH(CO_2Me)_2$

Thus, the 1,4-addition of the lactams **72** to nitroolefins provided access to the Michael adducts **73** with good stereoselectivities, which could be improved by recrystallization or chromatography. After reduction and protection of the amino group the γ-butyrolactams **74** were converted to the corresponding α-substituted lactams **75** in good overall yields (37–65%) and excellent diastereo- and enantiomeric excesses (*de, ee* > 96%) by removal of the chiral auxiliary through N—N cleavage (Scheme 1.1.21).

Since the trapping of the lactam enolate with electrophiles should not be limited to nitroolefins, extension to conjugate additions to alkenylsulfones, the ring opening of *N*-tosylaziridines, alkylation with functionalized halides, and silylation with chlorotrimethylsilane were explored.

The 1,4-addition of the *N*-dialkylaminolactams **76** to alkenylsulfones as Michael acceptors led to the desired products **77** in good yields but only with moderate diastereoselectivities (*de* = 38–41%). Only for **77a** (R = Me) the main diastereomer could be obtained with *de* > 96% after purification by HPLC (Scheme 1.1.22).

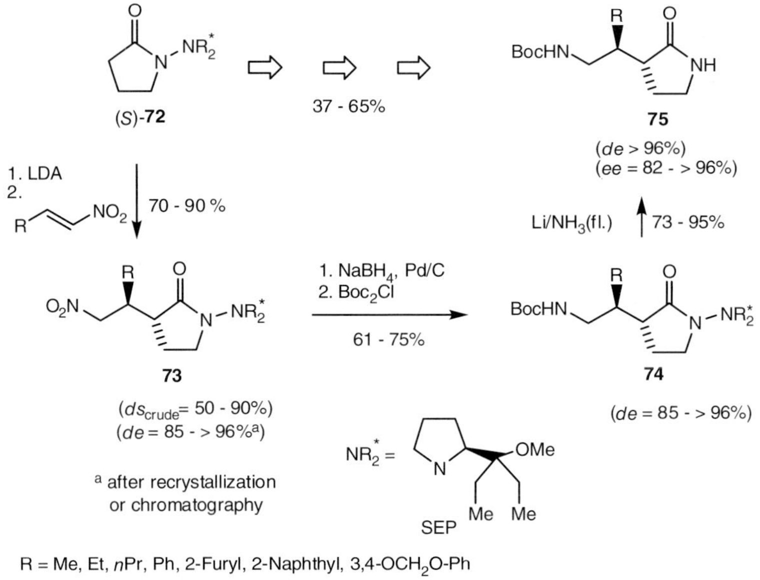

Scheme 1.1.21 Asymmetric Michael addition of lactams to nitroalkenes.

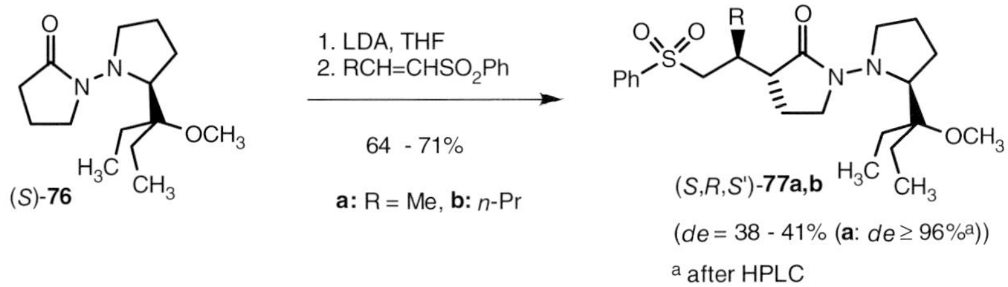

Scheme 1.1.22 Asymmetric Michael addition to alkenylsulfones.

1.1.5 Asymmetric Electrophilic α-Substitution of Lactones and Lactams

Based on our experience of asymmetric β-aminoethylations of lithiated SAMP hydrazones with *N*-tosylaziridines as electrophiles [67], the β-aminoethylation of lactams was explored, leading to the product **79a** in good yield and excellent diastereoselectivity (*de* > 96%). Cleavage of the chiral auxiliary again with lithium in liquid ammonia, which also caused the removal of the tosyl protecting group, led after BOC protection of the intermediate amino group to the protected aminolactam **80** with partial racemization.

Employing the same reaction conditions the addition of methyl bromoacetate, *t*-butyldimethylysilyloxyethyl bromide or chlorotrimethylsilane as electrophile afforded the corresponding α-functionalized *N*-dialkylaminolactams **79** and **81**. Auxiliary removal in the latter case to form **82** again led to a partial racemization (*ee* = 83%) (Scheme 1.1.23).

α,α-Disubstituted lactams are useful as key building blocks in the synthesis of natural and bioactive compounds. The six-membered derivatives especially have been used in the synthesis of alkaloids. Thus, employing our methodology, the lactams **83** were α-alkylated resulting in the mono-substituted lactams **84**, which were α-alkylated again following the same procedure. The lactams **85** could be achieved in this way with low to good yields and diastereomeric excesses *de* of 6–88%. Removal of the chiral auxiliary afforded the lactams **86** in yields of 64–88% and enantiomeric excesses *ee* of 50–88% (Scheme 1.1.24).

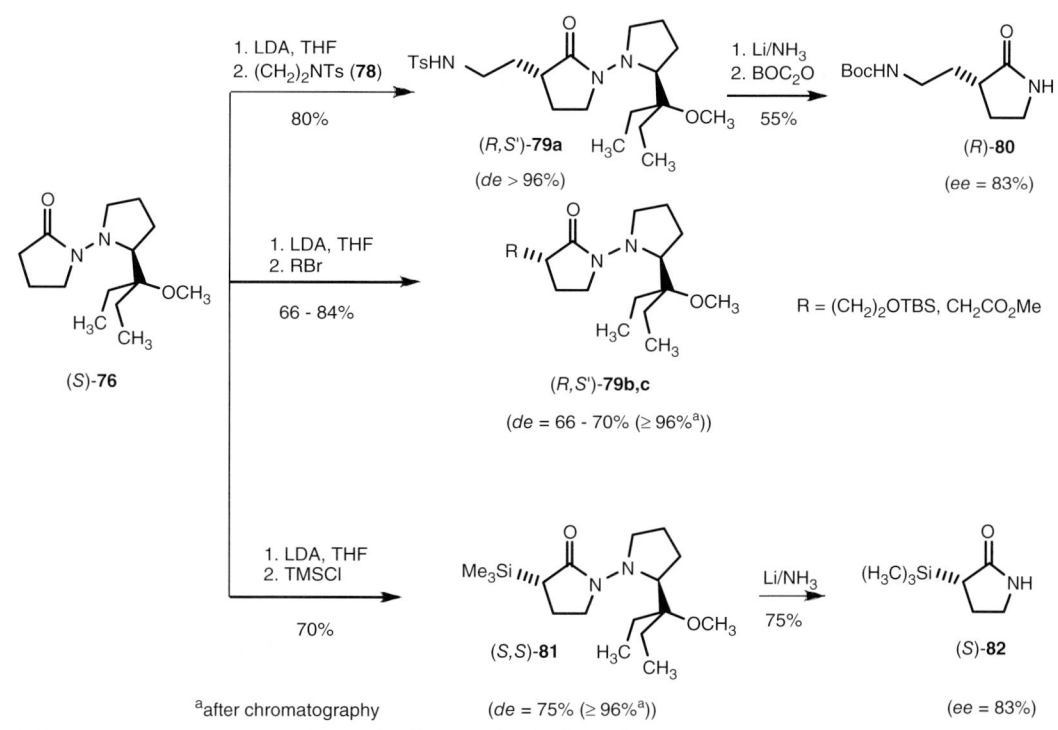

Scheme 1.1.23 Asymmetric synthesis of α-functionalized γ-butyrolactams.

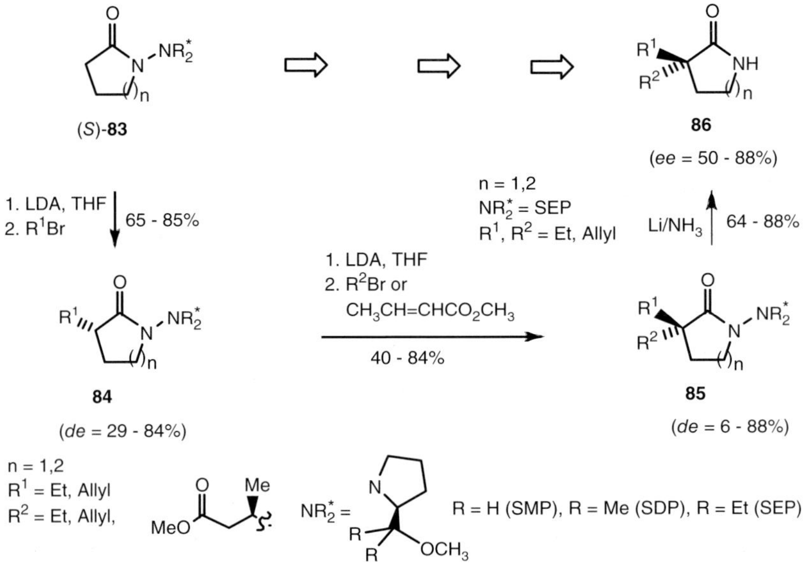

Scheme 1.1.24 Asymmetric synthesis of α,α-disubstituted lactams.

1.1.6
Asymmetric Synthesis of α-Phosphino Ketones and 2-Phosphino Alcohols

Parallel to the rapid development of homogeneous catalysis there has been a growing demand for the synthesis of functionalized phosphane ligands bearing an additional oxygen donor function to be used as hemilabile ligands. α-Phosphino ketones especially have been used successfully as ligands in the Shell Higher Olefin Process (SHOP) [68], whereas phosphino alcohols, for instance, have been used either directly or after further transformation to ethers and phosphites. Despite the utility of these important classes of chiral P-ligands for asymmetric catalysis, asymmetric syntheses had hardly been investigated at the beginning of our project [69].

In continuation of our efforts to explore the utility of the SAMP/RAMP hydrazone methodology, we developed the first asymmetric synthesis of α-phosphino ketones via formation of a carbon–phosphorus bond in the α-position to the carbonyl group [70]. The key step of this asymmetric C—P bond formation is the electrophilic phosphinylation of the ketone SAMP hydrazone **87**, giving rise to the borane-adduct of the phosphino hydrazone **88** with excellent diastereoselectivity (*de* = 95–98%). Since these phosphane–borane adducts are stable with respect to oxidation, the chemoselective cleavage of the chiral auxiliary by ozonolysis leading to the α-phosphino ketones (*R*)-**89** could be accomplished with virtually no racemization. Using RAMP as a chiral auxiliary, the synthesis of the enantiomer (*S*)-**89** was possible (Scheme 1.1.25).

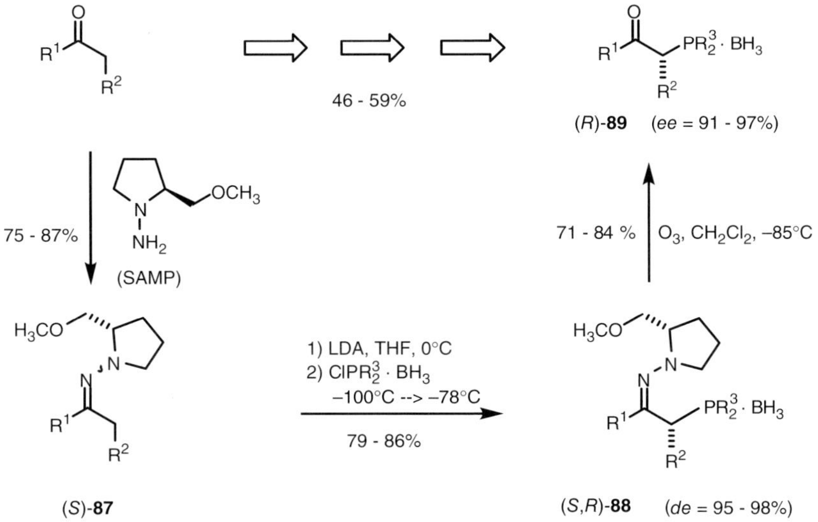

1.1.6 Asymmetric Synthesis of α-Phosphino Ketones and 2-Phosphino Alcohols

Scheme 1.1.25 Enantioselective synthesis of α-phosphino ketones.

In addition to the results described, enantioselective access to 2-phosphino alcohols could be accomplished, too [71]. Starting from a borane-protected α-phosphino aldehyde hydrazone **91** as the key intermediate and available by two different approaches, the enantioselective synthesis of the desired 2-phosphino alcohols **93** could be accomplished. Thus, the electrophilic phosphinylation of aldehyde hydrazones **90** (via route **I** with the chlorodiphenylphosphine–borane adduct; or via route **II** with chlorophosphines and subsequent phosphorus–boron bond formation) and the alkylation of phosphino acetaldehyde-SAMP hydrazones **92** (route **III**) was carried out (Scheme 1.1.26).

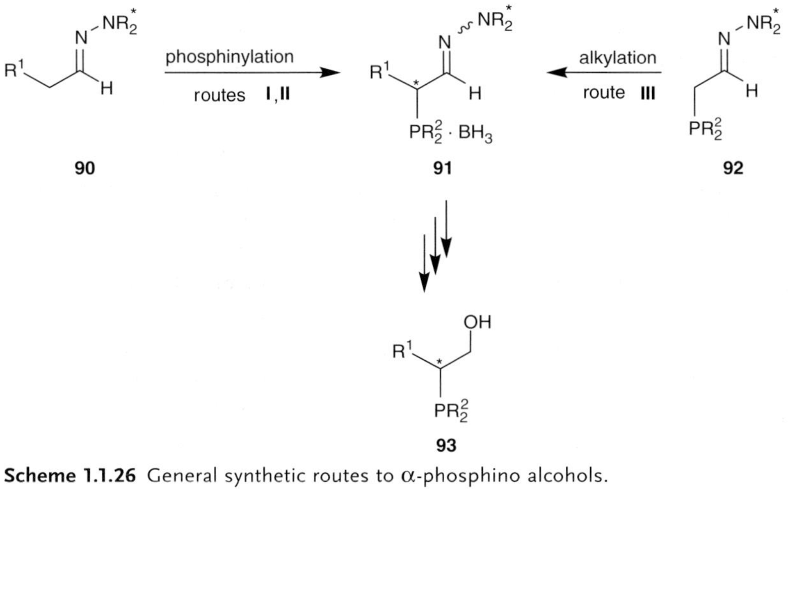

Scheme 1.1.26 General synthetic routes to α-phosphino alcohols.

Following this strategy, the azaenolates of the SAMP, RAMP, or SAEP aldehyde hydrazones **94** were trapped with the chlorodiphenylphosphine-borane adduct (route **I**) leading to the α-phosphino aldehyde hydrazones (*S,R*)-**96** in moderate yields (38–56%) and with good diastereomeric excesses (*de* = 73–80%) as *E/Z* mixtures with respect to the C=N double bond. Alternatively, trapping of the azaenolates with chlorodiphenylphosphine or chlorodiisopropylphosphine produced the products **96** in even better yields (45–75%) and diastereomeric excesses (*de* = 50–87%), predominantly as *Z*-isomers with regard to the C=N double bond. Finally, the *Z*-configured major diastereomer of the diphenylphosphanyl SAMP hydrazones could be isolated with high diastereoselectivity (*de* > 96%) by crystallization or separation of the diastereomers using HPLC.

An alternative access was achieved by alkylation of the α-diphenylphosphino acetaldehyde SAMP hydrazone **95**, yielding the hydrazone products **96** in good yields (60–63%) and good diastereomeric excesses (*de* = 68–71%) as *E/Z* mixtures, from which the major diastereomer was separated and purified by preparative HPLC. Ozonolysis and in-situ reduction with the borane–dimethyl sulfide complex of the aldehydes generated gave the air-stable borane-protected 2-diphenylphosphino alcohols **97** in good yields (67–83%). Reaction with DABCO afforded the unprotected 2-phosphino alcohols **98** in very good yields (85–91%) and excellent enantiomeric excesses (*ee* ≥ 96%) (Scheme 1.1.27).

1.1.7
Asymmetric Synthesis of 1,3-Diols and *anti*-1,3-Polyols

The development of highly stereoselective syntheses of 1,3-polyol chains has received considerable attention in recent years, mainly due to the growing interest in polyene macrolides as challenging synthetic targets with desirable pharmacological features [72]. Early research efforts explored the enantioselective reduction of β-keto esters and 1,3-diketones, the diastereoselective reduction of β-hydroxy ketones, and the Sharpless epoxidation of allylic alcohols with subsequent epoxide opening [73]. Evans et al. reported the synthesis of differentiated *anti*-1,3-diols by samarium-catalyzed intramolecular Tishchenko reduction of β-keto esters [74]. In 1991 Rychnovsky et al. developed a flexible access to *syn*- and *anti*-1,3-diols based on the epoxide opening of 1,2:4,5-diepoxypentane [75] and the efficient access to *syn*-1,3-diols based on the alkylation and reductive decyanation of 4-cyano-2,2-dimethyl-1,3-dioxanes [76]. In addition, the Lewis-acid-catalyzed addition of dialkylzinc reagents and allylic and propargylic organometallics to 4-acetoxy-1,3-dioxanes gives *anti*-1,3-diols in high yields [77]. The efficiency of these protocols has been demonstrated in the remarkable total syntheses of several members of the polyene-polyol macrolides [78]. An elegant approach by Harada and Oku toward enantiomerically pure *anti*-1,3-diols, which has been extended to *anti*-1,3-polyols, relies on the deracemization of *syn*-1,3-diols [79]. Brückner et al. introduced butyrolactones which can be transformed to *syn*-diols and -polyols and the usefulness of this approach was demonstrated in the total syntheses of *Tolypothrix* pentamethyl ethers [80]. Additionally, they developed a

1.1.7 Asymmetric Synthesis of 1,3-Diols and anti-1,3-Polyols

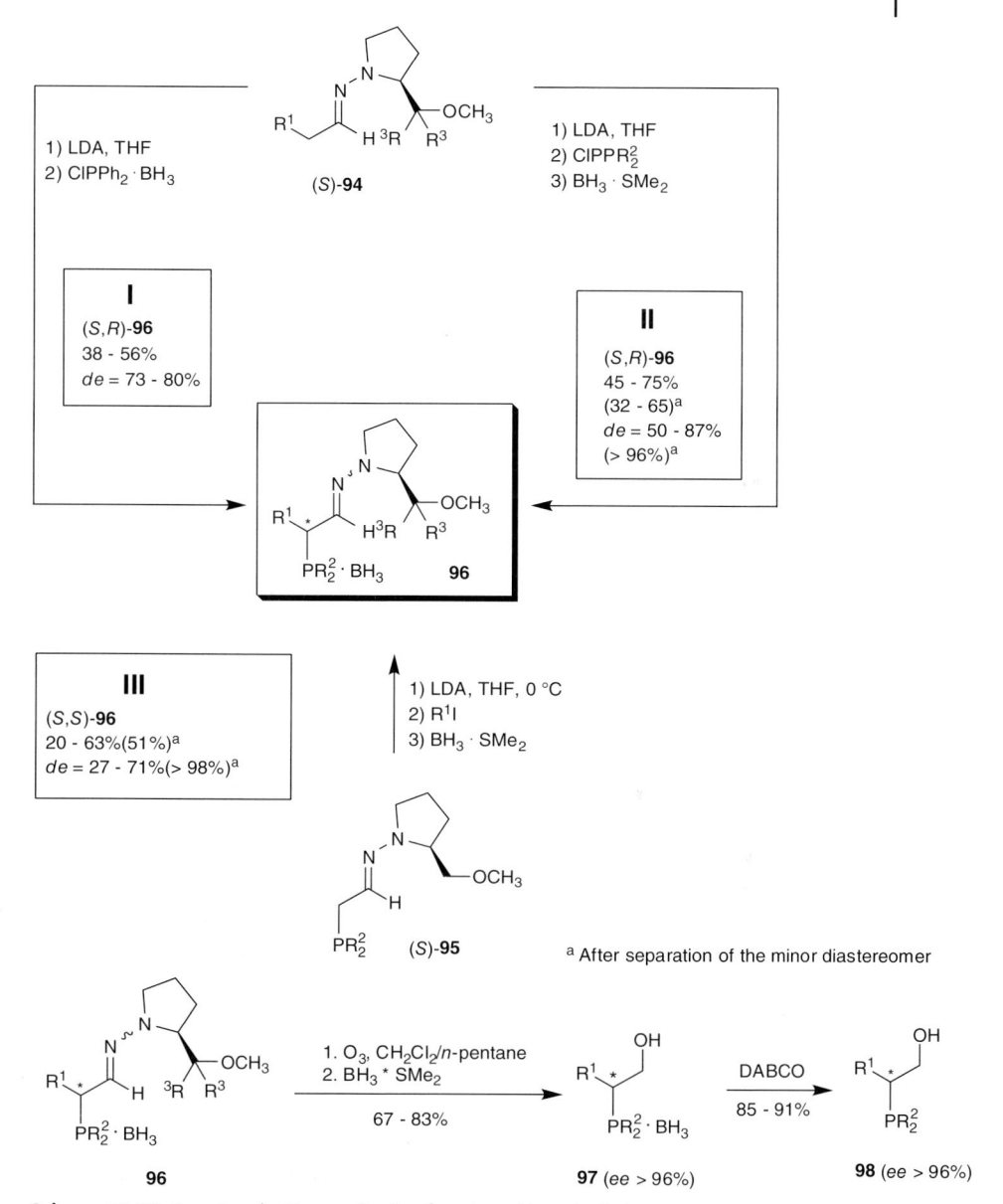

Scheme 1.1.27 Enantioselective synthesis of α-phosphino alcohols.

β-epoxy ketone building block, which can be transformed to either *syn-* or *anti-*1,3-diols depending on the epoxide opening and reduction sequence. The substitution pattern of these 1,3-diols allows a subsequent transformation after protecting-group manipulations [81].

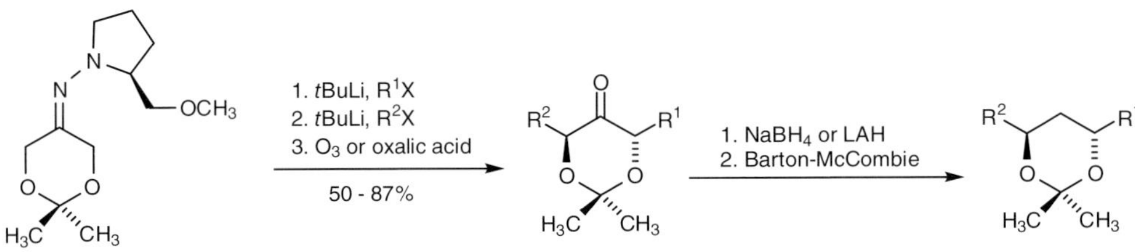

Scheme 1.1.28 Asymmetric synthesis of *anti*-1,3-diols employing the SAMP hydrazone method.

R¹,R² = Et, *i*Pr, Bn, CH$_2$(CH$_2$)TMS
R¹ = Me, R² = CH$_2$OBn; R¹ = Bn, R² = CH$_2$OBn

overall yield: 31 - 69%
($de \geq 98\%$, ee = 92 - 98%)

In the late 1980s we reported an efficient method for the asymmetric α-alkylation of 2,2-dimethyl-1,3-dioxan-5-one via its corresponding SAMP hydrazone [82]. This was later extended to many different classes of electrophiles [83], as for instance in the diastereo- and enantioselective α,α'-bisalkylation of **99** [84], demonstrating that these dialkylations occur with practically complete diastereo- and enantioselectivities to afford, after cleavage of the auxiliary 4,6-disubstituted 2,2-dimethyl-1,3-dioxan-5-ones **100** with the 4,6-substituents in a *trans* relationship. By reduction of the carbonyl group and subsequent deoxygenation of the hydroxy group according to Barton–McCombie a highly stereoselective method ($de \geq 98\%$, ee = 92–98%) for the synthesis of *anti*-1,3-diols **101** bearing a broad range of substitutents in good overall yield was developed [85] (Scheme 1.1.28).

Based on the results described above, an iterative asymmetric synthesis of protected *anti*-1,3-polyols was developed using the SAMP hydrazone of 2,2-dimethyl-1,3-dioxan-5-one **99** (Scheme 1.1.29) [86]. By employing the previously described reaction sequence, the protected diol **102** was synthesized in six steps in 69% yield. The use of BOMCl in the second alkylation allowed the conversion of the benzyl ether into the corresponding alcohol with calcium in liquid ammonia. A two-step conversion into the iodide via nucleophilic displacement of the corresponding nosylate gave **103**, which in turn could be used as a new electrophile in the iterative alkylation procedure. Sequential α,α'-bisalkylation of **99** with the iodide **103** and BOMCl led to the dioxanone **104** in 55% yield. Subsequent deoxygenation and conversion into the corresponding iodide gave **105**. Repetition of the reaction sequence with this electrophile finally afforded the protected all-*anti*-configured polyol **107** (*de, ee* > 96%).

1.1.8
Asymmetric Synthesis of α-Substituted Sulfonamides and Sulfonates

Sulfonamides are known to be versatile antibiotics since the discovery of the antibacterial activity of streptozone and sulfachrysoidine by Domagk et al. [87]. Among the very few efficient methods reported for their asymmetric synthesis,

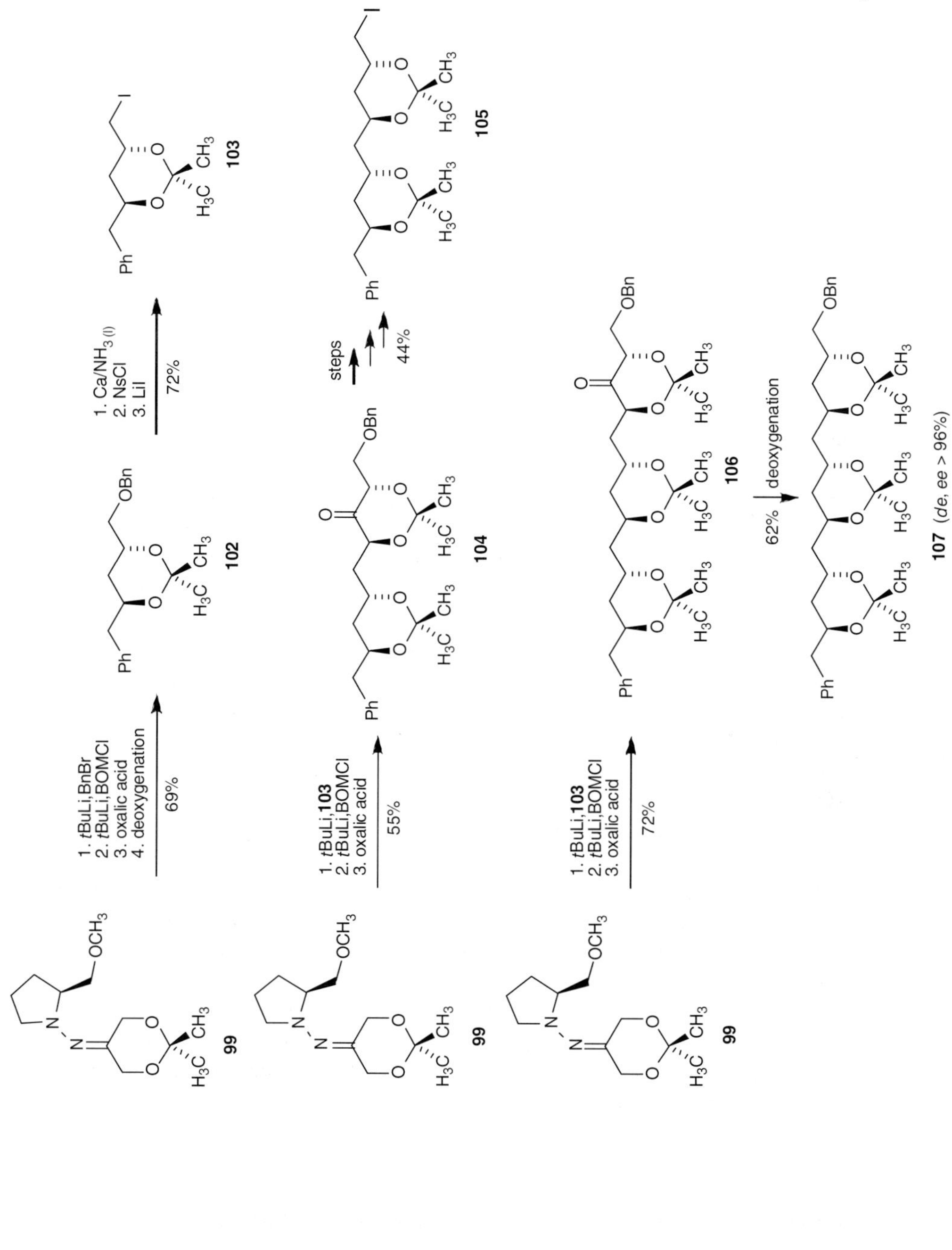

Scheme 1.1.29 Iterative asymmetric synthesis of protected *anti*-1,3-polyols.

Davis et al. [88] described the asymmetric synthesis of α-substituted primary sulfonamides involving the diastereoselective α-alkylation of N-sulfonylcamphorimine dianions, while Huart and Ghosez reported an enantioselective synthesis of bicyclic cyclopentenones via a stereoselective 1,4-addition of metallated enantiopure sulfonamides to cyclic enones [89].

In view of this background, we developed a new chiral auxiliary to allow for the first time the efficient asymmetric α-alkylation of sulfonamides [90]. After testing some amine auxiliaries mainly based on proline, which did not show high diastereoselectivities, we synthesized the 4-biphenyl-substituted 2,2-dimethyl-1,3-dioxan-5-amine **108** as a new auxiliary. The racemate obtained according to Erlenmeyer's phenylserine synthesis was resolved with tartaric acid to give both enantiomers.

(S,S)-**108**   (R,R)-**108**

With the new auxiliary in hand, the corresponding N-methylated tertiary sulfonamide **110** could be synthesized via the secondary sulfonamides **109** in very good yields. After metallation with BuLi/HMPA, the subsequent alkylation with aliphatic electrophiles R^2X (X = Br, I) gave the α-substituted sulfonamides **111** in good yields (67–86%) and high diastereomeric excesses (de = 83–94%). A major parameter that determines these relatively high asymmetric inductions is the steric hindrance of the electrophile attack by the "biphenyl wall" attached to the rigid dioxanone core. Diastereomerically pure sulfonamides (de > 98%) could be obtained by preparative HPLC purification. Finally, the efficient racemization-free cleavage of the auxiliary could be achieved using conc. hydrochloric acid (method A in the case of aromatic substituents) or conc. sulfuric acid in refluxing $CHCl_3$ (method B for nonaromatic starting material) (Scheme 1.1.30).

Enantiopure α-substituted sulfonic acids are typical structural units of several natural products and have been used as important building blocks and precursors of biologically interesting compounds. In general, they are obtained from the corresponding racemates by resolution using chiral amines [91]. The asymmetric synthesis of only a very few has been reported [92]. Since no general and practical method for the auxiliary-controlled asymmetric α-alkylation of metallated sulfonates using enantiopure alcohols as auxiliaries has been described, we decided to do so. Since enantiopure amines as chiral auxiliaries and these tertiary sulfonamides do not provide a solution (see Chapter 1.1.7) due to the harsh conditions required for their cleavage, potentially suitable alcohols have been investigated as chiral auxiliaries. Finally, the sugar derivative 1,2:5,6-di-O-isopropyliden-α-D-allofuranose **113** turned out to give the best results. Treatment of this allofuranose derivative with different sulfonyl chlorides provided access to various sulfonates

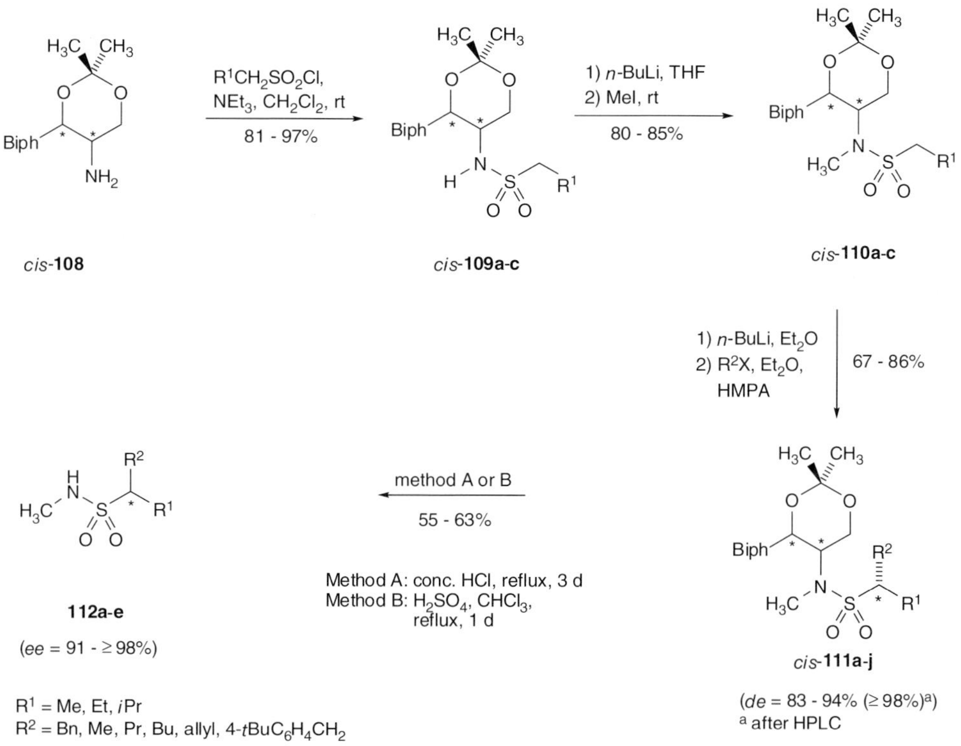

1.1.8 Asymmetric Synthesis of α-Substituted Sulfonamides and Sulfonates | 29

Scheme 1.1.30 Asymmetric synthesis of α-substituted N-methylsulfonamides.

R¹ = Me, Et, iPr
R² = Bn, Me, Pr, Bu, allyl, 4-tBuC$_6$H$_4$CH$_2$

114, which could be metallated. Subsequent reaction with various electropiles led to the α-substituted sulfonates **115** in excellent yields (93–98%) and high diastereoselectivities (*de* = 89–91%). In all cases, the diastereomerically pure sulfonates could be obtained by recrystallization (*de* > 98%). In order to cleave the chiral auxiliary, mild conditions had to be chosen. By employing a protocol reported by Lipshutz et al. [93] utilizing Pd(OAc)$_2$ in refluxing EtOH/H$_2$O, the sulfonic acids could be obtained. Alternatively, the removal of the auxiliary can be carried out with 2 mol% trifluoroacetic acid (TFA) in ethanol. In order to isolate the final products in a more accessible form, they were directly converted with diazomethane into the corresponding methyl sulfonates **116** in very good yields (77–98%) and as pure stereoisomers (*ee* > 98%). Finally, some of the methyl sulfonates could be converted to the corresponding α-alkylated sodium salts **117** without loss of enantiomeric purity (Scheme 1.1.31) [94].

Based on this gathered experience the diastereoselective alkylation of enantiopure α-lithiated sulfonates was extended to the Michael addition with aliphatic nitroolefins [95]. Thus the Michael adducts **118** could be achieved in excellent yields (84–99%) with high diastereoselectivities *de* of 80–88% (84 to >98% after recrystallization or chromatography). Cleavage of the chiral auxiliary and treatment with diazomethane furnished the *anti*-configured α,β-disubstituted γ-nitromethyl sulfonates **119** in overall yields of 41–70% and with excellent *de*- and *ee*-values (Scheme 1.1.32).

Scheme 1.1.31 Asymmetric synthesis of α-substituted sulfonates.

Scheme 1.1.32 Asymmetric 1,4-addition of lithiated sulfonates to nitroolefins.

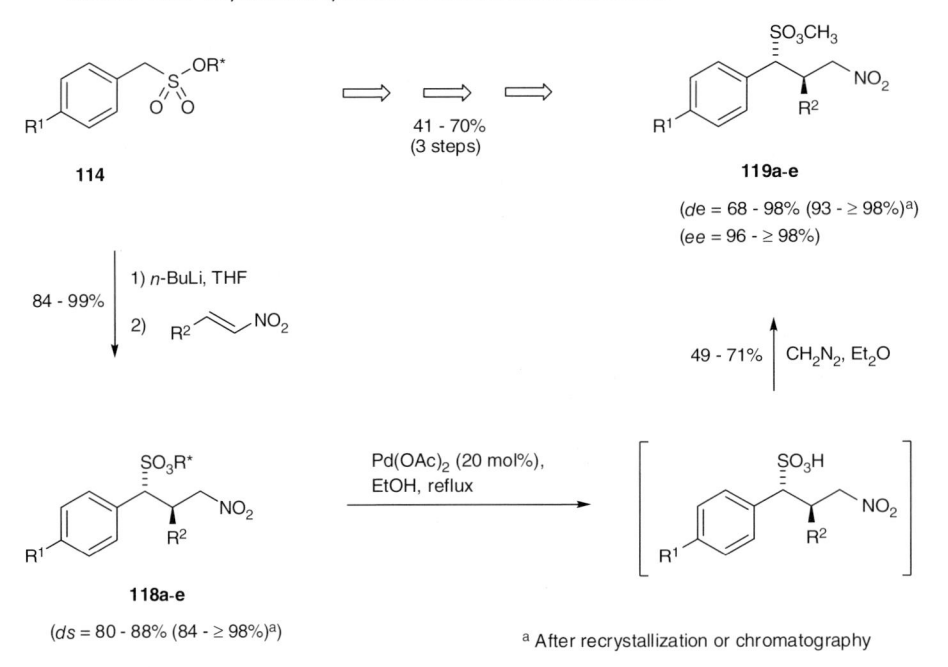

1.1.8 Asymmetric Synthesis of α-Substituted Sulfonamides and Sulfonates

Sultones are synthetically useful heterocycles, which react with a variety of nucleophiles with carbon–oxygen bond cleavage and thus behave as sulfoalkylating agents. Since to our knowledge no chiral auxiliary-controlled method for the asymmetric synthesis of sultones had been reported, we started a project to establish an efficient entry to α,γ-substituted γ-sultones via α-allylation of lithiated sulfonates bearing the allofuranose sugar as an auxiliary. In a first test example, the α-alkylated sulfonic acid sodium salt **120a** could be converted into the enantiopure sultone **121** in 45% yield and good diastereoselectivity (Scheme 1.1.33) [96].

Starting from the sulfonyl chlorides **122**, the lithiated chiral sulfonates **123** could be trapped with various alkylating reagents with high asymmetric inductions to afford the diastereomerically pure sulfonates (R)-**124** after recrystallization. The desired enantiopure γ-sultones (R,R)-**125** (de, ee ≥ 98%) were obtained by acid-catalyzed cleavage of the chiral auxiliary and subsequent diastereoselective ring closure of the intermediate sulfonic acid in one pot following a modified procedure using TFA instead of Pd(OAc)$_2$ (Scheme 1.1.34) [97].

By ring-opening of the new diastereo- and enantiomerically pure γ-sultones (R,R)-**125** a pathway to pharmacologically interesting sulfonic acid derivatives was opened proceeding via an S$_N$2 mechanism with inversion of configuration at the attacked γ-carbon atom. In this manner, we prepared the γ-alkoxy sulfonates (R,S)-**126** [98], the γ-hydroxy sulfonates (R,S)-**127** [99], and the γ-amino sulfonates (R,S)-**128** [100], all in excellent diastereo- and enantiomeric excesses of de, ee ≥ 98% (Scheme 1.1.35).

Furthermore, we have gained access to α,β-disubstituted γ-hydroxy and γ-amino sulfonate derivatives starting from the γ-nitro sulfonates **129** obtained via asymmetric Michael addition to nitroalkenes (Scheme 1.1.36) [101]. Cleavage of the chiral auxiliary with simultaneous conversion of the primary nitro group to the corresponding ester (R,R)-**130c** in a Meyer reaction was accomplished in 78% yield by refluxing the substrates in 2% TFA/EtOH. A chemoselective reduction of the ester group leaving the methyl sulfonate intact was achieved by treatment with DIBAL at low temperatures, leading to the desired α,β-substituted γ-hydroxy methylsulfonate (R,R)-**131**. The synthesis of α,β-substituted γ-amino sulfonates was carried out by reduction of the Michael adduct **129** with sodium borohydride

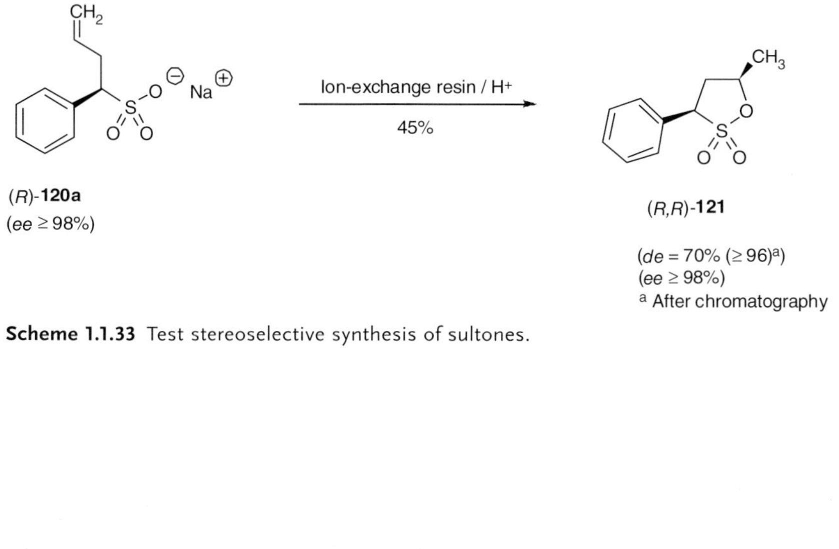

(R)-**120a**
(ee ≥ 98%)

Ion-exchange resin / H+
45%

(R,R)-**121**
(de = 70% (≥ 96)^a)
(ee ≥ 98%)
^{a} After chromatography

Scheme 1.1.33 Test stereoselective synthesis of sultones.

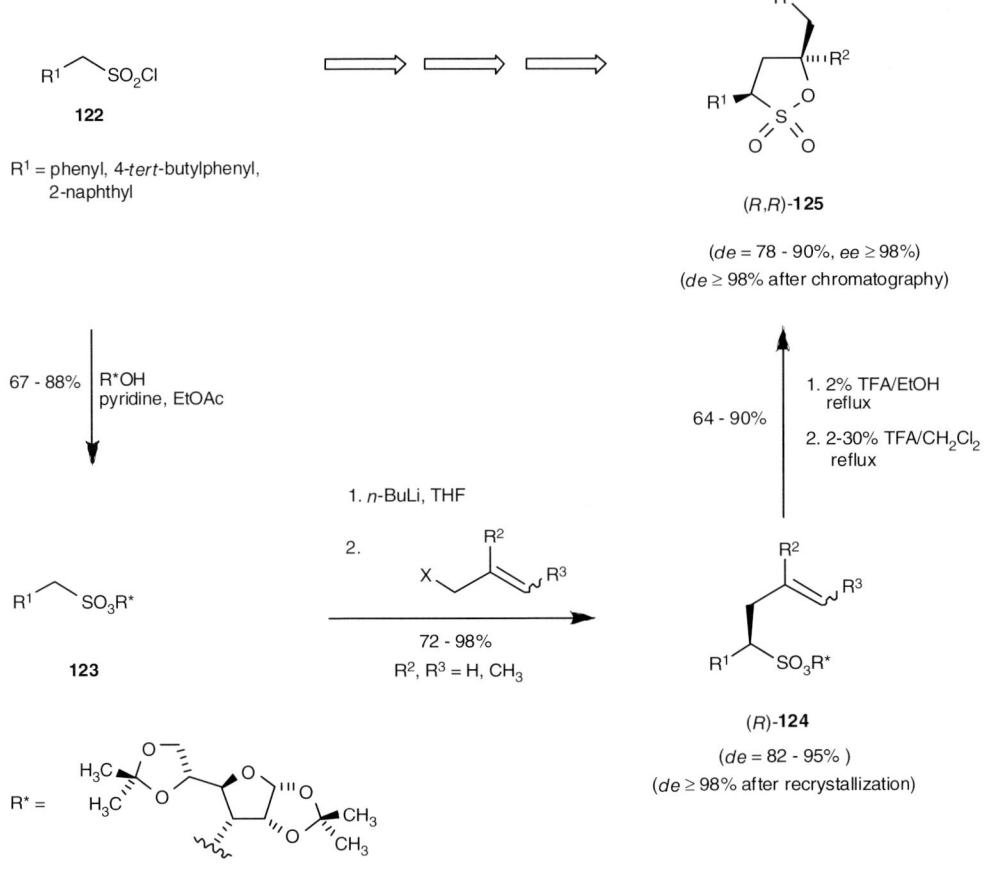

Scheme 1.1.34 Asymmetric synthesis of enantiopure γ-sultones.

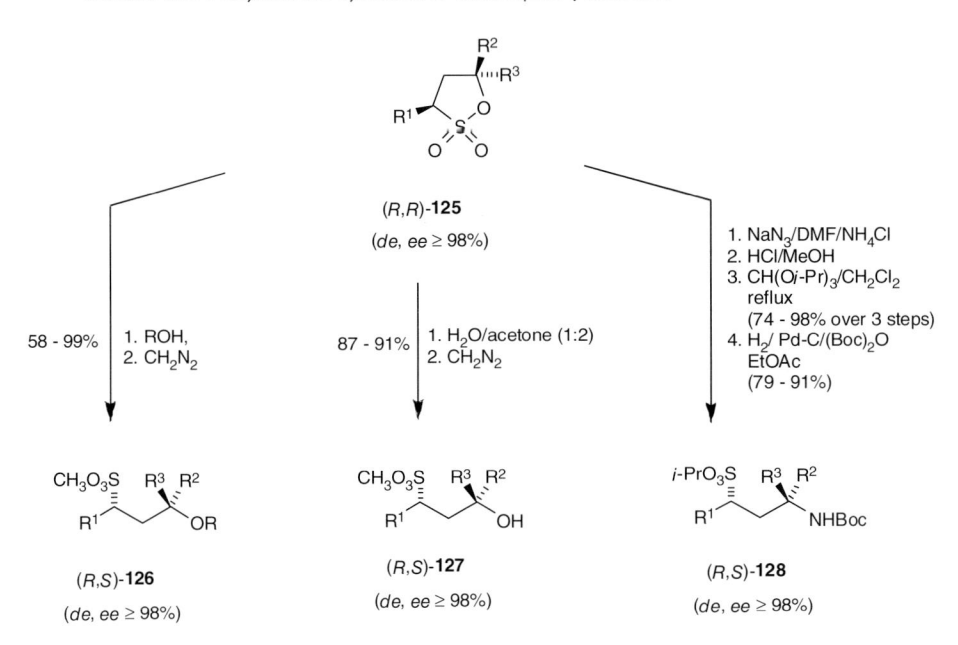

Scheme 1.1.35 Highly stereoselective synthesis of γ-alkoxysulfonates, γ-hyroxysulfonates, and γ-aminosulfonates.

R^1 = phenyl, 4-*tert*-butylphenyl, 2-naphthyl; R^2, R^3 = H, Me

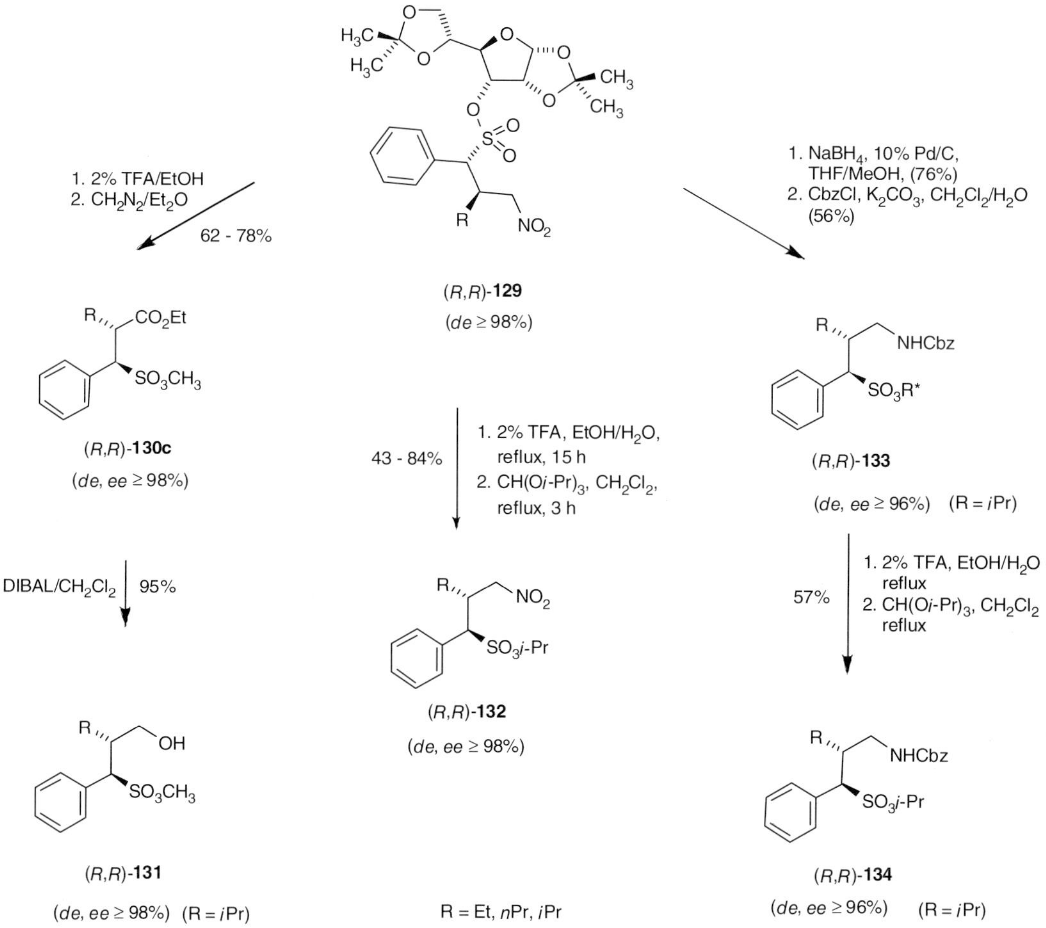

Scheme 1.1.36 Asymmetric synthesis of γ-hydroxy, γ-nitro, and γ-amino sulfonates.

in the presence of 10% Pd–C. After protection of the resulting amine, the N-Cbz-protected γ-amino sulfonate **133** was converted to the auxiliary free γ-amino sulfonate (R,R)-**134** (de, ee ≥ 96%). In addition, we demonstrated again that the sugar auxiliary of **129** can be removed with TFA under mild conditions to afford the γ-nitro sulfonates **132**.

In summary, with the novel reactions described, the door was opened to a rich new chemistry of optically active sulfonamides, sulfonates, and sultones.

References

1 D. Enders, R. W. Hoffmann, Asymmetrische Synthese, in *Chemie in unserer Zeit* **1985**, *19*, 177.
2 A. I. Meyers, *J. Org. Chem.* **2005**, *70*, 6137.
3 Review: A. Job, C. F. Janeck, W. Bettray, R. Peters, D. Enders, *Tetrahedron* **2002**, *58*, 225.
4 D. Enders, J. Adam, D. Klein, T. Otten, *Synlett* **2000**, 1371.

5 a) K. Mihami, Y. Itoh, M. Yamanaka, *Chem. Rev.* **2004**, *104*, 1; b) G. Sankar Lal, G. P. Pez, R. G. Syvret, *Chem. Rev.* **1996**, *96*, 1737.

6 a) D. Enders, M. Potthoff, G. Raabe, J. Runsink, *Angew. Chem.* **1997**, *109*, 2454; *Angew. Chem. Int. Ed. Engl.* **1997**, *36*, 2362; b) D. Enders, S. Faure, M. Potthoff, J. Runsink, *Synthesis* **2001**, 2307.

7 a) D. Enders. M. R. M. Hüttl, *Synlett* **2005**, 991; b) M. Marigo, D. Fielenbach, A. Braunton, A. Kjaersgaard, K. A. Joergensen, *Angew. Chem.* **2005**, *117*, 3769; *Angew. Chem. Int. Ed.* **2005**, *44*, 3703; c) D. D. Steiner, N. Nase, C. F. Barbas III, *Angew. Chem.* **2005**, *117*, 3772; *Angew. Chem. Int. Ed.* **2005**, *44*, 3706; d) T. D. Beeson, D. W. C. MacMillan, *J. Am. Chem. Soc.* **2005**, *127*, 8826.

8 C. Mannich, W. Krösche, *Arch. Pharm* **1912**, *250*, 647.

9 R. Kober, K. Papadopoulos, W. Miltz, D. Enders, W. Steglich, H. Reuter, H. Puff, *Tetrahedron* **1985**, *41*, 1963.

10 a) D. Enders, D. Ward, J. Adam, G. Raabe, *Angew. Chem.* **1996**, *108*, 1059; *Angew. Chem. Int. Ed. Engl.* **1996**, *35*, 981; b) D. Enders, J. Adam, S. Oberbörsch, D. Ward, *Synthesis* **2002**, 2737.

11 D. Enders, S. Oberbörsch, J. Adam, *Synlett* **2000**, 644.

12 D. Enders, S. Oberbörsch, *Synlett* **2002**, 471.

13 For recent organocatalytic Mannich reactions from our group, see: a) D. Enders, C. Grondal, M. Vrettou, G. Raabe, *Angew. Chem.* **2005**, *117*, 4147; *Angew. Chem. Int. Ed.* **2005**, *44*, 4079; b) D. Enders, M. Vrettou, *Synthesis* **2006**, 2155; c) D. Enders, C. Grondal, M. Vrettou, *Synthesis* **2006**, 3597.

14 a) D. Enders, W. Bettray, G. Raabe, J. Runsink, *Synthesis* **1994**, 1322; b) D. Enders, H. Wahl, W. Bettray, *Angew. Chem.* **1995**, *107*, 527; *Angew. Chem. Int. Ed. Engl.* **1995**, *34*, 453; c) D. Enders, J. Wiedemann, W. Bettray, *Synlett* **1995**, 369; d) D. Enders, J. Wiedemann, *Synthesis* **1996**, 1443; e) D. Enders, W. Bettray, J. Schankat, J. Wiedemann, in *Enantioselective Synthesis of β-Amino Acids* (Ed.: E. Juaristi), Wiley-VCH, New York **1997**, p. 187.

15 a) D. Enders, S. F. Müller, G. Raabe, *Angew. Chem.* **1999**, *111*, 212; *Angew. Chem. Int. Ed.* **1999**, *38*, 195; b) D. Enders, S. F. Müller, G. Raabe, *Synlett* **1999**, 741; c) D. Enders, S. F. Müller, G. Raabe, J. Runsink, *Eur. J. Org. Chem.* **2000**, 879; d) D. Enders, S. Wallert, *Synlett* **2002**, 304.

16 D. Seebach, A. K. Beck, D. J. Bierbaum, *Chem. Biodiversity* **2004**, *1*, 1111.

17 N. Sewald, *Angew. Chem.* **2003**, *115*, 5972; *Angew. Chem. Int. Ed.* **2003**, *42*, 5794, and references cited therein.

18 a) N. Asao, T. Uyehara, Y. Yamamoto *Tetrahedron* **1988**, *44*, 4173; b) T. Uyehara, N. Asao, Y. Yamamoto *J. Chem. Soc., Chem. Commun.* **1989**, 753; c) N. Asao, T. Uyehara, Y. Yamamoto *Tetrahedron* **1990**, *46*, 4563; d) Y. Yamamoto, N. Asao, T. Uyehara *J. Am. Chem. Soc.* **1992**, *114*, 5427; e) T. Uyehara, N. Shida, Y. Yamamoto *J. Org. Chem.* **1992**, *57*, 3139, 5049; f) I. Suzuki, H. Kin, Y. Yamamoto *J. Am. Chem. Soc.* **1993**, *115*, 10 139; g) N. Shida, C. Kabuto, T. Niwa, T. Ebata, Y. Yamamoto *J. Org. Chem.* **1994**, *59*, 4068.

19 F. Loydl, *Justus Liebigs Ann. Chem.* **1878**, *192*, 80.

20 T. Komnenos, *Justus Liebigs Ann. Chem.* **1883**, *218*, 145.

21 a) L. Claisen, *Justus Liebigs Ann. Chem.* **1883**, *218*, 170; b) L. Claisen, *Justus Liebigs Ann. Chem.* **1887**, *35*, 413.

22 a) A. Michael, *J. Prakt. Chem.* **1887**, *35*, 349; b) A. Michael, *Am. Chem. J.* **1887**, *9*, 112; c) A. Michael, O. Schulthess, *J. Prakt. Chem.* **1892**, *45*, 55; d) A. Michael, *Ber. Dtsch. Chem. Ges.* **1894**, *27*, 2126; e) A. Michael, *Ber. Dtsch. Chem. Ges.* **1900**, *33*, 3731.

23 a) L. Claisen, L. Crismer, *Justus Liebigs Ann. Chem.* **1883**, *218*, 129; b) C. Liebermann, *Ber. Dtsch. Chem. Ges.* **1893**, *26*, 1876; c) C. F. Koelsch, *J. Am. Chem. Soc.* **1943**, *65*, 437.

24 A. Berkessel, in *Methods of Organic Chemistry (Houben-Weyl)*, 4th ed., **1995**, Vol. E 21e, p. 4818, and references cited therein.

25 K. C. Nicolaou, C.-K. Hwang, M. E. Duggan, *J. Am. Chem. Soc.* **1989**, *111*, 6682.

26 A. G. M. Barrett, P. D. Weipert, D. Dhanak, R. K. Husa, S. A. Lebold, *J. Am. Chem. Soc.* **1991**, *113*, 9820.
27 a) A. Kamimura, N. Ono, *Tetrahedron Lett.* **1989**, *30*, 731; b) K. Hori, S. Higuchi, A. Kamimura, *J. Org. Chem.* **1990**, *55*, 5900.
28 W. Brade, A. Vasella, *Helv. Chim. Acta* **1990**, *73*, 1923.
29 a) D. Enders, A. Haertwig, G. Raabe, J. Runsink, *Angew. Chem.* **1996**, *108*, 2540–2542; *Angew. Chem. Int. Ed. Engl.* **1996**, *35*, 2388; b) D. Enders, A. Haertwig, G. Raabe, J. Runsink, *Eur. J. Org. Chem.* **1998**, 1771; c) D. Enders, A. Haertwig, J. Runsink, *Eur. J. Org. Chem.* **1998**, 1793.
30 a) Review: D. Enders, A. Saint-Dizier, M. I. Lannou, A. Lenzen, *Eur. J. Org. Chem.* **2006**, 29; b) D. Enders, L. Tedeschi, J. W. Bats, *Angew. Chem.* **2000**, *112*, 4774; *Angew. Chem. Int. Ed.* **2000**, *39*, 4605; c) L. Tedeschi, D. Enders, *Org. Lett.* **2001**, *3*, 3515.
31 a) J. G. Noltes, *Rec. Trav. Chim. Pays-Bas* **1965**, *84*, 782; b) J. Boersma, J. G. Noltes, *Rec. Trav. Chim. Pays-Bas* **1973**, *92*, 229.
32 T. Morita, Y. Okamoto, H. Sakurai, *Tetrahedron Lett.* **1978**, *28*, 2523.
33 a) D. Semenzin, G. Etemad-Moghadam, D. Albouy, O. Diallo, M. Koenig, M. *J. Org. Chem.* **1997**, *62*, 2414; b) D. Albouy, M. Laspéras, G. Etemad-Moghadam, M. Koenig, M. *Tetrahedron Lett.* **1999**, *40*, 2311.
34 a) H. Kabashima, H. Tsuji, H. Hattori, *Appl. Cat. A: General* **1997**, *165*, 319; b) B. M. Choudary, M. L. Kantam, C. V. Reddy, F. Figueras, F. *Tetrahedron* **2000**, *56*, 9357; c) K. Tanabe, W. F. Hölderich, *Appl. Cat. A: General* **1999**, *181*, 399; d) H. Hattori, *Chem. Rev.* **1995**, *95*, 537; e) Y. Ono, T. Baba, *Catalysis Today* **1997**, *38*, 321; f) A. Tungler, K. Fodor, *Catalysis Today* **1997**, *37*, 191; g) M. A. Barteau, *Chem. Rev.* **1996**, *96*, 1413.
35 D. Enders, J. P.- Shilvock, *Chem. Soc. Rev.* **2000**, *29*, 359.
36 D. Enders, J. Kirchhoff, P. Gerdes, D. Mannes, G. Raabe, J. Runsink, G. Boche, M. Marsch, H. Ahlbrecht, H. Sommer, *Eur. J. Org. Chem.* **1998**, 63, and references cited therein.
37 J. Chauffaille, E. Herbert, Z. Welvart, *J. Chem. Soc., Perkin Trans. 2* **1982**, 1645.
38 D. Enders, H. Lotter, N. Maigrot, J.-P. Mazaleyrat, Z. Welvart, *Nouv. J. Chim.* **1984**, *8*, 747.
39 a) H. M. Taylor, C. R. Hauser, *J. Am. Chem. Soc.* **1960**, *82*, 1790; b) E. Leete, *J. Org. Chem.* **1976**, *41*, 3438; c) J. D. Albright, F. J. McEvoy, *J. Org. Chem.* **1979**, *44*, 4597; d) H. Ahlbrecht, H.-M. Kompter, *Synthesis* **1983**, 645; e) M. Zervos, L. Wartski, J. Seyden-Penne, *Tetrahedron* **1986**, *42*, 4963.
40 a) D. Enders, P. Gerdes, H. Kipphardt, *Angew. Chem.* **1990**, *102*, 226; *Angew. Chem. Int. Ed. Engl.* **1990**, *29*, 179; b) D. Enders, D. Mannes, G. Raabe, *Synlett* **1992**, 837; c) D. Enders, J. Kirchhoff, D. Mannes, G. Raabe, *Synthesis* **1995**, 659; d) D. Enders, J. Kirchhoff, V. Lausberg, *Liebigs Ann.* **1996**, 1361. For recent applications, see: e) D. Enders, V. Lausberg, G. Del Signore, O. M. Berner, *Synthesis* **2002**, 515; f) D. Enders, G. Del Signore, O. M. Berner, *Chirality* **2003**, *15*, 510; g) D. Enders, M. Milovanovic, E. Voloshina, G. Raabe, J. Fleischhauer, *Eur. J. Org. Chem.* **2005**, 1984.
41 D. Enders, J. P. Shilvock, G. Raabe, *J. Chem. Soc. Perkin 1.* **1999**, 1617.
42 a) R. Noyori, *Asymmetric Catalysis in Organic Synthesis*, John Wiley & Sons, New York, 1994, p. 56; b) H. Takaya, T. Ohta, R. Noyori, in *Catalytic Asymmetric Synthesis* (Ed.: I. Ojima), VCH, New York, **1993**, Ch. 1.
43 T. F. Buckley III, H. Rapoport, *J. Am. Chem. Soc.* **1981**, *103*, 6157.
44 J. Jurczak, A. Golebiowski, *Chem. Rev.* **1989**, *89*, 149.
45 a) R. M. Jacobson, G. P. Lahm, J. W. Clader, *J. Org. Chem.* **1980**, *45*, 395; b) M. Reggelin, P. Tebben, D. Hoppe, *Tetrahedron Lett.* **1989**, *30*, 2915, 2919; c) J. M. Fang, W. C. Chon, G. H. Lee, S. M. Peng, *J. Org. Chem.* **1990**, *55*, 5515; d) S. Hünig, M. Schäfer, W. Schweeberg, *Chem. Ber.* **1993**, *126*, 205; e) C. N. Kirsten, M. Hern, T. H. Schrader, *J. Org. Chem.* **1997**, *62*, 6882.
46 S. Harada, T. Taguchi, N. Tabuchi, K. Narita, Y. Hanzawa, *Angew. Chem.* **1998**, *110*, 1796; *Angew. Chem. Int. Ed.* **1998**, *37*, 1696.

47 a) M. Braun, W. Hild, *Angew. Chem.* **1984**, *96*, 701; *Angew. Chem. Int. Ed. Engl.* **1984**, *23*, 723; b) H. Mahler, M. Braun, *Chem. Ber.* **1991**, *124*, 1379.

48 a) H. Ahlbrecht, C. Vonderheid, *Synthesis* **1975**, 512; b) H. Ahlbrecht, D. Liesching, *Synthesis* **1977**, 495.

49 C. J. Chang, J. M. Fang, L. F. Liao, *J. Org. Chem.* **1993**, *58*, 1754.

50 F. Pierre, D. Enders, *Tetrahedron Lett.* **1999**, *40*, 5301.

51 Reviews: a) H. Pielartzik, B. Irmisch-Pielartzik, T. Eicher, in *Houben-Weyl*, Vol. E5, Teil 1 (Ed.: J. Falbe), Thieme, Stuttgart, **1985**, p. 715; b) J. Mulzer, in *Comprehensive Organic Synthesis*, Vol. 6 (Eds.: B. M. Trost, I. Fleming, E. Winterfeldt), Pergamon, Oxford, **1991**, p. 323.

52 Reviews: a) J. Backes, in *Houben-Weyl*, Vol. E16, Teil 2 (Ed.: D. Klamann), Thieme, Stuttgart, **1991**, p. 31; b) L. Ghosez, J. Marchand-Brynaert, in *Comprehensive Organic Synthesis*, Vol. 5, (Eds.: B. M. Trost, I. Fleming, L. A. Paquette), Pergamon, Oxford, **1991**, p. 90.

53 Examples: a) A. I. Meyers, Y. Yamamoto, E. D. Mihelich, R. A. Bell, *J. Org. Chem.* **1980**, *45*, 2792; b) D. Enders, B. E. M. Rendenbach, *Chem. Ber.* **1987**, *120*, 1223; c) O. Zschage, D. Hoppe, *Tetrahedron* **1992**, *48*, 5657; d) Z.-M. Wang, X.-L. Zhang, K. B. Sharpless, S. C. Sinha, A. Sinha-Bagchi, E. Keinan, *Tetrahedron Lett.* **1992**, *33*, 6407.

54 T. Honda, N. Kimura, M. Tsubuki, *Tetrahedron: Asymmetry* **1993**, *4*, 1475.

55 C. Bolm, G. Schlinghoff, K. Weickardt, *Angew. Chem.* **1994**, *106*, 1944.

56 U. Gerlach, T. Haubenreich, S. Hünig, *Chem. Ber.* **1994**, *127*, 1981.

57 K. Nishide, A. Aramanta, T. Kamanaka, T. Inoue, M. Node, *Tetrahedron* **1994**, *50*, 8337.

58 C. W. Lee, H. Alper, *J. Org. Chem.* **1995**, *60*, 499.

59 a) I. Banssanne, C. Travers, J. Royer, *Tetrahedron: Asymmetry* **1998**, *9*, 797; b) T. Varea, M. Dufour, L. Miconin, C. Riche, A. Chiaroni, J.-C. Quirion, H.-P. Husson, *Tetrahedron Lett.* **1995**, *36*, 1035; c) L. Miconin, T. Varea, C. Riche, A. Chiaroni, J.-C. Quirion, H.-P. Husson, *Tetrahedron Lett.* **1994**, *35*, 2529.

60 J.-L. Matsuo, S. Kobayashi, K. Koga, *Tetrahedron Lett.* **1998**, *39*, 9723.

61 Y. L. Chung, D. Zhao, D. L. Hughes, J. M. McNamara, E. J. J. Grabowski, P. J. Reider, *Tetrahedron Lett.* **1995**, *36*, 7319.

62 B. Coates, P. J. Montgomery, P. J. Stevenson, *Tetrahedron* **1994**, *50*, 4025.

63 a) A. I. Meyers, G. P. Brengel, *Chem. Commun.* **1997**, 1; b) D. Romo, A. I. Meyers, *Tetrahedron* **1991**, *47*, 9503.

64 D. Enders, S. Brauer-Scheib, P. Fey, *Synthesis* **1985**, 393.

65 D. Enders, R. Gröbner, J. Runsink, *Synthesis* **1995**, 947.

66 a) D. Enders, R. Gröbner, G. Raabe, J. Runsink, *Synthesis* **1996**, 941; b) D. Enders, P. Teschner, G. Raabe, *Heterocycles* **2000**, *52*, 733; c) D. Enders, P. Teschner, G. Raabe, *Synlett* **2000**, 637; d) D. Enders, P. Teschner, G. Raabe, J. Runsink, *Eur. J. Org. Chem.* **2001**, 4463.

67 a) D. Enders, C. F. Janeck, *Synlett* **2000**, 641; b) D. Enders, C. F. Janeck, G. Raabe, *Eur. J. Org. Chem.* **2000**, 3337.

68 a) M. Peuckert, W. Keim, *Organometallics* **1983**, *2*, 594; b) W. Keim, *Chem.-Ing.-Tech.* **1984**, *56*, 850; c) E. R. Freitas, C. R. Gum, *Chem. Eng. Prog.* **1979**, *75*, 73.

69 a) S. D. Perera, B. L. Shaw, *J. Organomet. Chem.* **1991**, *402*, 133; b) D. A. Knight, D. J. Cole-Hamilton, D. C. Cupertino, *J. Chem. Soc., Dalton Trans.* **1990**, 3051; c) H. Brunner, A. Sicheneder, *Angew. Chem.* **1988**, *100*, 730; *Angew. Chem. Int. Ed. Engl.* **1988**, *27*, 718; d) H. B. Kagan, M. Tahar, J. C. Fiaud, *Tetrahedron Lett.* **1991**, *32*, 5959; e) T. Hayashi, Y. Uozomi, *Pure Appl. Chem.* **1992**, *64*, 1911; f) Y. Uozumi, A. Tanahashi, S.-Y. Lee, T. Hayashi, *J. Org. Chem.* **1993**, *58*, 1945; g) N. Sakai, S. Mano, K. Nozaki, H. Takaya, *J. Am. Chem. Soc.* **1993**, *115*, 7033.

70 D. Enders, T. Berg, G. Raabe, J. Runsink, *Helv. Chim. Acta* **1996**, *79*, 118.

71 a) D. Enders, T. Berg, *Synlett* **1996**, 796; b) D. Enders, T. Berg, G. Raabe, J. Runsink, *Liebigs Ann./Recueil* **1997**, 345.

72 a) S. Omura, H. Tanaka, in *Macrolide Antibiotics: Chemistry, Biology and Practice* (Ed.: S. Omura), Academic Press, New York, **1984**, p. 351; b) S. D. Rychnovsky, *Chem. Rev.* **1995**, *95*, 2021; c) C. Schneider, *Angew. Chem.* **1998**, *110*, 1445; *Angew. Chem. Int. Ed.* **1998**, *37*, 1375.

73 For a review of work up to 1990 see: T. Oishi, T. Nakata, *Synthesis* **1990**, 635.

74 D. A. Evans, A. H. Hoveyda, *J. Am. Chem. Soc.* **1990**, *112*, 6447.

75 S. D. Rychnovsky, G. Griesgraber, S. Zeller, D. Skalitzky, *J. Org. Chem.* **1991**, *56*, 5161.

76 S. D. Rychnovsky, S. S. Swenson, *J. Org. Chem.* **1997**, *62*, 1333, and references cited therein.

77 a) S. D. Rychnovsky, N. A. Powell, *J. Org. Chem.* **1997**, *62*, 6460; b) N. A. Powell, S. D. Rychnovsky, *Tetrahedron Lett.* **1998**, *39*, 3103.

78 a) S. D. Rychnovsky, R. C. Hoye, *J. Am. Chem. Soc.* **1994**, *116*, 1753; b) S. D. Rychnovsky, U. R. Khire, G. Yang, *J. Am. Chem. Soc.* **1997**, *119*, 2058; c) T. I. Richardson, S. D. Rychnovsky, *J. Am. Chem. Soc.* **1997**, *119*, 123 60.

79 T. Harada, T. Shintani, A. Oku, *J. Am. Chem. Soc.* **1995**, *117*, 123 46.

80 a) M. Menges, R. Brückner, *Liebigs Ann.* **1995**, 365; b) H. Priepke, S. Weigand, R. Brückner, *Liebigs Ann./Recueil* **1997**, 1635; c) H. Priepke, R. Brückner, *Liebigs Ann./Recueil* **1997**, 1645; d) S. Weigand, R. Brückner, *Liebigs Ann./Recueil* **1997**, 1657; e) S. Allerheiligen, R. Brückner, *Liebigs Ann./Recueil* **1997**, 1667.

81 S. Weigand, R. Brückner, *Synlett* **1997**, 225.

82 D. Enders, B. Bockstiegel, *Synthesis* **1989**, 493.

83 Review: D. Enders, M. Voith, A. Lenzen, *Angew. Chem.* **2005**, *117*, 1330; *Angew. Chem. Int. Ed.* **2005**, *44*, 1304.

84 D. Enders, W. Gatzweiler, U. Jegelka, *Synthesis* **1991**, 1137.

85 D. Enders, T. Hundertmark, C. Lampe, U. Jegelka, I. Scharfbillig, *Eur. J. Org. Chem.* **1998**, 2839.

86 D. Enders, T. Hundertmark, *Tetrahedron Lett.* **1999**, *40*, 4169.

87 Reviews: a) E. Meyle, *Pharmazie in unserer Zeit* **1984**, *6*, 177; b) R. Altstaedter, *Immunologie Aktuell* **2001**, *1*, 10.

88 F. A. Davis, P. Zhou, *J. Org. Chem.* **1993**, *58*, 4890.

89 C. Huart, L. Ghosez, *Angew. Chem.* **1997**, *109*, 627; *Angew. Chem. Int. Ed. Engl.* **1997**, *36*, 634.

90 a) D. Enders, C. R. Thomas, G. Raabe, J. Runsink, *Helv. Chim. Acta* **1998**, *81*, 1329; b) D. Enders, C. R. Thomas, N. Vignola, G. Raabe, *Helv. Chim. Acta* **2002**, *85*, 3657.

91 E. B. Evans, E. E. Mabbott, E. E. Turner, *J. Chem. Soc.* **1927**, 1159.

92 a) E. J. Corey, K. A. Cimprich, *Tetrahedron Lett.* **1992**, *33*, 4099; b) R. M. Lawrence, S. A. Biller, J. K. Dickson, J. V. H. Logan, D. R. Mignin, R. B. Sulsky, J. D. DiMarco, J. Z. Gougoutas, B. D. Beyer, S. C. Taylor, S. Lan, C. P. Ciosek, T. W. Harrity, K. G. Jolibois, L. K. Kunselman, S. A. Slusarchyk, *J. Am. Chem. Soc.* **1996**, *118*, 11 668; c) M. H. H. Nkunya, B. Zwanenburg, *Recl. Trav. Chim. Pays-Bas* **1983**, *102*, 461.

93 B. H. Lipshutz, D. Pollart, J. Monforte, H. Kotsuki, *Tetrahedron Lett.* **1985**, *26*, 4584.

94 a) D. Enders, N. Vignola, O. M. Berner, J. W. Bats, *Angew. Chem.* **2002**, *114*, 116; *Angew. Chem. Int. Ed.* **2002**, *41*, 109; b) D. Enders, N. Vignola, O. M. Berner, W. Harnying, *Tetrahedron* **2005**, *61*, 3231.

95 a) D. Enders, O. M. Berner, N. Vignola, J. W. Bats, *J. Chem. Soc. Chem. Commun.* **2001**, 2498; b) D. Enders, O. M. Berner, N. Vignola, W. Harnying, *Synthesis* **2002**, 1945.

96 N. Vignola, Dissertation, RWTH Aachen University, **2002**.

97 a) D. Enders, W. Harnying, N. Vignola, *Synlett* **2002**, 1727; b) D. Enders, W. Harnying, N. Vignola, *Eur. J. Org. Chem.* **2003**, 3939.

98 D. Enders, W. Harnying, *Arkivoc* **2004**(ii), 181.

99 D. Enders, W. Harnying, G. Raabe, *Synthesis* **2004**, 590.

100 D. Enders, W. Harnying, *Synthesis* **2004**, 2910.

101 D. Enders, J.-C. Adelbrecht, W. Harnying, *Synthesis* **2005**, 2962.

1.2
Asymmetric Synthesis of Natural Products Employing the SAMP/RAMP Hydrazone Methodology
Dieter Enders and Wolfgang Bettray

1.2.1
Introduction

The asymmetric synthesis of natural products and bioactive compounds in general is a central research area of organic chemistry and necessitates the development of an ever-increasing number of practical and highly stereoselective synthetic methods. Among the more general protocols in the field of stoichiometric asymmetric synthesis, the SAMP/RAMP hydrazone methodology developed in our group [1] has been shown to be a versatile instrument to provide access to a great variety of natural and bioactive compounds. In this chapter we describe some selected examples of such asymmetric syntheses carried out within the framework of the Sonderforschungsbereich 380.

1.2.2
Stigmatellin A

Stigmatellin A, **1** (Fig. 1.2.1), is a powerful inhibitor of the electron transport chain in mitochondria and chloroplasts and was first isolated by Höfle et al. in 1983 [2], together with the geometric isomer stigmatellin B (**2**) from the gliding bacterium *Stigmatella aurantiaca* [3]. The biological activity is based on points of attack on the one hand on the cytochrome bc_1-segment of the respiratory chain [4] and on the other hand on the reducing side of photosystem II as well as the

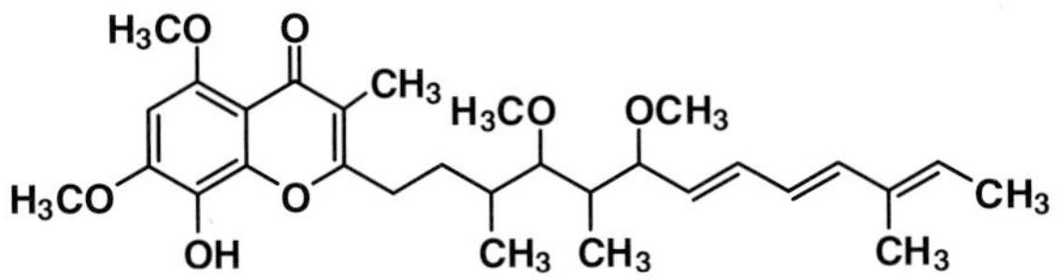

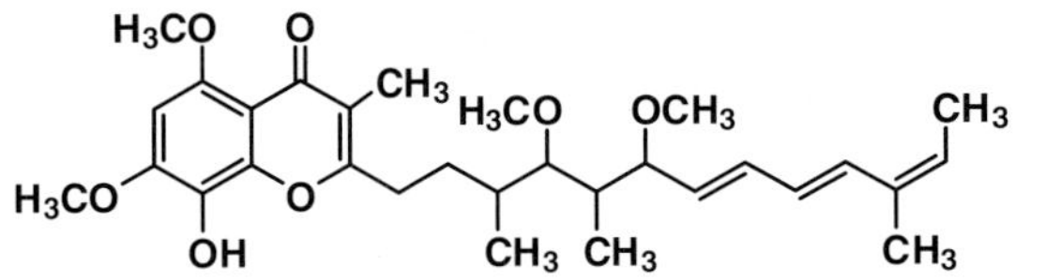

Fig. 1.2.1 Stigmatellin A and its geometric isomer stigmatellin B.

cytochrome b_6/f complex [5]. Investigations with derivatives of stigmatellin have shown the responsibility of the chromone system for the inhibition reaction [4]. It has been demonstrated that a change of either the 4-oxo or the 8-hydroxy function of the chromone part will have a considerable effect on the inhibition of the NADH oxidation in submitochondrial particles. Furthermore, it has been shown by Jagow et al. that alteration of the side chain or saturation of the C=C double bonds drastically affects the binding characteristics of stigmatellin A [6]. The structure of stigmatellin A was elucidated from ^{1}H-NMR, ^{13}C-NMR, MS, IR, UV, and CHN analysis data, although no information was obtained on the relative or absolute configuration until it was determined as (S,S,S,S) by our research group through chemical correlation by employing a combination of our SAMP/RAMP hydrazone method and the Evans *syn*-aldol protocol [7]. Taking this as a starting point, our research efforts led to the first highly diastereo- and enantioselective synthesis of stigmatellin A from achiral starting materials employing our SAMP/RAMP hydrazone methodology to create the stereogenic centers of the chain [8].

The retrosynthetic analysis of stigmatellin A (Scheme 1.2.1) led to the three subunits **A**, **B**, and **C** from which fragment **3** (**A**) had already been synthesized by Höfle et al. [2] without the protecting group on one of the hydroxyl groups, which is essential to target the right hydroxy group when coupling with the chiral fragment **B**. Using our SAMP/RAMP hydrazone methodology [1] and a *syn*-selective aldol reaction [9], access should be granted to the subunit **4** (**B**) as the most complicated building block. Since it is necessary to protect the two ends of the chain in different manners, we decided to use the benzyl group, easily removable

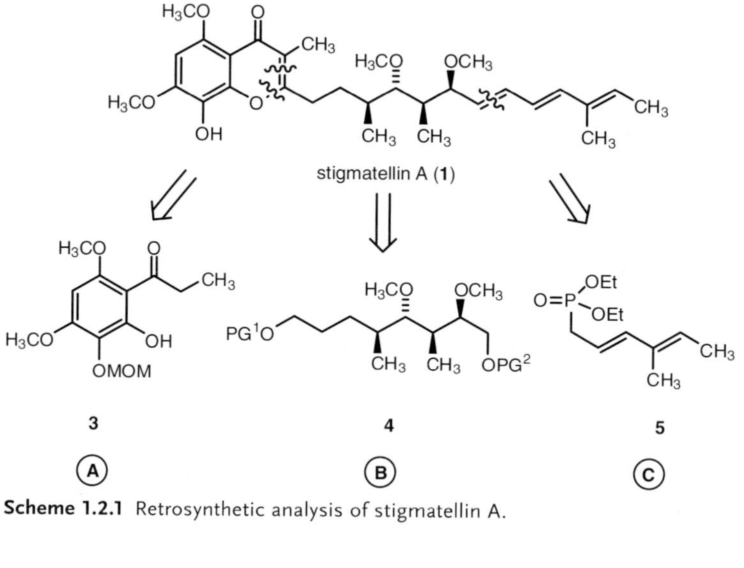

Scheme 1.2.1 Retrosynthetic analysis of stigmatellin A.

with hydrogen, on one side and the *p*-methoxyphenyl group, which should be removed with ceric ammonium nitrate (CAN), on the other. Subunit **5** (**C**) should be available starting from tiglinaldehyde and could be coupled to the chain by a Horner–Wadsworth–Emmons reaction.

Based on the work of Höfle et al., the synthesis of the aromatic ketone **9** could be accomplished by careful modification of the original reaction conditions with an improved overall yield of 63% starting from 3,5-dimethoxyphenol **6** (Scheme 1.2.2). After a first step of straightforward diacylation of 3,5-dimethoxyphenol with propionic acid in the presence of P_4O_{10} the conversion of the diketone to a monoester could be accomplished by a Baeyer–Villiger reaction with an excellent yield of 87%. Subsequent ester hydrolysis, followed finally by the protection of one hydroxyl group with chloromethyl methyl ether (MOMCl), gave the fragment **3** in virtually quantitative yield and with complete chemoselectivity. The second hydroxyl group is blocked by a hydrogen bond to the carbonyl function in the ortho position and does not react.

For the synthesis of the central portion **4** (**B**) of stigmatellin A we started with the alkylation of the (*S*)-1-amino-2-methoxymethylpyrrolidine (SAMP) hydrazone (*S*)-**13** with the iodide **12**, which was synthesized in two steps from 1-chloropropan-3-ol. The alkylation product (*S,S*)-**14** was obtained with virtually complete asymmetric induction (*de* > 98%) and was oxidatively cleaved to give the corresponding ketone [10]. The ketone was then subjected to a *syn*-selective titanium-mediated aldol reaction with benzyl-protected glycol aldehyde according to Evans et al. [11] to give the aldol adduct **15** with a *syn/anti* selectivity of 2 : 1. The envisaged aldol product (*R,R,S*)-**15** was isolated in 56% overall yield after HPLC-separation of the two isomers. *Anti*-selective reduction of the ketone group with tetramethylammonium triacetoxyborohydride according to Evans et al. [12] afforded the diol (*R,S,S,S*)-**16** in excellent yield (99%) and with complete induction

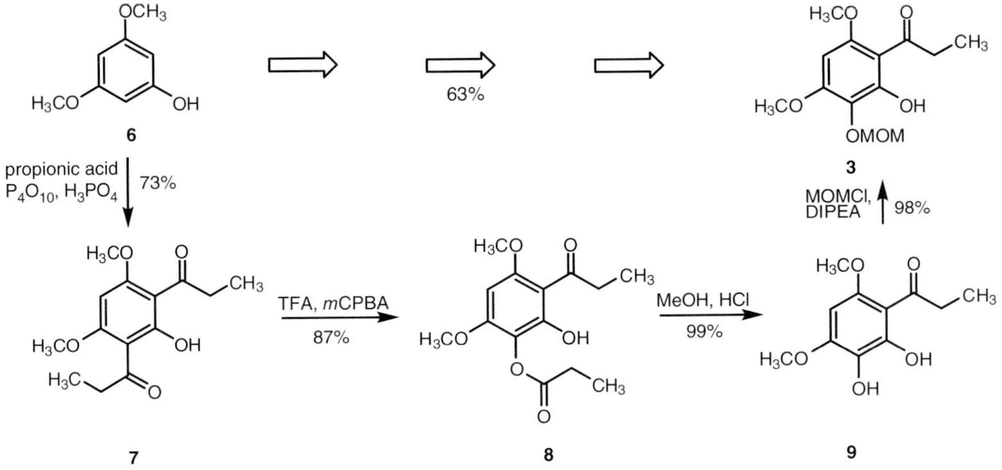

Scheme 1.2.2 Synthesis of fragment **3** (**A**).

at the newly generated stereogenic center. The conversion of the diol into the corresponding bismethyl ether was accomplished by reaction with potassium hydride/methyl iodide in the presence of 18-crown-6 [13], which gave the diastereomerically pure diether (R,S,S,S)-**17** in good yield (77%). In order to couple this fragment with the aromatic ketone **3**, the p-methoxyphenyl group was removed using ceric ammonium nitrate [14] in high yield (90%). Finally, oxidation of the hydroxy group with pyridinium dichromate (PDC) [15] gave the acid (S,S,S,R)-**19** in quantitative yield (Scheme 1.2.3).

In order to synthesize the diene subunit **5** of stigmatellin A (Scheme 1.2.4) we started from the commercially available tiglinaldehyde **20**, which was used in a Horner olefination to give the diene ester **21**. Selective reduction with diisobutylaluminum hydride (DIBAH) afforded the known alcohol **22** as a modification of a known procedure [16]. According to Corey et al. [17], the alcohol was converted into the bromide **23** by stirring with PBr$_3$ in ether at 0 °C. The phosphonate **5** was synthesized by a Michaelis–Arbuzov rearrangement [18] by heating the bromide **23** with triethylphosphite, which provided access to the product **5** in a good overall yield of 41% starting from tiglinaldehyde.

The final goal of the total synthesis of stigmatellin A was the coupling of the three building blocks **A**, **B**, and **C**. Keeping in mind the different methods described in the literature, we decided to generate the chromone systems with a long chain in the 2 or 3 position using the Baker–Venkataraman rearrangement [19]. As depicted in Scheme 1.2.5, the ester **24** was synthesized by conversion of the acid **19** in situ into the corresponding mixed anhydride with pivaloyl chloride, followed by reaction with the aromatic ketone **3** to give the product in 57% yield. The corresponding Baker–Venkataraman rearrangement of the ester gave the chromone **25** in good yield (75%). Deprotection of the primary alcohol with Pd/C under normal hydrogen pressure led to alcohol **26**, which opened access to the aldehyde **27** under modified Swern conditions. Finally, a Horner–Wadsworth–Emmons olefination gave stigmatellin A with an overall yield of 24% starting from acid **19**, and 7.4% starting from the SAMP hydrazone **13**. The synthetic stigmatellin A showed an optical rotation of $[\alpha]_D^{21} = +37.7$ ($c = 0.70$ in methanol), which is in perfect correlation with the natural product $[\alpha]_D^{21} = +38.5$ ($c = 2.3$ in methanol) [2].

1.2.3
Callistatin A

In 1997 (−)-callystatin A (Fig. 1.2.2), a potent cytotoxic polyketide, was isolated from the marine sponge *Callyspongia truncata* and structurally elucidated by Kobayashi et al. [20]; shortly afterward its absolute configuration was confirmed by the same authors by total synthesis [21]. The structure of (−)-callystatin A shows a polypropionate chain and a lactone ring connected to each other by two diene systems separated by two sp^3 carbon atoms (Fig. 1.2.2). Since this arrangement is structurally related to several antitumor antibiotics and due to the fact that only very small amounts can be isolated from natural sources, callistatin A has been

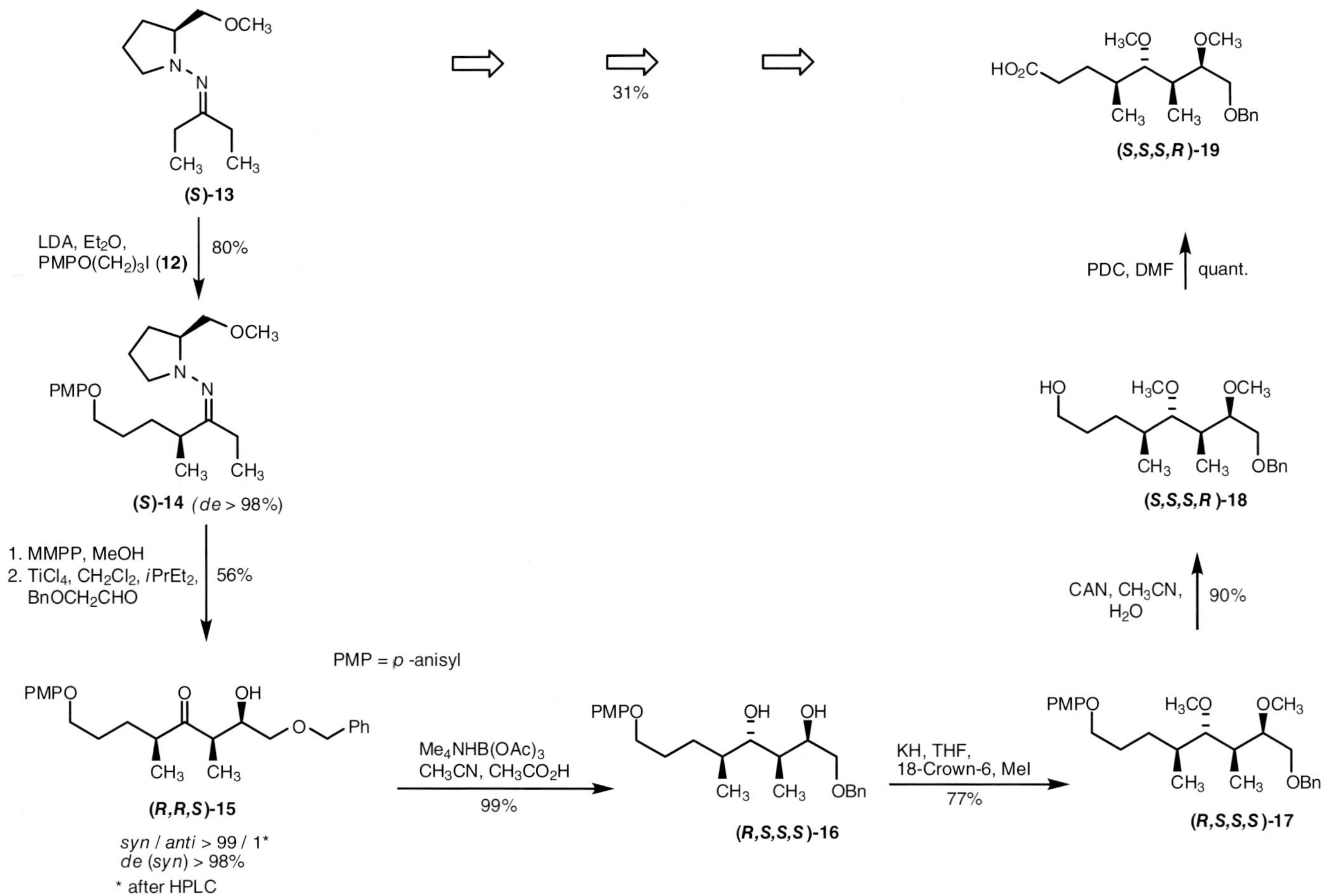

Scheme 1.2.3 Asymmetric synthesis of building block **19** (B).

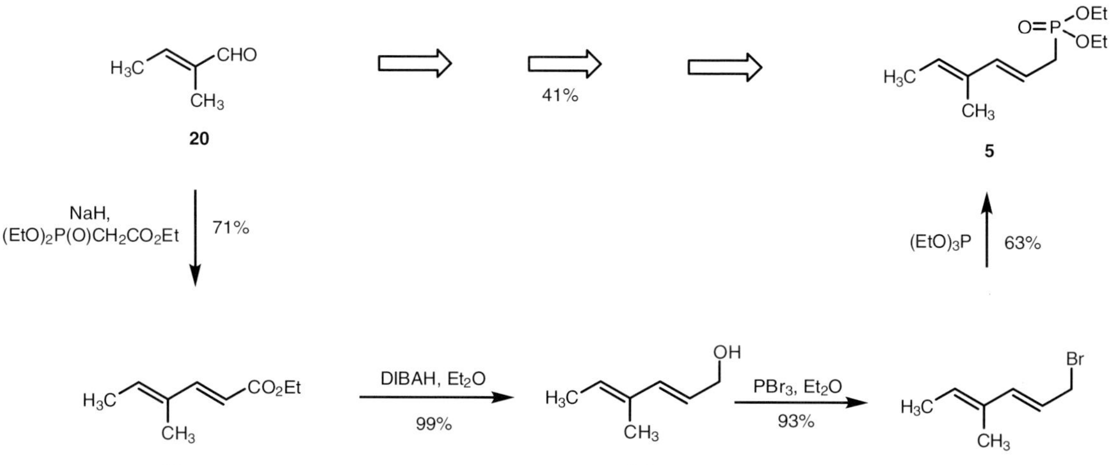

Scheme 1.2.4 Stereoselective synthesis of building block **5** (**C**).

a very attractive target for total synthesis. Just after the work of Kobayashi et al. another first new total synthesis was described by Crimmins et al. [22], who used a chiral auxiliary aldol methodology combined with an allylic Wittig reaction. In 2001 Smith et al. [23] reported the total synthesis of (−)-callystatin A using a combined Evans aldol methodology/Julia olefination approach, while Kalesse et al. [24] accomplished a total synthesis employing Heck and Wittig reactions to construct the C_1–C_{20} chain and performed a final diastereoselective aldol reaction in order to complete the synthesis. In 2002 Marshall et al. [25] reported a total synthesis of this polyketide in which the key attribute implicated the formation of the polypropionate chain by stereoselective addition of chiral allenylmetals to aldehydes. A feature shared by all the syntheses reported so far is that at least one of the stereogenic centers present in the skeleton of the natural product was introduced from a chiral-pool building block. Based on these research efforts and the challenging structure, we regarded the total synthesis of callistatin A as a unique opportunity to demonstrate the synthetic utility of our SAMP/RAMP hydrazone alkylation methodology for the introduction of several of the stereogenic centers, providing access to a flexible and convergent synthetic route to structural analogues with potentially improved biological activity [26]. Further total syntheses have been reported in the meantime by Lautens et al. [27], Kalesse et al. [28], Panek et al. [29] and Diaz et al. [30].

Retrosynthetic considerations reveal an approach (Scheme 1.2.6) in the first step based on disconnections of the C_6–C_7 and C_{12}–C_{13} double bonds. Those can be built up using highly *E*-selective Wittig olefinations between allyltributylphosphorous ylides derived from the corresponding allylic bromides **29** and **31** [31]. The aldehyde **30** is accessible from the keto ester **33**, which can be prepared in high enantiomeric purity by a biocatalytic enantioselective reduction of a

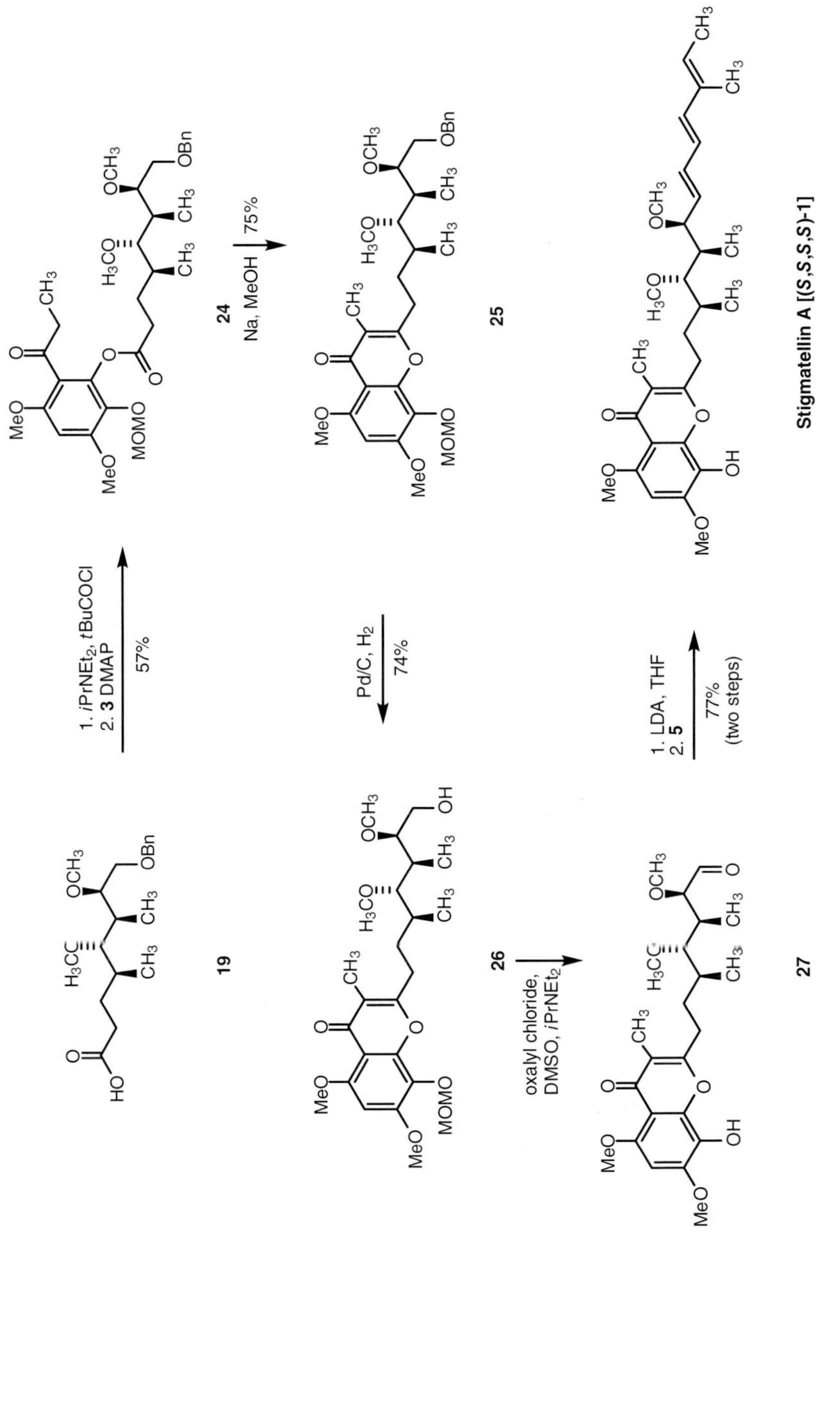

Scheme 1.2.5 Coupling of the building blocks.

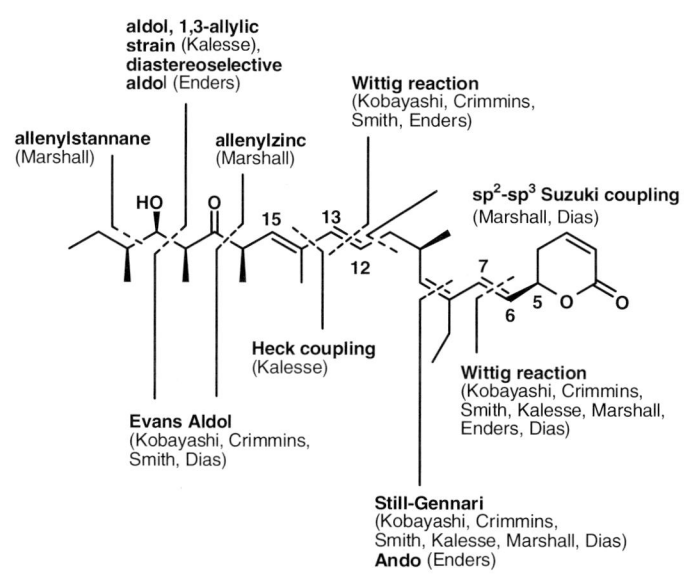

Fig. 1.2.2 Callistatin A and methods used for its total synthesis.

6-chloro-3,5-dioxohexanoate using methods developed in the course of our collaborative research [32]. A selective olefination of the functionalized aldehyde **34**, which can be prepared by asymmetric α-alkylation of the corresponding SAMP hydrazone, should open access to the bromide **31**. Finally, bromide **29** can be synthesized from the triol **32**, which should be obtained by means of a *syn*-selective aldol reaction between the enolate derived from the ketone **35** and the aldehyde **36**. The latter compounds can also be obtained as virtually single enantiomers by standard SAMP/RAMP hydrazone alkylations.

Based on our aim to use biological as well as chemical methods to synthesize bioactive compounds, we exploited the enantioselective biocatalytic reduction of 3,5-dioxocarboxylates in order to generate the stereogenic center in the key intermediate **30** (Scheme 1.2.7). We used the enantioselective enzymatic reduction developed by Müller et al. [32] and already mentioned. Treatment of *t*-butyl-6-chloro-3,5-dioxohexanoate with dried baker's yeast in a biphasic system (water/XAD-7 adsorber resin) furnished the regio- and enantioselectively reduced hydroxy keto ester **33** in 94% enantiomeric excess. By reduction of the remaining keto group with sodium borohydride and treatment with a catalytic amount of *p*-TsOH in refluxing toluene, lactonization together with elimination provided the α,β-unsaturated δ-lactone **37** in good yield.

In order to replace the chlorine atom by the desired hydroxy functionality, diisobutylaluminum hydride (DIBAL-H) reduction followed by an acid-catalyzed acetalization with isopropanol in refluxing benzene had to be performed to protect the lactone moiety as an acetal and provided the thermodynamically preferred α-anomer **38** together with 5–8% of the other epimer. The conversion of the

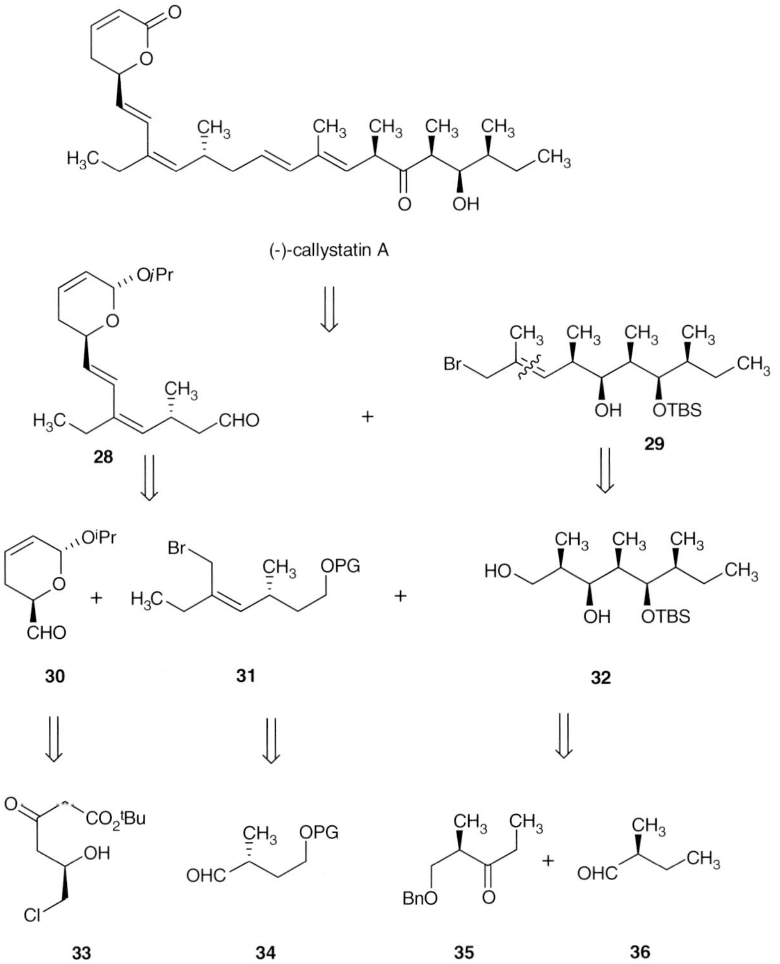

Scheme 1.2.6 Retrosynthetic analysis of (−)-callystatin A.

chlorine atom into a hydroxy group was achieved by a chloro–acetoxy substitution reaction of the protected intermediate **38** with tetrabutylammonium acetate (TBAA) in NMP, followed by saponification of the newly generated ester with potassium carbonate in MeOH. Finally, treatment of **39** under standard Swern oxidation conditions furnished the aldehyde **30**.

For the synthesis of the second required building block, the chiral allylic bromide **31**, we again used our SAMP/RAMP hydrazone methodology for the construction of the stereogenic center. The α-alkylation of the SAMP hydrazones of O-protected 4-hydroxybutanal **40a,b** with iodomethane led to the desired products **41** with very good yields and very high stereoselectivities and with the required

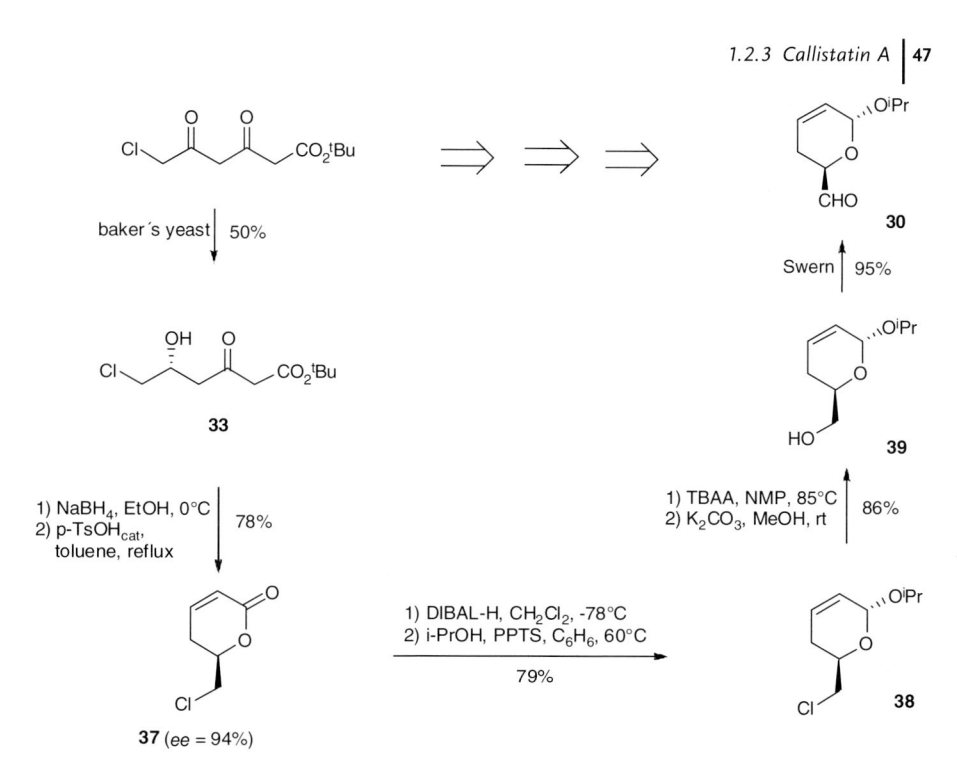

Scheme 1.2.7 Biocatalytic enantioselective synthesis of **30**.

(R) configuration at the newly created stereocenter. Removal of the auxiliary by ozonolysis provided access to the virtually enantiopure aldehydes **34a,b**. The following Z-selective generation of the double bond in the α,β-unsaturated esters **42a,b** was accomplished by a modified Horner–Wadsworth–Emmons reaction in good yields and with excellent diastereoselectivities (Z/E ratio = 34 : 1). Consecutive reduction afforded the corresponding allylic alcohols cleanly. At this point the enantiomeric excess could be determined accurately by GC analysis to be >98% for each compound. Finally, bromination employing CBr_4/PPh_3 in acetonitrile led to the allylic bromides **31a,b** (Scheme 1.2.8).

For the synthesis of the polypropionate fragment **29** we first had to provide the chiral ketone and aldehyde building blocks **35** and **36**. Here again our SAMP/RAMP hydrazone methodology turned out to be the best and most flexible way. The asymmetric α-alkylation of 3-pentanone via its RAMP hydrazone derivative **44** with benzyloxymethyl chloride (BOMCl) and subsequent cleavage of the auxiliary afforded ketone **35** in 96% *ee* and good yield (Scheme 1.2.9) [33]. In a similar approach, we prepared (S)- and (R)-2-methylbutanal **36** and *ent*-**36** by α-alkylation of butanal-RAMP or butanal-SAMP hydrazones **43** or *ent*-**43** with iodomethane, respectively. In order to generate the two new stereogenic centers a *syn*-selective aldol reaction between **35** and **36** was performed. With a Sn(II) mediated aldol

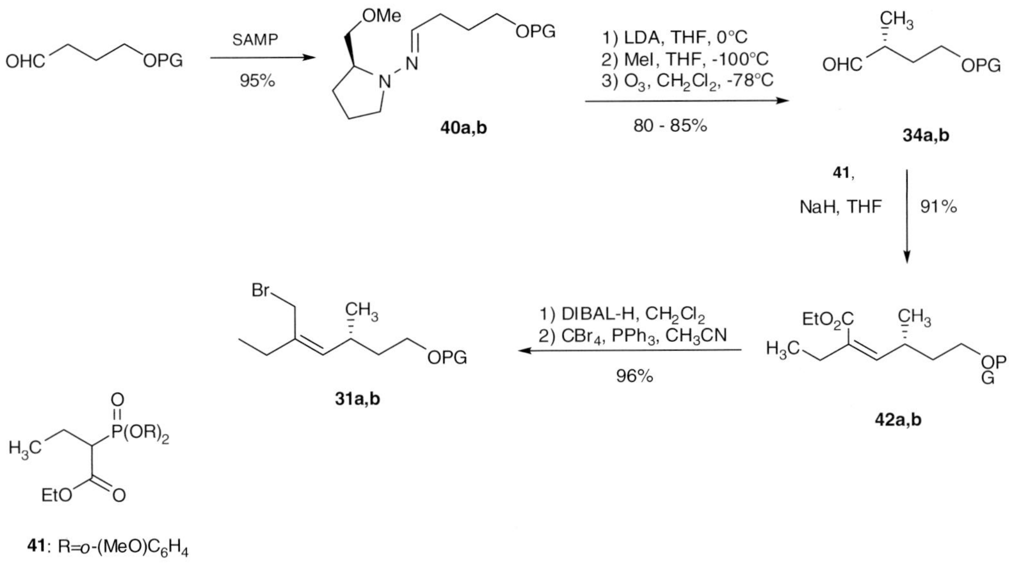

48 | *1.2 Natural Products Employing the SAMP/RAMP Hydrazone Methodology*

41: R=*o*-(MeO)C$_6$H$_4$

a: PG = TBS
b: PG = TBDPS

Scheme 1.2.8 Asymmetric synthesis of **31**.

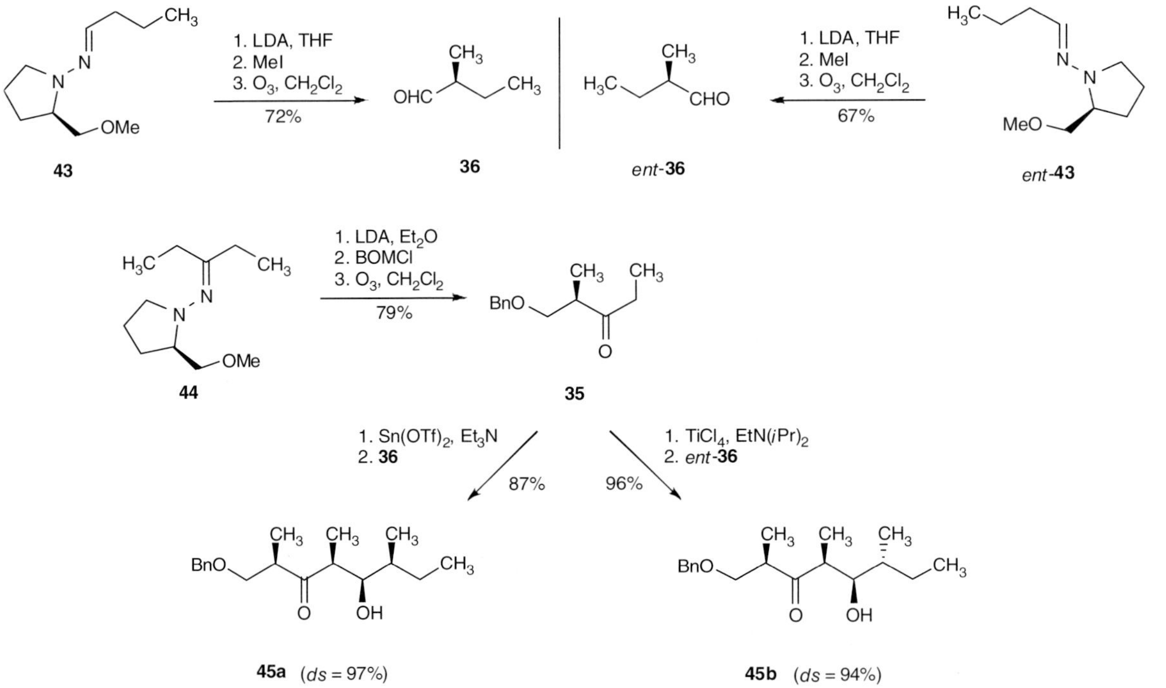

Scheme 1.2.9 Asymmetric synthesis of **45**.

reaction the product **45a** was obtained in excellent yield and 97% *ds*, whereas employing a titanium-mediated aldol reaction of **35** with aldehyde *ent*-**36** provided access to the aldol product **45b** in good yield and 94% *ds*. This matched/mismatched behavior of the substrates in the aldol reaction enabled us to access the natural product employing the adduct **45a** and its C-20 epimer using **45b**, which also constitutes the correct configuration of the ebelactone A side chain.

Continuing our total synthesis, the TBS-protected aldol adducts **45a,b** were subsequently debenzylated with hydrogen over Pd/C affording the β-hydroxy ketones **46a,b** with an enantiomeric excess $ee > 96\%$ (Scheme 1.2.10). Reduction with DIBAL-H in CH_2Cl_2 at low temperature yielded the stereopentads **32a,b** with excellent diastereoselectivities ($de = 91\%$ and 86% for **32a** and **32b**, respectively). Subsequent Swern oxidation of the primary hydroxy group followed by a Wittig reaction with $Ph_3P=C(CH_3)CO_2Et$ afforded the *E* isomers of α,β-unsaturated esters **47a,b** as the only detectable products. Because ester **47a** is a known compound we could confirm the proposed absolute configuration of this building block by comparison with the data reported in the literature [34]. Finally, DIBAL-H reduction of the ester moiety and selective bromination of the primary alcohol with CBr_4/PPh_3 furnished the desired polypropionate fragment **29a,b**.

In order to build up the skeleton of the target compounds, the allylic bromide **31b** was converted into the corresponding tributylphosphonium salt by treatment with PBu_3 in acetonitrile (Scheme 1.2.11). The Wittig reaction of the salt with

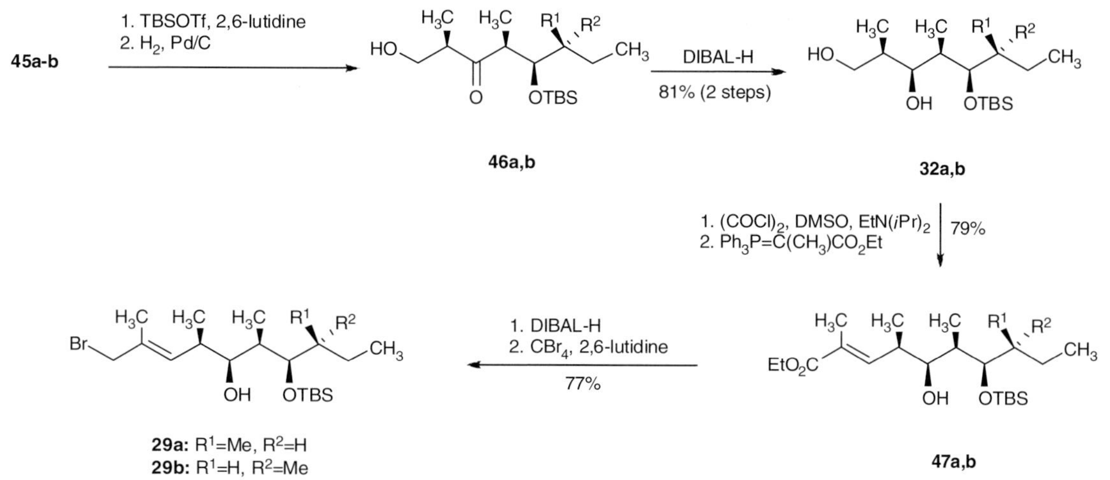

Scheme 1.2.10 Synthesis of the polypropionate fragment **29**.

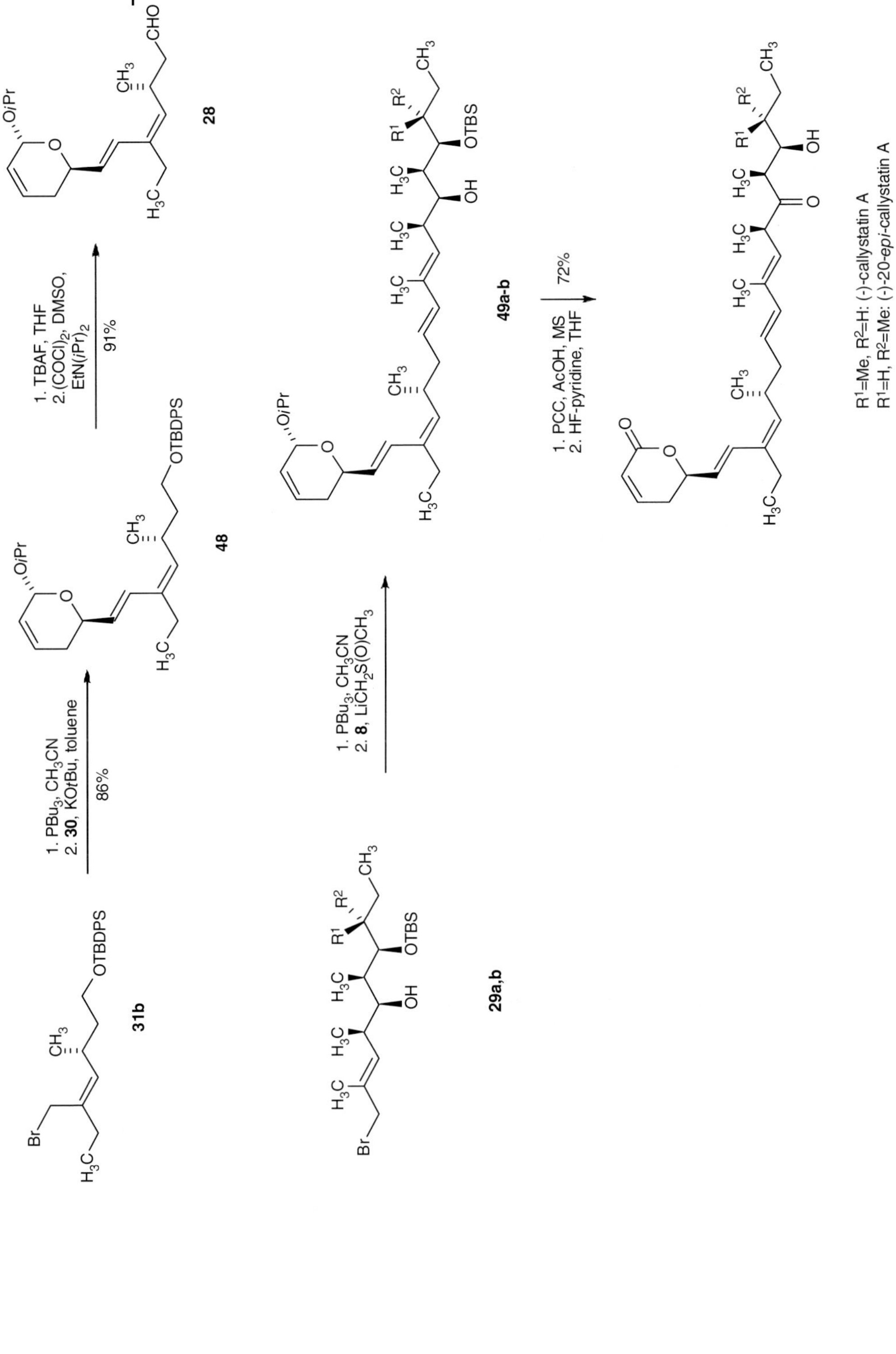

Scheme 1.2.11 Asymmetric synthesis of (−)-callystatin A and (−)-20-*epi*-callystatin A.

aldehyde **30** in the presence of KO*t*Bu afforded the left-hand side **48** of the target molecule in good yield and as a single *E* isomer with respect to the newly formed double bond. Deprotection of the *t*-butyldiphenylsilyl (TBDPS) ether followed by Swern oxidation yielded aldehyde **28** ready for the next Wittig coupling reaction with the polypropionate fragments **29a,b**. Thus using $LiCH_2S(O)CH_3$ as base the allyltributylphosphonium salts derived from **29a** and **29b** allowed the coupling with the aldehyde **28** to provide the pentaenes **49a,b** in good yield and again as single *E* isomers with respect to the newly generated double bond. The treatment of these pentaenes with PCC led to oxidation of the lactol moiety and the unprotected hydroxy group in the C_1 and C_{17} position, respectively. Finally the TBS ether was deprotected with HF–pyridine complex in THF to obtain the natural (−)-callystatin A and its 20-*epi* analogue.

1.2.4
Dehydroiridodiol(dial) and Neonepetalactone

In the cyclopentanoid monoterpene family, compounds containing an iridoid-structure exhibit various bioactivities in Nature. For example, dehydroiridodiol, isolated from dry leaves of the cat-attracting plant *Actinidia polygama* Miq., is known to be an attractant for the male adults of the Chrysopidae and shows activity in amounts as small as 10^{-4} μg [35]. Dehydroiridodial, a more oxidized product, was isolated as a pungent principle of *Actinidia polygama* Miq. and was characterized by T. Sakan et al. in 1978 [36].

Besides the synthesis of racemic dehydroiridodiol [37], some ex-chiral-pool syntheses using (*S*)-limonene have been described [38]. Dehydroiridodial was synthesized in the same manner [39]. Since the increasing number of cyclopentanoid natural products and their interesting biological activity has stimulated considerable interest in the synthesis of such compounds, we have used our methodology to provide a new asymmetric synthesis of dehydroiridodial, dehydroiridodiol, as well as analogues [40].

As shown in Scheme 1.2.12, the aldehyde or ketone SAMP hydrazones **50** were metallated using LDA to generate the desired azaenolate, and then TMEDA was added. Subsequent Michael addition with methyl-2-cyclopentenone carboxylate (**51**) resulted in a clean 1,4-addition leading to the desired adducts **52** in good yields.

The methyl group was introduced by a two-step procedure. Thus, the hydrazone Michael adducts **52** were converted into the enol pivaloates **53** in excellent yields and diastereomeric excesses (*de* > 96%) by treatment with pivaloyl chloride and triethylamine. After treatment with lithium dimethylcuprate the chiral auxiliary was removed by addition of 6N HCl in order to obtain the 5-substituted 2-methylcyclopentene carboxylate **54** in good yields and with excellent stereoselectivity (*de*, *ee* > 96%). Finally, the asymmetric synthesis of dehydroiridodiol (**55**, R^1 = Me, R^2 = H) and its analogues was accomplished by reduction of **54** with lithium aluminum hydride or L-selectride® leading to the desired products in excellent yields, diastereo- and enantiomeric excesses (*de*, *ee* > 96%).

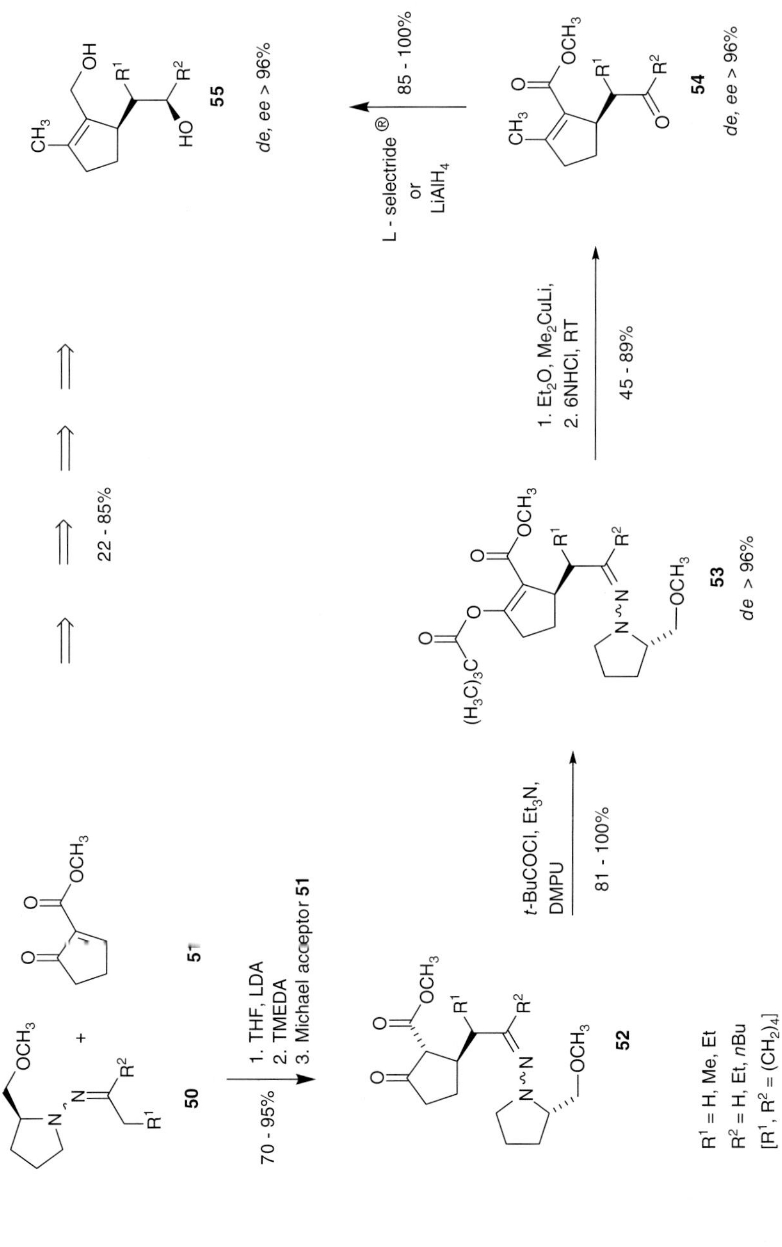

1.2 Natural Products Employing the SAMP/RAMP Hydrazone Methodology

R^1 = H, Me, Et
R^2 = H, Et, nBu
[R^1, R^2 = (CH$_2$)$_4$]

Scheme 1.2.12 Asymmetric synthesis of dehydroiridodiol.

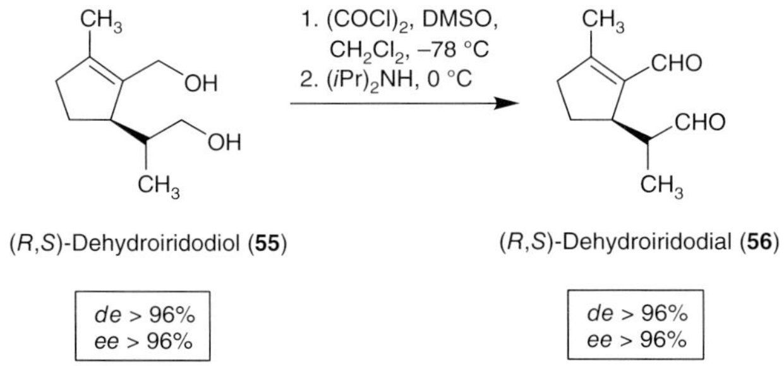

1.2.5 First Enantioselective Synthesis of Dendrobatid Alkaloids Indolizidine 209I and 223J

Scheme 1.2.13 Synthesis of dehydroiridodial.

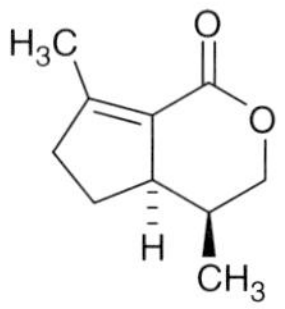

(R,S)-Dehydroiridodiol (**55**)
de > 96%
ee > 96%

Reagents: 1. (COCl)$_2$, DMSO, CH$_2$Cl$_2$, –78 °C; 2. (iPr)$_2$NH, 0 °C

(R,S)-Dehydroiridodial (**56**)
de > 96%
ee > 96%

(–)-neonepetalactone (**61**)

Fig. 1.2.3 Neonepetalactone.

Since the primary alcohol groups of dehydroiridodiol can be converted to the corresponding aldehydes, a pathway to dehydroiridodial (**56**) could be opened by Swern oxidation in excellent yield as well (Scheme 1.2.13).

Neonepetalactone, **61** (Fig. 1.2.3), a bioactive compound found to be quite attractive to cats [41], was isolated in 1965 from the leaves and galls of *Actinidia polygama* by T. Sakan et al. and its absolute configuration was determined in 1980 [41b]. As some syntheses of the racemic mixture or ex-chiral-pool syntheses had already been reported, we realized that our SAMP/RAMP hydrazone methodology would make it possible to develop a very short asymmetric synthesis of this bioactive δ-lactone.

Since neonepetalactone is closely related to the iridoids described above, the total synthesis followed the same procedure (Scheme 1.2.14) leading to the desired natural product in 55% overall yield and excellent *de* and *ee* values of >96%.

1.2.5
First Enantioselective Synthesis of Dendrobatid Alkaloids Indolizidine 209I and 223J

The skin of neotropical frogs of the Dendrobatidae family has been a rich source of bioactive compounds [42] such as the alkaloids epibatidine, batrachotoxins, histrionicotoxin, and indolizidines. The latter have been investigated due to their activity as noncompetitive blockers of the nicotinic receptor channels [43]. Because only small amounts are available and due to their pronounced physiological activity, these alkaloids are important targets for organic synthesis [44]. Since

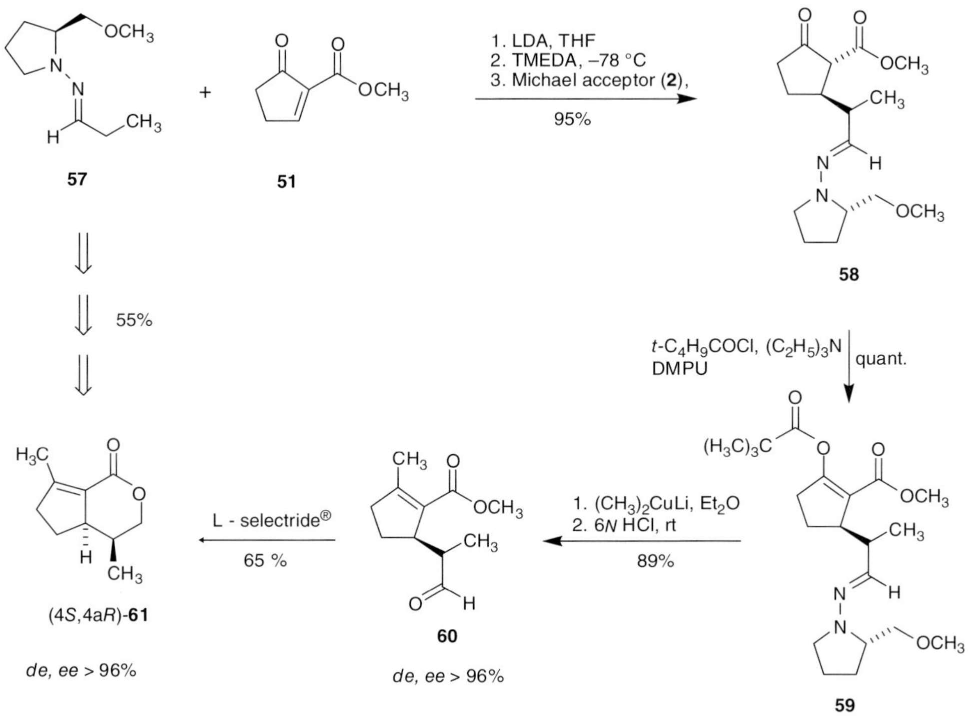

Scheme 1.2.14 Asymmetric synthesis of neonepetalactone.

they only occur in traces in the frog skin, the structural assignments turn out to be very difficult. The relative configuration of the two indolizidine alkaloids 209I [(5R,8R,8aS)-**62a**] and 223J [(5R,8R,8aS)-**62b**], detected in dendrobatid (209I, 223J) and mantelline frogs (223J), has been determined by GC analysis of a natural sample and comparison with racemic samples. The absolute configuration was assigned in accordance with those of other dendrobatid 5,8-disubstituted indolizidine alkaloids of known configuration.

In our group the diastereoselective 1,2-addition of organometallic reagents to aldehyde SAMP hydrazones was employed in the synthesis of several alkaloids and we have now extended our method to the efficient asymmetric synthesis of the poison-dart-frog indolizidine alkaloids 209I and 223J and their enantiomers via a common late-stage intermediate amino nitrile (5R,8R,8aS)-**63** [45]. This amino nitrile chemistry had previously been used by Polniaszek and Belmont in the first enantioselective total syntheses of 5,8-disubstituted indolizidine alkaloids [46]. They were able to prepare the indolizidines 205A (**65**) from **64** in one or two steps (Scheme 1.2.15).

Starting from the RAMP hydrazone (R)-**67**, which can easily be prepared from n-pentanal and RAMP in a multigram quantity, the electrophilic substitution (after metallation with LDA) using 2-(2-iodoethyl)-1,3-dioxolane (**68**) provided

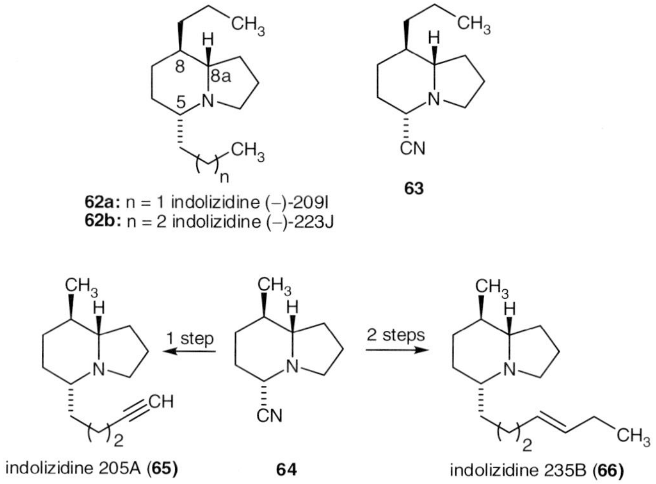

1.2.5 First Enantioselective Synthesis of Dendrobatid Alkaloids Indolizidine 209I and 223J

62a: n = 1 indolizidine (−)-209I
62b: n = 2 indolizidine (−)-223J

63

indolizidine 205A (**65**)   **64**   indolizidine 235B (**66**)

Scheme 1.2.15 Preparation of indolizidines.

access to the hydrazone **69** with a high diastereomeric excess (de = 90%) in 81% yield. 1,2-Addition of an organocerium reagent prepared from 3-t-butyldimethylsiloxypropyllithium and cerium(III) chloride to both epimers of the hydrazone **69** furnished hydrazine **70** in 82% yield as a 95 : 5 mixture of the (R,R,S)- and (R,S,S)-epimers (de = 90%), while no other stereoisomer could be detected, indicating that the complete asymmetric induction is exclusively controlled by the auxiliary. N–N bond cleavage with excess borane–tetrahydrofuran complex followed by protection of the crude amine with benzyl chloroformate converted the hydrazine **70** to the carbamate **71** in 86% yield. After removal of the silyl protecting group using tetra-n-butylammonium fluoride (TBAF) the amino alcohol **72** could be obtained quantitatively, followed by mesylation and ring closure reaction leading to the pyrrolidine **73** with diastereomeric excess de = 99% and a yield of 83%. The diastereomerically pure pyrrolidine **73** obtained after HPLC purification was then converted to the amino nitrile (5R,8R,8aS)-**63** by following standard procedures: that is, hydrogenating with Pd(OH)$_2$/charcoal leading to the crude pyrrolidino acetal **74**, which was converted by hydrolysis in the presence of potassium cyanide to **63** in 91% yield over two steps (Scheme 1.2.16).

Since the amino nitrile (5R,8R,8aS)-**63** is a key building block, we were now able to synthesize the indolizidines (−)-209I [(5R,8R,8aS)-**62a**] and (−)-223J [(5R,8R,8aS)-**62b**] and their C-5 epimers (5S,8R,8aS)-**62a,b**. Thus alkylation of **63** with n-propyl- and n-butyl bromide, respectively, resulted in the formation of amino nitriles **75a,b**, which underwent stereoselective reduction with sodium borohydride to furnish (−)-indolizidine 209I (**62a**, 88%, de = 96.9%) and (−)-indolizidine 223J (**62b**, 89%, de > 99%), both with an enantiomeric ee excess of 99%. Alternatively, a Bruylants reaction of (5R,8R,8aS)-**63** with n-propyl- and n-butylmagnesium bromide afforded (5S,8R,8aS)-**62a** (91%, de = 96.3%) and (5S,8R,8aS)-**62b** (87%, de = 96.0%), respectively (Scheme 1.2.17).

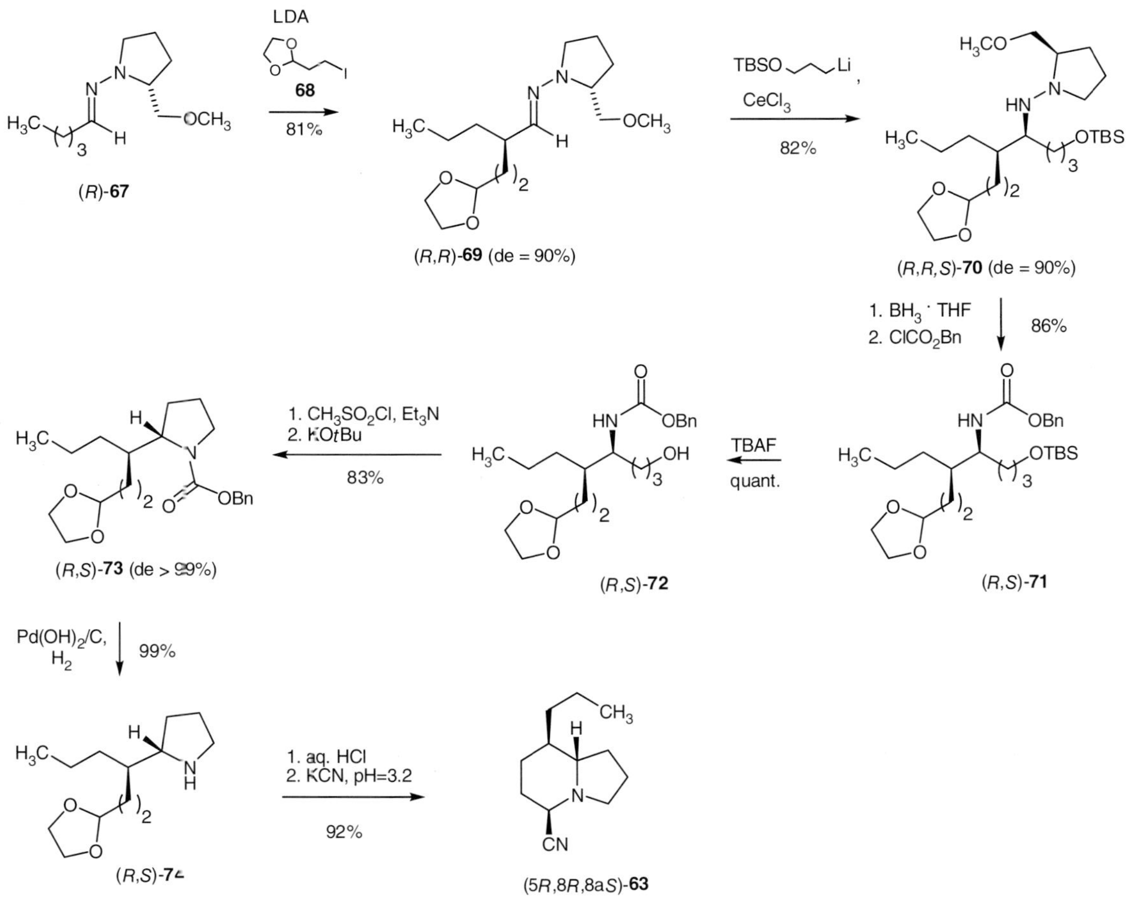

Scheme 1.2.16 Asymmetric synthesis of the amino nitrile **63**.

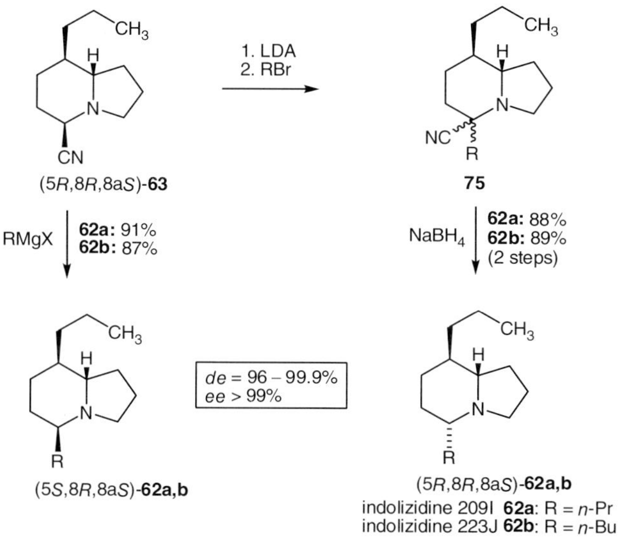

Scheme 1.2.17 Asymmetric synthesis of indolizidine 209I and 223J.

1.2.6
Efficient Synthesis of (2S,12′R)-2-(12′-Aminotridecyl)pyrrolidine, a Defense Alkaloid of the Mexican Bean Beetle

The Mexican bean beetle, *Epilachna varivestis* (Coccinellidae), discharges a blood droplet containing different defense alkaloids from its knee joints to defend itself from predators [47]. One of these alkaloids, (2S,12′R)-2-(12′-aminotridecyl)pyrrolidine [(S,R)-82] was extracted from adult bean beetles and finally characterized [48]. Using our methodology a new stereoselective synthesis starting from (R)-proline and generation of the stereogenic center at C-12′ by a diastereoselective 1,2-addition of methyllithium to an aldehyde SAMP hydrazone was developed [49].

Starting from the commercially available 11-bromoundecan-1-ol (**76**) the synthesis of the acetal **77** was performed by Swern oxidation and subsequent reaction with ethylene glycol. Treatment with triphenylphosphine yielded a phosphonium bromide, which was used in a Wittig olefination with (R)-N-benzylprolinal [(R)-**78**], derived from (R)-proline according to a literature procedure [50], to give the pyrrolidino acetal **79** in 80% yield. After conversion into the corresponding aldehyde the treatment with SAMP afforded the hydrazone **80**, to which methyllithium was added via 1,2-addition to the carbon–nitrogen double bond affording the hydrazine **81** as a single diastereomer (*de* ≥ 96%). In the final steps the double bond in (R,R,S)-**81** was hydrogenated using Pd(OH)$_2$/charcoal followed by treatment with an excess of borane–tetrahydrofuran complex, providing the defense alkaloid (S,R)-**82** in 61% yield over two steps and in a diastereo- and enantiomeric excess greater than 96% (Scheme 1.2.18).

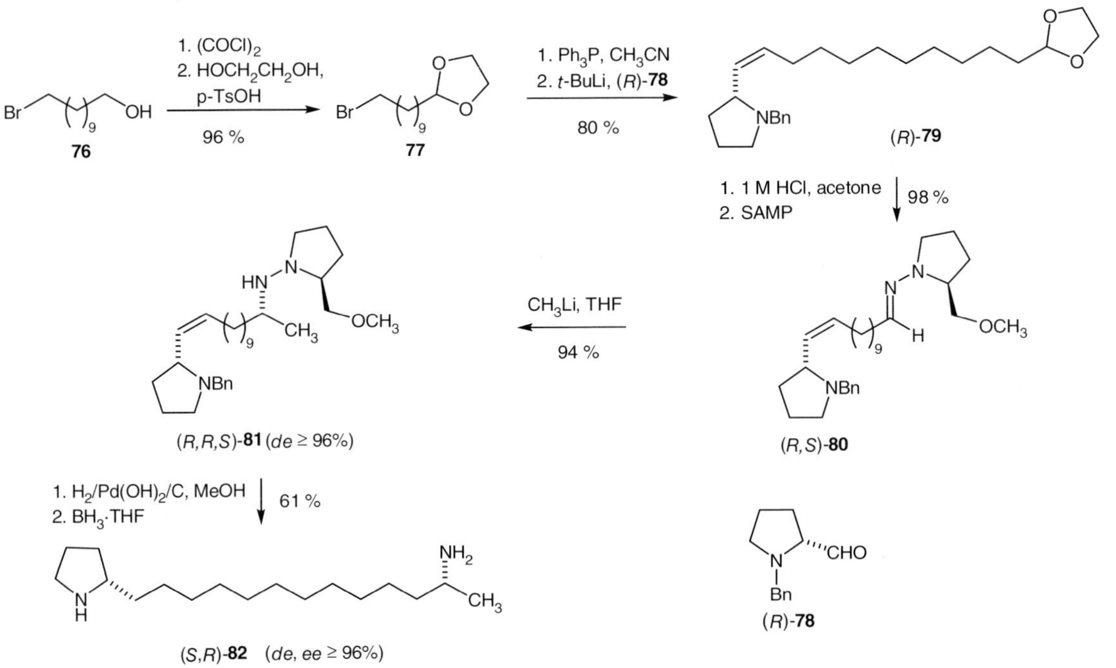

Scheme 1.2.18 Asymmetric synthesis of a defense alkaloid of the mexican bean beetle.

1.2.7
2-epi-Deoxoprosopinine

Hydroxylated piperidine alkaloids are found abundantly in living systems, exhibiting a wide range of potent physiological effects, such as the ability to mimic carbohydrates in a variety of enzymatic processes [51]. *Prosopis* alkaloids form a small subgroup of alkaloid lipids containing a 2,6-disubstituted 3-piperidinol framework with a long aliphatic appendage at the 6-position. At one end of these molecules is the polar head group with a configuration of the 1,3-diol unit similar to those in deoxynojirimycin, **83** (Fig. 1.2.4), a potent α-glucosidase I and II inhibitor [52], while the lipophilic tail resembles the membrane lipid sphingosine, **84**. Seven piperidine alkaloids, among them (+)-prosopinine **85** and (+)-prosophylline **86** have been isolated from the leaves of the West African savanna tree *Prosopis africana* Taub [53]. Due to their structural features mentioned above, these polysubstituted piperidine alkaloids and their deoxygenated derivatives (e. g. (+)-deoxoprosopinine, **87**) exhibit a wide variety of physiological properties, including analgesic, anesthetic, and antibiotic activity [54].

Since only a few asymmetric syntheses are known [55], we developed an efficient asymmetric synthesis of (S,S,R)-(+)-2-*epi*-deoxoprosopinine [(S,S,R)-**87**] employing the SAMP/RAMP hydrazone method as key steps [56]. Scheme 1.2.19 describes our retrosynthetic analysis, showing that the title alkaloid can be syn-

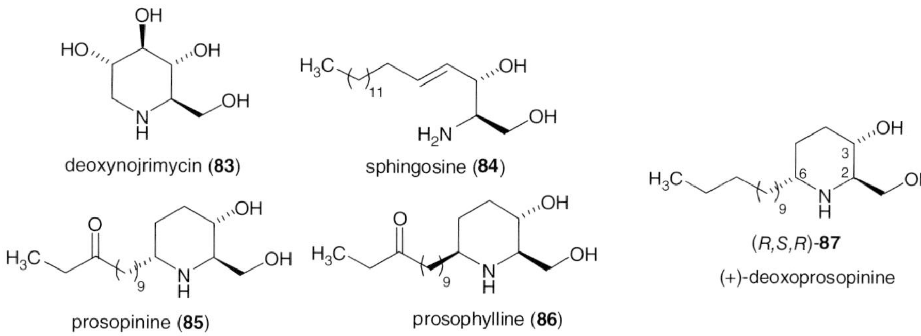

1.2.7 2-epi-Deoxoprosopinine

Fig. 1.2.4 *Prosopis* alkaloids **85–87** and the congeners **83** and **84**.

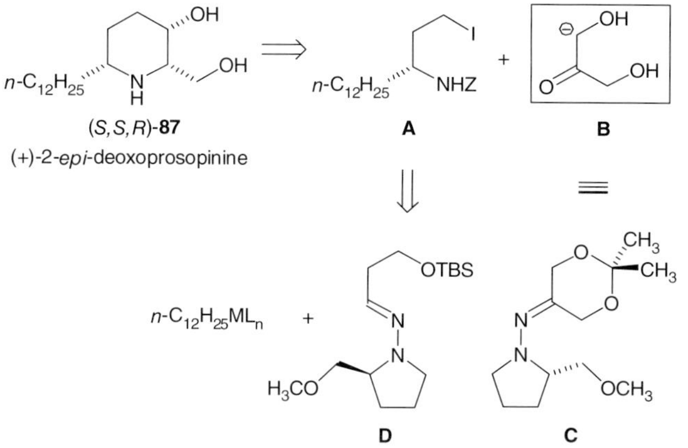

Scheme 1.2.19 Retrosynthetic analysis of (+)-2-*epi*-deoxoprosopinine.

thesized using the electrophilic iodo amine **A** and the dihydroxyacetone enolate synthon **B**. Both can be traced back to the SAMP hydrazones **C** and **D** and a dodecyl nucleophile.

Following these considerations, we started from 3-(*t*-butyldimethylsiloxy)-propanenitrile **88**, which provided access to the aldehyde **89** via reduction with diisobutylaluminum hydride, which in turn could be treated with SAMP to afford the hydrazone **90** in 72% yield (Scheme 1.2.20). The first key step of the synthesis was the 1,2-addition of the dodecyl nucleophile to the CN double bond of the hydrazone. Among various nucleophiles tested, a dodecylytterbium reagent, obtained from dodecyllithium and dry ytterbium(III) chloride, turned out to be the best one, furnishing the hydrazine **91** in 84% yield and excellent diastereomeric excess (*de* = 95%). In order to generate the corresponding amine, the borane–tetrahydrofuran complex was used for N–N bond cleavage without any racemization. The amine was protected as the carbamate using a two-phase system of water and chloroform with catalytic amounts of tetra-*n*-butylammonium iodide and Na_2CO_3 giving **93** in 79% yield. Finally, the desired electrophile could be

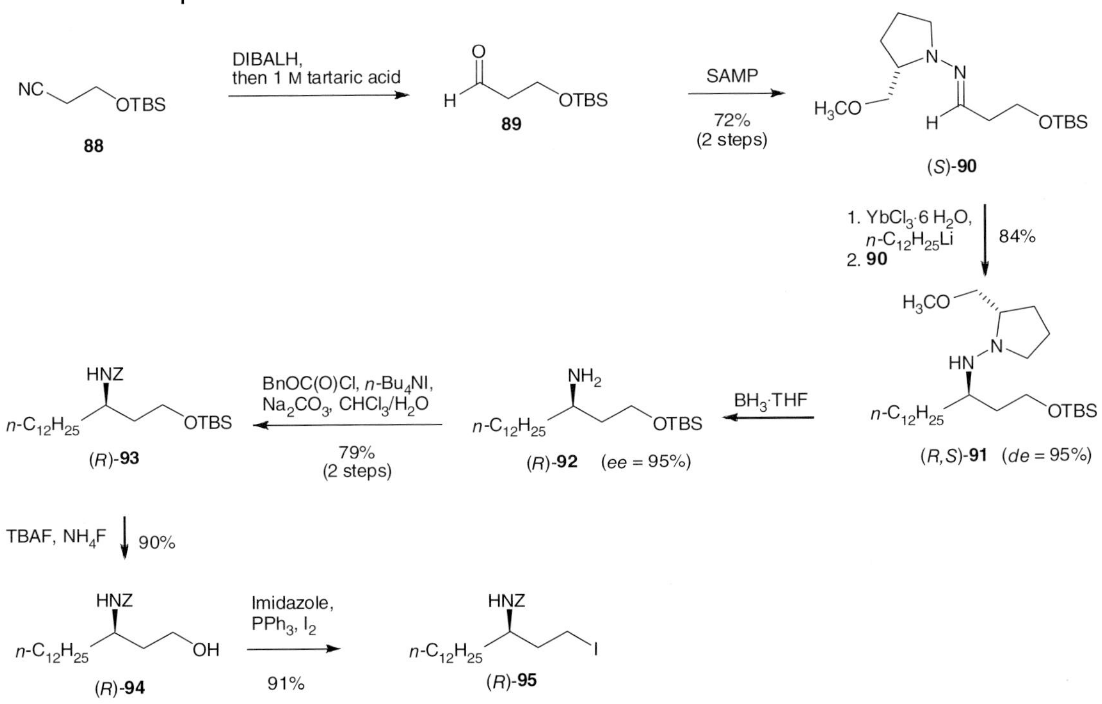

Scheme 1.2.20 Asymmetric synthesis of the building block **95**.

synthesized by conversion of **93** into the iodide **95** by cleavage of the silyl ether with an excess of TBAF in the presence of ammonium fluoride (yield 90%) and treatment of the alcohol with imidazole, triphenylphosphine, and elemental iodine (yield 91%).

Having the electrophile **95** to hand, the synthesis proceeded by metallation of the SAMP hydrazone **96** followed by alkylation with the iodide affording the α-alkylated hydrazone **97** with diastereomeric excess *de* = 95% (Scheme 1.2.21). The removal of the auxiliary proceeded smoothly with oxalic acid, leading to **98** in good yield (80%, two steps) and no epimerization without deprotection of the amino group.

In order to form the piperidine **99**, the removal of the protection group was performed in a domino reaction using catalytic hydrogenolysis over palladium on charcoal and cyclization under reductive amination conditions leading to the piperidines (*S*,*S*,*R*)-**99** and (*S*,*S*,*S*)-**99** in a ratio of 98 : 2 and an overall yield of 88% (Scheme 1.2.21). After chromatographic separation of the diastereomers the hydrolytic cleavage of the acetal group using an acid ion-exchange resin led to the (*S*,*S*,*R*)-(+)-2-*epi*-deoxoprosopinine [(*S*,*S*,*R*)-**87**] with a yield of 84% and diastereo- and enantiomeric excesses greater than 96%.

In conclusion, an efficient asymmetric synthesis of the prosopis alkaloid (+)-2-*epi*-deoxoprosopinine was successfully carried out in 11 steps to afford the target

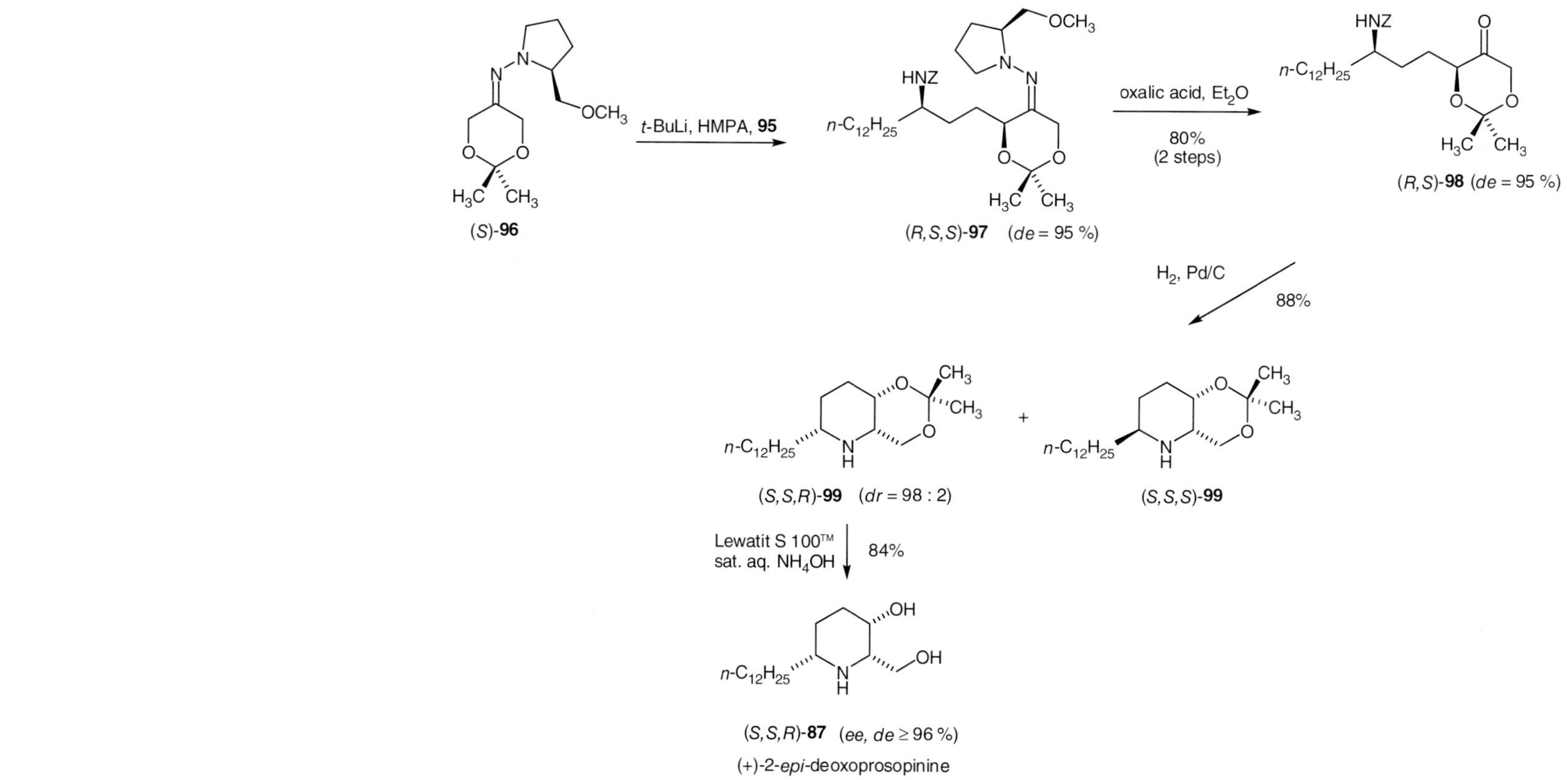

1.2.7 2-epi-Deoxoprosopinine

Scheme 1.2.21 Asymmetric synthesis of (+)-2-epi-deoxoprosopinine.

molecule in 23% overall yield and with excellent diastereo- and enantiomeric excesses.

1.2.8
Attenol A and B

In 1999 Uemura et al. isolated attenol A (**100**) and B (**101**) (Fig. 1.2.5), both marine natural products exhibiting a moderate cytotoxicity against P388 cells [57], from the Chinese *Pinna attenuata*. Since they are isomeric triols they differ only in the hydroxyl groups involved in the formation of the ketal functionality, so that this results in a 1,6-dioxaspiro[4.5]decane and a 6,8-dioxabicyclo[3.2.1]octane unit as the main structural feature of attenol A and B, respectively.

Due to their interesting bioactivity these products have attracted considerable interest as synthetic targets. The first total synthesis of attenol A and B has been carried out by Uemura, Suenaga et al. [58] followed by Eustache, Van de Weghe et al. [59].

Retrosynthetic considerations among our group have revealed the possibility of using our SAMP/RAMP hydrazone methodology in combination with the synthetic building block 2,2-dimethyl-1,3-dioxan-5-one. Thus starting from the 2,2-dimethyl-1,3-dioxan-5-one SAMP hydrazone **96**, successive alkylation with (2-bromoethoxy)-*t*-butyldimethylsilane and 5-bromopent-1-ene and consecutive cleavage with oxalic acid provided access to the disubstituted *trans*-2,2-dimethyl-1,3-dioxane **103** with diastereo- and enantiomeric excesses of *de*, *ee* ≥ 96% and in good yield (75%). Based on Barton and McCombie's results and our own experiences the keto group of **103** could be removed without racemization or epimerization via the xanthate **104** obtained in 96% yield over two steps. Subsequent reduction with Bu$_3$SnH and a catalytic amount of AIBN in toluene led to **105**. Iodination gave the iodide **106** (94%), which is the first of two electrophiles needed to be coupled by the Corey–Seebach reaction (Scheme 1.2.22).

As shown in Scheme 1.2.23, the synthesis of the second electrophile started from the corresponding SAMP hydrazone of 4-(4-methoxybenzyloxy)butyraldehyde (**107**), which was methylated with MeI yielding **108** in 86% yield and with *de* = 96%. Ozonolytic cleavage of the hydrazone moiety followed by Wittig olefination led to the enoate **109**, taking into account a small loss of enantiomeric purity (*ee* = 92%). Using the Sharpless asymmetric dihydroxylation a mixture of two

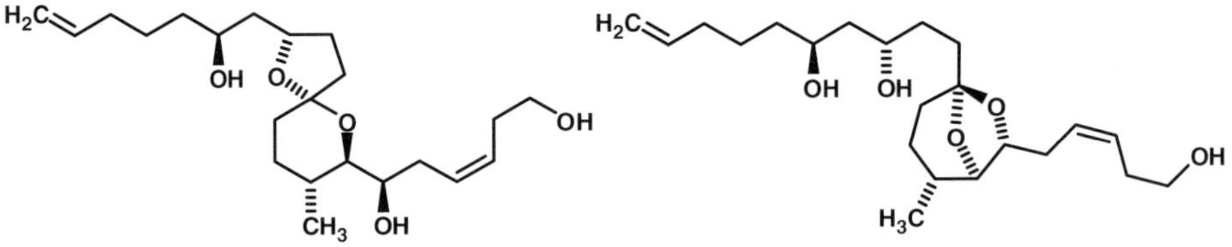

Fig. 1.2.5 Attenol A and B.

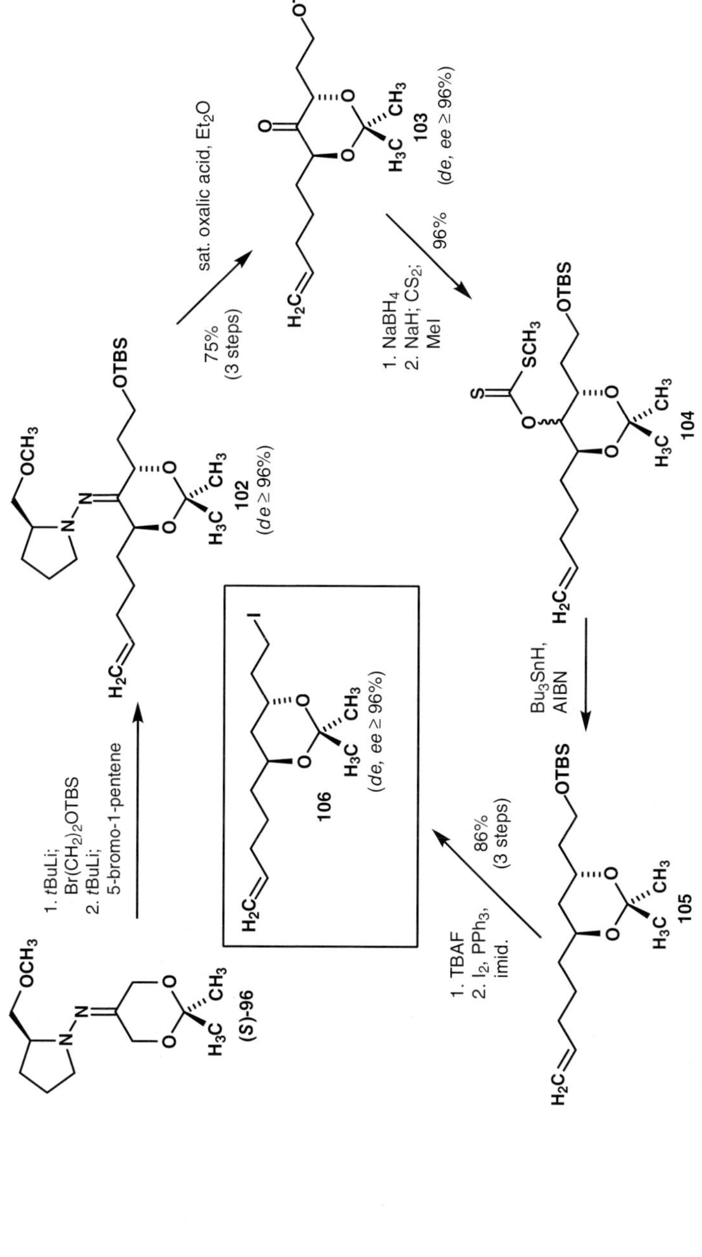

Scheme 1.2.22 Asymmetric synthesis of the iodide **106**.

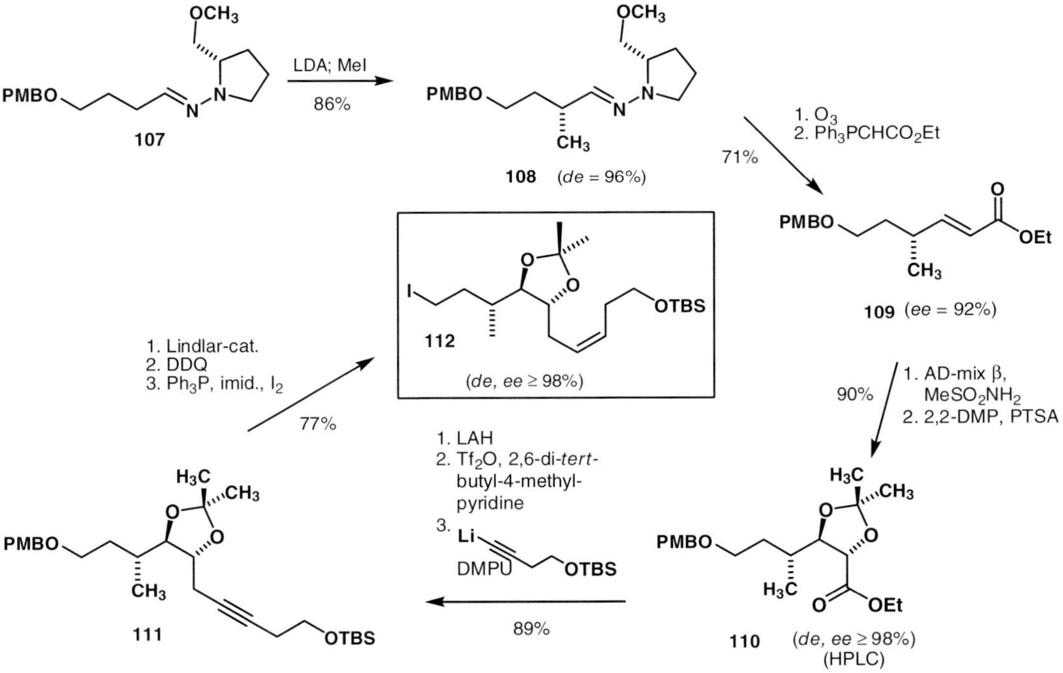

Scheme 1.2.23 Asymmetric synthesis of the iodide **112**.

diastereomers whose *de* corresponded with the *ee* of the unsaturated ester **109** could be achieved. After separation by HPLC of the minor diastereomer, **110** could be obtained in 90% yield with *de*, *ee* ≥ 98%. In a next step, reduction of the ester moiety and activation of the corresponding alcohol as its triflate followed by displacement with lithiated *t*-butylbut-3-ynyloxydimethylsilane gave alkyne **111** in 89% yield over two steps. Lindlar reduction, followed by the cleavage of the *p*-methoxybenzyl ether with 2,3-dichloro-5,6-dicyano-1,4-benzoquinone (DDQ) and iodination of the generated alcohol afforded **112** (77% yield over three steps) without loss of enantiomeric purity.

Finally, the synthesis of both attenols was accomplished by employing both electrophiles **106** and **112** as shown in Scheme 1.2.24. The alkylation of dithiane **113** with **106** proceeded cleanly in 96% yield. The second alkylation using **112** had to be carried out in the presence of HMPA and yielded **115** in 84% yield. Copper-mediated hydrolysis of the dithiane and *p*-toluenesulfonic acid catalyzed ketal formation finally gave a mixture of the title compounds **100** (57%) and **101** (9%), each over two steps. This synthesis is so far the shortest and most efficient one [60].

1.2.9
Asymmetric Synthesis of (+)- and (−)-Streptenol A

(+)-Streptenol A (**117**), a secondary metabolite of several Streptomyces species [61], is like the other streptenols B, C, and D a potent cholesterol biosynthesis

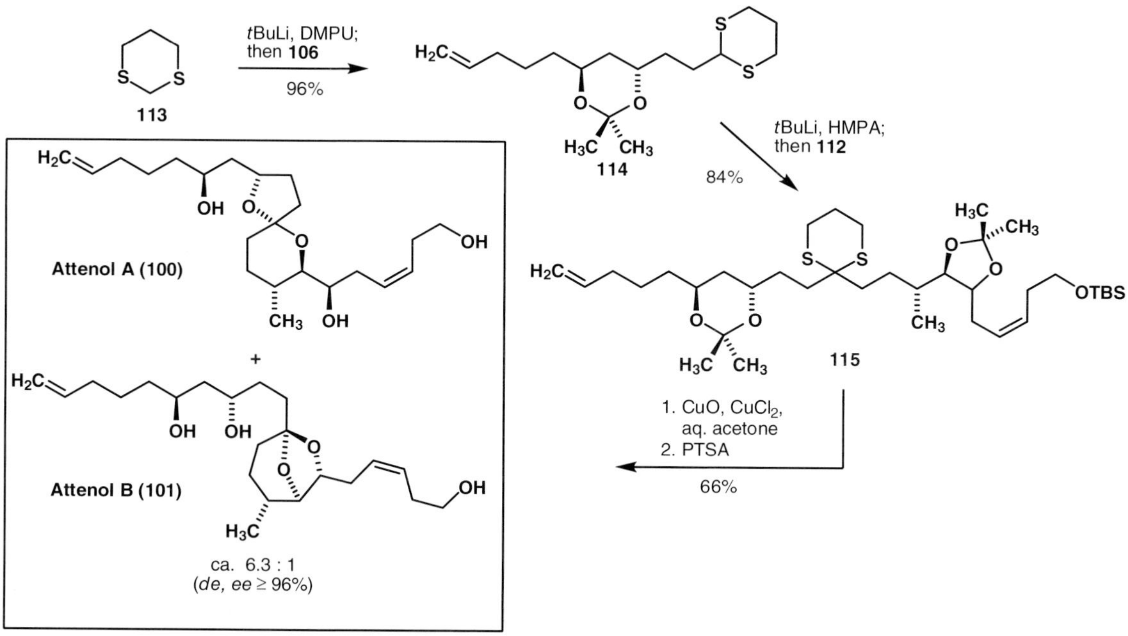

1.2.9 Asymmetric Synthesis of (+)- and (−)-Streptenol A

Scheme 1.2.24 Asymmetric synthesis of Attenol A and B.

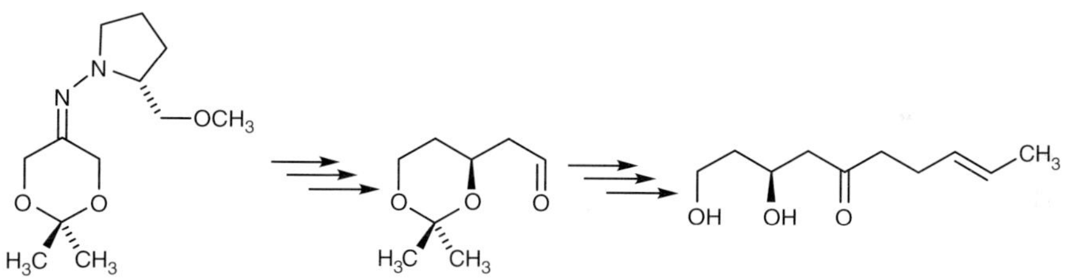

Scheme 1.2.25 Enantioselective pathway to streptenol A.

inhibitor [62] and in addition shows special anti-tumor and immuno-stimulating activity [63]. Several stereoselective syntheses have been reported [64], but the most general approach was developed by Blechert et al. using 2-(2,2-dimethyl-1,3-dioxan-4-yl)acetaldehyde **116** as the key building block in their syntheses of (±)-streptenol B, C, and D and of (−)-streptenol A [65]. Based on this previous work and on our protocol for the α-alkylation and deoxygenation of 2,2-dimethyl-1,3-dioxan-5-one RAMP/SAMP hydrazones **(R)-96** [66], access to both enantiomers of (+)-streptenol A, respectively, should be possible (Scheme 1.2.25) [67].

As already demonstrated in the previous natural product synthesis, the alkylation of 2,2-dimethyl-1,3-dioxan-5-one SAMP/RAMP hydrazones is a reliable tool with which to synthesize chiral 4-substituted 2,2-dimethyl-1,3-dioxan-5-ones in gram quantities and with high enantiomeric excesses [68]. Thus, after metallation of the RAMP hydrazone **(R)-96** the corresponding lithio azaenolate was alkylated with 2-bromo-1-*t*-butyldimethylsilyloxyethane, furnishing the α-alkylated hydrazone (*R*,*R*)-**118** with a diastereomeric excess of *de* ≥ 96%. Cleavage of the hydrazone with oxalic acid afforded (*R*)-**119** in 95% yield (two steps) and an enantiomeric excess *ee* of ≥96%. Reduction of the carbonyl group with sodium borohydride resulted in a diastereomeric mixture (*de* = 61%) of the corresponding alcohols **120**, which were converted to the xanthates **121** in 80% yield (two steps); this was followed by the Barton–McCombie deoxygenation to give the dioxane **122**. The primary alcohol **123** was obtained by deprotection of **122** with TBAF in 76% yield over the next two steps with a diastereomeric excess *de* > 96%. Finally, ruthenium-catalyzed oxidation with 7.5 mol% tetrapropylammonium perruthenate (TPAP) and *N*-methylmorpholine oxide (NMO) according to Ley et al. [69] gave (*S*)-2-(2,2-dimethyl-1,3-dioxan-4-yl)acetaldehyde [(*S*)-**116**] in 88% yield. This key building block of all members of the streptenol family was obtained in 51% overall yield starting from RAMP hydrazone (*R*)-**96**. The same reaction sequence employing the SAMP hydrazone (*S*)-**96** gave aldehyde (*R*)-**116** in 49% overall yield, thus undermining the reliability of this approach (Scheme 1.2.26).

Having both enantiomers of aldehyde **116** to hand, the final steps of the synthesis followed a slightly modified procedure of Blechert et al. [65], by which the Grignard-addition of (*E*)-5-magnesium bromopent-2-ene to the aldehyde (*S*)-**116** furnished the diastereomeric mixture (*S*/*R*,*S*)-**124** in 81% yield. Subsequent oxidation to the ketone **125** was performed with pyridinium chlorochromate (PCC) in combination with celite and ground molecular sieves in dichloromethane, giving **125** in 72% yield. The isopropylidene acetal was cleaved hydrolytically in the final step with trifluoroacetic acid to give access to the natural product (+)-streptenol A in 78% yield (Scheme 1.2.27). The enantiomer (−)-streptenol A was obtained in 37% overall yield starting from the aldehyde (*R*)-**117**.

1.2.10
Sordidin

The banana weevil *Cosmopolites sordidus* Germar is the most important worldwide insect pest of banana plants [70]. These long-lived weevils lay their eggs in the rhizome of the plant and their larvae damage the plant by weakening it due to feeding and tunneling in their rhizome, with the result that the rhizome snaps at ground level before the bunch is ripe. The male weevil produces a volatile pheromone, which was first found by Budenberg et al. in 1993 [71] and isolated and identified in 1995 by Ducrot et al. [72], who gave it the trivial name sordidin and who also carried out the first synthesis, employing a Baeyer–Villiger oxidation as the key step [73]. Further syntheses, which to our knowledge are based on the concept of ex-chiral-pool synthesis or led to the racemic mixture, were published by Ducrot [74], Oehlschlager [75], Mori [76], Kitching [77] and recently by Wardrop et al., who reported the synthesis of *rac*-7-*epi*-sordidin employing a regioselective rhodium(II)-catalyzed intramolecular diazocarbonyl C–H insertion as a key step [78]. Since the conversion of *rac*-7-*epi*-sordidin to *rac*-sordidin has been

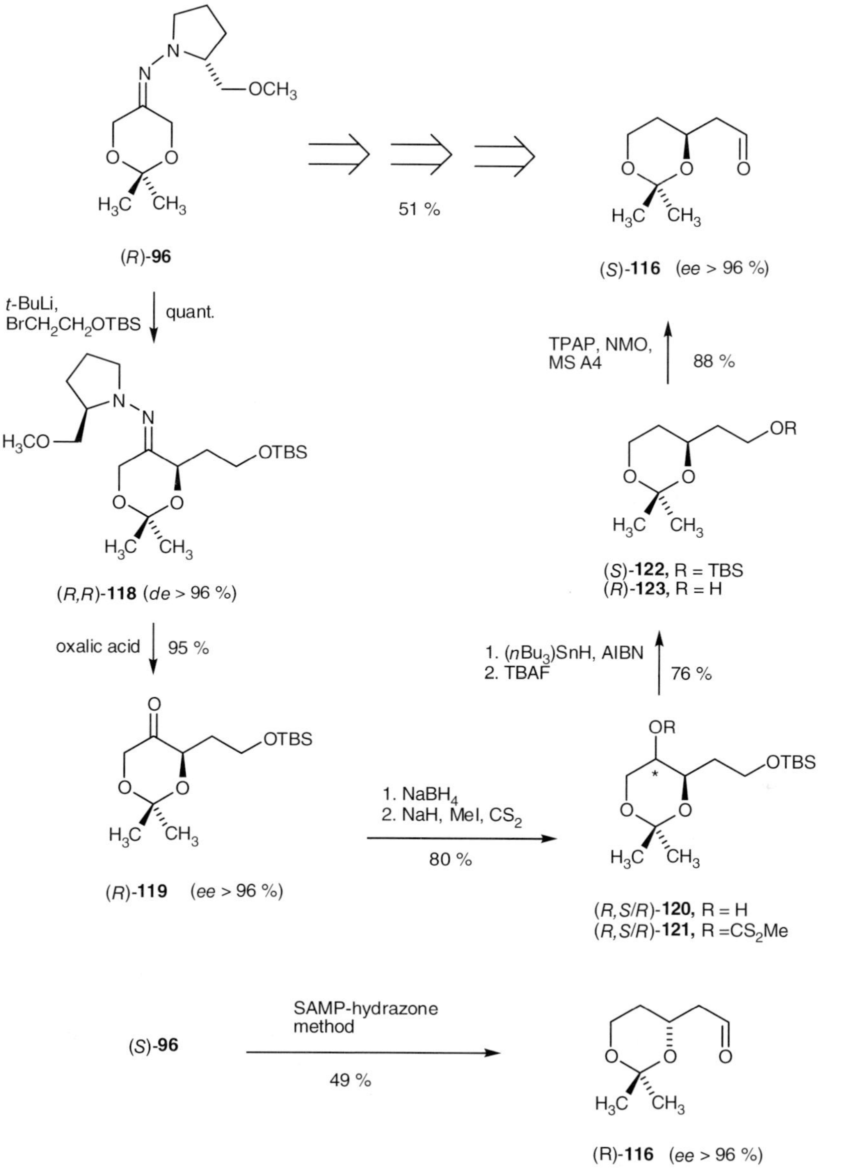

Scheme 1.2.26 Asymmetric synthesis of the key building block of streptenols.

discribed by Mori et al., the latter synthesis also constitutes a formal synthesis of rac-sordidin (**126**) [76].

Based on our SAMP/RAMP hydrazone methodology and using our dioxanone RAMP hydrazone **96** as a chiral 1,3-dihydroxyacetone equivalent, we have been

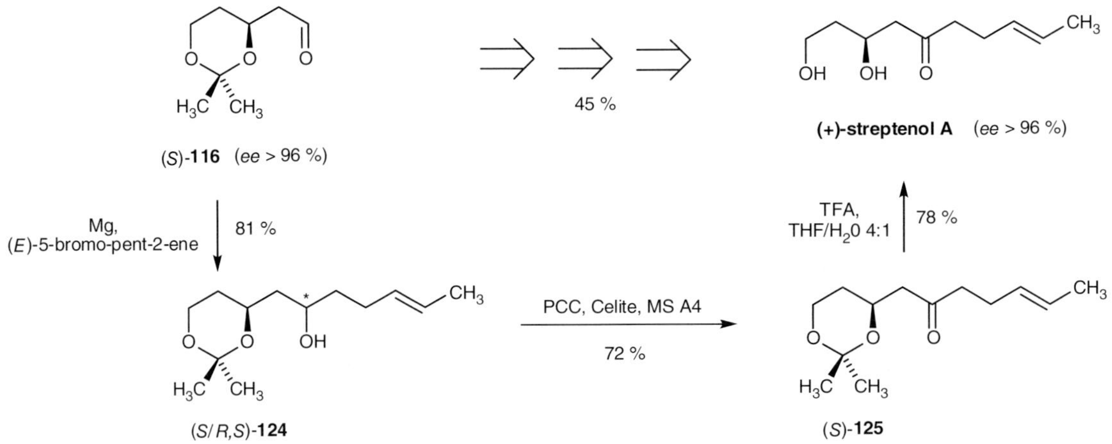

Scheme 1.2.27 Asymmetric synthesis of streptenol A.

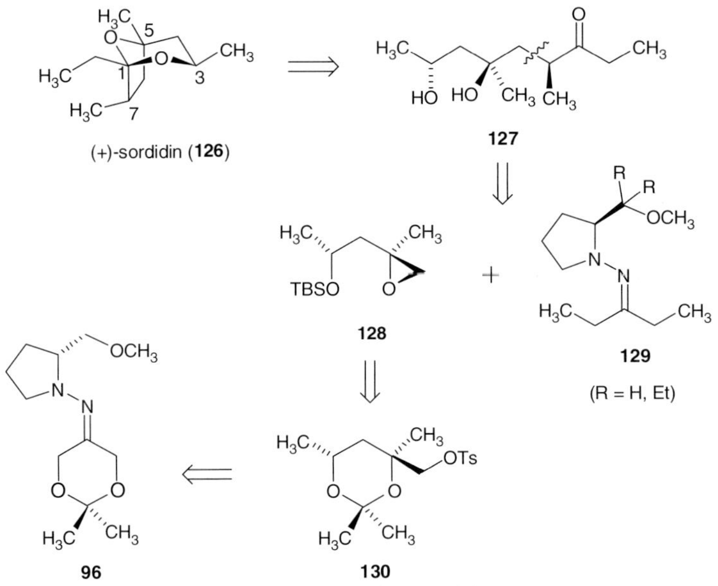

Scheme 1.2.28 Retrosynthetic analysis of (+)-sordidin.

able to establish the first asymmetric synthesis of (1*S*,3*R*,5*R*,7*S*)-(+)-sordidin (**126**), the main component of the pheromone and its epimer, 7-*epi*-(1*S*,3*R*,5*R*,7*R*)-(−)-sordidin (7-*epi*-**126**) [79].

As is shown in Scheme 1.2.28, our retrosynthetic analysis led us to the ketodiol **127**, providing access to the target pheromone after intramolecular acetalization. It was planned to prepare the dihydroxyketone **127** by diastereoselective ring opening of the epoxide **128** with either enantiomerically pure 3-pentanone SAMP

(R = H) or SAEP (R = Et) hydrazone **129** in one of the key steps of the synthesis. Finally, the oxirane **128** bearing the desired configuration can be synthesized starting from the RAMP hydrazone **96** via the corresponding the tosyl dioxane **130**.

As is depicted in Scheme 1.2.29, the epoxide **128** was synthesized starting from 2,2-dimethyl-1,3-dioxan-5-one RAMP hydrazone **(R)-96**, which was double-alkylated with methyl iodide at α- and α'-positions leading to the *trans*-dimethylated hydrazone **131** in 79% yield over two steps and excellent stereoselectivity (*de, ee* ≥ 96%) [68]. The quaternary stereocenter bearing the desired tertiary alcohol function was generated using benzyloxymethyl chloride (BOMCl) as the electrophile to trap the lithiated hydrazone **131**, providing the α-quaternary hydrazone **132** in very good yield (92%), excellent diastereomeric and enantiomeric excesses (*de, ee* ≥ 96%) and with the required *cis* relationship of the methyl substituents. The keto group of the trisubstituted dioxanone **133** generated by ozonolysis was removed by radical deoxygenation according to the Barton–McCombie protocol [80] via the alcohols **134** and the corresponding xanthate, leading after debenzylation to the dioxane **135** in excellent yield. After conversion to the tosylate, cleavage of the acetonide and protection of the secondary alcohol function as a TBS ether provided access to oxirane **128** by cyclization with NaH in 99% yield and in virtually diastereo- and enantiomerically pure form (*de, ee* ≥ 96%).

Finally, the oxirane **128** was opened with a large excess of the azaenolate of 3-pentanone SAEP hydrazone **129** in the presence of anhydrous LiCl additive according to a modified literature procedure for enolate oxirane ring openings [81]. The mixture of the crude product hydrazone **138** and the excess of hydrazone **129** was used directly in an acidic acetalization reaction utilizing aqueous 3N HCl in a biphasic system and gave a diastereomeric mixture of sordidin (**126**) and 7-*epi*-sordidin (7-*epi*-**126**) in 84% yield over two steps in a 1.5 : 1 ratio. Gratifyingly, we succeeded in separating the desired sordidin epimers by preparative gas chromatography. As a result, both could be obtained in diastereomerically pure form (sordidin: *de* ≥ 99%; 7-*epi*-sordidin: *de* ≥ 97%) and with a high enantiomeric excess for each epimer (*ee* ≥ 98%).

1.2.11
Prelactone B and V

The δ-lactone moiety can be found as a characteristic structural motif in a large number of natural products. Prelactones such as prelactone B (**139a**) and V (**139b**) (Fig. 1.2.6) have been isolated from various polyketide macrolide-producing microorganisms. Prelactone B was isolated from *Streptomyces griseus* by A. Zeeck et al. [82] in 1993 and is believed to be a product of the polyketide synthase (PKS) which is responsible for the synthesis of the macrolide itself. However, up to now no bioactivity has been determined, but it is used for the investigation of the mechanism of the PKS. Since only a few asymmetric syntheses have been reported [83], we regarded these natural lactones as suitable targets to demonstrate the synthetic utility of our SAMP/RAMP hydrazone methodology [84].

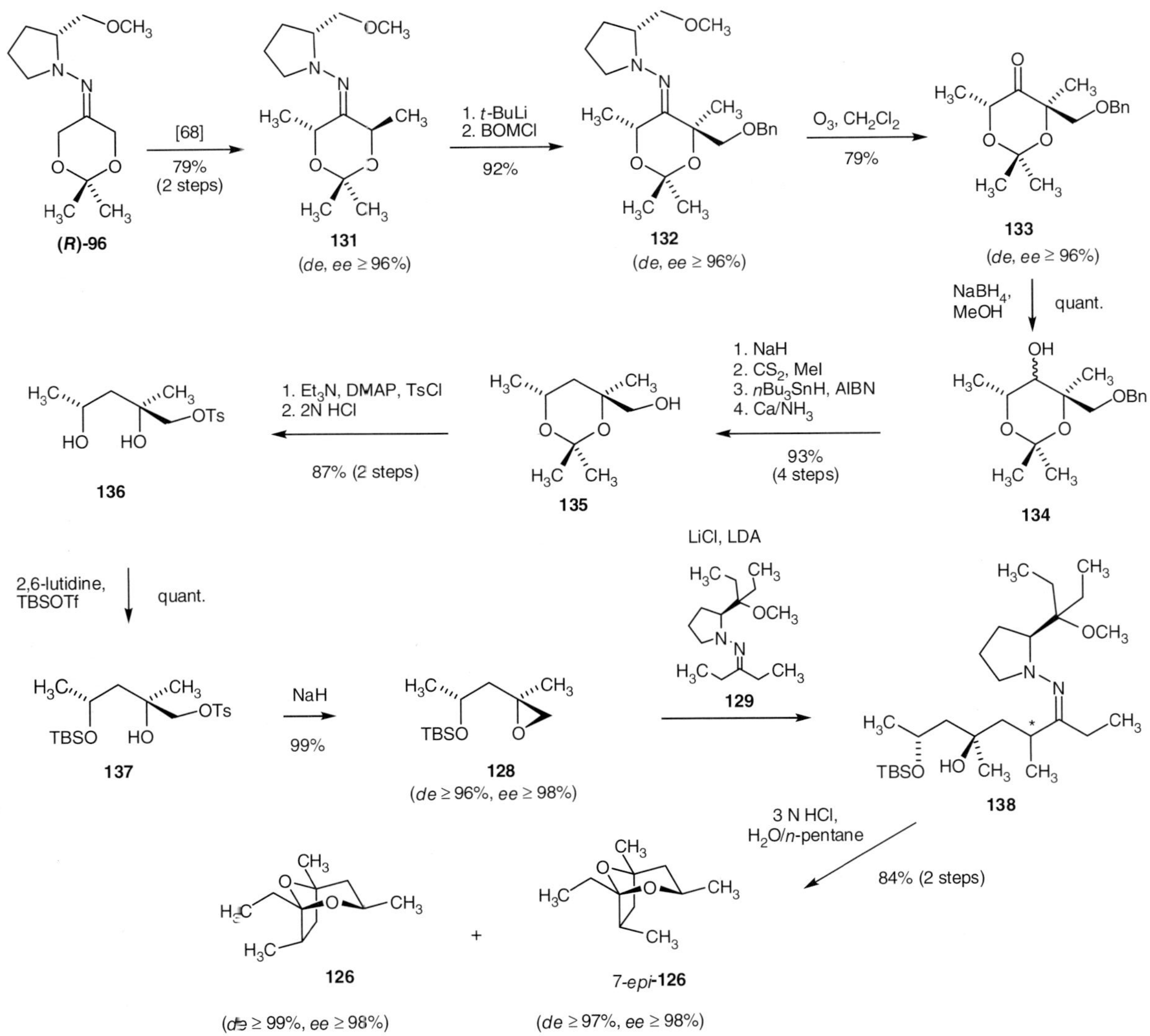

Scheme 1.2.29 Asymmetric synthesis of sordidin and 7-*epi*-sordidin.

1.2.11 Prelactone B and V

Fig. 1.2.6 Prelactones.

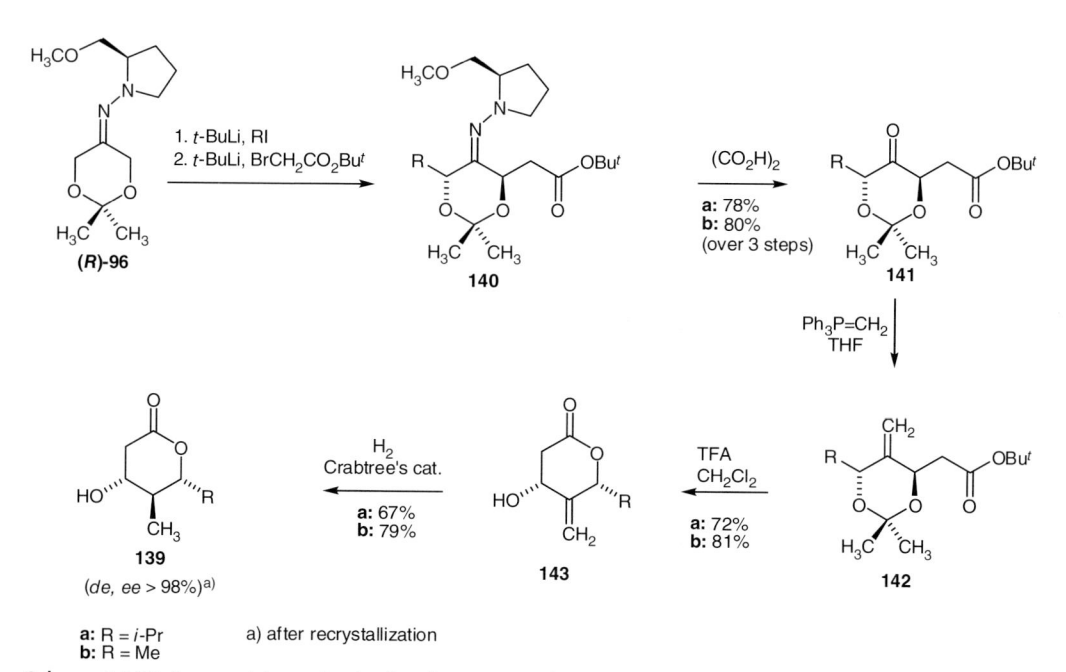

prelactone B (139a)

prelactone V (139)

Scheme 1.2.30 Asymmetric synthesis of prelactone B and V.

a: R = i-Pr
b: R = Me

a) after recrystallization

As starting material for our asymmetric syntheses of the δ-lactones we used the RAMP-hydrazone of 2,2-dimethyl-1,3-dioxane-5-one (**96**) already described. In a first step, alkylation with an alkyl iodide in the α-position was carried out employing standard conditions for this system, followed a second alkylation at the α′-position with bromo-t-butyl acetate. After removal of the chiral auxiliary the α,α′-disubstituted dioxanones **141** could be obtained in good yields of 78–80% over three steps. Subsequent Wittig reaction afforded the methylenated dioxanone derivatives **142**. Cleavage of the acetonide and cyclization to the lactone structure occurred by treatment with TFA and led to the δ-lactones **143**. The exocyclic double bond could be hydrogenated with Crabtree's catalyst [Ir(cod)(PCy$_3$)(py)]PF$_6$ [85], providing prelactone B (**139a**) and prelactone V (**139b**) in moderate yields and high diastereoselectivity. After recrystallization the single isomer of **139a** was obtained (de, ee > 98%) (Scheme 1.2.30).

References

1. A. Job, C. F. Janeck, W. Bettray, R. Peters, D. Enders, *Tetrahedron* **2002**, *58*, 2253.
2. a) G. Höfle, B. Kunze, C. Zorin, H. Reichenbach, *Liebigs Ann.* **1984**, 1883; b) B. Kunze, T. Kemmer, G. Höfle, H. Reichenbach, *J. Antibiotics* **1984**, *37*, 454; b) B. Kunze, G. Höfle, H. Reichenbach, *J. Antibiotics* **1987**, *40*, 258.
3. T. D. Brock, M. T. Madigan, *Biology of Microorganisms*, 6th ed., Prentice-Hall Englewood Cliffs, New Jersey **1991**.
4. G. Thierbach, B. Kunze, H. Reichenbach, G. Höfle, *Biochim. Biophys. Acta* **1984**, *765*, 227.
5. W. Oettmeier, D. Godde, B. Kunze, G. Höfle, *Biochim. Biophys. Acta* **1985**, *807*, 216.
6. T. Ohnishi, U. Brandt, G. von Jagow, *Eur. J. Biochem.* **1988**, *176*, 385.
7. D. Enders, S. Osborne, *J. Chem.Soc. Chem. Commun.* **1993**, 424.
8. D. Enders, G. Geibel, S. Osborne, *Chem. Eur. J.* **2000**, *6*, 1302.
9. a) D. A. Evans, J. S. Clark, R. Metternich, V. J. Novack, G. S. Sheppard, *J. Am. Chem. Soc.* **1990**, *112*, 866; b) D. A. Evans, F. Urpi, T. C. Somers, J. S. Clark, M. T. Bilodeau, *J. Am. Chem. Soc.* **1990**, *112*, 8215; c) D. A. Evans, D. L. Rieger, M. T. Bilodeau, F. Urpi, *J. Am. Chem. Soc.* **1991**, *113*, 1047.
10. a) D. Enders, A. Plant, *Synlett* **1990**, 725; b) D. Enders, A. Plant, *Synlett* **1994**, 1032; c) R. Fernandez, C. Gasch, J.-M. Lassaletta, J.-M. Llera, J. Vasquez, *Tetrahedron* **1993**, *34*, 141; d) D. Enders, A. Plant, D. Backhaus, U. Reinhold, *Tetrahedron* **1995**, *51*, 10 699.
11. a) D. A. Evans, J. S. Clark, R. Metternich, V. J. Novack, G. S. Sheppard, *J. Am. Chem. Soc.* **1990**, *112*, 866; b) D. A. Evans, F. Urpi, T. C. Somers, J. S. Clark, M. T. Bilodeau, *J. Am. Chem. Soc.* **1990**, *112*, 8215; c) D. A. Evans, D. L. Rieger, M. T. Bilodeau, F. Urpi, *J. Am. Chem. Soc.* **1991**, *113*, 1047.
12. D. A. Evans, K. T. Chapman, E. M. Carreira, *J. Am. Chem. Soc.* **1988**, *110*, 3560.
13. G. R. Scarlato, J. A. DeMattei, L. S. Chong, A. K. Ogawa, M. R. Lin, R. W. Armstrong, *J. Org. Chem.* **1996**, *61*, 6139.
14. R. Johansson, B. Samuelsson, *J. Chem. Soc. Perkin Trans.* **1984**, *1*, 2371.
15. a) E. J. Corey, G. Schmidt, *Tetrahedron Lett.* **1979**, *20*, 399; b) E. J. Corey, G. Schmidt, *Tetrahedron Lett.* **1980**, *21*, 731.
16. T. Schmidlin, W. Zürcher, C. Tamm, *Helv. Chim. Acta* **1981**, *64*, 235.
17. E. J. Corey, D. E. Cane, L. Libit, *J. Am. Chem. Soc.* **1971**, *93*, 7016.
18. a) A. K. Bhattacharya, G. Thyagarajan, *Chem. Rev.* **1981**, *81*, 415; b) L. Lombardo, R. J. K. Taylor, *Synthesis* **1978**, 131.
19. I. Hirao, M. Yamaguchi, M. Hamada, *Synthesis* **1984**, 1076.
20. a) M. Kobayashi, K. Higuchi, N. Murakami, H. Tajima, S. Aoki, *Tetrahedron Lett.* **1997**, *38*, 2859; b) N. Murakami, W. Wang, M. Aoki, Y. Tsutsui, K. Higuchi, S. Aoki, M. Kobayashi, *Tetrahedron Lett.* **1997**, *38*, 5533.
21. N. Murakami, W. Wang, M. Aoki, Y. Tsutsui, M. Sugimoto, M. Kobayashi, *Tetrahedron Lett.* **1998**, *39*, 2349.
22. M. T. Crimmins, B. W. King, *J. Am. Chem. Soc.* **1998**, *120*, 9084.
23. A. B. Smith III, B. M. Brandt, *Org. Lett.* **2001**, *3*, 1685.
24. M. Kalesse, M. Quitschalle, C. P. Khandavalli, A. Saeed, *Org. Lett.* **2001**, *3*, 3107.
25. a) J. A. Marshall, M. P. Bourbeau, *J. Org. Chem.* **2002**, *67*, 2751; J. A. Marshall, M. P. Bourbeau, *Org. Lett.* **2002**, *4*, 3931.
26. a) J. L. Vicario, A. Job, M. Wolberg, M. Müller, D. Enders, *Org. Lett.* **2002**, *4*, 1023; b) J. L. Vicario, A. Job, M. Wolberg, M. Müller, D. Enders, *Chem. Eur. J.* **2002**, *8*, 4272.
27. M. Lautens, T. A. Stammers, *Synthesis* **2002**, 1993.
28. M. Kalesse, K. P. Khandavalli, M. Quitschalle, A. Burzlaff, C. Kasper, T. Scheper, *Chem. Eur. J.* **2003**, *9*, 1129.

29 N. F. Langille, S. J. Panek, *Org. Lett* **2004**, *6*, 3203.
30 C. L. Dias, P. R. R. Meira, *J. Org. Chem.* **2005**, *70*, 4762.
31 R. Tamura, K. Saegusa, M. Kakihana, D. Oda, *J. Org. Chem.* **1988**, *53*, 2723.
32 a) M. Wolberg, W. Hummel, C. Wandrey, M. Müller, *Angew. Chem.* **2000**, *112*, 4476; *Angew. Chem., Int. Ed.* **2000**, *39*, 4306; b) M. Wolberg, W. Hummel, M. Müller, *Chem. Eur. J.* **2001**, *7*, 4562; c) A. Job, M. Wolberg, M. Müller, D. Enders, *Synlett* **2001**, 1796.
33 A. Job, R. Nagelsdiek, D. Enders, *Collect. Czech. Chem. Commun.* **2000**, *65*, 524.
34 J. A. Marshall, R. N. Fitzgerald, *J. Org. Chem.* **1999**, *64*, 4477.
35 S. B. Hyeon, S. Isoe, T. Sakan, *Tetrahedron Lett.* **1968**, 5325.
36 K. Yoshihara, T. Sakai, T. Sakan, *Chem. Lett.* **1978**, 433.
37 a) S. Isoe, T. Ono, S. B. Hyeon, T. Sakan, *Tetrahedron Lett.* **1968**, *51*, 5319; b) J. Wolinsky, D. Nelson, *Tetrahedron* **1969**, *25*, 3767; c) H. Kimura, S. Miyamoto, H. Shinkai, T. Kato, *Chem. Pharm. Bull.* **1982**, *30(2)*, 723; d) M. Nakayama, S. Ohira, S. Takata, K. Fukuda, *Chem. Lett.* **1983**, 147; e) K. Nozaki, K. Oshima, K. Utimoto, *J. Am. Chem. Soc.* **1987**, *109*, 2547.
38 T. Sakai, K. Nakajima, T. Sakan, *Tetrahedron* **1980**, *36*, 3115.
39 a) H. Takeshita, T. Hatsui, N. Kato, T. Masuda, H. Tagoshi, *Chem. Lett.* **1982**, 1153; b) F. Bellesia, F. Ghelfi, U. M. Pagnoni, A. Pinetti, *Tetrahedron Lett.* **1986**, *27*, 381; c) S. Uesato, Y. Ogawa, M. Doi, H. Inouye, *J. Chem. Soc. Chem. Commun.* **1987**, 1020.
40 a) D. Enders, A. Kaiser, *Liebigs Ann. / Recl.* **1997**, 485; b) D. Enders, A. Kaiser, *Heterocycles* **1997**, *46*, 631.
41 a) T. Sakan, S. Isoe, S. B. Hyeon, R. Katsumura, T. Maeda, J. Wolinsky, D. Dickerson, M. Slabaugh, D. Nelson, *Tetrahedron Lett.* **1965**, *46*, 4097; b) T. Sakai, K. Nakajima, K. Yoshihara, T. Sakan, S. Isoe, *Tetrahedron* **1980**, *36*, 3115.
42 a) N. Toyooka, K. Tanaka, T. Momose, J. W. Daly, H. M. Garraffo, *Tetrahedron* **1997**, *53*, 9553; b) For a comprehensive review on occurrence, synthesis and activity of alkaloids in amphibian skins, see: J. W. Daly, H. M. Garaffo, T. F. Spande, in *Alkaloids: Chemical and Biological Perspectives*, Vol. 13, Pelletier, S. W. (Ed.), Elsevier Science Ltd.: Oxford, **1999**; p. 1.
43 J. W. Daly, Y. Nishizawa, W. L. Padgett, T. Tokuyama, A. L. Smith, A. B. Holmes, C. Kibayashi, R. S. Aronstam, *Neurochem. Res.* **1991**, *16*, 1213.
44 a) J. P. Michael, *Nat. Prod. Rep.* **1999**, *16*, 675 and references cited therein; b) C. Kibayashi, S. Aoyagi, in *Studies in Natural Product Chemistry*; Vol. 19 (Part E), Atta-ur-Rahman (Ed.), Elsevier: Amsterdam, **1997**; p. 3; c) P. Michel, A. Rassat, J. W. Daly, T. F. Spande, *J. Org. Chem.* **2000**, *65*, 8908; d) S. Yu, W. Zhu, D. Ma, *J. Org. Chem.* **2005**, *70*, 7364.
45 a) D. Enders, C. Thiebes, *Synlett* **2000**, 1745; b) D. Enders, C. Thiebes, *Pure Appl. Chem.* **2001**, *73*, 573.
46 R. P. Polniaszek, S. E. Belmont, *J. Org. Chem.* **1991**, *56*, 4868.
47 P. Proksch, L. Witte, V. Wrag, T. Hartmann, *Entomol. Gener.* **1993**, *18*, 1.
48 A. B. Attygalle, S.-C. Xu, K. D. McCormick, J. Meinwald, C. L. Blankespoor, T. Eisner, *Tetrahedron* **1993**, *49*, 9333.
49 D. Enders, C. Thiebes, *Synthesis* **2000**, 510.
50 J. Deur, M. W. Miller, L. S. Hegedus, *J. Org. Chem.* **1996**, *61*, 2871.
51 For further information concerning hydroxylated piperidines, see: a) G. W. J. Fleet, L. E. Fellows, B. G. Winchester, Plagiarizing Plants: Aminosugars as a Class of Glucosidase Inhibitors, in *Bioactive Compounds from Plants*; Wiley: Chichester (Ciba Foundation Symposium 154), **1990**; p. 112; b) B. G. Winchester, G. W. Fleet, *J. Glycobiology* **1992**, *2*, 199; c) L. A. G. M. van den Broek, D. J. Vermaas, B. M. Heskamp, C. A. A. van Boeckel, M. C. A. A. Tan, J. G. M. Bolscher, H. L. Ploegh, F. J. van Kemenade, R. E. Y. de Goede, F. Miedema, *Recl. Trav. Chim. Pays-Bas* **1993**, *112*, 82; d) G. R. Cook, L. G. Beholz, J. R. Stille, *J. Org. Chem.* **1994**, *59*, 3575; e) N. Asano, R. J. Nash, R. J.

Molyneux, G. W. J. Fleet, *Tetrahedron: Asymmetry* **2000**, *11*, 1645.

52 a) N. Asano, M. Nishida, K. Oseki, H. Kizu, K. Matsui, *J. Med. Chem.* **1994**, *37*, 3701; b) N. Asano, A. Kato, H. Kizu, K. Matsui, Y. Shimada, I. Itoh, M. Baba, A. A. Watson, R. J. Nash, P. M. de Q. Lilley, D. J. Watkin, G. W. J. Fleet, *J. Med. Chem.* **1998**, *41*, 2565.

53 a) G. Ratle, B. C. Das, J. Yassi, Q. Khuong-Huu, R. Goutarel, *Bull. Soc. Chim. Fr.* **1966**, 2945; b) Q. Khuong-Huu, G. Ratle, X. Monseur, R. Goutarel, *Bull. Soc. Chim. Belg.* **1972**, *81*, 425; c) Q. Khuong-Huu, G. Ratle, X. Monseur, R. Goutarel, *Bull. Soc. Chim. Belg.* **1972**, *81*, 443.

54 a) Omnium Chimique S. A., French Patent FR1524395, **1968**; *Chem. Abstr.* **1969**, *71*, 91733w; b) P. Bourrinet, A. Quevauviller, *Ann. Pharm. Fr.* **1968**, *26*, 787; c) P. Bourrinet, A. Quevauviller, *Compt. Rend. Soc. Biol.* **1968**, *162*, 1138.

55 a) Y. Saitoh, Y. Moriyama, T. Takahashi, Q. Khuong-Huu, *Tetrahedron Lett.* **1980**, *21*, 75; b) K. Takao, Y. Nigawara, E. Nishino, I. Takagi, M. Izumi, T. Koji, K. Tandano, S. Ogawa, *Tetrahedron* **1994**, *50*, 5681; c) I. Kadota, M. Kawada, Y. Muramatsu, Y. Yamamoto, *Tetrahedron Lett.* **1997**, *38*, 7469; d) I. Kadota, M. Kawada, Y. Muramatsu, Y. Yamamoto, *Tetrahedron: Asymmetry* **1997**, *8*, 3887; e) D. L. Comins, M. J. Sandelier, T. A. Grillo, *J. Org. Chem.* **2001**, *66*, 6829; f) F.-X. Felpin, K. Boubekeur, J. Lebreton, *Eur. J. Org. Chem.* **2003**, 4518; g) Q. Wang, N. A. Sasaki, *J. Org. Chem.* **2004**, *69*, 4767; g) B.-G. Wie, J. Chen, P.-Q. Pei, *Tetrahedron* **2005**, *62*, 190.

56 D. Enders, J. H. Kirchhoff, *Synthesis* **2000**, 2099.

57 N. Takada, K. Suenaga, K. Yamada, S.-Z. Zheng, H.-S. Chen, D. Uemura, *Chem. Lett.* **1999**, 1025.

58 a) K. Suenaga, K. Araki, T. Senguko, D. Uemura, *Org. Lett.* **2001**, *3*, 527; b) K. Araki, K. Suenaga, T. Sengoku, D. Uemura, *Tetrahedron* **2002**, *58*, 1983.

59 P. Van de Weghe, D. Aoun, J.-G. Boiteau, J. Eustache, *Org. Lett.* **2002**, *4*, 4105.

60 D. Enders, A. Lenzen, *Synlett* **2003**, 2185.

61 W. Keller-Schierlein, D. Wuthier, H. Drautz, *Helv. Chim. Acta* **1983**, *66*, 1253.

62 S. Grabley, P. Hammann, H. Kluge, J. Wink, *J. Antibiotics* **1991**, *44*, 797.

63 S. Mizutani, H. Odai, T. Masuda, M. Iijima, M. Osono, M. Hamada, H. Naganawa, M. Ishizuka, T. Takeuchi, *J. Antibiotics* **1989**, *42*, 952.

64 a) M. Franck-Neumann, P. Bissinger, P. Geoffroy, *Tetrahedron Lett.* **1995**, *38*, 4469; b) M. Hayashi, H. Kaneda, N. Oguni, *Tetrahedron: Asymmetry* **1995**, *6*, 2511.

65 a) S. Blechert, H. Dollt, *Liebigs Ann.* **1996**, 2135; b) H. Dollt, P. Hammann, S. Blechert, *Helv. Chim. Acta* **1999**, *82*, 1111.

66 a) D. H. R. Barton, S. W. J. McCombie, *J. Chem. Soc. Perkin Trans.* **1975**, *1*, 1574; b) D. Enders, T. Hundertmark, C. Lampe, U. Jegelka, I. Scharfbillig, *Eur. J. Org. Chem.* **1998**, 2839.

67 D. Enders, T. Hundertmark, *Eur. J. Org. Chem.* **1999**, 751.

68 D. Enders, M. Voith, A. Lenzen, *Angew. Chem.* **2005**, *117*, 1330; *Angew. Chem. Int. Ed.* **2005**, *44*, 1304.

69 a) W. P. Griffith, S. V. Ley, G. P. Whitecombe, A. D. White, *J. Chem. Soc. Chem. Commun.* **1989**, 1625; b) S. V. Ley, J. Norman, W. P. Griffith, S. P. Marsden, *Synthesis* **1994**, 639.

70 H. E. Ostmark, *Annu. Rev. Entomol.* **1974**, *19*, 161.

71 W. J. Budenberg, I. O. Ndiege, F. W. Karago, *J. Chem. Ecol.* **1993**, 1905.

72 J. Beauhaire, P.-H. Ducrot, C. Malosse, D. Rochat, I. O. Ndiege, D. O. Otieno, *Tetrahedron Lett.* **1995**, *36*, 1043.

73 J. Beauhaire, P.-H. Ducrot, *Bioorg. Med. Chem.* **1996**, *4*, 413.

74 P.-H. Ducrot, *Synth. Comm.* **1996**, *26*, 3923.

75 I. O. Ndiege, S. Jayaraman, A. C. Oehlschlager, L. Gonzalez, D. Alpizar, M. Fallas, *Naturwissenschaften* **1996**, *83*, 280.

76 T. Nakayama, K. Mori, *Liebigs Ann.* **1997**, 1075.

77 M. T. Fletcher, C. J. Moore, W. Kitching, *Tetrahedron Lett.* **1997**, *38*, 3475.

78 a) D. J. Wardrop, R. E. Forslund, *Tetrahedron Lett.* **2002**, *43*, 737; b) D. J. Wardrop, R. E. Forslund, C. L. Landrie, A. I. Velter, D. Wink, B. Surve, *Tetrahedron: Asymmetry* **2003**, *14*, 929.
79 D. Enders, I. Breuer, A. Nühring, *Eur. J. Org. Chem.* **2005**, 2677.
80 a) W. Hartwig, *Tetrahedron* **1983**, *39*, 2609. b) W. B. Motherwell, D. Crich, in *Free Radical Chain Reactions in Organic Synthesis*, Academic Press: London, **1992**.
81 A. G. Myers, L. McKinstry, *J. Org. Chem.* **1996**, *61*, 2428.
82 K. U. Bindseil, A. Zeeck, *Helv. Chim Acta* **1993**, *76*. 150.
83 a) U. Hanefeld, A. M. Hooper, J. Staunton, *Synthesis* **1999**, 401; b) T. K. Chakraborty, S. Tapadar, *Tetrahedron Lett.* **2003**, *44*, 2541. For recent syntheses see: c) P. M. Pihko, A. Erkkila, *Tetrahedron Lett.* **2003**, *44*, 7607; d) A. G. Csaky, M. Mba, J. Plumet, *Synlett* **2003**, 2092; e) L. C. Dias, J. L. Steil, V. de A. Vasconcelos, *Tetrahedron: Asymmetry* **2004**, *15*, 147; f) J. S. Yadav, K. B. Reddy, G. Sabitha, *Tetrahedron Lett.* **2004**, *45*, 6475; g) V. K. Aggarwal, I. Bae, H.-Y. Lee, *Tetrahedron* **2004**, *60*, 9725; h) J. S. Yadav, M. S. Reddy, A. R. Prasad, *Tetrahedron Lett.* **2005**, *46*, 2133; i) A. A. Salaskar, N. V. Mayekar, A. Sharma, S. K. Nayak, A. Chattopadhyaya, S. Chattopadhyaya, *Synthesis* **2005**, 2777; j) S. Chandrasekaran, C. Rambabu, S. J. Prakash, *Tetrahedron Lett.* **2006**, *47*, 1213.
84 D. Enders, M. Haas, *Synlett*, **2003**, 2182.
85 R. H. Crabtree, M. W. Davis, *J. Org. Chem.* **1986**, *51*, 2655.

1.3
Asymmetric Synthesis Based on Sulfonimidoyl-Substituted Allyltitanium Complexes
Hans-Joachim Gais

1.3.1
Introduction

The phenylsulfonimidoyl group encompasses an almost unique combination of features (Scheme 1.3.1) including chirality, carbanion stabilization (Eqs. 1, 2), nucleofugacity (Eqs. 3, 4), Brønsted basicity (Eq. 5), nucleophilicity (Eq. 6), Lewis basicity (Eq. 7) and *ortho*-directed metallation (Eq. 8) [1]. Fine tuning of some of these features, including the acidity and nucleofugacity, can be achieved through choice of the substituent at the N atom (Fig. 1.3.1). For example, the acidity of *N,S*-dimethylphenylsulfoximine (pK_a = 33) increases strongly upon replacement of the *N*-methyl group by the tolylsulfonyl group (pK_a = 27.7) and upon methylation of the N atom (to pK_a = 14.4) [2]. The nucleofugacity of the sulfonimidoyl group increases strongly in the same direction.

We have developed asymmetric syntheses of isocarbacyclin [3] (Scheme 1.3.2) and cicaprost [4] (Scheme 1.3.3) featuring a Cu-mediated allylic alkylation of an allyl sulfoximine [5–7] and a Ni-catalyzed cross-coupling reaction of a vinyl sulfoximine [8–10], respectively, transformations that were both developed in our laboratories. The facile synthesis of an allyl sulfoximine by the addition–elimination–isomerization route aroused interest in the synthesis of sulfonimidoyl-substituted allyltitanium complexes of types **1** and **2** (Fig. 1.3.2) and their application as chiral heteroatom-substituted allyl transfer reagents [11].

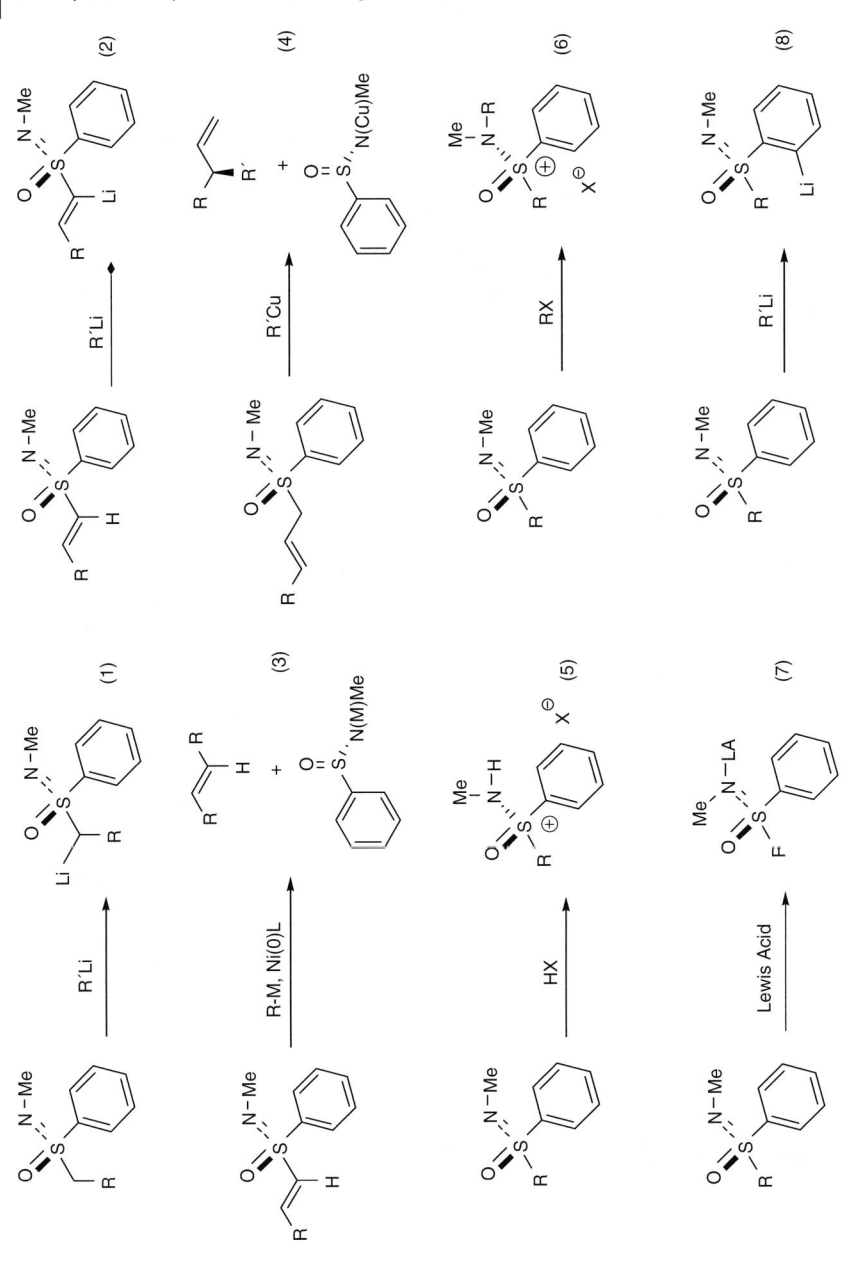

76 *1.3 Asymmetric Synthesis Based on Sulfonimidoyl-Substituted Allyltitanium Complexes*

Scheme 1.3.1 Chemical features of phenylsulfoximines.

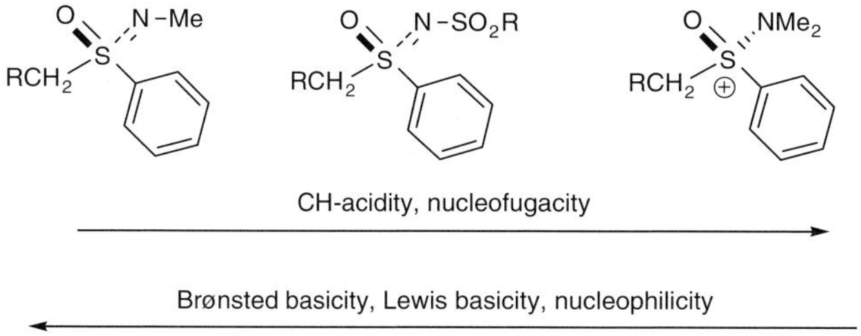

1.3.1 Introduction

Fig. 1.3.1 Fine-tuning of the properties of the sulfonimidoyl group through choice of the N-substituent(s).

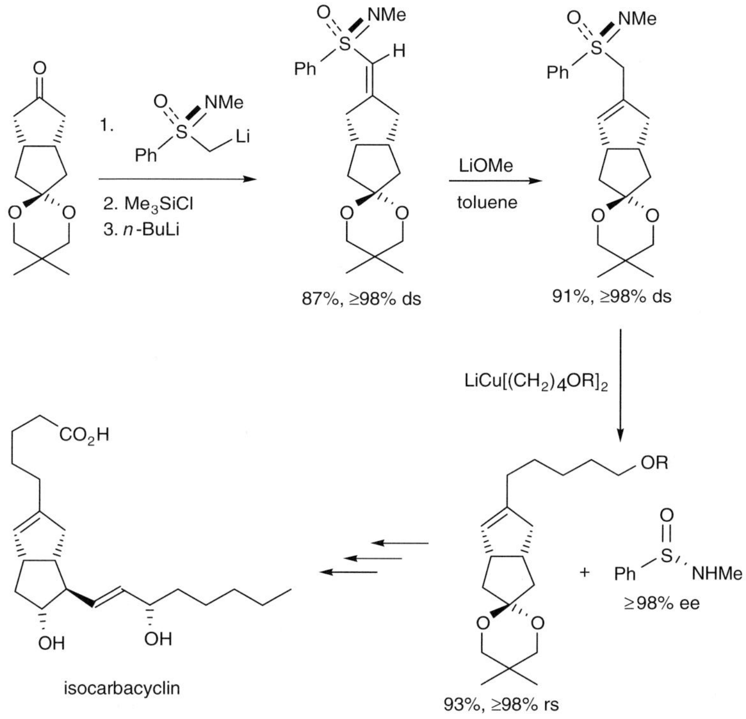

Scheme 1.3.2 Asymmetric synthesis of isocarbacyclin by the vinyl-allyl sulfoximine route.

We were particularly interested to see whether a regio- and stereoselective hydroxyalkylation and aminoalkylation of **1** and **2** with aldehydes and imino esters, perhaps by choice of the substituent X at the Ti atom, with formation of the corresponding sulfonimidoyl-substituted homoallyl alcohols **4–7** and the homoallyl amines **8–11** (Fig. 1.3.3) could be achieved. Reggelin et al. had already demonstrated that the sulfonimidoyl-substituted mono(allyl)titanium complexes **3**, the

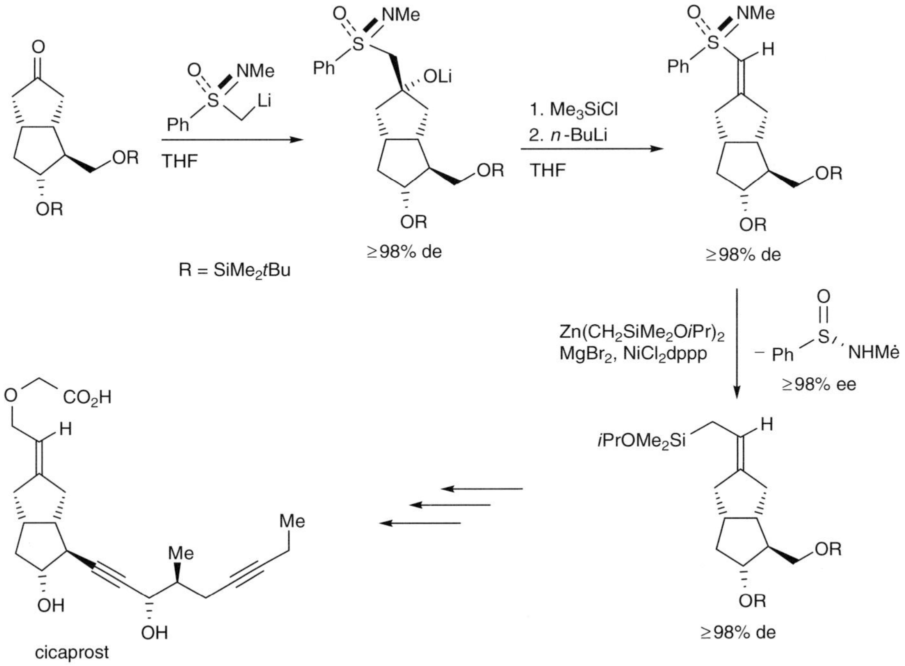

Scheme 1.3.3 Asymmetric synthesis of cicaprost by the vinyl sulfoximine route.

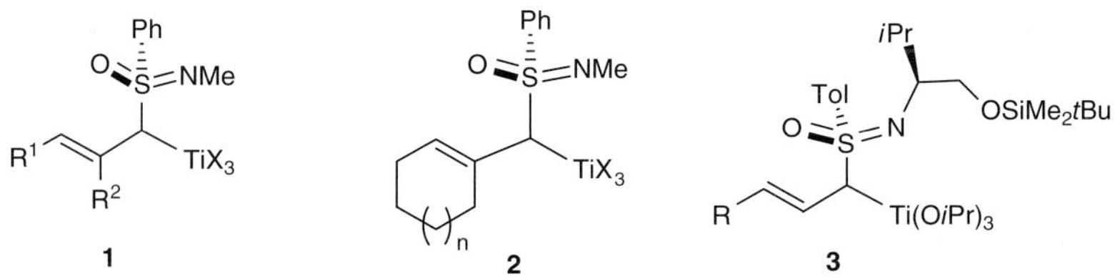

Fig. 1.3.2 Chiral sulfonimidoyl-substituted allyltitanium complexes.

N atom of which carries a chiral substituent, react with aldehydes with high selectivities with formation of homoallyl alcohols of type **4**, which were used in the asymmetric synthesis of complex molecules [12].

Because of high synthetic versatility of the sulfonimidoyl group, compounds of type **4–11** that carry a vinylic and allylic sulfonimidoyl group, respectively, could perhaps make excellent starting materials for the asymmetric synthesis of a number of interesting compounds including homopropargylic alcohols, 2,3-dihydrofurans, alkenyl oxiranes, unsaturated mono- and bicyclic proline derivatives, unsaturated mono- and bicyclic α-amino acids, and β-amino acids (vide infra). For example, vinyl sulfoximines (1) are excellent Michael acceptors, (2) they can be readily metallated at the α-position, and (3) their sulfonimidoyl group acts as

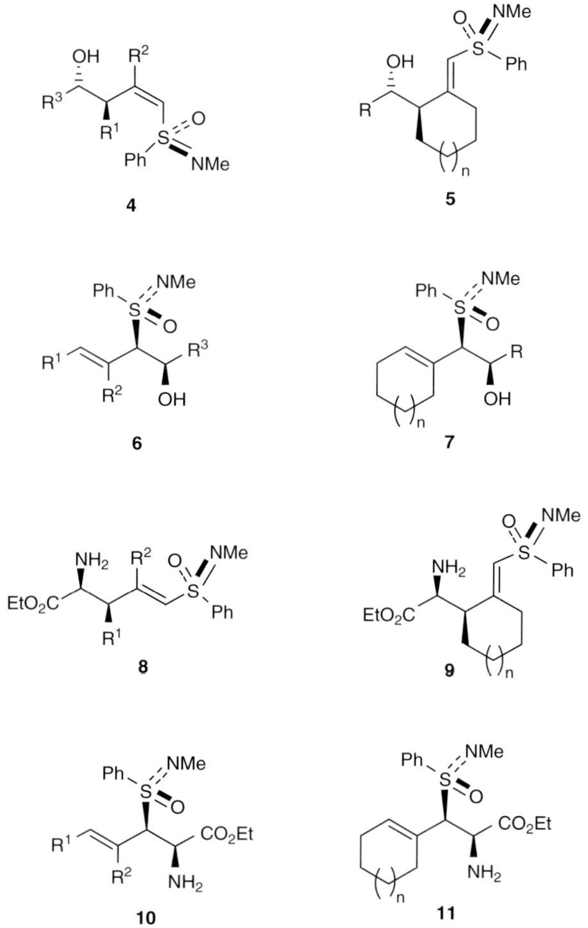

Fig. 1.3.3 Sulfonimidoyl-substituted homoallyl alcohols and homoallylamines.

a nucleofuge in transition metal catalyzed cross-coupling reactions with organometallics. Allyl sulfoximines in turn are amenable to S_N- and $S_{N'}$ reactions with copper organyls and are easily lithiated at the α-position. Finally the sulfonimidoyl group can be selectively activated and turned into an excellent nucleofuge through methylation at the N atom. Sulfonimidoyl-substituted allyltitanium complexes **1** and **2** could thus perhaps offer synthetic possibilities not attainable with the titanium complexes derived from lithiated allyl carbamates, which are synthetically highly valuable chiral heteroatom-substituted allyltitanium reagents for the allylation of aldehydes and ketones [11a].

The enantiopure acyclic and cyclic allyl sulfoximines **13** and **14**, respectively, required for the synthesis of the corresponding titanium complexes **1** and **2**, are available from sulfoximine **12** [13] and the corresponding aldehydes and cycloalkanones by the addition–elimination–isomerization route, which can be carried

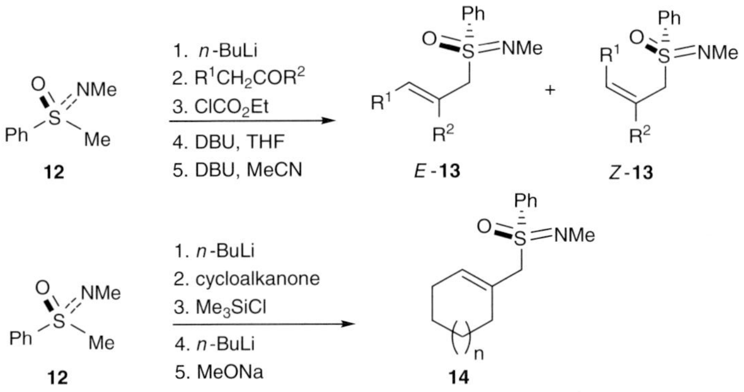

Scheme 1.3.4 Synthesis of acyclic and cyclic allyl sulfoximines by the addition–elimination–isomerization route.

out without the isolation of any intermediate (Scheme 1.3.4) [14]. The acyclic allyl sulfoximines are generally obtained as E/Z mixtures, the E isomer being by far the dominant one. In the case of R^1 = aryl and tBu, only the E isomer is obtained. The enantiopure sulfoximines **12** and ent-**12** can be obtained on a large scale through a highly efficient resolution with ω-camphor sulfonic acid, requiring no recrystallization, by the method of half-quantities [13], that is the use of 0.5 equiv. of the acid.

1.3.2
Hydroxyalkylation of Sulfonimidoyl-Substituted Allylltitanium Complexes

1.3.2.1 Sulfonimidoyl-Substituted Bis(allyl)titanium Complexes

The lithiation of the E-configured acyclic allyl sulfoximines E-**13** with n-BuLi gave the corresponding lithiated allyl sulfoximines E-**15** [15] which upon treatment with 1.1 equiv of ClTi(OiPr)$_3$ at −78 to 0 °C in THF furnished the bis(allyl)titanium complexes E-**16**, admixed with equimolar amounts of Ti(OiPr)$_4$, in practically quantitative yields (Scheme 1.3.5) [14, 16]. Surprisingly the bis(allyl)titanium complexes E-**16** together with Ti(OiPr)$_4$ and not the corresponding mono(allyl)titanium complexes were formed.

Reaction of the bis(allyl)titanium complexes E-**16** with saturated and unsaturated aldehydes at −78 °C in the presence of 1.1 equiv of ClTi(OiPr)$_3$ afforded the corresponding Z-anti-configured homoallyl alcohols **4** with ≥98% regioselectivity and ≥98% diastereoselectivity in good yields (Scheme 1.3.6) [14].

The treatment of the cyclic allyl sulfoximines **14** with n-BuLi gave the corresponding lithiated allyl sulfoximines **17**, which upon reaction with ClTi(OiPr)$_3$ afforded the cyclic bis(allyl)titanium complexes **18** and equimolar amounts of Ti(OiPr)$_4$ in practically quantitative yields [14]. The hydroxyalkylation of the cyclic complexes **18** with saturated and unsaturated aldehydes also proceeded with ≥98% regioselectivity and ≥98% diastereoselectivity at the γ-position and gave the corresponding Z-anti-configured homoallyl alcohols **5** in good yields [14].

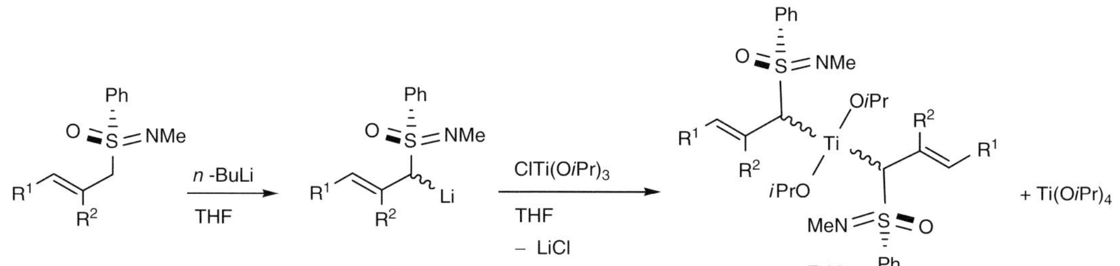

1.3.2 Hydroxyalkylation of Sulfonimidoyl-Substituted Allyltitanium Complexes

R^1 = Me, Et, iPr, cC_6H_{11}, tBu, Ph, Naphthyl; R^2 = H, Me, Et; R = iPr

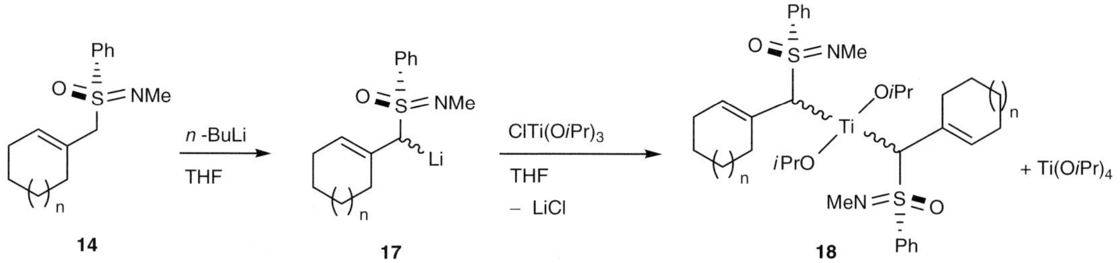

Scheme 1.3.5 Synthesis of cyclic and acyclic chiral sulfonimidoyl-substituted bis(allyl)titanium complexes.

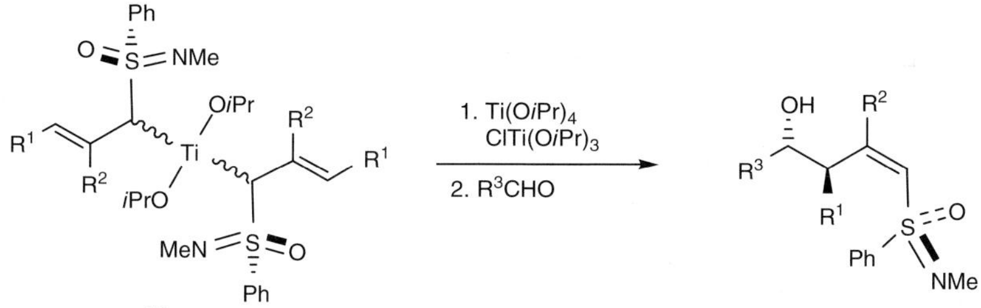

R^1 = Me, Et, iPr, cC_6H_{11}, tBu, Ph
R^2 = H, Me, Et
R^3 = Me, Et, iPr, Ar, CH=CHPh, C CPh, furanyl

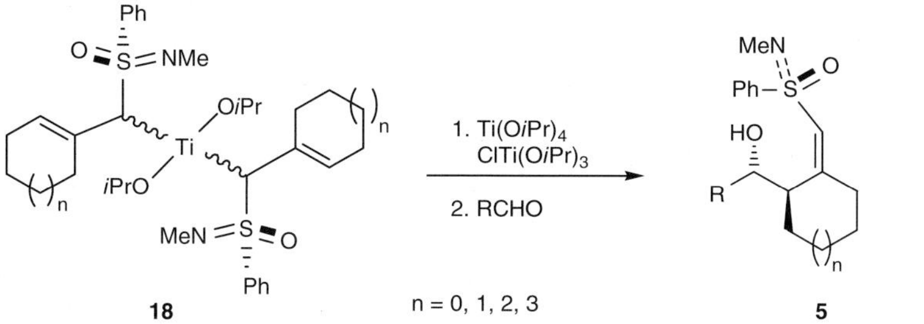

Scheme 1.3.6 Asymmetric synthesis of sulfonimidoyl-substituted homoallyl alcohols.

Reaction of the bis(allyl)titanium complexes **16** and **18** with aldehydes occurs in a step-wise fashion with intermediate formation of the corresponding mono(allyl)titanium complex containing the alcoholate derived from **4** and **5** as a ligand at the Ti atom. Then the mono(allyl)titanium complexes combine with a second molecule of the aldehyde. Both the bis(allyl)titanium complexes and the mixed mono(allyl)titanium complexes react with the aldehydes at low temperatures with high regio- and diastereoselectivities. Interestingly, control experiments revealed that for the reaction of the bis(allyl)titanium complexes with the aldehyde to occur the presence of Ti(OiPr)$_4$ is required, and for that of the intermediate mono(allyl)titanium complexes the addition of ClTi(OiPr)$_3$ is mandatory (vide infra).

1.3.2.2 Sulfonimidoyl-Substituted Mono(allyl)tris(diethylamino)titanium Complexes

The treatment of the lithiated allyl sulfoximines E-**15** with 1.1–1.2 equiv of ClTi(NEt$_2$)$_3$ at −78 to 0 °C in THF or ether afforded the corresponding mono(allyl)titanium complexes E-**19** in practically quantitative yields (Scheme 1.3.7) [14, 16]. Similarly the Z-configured complexes Z-**19** were obtained from the Z-configured allyl sulfoximines Z-**15**. Reaction of the titanium complexes E-**19** with aldehydes at −78 °C took place at the α-position and gave the corresponding homoallyl alcohols **6** with ≥98% diastereoselectivity in medium to good yields (Scheme 1.3.8) [14, 16].

However, a more detailed study of the reaction of the mono(allyl)titanium complexes E-**19** carrying different alkyl groups at the double bond with different aldehydes revealed in some cases the highly diastereoselective (≥98%) formation of significant amounts of the isomeric homoallyl alcohols **4** besides **6** (Table 1.3.1).

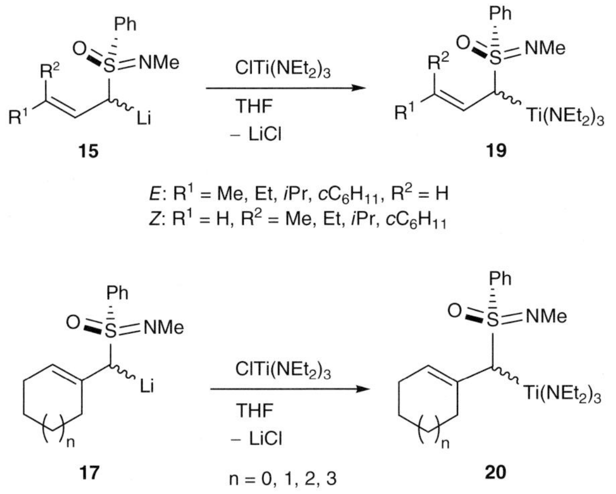

Scheme 1.3.7 Synthesis of chiral sulfonimidoyl-substituted mono(allyl)titanium complexes.

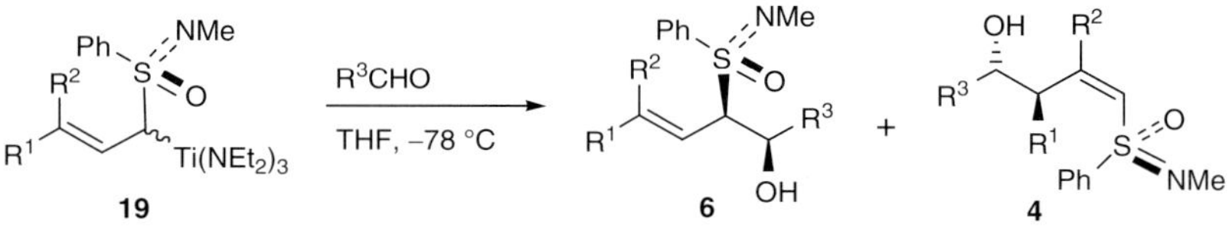

1.3.2 Hydroxyalkylation of Sulfonimidoyl-Substituted Allylltitanium Complexes

Scheme 1.3.8 Asymmetric synthesis of acyclic sulfonimidoyl-substituted homoallyl alcohols.

E: R^1 = Me, Et, iPr, cC_6H_{11}, R^2 = H
Z: R^1 = H, R^2 = Me, Et, iPr, cC_6H_{11}

Table 1.3.1 Reaction of the acyclic mono(allyl)titanium complexes E-19 with aldehydes.

Entry	E-19	Aldehyde	E-6:4	E-6, ds [%]	4, ds [%]
1	Me	MeCHO	1:1.5	≥98	92
2	Me	EtCHO	4.1:1	≥98	≥98
3	Me	EtCHO	7.2:1	≥98	≥98
4	Me	iPrCHO	23:1	≥98	≥98
5	Et	MeCHO	1.1:1	≥98	≥98
6	Et	iPrCHO	31:1	≥98	≥98
7	iPr	MeCHO	1.8:1	≥98	90
8	iPr	EtCHO	21:1	≥98	–
9	iPr	iPrCHO	≥100:1	≥98	–
10	iPr	PhCHO	≥100:1	≥98	–
11	cC_6H_{11}	iPrCHO	≥100:1	≥98	–
12	cC_6H_{11}	PhCHO	≥100:1	≥98	–

The regioselectivity of the hydroxyalkylation of E-19 is strongly dependent on the size of both the aldehyde and the substitutuent R^1 of the titanium complex. The α,γ-selectivity increases strongly with increasing size of both the aldehyde and R^1. From an inspection of Table 1.3.1 it seems that the size of the aldehyde has a stronger effect upon the α-selectivity than the size of R^1 of the titanium complex. In summary, a highly regioselective α-hydroxyalkylation of E-19 can generally be achieved except in those cases where both reactants carry small groups. It is noteworthy that, in general, the isomeric homoallyl alcohols **4** and **6** are both formed with high diastereoselectivity.

The titanation of Z-**15** with 1.1–1.2 equiv of ClTi(NEt$_2$)$_3$ at –78 to 0 °C in THF and reaction of the thus formed Z-configured titanium complexes Z-**19** with aldehydes gave rise to the isolation of the corresponding diastereopure Z-syn-configured homoallyl alcohols Z-**6** in medium to good yields (Table 1.3.2). Here, too, the regioselectivity of the hydroxyalkylation was studied by variation of the aldehyde and of the substituent R^2 of the titanium complex. An inspection

Table 1.3.2 Reaction of the acyclic mono(allyl)titanium complexes Z-**19** with aldehydes.

Entry	Z-**19**	Aldehyde	Z-**6**:**4**	E-**6**, ds [%]
1	Me	MeCHO	8.5:1	≥98
2	Me	EtCHO	≥100:1	≥98
3	Et	MeCHO	≥100:1	≥98
4	Et	iPrCHO	≥100:1	≥98
5	iPr	iPrCHO	≥100:1	≥98
6	iPr	PhCHO	≥100:1	≥98

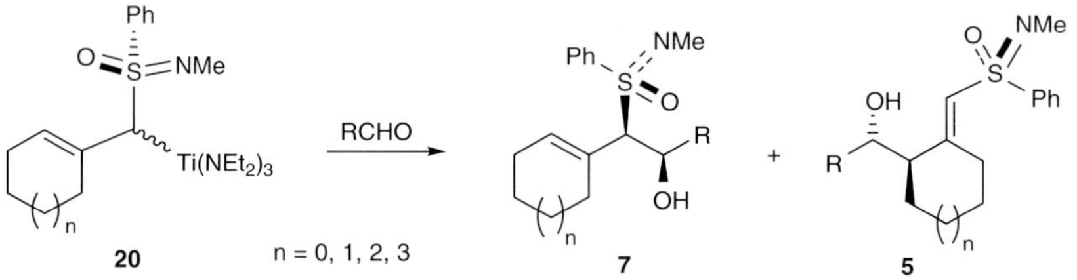

Scheme 1.3.9 Asymmetric synthesis of cyclic sulfonimidoyl-substituted homoallyl alcohols.

of Table 1.3.2 reveals that in the case of the Z-configured complexes Z-**19** the regioselectivity, which increases with increasing temperature, is significantly higher than in the case of the corresponding E-configured complexes E-**19**. Only in the reaction of the methyl-substituted complex E-**19** (R^1 = H, R^2 = Me) with ethanal was the formation of a mixture of Z-**6** (R^1 = H, R^2 = Me) and the E isomer of **4** (R^1 = Me) observed. In all cases investigated the Z-configured allyl alcohols were formed with high diastereoselectivity.

Selective α-hydroxyalkylation could be extended to the cyclic titanium complexes **20** as well [14]. Titanation of the lithiated allyl sulfoximine **17** with ClTi(NEt$_2$)$_3$ and the subsequent reaction of the corresponding mono(allyl)titanium complex **20** with aldehydes gave with medium to high α-regioselectivities and high diastereoselectivities the corresponding Z-syn-configured homoallyl alcohols **7** (Scheme 1.3.9, Table 1.3.3). An inspection of Table 1.3.3 reveals a similar dependency of the regioselectivity on the size of the aldehyde, as in the case of the hydroxyalkylation of the acyclic titanium complexes E-**19**. The α-selectivity strongly increases with increasing size of the aldehyde.

The structure of the sulfonimidoyl-substituted homoallyl alcohols **4**–**7** in solution and in the crystal is characterized by an intramoleculare hydrogen bond between the OH group and the N atom of the sulfonimidoyl group.

1.3.3 Aminoalkylation of Sulfonimidoyl-Substituted Allyltitanium Complexes

Table 1.3.3 Reaction of the cyclic mono(allyl)titanium complexes **20** ($n = 1$) with aldehydes.

Entry	Aldehyde	7:5	7, ds [%]	5, ds [%]
1	MeCHO	2.6:1	95	≥98
2	EtCHO	13:1	≥98	≥98
3	iPrCHO	22:1	≥98	≥98
4	PhCHO	≥100:1	95	≥98

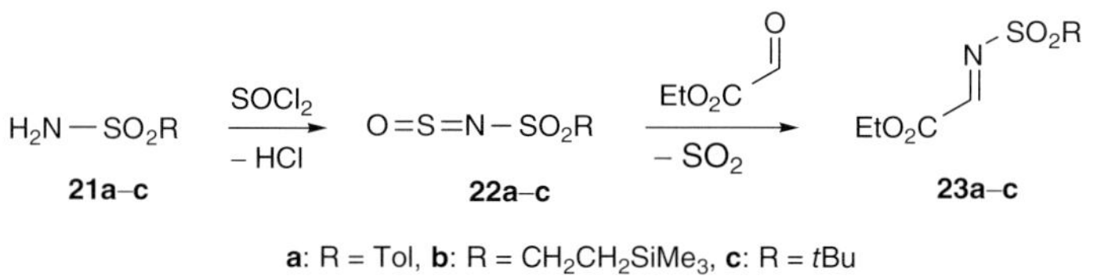

$$H_2N-SO_2R \xrightarrow[-HCl]{SOCl_2} O=S=N-SO_2R \xrightarrow[-SO_2]{EtO_2C-CHO} EtO_2C-CH=N-SO_2R$$

21a–c → **22a–c** → **23a–c**

a: R = Tol, **b**: R = CH$_2$CH$_2$SiMe$_3$, **c**: R = tBu

Scheme 1.3.10 Synthesis of N-sulfonyl imino esters.

1.3.3 Aminoalkylation of Sulfonimidoyl-Substituted Allyltitanium Complexes

1.3.3.1 Sulfonimidoyl-Substituted Bis(allyl)titanium Complexes

The N-sulfonyl imino esters **23a–c** (Scheme 1.3.10) were selected for the investigation of the aminoalkylation of the titanium complexes **16** and **18**. The N-tolylsulfonyl imino ester **23a** [17] has frequently been used in alkylation and allylation reactions. It is readily accessible from sulfonamide **21a** by the route shown in Scheme 1.3.10, which involves the treatment of sulfonamide **21a** with SOCl$_2$ with formation of thioisocyanate **22a** and its reaction with ethyl glyoxylate to give **23a** [17]. However, a major drawback associated frequently with the utilization of **23a** has been that the N-Tos group of the corresponding amine derivatives obtained through addition of RM to the double bond is difficult to remove [18]. Therefore the N-(trimethylsilyl)ethylsulfonyl imino ester **23b** and the N-t-butylsulfonyl imino ester **23c** were prepared in 67% and 93% overall yields, respectively, by the route depicted in Scheme 1.3.10 [19]. N-(Trimethylsilyl)ethyl sulfonyl- and N-t-butylsulfonyl-protected amines can generally be cleaved much more readily than N-tolylsulfonyl-protected amines [20].

The reaction of the acyclic bis(allyl)titanium complexes **16** with the imino esters **23a–c** in the presence of Ti(OiPr)$_4$ and ClTi(OiPr)$_3$ at low temperatures proceeded with ≥98% regioselectivity and ≥98% diastereoselectivity and gave the corresponding E-syn-configured unsaturated α-amino acid derivatives E-**24** in good yields (Scheme 1.3.11) [21, 22].

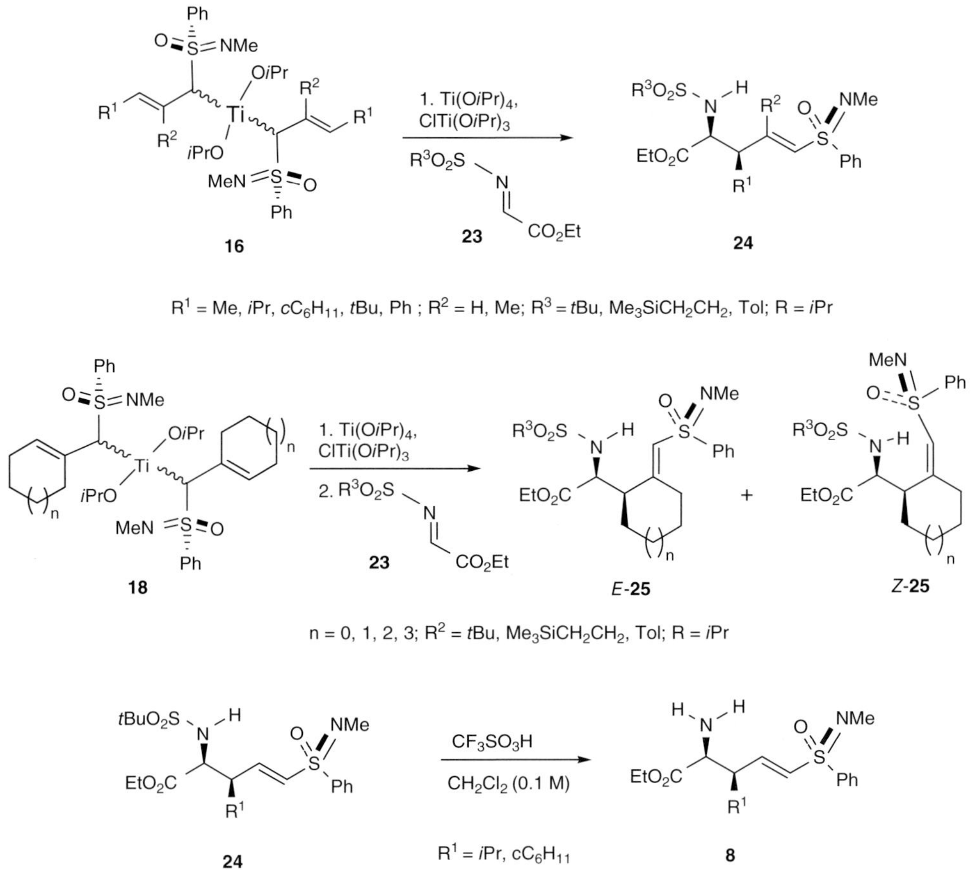

86 *1.3 Asymmetric Synthesis Based on Sulfonimidoyl-Substituted Allyltitanium Complexes*

R^1 = Me, iPr, cC_6H_{11}, tBu, Ph ; R^2 = H, Me; R^3 = tBu, Me$_3$SiCH$_2$CH$_2$, Tol; R = iPr

n = 0, 1, 2, 3; R^2 = tBu, Me$_3$SiCH$_2$CH$_2$, Tol; R = iPr

R^1 = iPr, cC_6H_{11}

Scheme 1.3.11 Asymmetric synthesis of unsaturated sulfonimidoyl-substiuted α-amino acids.

It is noteworthy that stereoselectivities were high, even with sterically demanding substituents at the double bond. Similarly, the treatment of the cyclic bis(allyl)titanium complexes **18** with the imino esters **23a–c** afforded the corresponding *E-syn*-configured cyclic unsaturated amino acid derivatives *E*-**25** and the *Z-syn*-configured isomers *Z*-**25** with ≥98% regioselectivity and ≥98% diastereoselectivity in good yields.

While a deprotection of the *N*-tolylsulfonyl amino acid derivatives **24** (R^3 = Tol) could not be effected the *N-t*-butylsulfonyl derivatives **24** (R^3 = *t*-butyl) were readily cleaved with CF$_3$SO$_3$H in anhydrous CH$_2$Cl$_2$ and afforded the amino acid derivatives **8** in high yields [21].

1.3.3.2 Sulfonimidoyl-Substituted Mono(allyl)tris(diethylamino)titanium Complexes
Surprisingly, the mono(allyl)titanium complexes **19** reacted with the imino ester **23c** also at the γ-position with high diastereoselectivities and gave the unsaturated

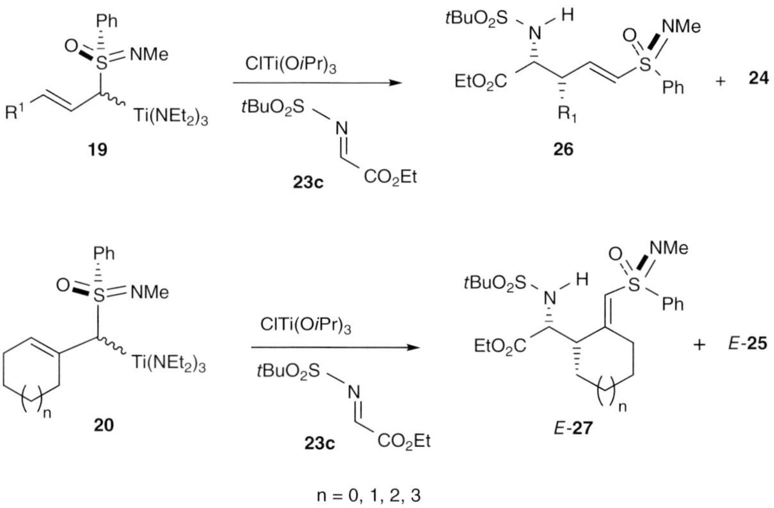

1.3.3 Aminoalkylation of Sulfonimidoyl-Substituted Allyltitanium Complexes

Scheme 1.3.12 Asymmetric synthesis of the α-amino acid derivatives **26** and *E*-**27**.

Table 1.3.4 Reaction of the acyclic mono(allyl)titanium complexes **19** with the imino ester **23c**.

R^1	26:24	26		24	
		Yield [%]	ds [%]	Yield [%]	ds [%]
*i*Pr	2:1	56	≥98	10	≥98
cC_6H_{11}	6:1	58	≥98	32	≥98

α-amino acid derivatives **26** and **24** in ratios of 2 : 1 and 6 : 1 (Scheme 1.3.12, Table 1.3.4) [23].

A similar observation was made in the case of the cyclic mono(allyl)titanium complexes **20**. The treatment of complexes **20** with the imino ester **23c** afforded the corresponding cyclic α-amino acids *E*-**27** and *E*-**25** in ratios ranging from 6 : 1 to 46 : 1 (Scheme 1.3.12, Table 1.3.5). The ratio of the two diastereomers is strongly dependent on the ring size of the titanium complex **20**, being the highest for the eight-membered cyclic complex ($n = 3$). Thus the mono(allyl)titanium complexes **19** and **20** and the corresponding bis(allyl)titanium complexes **16** and **18** react with the imino esters with similar regioselectivities but with different diastereoselectivities. Synthetically this is a most welcome result since both diastereomers *E*-**25** and *E*-**27**, having the opposite configuration at the two stereogenic C atoms, are accessible when starting from the allyl sulfoximines **14**, by choice of the titanation reagent.

Table 1.3.5 Reaction of the cyclic mono(allyl)titanium complexes **20** with the imino ester **23c**.

Entry	E-27:E-25	E-27		E-25	
		Yield [%]	ds [%]	Yield [%]	ds [%]
1	6:1	70	≥98	14	≥98
2	10:1	69	≥98	8	≥98
3	24:1	82	≥98	4	≥98
4	46:1	87	≥98	2	≥98

1.3.4
Structure and Reactivity of Sulfonimidoyl-Substituted Allyltitanium Complexes

1.3.4.1 Sulfonimidoyl-Substituted Bis(allyl)titanium Complexes

The bis(allyl)titanium complexes **16** and **18** react with both aldehydes and the N-sulfonyl imino esters **23a–c**, independently of the substitutents R^1 and R^2 and the size of the carbocyclic ring, with high diastereo- and regioselectivities at the γ-position, and give the *anti*-configured homoallyl alcohols **4** and **5** and the *syn*-configured homoallyl amines **24** and **25**, respectively. In contrast, the mono(allyl)titanium complexes **19** and **20** react with aldehydes with high diastereoselectivity at the α position to afford the regioisomeric *syn*-configured homoallyl alcohols **6** and **7**. Remarkably, the regioselectivity is characteristically dependent on the size of both substituents R^1 and R^2 and the ring size. While in the case of larger substituents and rings a highly regio- and diastereoselective α-hydroxyalkylation of **19** and **20** with formation of **6** and **7**, respectively, occurs, smaller substituents and rings result in both α- and γ-hydroxyalkylations (again, highly diastereoselective) with formation of **4/6** and **5/7**, respectively. Finally, the reactions of the mono(allyl)titanium complexes **19** and **20** with the N-sulfonyl imino esters are characterized by a dichotomy not in regio- but in diastereoselectivity. Complexes **19** and **20** react with high regioselectivity at the γ-position and deliver the *syn*-configured homoallyl amines **26** and *E*-**27** together with small amounts of diastereomers **24** and *E*-**25**, both, however, with high diastereoselectivities. These results strongly suggested that the Cα atoms of the titanium complexes **16**, **18**, **19**, and **20** are configurationally labile, resulting in the existence of equilibria of several reactive isomeric allyltitanium complexes. Therefore we carried out structural studies of the sulfonimidoyl-substituted allyltitanium complexes *E*-**16**, **18**, *E*-**19**, *Z*-**19**, and **20**. Surprisingly, despite their synthetic importance no direct information is available about the structure of other heteroatom-substituted allyltitanium complexes [2]. Structural studies of the phenyl-substituted bis(allyl)titanium complex **28** by X-ray crystal structure analysis and variable-

1.3.4 Structure and Reactivity of Sulfonimidoyl-Substituted Allyltitanium Complexes

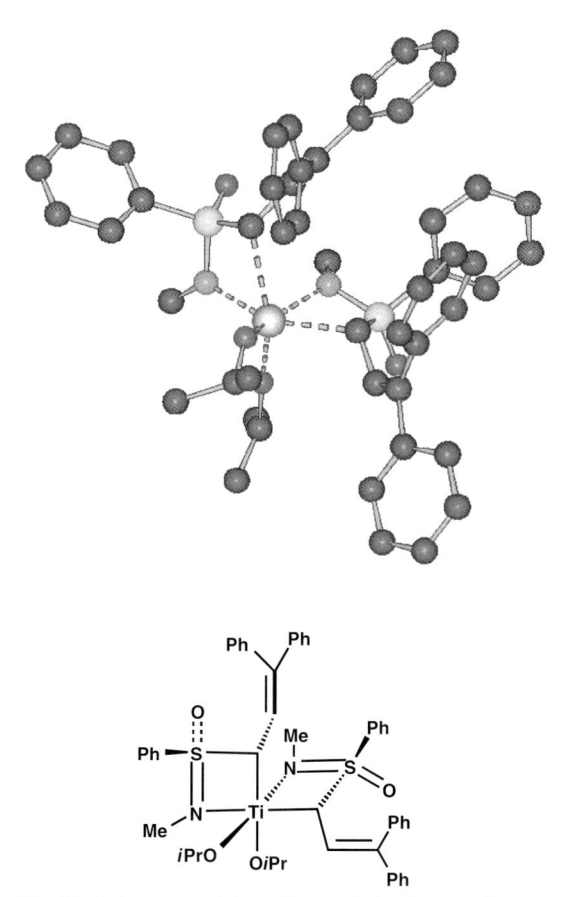

Fig. 1.3.4 Structure of the sulfonimidoyl-substituted bis(allyl)titanium complex **28** in the crystal. Selected bond lengths: Ti—C 244 and 229 pm, Ti—N 209 and 221 pm.

temperature NMR spectroscopy had revealed an octahedral complex with an η^2-coordination of the allyl sulfoximine ligands via the C and N atoms to the Ti atom (Fig. 1.3.4) [14]. NMR spectroscopy of **28** in toluene and THF showed a fast intramolecular exchange of the diastereotopic allyl sulfoximine and isopropoxy ligands, perhaps by a psendorotational mechanism [14].

According to NMR spectroscopy, complexes E-**16** have a low configurational stability of the Cα atoms, giving rise to the formation of a mixture of the diastereomeric complexes R_S,S_C-**16** and R_S,R_C-**16** (Schemes 1.3.13 and 1.3.14). Addition of the Lewis acids Ti(OiPr)$_4$ and ClTi(OiPr)$_3$ causes their coordination to one of the sulfonimidoyl groups and establishment of an equilibrium between the cor-

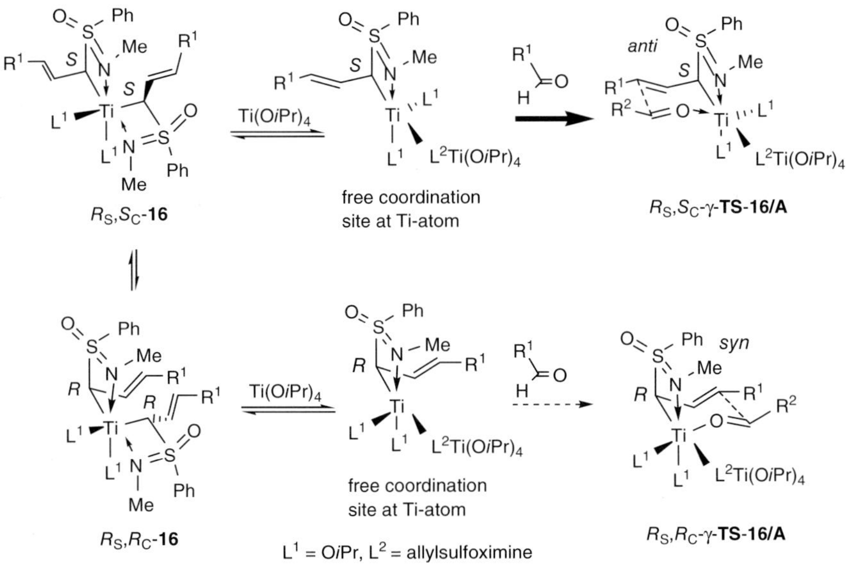

Scheme 1.3.13 Reactivity scheme for the reactions of the bis(allyl)titanium complexes **16** with aldehydes.

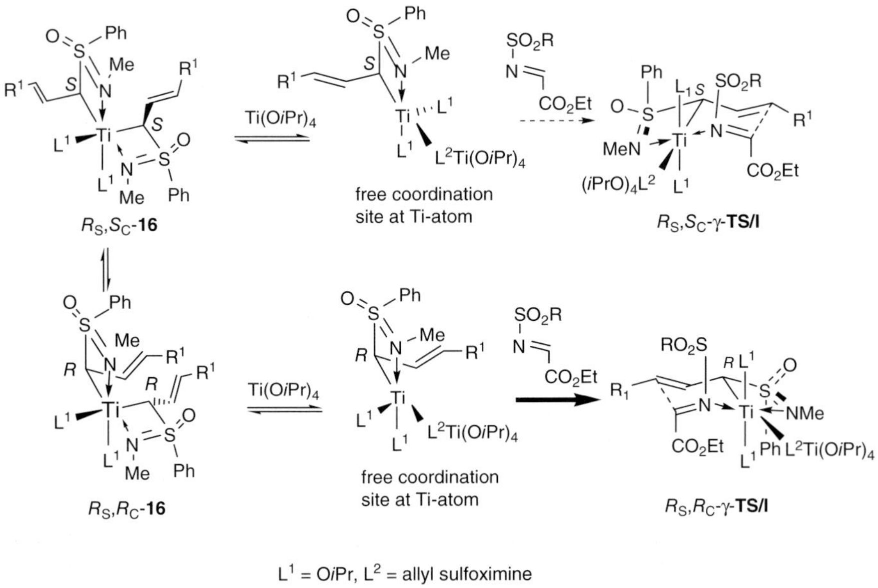

Scheme 1.3.14 Mechanistic scheme for the reactions of the bis(allyl)titanium complexes **16** with the imino esters **23**.

responding hexa- and penta-coordinated allyltitanium complexes. These features of complexes E-**16** and in particular the coordination of the Ti atom to the Cα atom and the N atom led to the proposal of reactivity schemes based on the assumption of (1) the operation of the Curtin–Hammett principle: that is, equilibration between R_S,S_C-**16** and R_S,R_C-**16** is faster than their reactions with the aldehydes and the imino esters, and (2) the attainment of a chair-like six-membered cyclic transition states featuring a coordination of the aldehyde molecule via the O atom and that of the imino ester molecule via the N atom to the Ti atom. It is proposed that, in the case of aldehydes, complexes R_S,S_C-**16** have a higher reactivity than complexes R_S,R_C-**16** because of the sterically less encumbered *anti* position of the S-phenyl group in transition state R_S,S_C-γ-**TS-16/A** as compared to R_S,R_C-γ-**TS-16/A**, the S-phenyl group of which is *syn* to the alkenyl group [1b, 12].

In the case of the reaction of **16** with the imino esters **23**, the situation is reversed. Here the complex R_S,R_C-**16** has the higher reactivity because in transition state R_S,R_C-γ-**TS-16/I** there is less steric interference between the S-phenyl group, the sulfonyl group, and the ligand L^1 than in transition state R_S,S_C-γ-**TS-16/I** derived from complexes R_S,S_C-**16**. It should be emphasized, however, that the transition states shown in Schemes 1.3.13 and 1.3.14 are only models, since their structures are not known.

1.3.4.2 Sulfonimidoyl-Substituted Mono(allyl)titanium Complexes

Variable-temperature NMR spectroscopy of the E-configured mono(allyl)titanium complexes E-**19** demonstrated the existence of a fast equilibrium between the S-configured C-titanium complex E-**19A**, the N-titanium ylide E-**19B**, and the R-configured C-titanium complex E-**19C** in ratios depending on the substituent R^1 (84 : 13 : 3 for R^1 = *i*Pr) (Scheme 1.3.15) [24]. In the case of the Z-configured complexes Z-**19** a fast equilibrium only between the S-configured C-titanium complex Z-**19A** and the N-titanium ylide Z-**19b** in ratios depending on the substituent R^2 (36 : 64 for R^2 = *i*Pr) was detected. Instrumental for the structural assignment of the equilibrium species were the magnitudes of $^1J(^{13}C,^1H)$ for the Cα atoms. In accordance with chemical evidence the establishment of an equilibrium between Z-**19** and E-**19** was not observed. In summary, complexes **19** are not configurationally stable at the Cα atom; even at low temperatures they experience a fast epimerization.

We had proposed the processes depicted in Scheme 1.3.15 in order to account for the epimerization and the existence of an equilibrium between the species detected by NMR spectroscopy at low temperatures. A 1,3-C,N shift of the Ti atom of the S-configured complex **19A** via transition state **TS-19A** generates ylide **19B1**. Rotation around the Cα—S bond of **19B1** leads to conformer **19B2** which suffers a 1,3-N,C shift with formation of the R-configured complex **19C**. Of particular interest are the N-titanium aminosulfoxonium allyl ylides **19B**. Although alkyl dimethylaminosulfoxonium ylides are valuable reagents for asymmetric synthesis, nothing is known about their structure, including the height of the Cα—S

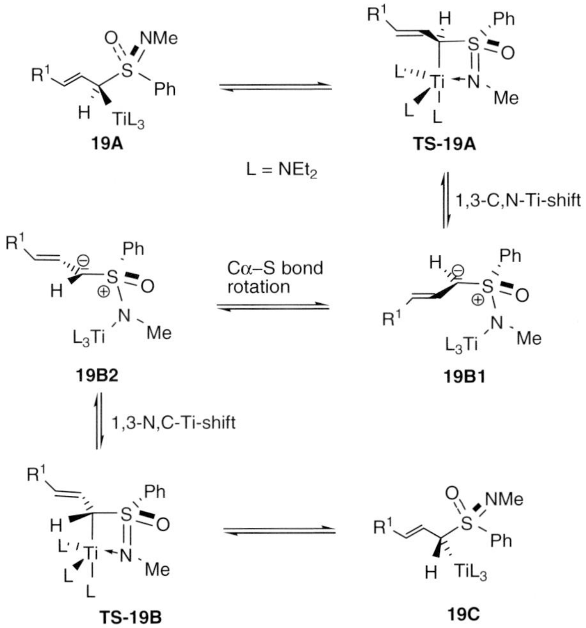

Scheme 1.3.15 Dynamics of the sulfonimidoyl-substituted mono(allyl)titanium complexes.

rotational barrier. Thus ab initio calculations (HF/6-31+G*, MP2/6-31+G*) of the methylene ylide **29** (Fig. 1.3.5) and the allyl ylide **30** (Fig. 1.3.6) were undertaken in order to gain information about the preferred Cα—S conformation and the rotational barrier [24]. The ab-initio calculations revealed short Cα—S bonds, a stabilization by both electrostatic interaction and negative hyperconjugation (n_C–σ^*_{SO} and n_C–σ^*_{SN}), and a low Cα—S rotational barrier ($\delta E_{C\alpha-S}$ = 10.2 kcal mol^{-1}, semi-rigid). Stabilization of the allyl ylide **30** through delocalization of n_C into σ_{SO}^* and π^* is of almost equal importance.

The ylides preferentially adopt Cα—S and Cα—N conformations in which the lone-pair orbital at the Cα atom is periplanar to the S=O bond, and that at the N atom is periplanar to the Cα—Ph bond. An important factor in the determination of the Cα—S and N—S energy minimum conformations of the ylides is the orthogonality of the lone pairs at the Cα and N atoms.

Because of the observation of a fast equilibrium between the C-titanium complexes **19A** and **19C** and the N-titanium ylides **19B**, the reactivity model depicted in Scheme 1.3.16 was proposed in order to account for the regio- and diastereoselectivity observed in the reaction of **19** with aldehydes. This model, which is based on the assumption of the operation of the Curtin–Hammett principle (that is, the reactions of **19A**, **19B**, and **19C** with aldehydes are significantly slower than their isomerization), features the six-membered cyclic chair-like transition states

1.3.4 Structure and Reactivity of Sulfonimidoyl-Substituted Allyltitanium Complexes

a)

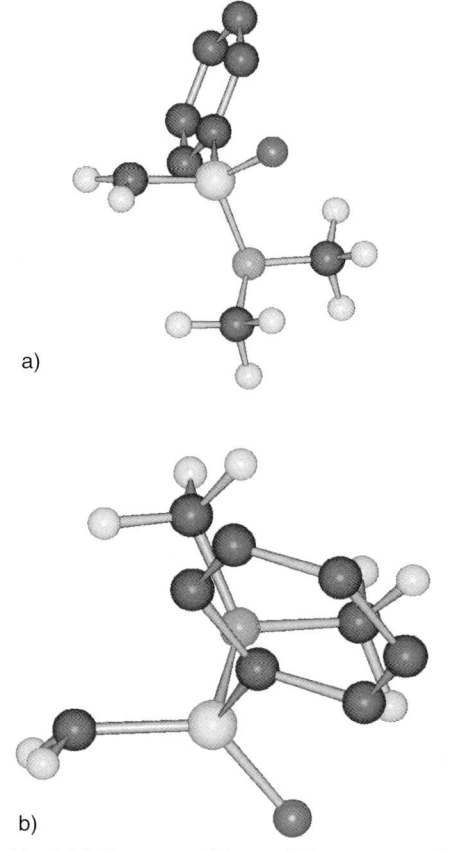

b)

Fig. 1.3.5 Structure of the methylene aminosulfoxonium ylide **29** according to ab-initio calculations. a) Rel. energy 0.0 kcal mol^{-1}; Cα—S 163 pm, σ ∠Cα 347°; b) rel. energy +1.7 kcal mol^{-1}; Cα—S 162 pm, σ ∠Cα 356°.

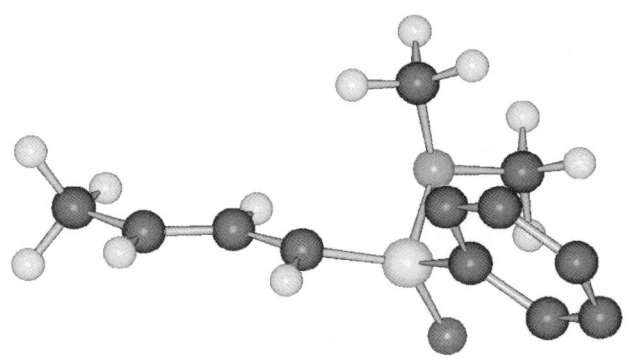

Fig. 1.3.6 Structure of the allyl aminosulfoxonium ylide **30** according to ab-initio calculations. Rel. energy 0.0 kcal mol^{-1}; Cα—S 163 pm, σ ∠Cα 356°.

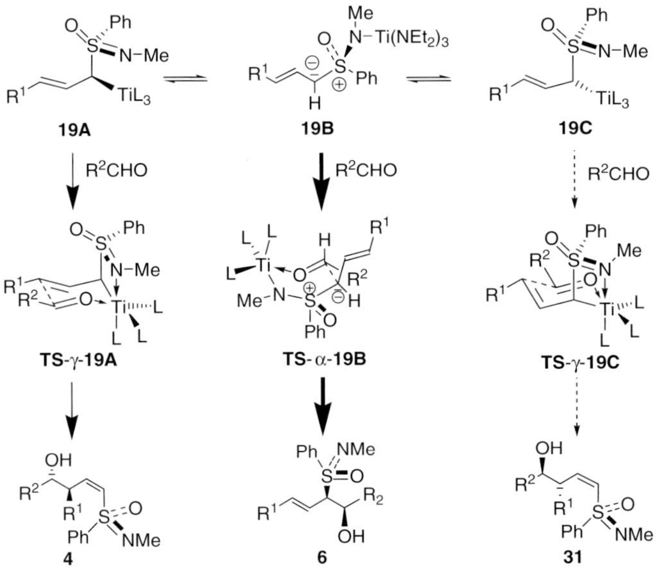

Scheme 1.3.16 Mechanistic scheme for the reactions of the sulfonimidoyl-substituted mono(allyl)titanium complexes with aldehydes.

TS-γ-19A and **TS-γ-19C** and the boat-like transition state **TS-α-19B** for the reactions of **19A**, **19C**, and **19B**, respectively. A key feature of **TS-γ-19A** and **TS-γ-19C** is the coordination of the sulfonimidoyl N atom to the Ti atom. Normally it is the N atom of the *N*-methylsulfonimidoyl group which has the highest Lewis and Brønsted basicity. The boat-like transition state **TS-α-19D** should be preferred over the corresponding chair-like transition state because of the Cα—S and S—N conformations in which the lone-pair orbitals at the Cα atom and the N atom are approximately periplanar to the S=O and Cα—Ph bonds, respectively. According to the ab-initio calculations of aminosulfoxonium ylides **29** and **30** this should allow for an efficient stabilization of the transition state by a twofold negative hyperconjugation, n_C–σ_{SO}* and n_N–σ_{SPh}*. Furthermore, in **TS-α-19B** the lone-pair orbitals at the Cα atom and the N atom are in an energetically favorable position orthogonal to each other. While **TS-α-19B** has a penta-coordinated Ti atom with a trigonal-bipyramidal geometry, **TS-γ-19A** and **TS-γ-19C** contain hexa-coordinated Ti atoms with an octahedral geometry. Decisive factors which will primarily determine the energy difference between **TS-γ-19A** and **TS-α-19B** and thus the ratio of **4** to **6** are the steric interactions between R^1 and R^2 in the first case and between R^2 and the alkenyl group in the second. While these interactions are presumably of the same magnitude for $R^1 = R^2 = Me$, the interaction between R^1 and R^2 in **TS-γ-19A** becomes more and more destabilizing with increasing size of the substituents compared with that between R² and the alkenyl group **TS-α-19B**. Thus with increasing size of the substituents, **TS-α-19B** should be energetically more and more preferred over **TS-γ-19A** and formation of **6** will dominate more and more. The *R*-configured *C*-titanium complex **19C** should react with aldehydes via transition state **TS-γ-19C** to give the diastereomer **28**. However, formation of this isomer has not been observed. This could be ascribed to an unfavorable steric interaction of the *endo*-positioned phenyl group with the aldehyde in transition state **TS-γ-19**. In **TS-γ-19A** the phenyl group is located in the sterically less demanding *exo* position.

1.3.5
Asymmetric Synthesis of Homopropargyl Alcohols

Enantio- and diastereomerically pure homopropargyl alcohols **38** constitute an interesting class of compounds, which have frequently served as building blocks in natural product synthesis (Scheme 1.3.17) [25].

The high synthetic utility of alcohols **38** stems from the fact that terminal alkynes are among the most versatile functional groups for the further elaboration of a carbon skeleton. Asymmetric synthesis of alcohols **38** from aldehydes with the concurrent formation of the two stereogenic C atoms has been accomplished mainly by two methods. The first features synthesis of chiral nonracemic allenylmetal compounds from the corresponding chiral nonracemic propargyl alcohols and addition of the former to aldehydes [26] and the second method in-

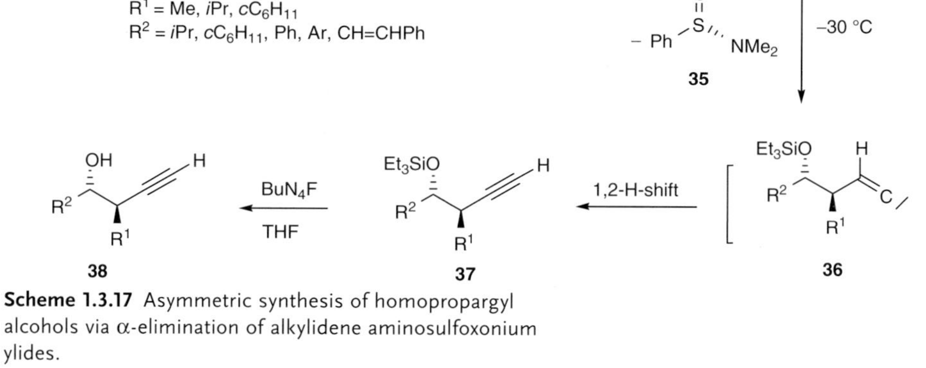

Scheme 1.3.17 Asymmetric synthesis of homopropargyl alcohols via α-elimination of alkylidene aminosulfoxonium ylides.

cludes the allylation of aldehydes with a chiral allylmetal reagent with formation of the corresponding homoallyl alcohol, which is then converted to **38** by a one-carbon homologation following conversion to the corresponding aldehyde [27]. While both methods are efficient, the enantio- and diastereoselectivities of the first one tend to be variable and the allylmetal method requires the oxidative cleavage of a double bond, which imposes restrictions upon R^1 and R^2.

An ideal starting material for the synthesis of alkynes **38** seemed to be the vinyl sulfoximines **32**, provided an elimination of sulfonamide could be achieved (Scheme 1.3.17). Methylation of sulfoximines **32** afforded the aminosulfoxonium salts **33** in practically quantitative yields. Treatment of salts **33** with the lithium amide at low temperatures gave the novel vinyl aminosulfoxonium ylides **34** which could be trapped with electrophiles at low temperatures. At elevated temperatures, however, ylides **34** suffered an elimination of sulfinamide **35** (≥98% ee) with the intermediate generation of the chiral alkylidene carbenes **36** (perhaps coordinated to THF molecules) which underwent a 1,2-H shift and gave the protected homopropargyl alcohols **37**. Cleavage of the silyl ethers **37** afforded the enantio- and diastereomerically pure homopropargyl alcohols **38** in good overall yields [28]. The synthesis of **38** from **32** not only gives testimony to the high synthetic versatility of the sulfonimidoyl group but also represents a new method for the generation of alkylidene carbenes [29]. The facile recycling of sulfinamide **35** with formation of sulfoximine **12** (≥98% ee) has already been demonstrated.

1.3.6
Asymmetric Synthesis of 2,3-Dihydrofurans

Highly substituted 2,3-dihydrofurans **44** (Scheme 1.3.18) would make particularly interesting starting materials for the asymmetric synthesis of tetrahydrofurans, structural motifs which can be found in many important natural products, including polyether antibiotics, lignans, and nucleosides [30]. Not only the activated double bond but also the vinylic silyl group of **44** should allow useful synthetic transformations.

Up to now the application of chiral 2,3-dihyrofurans in natural product synthesis has been hampered by the lack of general methods for their asymmetric synthesis [31]. The only more general route giving access to 2,3-dihydrofurans (carrying, however, no substituent at the 2-position) involves the reductive elimination of γ-butyrolactones [32]. These considerations led us to probe the generation of the silyloxy-substituted alkylidenes **42** from the aminosulfoxonium salts **40** and to study their propensity for 1,5-O—Si bond insertion [33]. Since a 1,2-methyl shift of **42** should be slower than the 1,2-H shift of **36**, the O—Si bond insertion of **42** was expected to be a likely process. However, very little was known at the beginning of our investigations about the reactivity of alkylidene carbenes of the **42** type in regard to the dependence on the substituents R^1 and R^2 of the competition between O—Si bond insertion, methyl migration, and 1,5-C—H bond insertion [34].

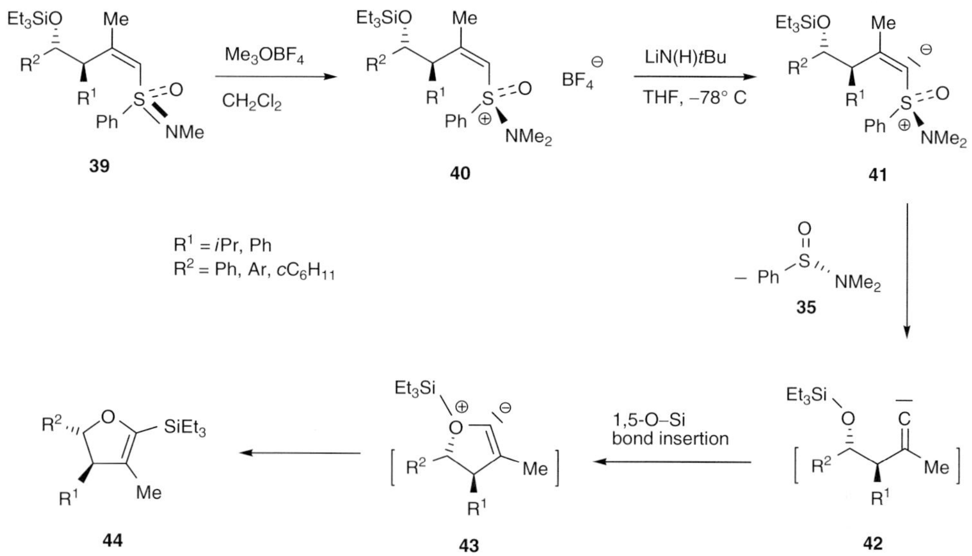

1.3.6 Asymmetric Synthesis of 2,3-Dihydrofurans | 97

Scheme 1.3.18 Asymmetric synthesis of monocyclic 2,3-dihydrofurans via O—Si bond insertion of alkylidene carbenes.

The methylation of the vinyl sulfoximines **39** with Me$_3$OBF$_4$ afforded in quantitative yields the aminosulfoxonium salts **40**. Treatment of salts **40** with LiN(H)t-Bu in THF first at −78 °C and then at room temperature gave the silyl-substituted 2,3-dihydrofurans **44** cleanly in good overall yields. This 2,3-dihydrofuran synthesis works equally well for derivatives carrying aliphatic and aromatic substituents, including sterically demanding ones. Besides the 2,3-dihydrofurans sulfinamide **35** with ≥98% ee was isolated in all cases in good yield [35].

When a high-yield synthesis of the enantio- and diastereomerically pure monocyclic 2,3-dihydrofurans **44** from the vinyl sulfoximines **39** had been developed, the synthesis of bicyclic 2,3-dihydrofurans of type **50** (Scheme 1.3.19) was investigated. Since a number of tetrahydrofuranoid natural products contain a fused bicyclic ring skeleton, the attainment of bicyclic 2,3-dihydrofurans of the **50** type would also be desirable.

The methylation of sulfoximines **45** with Me$_3$OBF$_4$ proceeded readily and gave the corresponding cyclic aminosulfoxonium salts **46** in quantitative yields. Upon treatment with LiN(H)t-Bu first at −78 °C and then at room temperature, salts **46** delivered the enantio- and diastereomerically pure bicyclic 2,3-dihydrofurans **50** cleanly in high overall yields. It is proposed that the reactions of the aminosulfoxonium salts **40** and **46** with the lithium amide at low temperatures afford the vinyl aminosulfoxonium ylides **41** and **47**, respectively. These alkylidene carbenoids eliminate sulfinamide **35** at higher temperatures with formation of the alkylidene carbenes **42** and **48**, respectively. Subsequently, the alkylidene

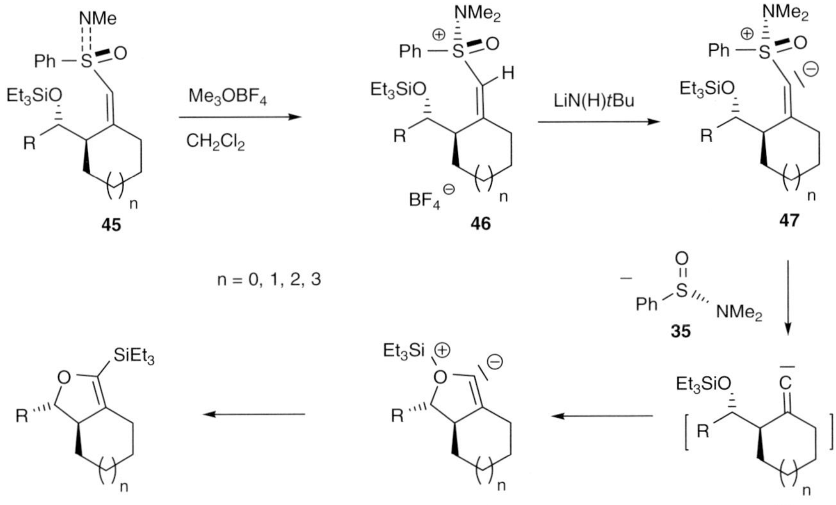

Scheme 1.3.19 Asymmetric synthesis of bicyclic 2,3-dihydrofurans via O—Si bond insertion of alkylidene carbenes.

carbenes preferentially undergo a 1,5-O—Si bond insertion rather than a 1,2-Me(CH$_2$R) shift to deliver the 2,3-dihydrofurans **44** and **50**, respectively. Thus the 1,5-O—Si bond insertion competes favorably with the 1,2-Me shift and the ring enlargement, irrespective of the substituents and the ring size. The O—Si bond insertion may occur either in a concerted or nonconcerted fashion. In the latter case the oxonium ylides **43** and **49** should be formed as intermediates, which then suffer a [1,2]-silyl migration. This scheme would be in accordance with the results of previous studies of acyclic siloxy alkylidene carbenes and of alkylidene carbenes in general [29, 33, 34]. Because of the different conformations the alkylidene carbenes **42** can adopt in regard to the C(sp^2)—C(sp^3) bond and the high propensity alkylidene carbenes normally show for C,H bond insertion, not only a 1,5-O,Si bond but also a 1,5-C,H bond insertion could in principle occur in this case. This was not observed, however, within the limits of the yields obtained.

1.3.7
Synthesis of Bicyclic Unsaturated Tetrahydrofurans

The ready deprotonation of the vinyl aminosulfoxonium salts **46** with a strong base at the α-position prompted a study of their reactions with weaker bases. It was speculated that in this case a vinyl–allyl isomerization of **46** might occur, followed by an intramolecular substitution of the allyl aminosulfoxonium salt

1.3.7 Synthesis of Bicyclic Unsaturated Tetrahydrofurans

52 with formation of an unsaturated bicyclic tetrahydrofuran of type **53** (Scheme 1.3.20).

Tetrahydrofurans **53** should be of synthetic interest because of the many natural products containing a bicyclic tetrahydrofuranoid skeleton or structural element [30]. The double bond would allow further synthetic transformations. While the chances of a regioselective isomerization of **46** to the allyl aminosulfoxonium salts **52** seemed to be good, the prospects of a cyclization of the latter with formation of **53** were considered to be less promising because of the poor nucleophilicity of the silyloxy group.

The treatment of **46** with 1,8-diazabicyclo [5.4.0] undec-7-ene (DBU) in CH_2Cl_2 at room temperature led to a highly regioselective migratory cyclization and gave the enantio- and diastereomerically pure bicyclic tetrahydrofurans **53** in good yields [35]. Besides **53**, sulfinamide **35** (≥98% ee) was isolated in good yields. It is proposed that the vinyl aminosulfoxonium salts **46** suffer a regioselective isomerization upon treatment with DBU with formation of the allyl aminosulfoxonium salts **52**. Although the mechanism of the isomerization of **46** is not known, the intermediate formation of the allyl aminosulfoxonium ylides **51** is much likely. Deprotonation of **46** at the γ-position is expected to be a facile process because of the stabilization of the negative charge of **51** by the aminosulfoxonium group and the double bond. For example N,N-dimethyl-S-methylphenylsulfoxonium tetrafluoroborate already has a pK_a-value of 15.5, and the double bond is expected to raise the acidity of **46** even further [2]. Cyclization of the allyl aminosulfoxonium salts **52** should lead primarily to the corresponding silyl tetrahydrofuranium salts [36] which are desilylated either by DBU, sulfinamide, or water

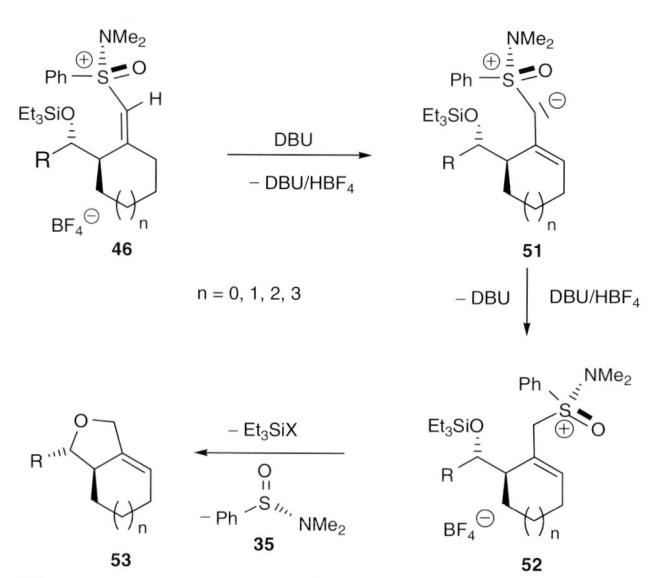

Scheme 1.3.20 Asymmetric synthesis of unsaturated bicyclic tetrahydrofurans via intramolecular substitution of allyl aminosulfoxonium salts.

upon aqueous workup, with formation of **53**. When considering the low reactivity of (for example) silyl ethers of 3-bromomethylbut-3-en-1-ol derivatives toward cyclization with formation of the corresponding 3-methylene tetrahydrofurans [37], the facile cyclization of **52** is noteworthy. These results show a high nucleofugacity of the allylic aminosulfoxonium group.

1.3.8
Asymmetric Synthesis of Alkenyloxiranes

Chiral alkenyl and cycloalkenyl oxiranes are valuable intermediates in organic synthesis [38]. Their asymmetric synthesis has been accomplished by several methods, including the epoxidation of allyl alcohols in combination with an oxidation and olefination [39a], the epoxidation of dienes [39b,c], the chloroallylation of aldehydes in combination with a 1,2-elimination [39f–h], and the reaction of S-ylides with aldehydes [39i]. Although these methods are efficient for the synthesis of alkenyl oxiranes, they are not well suited for cycloalkenyl oxiranes of the **56** type (Scheme 1.3.21). Therefore we had developed an interest in the asymmetric synthesis of the cycloalkenyl oxiranes **56** from the sulfonimidoyl-substituted homoallyl alcohols **7**. It was speculated that the allylic sulfoximine group of **7** could be stereoselectively replaced by a Cl atom with formation of corresponding chlorohydrins **55** which upon base treatment should give the cycloalkenyl oxiranes **56**. The feasibility of a Cl substitution of the sulfoximine group had been shown previously in the case of S-alkyl sulfoximines [40].

The treatment of sulfoximines **7** with $ClCO_2CH(Cl)Me$ at room temperature in CH_2Cl_2 readily afforded the chlorohydrins **55**, each with ≥98% ds in high yield. The substitution of the allylic sulfoximine group of **7** by the Cl atom occurred much faster than that of S-alkyl sulfoximines (vide infra). Remarkably, the substitution proceeded with retention of configuration. Presumably the chloroformate reacts with **7** with acylation at the N atom, thereby converting the sulfonimidoyl group into the corresponding allylic N-acyl aminosulfoxonium group, which is substituted by the Cl⁻ ion. Besides **55** sulfinamide **54** (≥98% ee) was obtained in good yield. The recycling of **54** with formation of sulfoximine **12** has already been carried out [40b]. The cyclization of chlorohydrins **55** with DBU gave the enantio- and diastereomerically pure cycloalkenyl oxiranes **56** in high

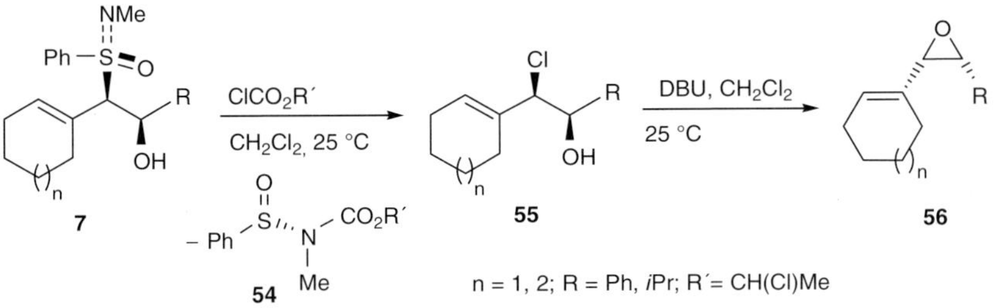

Scheme 1.3.21 Asymmetric synthesis of cycloalkenyl oxiranes via substitution of allyl sulfoximines.

1.3.8 Asymmetric Synthesis of Alkenyloxiranes

yields [40]. Having observed high selectivities in the synthesis of the cycloalkenyl oxiranes **56**, the possibility of an analogous synthesis of the alkenyl oxiranes **58** from the acyclic allyl sulfoximines **56** was investigated (Scheme 1.3.22).

The reaction of sulfoximine **6** with $ClCO_2CH(Cl)Me$ proceeded readily and gave, besides **54**, the chlorohydrins **57**, as mixtures of the *anti* and *syn* diastereomers in ratios of 78 : 22 and 73 : 27, respectively, in good yields. Treatment of the mixture of chlorohydrins with DBU furnished oxiranes **58** as mixtures of *trans* and *cis* isomers in ratios of 78 : 22 and 73 : 27, respectively, in good yields.

The sulfonimidoyl group acts in the asymmetric synthesis of the unsaturated oxiranes shown in Schemes 1.3.20 and 1.3.21 as both a chiral auxiliary and a nucleofuge. These results suggested the application of the sulfonimidoyl group as a chiral linker in solid-phase synthesis.

Synthesis of the polymer-bound allyl sulfoximine **60** was accomplished by the addition–elimination–isomerization route starting from the enantiomerically pure polymer-bound *N*-methyl-*S*-phenylsulfoximine **59**, which was prepared as previously described from Merrifield resin and sulfoximine **12** with a loading of 84% (Scheme 1.3.23) [42]. The successive treatment of resin **59** with *n*-BuLi in THF and with isovaleraldehyde furnished the corresponding polymer-bound lithium alcoholate, which upon reaction with $ClCO_2Me$ and DBU afforded the corresponding polymer-bound vinylic sulfoximine (not shown in Scheme 1.3.23), the isomerization of which with DBU in MeCN afforded sulfoximine **60**.

The loading of resin **60** and the success of each step of its synthesis was determined by an off-bead analysis involving the removal of the reaction products from the resin through an oxidative hydrolysis of the sulfoximine group at the S—N bond with formation of the corresponding sulfones by a method we have described recently [42].

Resin **60** was successively treated at −78 °C with *n*-BuLi, $ClTi(NEt_2)_3$, and benzaldehyde in THF to afford the polymer-bound homoallyl alcohol **61**. The cleavage of the linker group of **61** was achieved through treatment with $ClCO_2CH(Cl)Me$ in CH_2Cl_2 at room temperature. Thereby a mixture of chlorohydrins *syn*-**57** and *anti*-**57** in a ratio of 2 : 1 was obtained. These chlorohydrins were treated with DBU, which gave oxiranes *trans*-**58** and *cis*-**58** in a ratio of 2 : 1 in an overall yield of 34% based on sulfoximine **12**.

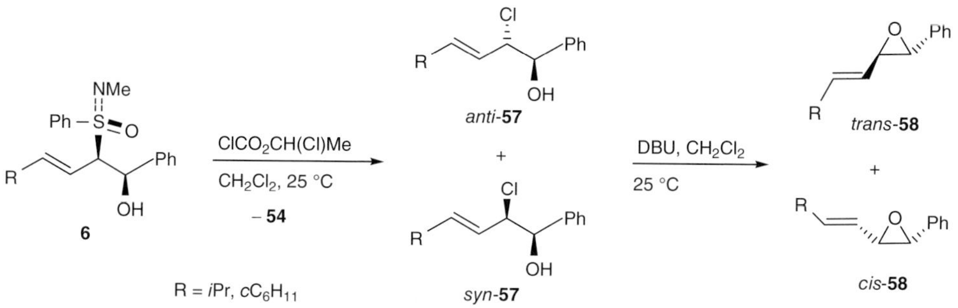

Scheme 1.3.22 Asymmetric synthesis of alkenyl oxiranes.

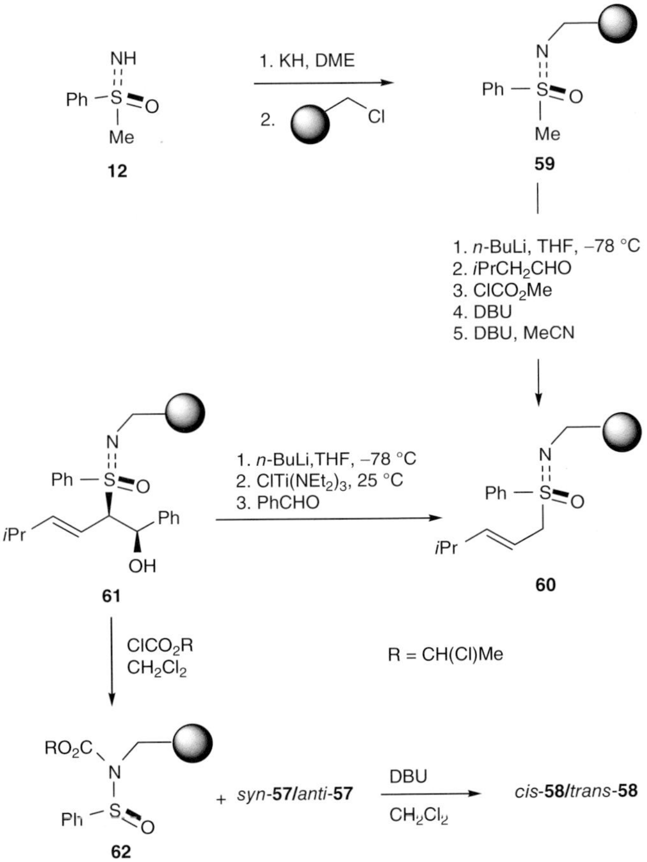

Scheme 1.3.23 Solid-phase asymmetric synthesis of alkenyl oxiranes by using the chiral sulfonimidoyl linker.

1.3.9
Asymmetric Synthesis of Unsaturated Mono- and Bicyclic Prolines

Proline derivatives have received much attention in recent years [43]. Peptide mimetics containing modified prolines are interesting probes for receptor studies and for the development of new drugs [44]. In addition, 3-substituted prolines are being currently studied as conformationally restricted α-amino acid analogues for the development of small-molecule drugs. This has led to intensive studies of the synthesis and biological activities of mono- and bicyclic proline derivatives. Of particular interest are 3,4-disubstituted prolines because of the occurrence of this substitution pattern as a core structure in the kainoid amino acids [45]. Because of their function as conformationally restricted L-glutamic acid analogues, the kanoid amino acids show neuroexcitatory properties and so are interesting probes for a study of neurological disorders such as Alzheimer's disease. Despite the considerable synthetic activity in the field of proline

1.3.9 Asymmetric Synthesis of Unsaturated Mono- and Bicyclic Prolines | 103

derivatives, there is a lack of methods for the asymmetric synthesis of 3,4-unsaturated prolines of the **63** and **64** type and of 4-methylene prolines of the **65** type (Fig. 1.3.7).

The acyclic sulfonimidoyl-substituted amino acids **24** were selected as starting material for the synthesis of the unsaturated prolines of type **63**. Because of the facile synthesis of the unsaturated bicyclic tetrahydrofurans **53** from the vinyl aminosulfoxonium salts **46** (cf. Scheme 1.3.20), it was speculated that upon treatment with a base the vinyl aminosulfoxonium salts **67** would experience a similar isomerization with formation of the allyl aminosulfoxonium salts **69**, which in turn could suffer an intramolecular substitution of the allylic aminosulfoxonium group (Scheme 1.3.24). The methylation of sulfoximines **24** with Me_3OBF_4 gave

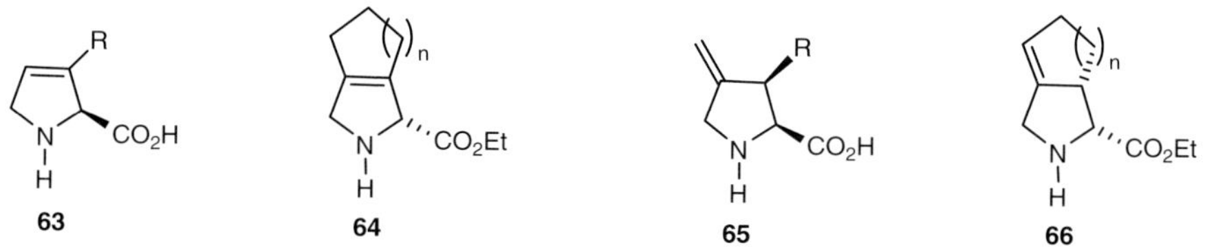

Fig. 1.3.7 Unsaturated mono- and bicyclic prolines.

R = Me, iPr, cC_6H_{11}, tBu, Ph

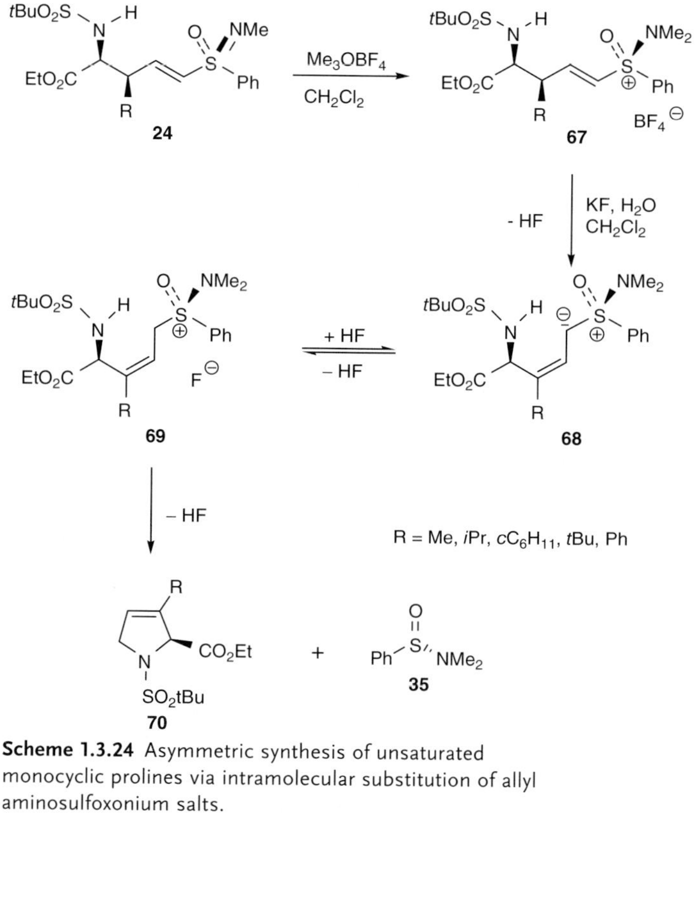

Scheme 1.3.24 Asymmetric synthesis of unsaturated monocyclic prolines via intramolecular substitution of allyl aminosulfoxonium salts.

the aminosulfoxonium salts **67** in practically quantitative yields. It was found eventually that the treatment of salts **67** with saturated aqueous KF in CH_2Cl_2 afforded the protected unsaturated prolines **70** in good overall yields [23, 46].

This method allows the synthesis of prolines of the **70** type bearing various substituents, including sterically demanding ones. It is assumed that the vinyl aminosulfoxonium salts **67** are deprotonated by the F^- ion (which enters the organic phase via anion exchange) at the γ-position with formation of the Z-configured allyl aminosulfoxonium ylid Z-**68**. Protonation of ylide Z-**68** at the α-position generates the allyl aminosulfoxonium salt Z-**69**, which suffers a F^- ion-induced cyclization. Presumably the corresponding E-configured species E-**68** and E-**69** are also present in equilibrium with their Z-isomers.

The extension of the synthesis of the unsaturated monocyclic prolines to that of bicyclic prolines **64** furnished some unexpected results.

The treatment with saturated aqueous NH_4Cl in CH_2Cl_2 of the cyclic aminosulfoxonium salt **71** derived from E-**27** resulted in a successive isomerization to the allyl aminosulfoxonium salt **72** and substitution of the latter by the Cl^- ion, and gave the allyl chloride **73** in good overall yield (Scheme 1.3.25). Both transformations caused by the Cl^- ion are notable. Cyclization of chloride **73** upon reaction with DBU afforded the bicyclic proline **74** in high yield. Interestingly the treatment of the aminosulfoxonium salts **71** with DBU furnished the isomeric bicyclic prolines **76**, the N-t-butylsulfonyl (N-Bus) derivatives of the bicyclic prolines of type **66**, in good yields. It is assumed that DBU causes a kinetically controlled isomerization of **71** with formation of the isomeric allyl aminosulfoxonium salts **75** while the Cl^- ion induces a thermodynamically controlled isomerization of the vinyl aminosulfoxonium salt. Formation of chloride ion containing ion pairs of the vinyl and allyl aminosulfoxonium salts in the organic phase via anion exchange seems to be an important factor in the isomerization and substitution.

The asymmetric synthesis of the 4-methylene prolines **65** was accomplished by starting from the methyl-substituted unsaturated amino acids **24** (Scheme 1.3.26) [23, 46]. The methylation of sulfoximines E/Z-**24** through treatment with Me_3OBF_4 in CH_2Cl_2 afforded the aminosulfoxonium salts E/Z-**77** in high yields. Isomerization of the vinyl aminosulfoxonium salts E/Z-**77** and cyclization of the allyl aminosulfoxonium salts **79** both proceeded readily upon treatment of the vinyl salts with aqueous KF in CH_2Cl_2 and furnished the corresponding cis-configured 3-substituted 4-methylene prolines **80** in medium to good yields. Both isomers, Z-**24** and E-**24**, afforded the proline derivative **80**, in addition to which sulfinamide **35** with ≥98% ee was isolated in each case in high yield. The functionalized proline derivative **80** (R = $(CH_2)_2OSitBuMe_2$) was prepared as a potential starting material for the synthesis of kainoid amino acids.

It is assumed that salts E/Z-**77** contained in the organic phase undergo an anion exchange with the F^- ion, which acts as a base and causes the isomerization to E/Z-**79**. Since the F^- ion is a weaker nucleophile but a stronger base than the Cl^- ion, cyclization of **79**, rather than substitution as in the case of **72**, takes place.

Finally, deprotection of the proline derivatives **70**, **76**, and **80** upon reaction with CF_3SO_3H in anhydrous CH_2Cl_2 gave prolines **63**, **66**, and **64**, respectively, in high yields (Scheme 1.3.27) [23, 46].

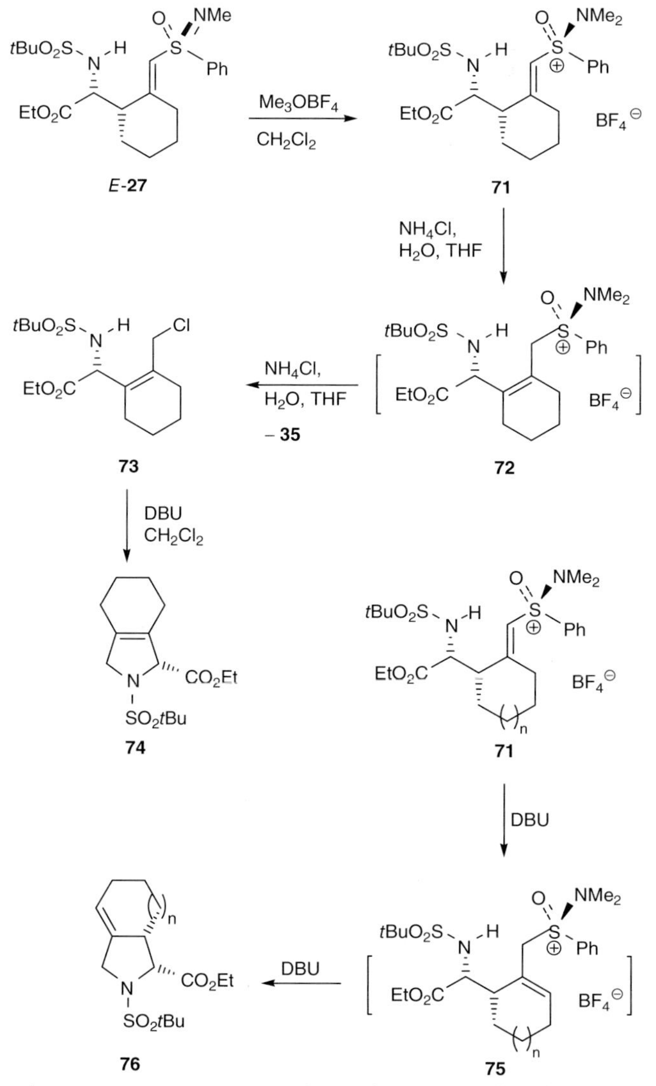

1.3.10 Asymmetric Synthesis of Bicyclic Amino Acids

Scheme 1.3.25 Asymmetric synthesis of unsaturated bicyclic pralines via inter- and intramolecular substitution of allyl aminosulfoxonium salts.

1.3.10
Asymmetric Synthesis of Bicyclic Amino Acids

The asymmetric synthesis of conformationally constrained α-amino acids and their incorporation into peptides are currently topics of great interest [47]. Much sought after are fused bicyclic α-amino acids having the N atom incorporated into a ring system [48]. Many of these amino acids display interesting pharmacological activities, and they have served as building blocks for the synthesis of pep-

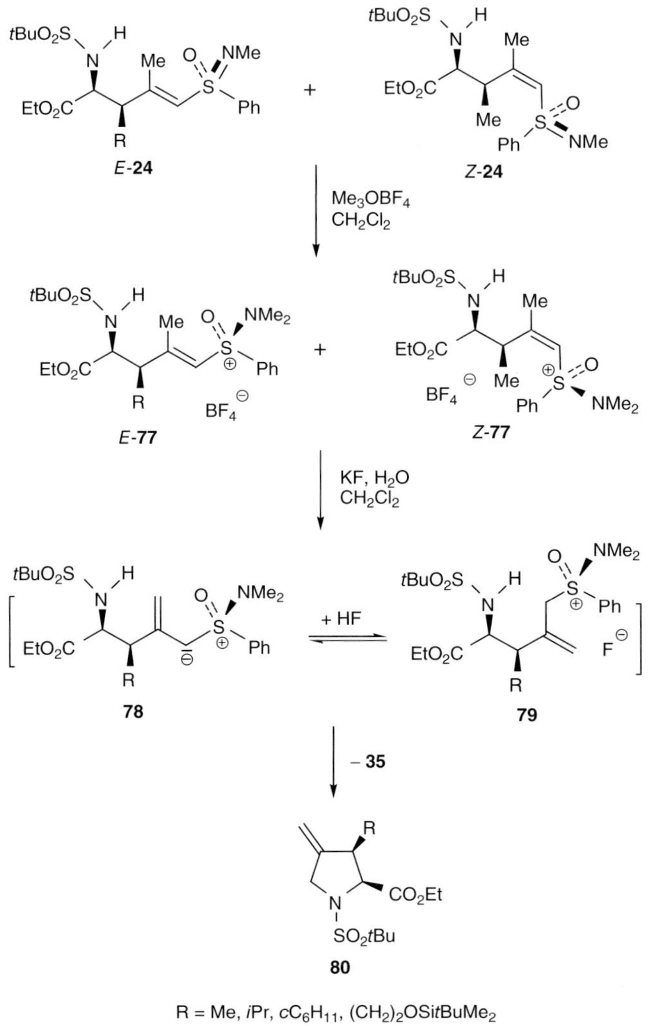

Scheme 1.3.26 Asymmetric synthesis of 4-methylene prolines via intramolecular substitution of allyl aminosulfoxonium salts.

tidomimetics. We were interested in the asymmetric synthesis of the bicyclic amino acids **82**, which contain a hexahydrocyclopenta[c]pyridine skeleton and carry, besides an enone moiety, a substituent at the α-position (Scheme 1.3.28). The bicyclic amino acids **82** and their derivatives should be interesting building blocks for the synthesis of peptidomimetics, and as bicyclic analogues of pipecolic acid they could exhibit interesting biological activities [49]. We envisioned a synthesis of **82** through a Pauson–Khand cycloaddition of the sulfonimidoyl-substituted enynes **77**, derived from the unsaturated amino acids **24**. Although

Scheme 1.3.27 Deprotection of *N-t*-butylsulfonyl prolines.

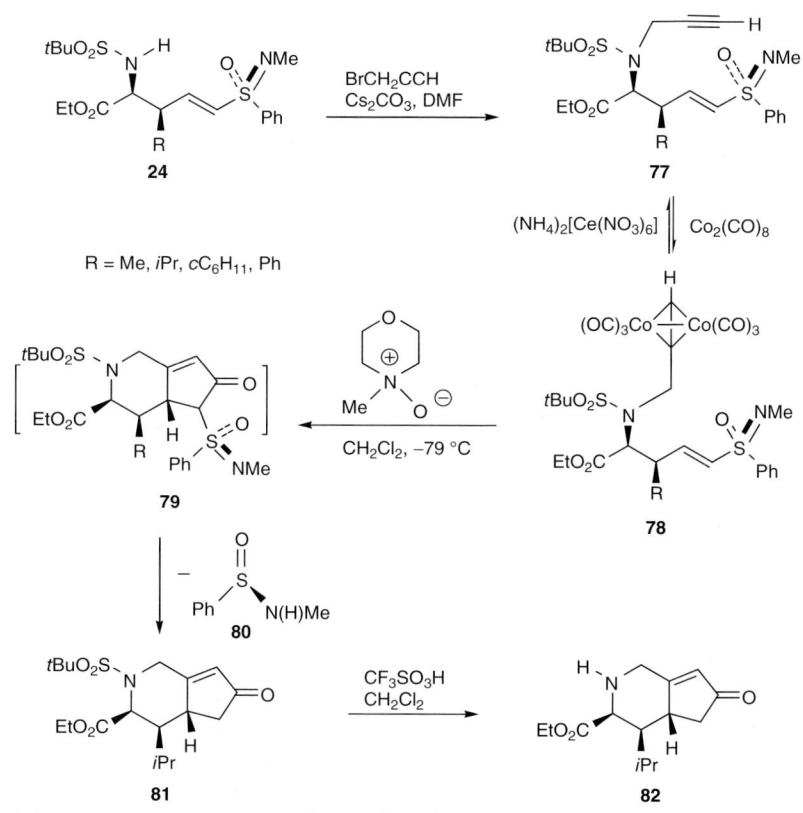

Scheme 1.3.28 Asymmetric synthesis of bicyclic amino acids via Pauson–Khand cycloaddition of vinyl sulfoximines.

intramolecular Pauson–Khand reactions involving vinyl sulfones and sulfoxides have been described [50], it was not clear whether vinyl sulfoximines of the **77** type, which carry the Lewis-basic sulfonimidoyl group, are also amenable to a [2+2+1]-cycloaddition.

Treatment of the amino acid derivatives **24** with propargyl bromide in DMF in the presence of $CsCO_3$ afforded the corresponding enyne derivatives **77** in high yields. The reaction of enynes **77** with $Co_2(CO)_8$ in THF led to their complete conversion to the corresponding cobalt complexes **78**, which were isolated in practically quantitative yields. Upon addition of N-morpholine-N-oxide (NMO) at –78 °C and warming the mixture to room temperature, complexes **78** gave the sulfonimidoyl free bicyclic amino acid derivatives **81** directly with high diastereoselectivity in good overall yields [51]. Thus the Pauson–Khand cycloaddition was accompanied by a reductive cleavage of the sulfonimidoyl group of the putative intermediate sulfoximines **79**. In all cases the R-configured sulfinamide **80** was isolated as a further reaction product. It is interesting to note that **80** is formed with inversion of configuration as a single enantiomer. The structural integrity of the cobalt complexes **78** was proven by chemical means in order to demonstrate that it is not this step which causes the reductive removal of the sulfonimidoyl group. Treatment of complex **78** with $(NH_4)_2Cl(NO_3)_6$ in acetone led to an almost quantitative recovery of the vinyl sulfoximine **77**.

To conclude the asymmetric synthesis of the new bicyclic amino acids the protected amino acid **81** was treated with CF_3SO_3H in CH_2Cl_2, affording the amino acid derivative **82** in high yield [51].

1.3.11
Asymmetric Synthesis of β-Amino Acids

β-Amino acids have attracted much attention because of their incorporation in naturally occurring peptides and their utilization as starting materials for the synthesis of β-lactams and β-peptides [52]. This has led to a strong interest in the asymmetric synthesis of β-amino acids. Currently of much interest are cyclic and acyclic β,β-disubstituted β-amino acids because of the expectation of unusual structural and pharmacological properties of the β-peptides derived therefrom. While numerous efficient asymmetric syntheses of β-substituted β-amino acids have been described, methods for the asymmetric synthesis of β,β-disubstituted β-amino acids are scarce. For example the well-established Arndt–Eistert homologation of α-amino acids fails in the case of β,β-disubstituted β-amino acids [52c]. The methods described for the asymmetric synthesis of β,β-disubstituted β-amino acids involve the addition of nucleophiles to N-sulfinylimines [53a–c], the hydrogenation of N-sulfinylimine-derived aziridines [53d], and the functionalization of aspartic acid derivatives [53e]. While the sulfinylimine method has been restricted to the synthesis of β-methyl-substiuted derivatives and seems to be limited because of a *syn* and *anti* isomerization of ketosulfinimines, the selectivities of the aziridine method tend to be variable, and the aspartic acid method is confined to special examples. Besides β,β-disubstituted derivatives, those β-amino

1.3.11 Asymmetric Synthesis of β-Amino Acids | 109

acids which carry a hydroxy group in the side chain are of particular interest. Hydroxy-β-amino acids occur as structural motifs in natural products, and they have been used as starting materials for the synthesis of biologically active compounds [54].

Therefore we were interested in the development of a method for the asymmetric synthesis of cyclic and acyclic δ-hydroxy-β-amino acids **83–85** (Fig. 1.3.8).

The synthesis of the protected δ-hydroxy β-amino acids **91** required a stereoselective amination of the vinyl sulfoximine **4** (Scheme 1.3.29). Carbamoylation of alcohols **4** with trichloroacetyl isocyanate afforded carbamates **86** in high yields. The treatment of carbamates **86** with *n*-BuLi resulted in a highly diastereoselective intramolecular addition of the deprotonated amino group to the activated double bond and, after protonation of the intermediate lithiated sulfoximine, gave the amino acid derivatives **87** in high yields. Substitution of the sulfonimidoyl group of **87** proceeded smoothly upon reaction with chloroformate and furnished chlorides **89** in high yields. Besides **89**, sulfinamide **88** (≥98% *ee*) was obtained in high yields. Conversion of sulfinamide **88** to the starting sulfoximine **12** (≥98% *ee*) has already been accomplished. Finally the N,O-diprotected amino acids **91**

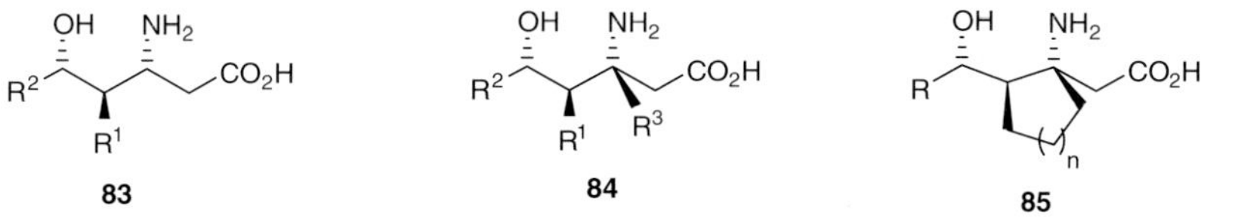

Fig. 1.3.8 Acyclic and cyclic δ-hydroxy-β-amino acids.

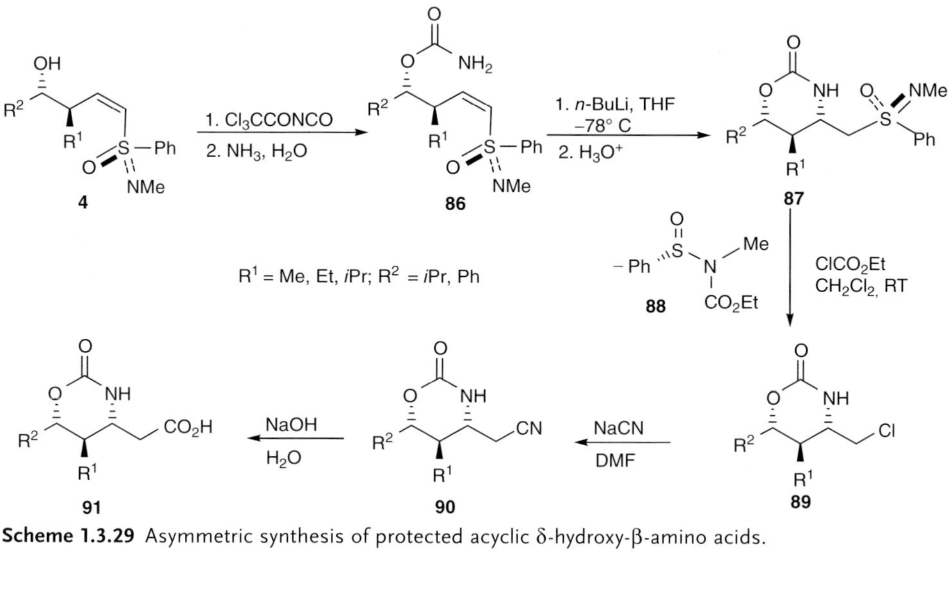

Scheme 1.3.29 Asymmetric synthesis of protected acyclic δ-hydroxy-β-amino acids.

were prepared through substitution of chlorides **89** with NaCN and hydrolysis of nitriles **90** in high overall yields [55].

Of particular interest are cyclic β,β-disubstituted β-amino acids of the **96** type (Scheme 1.3.30). Carbamoylation of alcohols **5** afforded carbamates **92** in high yields. Gratifyingly, cyclization of **92** delivered the bicyclic sulfoximines **93** with high diastereoselectivities in high yields. Replacement of the sulfonimidoyl group of **93** posed no problems despite its neopentyl-like position, and gave chlorides **94** in high yields. Reaction of chlorides **94** with NaCN and hydrolysis of the corresponding nitriles **95** finally gave the N,O-diprotected β-amino acids **92** in high overall yields [55].

The application of this method to the synthesis of the acyclic β,β-disubstituted β-amino acid **101** from the methyl-substituted vinyl sulfoximine **4** (Scheme 1.3.31) also posed no problems. Treatment of alcohol **4** with the isocyanate gave a mixture of the E- and Z-configured carbamates E-**97** and Z-**97** in a ratio of 80 : 20. Interestingly, under the conditions shown in Scheme 1.3.31 both isomeric vinyl sulfoximines gave the cyclic carbamate **98** with high diastereoselectivity. The facile conversion of sulfoximine **98** to chloride **99** again demonstrates the efficiency of the substitution of the sulfonimidoyl group of alkylsulfoximines by the chloroformate method. Reaction of chloride **99** with NaCN and the hydrolysis of nitrile **100** concluded the high-yield synthesis of the N,O-diprotected amino acid **101** [55].

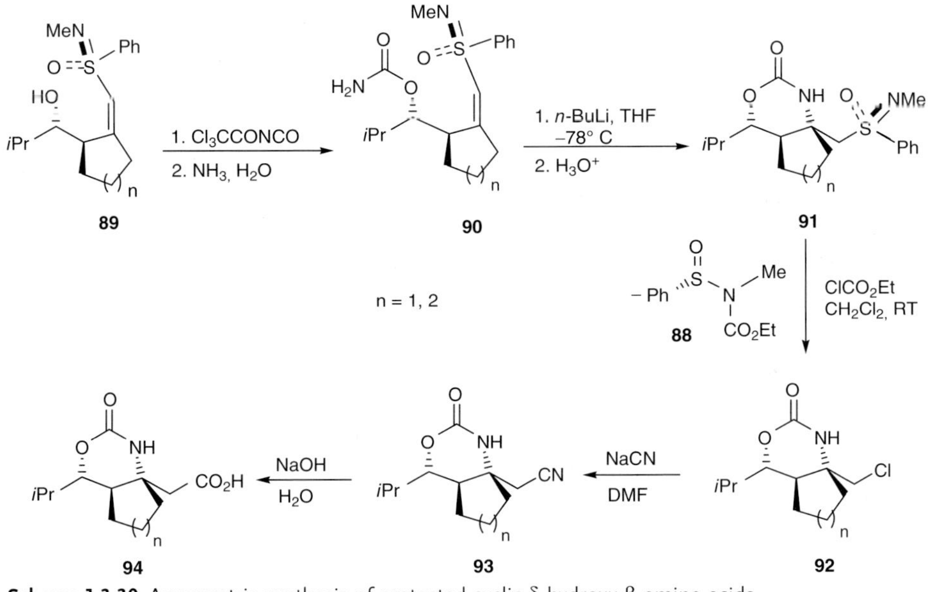

Scheme 1.3.30 Asymmetric synthesis of protected cyclic δ-hydroxy-β-amino acids.

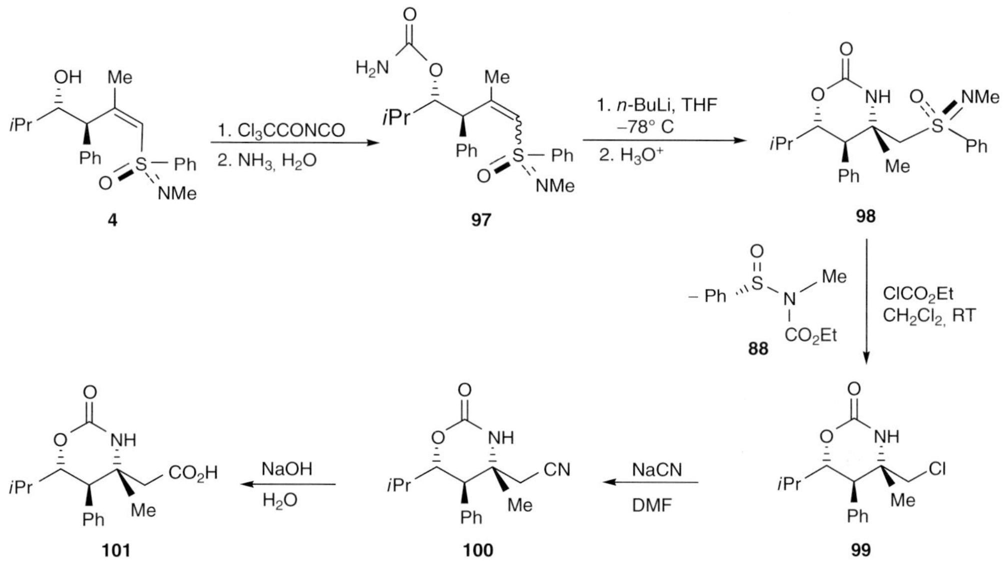

Scheme 1.3.31 Asymmetric synthesis of protected acyclic β,β-disubstiuted δ-hydroxy-β-amino acids.

1.3.12
Conclusion

Sulfonimidoyl-substituted mono(allyl)titanium and bis(allyl)titanium complexes, both of which are configurationally labile at the Cα atom, have emerged as valuable reagents in asymmetric synthesis. Their highly selective reaction with aldehydes and N-sulfonyl imino esters allows the attainment of enantio- and diastereomerically pure sulfonimidoyl-substituted homoallyl alcohols and unsaturated amino acids, which are valuable intermediates for the asymmetric synthesis of homopropargyl alcohols, 2,3-dihydrofurans, bicyclic unsaturated tetrahydrofurans, cycloalkenyl oxiranes, unsaturated monocyclic and bicyclic prolines, bicyclic unsaturated α-amino acids, and hydroxy-substituted cyclic and acyclic β-amino acids and β,β-disubstituted β-amino acids. The high synthetic versatility of the sulfonimidoyl group stems from its utilization as a chiral carbanion-stabilizing nucleofuge [56].

Acknowledgements

This study was supported by the Deutsche Forschungsgemeinschaft (SFB 380, "Asymmetric Synthesis with Chemical and Biological Methods," and GK 440 "Methods in Asymmetric Synthesis") and the Grünenthal GmbH, Aachen. The work described here would not have been possible without the dedicated efforts of my coworkers who are named in the references and to whom I am grateful for their contributions.

References

1 a) S. G. Pyne, *Sulfur Reports* **1999**, *21*, 281–334; b) M. Reggelin, C. Zur, *Synthesis* **2000**, 1–64.
2 a) F. G. Bordwell, J. C. Branca, C. R. Johnson, N. R. Vanier, *J. Org. Chem.* **1980**, *45*, 3884–3889; b) F. G. Bordwell, *Acc. Chem. Res.* **1988**, *21*, 456–463.
3 J. Bund, H.-J. Gais, E. Schmitz, I. Erdelmeier, G. Raabe, *Eur. J. Org. Chem.* **1998**, 1319–1335.
4 M. Lerm, H.-J. Gais, K. Cheng, C. Vermeeren, *J. Am. Chem. Soc.* **2003**, *125*, 9653–9667.
5 J. Bund, H.-J. Gais, I. Erdelmeier, *J. Am. Chem. Soc.* **1992**, *113*, 1442–1444.
6 M. Scommoda, H.-J. Gais, S. Bosshammer, G. Raabe, *J. Org. Chem.* **1996**, *61*, 4379–4390.
7 H.-J. Gais, H. Müller, J. Bund, M. Scommoda, J. Brandt, G. Raabe, *J. Am. Chem. Soc.* **1995**, *117*, 2453–2466.
8 H.-J. Gais, G. Bülow, *Tetrahedron Lett.* **1992**, *33*, 465–468.
9 H.-J. Gais, G. Bülow, *Tetrahedron Lett.* **1992**, *33*, 461–464.
10 H.-J. Gais, H. Müller, J. Decker, R. Hainz, *Tetrahedron Lett.* **1995**, *36*, 7433–7436.
11 a) D. Hoppe, in *Stereoselective Synthesis, Methods of Organic Chemistry* (Houben-Weyl) (Eds.: G. Helmchen, R. W. Hoffmann, J. Mulzer, E. Schaumann), Thieme, Stuttgart, **1996**, E21, Vol. 3, pp. 1551–1602; b) F. Sato, H. Urabe, S. Okamoto, *Chem. Rev.* **2000**, *100*, 2835–2886; c) S. E. Denmark, N. G. Almstead, in *Modern Carbonyl Chemistry* (Ed.: J. Otera), Wiley-VCH, Weinheim, **2000**, pp. 293–394; d) J. Szymoniak, C. Moise, in *Titanium and Zirconium in Organic Synthesis* (Ed.: I. Marek), Wiley-VCH, Weinheim, **2002**, pp. 451–474; e) K. Mikami, Y. Matsumoto, T. Spiono, in *Science of Synthesis, Methods of Molecular Transformations* (Houben-Weyl) (Eds.: T. Imamoto, R. Noyori), Thieme, Stuttgart, **2003**, Vol. 2, pp. 457–679.
12 a) M. Reggelin, H. Weinberger, M. Gerlach, R. Welcker, *J. Am. Chem. Soc.* **1996**, *118*, 4765–4777; b) M. Reggelin, H. Weinberger, T. Heinrich, *Liebigs Annalen/Recueil.* **1997**, *9*, 1881–1886; c) M. Reggelin, T. Heinrich, *Angew. Chem.* **1998**, *110*, 3005–3008; *Angew. Chem., Int. Ed.* **1998**, *37*, 2883–2886; d) M. Reggelin, M. Gerlach, M. Vogt, *Eur. J. Org. Chem.* **1999**, *5*, 1011–1031.
13 J. Brandt, H.-J. Gais, *Tetrahedron: Asymmetry* **1997**, *8*, 909–912 and references cited therein.
14 H.-J. Gais, R. Hainz, H. Müller, P. R. Bruns, N. Giesen, G. Raabe, J. Runsink, S. Nienstedt, J. Decker, M. Schleusner, J. Hachtel, R. Loo, C.-W. Woo, P. Das, *Eur. J. Org. Chem.* **2000**, 3973–4009.
15 H.-J. Gais, D. Lenz, G. Raabe, *Tetrahedron Lett.* **1995**, *36*, 7437–7440.
16 R. Hainz, H.-J. Gais, G. Raabe, *Tetrahedron: Asymmetry* **1996**, *7*, 2505–2508.
17 a) R. Albrecht, G. Kresze, *Chem. Ber.* **1965**, *98*, 1431–1434; b) D. M. Tschaen, E. Turos, S. Weinreb, *J. Org. Chem.* **1984**, *49*, 5058–5064.
18 B. Nyasse, L. Grehn, U. Ragnarsson, *J. Chem. Soc., Chem. Commun.* **1997**, 1017–1018.
19 M. Schleusner, S. Koep, M. Günter, S. K. Tiwari, H.-J. Gais, *Synthesis* **2004**, *6*, 967–969.
20 P. Sun, S. M. Weinreb, M. Shang, *J. Org. Chem.* **1997**, *62*, 8604–8608.
21 M. Schleusner, H.-J. Gais, S. Koep, G. Raabe, *J. Am. Chem. Soc.* **2002**, *124*, 7789–7800.
22 S. Koep, H.-J. Gais, G. Raabe, *J. Am. Chem. Soc.* **2003**, *125*, 13243–13251.
23 S. K. Tiwari, H.-J. Gais, G. S. Babu, L. R. Reddy, F. Köhler, M. Günter, S. Koep, V. B. R. Iska, *J. Am. Chem. Soc.* **2006**, *128*, 7360.
24 H.-J. Gais, P. R. Bruns, G. Raabe, R. Hainz, M. Schleusner, J. Runsink, G. S. Babu, *J. Am. Chem. Soc.* **2005**, *127*, 6617–6631.
25 a) J. A. Marshall, M. Yanik, M. *Org. Lett.* **2000**, *2*, 2173–2175; b) S. J. O'Malley, J. L. Leighton, *Angew. Chem.* **2001**, *113*, 2999–3001; *Angew. Chem., Int. Ed.* **2001**, *40*, 2915–2917; c) J. A. Marshall, M. P. Bourbeau, *J. Org. Chem.* **2002**, *67*, 2751–2754.
26 a) J. A. Marshall, *Chem. Rev.* **1996**, *96*, 31–48; b) J. A. Marshall, *Chem. Rev.*

2000, *100*, 3163–3186; c) J.-F. Poisson, J. F. Normant, *J. Org. Chem.* **2000**, *65*, 6553–6560; d) J. A. Marshall, H. R. Chobanian, M. Yanik, *M. Org. Lett.* **2001**, *3*, 3369–3372.

27 a) S. E. Denmark, N. B. Almstead, in *Modern Carbonyl Chemistry* (Ed.: J. Otera), Wiley-VCH, Weinheim, Germany, **2000**; pp. 299–401; b) S. R. Chemler, W. R. Roush, in *Modern Carbonyl Chemistry* (Ed.: J. Otera), Wiley-VCH, Weinheim, Germany, **2000**; pp. 403–490.

28 L. R. Reddy, H.-J. Gais, C.-W. Woo, G. Raabe, *J. Am. Chem. Soc.* **2002**, *124*, 10427–10434.

29 R. Knorr, *Chem. Rev.* **2004**, *104*, 3795–3849.

30 a) O. R. Gottlieb, in *New Natural Products and Plant Drugs with Pharmacological, Biological, or Therapeutic Activity*, Springer Verlag, Berlin, **1977**, p. 227; b) R. S. Ward, *Tetrahedron* **1990**, *46*, 5029–5041; c) B. M. Fraga, *Nat. Prod. Rep.* **1992**, *9*, 217–241; d) A. T. Merrit, S. V. Ley, *Nat. Prod. Rep.* **1992**, *9*, 243–287; e) C. J. Moody, M. Davies, *Stud. Nat. Prod. Chem.* **1992**, *10*, 201–239; f) U. Koert, *Synthesis* **1995**, 115–132; g) R. Benassi, in *Comprehensive Heterocyclic Chemistry II* (Eds.: A. R. Katritzky, C. W. Rees, E. F. V. Scrivan, C. W. Bird), Elsevier, Oxford, **1996**; Vol. 2, pp. 259–295; h) S. S. C. Koch, A. R. Chamberlin, *Stud. Nat. Prod. Chem.* **1995**, *16*, 687–725; i) R. S. Ward, *Nat. Prod. Rep.* **1999**, *16*, 75–96.

31 a) M. Bottex, M. Cavicchioli, B. Hartmann, N. Monteiro, G. Balme, *J. Org. Chem.* **2001**, *66*, 175–179; b) S. Ma, W. Gao, *Synlett* **2002**, 65–68; c) V. Calo, F. Scordari, A. Nacci, E. Schingaro, L. D´Accolti, A. Monopoli, *J. Org. Chem.* **2003**, *68*, 4406–4409.

32 a) K. Jarowicki, P. Kocienski, S. Norris, M. O'Shea, M. Stocks, *Synthesis* **1995**, 195–198; b) J. A. Walker II; J. J. Chen, D. S. Wise, L. B. Townsend, *J. Org. Chem.* **1996**, *61*, 2219–2221; c) V. Fargeas, P. Le Ménez, I. Berque, J. Ardisson, A. Pancrazi, *Tetrahedron* **1996**, *52*, 6613–6634; d) Z.-X. Wang, L. I. Wiebe, E. De Clercq, J. Balzarini, E. E. Knaus, *Can. J. Chem.* **2000**, *78*, 1081–1088; e) R. B. Chhor, B. Nosse, S. Sörgel, C. Böhm, M. Seitz, O. Reiser, *Chem. Eur. J.* **2003**, *9*, 260–270.

33 a) K. Miwa, T. Aoyama, T. Shioiri, *Synlett* **1994**, 461–462; b) S. Kim, C. M. Cho, *Tetrahedron Lett.* **1995**, *36*, 4845–4848; c) K. S. Feldman, M. L. Wrobleski, *J. Org. Chem.* **2000**, *65*, 8659–8668.

34 W. Kirmse, *Angew. Chem.* **1997**, *109*, 1212–1218; *Angew. Chem., Int. Ed. Engl.* **1997**, *36*, 1164–1170.

35 H.-J. Gais, L. R. Reddy, G. S. Babu, G. Raabe, *J. Am. Chem. Soc.* **2004**, *126*, 4859–4864.

36 J. S. Hrkach, K. Matyjaszewski, *Macromolecules* **1990**, *23*, 4042–4046.

37 a) D. C. Russell, C. J. Forsyth, *J. Org. Chem.* **1997**, *62*, 5672–5673; b) L. A. Paquette, G. D. Bennett, M. B. Isaac, A. Chhatriwalla, *J. Org. Chem.* **1998**, *63*, 1836–1845; c) T.-P. Loh, H.-Y. Song, *Synlett* **2002**, *12*, 2119–2121.

38 a) J. A. Marshall, *Chem. Rev.* **1989**, *89*, 1503–1511; b) C. Hertweck, W. Boland, *Rec. Res. Develop. Org. Chem.* **1999**, *3*, 219–235; c) C. Courillon, S. Thorimbert, M. Malacria, in *Handbook of Organopalladium Chemistry for Organic Synthesis* (Ed.: E. Negishi), Wiley, New York, **2002**, Vol. II, pp. 1795–1810.

39 a) R. A. Johnson, K. B. Sharpless, in *Catalytic Asymmetric Synthesis* (Ed.: I. Ojima), Wiley-VCH, New York, **2000**, pp. 231–280; b) S. Chang, N. H. Lee, E. N. Jacobsen, *J. Org. Chem.* **1993**, *58*, 6939–6941; c) S. Chang, R. M. Heid, E. N. Jacobsen, *Tetrahedron Lett.* **1994**, *35*, 669–672; d) K. G. Rasmussen, D. S. Thomsen, K. A. Jørgensen, *J. Chem. Soc., Perkin Trans. 1* **1995**, 2009–2017; e) M. Frohn, M. Dalkiewicz, Y. Tu, Z.-X. Wang, Y. Shi, *J. Org. Chem.* **1998**, *63*, 2948–2953; f) S. Hu, S. Jayaraman, A. C. Oehlschlager, *J. Org. Chem.* **1999**, *64*, 2524–2526; g) C. Hertweck, W. Boland, *J. Org. Chem.* **2000**, *65*, 2458–2463; h) M. Bandini, P. G. Cozzi, P. Melchiorre, S. Morganti, A. Umani-Ronchi, *Org. Lett.* **2001**, *3*, 1153–1155; i) J. Zanardi, D. Lamature, S. Miniere, V. Reboul, P. Metzner, *J. Org. Chem.* **2002**, *67*, 9083–9086.

40 a) R. Loo, Ph. D. Thesis, RWTH Aachen, **1999**; b) D. Roder, Ph. D. Thesis, RWTH Aachen, **2004**.
41 H.-J. Gais, G. S. Babu, M. Günter, P. Das, *Eur. J. Org. Chem.* **2004**, 1464–1473.
42 J. Hachtel, H.-J. Gais, *Eur. J. Org. Chem.* **2000**, 1457–1465.
43 a) S. Liao, M. Shenderovich, K. E. Köver, Z. Zhang, K. Hosohata, P. Davis, F. Porreca, H. I. Yamamura, V. J. Hruby, *J. Pept. Res.* **2001**, *57*, 257–276; b) S. Hanessian, G. McNaughton-Smith, H.-G. Lombart, W. D. Lubell, *Tetrahedron* **1997**, *53*, 12789–12854; c) R. Galeazzi, G. Mobili, M. Orena, *Curr. Org. Chem.* **2004**, *8*, 1799–1829.
44 a) N. Pellegrini, M. Schmitt, S. Guery, J.-J. Bourguignon, *Tetrahedron Lett.* **2002**, *43*, 3243–3246; (b) I.-L. Lu, S.-J. Lee, H. Tsu, S.-Y. Wu, K.-H. Kao, Y.-Y. Chien, Y.-S. Chen, J.-H. Cheng, C.-N. Cheng, T.-W. Chen, S.-P. Chang, X. Chen, W.-T. Jiaang, *Bioorg. Med. Chem. Lett.* **2005**, *15*, 3271–3275.
45 a) A. F. Parsons, *Tetrahedron* **1996**, *52*, 4149–4174; b) M. G. Moloney, *Nat. Prod. Rep.* **1998**, *15*, 205–219; c) M. G. Moloney, *Nat. Prod. Rep.* **1999**, *16*, 485–498.
46 S. K. Tiwari, A. Schneider, S. Koep, H.-J. Gais, *Tetrahedron Lett.* **2004**, *45*, 8343–8346.
47 a) R. M. Williams, *Synthesis of Optically Active χ-Amino Acids*; Pergamon, Oxford, **1989**; b) R. O. Duthaler, *Tetrahedron* **1994**, *50*, 1539–1650; c) F. P. J. T. Rutjes, L. B. Wolf, H. E. Schoemaker, *J. Chem. Soc., Perkin Trans. 1*, **2000**, 4197–4212; d) G. C. Barrett, J. S. Davies, *Amino Acids, Peptides, and Proteins*, The Royal Society of Chemistry, Cambridge, **2002**; Vol. 33.
48 a) R. Chinchilla, L. R. Falvello, N. Galindo, C. Nájera, *J. Org. Chem.* **2000**, *65*, 3034–3041; b) N. Valls, M. López-Canet, M. Vallribera, J. Bonjoch, *Chem. Eur. J.* **2001**, *7*, 3446–3460; c) S. Tanimori, K. Fukubayashi, M. Kirihata, *Tetrahedron Lett.* **2001**, *42*, 4013–4016; d) S. Kotha, N. Sreenivasachary, *Eur. J. Org. Chem.* **2001**, 3375–3383; e) R. Millet, J. Domarkas, P. Rombaux, B. Rigo, R. Houssin, J.-P. Hénichart, *Tetrahedron Lett.* **2002**, *43*, 5087–5088; f) W. Maison, G. Adiwidjaja, *Tetrahedron Lett.* **2002**, *43*, 5957–5960; g) J. Zhang, C. Xiong, W. Wang, J. Ying, V. J. Hruby, *Org. Lett.* **2002**, *4*, 4029–4032; h) N. Valls, M. Vallribera, S. Carmeli, J. Bonjoch, *Org. Lett.* **2003**, *5*, 447–450; i) K. Hattori, R. B. Grossman, *J. Org. Chem.* **2003**, *68*, 1409–1417.
49 a) F. Couty, *Amino Acids* **1999**, *16*, 297–320; b) K. C. M. F. Tjen, S. S. Kinderman, H. E. Schoemaker, H. Hiemstra, F. P. J. T. Rutjes, *Chem. Commun.* **2000**, 699–700; c) J. Barluenga, M. A. Fernández, F. Aznar, C. Valdés, *Tetrahedron Lett.* **2002**, *43*, 8159–8163; d) S. Carbonnel, Y. Troin, *Heterocycles* **2002**, *57*, 1807–1830; e) H. Huang, T. F. Spande, J. S. Panek, *J. Am. Chem. Soc.* **2003**, *125*, 626–627; f) K. Partogyan-Halim, L. Besson, D. J. Aitken, H.-P. Husson, *Eur. J. Org. Chem.* **2003**, 268–273.
50 Rivero, M. R.; Adrio, J.; Carretero, J. C. *Eur. J. Org. Chem.* **2002**, 2881–2889.
51 M. Günter, H.-J. Gais, *J. Org. Chem.* **2003**, *68*, 8037–8041.
52 a) G. Cardillo, C. Tomassini, *Chem. Soc. Rev.* **1996**, *25*, 117–128; b) *Enantioselective Synthesis of ß-Amino Acids*, (Eds.: E. Juaristi, V. A. Soloshonok), Wiley-VCH, New York, **2005**; c) E. Juaristi, H. Lopéz-Ruiz, *Curr. Med. Chem.* **1999**, *6*, 983–1004; c) S. Abele, D. Seebach, *Eur. J. Org. Chem.* **2000**, 1–15.
53 a) D. H. Hua, S. W. Miao, J. S. Chen, S. Iguchi, *J. Org. Chem.* **1991**, *56*, 4–6; b) F. A. Davis, R. T. Reddy, R. E. Reddy, *J. Org. Chem.* **1992**, *57*, 6387–6389; c) T. P. Tang, J. A. Ellman, *J. Org. Chem.* **2002**, *67*, 7819–7832; d) F. A. Davis, J. Deng, Y. Zhang, R. C. Haltiwanger, *Tetrahedron* **2002**, *58*, 7135–7143; e) E. Juaristi, M. Balderas, H. López-Ruiz, V. M. Jiménez-Pérez, M. L. Kaiser-Carril, Y. Ramírez-Quirós, *Tetrahedron: Asymmetry* **1999**, *10*, 3493–3505.
54 a) M. Hirama, S. Itô, *Heterocycles* **1989**, *28*, 1229–1247; b) T. Kitazume, T. Kobayashi, T. Yamamoto, T. Yamazaki, *J. Org. Chem.* **1987**, *52*, 3218–3223; c) J. de Blas, J. C. Carretero, E. Domínguez, *Tetrahedron Lett.* **1994**, *35*, 4603–4606; d) N. Asao, T. Shimada, T. Sudo, N. Tsukada, K. Yazawa, Y. S. Gyoung, T. Uyehara, Y. Yamamoto, *J. Org. Chem.*

1997, *62*, 6274–6282; e) S. G. Davies, O. Ichihara, *Tetrahedron Lett.* **1999**, *40*, 9313–9316. G. Delle Monache, D. Misiti, P. Salvatore, G. Zappia, *Tetrahedron: Asymmetry* **2000**, *11*, 1137–1149; f) G. Guanti, A. Moro, E. Narisano, *Tetrahedron Lett.* **2000**, *41*, 3203–3207; g) D. Seebach, A. Jacobi, M. Rueping, K. Gademann, M. Ernst, B. Jaun, *Helv. Chim. Acta* **2000**, *83*, 2115–2140; h) F. Schweizer, A. Lohse, A. Otter, O. Hindsgaul, *Synlett* **2001**, *9*, 1434–1436; i) G. V. M. Sharma, V. G. Reddy, A. S. Chander, K. R. Reddy, *Tetrahedron: Asymmetry* **2002**, *13*, 21–24; j) C. Palomo, M. Oiarbide, A. Landa, M. C. González-Rego, J. M. Garcia, A. González, J. M. Odriozola, M. Martín-Pastor, A., Linden, *J. Am. Chem. Soc.* **2002**, *124*, 8637–8643.

55 H.-J. Gais, R. Loo, D. Roder, P. Das, G. Raabe, *Eur. J. Org. Chem.* **2003**, 1500–1526.

56 The sulfoximine group is depicted in the Schemes for the sake of simplicity with SO- and SN-double bonds. However, ab initio calculation (Ref. [23, 24] and P. S. Kumar, P. V. Bharatam, *Tetrahedron* **2005**, *61*, 5633–5639; C. P. R. Hackenberger, G. Raabe, C. Bolm, *Chem. Eur. J.* **2004**, *10*, 2942–2952; E. Voloshina, C. Bolm, J. Fleischhauer, I. Atodiresei, G. Raabe, unpublished results) give no evidence for the existence of a double bond. Thus a correct structural representation of the sulfoximine group would be the one containing SO- and SN-single bonds with two positive charges at the S-atom and a negative charge at each the O- and N-atom.

1.4
The "Daniphos" Ligands: Synthesis and Catalytic Applications
Albrecht Salzer and Wolfgang Braun

1.4.1
Introduction

Enantioselective homogeneous catalysis remains one of the most challenging topics in organic and organometallic research. Improvements are being made continually, largely due to the development of new families of ligands. The Nobel committee recognized this by awarding the 2001 Nobel Prizes in Chemistry for achievements in this field [1].

While searching for novel catalysts and ligands, chemists have used all kinds of organic backbones, natural as well as unnatural. While earlier research focused on ligands derived from the "chiral pool", e.g., tartaric acid, terpenes, amino acids, and sugars, the emphasis has shifted somewhat toward other ligand systems not derived from natural sources. This is largely due to the enormous success of such ligand families as BINAP and its derivatives, the DUPHOS ligands [2, 3], and the "Josiphos" family of ligands [4, 5]. It should be noted that these ligands contain unusual elements of chirality, such as planar and axial chirality. The Josiphos ligands are also special in that they incorporate a transition metal-containing unit as the central core around which the ligand framework is constructed. This transition metal itself is not involved in the catalysis, but is part of the scaffold to which the ligand functions are attached. The function of the metal is to create a chiral environment whose spatial and dynamic properties could not be achieved by an organic framework alone.

We have developed a similar class of metal complexes, labeled "Daniphos", which are based on chiral arenechromiumtricarbonyl complexes [6–14]. In many aspects these ligands are similar to the Josiphos ligands, but they differ both in steric and in electronic properties. They also match the ferrocene ligands in enantioselectivity in a number of catalytic applications. The major advantage of the chromium system is its easy availability in optically active form without any need for resolution, as the organic ligand itself, phenylethylamine, is available commercially as both enantiomers and with a variety of functional derivatives. Furthermore the simple and straightforward synthetic route to the Daniphos ligands allows the construction of these diphosphines in a modular fashion, a concept that has attracted considerable attention since the mid 1990s.

1.4.2
General Synthesis

The general synthesis of the Daniphos ligands starting from enantiomerically pure [(R)-1-(phenylethyl)dimethylamine]chromiumtricarbonyl **1**, is depicted in Scheme 1.4.1 [15]. A directed *ortho*-metallation (DOM) and subsequent quench with a chlorophosphine leads to an enantiomerically pure planar-chiral complex, which after chlorination using ACE chloride (1-chloroethyl chloroformate) is transformed into the desired diphosphine by a nucleophilic substitution without any loss of optical purity (Scheme 1.4.1) [6, 10].

One of the key steps in this reaction sequence is the *ortho*-functionalization, which introduces planar chirality in a diastereospecific manner (the corresponding diastereomer is not detectable by ^{31}P NMR spectroscopy). This can be ascribed to the preferred conformation that the side chain of complex **1** adopts: due to the steric bulk of the chromiumtricarbonyl unit the free rotation around the α-*ipso* bond is hindered and the smallest substituent, the hydrogen atom, stays synperiplanar to the chromium tripod, leaving the NMe$_2$ group above the plane of the arene ring in close proximity to the pro-*R*-*ortho*-hydrogen. Addition of *t*-BuLi now leads to pre-coordination of the lithium atom by the nitrogen atom and therefore the pro-*R*-*ortho*-hydrogen is removed exclusively. This explanation for the extraordinary diastereoselectivity was anticipated by Heppert employing NOE-NMR ex-

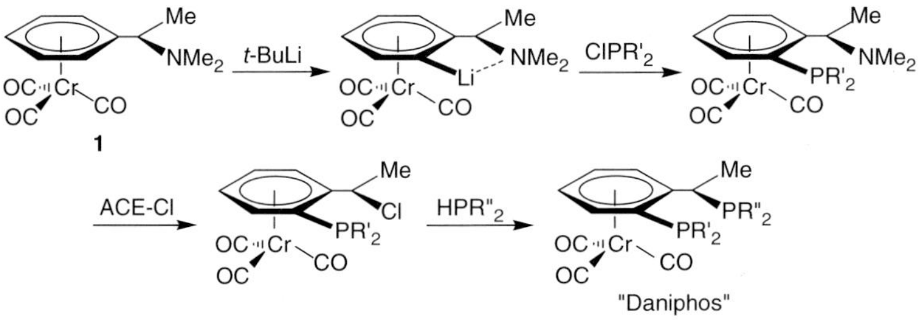

Scheme 1.4.1 General synthesis of the Daniphos ligands.

periments in solution [16]. Now we were able to prove this for the solid state as well by an X-ray structure of complex 1 (Fig. 1.4.1) [15].

As can be clearly seen from the image, the results for the solid state match those obtained for solutions: the stereogenic side-chain adopts a conformation in which the nitrogen atom is located above the plane of the arene ring in close proximity to the hydrogen atom which is to be abstracted, thus explaining the unique diastereoselectivity of the deprotonation. We have calculated a dihedral angle of 64° between the planes defined by nitrogen, *ipso*-C8, α-C7, and *ortho*-C13. This indicates an ideal setup for the abstraction of the desired pro-*R* proton. The chromiumtricarbonyl moiety also facilitates the two consecutive nucleophiic substitutions of the reaction pathway, which both proceed with retention of configuration, by shielding the bottom of the molecule sterically and also by stabilizing the benzylic carbocationic intermediate [6, 7].

As mentioned above, this synthetic route enables a modular approach to ligand design. The concept of modularity, as introduced for the Josiphos ligands by Togni [17], is sketched schematically in Fig. 1.4.2 for the Daniphos ligands.

While the backbone of the ligand provides the stereochemical information, the two different donor groups (**D**$_x$) can be introduced independently of one another in consecutive steps and therefore make a fine-tuning of the ligand's steric and electronic features possible. So with a stock of electro- and nucleophiles at hand an enormous number of ligands is accessible by employing the same synthetic procedure and workup, while the synthesis is not limited to phosphines; other functionalities such as Hal, C, Si, S, O, and N can also be introduced [7b–d,i].

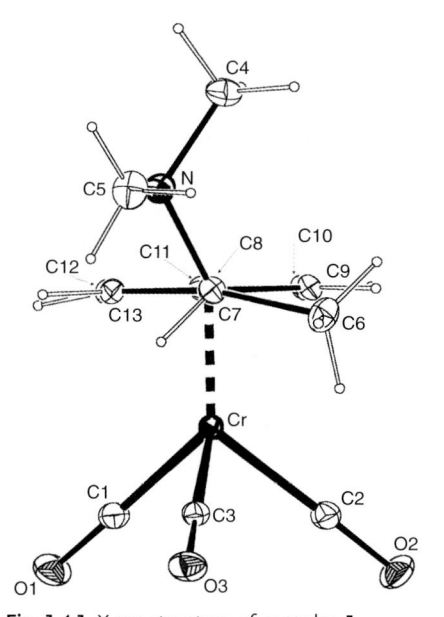

Fig. 1.4.1 X-ray structure of complex **1**.

Electrophilic synthon D₁   D₂ **Nucleophilic synthon**

Ligand backbone: Source of stereochemical information

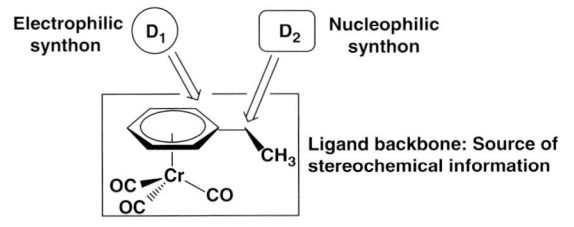

D: Fine tuning of steric and electronic features of the ligand

Fig. 1.4.2 The concept of modularity.

This modularity is an intrinsic advantage for an industrially successful ligand as it allows the rapid synthesis and screening of a huge number of ligands. A special efficiency in time and cost can be achieved when the *ortho* group in the synthesis is kept the same and only the group in the α-position is varied ("late stage modularity"). This is underlined by the success of the Josiphos-derived Xyliphos ligand, which is employed in the industrial synthesis of the herbicide Metolachlor®, to date the largest enantioselective homogeneous transition-metal catalyzed process (>10 000 t y^{-1}). Nevertheless the use of chromiumtricarbonyl ligands has another inherent advantage as other commercially available phenylethylamines, such as those bearing additional substituents on the arene ring (e.g., *t*-Bu, MeO, Me, and Cl groups) or indane and tetralin derivatives, may be used, whereas Josiphos is somewhat limited in this respect because Ugi's amine is used as the starting material.

In this manner a large ligand library of planar-chiral diphosphines of great structural variety has been accumulated, mainly comprising the following derivatives (for the nomenclature system, see below): PPh₂/PPh₂ **2**, P(O)Ph₂/PPh₂ **2a**, PPh₂/PCy₂ **3**, PPh₂/P(*t*-Bu)₂ **4**, PPh₂/P(*i*-Bu)₂ **5**, PPh₂/PbA **6**, PPh₂/PbB **7**, (*S*,*S*)-PPh₂/(*R*,*R*)-DMP **8a**, (*R*,*R*)-PPh₂/(*R*,*R*)-DMP **8b**, PPh₂/P(Cyclpent)₂ **9**, PPh₂/P(*p*-FPh)₂ **10**, P(*i*-Pr)₂/PPh₂ **11**, PCy₂/PCy₂ **12**, PPh₂/P(*o*-Tol)₂ **13**, PPh₂/P(*m*-Tol)₂ **14**, PPh₂/P(*p*-Tol)₂ **15**, PPh₂/PXyl₂ **16**, PXyl₂/PPh₂ **17**, PXyl₂/PXyl₂ **18**, P(*o*-Tol)₂/PPh₂ **19**, P(*o*-Tol)₂/PCy₂ **20**, P(*o*-Tol)₂/(*R*,*R*)-DMP **21**, P(*o*-Tol)₂/P(*o*-Tol)₂ **22**, *t*-Bu/PPh₂/PCy₂ **23**, Me/PPh₂/PPh₂ **24**, Me/PPh₂/PCy₂ **25**, Me/PPh₂/P(*t*-Bu)₂ **26**, MeO/PPh₂/PPh₂ **27**, MeO/PPh₂/PCy₂ **28**, MeO/PPh₂/P(*t*-Bu)₂ **29**, Cl/PPh₂/PPh₂ **30**, Cl/PPh₂/PCy₂ **31**, Cl/PPh₂/P(*t*-Bu)₂ **32**, Ind/PPh₂/PPh₂ **33**, Ind/PPh₂/PCy₂ **34**, Ind/PPh₂/P(*t*-Bu)₂ **35**. Representative possible structural motifs are depicted in Fig. 1.4.3.

For the further discussion of the catalytic applications it is important to note the acronym system which we apply to designate each ligand easily and unequivocally. According to the general label R/PR′₂/PR″₂, R denotes the substituent in the *para*-position to the chiral α-chain (when R = H, it is omitted), PR′₂ stands for the group in the *ortho*-position and finally PR″₂ for the phosphino group in the α-chain itself (see Fig. 1.4.4). The abbreviations "PbA" and "PbB" refer to the regioisomeric phobane skeletons (PbA = phosphabicyclo[3.3.1]nonane, PbB = phosphabicyclo[4.2.1]nonane), "Ind" stands for indane and the (*R*,*R*)-2,5-dimethylphospholane unit is abbreviated to "(*R*,*R*)-DMP".

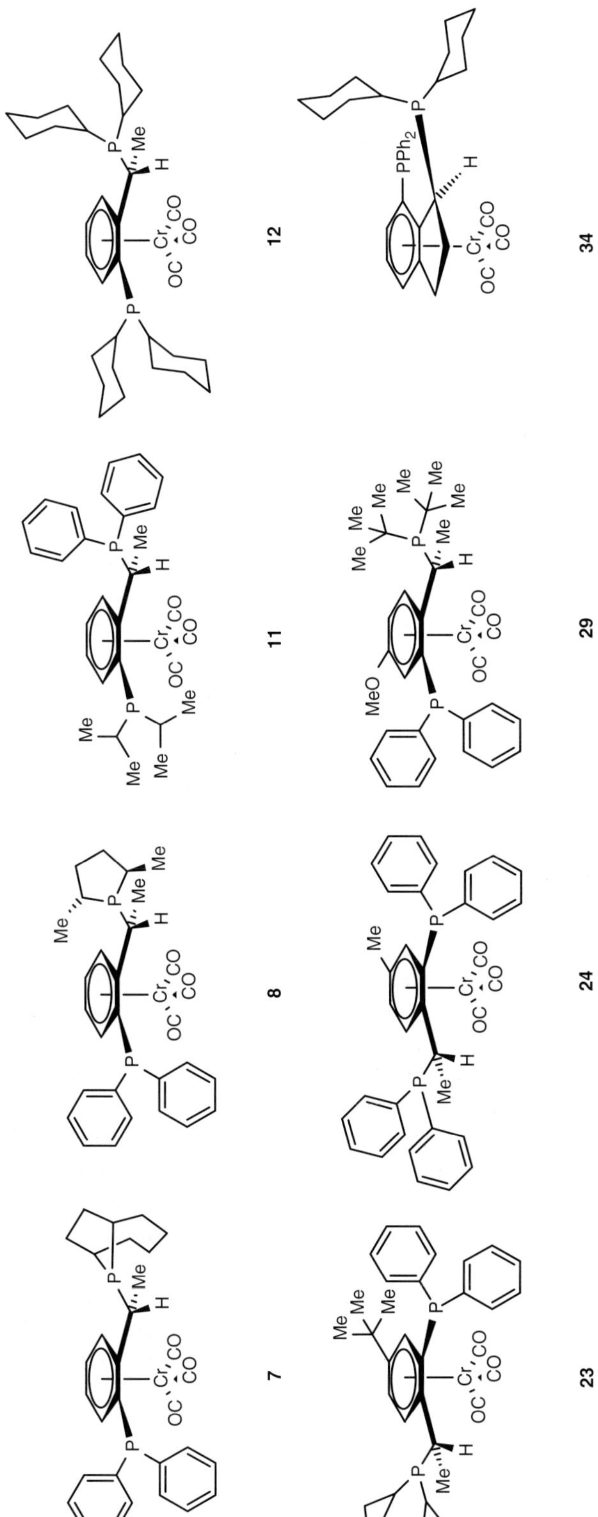

Fig. 1.4.3 A representative set of structural variations from our library.

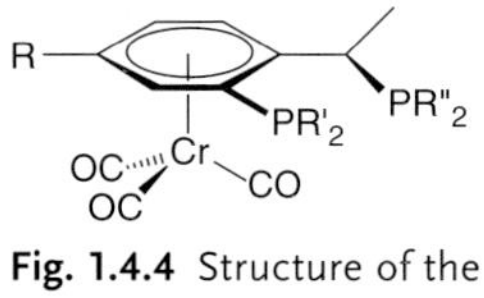

Fig. 1.4.4 Structure of the general acronym labeled R/PR′$_2$/PR″$_2$.

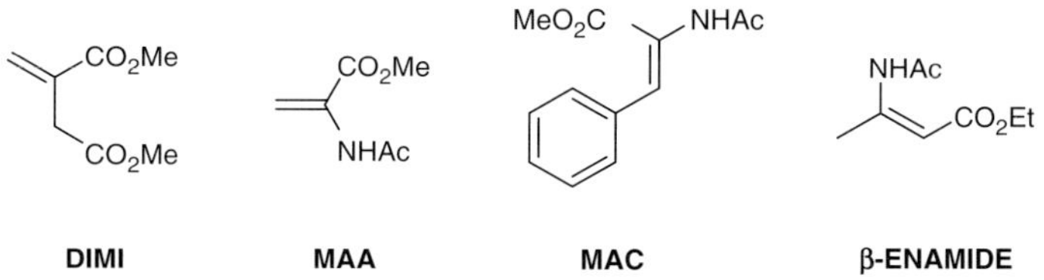

DIMI   **MAA**   **MAC**   **β-ENAMIDE**

Fig. 1.4.5 Substrates for the hydrogenation of C=C double bonds.

1.4.3
Applications in Stereoselective Catalysis

1.4.3.1 Enantioselective Hydrogenations

As enantioselective hydrogenations of prochiral substrates are undoubtedly the most common applications of chiral diphosphine ligands, a broad screening of our ligands was undertaken with some commonly used standard substrates. As substrates for the hydrogenation of C=C double bonds dimethyl itaconate (DIMI), methyl 2-acetamidoacrylate (MAA), methyl acetamidocinnamate (MAC) as an α-amino acid precursor, and ethyl (Z)-3-acetamidobutenoate (β-ENAMIDE) as a β-amino acid precursor were chosen (see Fig. 1.4.5).

The reactions were carried out with rhodium catalyst precursors at a hydrogen pressure of 1 bar, a temperature of 25 °C and a substrate/catalyst ratio of 100 : 1 with methanol as the solvent (ethanol was used in case of the β-enamide). It was found that for this kind of reaction the enantioselectivities varied over a broad range, strongly depending on the substitution pattern employed, thus to this extent justifying the concept of modularity in ligand design. It showed that comparatively electron-rich ligands with additional electron-donating groups on the arene ring and bulky, electron-rich phosphine groups such as bicyclohexylphosphine and (R,R)-dimethylpholane on the α-chain gave the best results. The highest enantioselectivities were obtained for DIMI with ligands Me/PPh$_2$/PCy$_2$ **25** (79.2% ee) and Me/PPh$_2$/PCy$_2$ **23** (77.9% ee), for MAA also with ligands Me/PPh$_2$/PCy$_2$ **25** (90.8% ee) and Me/PPh$_2$/PCy$_2$ **23** (91.2% ee), for MAC with ligands PPh$_2$/PbB **7** (87.3% ee) and **25** (81.6% ee), and for the β-enamide with ligands PPh$_2$/PbB **7** (83.1% ee) and (S,S)-PPh$_2$/(R,R)-DMP **8a** (85.1% ee). A detailed description of the experimental results can be found in Ref. [18].

As an example of a substrate with a C=O double bond, ethyl pyruvate was chosen and tested with a number of candidates from our ligand stock. The reaction is outlined in Scheme 1.4.2 and the results are summarized in Table 1.4.1 [18].

Although it is clear that the enantioselectivities obtained so far are somewhat low, it has to be taken into account that these are only the results of a first screening and hence they can be considered encouraging for a further optimization.

The MEA imine, an intermediate in the synthesis of the well-known Ciba–Geigy/Syngenta herbicide Metolachlor (DUAL MAGNUM®), still remains a challenging substrate for asymmetric hydrogenation and gives a good insight into the performance of a ligand in the hydrogenation of a C=N double bond. Therefore we decided to take it as our candidate of choice for such an imine substrate, and we screened a number of our ligands in this reaction (Scheme 1.4.3). Table 1.4.2 gives an overview of the results obtained [18].

The benchmark ligand for this reaction, which is also the one used for the industrial-scale reaction, is the Xyliphos ligand, a ferrocene-based diphosphine bearing the same stereochemical features and substitution pattern as ligand

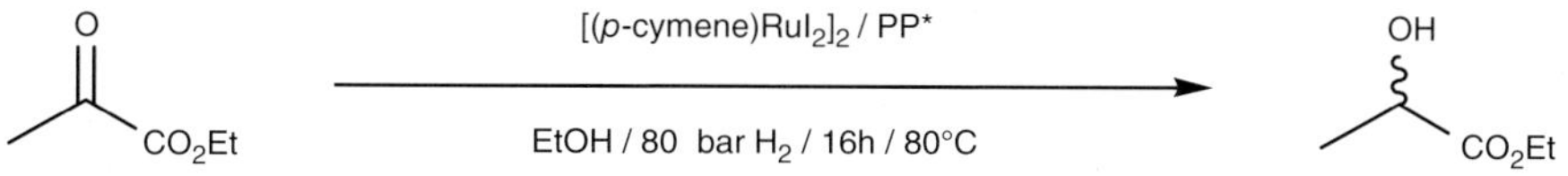

Scheme 1.4.2 Hydrogenation of ethyl pyruvate.

Table 1.4.1 Results of the hydrogenation of ethyl pyruvate.

Ligand	8b	3	12	23	4	10
ee [%]	15.6	35.0	42.2	2.7	30.1	32.3
Conversion [%]	5	6	65	9	88	3
c.y. [%]	3	6	24	7	87	3
Configuration	R	R	R	S	R	R

a c.y. = chemical yield.

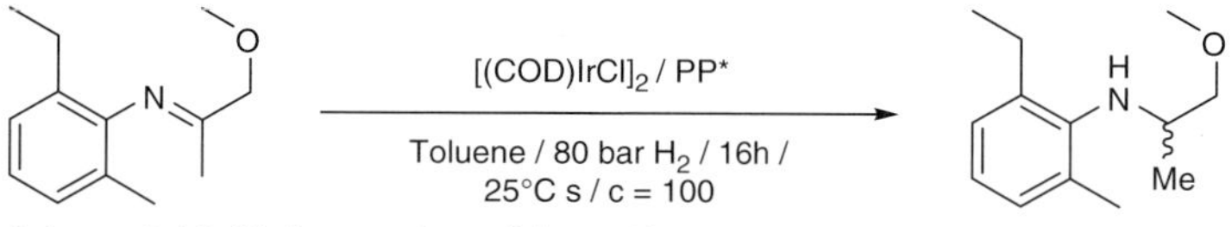

Scheme 1.4.3 Hydrogenation of the MEA imine.

Table 1.4.2 Results of the hydrogenation of the MEA imine.

Ligand	3	7	8a	8b	10	11	16	18	23	25
ee [%]	63.9	48.5	61.6	81.8	65.7	42.5	21.6	44.7	63.1	54.7
Conversion [%]	60	30	65	96	90	37	34	87	63	60
c.y. [%]	60	30	65	96	90	37	34	87	63	60
Configuration	S	S	R	S	S	S	S	S	R	R

$PPh_2/PXyl_2$ **16**. During this survey Xyliphos gave 81.6% ee at a conversion of 97%, compared to which the results obtained with the chromiumtricarbonyl analogues are somewhat lower with most of the ligands, of course. Nonetheless some of them produced quite encouraging results. Compounds (S,S)-PPh_2/(R,R)-DMP **8a**, PPh_2/PCy_2 **3**, t-Bu/PPh_2/PCy_2 **23**, and $PPh_2/P(p$-$FPh)_2$ **10** gave selecivities higher than 60% ee, while it is noteworthy that the comparatively electron-poor ligand **10**, which carries a *para*-fluoro substituent, showed a conversion of 90% after 1 h. Especially encouraging are the results for ligand **8b** (81.8% ee, 96% conversion). Diphosphine (R,R)-PPh_2/(R,R)-DMP **8b** almost exactly matches the performance of the benchmark ligand Xyliphos, but it should be noted that the reactions are still unoptimized.

1.4.3.2 Diastereoselective Hydrogenation of Folic Acid Ester

L-Tetrahydrofolic acid is a versatile intermediate for the manufacture of various folates, e.g., L-leucovorin [19], which is used in cancer therapy, or Metafolin®, which is used as a vitamin in functional food. To our knowledge optically pure L-tetrahydrofolic acid is still obtained by repeated fractional crystallization from an equimolar mixture of diastereoisomers formed by nondiastereoselective hydrogenation of folic acid. In order to increase the yield of l-tetrahydrofolic acid and to avoid recrystallization steps, we checked the utility of our ligand for the diastereoselective hydrogenation of folic acid dimethyl ester benzenesulfonate (Scheme 1.4.4).

In this reaction the diazadiene system of the pteridine ring of the substrate is hydrogenated, giving a new stereogenic center at position 6. The distance of the stereogenic center of the glutamic acid residue is too far from the newly formed chiral center and therefore it does not cause any measurable stereochemical induction in the hydrogenation and the outcome of the reaction is governed by the chiral ligand alone. Quite a few experiments have been undertaken so far to reach a reasonable induction, but the best results achieved do not exceed some 40% *de*, where the desired product is the (S,S)-diastereomer [20]. In the course of the reaction a significant percentage of an undesired cleavage product is observed in varying amounts ($ABGAMe_2$, aminobenzoylglutamic acid dimethyl ester) depending on the ligand used, diminishing the yield of the valuable tetrahydrofolic acid dimethyl ester ($THFMe_2$).

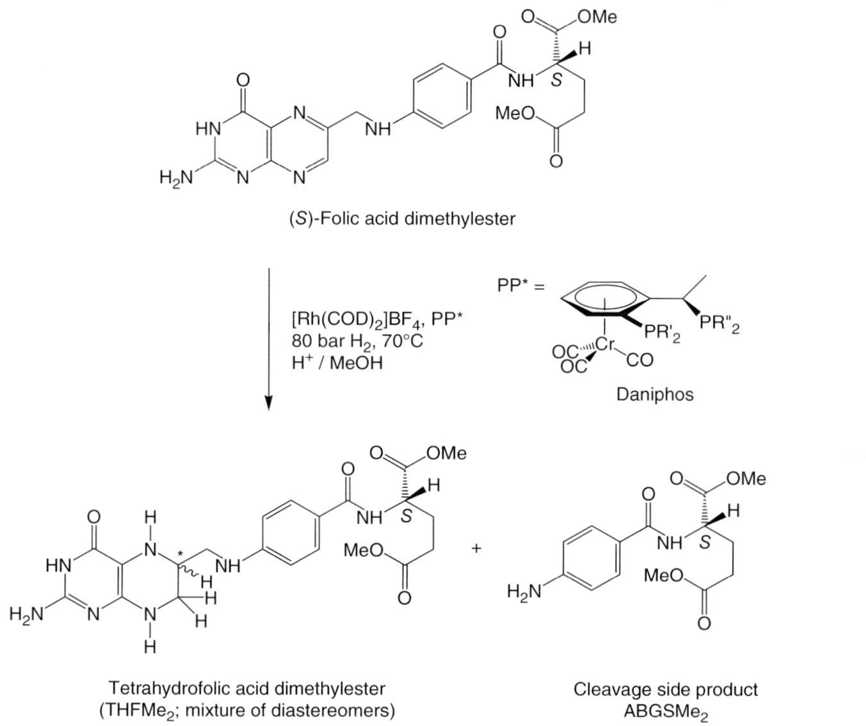

1.4.3 Applications in Stereoselective Catalysis | 123

Scheme 1.4.4 Hydrogenation of folic acid dimethyl ester benzenesulfonate.

A number of candidates from our ligand library have been employed in this survey, bearing aliphatic as well as aromatic substituents on the phosphorus donor atoms, together with the benchmark ligands Josiphos and BINAP, which were included in this examination for comparison purposes. The results are collected in Table 1.4.3.

From the results achieved with ligand **3** it can be seen that the use of the other enantiomer of the ligand results in a reversal of the configuration in the product together with (nearly) the same induction, as is to be expected. Moreover, it is remarkable that all derivatives that carry only aromatic substituents on the phosphorus donor atoms deliver preferentially the undesired 6R,S diastereomer (indicated by the minus sign). In contrast to this, the desired isomer is formed in excess when the ligands contain aliphatic side chains on the donor centers as well (it should be kept in mind that all ligands are configured R,R). Here an especially high stereochemical induction and chemical yield of tetrahydrofolic acid ester is achieved in the case of the P(i-Pr)$_2$/PPh$_2$ derivative **11**, which matches the best

Table 1.4.3 Results of the hydrogenation of folic acid dimethyl ester.

Ligand	Form	de [%]	c.y. [% THFMe$_2$]	c.y. [% ABGAMe$_2$]
3	S,S	−14.0	70.1	26.9
3	R,R	17.0	64.0	18.6
11	R,R	42.4	81.5	16.7
5	R,R	8.0	69.5	16.8
13	R,R	−31.6	87.9	7.2
14	R,R	−21.6	77.6	21.0
15	R,R	−20.8	80.6	9.9
19	R,R	−25.6	63.0	20.9
22	R,R	−35.0	76.1	21.7
16	R,R	−17.4	71.3	14.9
17	R,R	−27.8	83.1	14.2
18	R,R	−32.0	79.8	10.4
Josiphos	S,S	22	60	unknown
BINAP	R	46	90	unknown

selectivities known to date when employing ligands such as BINAP. So one might conclude that a sterically demanding, Lewis-basic group in the *ortho* position influences the reaction in the desired direction [21].

1.4.3.3 Enantioselective Isomerization of Geranylamine to Citronellal

Enantiomerically pure citronellal in both of its antipodal forms has outstanding importance as a key intermediate for the production of fine chemicals, especially for the production of fragrances and flavors. In this respect the isomerization of diethylgeranylamine to (R)-citronellal enamine in the presence of Rh^I/(S)-BINAP is an exceptional industrial process, for instance as one of the key steps of the Takasago process for the production of (−)-menthol [22]. In the search for alternatives for this process, both Josiphos and Daniphos derivatives were evaluated (Scheme 1.4.5) [23].

Both classes of ligands gave very good results, the Daniphos ligands having higher enantioselectivities but slower kinetics (see Table 1.4.4).

1.4.3.4 Nucleophilic Asymmetric Ring-Opening of Oxabenzonorbornadiene

While the discovery of new reactions for stereoselective organometallic catalysis is somewhat rare, it is especially noteworthy that Lautens and coworkers have recently reported on the development of a nucleophilic asymmetric ring-opening (ARO) reaction of oxabenzonorbornadienes [24], achieving excellent results with different substrates and various kinds of nucleophiles. It reveals an interesting new application for chiral diphosphine ligands. Lautens had successfully employed the ferrocene-based Josiphos class of ligands, which are structurally related to ours, so we found it challenging to test the Daniphos ligands with one of these substrates as well.

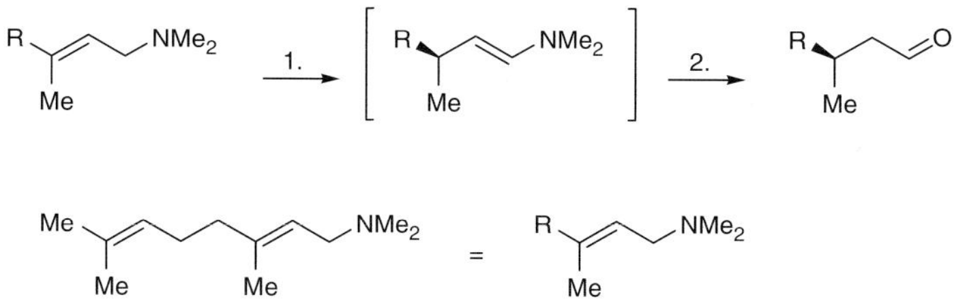

Scheme 1.4.5 Enantioselective isomerization of geranylamine to citronellal.

1. [CODRhPP*COD]Tf 0.25 mol-%, THF, 66°C; 2. AcOH, H$_2$O, Et$_2$O

Table 1.4.4 Results of the isomerization of geranylamine.

L*	Rest/No.	Conv./%	ee/%	Prod.-Conf.
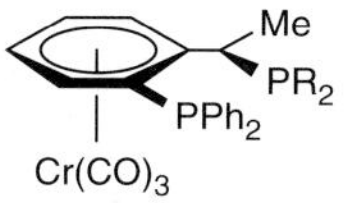 Cr(CO)$_3$ arene-PR$_2$/PPh$_2$ (Me)	Cy 3	99	94	S
	t-Bu 4	99	96	S
	Ph (1 mol%) 2	44	76	S
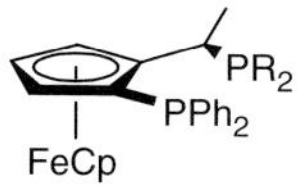 FeCp-PR$_2$/PPh$_2$	Cy	99	78	S
	t-Bu	99	92	S
(R)-BINAP	–	99	97	S

We applied a selection of our ligands in the asymmetric ring-opening of oxabenzonorbornadiene using methanol as the nucleophile. The rhodium-catalyzed reaction proceeds at a substrate/ligand ratio of 100 : 1 with a reaction time of 5 h at 80 °C. The Josiphos ligand ("Fc/PPh$_2$/PCy$_2$") was also included for purposes of comparison. The reaction scheme is depicted in Scheme 1.4.6, and Table 1.4.5 summarizes the results obtained [25].

The enantioselectivities achieved (Table 1.4.5) range from moderate to very good, with the best value of 97.5% ee obtained with ligand 4. Taking into account that the ligands used have the same backbone and differ only in the nature of the second donor group in the α-chain, it appears noteworthy that the values vary over a considerable range. This shows the sensitivity of the reaction toward the steric and electronic properties of the ligands. While Josiphos and the structurally equivalent ligand 3 give almost the same selectivities, it was found that the chromium-based ligand performs faster in this reaction.

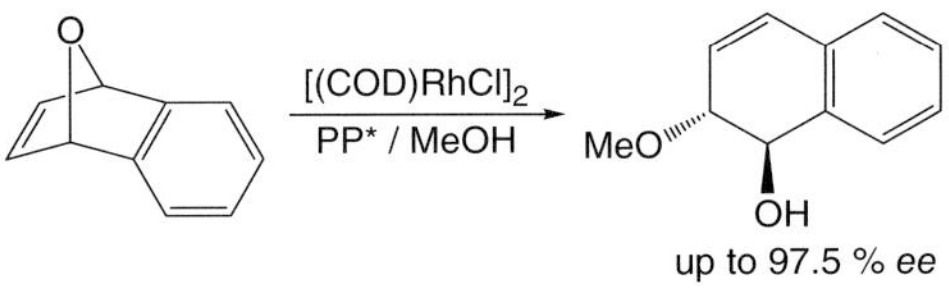

Scheme 1.4.6 Asymmetric ring-opening of oxabenzonorbornadiene.

Table 1.4.5 Results of the ARO reaction.

Ligand (ortho/α)	Yield [%]	ee [%]
PPh_2/PPh_2 **2**	69	80.3
$PPh_2/P(p-FPh)_2$ **10**	10	50.9
$PPh_2/P(i-Bu)_2$ **5**	20	75.0
$PPh_2/P(t-Bu)_2$ **4**	59	97.5
$PPh_2/PCyclpent_2$ **9**	79	84.7
PPh_2/PCy_2 **3**	53	93.1
$Fc/PPh_2/PCy_2$, "Josiphos"	36	92.4

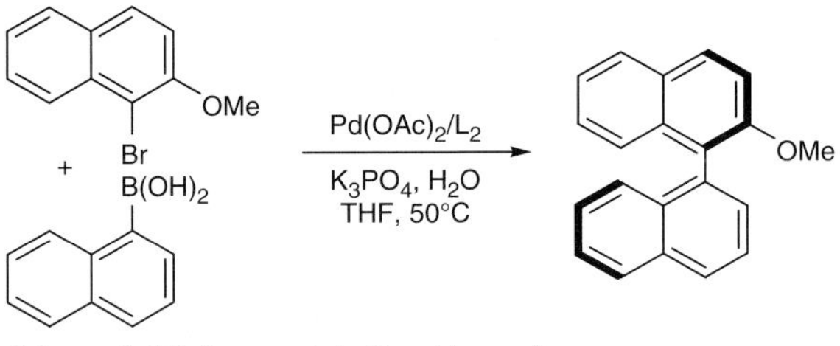

Scheme 1.4.7 Asymmetric Suzuki coupling.

1.4.3.5 Enantioselective Suzuki Coupling

There are as yet few examples of enantioselective Suzuki couplings [26–28]. Cammidge and Crépy [27] used planar-chiral ferrocene derivatives with PPh_2/NMe_2 ligand functions which gave up to 86% *ee*. We have performed experiments using both ferrocene and arenechromium bidentate ligands; the reaction and the results are shown in Scheme 1.4.7 and Table 1.4.6 respectively [29].

The yield for the chromium complex was lower, but the enantioselectivity was higher than for the corresponding ferrocene catalyst.

1.4.3.6 Asymmetric Hydrovinylation

Asymmetric hydrovinylation has been pioneered by Bogdanovic [30] and Wilke [31] using nickel catalysts. Of special interest is the reaction between vinylarenes and ethylene, as enantioselective codimerization provides a convenient route to

Table 1.4.6 Results of the asymmetric Suzuki coupling.

Ligand	Yield [%]	ee [%]
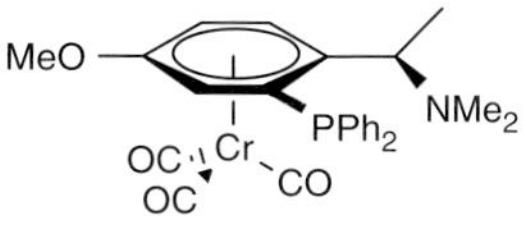 MeO–(arene)Cr(CO)$_2$(PPh$_2$)–CH(CH$_3$)–NMe$_2$	27	87
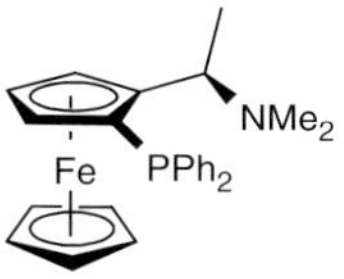 Fe(Cp)(Cp–PPh$_2$)–CH(CH$_3$)–NMe$_2$	47	63

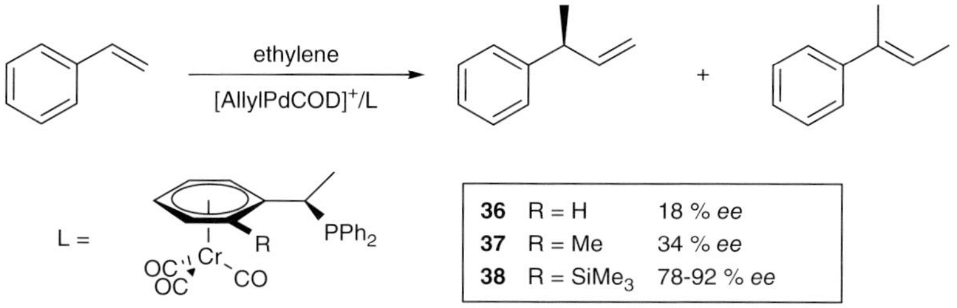

Scheme 1.4.8 Asymmetric hydrovinylation.

styrene + ethylene →[[AllylPdCOD]$^+$/L] branched product + isomerized product

L = (arene)Cr(CO)$_2$(PPh$_2$)(R)–CH(CH$_3$)

36	R = H	18 % ee
37	R = Me	34 % ee
38	R = SiMe$_3$	78–92 % ee

2-arylpropionic acids. A very efficient system was found using a ligand derived from myrtenal, which at −70 °C gave enantioselectivities of >95% *ee* [31]. A catalytic system based on palladium was recently reported by Vogt et al. for the same reaction using the diastereomerically pure *t*-butyl(methyl-*O*)phenylphosphinite possessing a stereogenic P atom, which gave *ee* values up to 87% [32]. We then tested monophosphines with organometallic backbones. After initial success with monophosphines based on tri(carbonyl)iron complexes, we also investigated the monophosphines **36–38**. Even at room temperature, high activity and selectivity toward the codimers were observed with all ligands (Scheme 1.4.8).

Stability, activity and chemo- and enantioselectivity increased with increasing steric demand of the *ortho* substituent R. Introduction of the trimethylsilyl group at this position (ligand **38**) therefore resulted in an excellent enantioselective system which belongs among the best Pd catalysts described so far for asymmetric hydrovinylation. Almost 70% conversion was observed within 15 min. The product was obtained in 78.5% *ee* and only a small amount of the isomerization products was detected in the reaction mixture. However, at higher conversions, isomerization of the product to the internal achiral olefin took place. Therefore, after 0.5 h and at complete conversion, selectivity toward 3-phenylbut-1-ene had dropped to 48.5%; but even at these high conversions chemoselectivity toward the codimers remained very good (98%). The consecutive isomerization reaction goes along with a kinetic resolution, due to which the *ee* of the product rises to 92% within 0.5 h. Another feature is the remarkable stability of the catalytic system. No Pd(black) formation was observed after the reaction, which is quite unusual when a monodentate phosphine ligand is being used. The steric bulk of ligands **36–38** probably prevents the formation of binuclear palladium species, an observation similar to those recently made by Buchwald with his biphenyl-derived monophosphines (X-PHOS) [33]. The increase in stability and activity of the catalytic system going from **37** to **38** conforms with this explanation [7].

1.4.3.7 Allylic Sulfonation

Whereas Pd-catalyzed asymmetric allylic substitution reactions, with carbon as well as with heteronucleophiles, are widespread in stereoselective catalysis, it seems unusual that sulfur nucleophiles are less commonly used. Therefore we tested our ligands in such a reaction. We employed ligands **2** and **3** successfully in the reaction of racemic 3-methoxycarbonyloxyhept-4-ene with lithium *t*-butylsulfinate in the presence of 1.5 mol% of Pd_2dba_3 and 4.5 mol% of the ligands. In all cases full conversion was achieved, but with marked differences in the product selectivities (Scheme 1.4.9, Table 1.4.7).

The highest *ee* value (57%) was obtained with ligand **2**, but here again there should be considerable scope for improvement by variation of the steric and electronic properties of the two ligand functions [8].

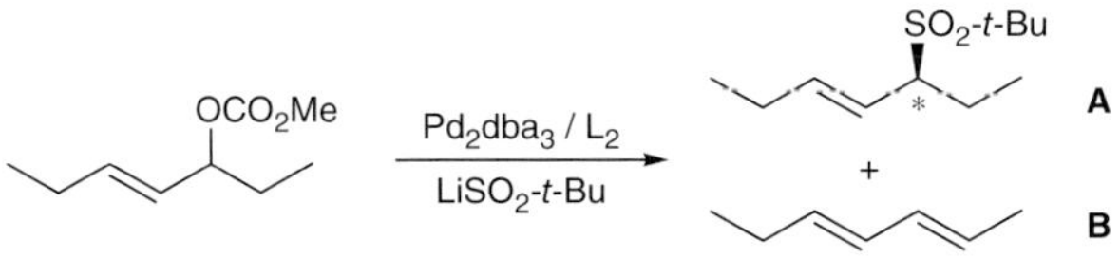

Scheme 1.4.9 Allylic sulfonation with the Daniphos ligands.

Table 1.4.7 Results of the allylic sulfonation.

Ligand (ortho/α)	S (A/B)	Conversion [%]	ee [%]
PPh_2/PPh_2 **2**	73:27	100	57
PPh_2/PCy_2 **3**	15:85	100	14
$PPh_2/P(t\text{-}Bu)_2$ **4**	100:0	100	11

1.4.4
Conclusion

We have shown that a large library of mono- and bidentate ligands based on arenechromiumtricarbonyl complexes can be derived from optically active phenylethylamine and its derivatives. The complexes are very similar to the well-known "Josiphos" ligands but differ in both steric and electronic properties. The "Daniphos" ligands match the corresponding "Josiphos" ligands closely in enantioselectivity in a number of different catalytic applications. It must be acknowledged, however, that the ferrocene ligands, due to their industrial applications, are currently better established and more thoroughly studied. An interesting feature of both ligand classes is the fact that for most catalytic applications a different ligand gives the highest *ee* values, therefore underlining the importance of having a modular ligand system.

Structural data confirm that the "Daniphos" ligands are readily adaptable to most ligand environments. A major advantage of the chromium system is the commercial availability of phenylethylamine and its derivatives in both enantiomeric forms. In contrast to the cyclopentadienyl ring, arenes have an almost unlimited potential for controlled substitutional variation, accessible through standard reaction protocols.

References

1 a) W. S. Knowles, *Adv. Synth. Catal.* **2003**, *345 (1–2)*, 3; b) R. Noyori, *Adv. Synth. Catal.* **2003**, *345 (1–2)*, 15; c) W. S. Knowles, *Angew. Chem. Int. Ed.* **2002**, *41(12)*, 1998; d) R. Noyori, *Angew. Chem. Int. Ed.* **2002**, *41(12)*, 2008; e) K. B. Sharpless, *Angew. Chem. Int. Ed.* **2002**, *41(12)*, 2024.

2 M. J. Burk, in: *Handbook of Chiral Ligands*, Marcel Dekker, New York, **1999**, p. 339.

3 M. J. Burk, *Acc. Chem. Res.* **2000**, *33*, 363.

4 H. U. Blaser, *Adv. Synth. Catal.* **2002**, *344*, 17.

5 H. U. Blaser, W. Brieden, B. Pugin, F. Spindler, M. Studer, A. Togni, *Top. Catal.* **2002**, *19*, 3.

6 U. Englert, A. Salzer, D. Vasen, *Tetrahedron: Asymmetry* **1998**, *9*, 1867.

7 U. Englert, R. Haerter, D. Vasen, A. Salzer, E. B. Eggeling, D. Vogt, *Organometallics* **1999**, *18*, 4390.

8 D. Vasen, A. Salzer, F. Gerhards, H. J. Gais, R. Stürmer, N. H. Bieler, A. Togni, *Organometallics* **2000**, *19*, 539.

9 A. Salzer, *Coord. Chem. Rev.* **2003**, *242*, 59.

10 W. Braun, A. Salzer, H.-J. Drexler, A. Spannenberg, D. Heller, *J. Chem. Soc., Dalton Trans.* **2003**, 1606.

11 U. Englert, C. Hu, A. Salzer, E. Alberico, *Organometallics* **2004**, *23(23)*, 5419.

12 D. Totev, A. Salzer, D. Carmona, L. A. Oro, F. J. Lahoz, I. T. Dabrinovitch, *Inorg. Chim. Acta* **2004**, *357(10)*, 2989.

13 W. Braun, B. Calmuschi, J. Haberland, W. Hummel, A. Liese, T. Nickel, O. Stelzer, A. Salzer, *Eur. J. Inorg. Chem.* **2004**, *11*, 2235.

14 W. Braun, B. Calmuschi, H.-J. Drexler, U. Englert, D. Heller, A. Salzer, *Acta Crystallogr. Sect. C*, **2004**, *60*, 532.

15 B. Calmuschi, W. Braun, A. Salzer, *Acta Crystallogr. Sect. E* **2005**, *61*, 828.

16 J. A. Heppert, J. Aube, M. E. Thomas-Miller, M. L. Milligan, F. Takusagawa, *Organometallics* **1990**, *9(3)*, 727.
17 a) A. Togni, C. Breutel, A. Schnyder, F. Spindler, H. Landert, A. Tijani, *J. Am. Chem. Soc.* **1994**, *116*, 4062; b) H. C. L. Abbenhuis, U. Burckhardt, V. Gramlich, A. Togni, A. Albinati, B. Müller, *Organometallics* **1994**, *13*, 4481; c) U. Burckhardt, L. Hintermann, A. Schnyder, A. Togni, *Organometallics* **1995**, *14*, 5415.
18 W. Braun, A. Salzer, F. Spindler, E. Alberico, *Appl. Cat. A: General* **2004**, *274(1–2)*, 191.
19 V. Groehn, R. Moser, *Pteridines*, **1999**, *10*, 95.
20 a) H. Brunner, C. Huber, *Asymm. Cat.*, **1992**, *67*, 2085; b) H. Brunner, P. Bublak, M. Helget, *Asymm. Cat.*, **1996**, *105*, 55; c) BASF Patent EP 0551642A1, **1992**; d) H. Brunner, S. Rosenboem, *Monatsh. Chemie*, **2000**, *131*, 1271; e) H. Brunner, C. Huber, *Chem. Ber.* **1992**, *125*, 2085; f) H. Brunner, P. Bublak, M. Helget, *Chem. Ber.* **1997**, *127*, 55; g) V. Groehn, R. Moser, B. Pugin, *Adv. Synth. Catal.* **2005**, *347*, 1855.
21 W. Braun, B. Calmuschi-Cula, A. Salzer, V. Groehn, *J. Organomet. Chem.* **2006**, *691*, 2263.
22 R. Noyori, *Angew. Chem.* **2002**, *114(12)*, 2108; R. Noyori, *Angew. Chem. Int. Ed.* **2002**, *41(12)*, 2008.
23 C. Chapuis, M. Barthe, J. Y. de Saint Laumer, *Helv. Chim. Acta* **2001**, *84*, 230.
24 M. Lautens, K. Fagnou, S. Hiebert, *Acc. Chem. Res.* **2003**, *36*, 48.
25 W. Braun, W. Müller, B. Calmuschi, A. Salzer, *J. Organomet. Chem.* **2005**, *690(5)*, 1166.
26 S.L. Buchwald, J. Yin, *J. Am. Chem. Soc.* **2000**, *122*, 12 051.
27 N. Cammidge, K.V.L. Crépy, *Chem. Commun.* **2000**, 1723.
28 A.S. Castanet, F. Colobert, P.-E. Brontin, M. Obringer, *Tetrahedron: Asymmetry* **2002**, *13*, 659.
29 A. Grosse-Boewing, Diploma Thesis, RWTH Aachen, **2002**.
30 B. Bogdanovic, *Adv. Organomet. Chem.* **1979**, *17*, 105.
31 G. Wilke, *Angew. Chem.* **1988**, *100*, 189; *Angew. Chem. Int. Ed. Engl.* **1988**, *27*, 185.
32 R. Bayersdörfer, B. Ganter, U. Englert, W. Keim, D. Vogt, *J. Organomet. Chem.* **1998**, *552*, 187.
33 E. R. Strieter, D. G. Blackmond, S. L. Buchwald, *J. Am. Chem. Soc.* **2003**, *125(46)*, 13 978.

1.5
New Chiral Ligands Based on Substituted Heterometallocenes
Christian Ganter

1.5.1
Introduction

The development of new chiral ligand systems remains a real challenge with particular importance for asymmetric catalysis with transition metal complexes. Different approaches have been used to create chiral ligand structures. The modification of starting compounds from the chiral pool represents one possibility and molecules like DIOP or taddol which are derived from tartaric acid may serve as examples. Similarly, oxazoline derivatives are powerful building blocks for the assembly of chiral ligands and can be obtained easily from naturally occurring amino acids with stereogenic centers. However, axially chiral systems featuring a biaryl skeleton are not obtained directly from chiral-pool precursors although the latter substances may be used during the synthesis as resolving agents to obtain enantiomerically pure compounds. The binaphthyl backbone has yielded particularly successful ligands such as BINAP and related derivatives [1]. The chiral phospholane ring present in the DUPHOS [2] ligand family is another example of a chiral donor function which is not directly related to a chiral-pool precursor.

Metal complexes themselves may also serve as ligands, provided they carry suitable donor functions. Ferrocene [3, 4] and (arene)chromiumtricarbonyl [5–7] derivatives have been studied in great detail (for recent developments see, for example, Chapter 1.4). Both systems feature π-bound carbocyclic ligands (Cp or benzene) carrying two donor groups. If the two groups are different a planar-chiral compound is obtained. Both the ferrocene as well as the (arene)chromium systems may be synthesized according to a highly modular approach which allows the preparation of a huge number of ligands with different combinations of the two donor groups. Ferrocene-based ligands ("Josiphos" ligands) have found applications in industrial-scale enantioselective catalytic processes.

In recent years we have developed a new type of planar-chiral ligand system which is based on a phosphaferrocene skeleton equipped with an additional donor function Y [8]. This structure is closely related to the well-known ferrocene-type ligands in that a CH unit of the latter has been replaced by a P atom. The phosphaferrocene moiety serves as both a chiral metallocene-type backbone and as a donor group via the phosphorus atom lone pair. These new ligands are unique in their topological architecture and show interesting ligand properties.

1.5.2
General Properties of Phosphaferrocenes

The motivation to use π-coordinated heterocycles as ligand components came from the successful performance of ferrocene derivatives as mentioned in Section 1.5.1. The phosphaferrocene system was chosen from several candidates for a couple of reasons. Compared to complexes of other heterocycles, the synthesis of phosphaferrocenes is quite straightforward, relying mainly on procedures developed by Mathey, who carried out fundamental studies starting as early as 1977 [9]. From Mathey's work it is also known that phosphaferrocenes can act as ligands by virtue of their electron lone pair located at the P atom. While the commonly used trialkyl- and triarylphosphines are good σ-donors and moderate π-acceptors, the reverse is true for the phosphaferrocene, which acts as an electron-poor good acceptor ligand much more like a phosphite than a phosphine. Introduction of a suitable donor group Y into the 2-position of the phospholyl ring yields a planar-chiral chelate ligand (Figure 1.5.1). A metal coordinated to the P atom in such a substituted phosphaferrocene is close to the source of chirality and a high level of stereochemical discrimination may be expected for a reaction proceeding at that metal atom.

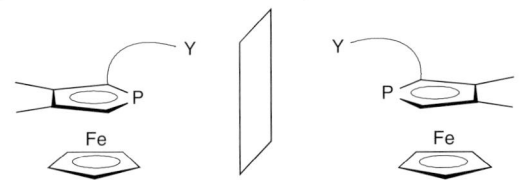

Fig. 1.5.1 The planar-chiral donor-substituted phosphaferrocene chelate ligand.

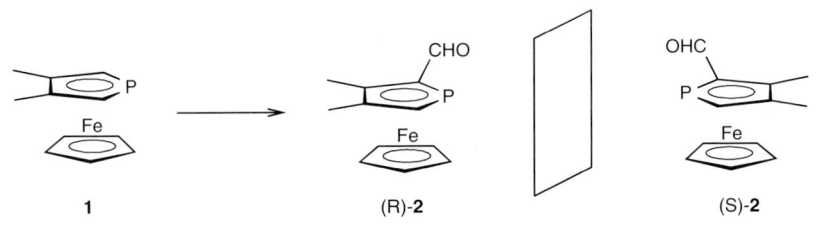

Scheme 1.5.1 Two enantiomers of aldehyde **2**.

1.5.3
Synthesis of Phosphaferrocenes

The parent compound 3,4-dimethylphosphaferrocene **1** which serves as starting material for chelate ligands is readily obtained in multigram quantities from 1-*t*-butyl-3,4-dimethylphosphole and $[CpFe(CO)_2]_2$ in a thermal reaction. It has been shown recently that this reaction proceeds via a nonradical pathway [10]. Another approach is the reaction of a phospholide anion with a reagent delivering a $CpFe^+$ synthon [9]. In contrast to ferrocene, the deprotonation of **1** in the α-position of the phosphole ring is unfortunately not possible, precluding the introduction of a substituent by treatment of a metallocenyl anion with an electrophile. Yet classical electrophilic aromatic substitutions are possible and proceed selectively at the phospholyl ring. The Vilsmeyer formylation was therefore employed as the key step for introducing a synthetically versatile formyl group into the 2-position of the phosphole ring (Scheme 1.5.1) [11].

The racemic mixture of aldehyde **2** was thus obtained in 65% yield and transformation of the aldehyde function allowed a large number of different chelate ligands to be prepared. An efficient separation method was developed to obtain the enantiomerically pure aldehydes (*R*)-**2** and (*S*)-**2**, respectively [12]. Thus, treatment of the racemic aldehyde with enantiomerically pure *trans*-1,2-bis(methylamino)cyclohexane yields the diastereomeric aminals, which can be separated easily by column chromatography on silica. Acidic hydrolysis of the diastereomers gives the pure enantiomers of **2** in 95% yield. The absolute configuration of the (*S*)-enantiomer was determined by X-ray diffraction (Fig. 1.5.2). The structure features a coplanar arrangement of the formyl group with the phospholyl ring and a *cis* orientation of the formyl O and the phospholyl P atoms.

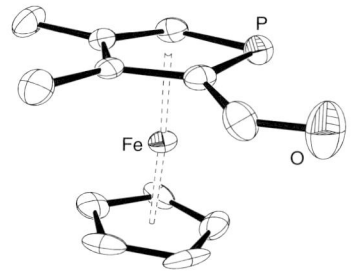

Fig. 1.5.2 Molecular structure of (S)-**2**.

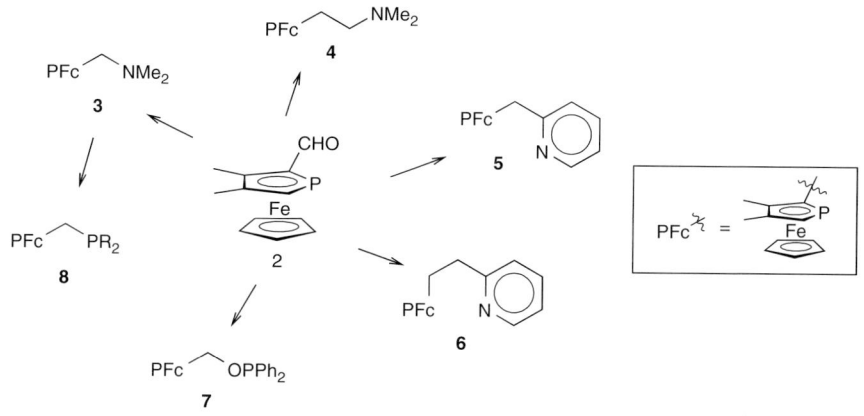

Scheme 1.5.2 A variety of bidentate ligands derived from aldehyde **2**.

All the derivatives prepared from aldehyde **2** can be obtained as pure enantiomers if desired, e.g., for use as ligands in asymmetric catalysis. Investigations of diastereoselective processes can of course be carried out using racemic phosphaferrocene compounds.

1.5.4
Preparation of Bidentate P,P and P,N Ligands

Nucleophilic additions to the formyl group of aldehyde **2** give access to a range of potential bidentate ligands. Representative examples are depicted in Scheme 1.5.2. By choice of the appropriate reaction protocol, P,N ligands with alkylamino or pyridyl donor groups are available in which the linker group to the phosphaferrocene nucleus has a variable length. For example, reductive amination leads directly to the aminomethyl complex **3** and the analogous aminoethyl compound **4** is available via the condensation of the aldehyde with CH_3NO_2 and subsequent reduction [11]. Similarly, the addition of 2-lithiopyridine or 2-(lithiomethyl)pyridine gives access to the pyridine derivatives **5** and **6** after reduction of the alcohols formed initially [13].

Reduction of the aldehyde **2** and treatment of the alcohol with PPh$_2$Cl gives the phosphite **7** [11], and the phosphinomethyl compounds **8** can be obtained from the amino derivative **3** by reaction with the secondary phosphine HPR$_2$ [14]. The availability of different compounds is significantly enhanced by homologation of the aldehyde **2** with a methylene group: a Wittig reaction with Ph$_3$P=CHOCH$_3$ gives the enol-ether **9** which is then hydrolyzed to the extended aldehyde **10** (Scheme 1.5.3). Reactions similar to those carried out with **2** can be used to transform aldehyde **10** into bidentate ligands [14].

All the reactions described above rely on the electrophilic reactivity of the carbonyl C atom of the aldehyde **2**. Much effort was devoted to the development of phosphaferrocenes with nucleophilic reactivity at this position. For example, transformation of the formyl group into a halomethyl function would pave the way for the preparation of Grignard or lithium derivatives by halogen–metal exchange. However, all attempts to do this were unsuccessful.

Quite a number of different metal complexes were prepared with the bidentate ligands described above. The common characteristics of those complexes will be described for illustrative examples.

Reaction of aminoethyl ligand **4** with [Cp*RuCl]$_4$ in THF gave a half-sandwich complex **11** in quantitative yield in which the ligand forms a seven-membered chelate ring with the Ru center. The stereogenic Ru atom is formed diastereoselectively, as revealed by NMR spectroscopy. The efficient control of the metal configuration by the chiral phosphaferrocene ligand reflects the avoidance of steric interference between the two bulky groups Cp* and CpFe (Fig. 1.5.2). In

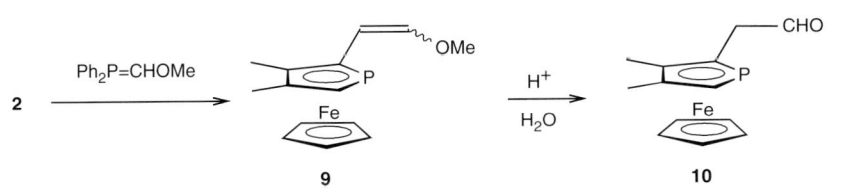

Scheme 1.5.3 Homologation of aldehyde **2**.

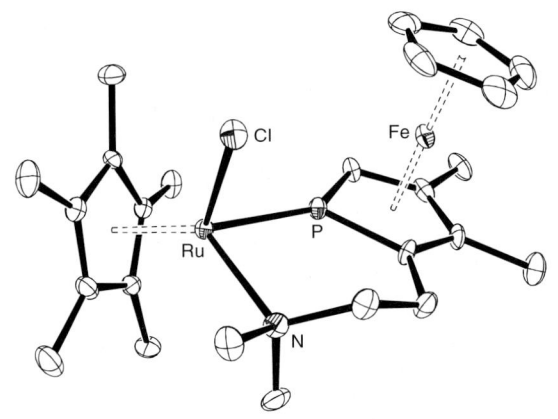

Fig. 1.5.2 Molecular structure of complex **11**.

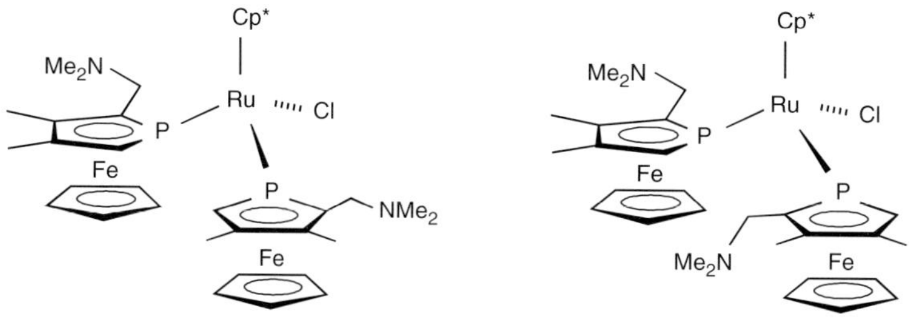

Scheme 1.5.4 Two diastereomeric bis(ligand)complexes **12**.

solution the NMe$_2$ group shows a hemilabile coordination to the Ru atom; only one ^{1}H NMR signal is observed for the two diastereotopic methyl groups at room temperature, which is split into two separate resonances when the temperature is lowered [11].

In contrast to the chelate formation with ligand **4** the reaction of the aminomethyl ligand **3** follows another path and no chelate complex is formed. Instead two molecules of ligand **3** are P-coordinated to the Ru atom while the aminomethyl side-chains remain uncoordinated. Two diastereomeric products were observed, a *meso* complex **12a** and a C_1-symmetric diastereomer **12b**, which has two molecules of **3** with identical configurations coordinated to ruthenium (Scheme 1.5.4) [11].

These results are noteworthy, because five-membered chelate complexes are usually thought to be particularly stable. In our case, however, only the formation of a six-membered chelate **11** is found, while a five-ring complex is strictly avoided –the bis(ligand) complexes **12** are obtained instead. The special geometry of the phosphaferrocene moiety provides an explanation for this behavior. The phospholyl ring deviates significantly from a regular five-ring arrangement (intra-ring angle 108°) with a compressed P-C-P angle of ca. 90° and expanded P-C-C angles of ca. 112°. Furthermore, the P atom lone pair is located in the phospholyl ring plane, pointing radially away from the ring. The overall result of these structural features is that only the extended ligand **5** can form a chelate with the Ru fragment without being severely strained. The bite angle of the short ligand **4** is too small for a chelating coordination and the bis(ligand) complex is formed instead.

The different features of ligands leading to five- and six-membered chelate complexes are also evident from comparison of the two Mo(CO)$_4$ complexes **13** and **14** (Fig. 1.5.3), which are both obtained from the respective P,P ligand, PFcCH$_2$PCy$_2$ or PFc(CH$_2$)$_2$PPh$_2$, and (norbornadiene)Mo(CO)$_4$ [14]. Here, chelate formation is observed in both cases. However, the five-membered complex **13** clearly features the more strained structure and shows more pronounced deviations from the idealized geometry of an octahedral complex with a P-Mo-P angle of 76°, as compared to complex **14** where the P-Mo-P angle is 82°. In both complexes, the Mo bond to the phospholyl P atom is ca. 7 pm shorter than the bond

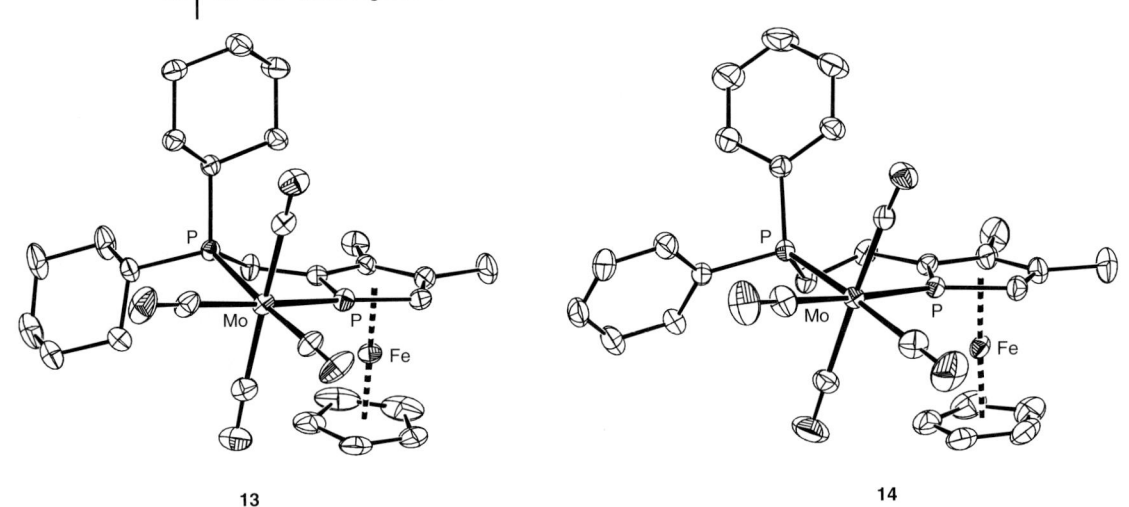

Fig. 1.5.3 Molecular structures of complexes **13** and **14**.

to the phosphanyl P atom. It is a general observation that the phosphaferrocene has quite short bonds to P-coordinated metal fragments, thereby resembling phosphites P(OR)$_3$ much more closely than trialkyl- or triarylphosphines PR$_3$.

This is also true for the donor/acceptor properties, because the phosphaferrocene is a reasonable π-acceptor due to the large p$_z$ orbital contribution of the P to the LUMO of the molecule [15].

1.5.5
Modification of the Backbone Structure

In order to gain access to a greater variety of ligand structures the introduction of substituents into the backbone was investigated. For this purpose, the aldehyde **2** was reacted with different Grignard or organolithium reagents leading to the respective secondary alcohols, which in turn could be transformed by replacement of the OH group with a suitable donor group. The transformation of the methyl-substituted compound **15** was studied in detail and revealed interesting results [16]. First, the addition of MeMgI to **2** proceeds diastereoselectively giving only one diastereomeric alcohol **15a** (Scheme 1.5.5). The selectivity is assumed to arise from an *exo* attack by the Grignard reagent on the aldehyde **2**, where the O and P atoms are in an *s-cis* conformation. The exchange of the OH group of the alcohol for a PR$_2$ function is carried out by protonation of **15a** with HBF$_4$ and subsequent addition of a secondary phosphine HPR$_2$ followed by aqueous workup under basic conditions; this gives the P,P ligands **16a** in excellent yields. Interestingly, it was found that the transformation **15** → **16** occurs with complete inversion on the former carbonyl C atom.

This is in contrast to the situation with analogous ferrocene compounds, where the substitution of OH with nucleophiles under the same conditions proceeds

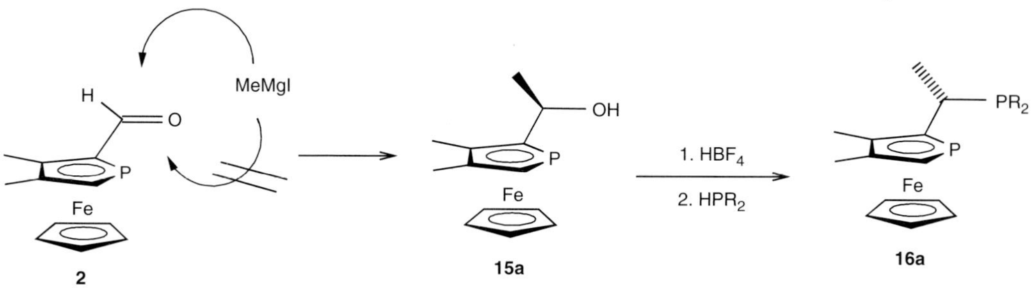

1.5.5 Modification of the Backbone Structure

Scheme 1.5.5 Diastereoselective transformation of aldehyde **2**.

with retention of configuration [17]. A mechanistic scheme was proposed to rationalize the stereochemical course of the reaction. In analogy to the related ferrocene chemistry, it is assumed that a water molecule is released from the protonated alcohol *trans* to the sterically demanding CpFe fragment. Thus, the diastereomerically pure alcohol **15a** gives the fulvene-like cationic intermediate (*E*)-**17** with an *E* arrangement of the P atom and the Me group (Scheme 1.5.6). In the absence of a nucleophilic reagent, the cationic species (*E*)-**17** isomerizes smoothly to the thermodynamically more stable cation (*Z*)-**17**, which, upon *exo* attack of an HPR$_2$ nucleophile, gives rise to the formation of the substitution product **16** with inverted configuration at the stereogenic center as compared to the starting material **15**. The *E* configuration of the kinetic product (*E*)-**17** and the activation barrier for its conversion to the thermodynamic product (*Z*)-**17** could be deduced by NMR methods and was confirmed by the X-ray crystal structure determination of a reaction product [16].

Another possibility for creating a new ligand structure is to connect two phosphaferrocene molecules via their formyl functional groups. The reductive coupling of carbonyl compounds with low-valent Ti reagents – the McMurry reaction – is a technique that has found wide application in organic synthesis [18].

Depending on the reaction conditions, either olefinic or pinacol-type products can be obtained. Thus, treatment of enantiopure aldehyde (*R*)-**2** with TiCl$_3$(DME)$_{1.5}$ and Zn/Cu in DME at elevated temperature gives the 1,2-bis(phosphaferrocenyl) ethene **18** as a mixture of the *E*- and *Z* isomers in a ratio of 8:1 (Scheme 1.5.7); they can be separated by chromatography.

The *E* and *Z* configurations were assigned by determining the $^3J_{HH}$ coupling constant between the two chemically equivalent olefinic protons from the ^{13}C satellites in the $^1H\{^{31}P\}$ NMR spectrum, and has been confirmed by X-ray analysis of both isomers [19].

From a stereochemical point of view, the pinacol-type coupling is much more interesting, because two new stereogenic centers are established at the former carbonyl C atoms. When the coupling reactions described above were carried out with (*R*)-**2** at 0 °C the pinacol product **19** (Fig. 1.5.4) was obtained as one out of three possible stereoisomers.

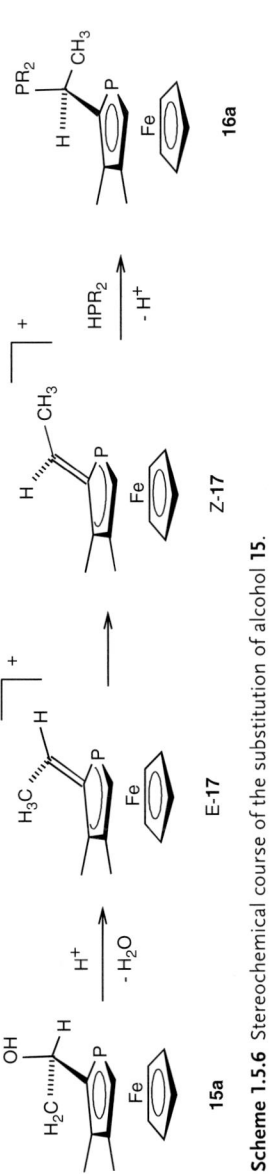

Scheme 1.5.6 Stereochemical course of the substitution of alcohol **15**.

Scheme 1.5.7 McMurry-coupling of aldehyde **2** yielding olefinic products.

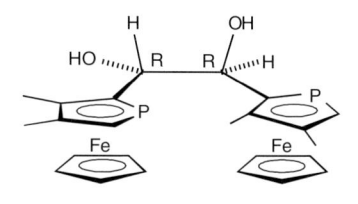

Fig. 1.5.4 Pinacol **19** obtained by coupling (R)-**2** by the McMurry reaction at 0 °C.

Racemic aldehyde **2** under similar reaction conditions yielded two diastereomeric pinacols; the configuration of all the products was assigned by X-ray diffraction. On the basis of these results a mechanistic scheme could be proposed to explain the stereochemical course of the coupling reactions [19]. It is assumed that the reaction proceeds via an intermediate titanaoxirane by insertion of a molecule of aldehyde **2** into the Ti—C bond – a proposal that has been put forward recently on the basis of DFT calculations [20]. The pinacol-type coupling products could be employed as bidentate P,P chelate ligands to $Mo(CO)_4$ fragments; however analogous experiments carried out with the olefinic derivatives **18** were unsuccessful [19].

1.5.6
Cp–Phosphaferrocene Hybrid Systems

The condensation of aldehyde **2** with cyclopentadiene yields the fulvene **20** which can be transformed to the respective cyclopentadienyl anion **21** by addition of a hydride anion (Scheme 1.5.8). The anion **21** opens up the possibility of the formation of sandwich and half-sandwich type of complexes with suitable metal fragments.

For example, treatment of anion **21** with anhydrous $FeCl_2$ gives the 1,1′-disubstituted ferrocene **22** with a phosphaferrocenylmethyl group attached to each Cp ring (Scheme 1.5.9). Because the aldehyde employed was a racemic

Scheme 1.5.8 Formation of the Pfc-functionalized Cp anion **21**.

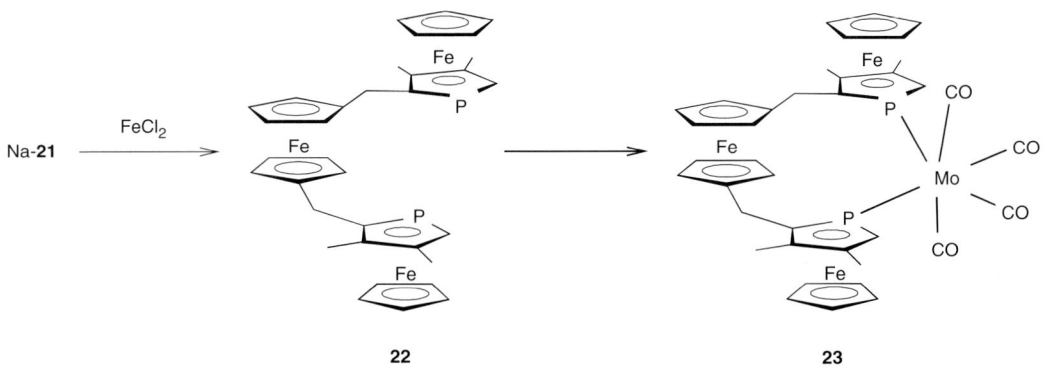

Scheme 1.5.9 Formation and use of the bidentate trimetallocene ligand **22**.

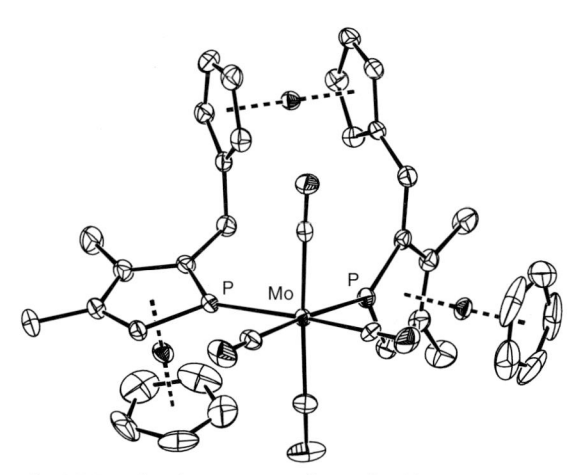

Fig. 1.5.5 Molecular structure of complex **23**.

mixture, complex **22** was obtained as a mixture of rac-(R*R*) and meso-(RS) isomers in a ratio of 1 : 1 as determined from the ^{31}P NMR spectra.

The propensity of ferrocene **22** to act as a bidentate P,P ligand was manifested by the preparation of the molybdenumcarbonyl complex **22** · Mo(CO)$_4$, **23**, which was characterized by X-ray structure analysis (Fig. 1.5.5) [21].

Reaction of the anion **21** with Cp or Cp* metal fragments provides further metallocene-type complexes with a pendant phosphaferrocene side-chain. For example, the reaction of the thallium derivative Tl · **21** with [Cp*RhCl$_2$]$_2$ yields the cationic pentamethylrhodocenium **24** as its chloride (Scheme 1.5.10). This is an interesting species because it is a chiral water-soluble P ligand. The chloride anion can be exchanged by PF$_6^-$ to make the compound more soluble in organic solvents.

As an overall cationic species, coordination of complex **24** to metals is feasible, as shown in the reaction with W(CO)$_5$(THF) to produce the respective W(CO)$_5$ complex. The P coordination is evident from the tungsten satellites in the ^{31}P NMR spectrum [22].

The Cp anion **21** could also be used to prepare half-sandwich complexes. Ruthenium compounds of the type [(Cp-L)Ru(PR$_3$)X] with stereogenic Ru atoms have attracted considerable attention during the last years. In order to control the metal configuration in a diastereoselective complexation reaction, several chiral Cp-L ligands were designed which resulted in de values of 18–83% in the subsequent complexation reactions [23]. When the anion **21** was treated with [(PPh$_3$)$_3$RuCl$_2$] in toluene at 90 °C over night, the diastereomeric halfsandwich complexes **25a** and **25b** (R = H) were formed in a ratio of 95 : 5. This ratio was established for the crude product by ^{31}P NMR spectroscopy and reflects therefore the selectivity of the complexation reaction itself, unaffected by workup manipulations. The high selectivity could be further increased by replacement of the Cp ring by the more sterically demanding Cp* ligand in the phosphaferrocene.

Thus, with anion **26** under similar reaction conditions only one diastereomeric half-sandwich complex, **27a** (R = Me), was observed in the crude product, isolated in high yield after chromatographic workup, and characterized by X-ray diffraction. The relative configuration is such that the two bulky groups – the PPh$_3$ ligand and the Cp*Fe moiety – avoid steric interference (Fig. 1.5.6) [24].

A couple of subsequent reactions which were carried out with the Cp* complex **27** all proceeded diastereoselectively, presumably with retention at Ru. These included exchange of the chloride for an iodide ligand under retention as determined by X-ray diffraction, reaction with NaOMe in MeOH to give the neutral monohydride as a single diastereomer, and removal of the chloride with AgBF$_4$

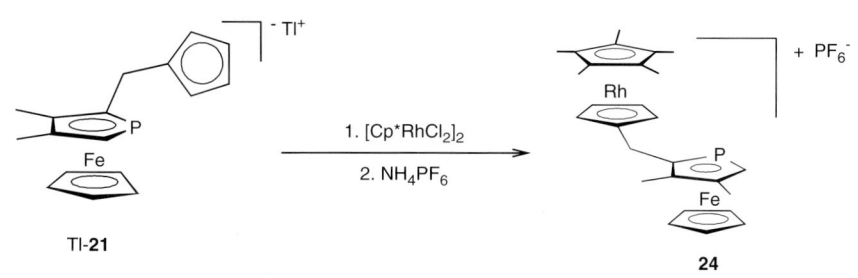

Scheme 1.5.10 Synthesis of the cationic phosphaferrocene derivative **24**.

21, R = H
26, R = Me

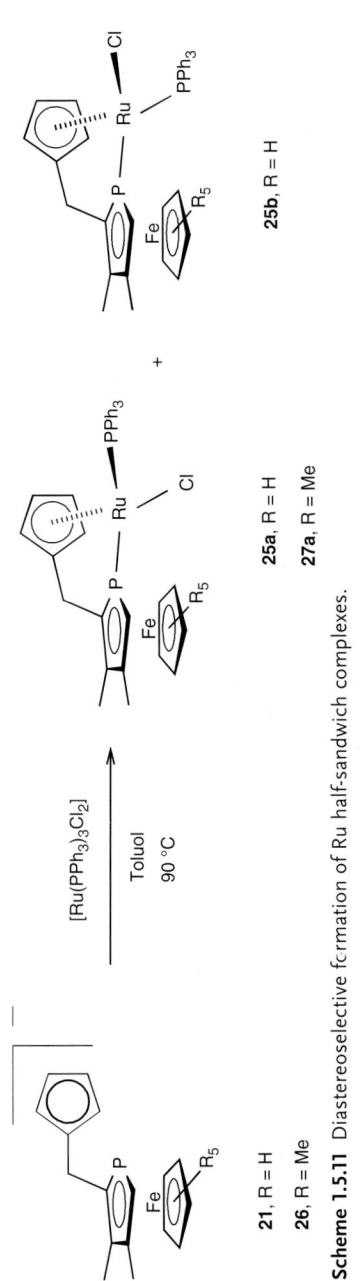

25a, R = H
27a, R = Me

25b, R = H

Scheme 1.5.11 Diastereoselective formation of Ru half-sandwich complexes.

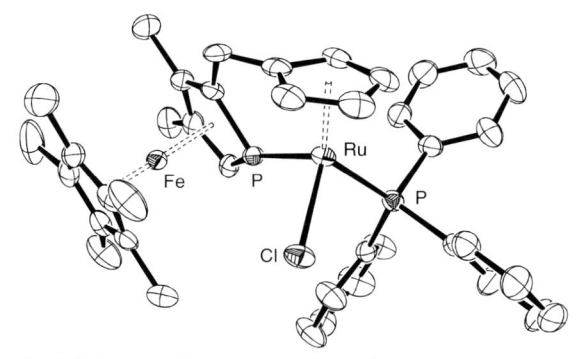

Fig. 1.5.6 Molecular structure of complex **27a**.

and treatment with H_2 to give the cationic dihydrogen complex. The nature of the compound as a dihydrogen complex was evident from the short T_1 relaxation time of 35 ms.

The high level of stereocontrol in the formation of complexes **25** and **27** suggests that compounds of this type may be useful as chiral catalysts. Indeed, several examples of enantioselective catalytic reactions carried out with half-sandwich complexes have been published recently [23, 25]. However, it seemed desirable to have access to complexes of the $[21 \cdot Ru(solv)_2]^+$ type, which have two easily removable solvent molecules coordinated to the central metal, in order to provide coordination sites for a substrate to be transformed. Although the chloride ligand could be easily removed from **23** and **25** all attempts to strip off the PPh_3 were unsuccessful. Therefore a new reaction scheme was developed which precluded the use of phosphine ligands, and the bis(acetonitrile) complex **28** could be obtained in a multi-step protocol via the Tl salt Tl · **21** (Scheme 1.5.12) [26].

The acetonitrile molecules are labile and can be exchanged for other mono- or bidentate ligands such as pyridine or tmeda. With N-(2-dimethylamino)ethylmorpholine two diastereomers are obtained in excellent yield in a ratio of 3 : 2 (*de* 20%), one of which could be characterized by X-ray diffraction analysis.

Half-sandwich complexes of rhenium and manganese have been investigated thoroughly regarding their synthetic potential and particularly their stereochemical properties [27]. The Cp-phosphaferrocene ligand **21** could be used to obtain complexes with Re and Mn as well. Treatment of the bromides $(CO)_5MBr$ (M = Mn, Re) with the thallium derivative Tl · **21** gave the tricarbonyl complexes **29** in excellent yield. Irradiation led to the release of one CO and intramolecular coordination of the phosphaferrocene donor group to the metal, giving the dicarbonyls **30** (Scheme 1.5.13). The usual features for P coordination are observed in the NMR spectra: a pronounced downfield shift in the ^{31}P spectrum upon coordination (**29**(Mn): −77 ppm, **30**(Mn): +72 ppm) and a reduction of the $^2J_{PH}$ coupling for the α-phospholyl proton in the ^1H NMR spectrum from 36 Hz in the free ligand to 32 Hz in the coordinated form. The X-ray structure of complex **30** (M = Mn) shows two CO groups in quite different environments, one being in a

Scheme 1.5.12 Multi-step synthesis of the bis(acetonitrile) complex **28**.

29, M = Mn, Re

30, M = Mn, Re

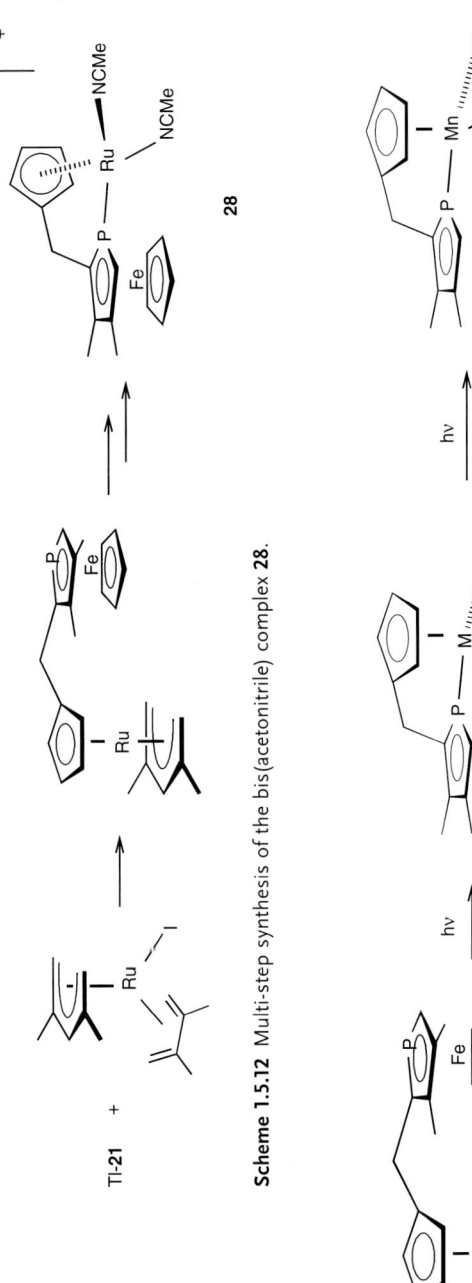

Scheme 1.5.13 Mn and Re half-sandwich complexes.

syn and one in an *anti* arrangement with the CpFe fragment when the complex is viewed along the Mn—P bond. Irradiation of the dicarbonyl **30** (M = Mn) in the presence of 1 equiv of PPh$_3$ released a further CO group and produced the chiral-at-metal complex **31** as a 11 : 1 mixture of diastereomers. Interestingly, the formation of the analogous monocarbonyl derivative with Re was not successful [28].

The X-ray diffraction analysis of the major isomer revealed that the CO *anti* to the CpFe fragment in complex **31** had been replaced by the bulky PPh$_3$ ligand (Fig. 1.5.7).

1.5.7
Catalytic Applications

Selected derivatives of the ligands and complexes described above have been tested in catalytic applications. Early tests with bidentate P,P or P,N ligands such as **4**, **6**, and **8** in Rh-catalyzed asymmetric hydrogenation were disappointing, with *ee* values below 20%. However, as was demonstrated mainly by the Fu group, phosphaferrocene derivatives do have the potential for successful applications in asymmetric catalytic reactions, provided the phosphaferrocene is endowed with sufficient steric bulk. Examples are depicted in Fig. 1.5.8: the Cp* derivative **32**,

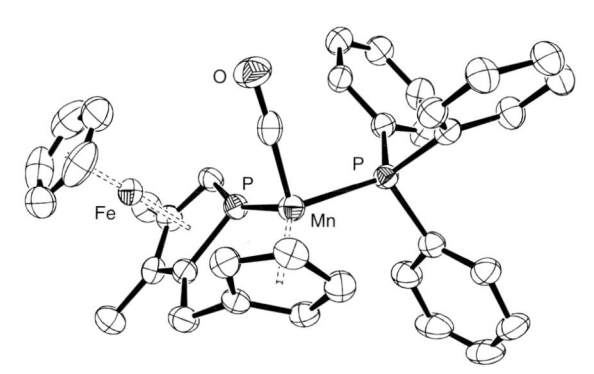

Fig. 1.5.7 Molecular structure of complex **31**.

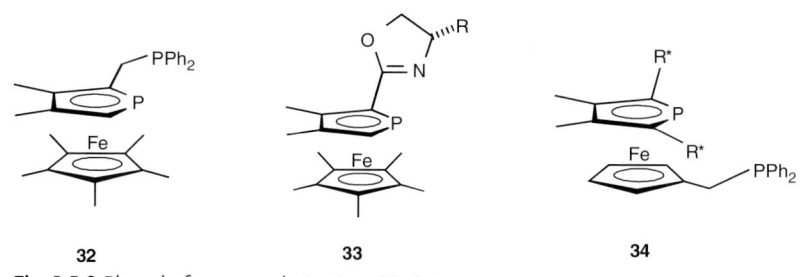

Fig. 1.5.8 Phosphaferrocene derivatives **32**–**34**.

closely related to ligand **8**, provides high enantioselectivities in asymmetric hydrogenation reactions [29] as well as in the Rh-catalyzed enantioselective isomerization of allylic alcohols [30]. The oxazoline derivatives **33**, also developed by Fu, were applied successfully in asymmetric allylic substitutions as well as in [3 + 2] cycloadditions [31]. The 2,5-di((−)-menthyl)-substituted derivative **34**, which has the additional donor function attached to the Cp ring, was prepared in the Hayashi group and gave high enantioselectivities in asymmetric allylic substitutions [32].

The bis(phosphaferrocene) **22** gave good results in the hydroformylation of α-olefins regarding activity and *n/iso* selectivity [33].

1.5.8
Conclusion

A new family of chiral multidentate ligands has been developed featuring a phosphaferrocene moiety as a common structural unit. This heterometallocene fragment is characterized by uncommon steric and electronic properties as compared to the vast majority of chiral phosphine ligands. Some of the factors determining the stereochemical course of transformations of the phosphaferrocene group could be elucidated. Promising results were obtained with the new ligands in homogeneous catalytic reactions.

References

1 R. Noyori, *Angew. Chem. Int. Ed.* **2002**, *41*, 2008.
2 M. J. Burk, *Acc. Chem. Res.* **2000**, *33*, 363.
3 R. C. J. Atkinson, V. C. Gibson, N. J. Long, *Chem. Soc. Rev.* **2004**, *33*, 313.
4 H. U. Blaser, W. Brieden, B. Pugin, F. Spindler, M. Studer, A. Togni, *Top. Catal.* **2002**, *19*, 3.
5 A. Salzer, *Coord. Chem. Rev.* **2003**, *242*, 59.
6 W. Braun, B. Calmuschi, J. Haberland, W. Hummel, A. Liese, T. Nickel, O. Stelzer, A. Salzer, *Eur. J. Inorg. Chem.* **2004**, 2235.
7 U. Englert, C. Hu, A. Salzer, E. Alberico, *Organometallics* **2004**, *23*, 5419.
8 For a short review see: C. Ganter, *J. Chem. Soc., Dalton Trans.* **2001**, 3541.
9 For comprehensive reviews see: a) F. Mathey, *Angew. Chem.* **2003**, *115*, 1616; *Angew. Chem. Int. Ed.* **2003**, *42*, 1578; b) F. Mathey, *J. Organomet. Chem.* **2002**, *646*, 15; c) K. B. Dillon, F. Mathey, J. F. Nixon, *Phosphorus: The Carbon Copy*, Wiley, Chichester, **1998**; d) F. Mathey, *Coord. Chem. Rev.* **1994**, *137*, 1.
10 J. Bitta, S. Fassbender, G. Reiss, C. Ganter, *Organometallics* **2006**, *25*, 2394.
11 C. Ganter, L. Brassat, C. Glinsböckel, B. Ganter, *Organometallics* **1997**, *16*, 2862.
12 C. Ganter, L. Brassat, B. Ganter, *Tetrahedron: Asymmetry* **1997**, *8*, 2607.
13 C. Ganter, C. Glinsböckel, B. Ganter, *Eur. J. Inorg. Chem.* **1998**, 1163.
14 C. Ganter, L. Brassat, B. Ganter, *Chem. Ber./Recueil* **1997**, *130*, 1771.
15 a) G. Frison, F. Mathey, A. Sevin, *J. Phys. Chem. A* **2002**, *106*, 5653; b) N. M. Kostic, R. F. Fenske, *Organometallics* **1983**, *2*, 1008.
16 L. Brassat, B. Ganter, C. Ganter, *Chem. Eur. J.* **1998**, *4*, 2148.
17 Comprehensive review: G. Wagner, R. Herrmann, in *Ferrocenes: Homogenous Catalysis, Organic Synthesis, Materials Science*, A.Togni, T. Hayashi (Eds.), VCH, Weinheim, **1995**.

18 Review: A. Fürstner, B. Bogdanovic, *Angew. Chem.* **1996**, *108*, 2582; *Angew. Chem. Int. Ed.* **1996**, *35*, 2442.
19 O. Agustsson, U. Englert, C. Hu, T. Marx, L. Wesemann, C. Ganter, *Organometallics* **2002**, *21*, 2993.
20 M. Stahl, U. Pidun, G. Frenking, *Angew. Chem.* **1997**, *109*, 2308; *Angew. Chem. Int. Ed.* **1997**, *36*, 2234.
21 C. Ganter, C. Kaulen, U. Englert, *Organometallics* **1999**, *18*, 5444.
22 J. Bitta, H. Willms, C. Ganter, unpublished results.
23 Review: C. Ganter, *Chem. Soc. Rev.* **2003**, *32*, 130.
24 C. Kaulen, C. Pala, C. Hu, C. Ganter, *Organometallics* **2001**, *20*, 1614.
25 a) M. Ito, M. Hirakawa, K. Murata, T. Ikariya, *Organometallics* **2001**, *20*, 379; b) Y. Morisaki, T. Kondo, T. Mitsudo, *Organometallics* **1999**, *18*, 4742; c) B. M. Trost, P. L. Fraisse, Z. T. Ball, *Angew. Chem.* **2002**, *114*, 1101; *Angew. Chem. Int. Ed.* **2002**, *41*, 1059; d) Y. Matsushima, K. Onitsuka, T. Kondo, T. Mitsudo, S. Takahashi, *J. Am. Chem. Soc.* **2001**, *123*, 10 405; e) B. M. Trost, B. Vidal, M. Thommen, *Chem. Eur. J.* **1999**, *5*, 1055; f) C. Standfest-Hauser, C. Slugovc, K. Mereiter, R. Schmid, K. Kirchner, L. Xiao, W. Weissensteiner, *J. Chem. Soc., Dalton Trans.* **2001**, 2989.
26 L. Jekki, C. Pala, B. Calmuschi, C. Ganter, *Eur. J. Inorg. Chem.* **2005**, 745.
27 a) J. A. Gladysz, B. J. Boone, *Angew. Chem.* **1997**, *109*, 566; *Angew. Chem. Int. Ed.* **1997**, *36*, 550; b) M. Otto, B. J. Boone, A. M. Arif, J. A. Gladysz, *J. Chem. Soc., Dalton Trans.* **2001**, 1218.
28 J. Bitta, S. Fassbender, G. Reiss, W. Frank, C. Ganter, *Organometallics* **2005**, *24*, 5176.
29 S. Qiao, G. C. Fu, *J. Org. Chem.* **1998**, *63*, 4168.
30 K. Tanaka, S. Qiao, M. Tobisu, M. M.-C. Lo, G. C. Fu, *J. Am. Chem. Soc.* **2000**, *122*, 9870.
31 a) R. Shintani, M. M.-C. Lo, G. C. Fu, *Org. Lett.* **2000**, *2*, 3695; b) R. Shintani, G. C. Fu, *J. Am. Chem. Soc.* **2003**, *125*, 10 778; c) R. Shintani, G. C. Fu, *Angew. Chem.* **2003**, *115*, 4216; *Angew. Chem. Int. Ed.* **2003**, *42*, 4082.
32 M. Ogasawara, K. Yoshida, T. Hayashi, *Organometallics* **2001**, *20*, 3913.
33 W. Ahlers, T. Mackewitz, M. Röper, F. Mathey, C. Ganter, B. Breit, Ger. Offen. DE19 921 730, **2000**.

2
Catalytic Asymmetric Synthesis

2.1
Chemical Methods

2.1.1
Sulfoximines as Ligands in Asymmetric Metal Catalysis
Carsten Bolm

2.1.1.1 Introduction

Sulfoximines are monoaza analogues of sulfones, the first representative of which was discovered at the end of the 1940s when wheat that had been bleached with NCl_3, and fed to dogs caused them to develop a disease called canine hysteria [1]. After isolation of the toxic factor and determination of the structure of the bioactive component, a compound **1** with an as-yet unknown structural feature at sulfur was identified. Apparently, sulfoximine **1** stemmed from the amino acid methionine, and it was remarkable how the modification of the sulfur unit affected the biological activity of this natural product. After extensive biological studies it became apparent that sulfoximines such as **1** were glutamine synthetase inhibitors [2], and subsequently derivative **2** (L-buthionine (S,R)-sulfoximine, BSO) attracted significant attention in the medical community due to its specific inhibitory effects on γ-glutamylcysteine synthetase. The importance of BSO for tumor cell drug resistance relates to the reduction in the glutathione biosynthesis, which then enhances the potency of cytotoxic agents [3].

Several other studies revealed alternative biological effects of sulfoximines. For example, Mock found **3** to be a transition-state analogue of carboxypeptidase

Asymmetric Synthesis with Chemical and Biological Methods. Edited by Dieter Enders and Karl-Erich Jaeger
Copyright © 2007 WILEY-VCH Verlag GmbH & Co. KGaA, Weinheim
ISBN 978-3-527-31473-7

A [4], and Tsujihara demonstrated the antifungal activity of sulfoximine 4 [5]. In a collaboration with Weinhold [6] we discovered an unusual cleavage pattern of pseudotripeptide 5 when it was treated with peptidase A and attributed this effect to the presence of the central acylated β-carboxysulfoximidoyl unit, which we had studied separately previously [7].

The amide-like linkage between the sulfoximine nitrogen and the natural α-amino acid of 5 revealed interesting properties, which were investigated spectroscopically and theoretically in collaboration with Raabe and Fleischhauer [8].

2.1.1.2 Development of Methods for Sulfoximine Modification

Since the 1970s, sulfoximines have been widely used in organic synthesis [9]. In particular, the studies by Johnson on their application as chiral auxiliaries for asymmetric synthesis stimulated subsequent investigations by several researchers, including Gais, Harmata, Pyne, Craig, and Reggelin, just to mention a few [10, 11]. In all of those cases stoichiometric quantities of sulfoximines were applied and commonly diastereoselective transformations were studied. With the objective of using sulfoximines as ligands in asymmetric catalysis, we initiated a research program, which resulted in the first publication on this topic in 1992 [12]. For the first time we demonstrated that a catalytic amount of a sulfoximine could be applied in enantioselective C–C bond formation, resulting in enantioselectivity which reached the 80–90% *ee* level. Subsequently, we investigated other metal-catalyzed reactions, including, for example, asymmetric ketone and imine reductions, enantioselective cyanohydrin formations, and asymmetric allylic substitution reactions [13, 14]. Most of the sulfoximines used at that time (such as compounds **6–8**) were based on optically active methylphenyl sulfoximine (**9**), which is readily available in both enantiomeric forms by way of well-established, standard protocols [15].

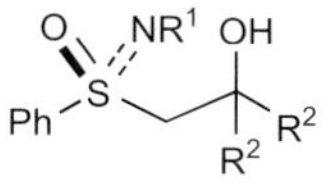

2.1.1 Sulfoximines as Ligands in Asymmetric Metal Catalysis | 151

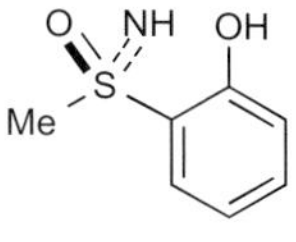

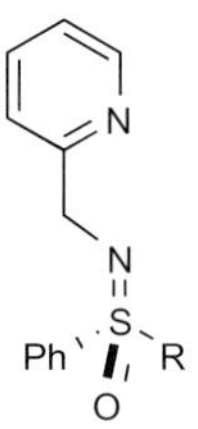

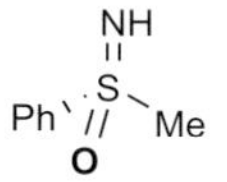

In general, sulfoximines are accessible by various routes, and most of them involve sulfur oxidation/imination sequences. For example, enantiopure **9** is commonly prepared starting from sulfide **10**, which is oxidized with hydrogen peroxide (under acidic conditions) giving sulfoxide **11** (Scheme 2.1.1.1). Subsequent imination of **11** with a mixture of sodium azide and sulfuric acid affords sulfoximine **9** as a racemate. Enantiomer resolution can then be achieved with camphorsulfonic acid, leading to both enantiomers of **9** with high efficiency [15]. Alternatively, many sulfoximine syntheses start from enantiopure sulfoxides [16, 17], which can be stereospecifically iminated with O-mesitylenesulfonylhydroxylamine (MSH) [18], as shown for the synthesis of sulfoximine (R)-**13** in Scheme 2.1.1.2.

Both protocols work well on a laboratory scale: preparations of larger quantities of sulfoximines are, however, hampered by the toxicity and potential danger of the iminating agents (hydrazoic acid formed in situ and MSH, in the first and the second approach, respectively) [19]. On this basis, metal-catalyzed iminations have recently caught the attention of the synthetic community [11a, 20]. Several metal salts and complexes proved applicable to the catalysis, and the iminating agents could be widely modified. In order to further improve the accessibility of sulfoximines, we studied alternative iminating reactions and found rhodium diacetate to be an excellent catalyst for the nitrogen-transfer reaction to sulfoxides

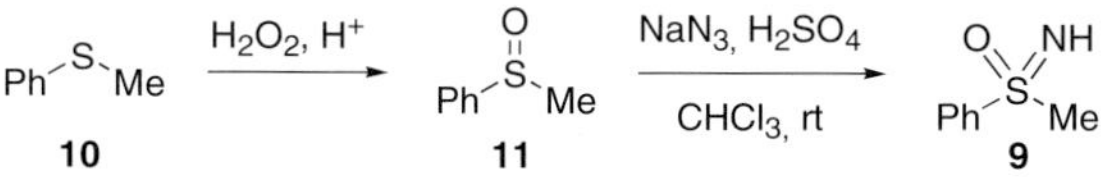

Scheme 2.1.1.1

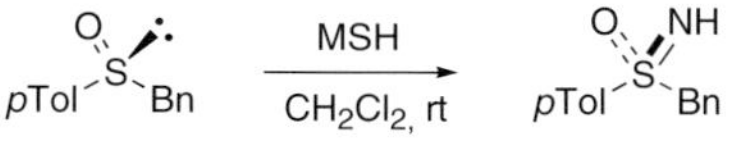

Scheme 2.1.1.2

[21]. A wide variety of sulfonyl or carboxyl amides can be applied as nitrogen sources and nontoxic iodobenzene diacetate serves as a mild oxidant. Furthermore, the reaction conditions (e.g., room temperature) are mild, and the imination proceeds with retention of configuration at the stereogenic sulfur (as shown in Scheme 2.1.1.3 for the conversion of (R)-**14** into (R)-**15**). Most importantly, sulfoximine derivatives with easy-to-cleave protective groups at the sulfoximine nitrogen are obtainable, which allow access to the synthetically important "free" NH-sulfoximines (vide infra).

Although the development of the rhodium catalysis route represented significant progress in the accessibility of sulfoximines, the high cost of the metal complex still hampered its large-scale application. Stimulated by recent reports from Cui and He on silver-catalyzed aziridinations of olefins and nitrene insertions into C–H bonds [22], we began to investigate the catalytic application of silver salts to the imination reaction of sulfoxides. Gratifyingly, we found that a simple combination of silver acetate and 4,4′,4″-tri-t-butyl-2,2′ : 6′,2″-terpyridine (4,4′,4″-tBu₃tpy) served to afford a wide range of synthetically interesting N-protected sulfoximines **16–21** (Scheme 2.1.1.4) [23].

Again, sulfoxide imination is stereospecific and "free" NH-sulfoximines such as **9** may be obtained with retention of stereochemistry by cleavage of the nosyl protecting group (Scheme 2.1.1.5) [23].

During the optimization of the reaction conditions it was noticed that at higher temperatures the imination proceeded even in the absence of the metal. Subse-

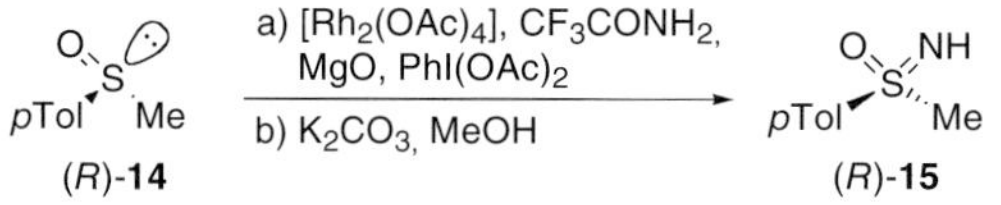

Scheme 2.1.1.3

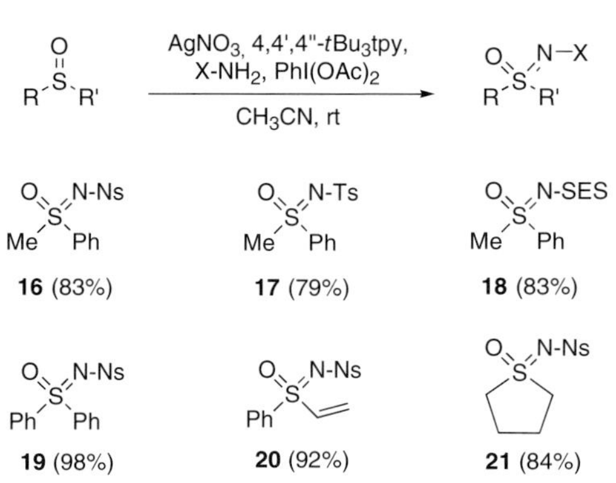

Scheme 2.1.1.4

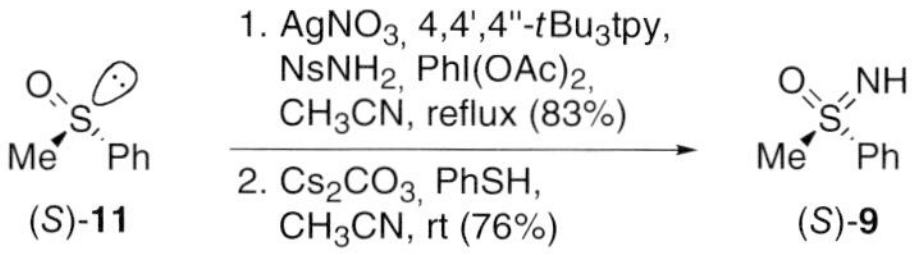

Scheme 2.1.1.5

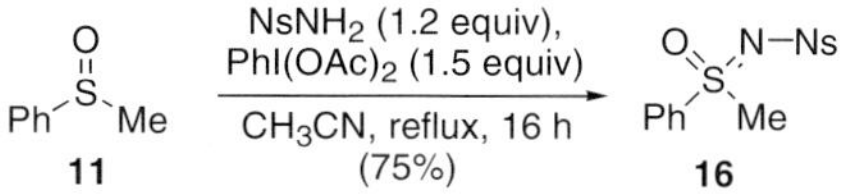

(S)-11 → (S)-9
1. AgNO$_3$, 4,4',4''-tBu$_3$tpy, NsNH$_2$, PhI(OAc)$_2$, CH$_3$CN, reflux (83%)
2. Cs$_2$CO$_3$, PhSH, CH$_3$CN, rt (76%)

Scheme 2.1.1.6

11 → 16
NsNH$_2$ (1.2 equiv), PhI(OAc)$_2$ (1.5 equiv), CH$_3$CN, reflux, 16 h (75%)

quently this observation led to the development of a metal-free sulfoxide-to-sulfoximine conversion [24]. Although partial racemization occurs under these conditions, the protocol is still synthetically useful, since it avoids the application of both toxic iminating agents (vide supra) and high-cost metal catalysts [25, 26].

The metal-free imination, as well as the Rh and Ag catalyses, can be applied to the conversion of sulfides into sulfilimines such as **22** and **23** [24]. Unfortunately, however, double iminations leading to sulfondiimides (N-protected forms of **24**) did not occur, and compounds such as **25** [27] still require the intermediacy of sulfondiimide **24**. Thus, its preparation was achieved by treatment of the corresponding sulfide with a mixture of ammonia and t-butyl hypochlorite [28].

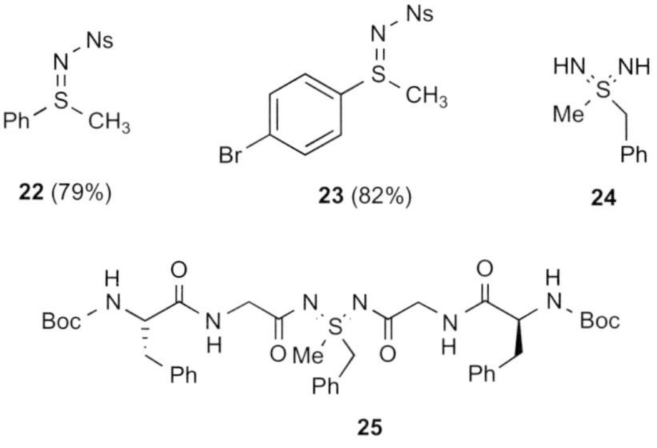

22 (79%)   **23** (82%)   **24**

25

As mentioned above, enantiopure methylphenyl sulfoximine (**9**) served as a valuable starting material in several initial ligand syntheses, and generally approaches toward desired ligand structures could rely only on literature knowledge or our own expertise. Expansion of the range of such sulfoximines to allow a reliable and straightforward alternation of the ligand structure was also considered important. We therefore initiated a study focused on varying all four substitutents of the sulfoximine core (Fig. 2.1.1.1).

During the investigation of the synthesis and use of sulfilimines (such as **22** and **23**) [21, 23, 24] and sulfondiimides (of the **25** type) [27], it became apparent that appropriately substituted nonsymmetric derivatives of these compounds were much more difficult to prepare in enantiopure form than their sulfoximine counterparts. It was therefore decided to retain the sulfur/oxygen moiety and to concentrate further efforts exclusively on the synthesis and application of sulfoximines.

Functionalization of the alkyl substituent α to the sulfur atom is well established in the literature and usually involves the intermediacy of metallated species, which may be obtained by deprotonation at this position with strong base such as *n*-butyllithium or lithio amide bases [29]. Metal-catalyzed C-arylations of sulfoximines were, however, unknown before our work, although they could be synthetically valuable processes for the preparation of optically active benzylphenyl sulfoximines. Furthermore, such an approach was attractive, since a single enantiopure starting material (having, for example, a methyl substituent as in **26**) could be converted into a range of products **27** by applying various aryl sources as coupling partners (Scheme 2.1.1.7).

The initial observation indicating that such C-arylation was indeed feasible stemmed from an attempted palladium-catalyzed double N-arylation reaction (for details of this type of coupling reaction, see below). With the objective of preparing C_2-symmetric bissulfoximine **28**, dibromonaphthalene **29** was treated with an excess of sulfoximine **9** under standard coupling conditions {using [Pd$_2$dba$_3$]/BINAP as a catalyst in the presence of a base} and, to our surprise, cyclic sulfoximine **30** was obtained in excellent yield (Scheme 2.1.1.8) [30].

Apparently, the first N-arylation had proceeded well, but then an intramolecular ring closure occurred faster than the second coupling reaction, thereby convert-

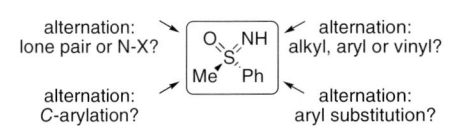

Fig. 2.1.1.1 Schematic representation of the possible modifications of the four substituents on the sulfoxime starting material.

Scheme 2.1.1.7

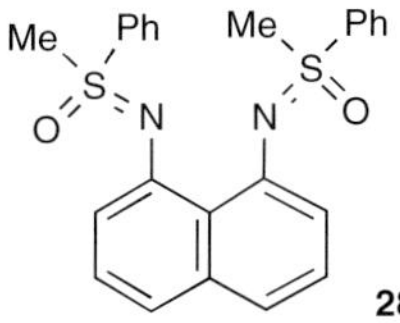

2.1.1 Sulfoximines as Ligands in Asymmetric Metal Catalysis | 155

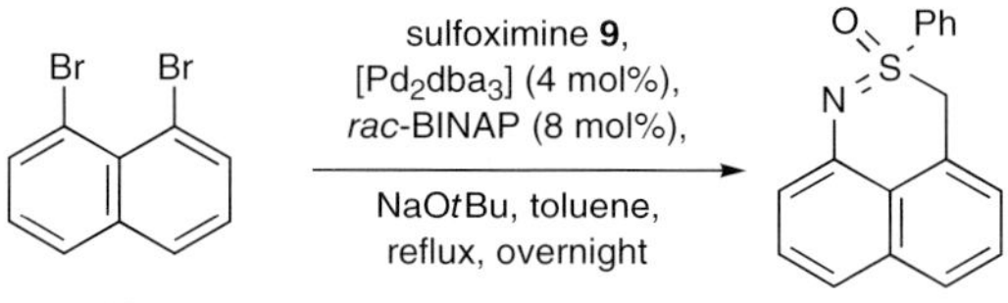

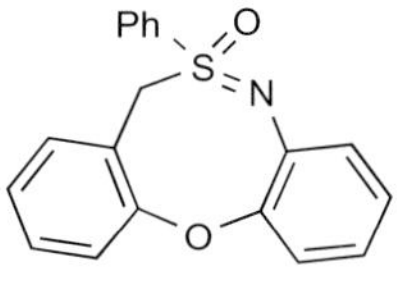

Scheme 2.1.1.8

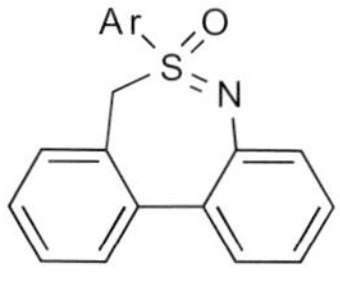

ing the sulfoximine methyl group into the benzylic methylene of the final product. This (unexpected) palladium-catalyzed intramolecular C-arylation reaction was found to be fairly general, yielding a range of interesting sulfoximine-based heterocycles such as **31** (with Ar = Ph or Tol) and **32** in excellent yields (up to 99%) [30]. Subsequent studies showed that heterocycles **35** and **36** could be obtained in an analogous manner, starting from N-benzylated and N-acylated sulfoximines **33** and **34**, respectively (Scheme 2.1.1.9) [31].

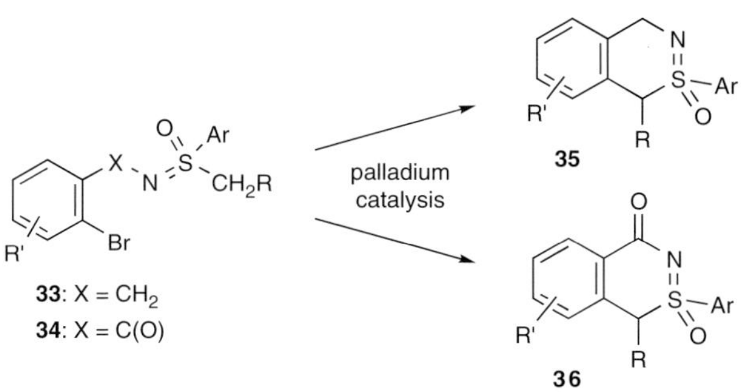

Scheme 2.1.1.9

Subsequently, alternative syntheses of sulfoximine-containing heterocycles were studied, and one such approach was based on Grubbs' olefin metathesis reaction. Using the ruthenium carbene complex **39** as catalyst, a wide range of heterocycles of the **38** type became available in good to excellent yield starting from the appropriately functionalized (doubly unsaturated) sulfoximine derivatives, such as **37** (Scheme 2.1.1.10) [32].

After the successful demonstration of the potential of *intra*molecular C-arylation reactions of sulfoximines, we returned to our initial goal: the preparation of various benzylphenyl sulfoximine derivatives by *inter*molecular C-arylation (vide supra). Unexpectedly, this project proved more challenging than expected [33]. All attempts to directly arylate simple methylphenyl sulfoximines **26** having various N-protective groups were unsuccessful and led to undesired products in only small quantities. Finally, hypothesizing that the pK_a of the starting material had to be lowered, *N*-benzoyl β-sulfoximidoyl carboxy ester **40** was applied; gratifyingly, with this starting material the α-arylation proceeded well, giving a wide range of products **41** (Scheme 2.1.1.11) [34]. A combination of Pd(OAc)$_2$ and PCy$_3$ as a catalyst, sodium *t*-butoxide as the base, and dioxane as the solvent proved best.

Both, the N-protection by the benzoyl group and the presence of the carboxy ester were essential for the success of the α-arylation. On one hand, they activated the substrate, and on the other, they could be cleaved selectively (reductively and hydrolytically, respectively), as demonstrated in the conversion of ester **42** into *NH*-benzylmethyl sulfoximine **44** via compound **43** (Scheme 2.1.1.12) [34].

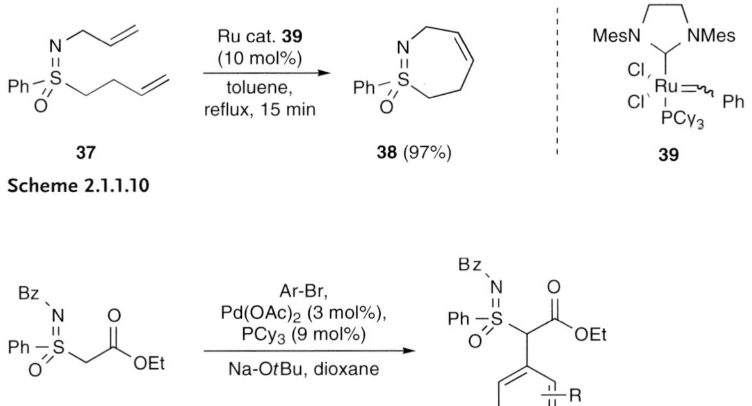

Scheme 2.1.1.10

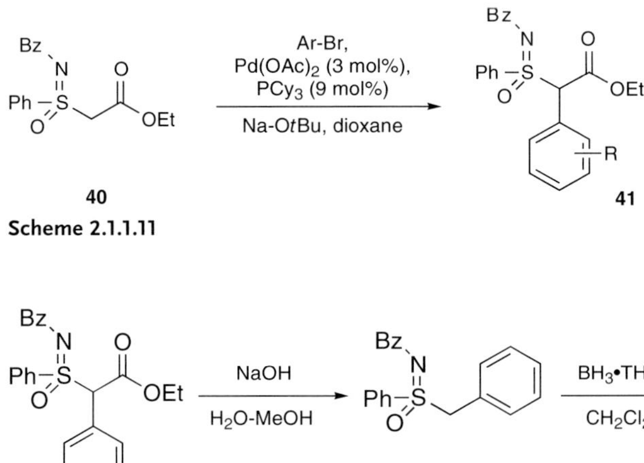

Scheme 2.1.1.11

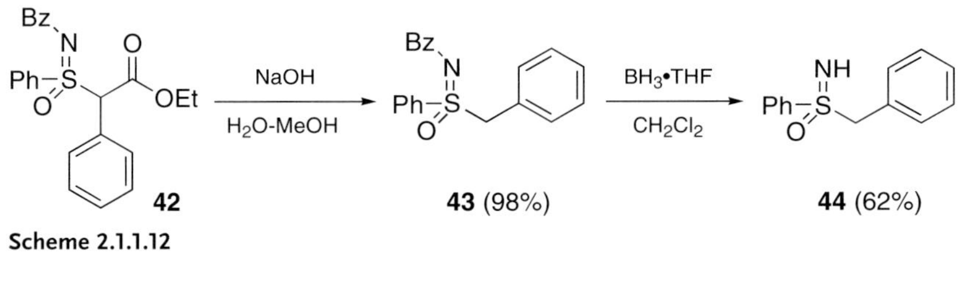

Scheme 2.1.1.12

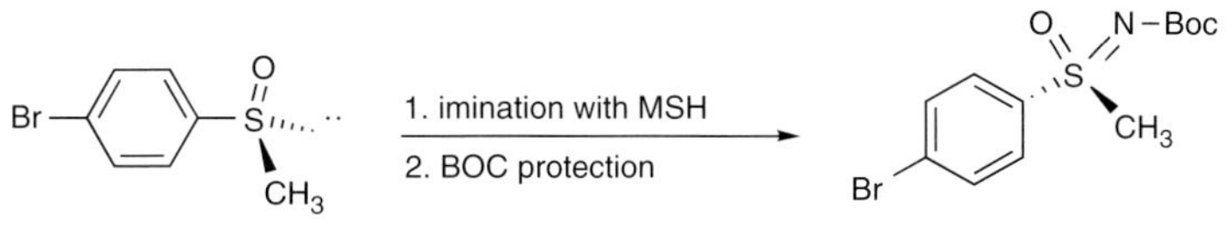

Scheme 2.1.1.13

Having examined the alkyl substituent of the sulfoximine, we next investigated the modification of the sulfoximine aryl group. Although *ortho*-metallations with strong lithio bases (using the sulfoximine moiety as a directing group) have been described [11b, 35], they are commonly restricted to a narrow range of well-defined substrates which have a rather limited array of substituents (such as *t*-butyl for the alkyl group). With respect to their application as ligands in catalysis, however, the aryl substitution site of the sulfoximines should be kept as unhindered as possible. On that basis and with the hope that the electronic effects of remote substitutents could be utilized, sulfoximine **46**, having a bromo substituent in the *para* position of the sulfoximine aryl group (and an easy-to-cleave *t*-butoxycarbonyl (BOC) group at the nitrogen) became the target of choice. The synthesis of the corresponding sulfoxide (S)-**45** was well established by Naso [36] and, as we found, its imination with MSH proceeded smoothly giving (S)-**46** stereospecifically (after *N*-BOC protection) in high yield (86%; Scheme 2.1.13) [37].

Hartwig–Buchwald, Suzuki, and Stille type cross-coupling reactions with key intermediate **46** led to a wide range of substituted sulfoximines such as **47–49** [37]. In order to demonstrate the synthetic utility of the resulting products, pseudo tripeptide **50** was prepared from a related intermediate.

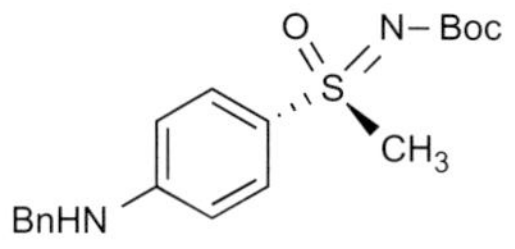

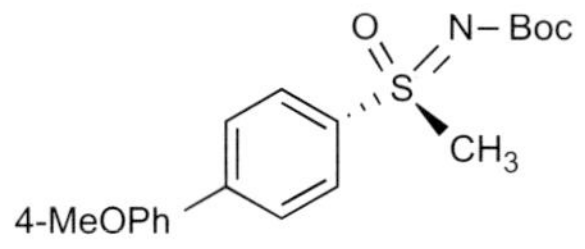

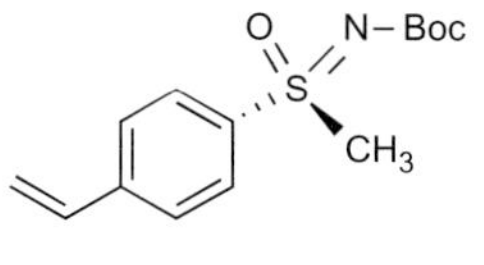

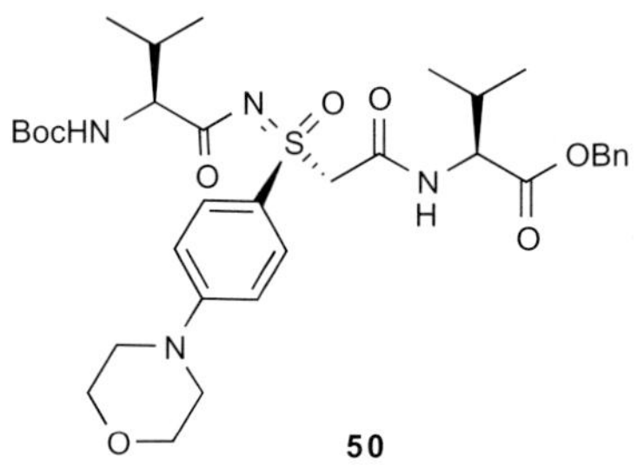

As pointed out already, "free" *NH*-sulfoximines were considered to be the most useful synthetically, because they allowed a variety of post-modifications at the sulfoximine nitrogen. Although at the outset of our studies the binding site of metals (in the sense of metal ions, metal complexes, and suchlike) was uncertain and it was difficult to predict which heteroatom of the sulfoximine (or even both) would coordinate [38, 39], we expected the sulfoximine nitrogen to play an important role in the activity and the stereoselectivity of a given catalytic system.

Several modifications (including alkylation, acylation, silylation, phosphorylation and so on) of the sulfoximine nitrogen had been known [9], but for the preparation of novel chiral ligands and target-directed synthesis an improvement and extension of this chemistry appeared desirable. For example, in 1996 we had already tried to connect two sulfoximines **51** by an ethylene bridge to give (after further functionalizations) salen-type bissulfoximine **52** (Scheme 2.1.1.14) [40]. However, since simple N-alkylations of sulfoximines (with the exception of the methylation under Eschweiler–Clark conditions) are not easy to achieve due to the low nucleophilicity of the deprotonated intermediates [41], we had to spend considerable time and effort on the synthesis of **52** (which later turned out to lead to racemates when applied to attempted asymmetric catalyses). Finally, the key intermediate **54** on the way to bissulfoximine **52** was obtained by an acylation/reduction sequence (using oxalyl chloride and boranes, respectively), as generally depicted for the conversion of **53** into **55** in Scheme 2.1.1.15 [40].

Later, this reaction sequence was further optimized [42], and also shown to be useful for the synthesis of other N-alkylated sulfoximines, including those (such as **57**) prepared from amino acids (Scheme 2.1.1.16).

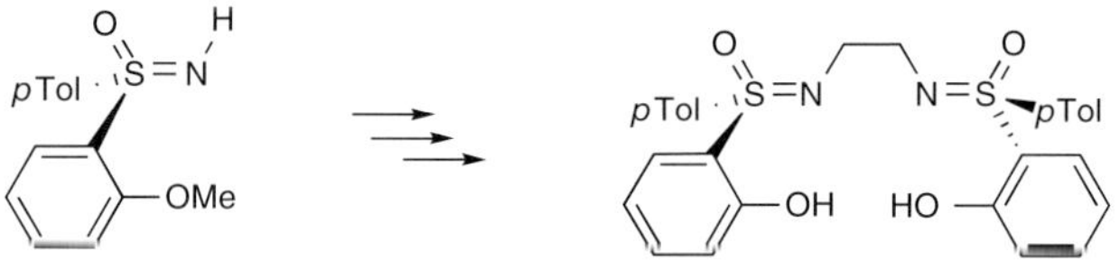

Scheme 2.1.1.14

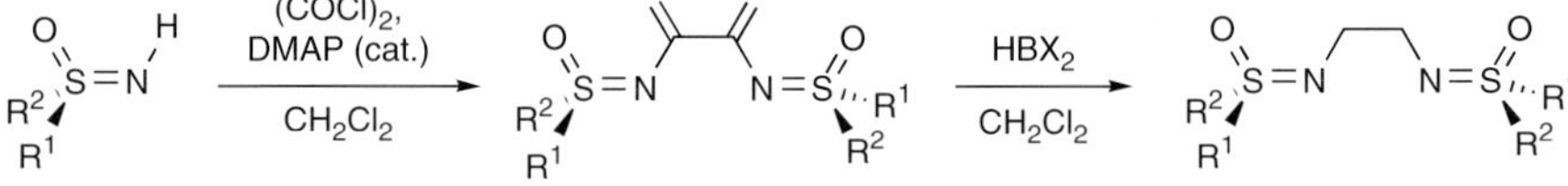

Scheme 2.1.1.15

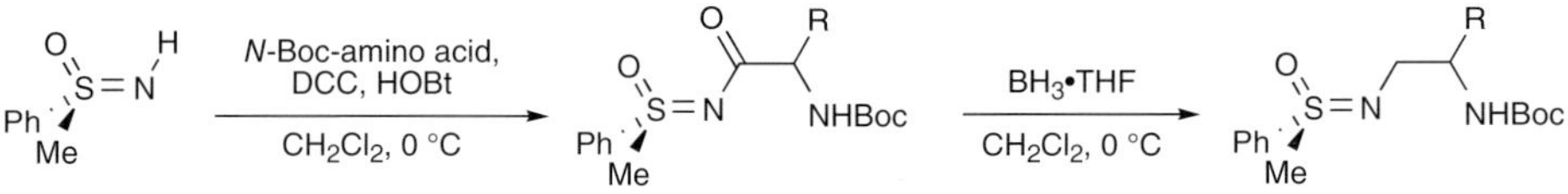

Scheme 2.1.1.16

2.1.1 Sulfoximines as Ligands in Asymmetric Metal Catalysis

Whereas there was precedence in the literature for N-alkylation reactions of sulfoximines, the analogous N-arylation was entirely unknown before our work. Commonly, N-aryl sulfoximines had to be prepared by oxidation of N-aryl sulfimines with permanganate or peracid, or by a multi-step reaction sequence starting from N-aryl sulfonimidoyl chlorides, which were treated with an organometallic reagent such as an alkyl- or aryllithium [43]. In order to avoid the obvious disadvantages of these strategies (such as the insufficient functional group tolerance and the lengthy and hazardous preparation of the sulfonimidoyl halide precursors from the corresponding sulfinamides using t-butyl hypochlorite) we decided to initiate a study on the applicability of Buchwald–Hartwig-type coupling reactions [44] to the N-arylation of readily accessible NH-sulfoximines.

To our delight we found that palladium-catalyzed cross-couplings between sulfoximines and aryl halides (or aryl nonaflates) work very well, affording N-arylated products in high yield [45]. Generally, Buchwald's Pd(OAc)$_2$/BINAP catalyst system (with Cs$_2$CO$_3$ and toluene at 100 °C) was applied, and thereby a variety of N-aryl sulfoximines 58 – including enantiopure ones – were prepared (Scheme 2.1.1.17).

Despite the fact that *ortho*-substituted aryl halides coupled well, a double sulfoximination of 59 with an excess of 9 to give bissulfoximine 60 was impossible under the conditions mentioned above (Scheme 2.1.1.18). Instead, monoarylated product 61 was obtained in good yield (75%) [45b].

Generally speaking, 61 (like 60) also appeared attractive for ligand syntheses, since it could be envisaged that subsequent palladium catalyses could activate the

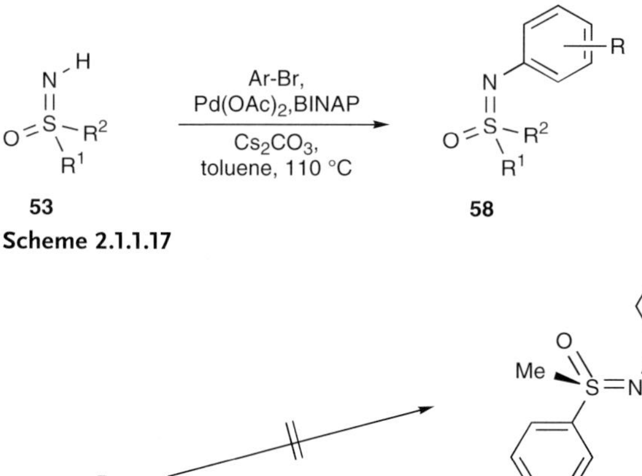

Scheme 2.1.1.17

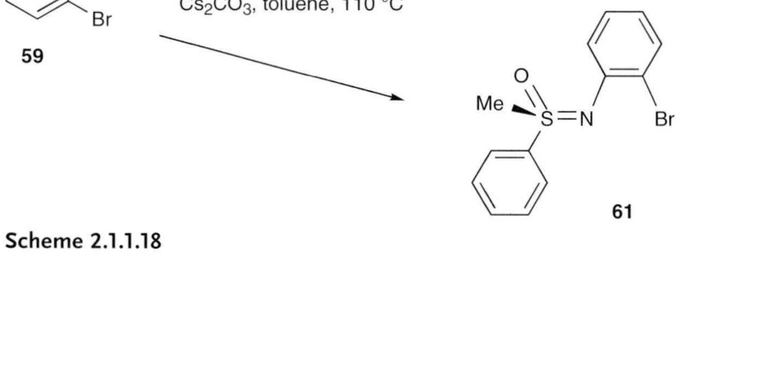

Scheme 2.1.1.18

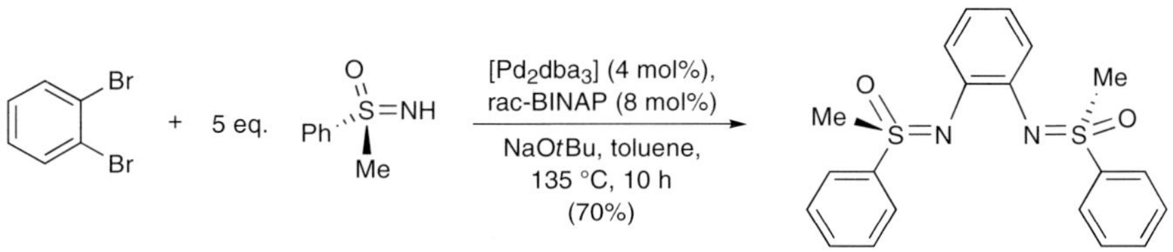

Scheme 2.1.1.19

59 + 5 eq. (S)-9 → [Pd₂dba₃] (4 mol%), rac-BINAP (8 mol%), NaOtBu, toluene, 135 °C, 10 h (70%) → (S,S)-60

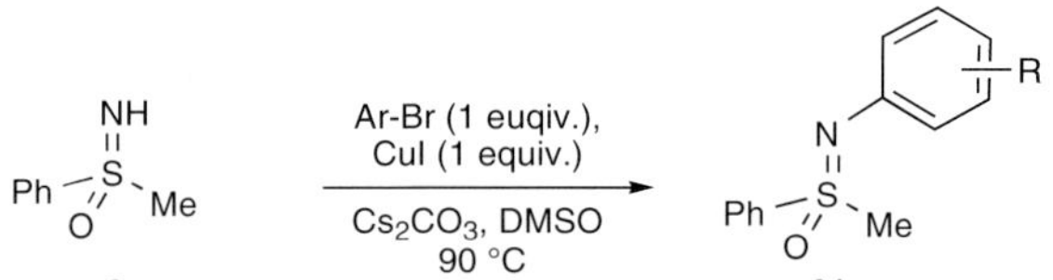

Scheme 2.1.1.20

9 + Ar-Br (1 equiv.), CuI (1 equiv.), Cs₂CO₃, DMSO, 90 °C → 61

remaining aryl bromide bond and lead to amino, phosphino, and similar substitutions. All approaches toward this goal, however, remained unsuccessful at this stage [46]. Very soon after these findings, a publication by Diver on the bis-amination of *ortho*-dibromobenzene [47] caught our attention. There, a different reagent combination was applied, and in fact, by changing the catalyst and conditions [from Pd(OAc)₂/BINAP, Cs₂CO₃, toluene, 100 °C] to [Pd₂dba₃]/BINAP and NaOt-Bu in toluene at 135 °C, to our great surprise, the doubly substituted bissulfoximine **60** was afforded in 70% yield (Scheme 2.1.1.19) [48, 49].

The double coupling with (S)-**9** afforded C_2-symmetric (S,S)-**60** in enantiopure form, which was then applied as a ligand in hetero-Diels–Alder reactions (vide infra) [48]. Although most arylations of sulfoximines proceeded well, some substrates proved difficult to convert or the product could not be purified because of heavy metal contamination; as a result, the yield of some arylations was low. Furthermore, taking into account the high price of the ligand – although it was applied as a racemate – and the cost of the noble metal, the coupling protocol needed improvement in terms of the choice of reagent. Recently, several groups have reported effective copper-mediated (and -catalyzed) cross-couplings [50], and therefore we investigated the application of such systems to the chemistry of sulfoximines. To our delight, our expectations were fulfilled and, indeed, copper salts proved useful for the N-arylation of sulfoximines [51–54]. First, a stoichiometric version using copper(I) iodide in combination with cesium carbonate (in DMSO at 90 °C) was developed [51]. Various aryl bromides coupled with sulfoximines such as **9**, smoothly affording the corresponding arylated products in high yield (Scheme 2.1.1.20). In some cases, the results achieved were noticeably better than with the palladium catalysts.

Furthermore, this copper system was interesting, since it changed the selectivity of the coupling reaction with **29**. Thus, by application of the copper(I) iodide

2.1.1 Sulfoximines as Ligands in Asymmetric Metal Catalysis | 161

version in the cross-coupling between dibromoarene **29** and **9**, bissulfoximine **28** became accessible (Scheme 2.1.1.21), whereas it could not be prepared by the palladium catalysis described earlier (where only heterocyclic **30** was formed) [51].

Subsequently, a copper-*catalyzed* cross-coupling [with substoichiometric amounts of copper(I) iodide and N,N'-dimethylethylenediamine (DMEDA)] between aryl halides and sulfoximines was developed [52]. In this case, both aryl bromides and aryl iodides reacted well. For the conversion of the former substrates an in-situ copper-catalyzed "aryl Finkelstein reaction" [53] had to be performed first, as shown in Scheme 2.1.1.22 for the preparation of **64** starting from bromobenzene (**62**).

Finally, we demonstrated that the N-arylation of sulfoximines could also be achieved using arylboronic acids as an aryl source [54]. This protocol (with the formally *umgepolte* reagents **65**; Scheme 2.1.1.23) has the advantages that the

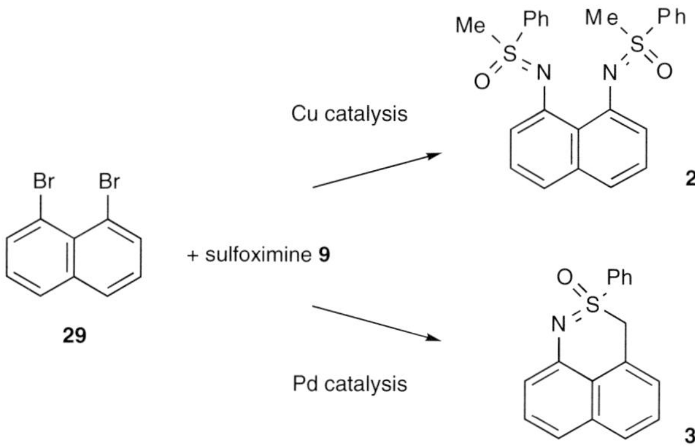

Scheme 2.1.1.21

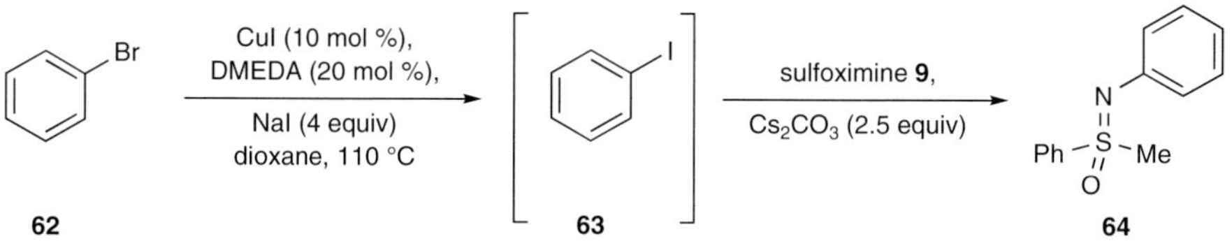

Scheme 2.1.1.22

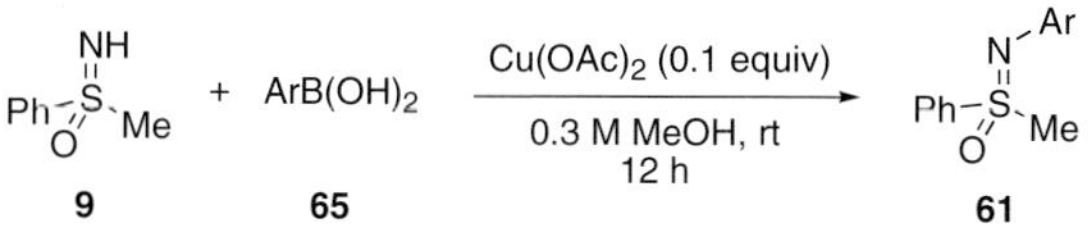

Scheme 2.1.1.23

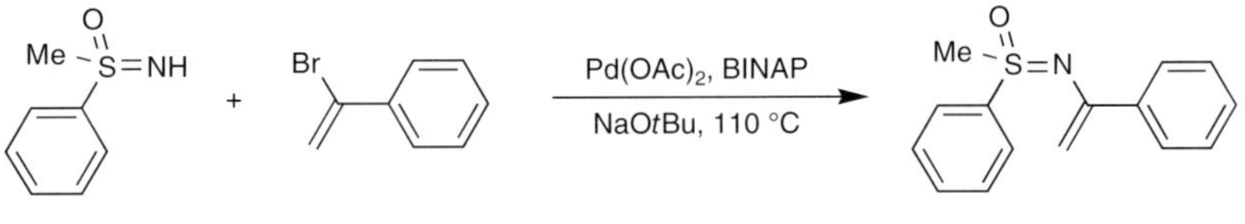

Scheme 2.1.1.24

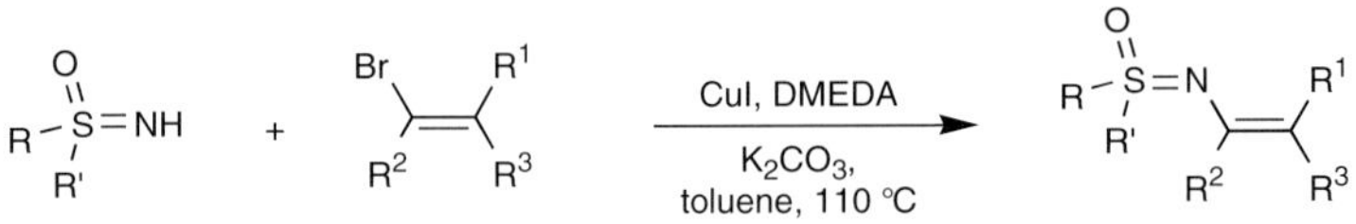

Scheme 2.1.1.25

cross-coupling proceeds at room temperature and that no additional base is needed. Interestingly, and in contrast to the preceding protocols, copper(II) acetate is now the copper salt of choice and methanol the best solvent.

As an extension of the coupling chemistry discussed above, we also investigated the N-vinylation of sulfoximines. Metal-catalyzed reactions of this type have recently been introduced for the synthesis of enol ethers and enamides, which are important structural units in a number of biologically interesting natural products [55]. Gratifyingly we found that both palladium and copper catalysts were applicable for such C–N coupling reactions, leading to vinylated sulfoximines such as **67** in good yield (Scheme 2.1.1.24) [56, 57].

It is noteworthy that unsaturated products such as **69** (Scheme 2.1.1.25) were found to be rather unstable, which might explain why they have only rarely been described in the literature [58].

2.1.1.3 Sulfoximines as Ligands in Asymmetric Metal Catalysis

As mentioned above, several asymmetric metal catalyses using sulfoximines as ligands have been investigated in recent years, with most of them focused on the formation of new C–C bonds. With the early metal/sulfoximine catalysts enantioselectivities above 90% were rare, and seldom could they seriously compete with existing systems for a given catalysis. In 2001, however, a turning point was reached, and excellent results in terms of both activity and enantioselectivity were achieved. During an investigation of the catalytic effects of copper(II) complexes bearing C_2-symmetric sulfoximine **60** as a ligand, a product **72** with an *ee* of 98% was obtained in a hetero-Diels–Alder reaction between cyclohexadiene **70** and ethyl glyoxylate **71** (Scheme 2.1.1.26). Furthermore it was found that keto diester **73** reacted equally well (giving **74** with an excellent *ee*) and that the catalyst loading could be reduced (to 1 mol%) without significantly affecting the *ee* [48].

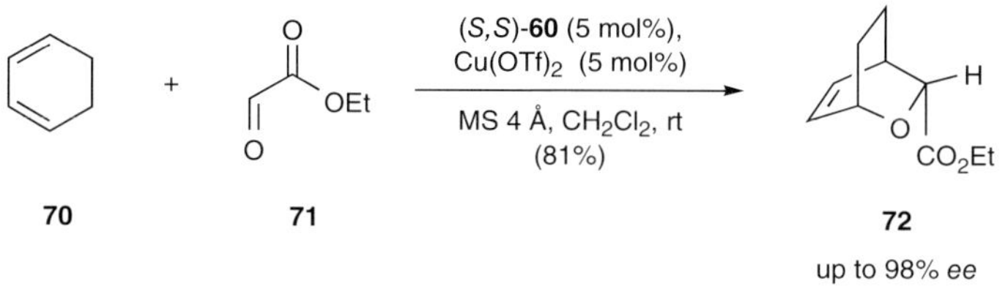

Scheme 2.1.1.26

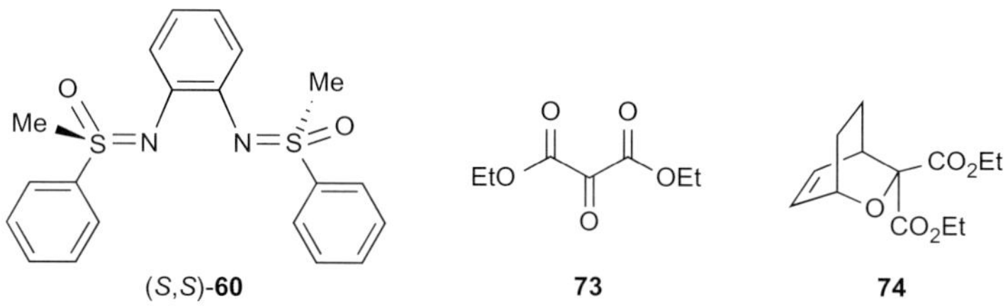

In this type of hetero-Diels–Alder reaction ethylene-bridged bissulfoximines **55** were also applied as ligands, giving bicyclic products with up to 99% *ee* [59, 60]. It is noteworthy that these results compared very well with those obtained using the well-established Cu(II)/BOX-based systems, which were investigated by Jørgensen and afforded cycloaddition products with up to 97% *ee* [61].

After this first successful application of bissulfoximine **60**, we investigated its use in other copper-catalyzed cycloaddition reactions. For example, Diels–Alder reactions between cyclopentadiene **75** and dienophile **76** had been studied with Cu(II)/BOX systems before, and it was shown by Evans that almost complete enantioselectivities in the fomation of **77** could be achieved [62]. The attempted use of **55** or **60**, however, gave only unsatisfactory results (up to 83% *ee*). Significant modification of the ligand structure and a major adjustment to the reaction conditions were required until we were able to demonstrate that in this reaction also the use of a sulfoximine ligand on a copper(II) salt could lead to *ee* values above 90%. In this case, bissulfoximine **78** was found to be the best ligand, and the previously used copper triflate had to be replaced by copper(II) perchloride (Scheme 2.1.1.27). Furthermore, the solvent (CHCl$_3$) and the temperature (−60 °C) had to be well chosen to allow **77** to be produced with 93% *ee* (*endo/exo* ratio: 89 : 11) [63]. Particularly remarkable was the pronounced positive effect of the *ortho*-methoxy group on the ligand structure.

An interpretation of the results and a hypothesis on potential intermediates was difficult at this stage, since no defined complexes could be obtained. The first identification of species present in solution during the catalysis of the Diels–Alder reaction was achieved in a collaboration between our group and three physical chemists, Bertagnolli, Gescheidt, and Schweiger [64]. Using various techniques

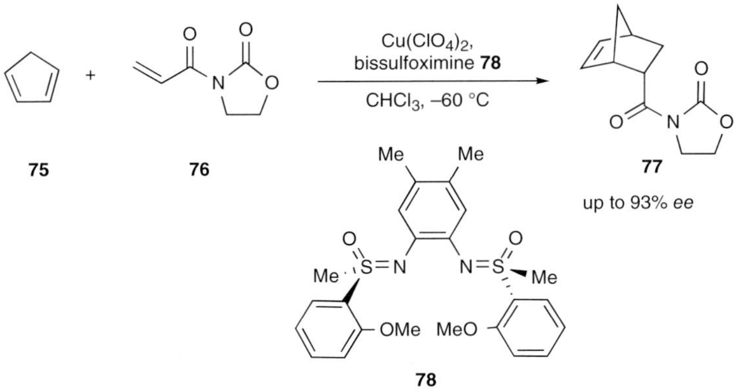

Scheme 2.1.1.27

including EXAFS, CW-EPR, HYSCORE, pulsed ENDOR, and UV–Vis spectroscopy they were able to find a copper(II)/ligand/substrate complex which was formed under the reaction conditions. The data showed that a tetragonally distorted complex was obtained, in which the ligand was bound to the copper(II) center through the imine nitrogens. Furthermore, substrate **76** interacts through the carbonyl oxygens, and a triflate anion binds with an oxygen atom in the axial position.

During the data optimization a schematic representation of a species was transmitted, which is shown here as structure **79**. It differed from the final result, but proved most stimulating for our future ligand design.

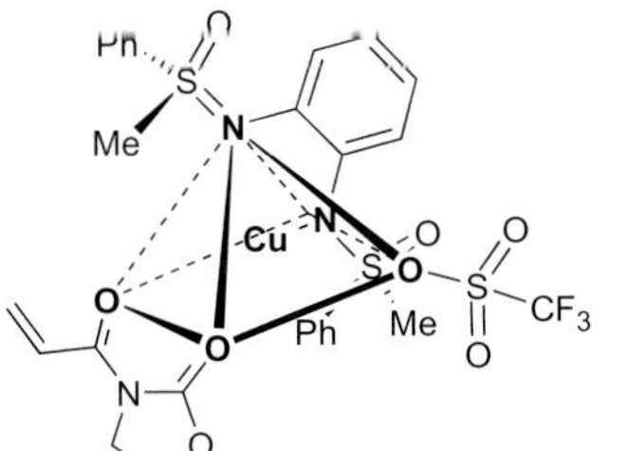

From this initially proposed (incorrect) arrangement **79** it was apparent that the bissulfoximine ligand had two unequal binding sites. The two sulfoximine nitrogens differed in their position! This meant that the ligand had lost its C_2-symmetry and that it would also be worthwhile to study C_1-symmetric sulfoximines that could – perhaps – lead to high enantioselectivities as well.

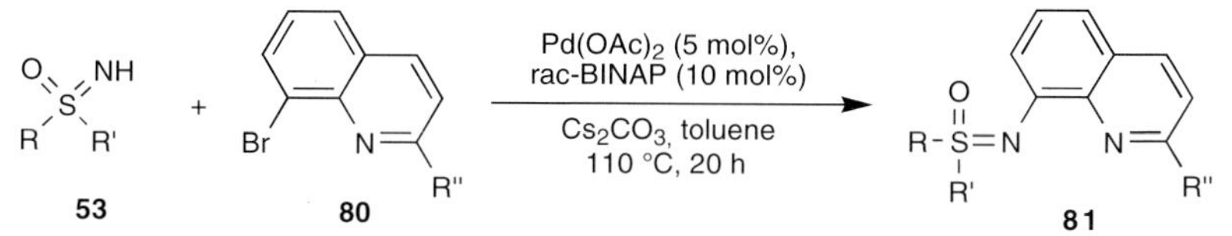

2.1.1 Sulfoximines as Ligands in Asymmetric Metal Catalysis | 165

Scheme 2.1.1.28

Based on this concept, simple monosulfoximines such as **81** were prepared and tested [65]. Again, the palladium-catalyzed N-arylation of the sulfoximines **53** with the corresponding aryl halides **80** (Scheme 2.1.1.28) allowed rapid access to a wide range of such compounds.

The results of the hetero-Diels–Alder reaction between **70** and **71** then confirmed our hypothesis. Even with these C_1-symmetric "easy-to-come-by" sulfoximines, cycloaddition product **72** was formed with up to 96% ee [65].

To our delight, it was also possible at this stage to isolate a 1 : 1 complex composed of $CuCl_2$ and sulfoximine **82** [65]. Its X-ray crystal structure proved most instructive in interpreting the observed effects and stereochemical results. Thus, by retaining the overall spatial arrangement, removing both chlorides from copper, and attaching substrate **71** (in an orientation that minimized steric effects with the ligand methoxy group), a formally dicationic arrangement **83** resulted. From this it became apparent that the experimentally observed enantiomer of **72** had to be formed, if the diene approached from the less hindered "eastern" side of the complex. Furthermore, the developing diene–carbonyl interaction should not be hampered by an enlargement of substituent R" (see general structure **81**). If the sulfoximine methyl group (R' in **81**) was (formally) increased in size (by having ethyl, isopropyl or similar substituents at that position), the coordination of the carbonyl component would become more difficult. Furthermore, the aryl methoxy group blocked the top face of the complex and helped in orienting both substrates [66]

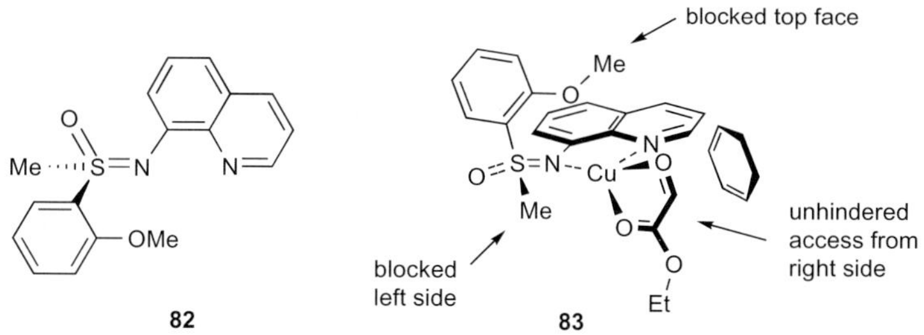

The success of copper-catalyzed cycloaddition reactions yielding products with up to 99% ee stimulated further investigations in this regard. Both of the key ligand structure types represented by C_2-symmetric sulfoximine **60** and C_1-symmetric **82** had a two-carbon distance between the two coordination nitrogens.

By keeping this successful array, it became of interest to see what effect a further modification of the nature of the second (non-sulfoximine) nitrogen would have on the activity and enantioselectivity of a given copper complex. Consequently, C_1-symmetric compounds **85** having an aniline nitrogen became the targets of choice. Starting from 1-bromo-2-nitrobenzene **84**, a wide range of variously substituted sulfoximines **85** were available through a straightforward palladium-catalyzed cross-coupling/nitro group reduction/reductive alkylation reaction sequence (Scheme 2.1.1.29) [67, 68].

Additionally, another route to copper-catalyzed C–C bond formation, the Mukaiyama-type aldol reaction between enol ether **86** and ketone **87** to give ester **88** (Scheme 2.1.1.30), was investigated [69]. This C–C bond-forming reaction is particularly interesting, since **88** has a stereogenic center at a quaternary carbon, and those compounds are generally difficult to prepare [70]. As a result, sulfoximine **89** was identified as the ligand of choice, which allowed the preparation of **88** with up to 99% *ee* [67]. In subsequent studies the full scope of substrates as well as the effect of related ligand structures and variations of the reaction conditions were investigated [68].

As an interesting challenge we examined an extension of this procedure to its vinylogous analogue [71]. Commonly, reactions between "extended dienolates" such as **90** and carbonyl compounds have focused on the use of aldehydes as an acceptor component. In contrast, we decided to investigate catalytic asymmetric

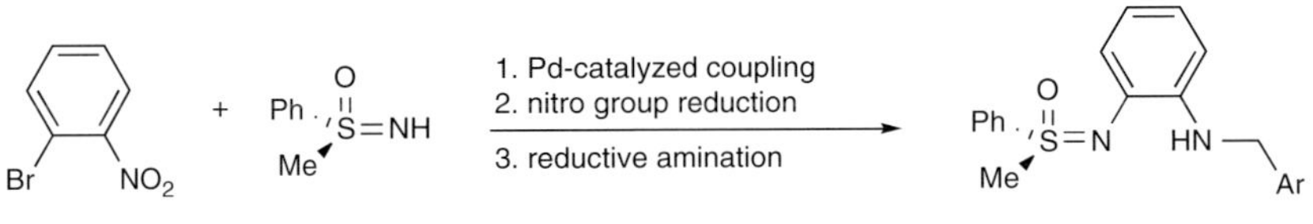

Scheme 2.1.1.29

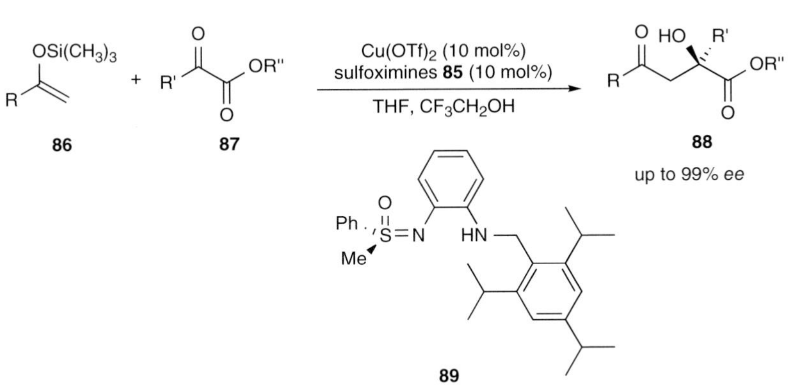

Scheme 2.1.1.30

addition to pyruvates **91** [72]. To our delight, sulfoximine **89** also proved applicable in this case (Scheme 2.1.1.31). Among a series of ten chelating ligands (and using a substrate with TMDMS as silyl group, R′ = *t*-butyl, and R″ = isopropyl), this particular sulfoximine gave the highest enantioselectivities, reaching up to 99% *ee* [72].

The potential of aminosulfoximines **85** to be used as ligands in copper-catalyzed carbonyl–ene reactions was also examined [73, 74]. Interestingly, in this case, unsymmetrically substituted **95** gave the best results (up to 91% *ee* for the synthesis of **94** with R = Me; Scheme 2.1.1.32) [75].

At this stage, sulfoximines had been shown to function as chiral ligands for various palladium and copper catalysts which led to enantioselectivities of >95% *ee* in various reactions. Furthermore, most of those catalyzed reactions were C–C bond formations. Obvious questions were, therefore, whether sulfoximines could also be applied in combination with other metals and whether reductions and oxidations could be catalyzed as well. A structural comparison of the sulfoximines leading to high *ee* values such as **55**, **60**, **81**, and **85** revealed that all of them had a two-carbon distance between the two coordinating atoms (which were all nitrogen in these cases).

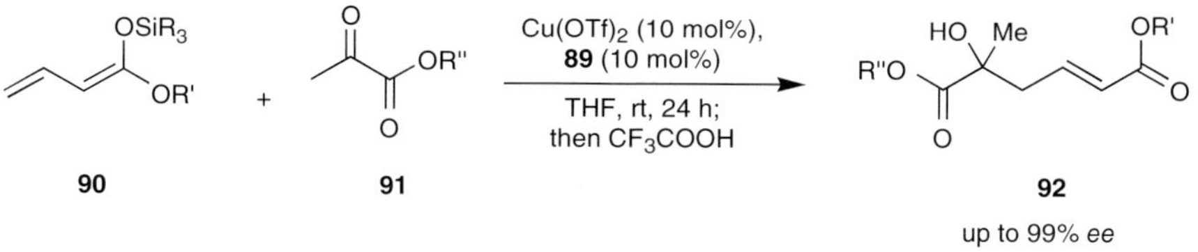

Scheme 2.1.1.31

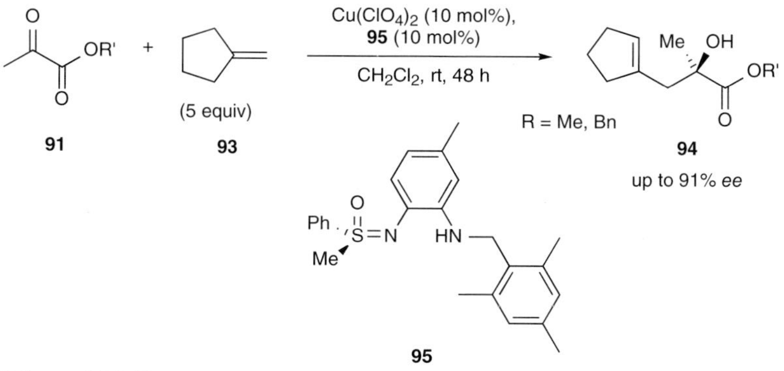

Scheme 2.1.1.32

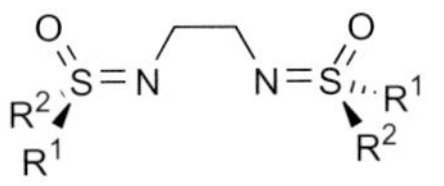

55
allylic alkylation
reactions

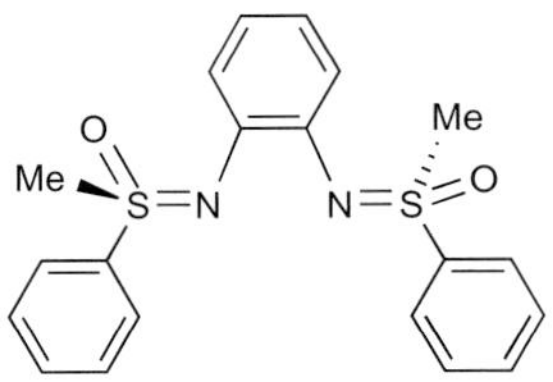

60
hetero-Diels-
Alder reactions

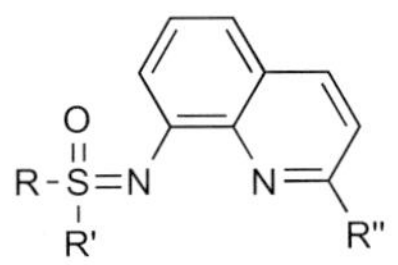

81
hetero-Diels-
Alder reactions

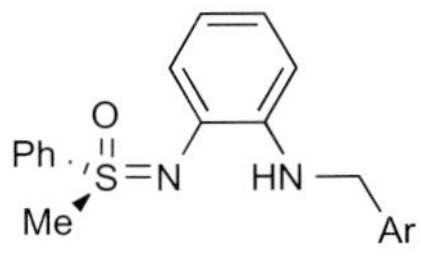

85
Mukaiyama-type
aldol reactions
and vinylogous
aldol reactions

Consequently, compounds such as **96** having a suitable donor atom X in the appropriate position became desirable targets. In particular, phosphino sulfoximine **97** appeared attractive because other P,N-ligands had already been applied successfully in various catalyses [76, 77].

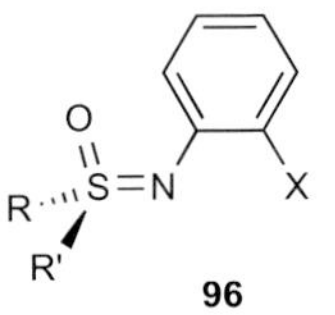

96
(X = donor atom)

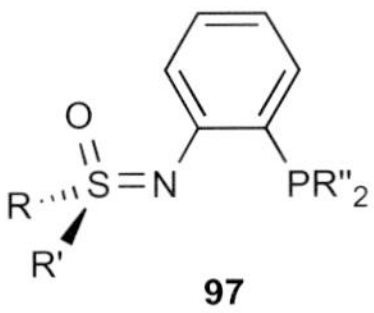

97

The synthesis of phosphino sulfoximine **97** relied significantly on the successful development of methods pursued in parallel in our group. Whereas palladium-catalyzed cross-couplings between **53** and **98** proceeded in low yield, the copper catalysis with a combination of copper(I) iodide and cesium acetate worked well, affording **99** in up to 83% yield [78]. The resulting phosphine oxides **99** were then reduced to the corresponding phosphines **97** using a mixture of trichlorosilane and triethylamine (Scheme 2.1.1.33).

In order to test the capability of the novel P,N-ligands **97** in asymmetric catalyses, a challenging reductive transformation, the enantioselective hydrogenation of acyclic imines [79], was chosen. Based on previous work [76], iridium was

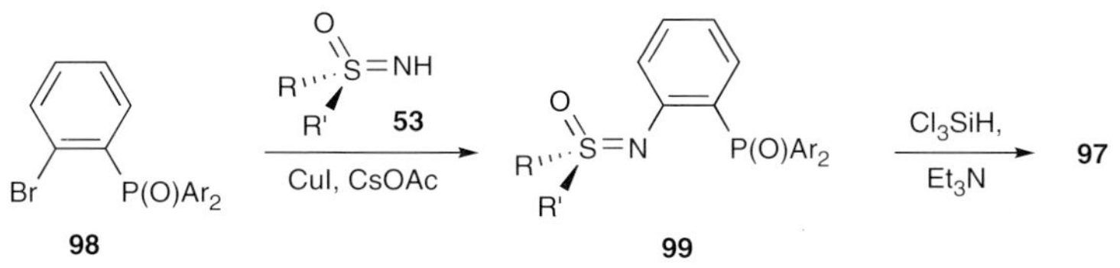

Scheme 2.1.1.33

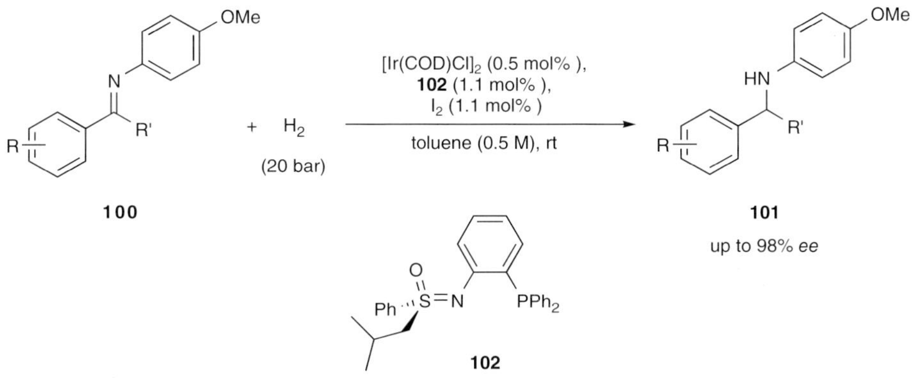

Scheme 2.1.1.34

selected as the metal for the catalyst. To our delight we found that the phosphino sulfoximines **97** were not only applicable in these reactions, they also led to products with excellent enantioselectivities [78]. An optimization study revealed that **102** was the best ligand in these reactions, affording protected amines **101** with up to 98% *ee* (Scheme 2.1.1.34).

Quite a wide range of substrates **100** could be converted into products **101** with high *ee* values; since it is known that the N-protecting group of **101** can easily be cleaved, the approach represents a formal synthesis of optically active amines. It remains to be seen if this iridium/sulfoximine combination also opens up an alternative access to industrially relevant products such as the herbicide (S)-metolachlor produced by Syngenta [80].

As an extension of these studies on the use of sulfoximines in asymmetric reductions, BINOL-derived phosphino sulfoximines of the **105** type were tested in both rhodium-catalyzed hydrogenations (yielding optically active diesters **104** or amino acid derivatives; Scheme 2.1.1.35) and palladium-catalyzed allylic alkylations (not shown) in collaboration with Reetz and Gais [81, 82]. Here, enantioselectivities of up to ≥99 and 66% *ee*, respectively, were achieved.

Depending on the substrate, matched/mismatched combinations between the stereogenic elements at BINOL and the sulfoximine were observed. In the conver-

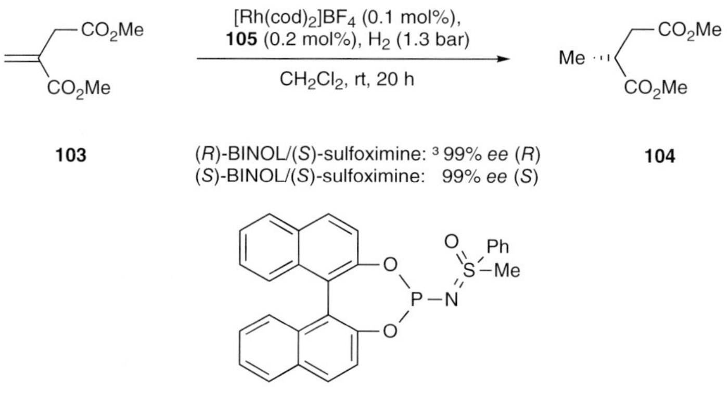

Scheme 2.1.1.35

Reaction of **103** → **104**: [Rh(cod)$_2$]BF$_4$ (0.1 mol%), **105** (0.2 mol%), H$_2$ (1.3 bar), CH$_2$Cl$_2$, rt, 20 h.

(R)-BINOL/(S)-sulfoximine: [3] 99% ee (R)
(S)-BINOL/(S)-sulfoximine: 99% ee (S)

sion of **103** to **104** these were less pronounced but, for example, in the synthesis of N-acetyl-1-phenylethylamine (by hydrogenation of the corresponding enamide) a difference in ee of 12% resulted [81]. In all cases the absolute configuration of the product was determined by the BINOL fragment of the ligand, and generally – although not in all cases – the stereochemistry at sulfur was of minor influence. Similar trends were observed in the palladium-catalyzed allylic alkylation reactions.

2.1.1.4 Conclusions

Since the mid-1990s, sulfoximines have been shown to be excellent ligands for asymmetric metal catalysis. Whereas initially the resulting enantioselectivities were only moderate to good, more recently copper-, palladium-, and iridium-based systems have been discovered which afford products with excellent ee values and which can compete very well with established ligands having similar N,N- or P,N-chelating units. Since both enantiomers of the ligands are equally accessible, the absolute configurations of the products can be chosen at will. Furthermore it is noteworthy that a significant range of C–C bond-forming reactions can now be catalyzed and that the reaction scope has been expanded into the area of asymmetric hydrogenations. Still, the synthesis of sulfoximines requires attention: higher selectivities and milder reaction conditions for their preparation (for example, in the sulfide oxidation or the sulfoxide imination) would be desirable. Nevertheless, significant progress has also been made in these fields, and both catalytic studies and method development benefit from being pursued simultaneously. The future will show whether further applications of sulfoximines – as ligands in asymmetric catalysis and perhaps even in entirely unrelated areas [83] – can be found [84], and whether these sulfur reagents will then be recognized as "privileged" compounds.

References

1. a) E. Mellanby, *Br. Med. J.* **1946**, *2*, 885; b) E. Mellanby, *Br. Med. J.* **1947**, *3*, 288; c) T. Morau, *Lancet* **1947**, 289; d) H. R. Bentley, E. E. McDermott, J. Pace, J. K. Whitehead, T. Moran, *Nature* **1949**, *163*, 675; e) H. R. Bentley, E. E. McDermott, J. Pace, J. K. Whitehead, T. Moran, *Nature* **1950**, *165*, 150; f) H. R. Bentley, E. E. McDermott, J. Pace, J. K. Whitehead, *Nature* **1950**, *165*, 735; g) H. R. Bentley, E. E. McDermott, T. Moran, J. Pace, J. K. Whitehead, *Proc. Roy. Soc. B* **1950**, *137*, 402.
2. a) R. A. Ronzio, W. B. Rowe, S. Wilk, A. Meister, *Biochemistry* **1969**, *8*, 2670; b) W. B. Rowe, R. A. Ronzio, A. Meister, *Biochemistry* **1969**, *8*, 2674; c) J. M. Manning, S. Moore, W. B. Rowe, A. Meister, *Biochemistry* **1969**, *8*, 2681.
3. For applications of sulfoximines in medical studies, see: a) A. C. Barnes, P. W. Hairine, S. S. Matharu, P. J. Ramm, J. B. Taylor, *J. Med. Chem.* **1979**, *22*, 418; b) R. D. Dillard, T. T. Yen, P. Stark, D. E. Pavey, *J. Med. Chem.* **1980**, *23*, 717; c) C. H. Levenson, R. B. Meyer, Jr., *J. Med. Chem.* **1984**, *27*, 228; d) P. J. Harvison, T. I. Kalman, *J. Med. Chem.* **1992**, *35*, 1227; e) M. Morimoto, T. Ashizawa, N. Nakamizo, Y. Otsuji, *J. Heterocyclic Chem.* **1992**, *29*, 1133; f) R. E. Dolle, D. McNair, *Tetrahedron Lett.* **1993**, *34*, 133. Reviews: g) A. Meister, *Biochem. Biophys. Acta* **1995**, 35; h) H. H. Bailey, *Chem.–Biol. Interact.* **1998**, *111–112*, 239; i) E. Obrador, J. Navarro, J. Mompo, M. Asensi, J. A. Pellicier, J. M. Estrela, *BioFactors* **1998**, 8, 23; j) M. E. Anderson, *Chem.–Biol. Interact.* **1998**, *111*, 1; k) L. L. Muldoon, L. S. L. Walker-Rosenfeld, C. Hale, S. E. Purcell, L. C. Bennett, E. A. Neuwelt *J. Pharmacol. Exp. Ther.* **2001**, *296*, 797; l) see also M. Kahraman, S. Sinishtaj, P. M. Dolan, T. W. Kensler, S. Peleg, U. Saha, S. S. Chuang, G. Bernstein, B. Korczak, G. H. Posner, *J. Med. Chem.* **2004**, *47*, 6854 and references therein.
4. a) W. L. Mock, J.-T. Tsay, *J. Am. Chem. Soc.* **1989**, 111, 4467; b) W. L. Mock, J. Z. Zhang, *J. Biol. Chem.* **1991**, *266*, 6393.
5. a) H. Kawanishi, H. Morimoto, T. Nakano, T. Watanabe, K. Oda, K. Tsujihara, *Heterocycles* **1998**, *49*, 181; b) for insecticidal activity, see: Y. Zhu, R. B. Rogers, J. X. Huang, Patent Appl. US 2005/0228027 A1 (Dow Agroscience), Oct. 13, **2005**.
6. C. Bolm, D. Müller, C. Dalhoff, C. P. R. Hackenberger, E. Weinhold, *Bioorg. Med. Chem. Lett.* **2003**, *13*, 3207.
7. a) C. Bolm, J. D. Kahmann, G. Moll, *Tetrahedron Lett.* **1997**, *38*, 1169; b) C. Bolm, G. Moll, J. D. Kahmann, *Chem. Eur. J.* **2001**, *7*, 1118; c) C. Bolm, D. Müller, C. P. R. Hackenberger, *Org. Lett.* **2002**, *4*, 893; d) see also: H. Tye, C. L. Skinner, *Helv. Chim. Acta* **2002**, *85*, 3272.
8. a) C. P. R. Hackenberger, G. Raabe, C. Bolm, *Chem. Eur. J.* **2004**, *10*, 2942; b) C. Bolm, J. Fleischhauer, G. Raabe, E. Voloshina, submitted for publication; c) see also: P. Kumar, P. V. Bharatam, *Tetrahedron* **2005**, *61*, 5633.
9. Reviews: a) C. R. Johnson, *Aldrichim. Acta* **1985**, *18*, 3; b) C. R. Johnson, *Acc. Chem. Res.* **1973**, *6*, 341; c) C. R. Johnson, in *Comprehensive Organic Chemistry* (Eds.: D. Barton, W. D. Ollis), Pergamon Press, Oxford, **1979**, *3*, 223; d) C. R. Johnson, M. R. Barachyn, N. A. Meanwell, C. J. Stark, Jr., J. R. Zeller, *Phosphorus Sulfur* **1985**, *24*, 151; e) S. L. Huang, D. Swern, *Phosphorus Sulfur* **1976**, *1*, 309; f) S. G. Pyne, *Sulfur Rep.* **1992**, *12*, 57; g) M. Reggelin, C. Zur, *Synthesis* **2000**, 1.
10. For the use of sulfoximines as chiral auxiliaries, see for example: a) M. Haiza, J. Lee, J. K. Snyder, *J. Org. Chem.* **1990**, *55*, 5008; b) B. M. Trost, R. T. Matuoka, *Synlett* **1992**, 27; c) S. G. Pyne, Z. Dong, B. W. Skelton, A. H. White, *J. Org. Chem.* **1997**, *62*, 2337; d) M. Reggelin, T. Heinrich, *Angew. Chem.* **1998**, *110*, 3005; *Angew. Chem. Int. Ed.* **1998**, *37*, 2883; e) S. Bosshammer, H.-J. Gais, *Synthesis* **1998**, 919; f) L. A. Paquette, Z. Gao, Z. Ni, G. F. Smith, *J. Am. Chem. Soc.* **1998**, *120*, 2543; g) M. Harmata, M. Kahraman, D. E. Jones, N. Pavri, S. E. Weatherwax, *Tetrahedron* **1998**, *54*, 9995;

h) M. Harmata, N. Pavri, *Angew. Chem.* **1999**, *111*, 2577; *Angew. Chem. Int. Ed.* **1999**, *38*, 2419; i) R. R. Reddy, H.-J. Gais, C.-W. Woo, G. Raabe, *J. Am. Chem. Soc.* **2002**, *124*, 10 427; j) S. Koep, H.-J. Gais, G. Raabe, *J. Am. Chem. Soc.* **2003**, *125*, 13 243; k) H.-J. Gais, G. S. Babu, M. Günter, P. Das, *Eur. J. Org. Chem.* **2004**, 1464; l) M. Harmata, X. Hong, C. L. Barnes, *Tetrahedron Lett.* **2003**, *44*, 7261; m) M. Harmata, X. Hong, *J. Am. Chem. Soc.* **2003**, *125*, 5754; n) M. Harmata, X. Hong, *Org. Lett.* **2005**, *7*, 3581; o) D. Craig, F. Grellepois, A. J. P. White, *J. Org. Chem.* **2005**, *70*, 6827.

11 For contributions from our group in this area, see: a) C. Bolm, K. Muñiz, N. Aguilar, M. Kesselgruber, G. Raabe, *Synthesis* **1999**, 1251; b) C. Bolm, M. Kesselgruber, K. Muñiz, G. Raabe, *Organometallics* **2000**, *19*, 1648; c) G. Süss-Fink, G. Rheinwald, H. Stoeckli-Evans, C. Bolm, D. Kaufmann, *Inorg. Chem.*, **1996**, *35*, 3081; d) V. Ferrand, C. Gambs, N. Derrien, C. Bolm, H. Stoeckli-Evans, G. Süss-Fink, *J. Organomet. Chem.* **1997**, *549*, 275.

12 C. Bolm, M. Felder, J. Müller, *Synlett* **1992**, 439.

13 a) C. Bolm, M. Felder, *Tetrahedron Lett.* **1993**, *34*, 6041; b) C. Bolm, J. Müller, G. Schlingloff, M. Zehnder, M. Neuburger, *J. Chem. Soc., Chem. Commun.* **1993**, 182; c) C. Bolm, A. Seeger, M. Felder, *Tetrahedron Lett.* **1993**, *34*, 8079; d) C. Bolm, M. Felder, *Synlett* **1994**, 655; e) C. Bolm, J. Müller, *Tetrahedron* **1994**, *50*, 4355; f) C. Bolm, P. Müller *Tetrahedron Lett.* **1995**, *36*, 1625; g) C. Bolm, P. Müller, *Acta Chem. Scand.* **1996**, *50*, 305; h) C. Bolm, D. Kaufmann, M. Zehnder, M. Neuburger, *Tetrahedron Lett.* **1996**, *37*, 3985; i) C. Bolm, N. Derrien, A. Seger, *Synlett* **1996**, 386.

14 For recent reviews summarizing applications of sulfoximines as ligands in asymmetric catalysis, see: a) H. Okamura, C. Bolm, *Chem. Lett.* **2004**, *33*, 482; b) M. Harmata, *Chemtracts* **2003**, *16*, 660.

15 a) R. Fusco, F. Tericoni, *Chim. Ind. (Milan)* **1965**, *47*, 61; b) C. R. Johnson, C. W. Schroeck, *J. Am. Chem. Soc.* **1973**, *95*, 7418; c) C. S. Shiner, A. H. Berks, *J. Org. Chem.* **1988**, *53*, 5542; d) K. Mori, F. Toda, *Chem. Lett.* **1988**, 1997; e) for an improved protocol, which was applied in most of our studies, see: J. Brandt, H.-J. Gais, *Tetrahedron: Asymmetry* **1997**, *6*, 909.

16 For reports on catalyzed asymmetric sulfide oxidations from our group, see: a) C. Bolm, F. Bienewald, *Angew. Chem.* **1995**, *107*, 2883; *Angew. Chem. Int. Ed. Engl.* **1995**, *34*, 2640; b) C. Bolm, G. Schlingloff, F. Bienewald, *J. Mol. Catal. A: Chem.* **1997**, *117*, 347; c) C. Bolm, F. Bienewald, *Synlett* **1998**, *34*, 1327; d) C. Bolm, *Coord. Chem. Rev.* **2003**, *237*, 245; e) C. Bolm, O. A. G. Dabard, *Synlett* **1999**, 360; f) J. Legros, C. Bolm, *Angew. Chem.* **2003**, *115*, 5645; *Angew. Chem. Int. Ed.* **2003**, *42*, 5487; g) J. Legros, C. Bolm, *Angew. Chem.* **2004**, *116*, 4321; *Angew. Chem. Int. Ed.* **2004**, *43*, 4225; h) A. Korte, J. Legros, C. Bolm, *Synlett* **2004**, 2397; i) see also in: C. Bolm, J. Legros, J. LePaih, L. Zani, *Chem. Rev.* **2004**, *104*, 6217.

17 For reviews on the preparation and use of chiral sulfoxides, see: a) H. B. Kagan, T. Luukas in *Transition Metals for Organic Synthesis*, Vol. 2 (Eds.: M. Beller, C. Bolm), Wiley-VCH, Weinheim, **2004**, p. 479; b) C. Bolm, K. Muñiz, J. P. Hildebrand in *Comprehensive Asymmetric Catalysis* (Eds.: E. N. Jacobsen, A. Pfaltz, H. Yamamoto), Springer-Verlag, Berlin, **1999**, p. 697; c) H. B. Kagan in *Catalytic Asymmetric Synthesis*, 2nd ed. (Ed.: I. Ojima), Wiley-VCH, New York, **2000**, p. 327; d) J.-E. Bäckvall, in *Modern Oxidation Methods* (Ed.: J.-E. Bäckvall), VCH-Wiley, Weinheim, **2004**, p. 193; e) E. N. Prilezhaeva, *Russ. Chem. Rev.* **2000**, *69*, 367; f) E. N. Prilezhaeva, *Russ. Chem. Rev.* **2001**, *70*, 897; g) I. Fernández, N. Khiar, *Chem. Rev.* **2003**, *103*, 3651; h) M. C. Carreño, *Chem. Rev.* **1995**, *95*, 1717.

18 a) Y. Tamura, J. Minamikawa, K. Sumoto, S. Fujii, M. Ikeda, *J. Org. Chem.* **1973**, *38*, 1239; b) C. R. Johnson, R. A. Kirchhoff, H. G. Corkins, *J. Org. Chem.* **1974**, *39*, 2458; c) Y. Tamura, H. Matushima, J. Minamikawa, M. Ikeda, K. Sumoto, *Tetrahedron* **1975**, *31*, 3035;

d) M. Fieser, L. F. Fieser, *Reagents for Organic Synthesis*, Vol. 5, John Wiley, New York, **1975**, p. 430.

19 For the toxicity of NaN$_3$: *Merck Index*, 12th ed. (Ed.: S. Budavari), Merck & Co., Whitehouse Station, NJ, **1996**, p. 451.

20 Cu salts: a) J. F. K. Müller, P. Vogt, *Tetrahedron Lett.* **1998**, *39*, 4805; b) E. Lacôte, M. Amatore, L. Fensterbank, M. Malacria, *Synlett* **2002**, 116; c) S. Cren, T. C. Kinahan, C. L. Skinner, H. Tye, *Tetrahedron Lett.* **2002**, *43*, 2749; d) C. S. Tomooka, E. M. Carreira, *Helv. Chim. Acta* **2003**, *85*, 3773; e) H. Takada, K. Ohe, S. Uemura, *Angew. Chem.* **1999**, *111*. 1367; *Angew. Chem. Int. Ed.* **1999**, *38*, 1288. Fe salts: f) T. Bach, C. Körber, *Tetrahedron Lett.* **1998**, *39*, 5015; g) T. Bach, C. Körber, *C. Eur. J. Org. Chem.* **1999**, *64*, 1033. Mn complexes: h) H. Nishikori, C. Ohta, E. Oberlin, R. Irie, T. Katsuki, *Tetrahedron* **1999**, *55*, 13937; i) C. Ohta, T. Katsuki, *Tetrahedron Lett.* **2001**, *42*, 3885. Ru complexes: j) M. Murakami, T. Uchida, T. Katsuki, *Tetrahedron Lett.* **2001**, *42*, 7071; k) Y. Tamura, T. Uchida, T. Katsuki, *Tetrahedron Lett.* **2003**, *44*, 3301; l) M. Murakami, T. Uchida, B. Saito, T. Katsuki, *Chirality* **2003**, *15*, 116; m) T. Uchida, Y. Tamura, M. Ohba, T. Katsuki, T. *Tetrahedron Lett.* **2003**, *44*, 7965.

21 H. Okamura, C. Bolm, *Org. Lett.* **2004**, *6*, 1305.

22 (a) Y. Cui, C. He, *C. J. Am. Chem. Soc.* **2003**, *125*, 16 202; b) Y. Cui, C. He, *Angew. Chem.* **2004**, *116*. 4306; *Angew. Chem. Int. Ed.* **2004**, *43*, 4210.

23 G. Y. Cho, C. Bolm, *Org. Lett.* **2005** *7*, 4983.

24 G. Y. Cho, C. Bolm, *Tetrahedron Lett.* **2005**, *46*, 8007.

25 For another metal-free approach toward sulfoximines, which utilizes electrochemistry, see: a) T. Siu, C. J. Picard, A. K. Yudin, *J. Org. Chem.* **2005**, *70*, 932; b) T. Siu, A. K. Yudin, *Org. Lett.* **2002**, *4*, 1839; for related work see also: c) L. B. Krasnova, R. M. Hili, O. V. Chernoloz, A. K. Yudin, *Arkivoc* **2005**, *iv*, 26.

26 For a sulfoxide imination with a 3-acetoxyaminoquinazolinone (which requires the use of toxic Pb(OAc)$_4$ for its preparation), see: S. Karabuga, C. Kazaz, H. Kilic, S. Ulukanli, A. Celik, *Tetrahedron Lett.* **2005**, *46*, 5225.

27 J. R. Dehli, C. Bolm, *Synthesis* **2005**, 105.

28 a) Review: M. Haake, in *Topics in Sulfur Chemistry*, Vol. 1 (Ed.: A. Senning), Thieme, Stuttgart, **1976**, 185; b) see also references in: W. E. Diederich, M. Haake, *J. Org. Chem.* **2003**, *68*, 381.

29 For the use of ZnEt$_2$ as a base and structural studies of the resulting sulfoximine-containing zinc enolates, see: C. Bolm, J. Müller, M. Zehnder, M. A. Neuburger, *Chem. Eur. J.* **1995**, *1*, 312.

30 C. Bolm, M. Martin, L. Gibson, *Synlett* **2002**, 832.

31 C. Bolm, H. Okamura, M. Verrucci, *J. Organometal. Chem.* **2003**, *687*, 444.

32 C. Bolm, H. Villar, *Synthesis* **2005**, 1421.

33 a) For a review on palladium-catalyzed χ-arylations of carbonyl compounds and nitriles, see: D. A. Culkin, J. K. Hartwig, *Acc. Chem. Soc.* **2003**, *36*, 234. For recent examples, see: b) H. Muratake, M. Natsume, *Angew. Chem.* **2004**, *116*, 4746; *Angew. Chem. Int. Ed.* **2004**, *43*, 4646; c) H. Muratake, M. Natsume, H. Nakai, *Tetrahedron* **2004**, *60*, 11783; d) J. Chae, J. Yun, S. L. Buchwald, *Org. Lett.* **2004**, *6*, 4809; e) M. C. Willis, G. N. Brace, I. P. Holmes, *Angew. Chem.* **2005**, *117*, 407; *Angew. Chem. Int. Ed.* **2005**, *44*, 403.

34 G. Y. Cho, C. Bolm, *Org. Lett.* **2005**, *70*, 1351.

35 S. Gaillard, C. Papamicael, G. Dupas, F. Marsais, V. Levachet, *Tetrahedron* **2005**, *61*, 8138.

36 a) M. A. M. Capozzi, C. Cardellicchio, F. Naso, G. Spina, P. Tortorella, *J. Org. Chem.* **2001**, *66*, 5933; b) for a review, see: M. A. M. Capozzi, C. Cardellicchio, F. Naso, *Eur. J. Org. Chem.* **2004**, 1855.

37 G. Y. Cho, H. Okamura, C. Bolm, *J. Org. Chem.* **2005**, *70*, 2346.

38 For an early study of the coordination behavior of sulfoximine–copper complexes, see: M. Zehnder, C. Bolm, S. Schaffner, D. Kaufmann, J. Müller, *Liebigs Ann.* **1995**, 125.

39 For an X-ray crystal structure of a bissulfoximine–copper(I) olefin complex,

see: M. Harmata, S. K. Gosh, C. L. Barnes, *J. Supramol. Chem.* **2002**, *2*, 349.
40 C. Bolm, F. Bienewald, H. Harms, *Synlett* **1996**, 775.
41 C. R. Johnson, C. W. Schroeck, J. R. Shanklin, *J. Am. Chem. Soc.* **1973**, *95*, 7424.
42 C. Bolm, C. P. R. Hackenberger, O. Simic, M. Verrucci, D. Müller, F. Bienewald, *Synthesis* **2002**, 879.
43 a) P. Claus, W. Rieder, P. Hofbauer, E. Vilsmaier, *Tetrahedron* **1975**, *31*, 505; b) S.-L. Huang, D. Swern, *J. Org. Chem.* **1979**, *44*, 2510; c) R. W. Heintzelman, R. B. Bailey, D. Swern, *J. Org. Chem.* **1976**, *41*, 2207; d) C. R. Johnson, K. G. Bis, J. H. Cantillo, N. A. Meanwell, M. F. D. Reinhard, J. R. Zeller, G. P. Vonk, *J. Org. Chem.* **1983**, *48*, 1; e) M. Harmata, *Tetrahedron Lett.* **1989**, *30*, 437; f) M. Harmata, R. J. Claassen II, *Tetrahedron Lett.* **1991**, *32*, 6497.
44 a) A. R. Muci, S. L. Buchwald, *Top. Curr. Chem.* **2002**, *219*, 131; b) J. F. Hartwig, in *Modern Amination Methods* (Ed.: A. Ricci), Wiley-VCH, Weinheim, **2000**, p. 195; c) J. F. Hartwig, in: *Handbook of Organopalladium Chemistry for Organic Synthesis* (Ed.: E. Negishi), Wiley-Interscience, New York, **2002**, p. 1051.
45 a) C. Bolm, J. P. Hildebrand, *Tetrahedron Lett.* **1998**, *39*, 5731; b) C. Bolm, J. P. Hildebrand, *J. Org. Chem.* **2000**, *65*, 169; c) C. Bolm, J. P. Hildebrand, J. Rudolph, *Synthesis* **2000**, 911.
46 J. P. Hildebrand, C. Bolm, unpublished results.
47 F. M. Rivas, U. Riaz, S. T. Diver, *Tetrahedron: Asymmetry* **2000**, *11*, 1703.
48 C. Bolm, O. Simic, *J. Am. Chem. Soc.* **2001**, *123*, 3830.
49 Subsequently, palladium-catalyzed couplings between aryl halides and sulfoximines were also used by Gais, Harmata, and others. For example, see: M. Harmata, S. K. Ghosh, *Org. Lett.* **2001**, *3*, 3321 and citations given in Ref. [14b].
50 a) S. V. Ley, A. W. Thomas, *Angew. Chem.* **2003**, *115*, 5558; *Angew. Chem. Int. Ed.* **2003**, *42*, 5400; b) I. P. Beletskaya, A. V. Cheprakov, *Coord. Chem. Rev.* **2004**, *248*, 2337.
51 C. Bolm, G. Y. Cho, P. Rémy, J. Jansson, C. Moessner, *Org. Lett.* **2004**, *6*, 3293.
52 J. Sedelmeier, C. Bolm, *J. Org. Chem.* **2005**, *70*, 6904.
53 A. Klapars, S. L. Buchwald, *J. Am. Chem. Soc.* **2002**, *124*, 14 844.
54 C. Moessner, C. Bolm, *Org. Lett.* **2005**, *7*, 2667.
55 For a review, see: J. R. Dehli, J. Legros, C. Bolm, *Chem. Commun.* **2005**, 973.
56 J. R. Dehli, C. Bolm, *J. Org. Chem.* **2004**, *69*, 8518.
57 J. R. Dehli, C. Bolm, *Adv. Synth. Catal.* **2005**, *347*, 239.
58 a) H. G. Bonacorso, S. R. T. Bittencourt, R. V. Lourega, A. F. C. Flores, N. Zanatta, M. A. P. Martins, *Synthesis* **2000**, 1431; b) M. Ikeda, H. Tsubouchi, M. Tsunekawa, H. Kondo, Y. Tamura, *Chem. Pharm. Bull.* **1984**, *32*, 3028; c) M. Ikeda, H. Tsubouchi, M. Tsunekawa, H. Kondo, Y. Tamura, *Heterocycles* **1983**, *20*, 2185; c) Y. Tamura, S. M. Bayomi, M. Tsunekawa, M. Ikeda, *Chem. Pharm. Bull.* **1979**, *27*, 2137.
59 C. Bolm, M. Verrucci, O. Simic, C. P. R. Hackenberger, *Adv. Synth. Catal.* **2005**, *347*, 1696.
60 Ethylene-bridged bissulfoximines 55 have also been applied in palladium-catalyzed asymmetric allylic alkylations affording products with up to 98% *ee*; see: C. Bolm, O. Simic, M. Martin, *Synlett* **2001**, 1878.
61 a) M. Johannsen, K. A. Jørgensen, *J. Org. Chem.* **1995**, *60*, 5757; b) M. Johannsen, K. A. Jørgensen, *J. Chem. Soc., Perkin Trans. 2* **1997**, 1183; c) M. Johannsen, K. A. Jørgensen, *Tetrahedron* **1996**, *52*, 7321; d) M. Johannsen, S. Yao, A. Graven, K. A. Jørgensen, *Pure Appl. Chem.* **1998**, *70*, 1117.
62 a) D. A. Evans, T. Rovis, J. S. Johnson, *Pure Appl. Chem.* **1999**, *71*, 1407; b) J. S. Johnson, D. A. Evans, *Acc. Chem. Res.* **2000**, *33*, 325.
63 C. Bolm, M. Martin, O. Simic, M. Verrucci, *Org. Lett.* **2003**, *5*, 427.
64 C. Bolm, M. Martin, G. Gescheidt, C. Palivan, D. Neshchadin, H. Bertagnolli, M. P. Feth, A. Schweiger, G. Mitrikas, J. Harmer, *J. Am. Chem. Soc.* **2003**, *125*, 6222.

65 C. Bolm, M. Verrucci, O. Simic, P. G. Cozzi, G. Raabe, H. Okamura, *Chem. Commun.* **2003**, 2826.

66 Based on this data interpretation it was also possible to suggest a new ligand structure (having an O-*i*Pr substituent instead of the OMe group at the arene), which led to an additional increase in ee.

67 M. Langner, C. Bolm, *Angew. Chem.* **2004**, *116*, 6110; *Angew. Chem. Int. Ed.* **2004**, *43*, 5984.

68 M. Langner, P. Rémy, C. Bolm, *Chem. Eur. J.* **2005**, *11*, 6254.

69 For previous studies on the use of copper catalysts, see: a) D. A. Evans, D. W. C. MacMillan, K. R. Campos, *J. Am. Chem. Soc.* **1997**, *119*, 10859; b) D. A. Evans, M. C. Kozlowski, C. S. Burgey, D. W. C. MacMillan, *J. Am. Chem. Soc.* **1997**, *119*, 7843; c) D. A. Evans, C. S. Burgey, M. C. Kozlowski, S. W. Tregay, *J. Am. Chem. Soc.* **1999**, *121*, 686.

70 a) I. Denissova, L. Barriault, *Tetrahedron* **2003**, *59*, 10105; b) J. Christoffers, A. Mann, *Angew. Chem.* **2001**, *113*, 4725; *Angew. Chem. Int. Ed.* **2001**, *40*, 4591; c) E. J. Corey, A. Guzman-Perez, *Angew. Chem.* **1998**, *110*, 402; *Angew. Chem. Int. Ed.* **1998**, *37*, 388; d) K. Fuji, *Chem. Rev.* **1993**, *93*, 2037; e) S. F. Martin, *Tetrahedron* **1998**, *36*, 419.

71 For a recent review on this topic, see: S. E. Denmark, J. R. Heemstra, Jr., G. L. Beutner, *Angew. Chem.* **2005**, *1137* 4760; *Angew. Chem. Int. Ed.* **2005**, *44*, 4682.

72 P. Rémy, M. Langner, C. Bolm, *Org. Lett.* **2006**, *8*, 1209.

73 M. Langner, P. Rémy, C. Bolm, *Synlett* **2005**, 781.

74 For previous examples, see: a) K. Maruoka, Y. Hoshino, T. Shirasaki, H. Yamamoto, *Tetrahedron Lett.* **1988**, *29*, 3967; b) K. Mikami, M. Terada, T. Nakai, *J. Am. Chem. Soc.* **1990**, *112*, 3949; c) K. Mikami, *Pure Appl. Chem.* **1996**, *68*, 639; d) D. A. Evans, S. W. Tregay, C. S. Burgey, N. A. Paras, T. Vojkovsky, *J. Am. Chem. Soc.* **2000**, *122*, 7936.; e) Y. Yuan, X. Zhang, K. Ding, *Angew. Chem.* **2003**, *115*, 5636; *Angew. Chem. Int. Ed.* **2003**, *42*, 5478; f) H. Guo, X. Wang, K. Ding, *Tetrahedron Lett.* **2004**, *45*, 2009. Reviews: g) K. Mikami, M. Shimizu, *Chem. Rev.* **1992**, *92*, 1021; h) K. Mikami, M. Terada, in *Comprehensive Asymmetric Catalysis*, Vol. III (Eds.: E. N. Jacobsen, A. Pfaltz, H. Yamamoto), Springer, Berlin, **1999**, p. 1143; i) L. C. Dias, *Curr. Org. Chem.* **2000**, *4*, 305.

75 Recently, we found that copper complexes bearing sulfoximines such as 85 were also capable of catalyzing chlorinations of β-keto esters. G. Y. Cho, M. Carill, C. Bolm, to be published.

76 For examples of the use of P,N-ligands in asymmetric catalysis, see: a) D. Xiao, X. Zhang, *Angew. Chem.* **2001**, *113*, 3533; *Angew. Chem. Int. Ed.* **2001**, *40*, 3425; b) Y. Chi, Y.-G. Zhou, X. Zhang, *J. Org. Chem.* **2003**, *68*, 4120; c) P. Schnider, G. Koch, R. Prétôt, G. Wang, F. M. Bohnen, C. Krüger, A. Pfaltz, *Chem. Eur. J.* **1997**, *3*, 887; d) S. Kainz, A. Brinkmann, W. Leitner, A. Pfaltz, *J. Am. Chem. Soc.* **1999**, *121*, 6421; e) F. Menges, A. Pfaltz, *Adv. Synth. Catal.* **2002**, *344*, 40; f) P. G. Cozzi, F. Menges, S. Kaiser, *Synlett* **2003**, 833; g) A. Trifonova, J. S. Diesen, C. J. Chapman, P. G. Andersson, *Org. Lett.* **2004**, *6*, 3825; h) M. Solinas, A. Pfaltz, P. G. Cozzi, W. Leitner, *J. Am. Chem. Soc.* **2004**, *126*, 16 142.

77 For previous applications of phosphino-substituted sulfoximines, which only led to low enantioselectivities, see: a) Ref. [12]; b) J. Müller, Ph.D. Thesis, Universität Basel, **1993**; c) T. C. Kinahan, H. Tye, *Tetrahedron: Asymmetry* **2001**, *12*, 1255.

78 C. Moessner, C. Bolm, *Angew. Chem.* **2005**, *117*, 7736; *Angew. Chem. Int. Ed.* **2005**, *44*, 7564.

79 Reviews on asymmetric imine hydrogenations: a) F. Spindler, H.-U. Blaser, in *Transition Metals for Organic Synthesis*, 2nd ed. (Eds.: M. Beller, C. Bolm), Wiley-VCH, Weinheim, **2004**; Vol. 2, p. 113; b) H.-U. Blaser, F. Spindler, in *Comprehensive Asymmetric Catalysis* (Eds.: E. N. Jacobsen, A. Pfaltz, H. Yamamoto); Springer, Berlin, **1999**; Vol. 1, p. 247; c) H.-U. Blaser, C. Malan, B. Pugin, F. Spindler, H. Steiner, M. Studer, *Adv. Synth. Catal.* **2003**, *345*, 103;

d) M. J. Palmer, M. Wills, *Tetrahedron: Asymmetry* **1999**, *10*, 2045.
80 a) F. Spindler, H.-U. Blaser, *Adv. Synth. Catal.* **2001**, *343*, 68; b) H.-U. Blaser, H.-P. Buser, R. Häusel, H.-P. Jalett, F. Spindler, *J. Organomet. Chem.* **2001**, *621*, 34; c) H.-U. Blaser, H.-P. Buser, H.-P. Jalett, B. Pugin, F. Spindler, *Synlett* **1999**, 867; d) H.-U: Blaser, H.-P. Buser, K. Coers, R. Hanreich, H.-P. Jalett, E. Jelsch, B. Pugin, H. D. Schneider, F. Spindler, A. Wegmann, *Chimia* **1999**, *53*, 275; e) H.-U. Blaser, F. Spindler, in *Comprehensive Asymmetric Catalysis* (Eds.: E. N. Jacobsen, A. Pfaltz, H. Yamamoto), Springer, Berlin **1999**, Vol. 3, p. 1427.
81 M. T. Reetz, O. G. Bondarev, H.-J. Gais, C. Bolm, *Tetrahedron Lett.* **2005**, *46*, 5643.
82 For examples of asymmetric catalyses with BINOL-derived monodentate phosphites, phosphonites, and phosphoramidites, see: a) M. T. Reetz, G. Mehler, *Angew. Chem.* **2000**, *112*, 4047; *Angew. Chem. Int. Ed.* **2000**, *39*, 3889; b) M. T. Reetz, T. Sell, *Tetrahedron Lett.* **2000**, *41*, 6333; c) C. Claver, E. Fernandez, A. Gillon, K. Heslop, D. J. Hyett, A. Martorell, A. G. Orpen, P. G. Pringle, *Chem. Commun.* **2000**, 961; d) M. van den Berg, A. J. Minnaard, E. P. Schudde, J. van Esch, A. H. M. de Vries, J. G. de Vries, B. L. Feringa, *J. Am. Chem. Soc.* **2000**, *122*, 11 539.
83 For a recent application of sulfoximines in material sciences, see: P. Kirsch, M. Lenges, D. Kühne, K.-P. Wanczek, *Eur. J. Org. Chem.* **2005**, 797.
84 A. K. Bolm, to be published.

2.1.2
Catalyzed Asymmetric Aryl Transfer Reactions
Carsten Bolm

2.1.2.1 Introduction

In early 1990 we began to work on catalyzed carboligations using organozinc reagents [1]. The reactions commonly involved diethylzinc **2**, which was added enantioselectively to aldehydes **1** affording secondary alcohols **3**. Initially, bipyridine **4** [2–6], pyridine **5** [3], and sulfoximines **6** [7] were applied as catalysts, which in some cases led to enantioselectivities greater than 95% *ee* [8, 9].

In the late 1990s the focus of our work changed in two respects. First, we started to use planar-chiral ferrocene **9** as a catalyst [10, 11], and second, instead of applying the well-investigated di*alkyl*zinc reagents [12], we began to explore reac-

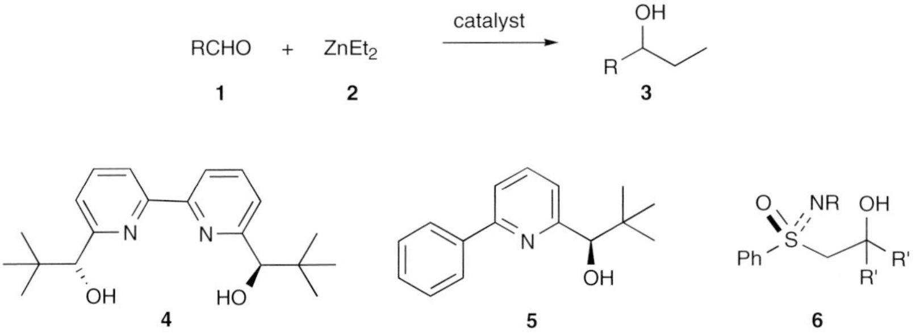

Scheme 2.1.2.1

tions with di*aryl*zincs such as **7** (Scheme 2.1.2.2) [13, 14]. The latter organometallics are the more reactive of the two species when involved in addition reactions with carbonyl compounds, which makes their control in asymmetric catalysis more challenging.

2.1.2.2 Catalyst Design

The development of ferrocene **9** was part of our studies on planar-chiral compounds, which also involved the synthesis of other scaffolds such as chromium-tricarbonyl arenes [15], sulfoximidoyl ferrocenes [16], and [2.2]paracyclophanes [17]. In aryl transfer reactions, however, ferrocene **9** proved to be the best catalyst in this series, and it is still used extensively today.

The synthesis of ferrocene **9** relied on chemistry introduced by Sammakia, Uemura, and Richards [18]. They had shown that 2-ferrocenyl oxazoline **10** derived from *t*-leucine could be selectively deprotonated and trapped with electrophiles to afford *ortho*-functionalized planar-chiral products **11** with excellent diastereoselectivities (Scheme 2.1.2.3). Following this strategy, **9** became accessible in a highly straightforward manner by trapping the lithiated intermediate derived from **10** with benzophenone [10, 11].

With other electrophiles, ferrocenes **12** and **13** could be obtained, bearing a selenium group [19] or a silanol moiety [20], respectively, in the *ortho* position. Those compounds proved to be catalytically active as well, and in particular **13** was of interest, since – to the best of our knowledge – it was the first silanol ever used as a chiral ligand in asymmetric catalysis. Details of this study will be discussed below.

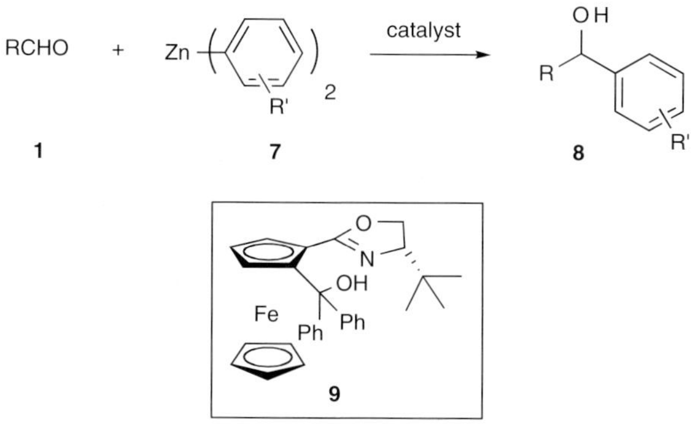

Scheme 2.1.2.2

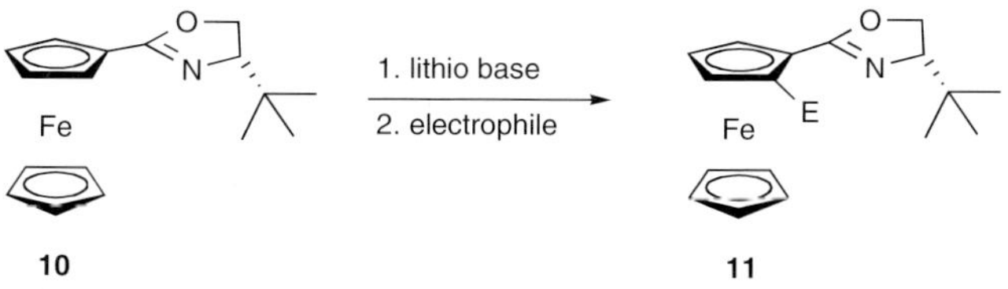

Scheme 2.1.2.3

12

13

The same synthetic strategy as in the synthesis of planar-chiral ferrocenes was applied to the preparation of rheniumtricarbonyl **14**, which has also been studied as a catalyst in aryl transfer reactions [21]. Subsequently, this chemistry has been extended, and various catalytic applications of cyrhetrenes **15**, **16** (AAPhos), and related derivatives have recently been demonstrated [22].

14

15

16 (AaPhos)

Ortho-directed metallations also allowed the synthesis of ferrocene-based hydroxamic acids such as **17** and **18** [23] as well as the preparation of planar-chiral carbenes **19** and **21** (which were trapped as chromium and rhodium complexes, **20** and **22**, respectively) [24]. In this context it is noteworthy that **19** was the first carbene with planar chirality ever reported.

17
a: R = 1-adamantyl
b: R = Bn

18

19

20

21

22

Ferrocenes of type **11** (as well as cyrhetrenes such as **14**) are characterized as having two elements of chirality: a stereogenic center at the oxazoline ring and a plane of chirality due to the two *ortho* substituents on the ferrocene core.

Consequently, matched/mismatched cases [25] can result, and indeed our investigations on cooperative effects of stereogenic elements in such systems revealed **9** be the matched case and **23** (which is also easily prepared by following a directed deprotonation–silylation–deprotonation–trapping–desilylation sequence [11]) to be the mismatched case in diethylzinc additions to aldehydes [26]. Later, these investigations were extended to more complex systems such as **24** [27], but ferrocene **9** still remains superior to all other compounds.

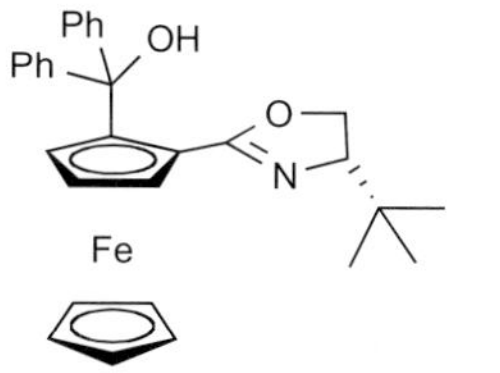

23

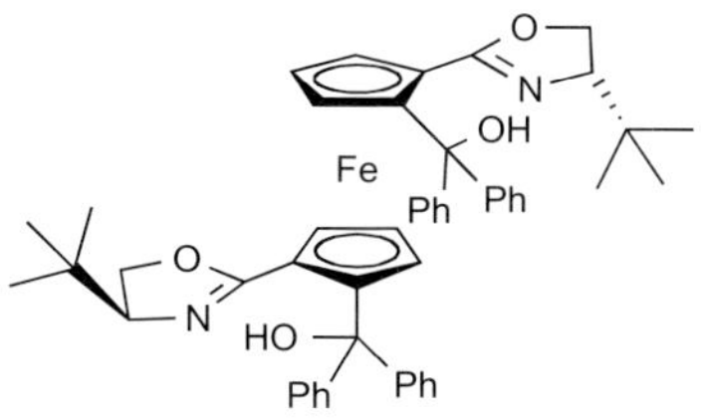

24

A study of the catalytic use of diastereomeric mixtures of catalysts in additions of dimethylzinc to benzaldehyde, in which combinations of ferrocenes **9** and **23** were applied, showed a strong kinetic preference for **9** over **23**. Thus, even with **23** in large excess, the stereochemical result was mainly determined by ferrocene **9** [28].

Later, the oxazolines **25** were examined to study the effects of matched/mismatched combinations of stereogenic centers on catalyzed aryl transfer reactions to aldehydes. Of these mandelic acid-derived catalysts, **25b** gave the best results in terms of enantioselectivity (up to 35% *ee*), while diastereomer (*R*,*S*)-**25b** proved to be superior to (*S*,*S*)-**25b** with respect to catalyst activity [29]. With both compounds, the absolute configuration of the product was determined by the oxazoline moiety.

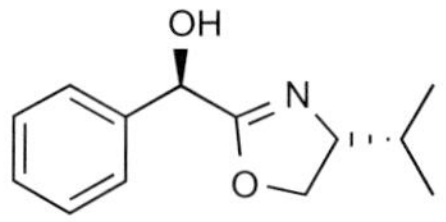

(*R*,*R*)-**25a** 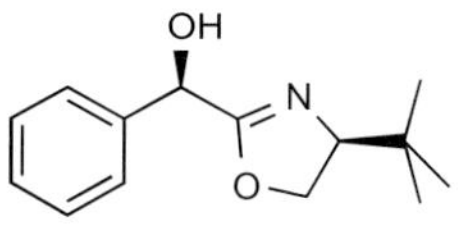   (*R*,*S*)-**25b** 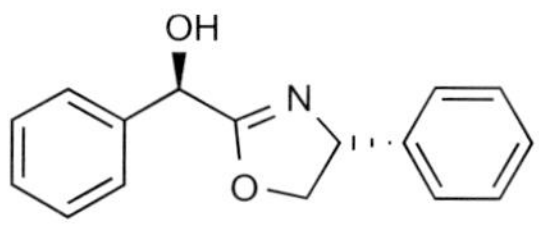   (*R*,*R*)-**25c**

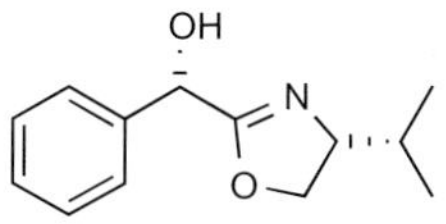

(*S*,*R*)-**25a** 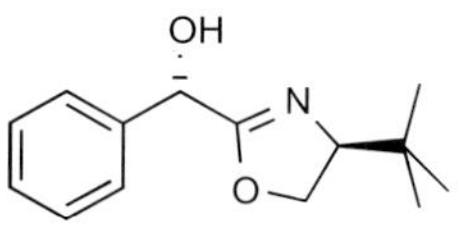   (*S*,*S*)-**25b** 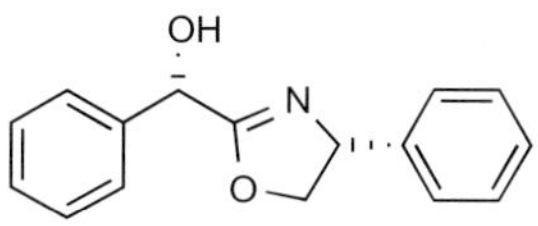   (*S*,*R*)-**25c**

The use of hydroxy oxazolines **26** in catalytic aryl transfer reactions was investigated subsequently (specifically, addition of BPh$_3$ to *p*-methoxybenzaldehyde in the presence of dimethyl poly(ethylene glycol) (DiMPEG); for details of this protocol, vide infra). The highest enantioselectivity (81% *ee*) was achieved with (*S*)-**26a**, which has a *t*-butyl substituent on the oxazoline group and two 2-methylphenyl substituents on the hydroxyl-bearing carbon [30a].

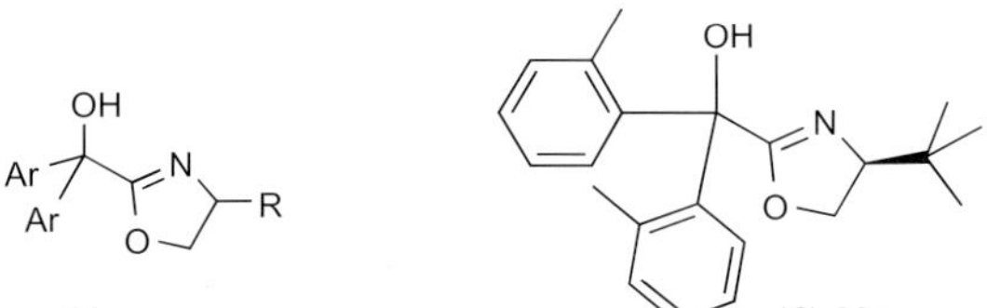

The use of amino alcohol derivatives with a cyclohexyl backbone was reported recently [30b].

2.1.2.3 Catalyzed Aryl Transfer Reactions

The synthesis of optically active diaryl methanols **27** is of particular synthetic value, since their core structure represents a molecular scaffold which is relevant to numerous biologically active compounds and pharmaceuticals possessing antihistaminic, anticholinergic, local-anesthetic, and laxative properties. Examples include neobenodine **28**, orphenadrine **29**, and carbinoxamine **30** [31].

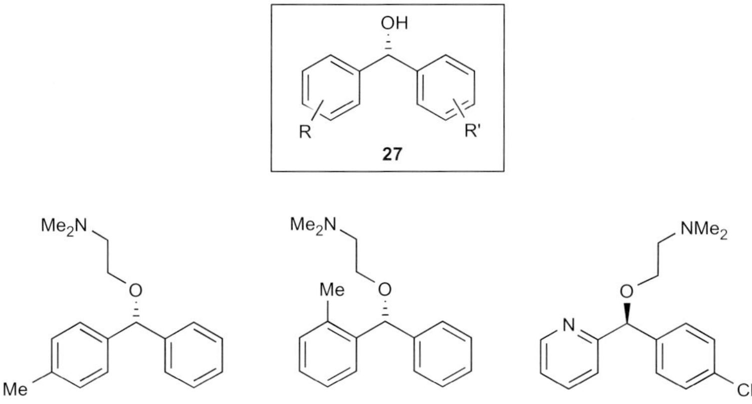

Furthermore, it was recently shown, that enantiopure diaryl methanols can undergo stereospecific S_N2-type substitution reactions with enolates at the (di)benzylic carbon affording 1,1-diarylalkyl derivatives (**31** → **32** → **33**; Scheme 2.1.2.4). This discovery widened the preparative applicability of diaryl methanols significantly, and was used by Merck in the synthesis of selective PDE-IV inhibitors **34** and **35** [32].

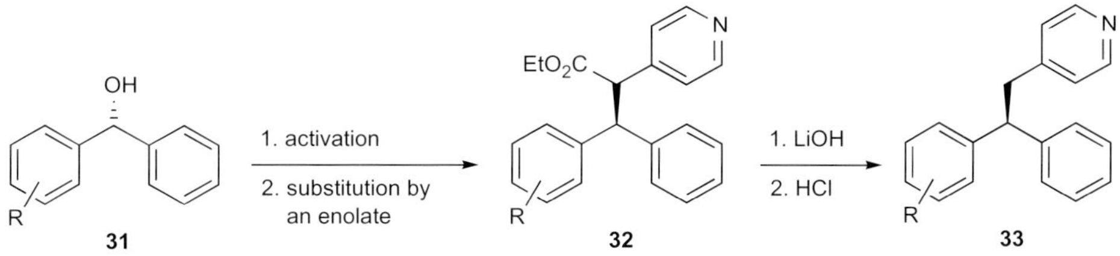

Scheme 2.1.2.4

Two strategies for the synthesis of enantiomerically enriched diaryl methanols **27** are apparent: first, asymmetric reductions of the corresponding diaryl ketones **36** [33], and, second, enantioselective aryl transfer reactions to the respective benzaldehyde derivatives **37** (Scheme 2.1.2.5) [34, 35].

Before our work, both approaches had already been investigated, but severe limitations were encountered. For example, to achieve high enantioselectivities in asymmetric reductions following the protocols developed by Corey and Noyori [33], the two aryl groups of **36** have to be very different in terms of their electronic or steric nature. In the alternative approach toward **27** starting from benzaldehyde **37**, only moderate to good enantioselectivities had been observed. In this case, the most significant work originated from Soai, Fu, and Pu and involved the use of various phenylzinc reagents [34].

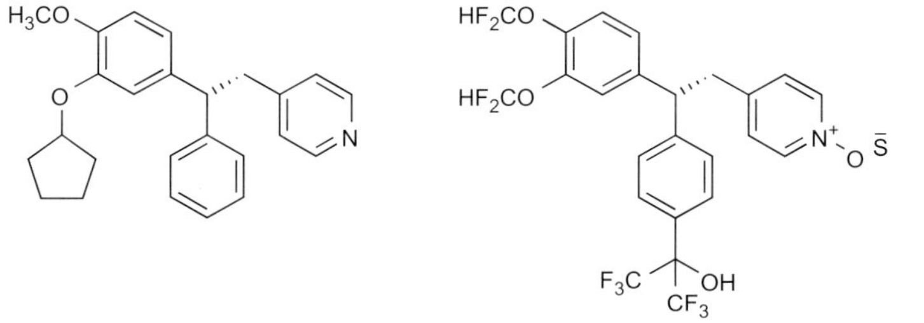

34 (CDP-840)

35

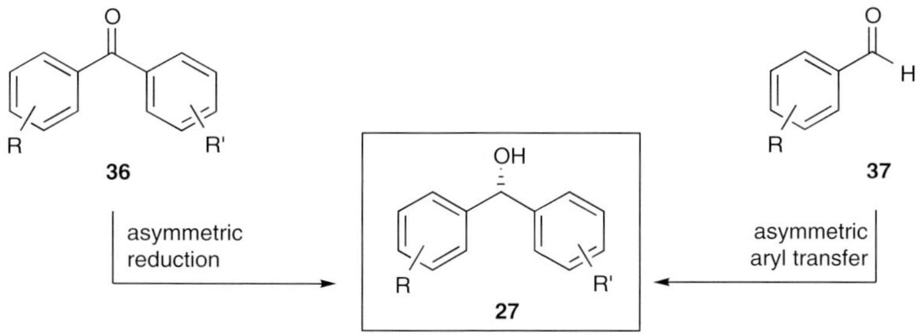

Scheme 2.1.2.5

Our first results in this field were published in 1999 [13], when we reported that ferrocene **9** was capable of catalyzing the enantioselective addition of diphenylzinc to aromatic aldehydes, affording diaryl methanols with good to high enantioselectivities. For example, from the reaction between *p*-chlorobenzaldehyde **37a** and diphenylzinc **7** in the presence of 5 mol% ferrocene **9** (at 0 °C), product **27a** was obtained with 82% *ee* (Scheme 2.1.2.6).

Increasing the catalyst loading to 10 mol% and performing the reaction at −20 °C raised the *ee* of **27a** to 90%. Of particular note is the observation that ferrocene carbaldehyde gave the best enantioselectivity in such addition reactions (≥96% *ee*).

An important issue was the fact that isolated diphenylzinc had to be used in this protocol and that samples of this arylzinc reagent prepared in situ (and presumably still containing lithium or magnesium salts) led to products with lower enantioselectivities. Thus, although pure diphenylzinc was commercially available, its high price was expected to hamper large-scale applications of this catalyzed asymmetric phenyl transfer process [36, 37].

A major breakthrough was achieved in 2000 [38]. In an attempt to render the aryl transfer from diphenylzinc more selective by modifying the nature of the aryl source, we tested the applicability of mixtures of zinc reagents. For example, diethylzinc **2** was added to diphenylzinc **7**, and various ratios of these organometallics were then tested in the phenyl transfer reaction to benzaldehydes (Scheme 2.1.2.7). Conceptionally, we hoped to achieve a better control of the asymmetric catalysis, either by making the catalytic process more efficient or by suppressing the uncatalyzed (background) reaction, which reduced the enantiomeric excess of the product by concurrent formation of the racemate. To our delight this

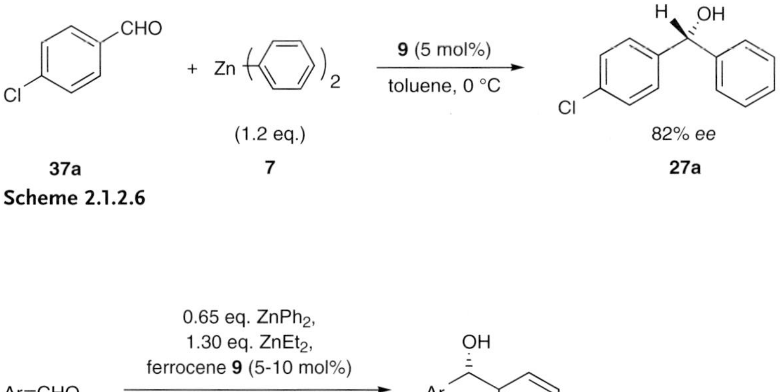

Scheme 2.1.2.6

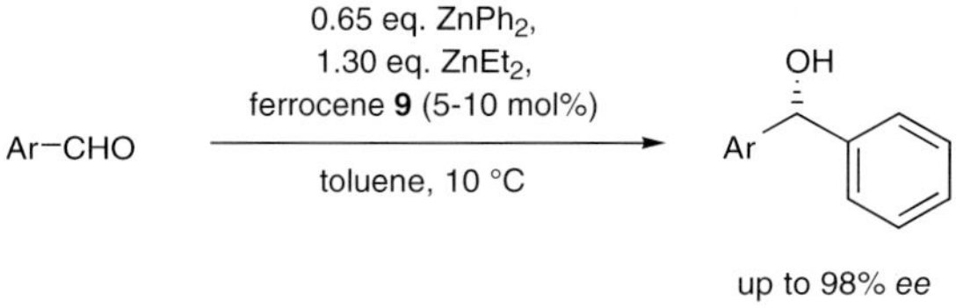

Scheme 2.1.2.7

hypothesis was confirmed and we found that the use of such modified (mixed) zinc reagents led to significantly improved enantioselectivities (of up to 98% *ee*) [38–40].

It is important to mention that the catalysis with this modified arylzinc reagent not only leads to improved enantioselectivity (at high product yield), but also that in this process *sub*stoichiometric quantities of diphenylzinc could be applied. This also meant that now both of the phenyl groups could be activated and transferred to the aldehydes. A reaction profile obtained by FT-IR studies revealed that the modification of the zinc reagent had a significant effect on its reactivity [41].

Representative results are shown in Fig. 2.1.2.1. Both reactions with mixtures of diphenylzinc and diethylzinc and those with pure diphenylzinc are catalyzed by ferrocene **9** as indicated by the comparisons between curves A and C as well as B and D, respectively. Apparently the reaction becomes slower when the mixture of the zinc reagents is applied (curve A versus curve B). We presume that this effect is – as hoped – due to a less pronounced background reaction as well as to a modification of the aryl source (potentially PhZnEt or complexes thereof). Consequently, a better control of the enantioselectivity at the expense of the reaction rate is observed.

Use of ferrocene **9** in combination with the mixed zinc reagent ($ZnPh_2/ZnEt_2$ = 1 : 2) allowed a wide range of diarylmethanols **27** to be prepared in a highly enantioselective manner [38]. A selection of products is shown in Scheme 2.1.2.8.

With the aim of further improving the catalyst system, other metallocenes were tested as catalysts. Neither diselenide **12** [19] nor doubly substituted **24** [27] led to better results. More interesting was the use of ferrocene **13a** [20] since – to the

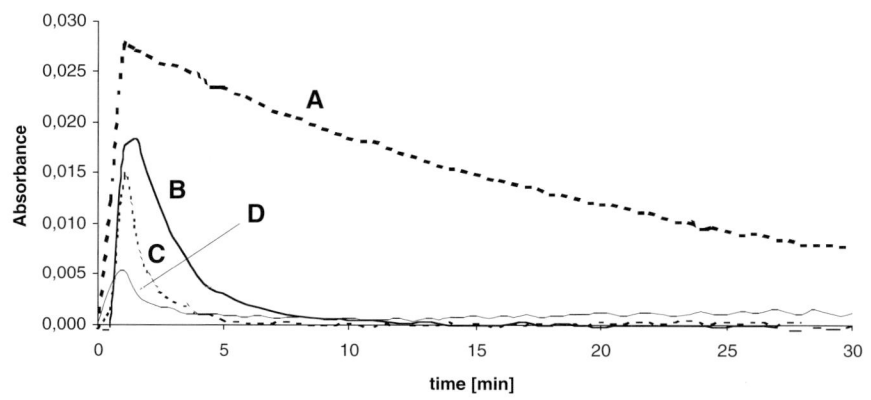

Fig. 2.1.2.1 Reaction profile of the phenylzinc addition to *p*-chlorobenzaldehyde **37a**, as determined by FT-IR. Curve A: $ZnPh_2$ (0.65 equiv), $ZnEt_2$ (1.3 equiv), toluene, rt. Curve B: $ZnPh_2$ (1.50 equiv), toluene, rt. Curve C: $ZnPh_2$ (0.65 equiv), $ZnEt_2$ (1.3 equiv), ferrocene **9** (10 mol%), toluene, rt. Curve D: $ZnPh_2$ (1.50 equiv), ferrocene **9** (10 mol%), toluene, rt. All reactions were carried out with 0.25 mmol *p*-chlorobenzaldehyde **37a** in 3 mL toluene.

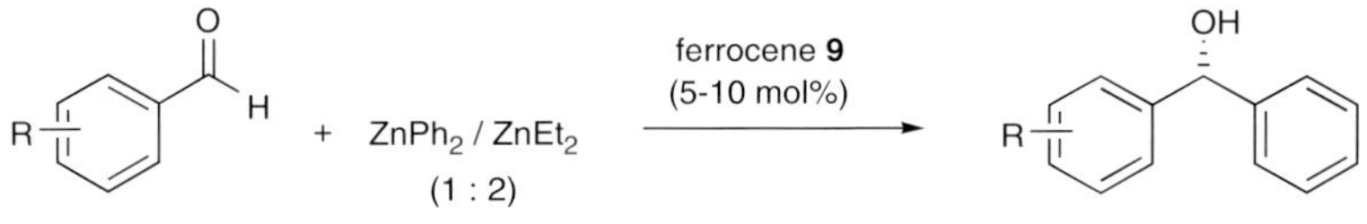

Scheme 2.1.2.8

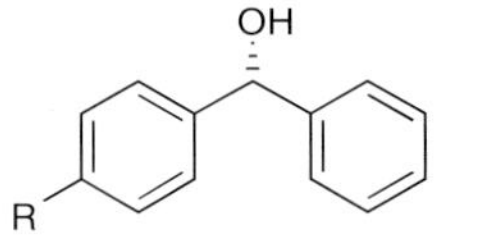

R	ee (%)
MeO	98
Me	98
Cl	97

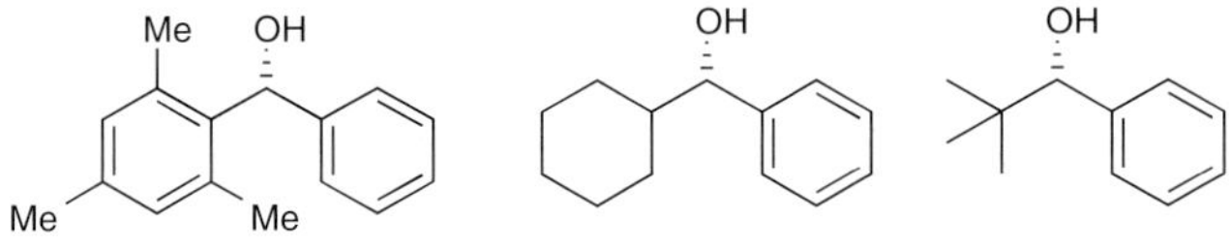

92% ee   86% ee   94% ee

best of our knowledge – it was the first chiral silanol to be applied in an asymmetric catalysis [42]. With 10 mol% of this catalyst the phenyl transfer (from the 1:2 mixture of ZnPh$_2$ and ZnEt$_2$) onto aldehyde **37a** afforded **27a** smoothly in 91% ee with 82% yield [20, 43].

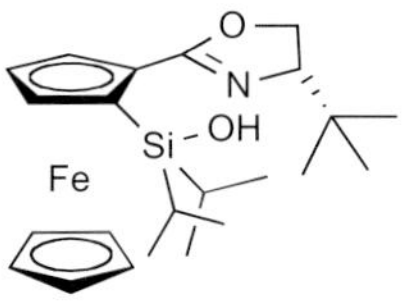

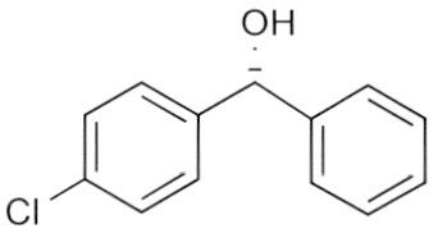

The most significant effect on the aryl transfer was observed when the metallocene backbone was changed [21]. Both ee and catalyst turnover were affected, and even with a reduced catalyst loading excellent enantioselectivities were observed in some cases. Three examples will illustrate details of the effects observed using the most effective catalyst, which was obtained from rheniumtricarbonyl **14**. In the formation of **27a**, use of 10 mol% cyrhetrene **14** afforded the product with 98% ee. Compared to the result obtained with ferrocene **9** (97% ee) under identical conditions, the ee value achieved with **14** revealed a slight improvement in the enantioselectivity. Remarkably (and in contrast to a catalysis with ferrocene **9**) almost the same ee was obtained in an experiment with only 2 mol% **14** (formation of **27a** with 96% ee). In the addition to 2-furylcarbaldehyde neither the change from cyrhetrene **14** to ferrocene **9** nor the reduction of the catalyst loading (of **14**) had an effect on the enantioselectivity of the aryl transfer (95% *ee* in all cases). In the formation of sterically hindered **27c**, cyrhetrene **14** showed a much better result than ferrocene **9** (98% versus 92% *ee*, respectively) but unfortunately, in this case, use of 2 mol% **14** led to the product with a significantly lower *ee* (80%) [21].

	27a % ee	27b % ee	27c % ee
10 mol% of cyrhetrene **14**	98	95	98
10 mol% of ferrocene **9**	97	95	92
2 mol% of cyrhetrene **14**	96	95	80

In an alternative approach, polymer-supported **38** (attached to trityl chloride resin) and MPEG-bound **39** were prepared and tested in the aryl transfer [44]. The intention of this study was to improve the efficiency of the system by recycling and reuse of the catalyst [45–47]. To our surprise we found that heterogeneous catalyst **38** performed well in terms of product yield, but unfortunately the diarylmethanols were racemic. In contrast, MPEG-supported ferrocene **39** gave very positive results (a quantitative yield and up to 97% *ee* in the formation of **27a**), and the recovery of the ferrocene (with an MPEG tether having a molecular weight of 5000) was facile. Furthermore, the excellent enantioselectivity was retained throughout successive addition reactions, so that even after consecutive use of **39** in five catalytic cycles, **27a** was obtained with an excellent *ee* of 95% [44].

Although in the studies summarized so far excellent yields and enantioselectivities could be achieved in phenyl transfer reactions to aldehydes (affording diarylmethanols), large-scale applications of such processes were hampered by the high cost of the aryl source, diphenylzinc. In subsequent investigations we therefore focused our attention on the search for new aryl transfer agents. One of them was triphenylborane, which is quite inexpensive compared to diphenylzinc and (commercially) available in large quantities. To our delight we found that in many cases similar or even better results (yields and *ee* values) were obtained compared to the original system and that the use of triphenylborane also led to a broad range of secondary alcohols with high enantioselectivities in good to excellent yields [48]. Furthermore, remarkable enantioselectivities were achieved in reactions with aliphatic aldehydes. Applying a modified workup (extraction with dilute acetic acid) allowed the removal of residual boron compounds from the crude product and avoided any tedious chromatographical separation of **27** from those by-products. Some results obtained in gram-scale syntheses are shown in Scheme 2.1.2.9. A slight decrease in *ee* from 98% (on the original 0.25 mmol scale) to 95% *ee* (on a gram scale) had to be accepted in the synthesis of **27d**. However, *ortho*-substituted diarylmethanol **27e**, which had been difficult to prepare in good yield before, was now obtained in 97% yield [48].

Another limitation of the protocols described so far related to the aryl moiety itself. Until this stage only phenyl groups had been transferred onto (aromatic) aldehydes, which significantly restricted the substrate scope. In order to expand the applicability of the process, alternative aryl sources had to be found. This problem was solved in 2002, when we discovered that simple boronic acids **40** could also be applied in the transmetallation process [49]. For the first time the catalyzed asymmetric aryl transfer allowed a high degree of flexibility in the substitution pattern of the substrates. A wide variety of aryl boronic acids is commercially available; alternatively, specific derivatives can easily be synthesized, selected examples of which (with benzaldehyde **37b** as the aryl acceptor) are shown in Scheme 2.1.2.10.

Since the starting materials could be freely chosen, another interesting aspect arose. It was possible to use a single catalyst for the preparation of both enantiomers, as illustrated in Scheme 2.1.2.11 for the synthesis of (*R*)-**27a** and (*S*)-**27a** using ferrocene **9** as the catalyst [49].

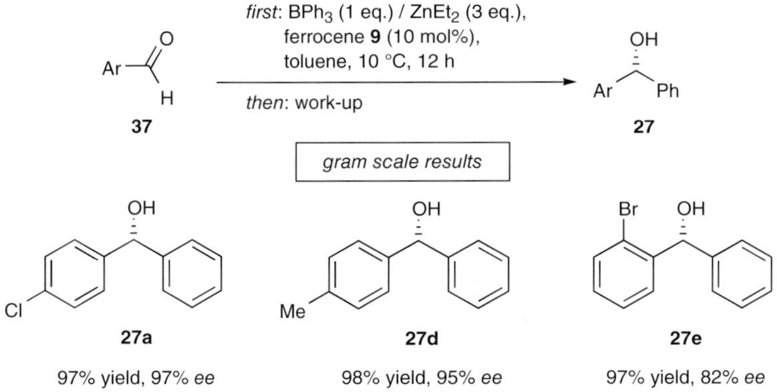

Scheme 2.1.2.9

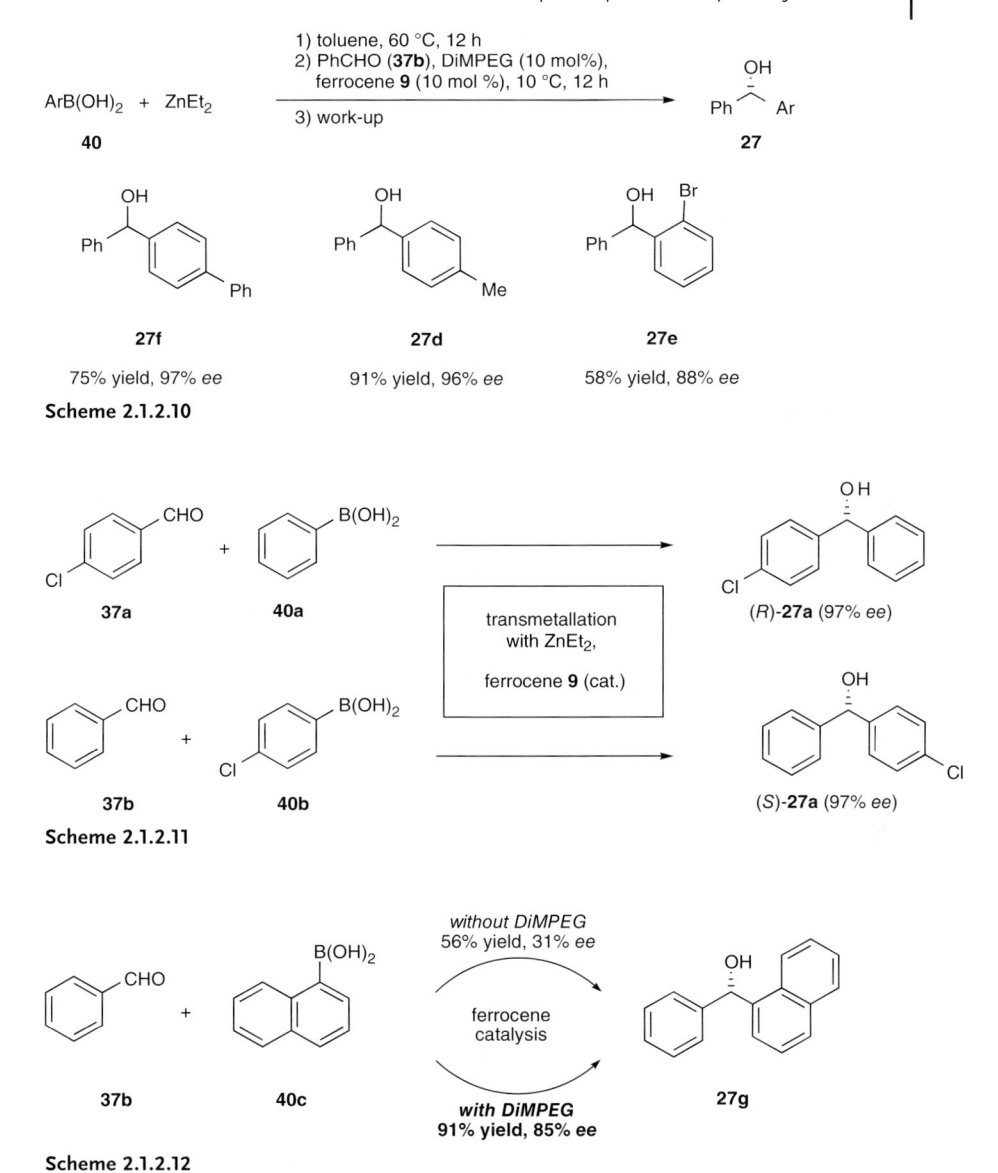

2.1.2 Catalyzed Asymmetric Aryl Transfer Reactions

ArB(OH)$_2$ + ZnEt$_2$
40

1) toluene, 60 °C, 12 h
2) PhCHO (**37b**), DiMPEG (10 mol%), ferrocene **9** (10 mol %), 10 °C, 12 h
3) work-up

→ Ph–CH(OH)–Ar
27

27f
75% yield, 97% ee
(Ph–CH(OH)–C$_6$H$_4$–Ph)

27d
91% yield, 96% ee
(Ph–CH(OH)–C$_6$H$_4$–Me)

27e
58% yield, 88% ee
(Ph–CH(OH)–C$_6$H$_4$–Br, ortho)

Scheme 2.1.2.10

4-Cl-C$_6$H$_4$-CHO (**37a**) + PhB(OH)$_2$ (**40a**) → transmetallation with ZnEt$_2$, ferrocene **9** (cat.) → (R)-**27a** (97% ee)

PhCHO (**37b**) + 4-Cl-C$_6$H$_4$-B(OH)$_2$ (**40b**) → (S)-**27a** (97% ee)

Scheme 2.1.2.11

PhCHO (**37b**) + 1-naphthyl-B(OH)$_2$ (**40c**) → ferrocene catalysis → **27g**

without DiMPEG: 56% yield, 31% ee
with DiMPEG: 91% yield, 85% ee

Scheme 2.1.2.12

The success of the asymmetric aryl transfer from boronic acids also relied on another important phenomenon. The presence of catalytic amounts (10 mol%) of DiMPEG (MW = 2000) increased the enantioselectivity of the process significantly [49]. For example, without DiMPEG the reaction between benzaldehyde **37b** and 1-naphthylboronic acid **40c** gave the corresponding diarylmethanol **27g** with 31% ee in 56% yield, whereas in the presence of the polyether, **27g** was obtained in 85% ee and 91% yield (Scheme 2.1.2.12) [49].

Although the precise reason for the "MPEG-effect" [50] is still unknown, we assume that catalytically active achiral metal species are inactivated by trapping with the polyether and that the catalyzed asymmetric reaction path thereby becomes more dominant. This phenomenon had already been observed in aryl transfer reactions catalyzed by MPEG-bound ferrocene **39** [44], and it can be used to perform catalyses with lower catalyst loadings [50].

This protocol has attracted much attention recently, and several groups (those of Chan, Katsuki, Braga, and Zhao) have applied their catalysts to aryl transfer reactions from boronic acids (or derivatives thereof) [51].

In collaboration with Dahmen and Lormann from cynora GmbH, we then developed an automated, high-throughput, screening approach for the determination of additive effects in organozinc addition reactions to aldehydes [52, 53]. With a catalyst derived from N,N-dibutylnorephedrine (dbne, **41**) an improvement of 20% ee over the catalyzed reaction ($BPh_3/ZnEt_2$ variant) in the absence of an additive was observed in phenyl addition to 2-bromobenzaldehyde **37c**. Both the type of additive as well as its amount affected the catalysis [54]. As was found previously [49, 50], polyethyleneglycol derivatives improved the enantioselectivity, and DiMPEG 2000 was the best additive. Imidazole also had a strong effect, and it reversed the absolute configuration of the product [52, 55].

Briefly, an alternative route toward optically active diarylmethanols relying on rhodium catalysis and with arylboronic acids as aryl sources was investigated [56, 57]. The basis for this approach stemmed from work by Miyaura, who achieved 41% ee in the formation of **27g** using MeO–MOP as a ligand [58, 59]. Later, Fürstner demonstrated the applicability of (achiral) carbenes in this reaction [60]. In our study, planar-chiral imidazolium salts were used as precursors for N-heterocyclic carbene (NHC) ligands [61, 62], and with the pseudo-*ortho*-disubstituted [2.2]paracyclophane (S,R_p)-**42** bearing a phosphinoyl substituent on the paracyclophane backbone [63], an ee of 38% was achieved in the formation of diarylmethanol **27g** (Scheme 2.1.2.13) [56].

Obviously, this rhodium-catalyzed diarylmethanol formation with boronic acids still requires significant optimization in order to be competitive with the forementioned existing methodology using phenylzinc reagents, but these early results appear particularly promising.

With the aim of improving understanding of the mechanism of the ferrocene-catalyzed aryl transfer reactions from organozincs to aldehydes, DFT calculations were performed in close collaboration with Norrby [64]. As a result, the experimentally observed higher reactivity of the phenyl transfer to aldehydes compared

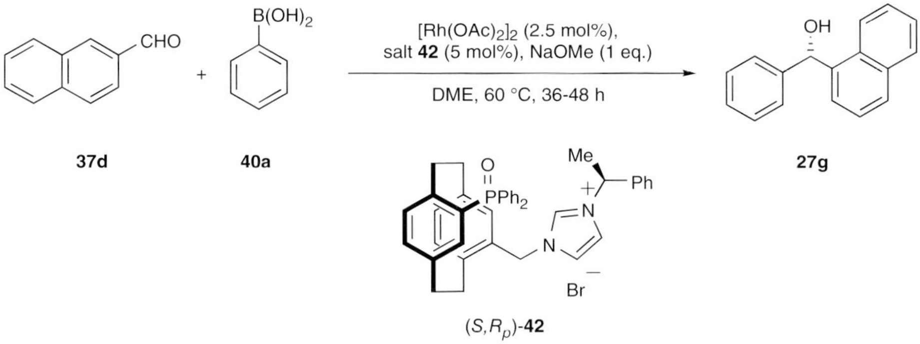

2.1.2 Catalyzed Asymmetric Aryl Transfer Reactions | 189

Scheme 2.1.2.13

to the analogous alkyl transfer could be rationalized. In the former case the π system of the phenyl group allows a simultaneous overlap with both the zinc and the reacting carbonyl, which leads to a substantially lower energy of the transition state compared to that for the alkyl group transfer. Subsequently, various transition states of the phenyl transfer were calculated, taking into account the presence of a number of catalyst/aldehyde adducts with varying combinations of phenyl and ethyl groups at the zinc atoms. This led to a concise picture, which explained theoretically the increased enantioselectivity observed for the mixed zinc species [64].

Based on the successful development of the catalytic formation of diarylmethanols by arylzinc additions to aldehydes, we also considered asymmetric aryl-to-imine transfer reactions, which would lead to (protected) diarylmethylamines **44** (Scheme 2.1.2.14). In the light of such compounds being important as intermediates in syntheses of biologically active products [65] and considering the fact that at the outset of our investigation no efficient asymmetric catalysis toward **44** was known [66], the search for such a process appeared particularly attractive. Furthermore it was realized that solving this problem could also lead to an alternative synthesis of enantiopure cetirizine hydrochloride (**45**), which is a commercially important, nonsedative antihistamine agent [67, 68].

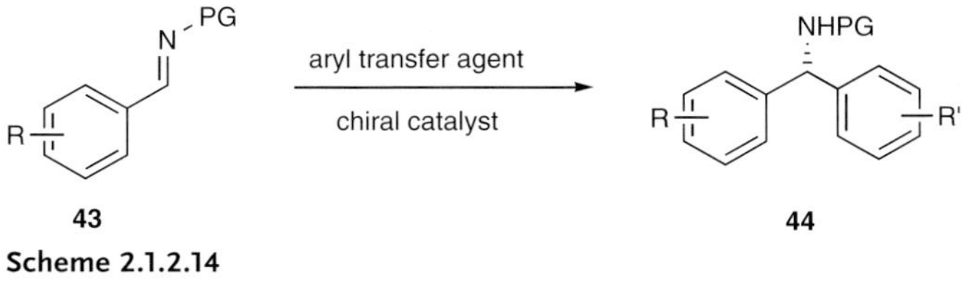

Scheme 2.1.2.14

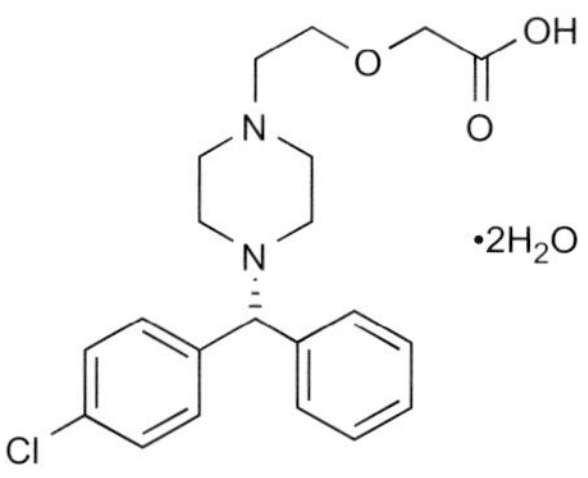

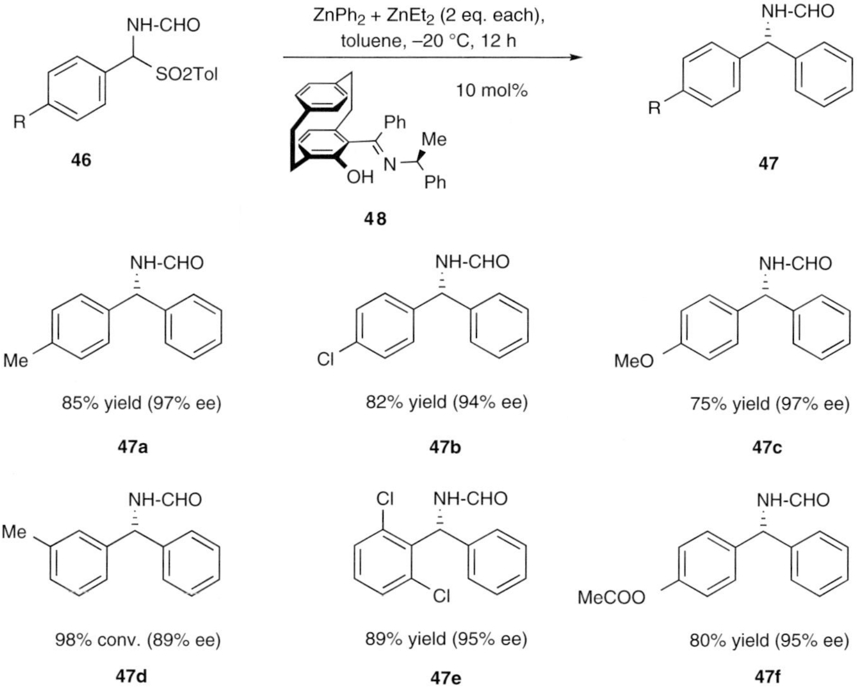

Scheme 2.1.2.15

After screening several compounds, including ferrocene **9** or cyrhetrene **14**, which had previously been applied in the aryl transfer reactions to aldehydes and led to excellent enantioselectivities, we found, in collaboration with Bräse, that only [2.2]paracyclophane **48** was capable of forming a catalyst system (Scheme 2.1.2.15). Subsequently, the substrate scope was evaluated, and several diarylmethylamines **47** were obtained with excellent enantioselectivities (up to 97% *ee*) [69, 70].

For the synthesis of cetirizine hydrochloride **45**, diarylmethylamine **47b** was of particular importance, since deprotection under acidic conditions led to **49**, which is a known intermediate for the synthesis of the desired antihistamine agent. As

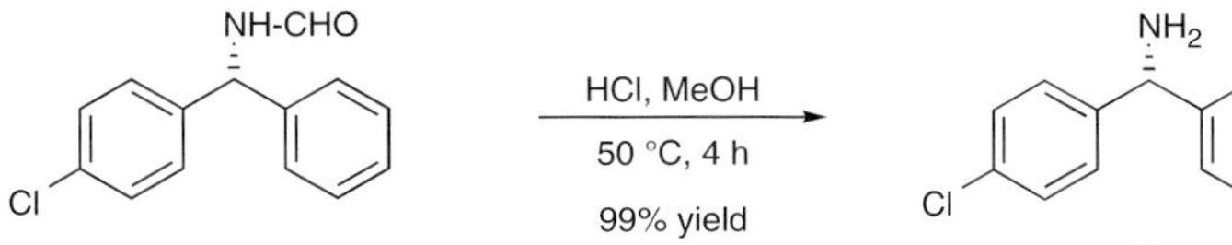

Scheme 2.1.2.16

shown in Scheme 2.1.2.16, the deformylation proceeded in excellent yield and without any detectable epimerization at the stereogenic center [69].

So far, attempts to expand the substrate scope further by modifying the aryl transfer agent have remained unsuccessful. Thus, imine addition reactions with arylzinc species other than the one prepared in situ by mixing diphenylzinc and diethylzinc still deserve attention, and will be developed in the near future.

References

1 C. Bolm, M. Zehnder, D. Bur, *Angew. Chem.* **1990**, *102*, 206; *Angew. Chem. Int. Ed. Engl.* **1990**, *29*, 205.

2 Review: C. Bolm, in *Advances in Organic Synthesis via Organometallics* (Eds.: R. W. Hoffmann, K. H. Dötz), Vieweg Verlag, Wiesbaden, **1991**, p. 223.

3 C. Bolm, M. Ewald, M. Felder, G. Schlingloff, *Chem. Ber.* **1992**, *125*, 1169; b) C. Bolm, G. Schlingloff, K. Harms, *Chem. Ber.* **1992**, *125*, 1191; c) for a hyperbranched version of pyridine **5** and its use in catalysis, see: C. Bolm, N. Derrien, A. Seger, *Synlett* **1996**, 386.

4 For a nickel-catalyzed conjugate addition, see: a) C. Bolm, M. Ewald, *Tetrahedron Lett.* **1990**, *31*, 5011; b) C. Bolm, *Tetrahedron: Asymmetry* **1991**, *2*, 701; c) C. Bolm, M. Ewald, M. Felder, *Chem. Ber.* **1992**, *125*, 1205, 1781.

5 For the use of such bipyridines in other applications, see: a) supramolecular chemistry: W. Zarges, J. Hall, J.-M. Lehn, C. Bolm, *Helv. Chim. Acta* **1991**, *74*, 1843; asymmetric catalysis: b) S. Ishikawa, T. Hamada, K. Manabe, S. Kobayashi, *J. Am. Chem. Soc.* **2004**, *126*, 12 236; c) S. Azoulay, K. Manabe, S. Kobayashi, *Org. Lett.* **2005**, *7*, 4593; d) S. Kobayashi, T. Ogina, H. Shimizu, S. Ishikawa, T. Hamada, K. Manabe, *Org. Lett.* **2005**, *7*, 4729; e) C. Schneider, A. R. Shreekanth, E. Mai, *Angew. Chem.* **2004**, *116*, 5809; *Angew. Chem. Int. Ed.* **2004**, *43*, 5691; crown ethers: f) H.-L. Kwong, W.-S. Lee, H.-F. Ng, W.-H. Chiu, W.-T. Wong, *J. Chem. Soc., Dalton Trans.* **1998**, 1043; g) H.-L. Kwong, K.-M. Lau, W.-S. Lee, W.-T. Wong, *New J. Chem.* **1999**, *23*, 629; h) C.-S. Lee, P.-F. Teng, W.-L. Wong, H.-L. Kwong, A. S. C. Chan, *Tetrahedron* **2005**, *61*, 7924; i) aldol additions: S. E. Denmark, Y. Fan, *J. Am. Chem. Soc.* **2002**, *124*, 4233; j) S. E. Denmark, Y. Fan, M. D. Eastgate, *J. Org. Chem.* **2005**, *70*, 5235.

6 For a related type of C_2-symmetric bipyridine, see: C. Bolm, M. Ewald, M. Zehnder, M. A. Neuburger, *Chem. Ber.* **1992**, *125*, 453.

7 a) C. Bolm, J. Müller, G. Schlingloff, M. Zehnder, M. Neuburger, *J. Chem. Soc., Chem. Commun.* **1993**, 182; b) C. Bolm, J. Müller, *Tetrahedron* **1994**, *50*, 4355; c) see also: C. Bolm, M. Felder, J. Müller, *Synlett* **1992**, 439.

8 For the use of ROMP polymers in reactions of $ZnEt_2$, see: C. Bolm, C. L. Dinter, A. Seger, H. Höcker, J. Brozio, *J. Org. Chem.* **1999**, *64*, 5730.

9 For the application of fluorinated catalysts, see: J. K. Park, H. G. Lee, C. Bolm, B. M. Kim, *Chem. Eur. J.* **2005**, *11*, 945.

10 C. Bolm, K. Muñiz-Fernandez, A. Seger, G. Raabe, *Synlett* **1997**, 1051.

11 C. Bolm, K. Muñiz-Fernandez, A. Seger, G. Raabe, K. Günther, *J. Org. Chem.* **1998**, *63*, 7860.

12 For general reviews of this reaction, see: a) K. Soai, S. Niwa, *Chem. Rev.* **1992**, *92*, 833; b) L. Pu, H.-B. Yu, *Chem. Rev.* **2001**, *101*, 757; c) K. Soai, T. Shibata, in *Comprehensive Asymmetric Catalysis*, Vol. 2 (Eds.: E. N. Jacobsen, A. Pfaltz, H. Yamamoto), Springer, Berlin, **1999**, p. 911.

13 C. Bolm, K. Muñiz, *Chem. Commun.* **1999**, 1295.

14 For a general review of catalyzed asymmetric aryl transfer reactions, see: C. Bolm, J. P. Hildebrand, K. Muñiz, N. Hermanns, *Angew. Chem.* **2001**, *113*, 3382; *Angew. Chem. Int. Ed.* **2001**, *40*, 3284.

15 a) C. Bolm, K. Muñiz, C. Ganter, *New J. Chem.* **1998**, *22*, 1371; b) review: C. Bolm, K. Muñiz, *Chem. Soc. Rev.* **1999**, *28*, 51.

16 a) C. Bolm, K. Muñiz, N. Aguilar, M. Kesselgruber, G. Raabe, *Synthesis* **1999**, 1251; b) C. Bolm, M. Kesselgruber, K. Muñiz, G. Raabe, *Organometallics* **2000**, *19*, 1648.

17 a) C. Bolm, T. Kühn, *Synlett* **2000**, 899; b) C. Bolm, K. Wenz, G. Raabe, *J. Organomet. Chem.* **2002**, *662*, 23; c) C. Bolm, T. Focken, G. Raabe, *Tetrahedron: Asymmetry* **2003**, *14*, 1733; d) T. Focken, G. Raabe, C. Bolm, *Tetrahedron: Asymmetry* **2004**, *15*, 1693.

18 a) T. Sammakia, H. A. Latham, D. R. Schaad, *J. Org. Chem.* **1995**, *60*, 10; b) C. J. Richards, T. Damalidis, D. E. Hibbs, M. B. Hursthouse, *Synlett* **1995**, 74; c) C. J. Richards, A. W. Mulvaney, *Tetrahedron: Asymmetry* **1996**, *7*, 1419; d) Y. Nishibayashi, S. Uemura, *Synlett* **1995**, 79.

19 C. Bolm, M. Kesselgruber, A. Grenz, N. Hermanns, J. P. Hildebrand, *New J. Chem.* **2001**, *25*, 13.

20 S. Özçubukçu, F. Schmidt, C. Bolm, *Org. Lett.* **2005**, *7*, 1407.

21 C. Bolm, M. Kesselgruber, N. Hermanns, J. P. Hildebrand, G. Raabe, *Angew. Chem.* **2001**, *113*, 1536; *Angew. Chem. Int. Ed.* **2001**, *40*, 1488.

22 a) C. Bolm, L. Xiao, M. Kesselgruber, *Org. Biomol. Chem.* **2003**, *1*, 145; b) C. Bolm, L. Xiao, L. Hintermann, T. Focken, G. Raabe *Organometallics* **2004**, *23*, 2362; c) R. Stemmler, C. Bolm, *J. Org. Chem.* **2005**, *70*, 9925.

23 C. Bolm, T. Kühn, *Isr. J. Chem.* **2001**, *41*, 263.

24 a) C. Bolm, M. Kesselgruber, G. Raabe, *Organometallics* **2002**, *21*, 707; b) Y. Yuan, G. Raabe, C. Bolm, *J. Organomet. Chem.* **2005**, *690*, 5747. For related C_2-symmetric carbenes, see: c) D. Broggini, A. Togni, *Helv. Chim. Acta* **2002**, *85*, 2518; d) S. Gischig, A. Togni, *Organometallics* **2004**, *23*, 2479.

25 For early investigations on matched/mismatched effects in asymmetric metal catalysis, see: a) S. D. Pastor, A. Togni, *J. Am. Chem. Soc.* **1989**, *111*, 2333; b) A. Togni, S. D. Pastor, *J. Org. Chem.* **1990**, *55*, 1649; c) S. D. Pastor, A. Togni, *Helv. Chim. Acta* **1991**, *74*, 905.

26 K. Muñiz, C. Bolm, *Chem. Eur. J.* **2000**, *6*, 2309.

27 C. Bolm, N. Hermanns, M. Kesselgruber, J. P. Hildebrand, *J. Organomet. Chem.* **2001**, *624*, 157.

28 C. Bolm, K. Muñiz, J. P. Hildebrand, *Org. Lett.* **1999**, *1*, 491.

29 C. Bolm, L. Zani, J. Rudolph, I. Schiffers, *Synthesis* **2004**, 2173.

30 a) C. Bolm, F. Schmidt, L. Zani, *Tetrahedron: Asymmetry* **2005**, *16*, 1367; b) I. Schiffers, T. Rantanen, F. Schmidt, W. Bergmans, L. Zani, C. Bolm, *J. Org. Chem.* **2006**, *71*, 2320.

31 a) K. Meguro, M. Aizawa, T. Sohda, Y. Kawamatsu, A. Nagaoka, *Chem. Pharm. Bull.* **1985**, *33*, 3787; b) F. Toda, K. Tanaka, K. Koshiro, *Tetrahedron: Asymmetry* **1991**, *2*, 873; c) S. Stanev, R. Rakovska, N. Berova, G. Snatzke, *Tetrahedron: Asymmetry* **1995**, *6*, 183; d) M. Botta, V. Summa, F. Corelli, G. Di Pietro, P. Lombardi, *Tetrahedron: Asymmetry* **1996**, *7*, 1263; e) E. Regel, W. Draber, K. H. Büchel, M. Plempel, German Patent 2 461 406, **1976**; f) P. A.

J. Janssen, French Patent 2 014 487, **1970**; g) for a classification, see: M. Murcia-Soler, F. Pérez-Giménez, F. García-March, M. T. Salabert-Salvador, W. Díaz-Víllanueva, M. J. Castro-Bleda, *J. Chem. Inf. Comput. Sci.* **2003**, *43*, 1688.

32 a) Y. Bolshan, C.-Y. Chen, J. R. Chilenski, F. Gosselin, D. J. Mathre, P. D. O'Shea, A. Roy, R. D. Tillyer, *Org. Lett.* **2004**, *6*, 111; b) M. C. Hillier, J.-N. Desrosiers, J.-F. Marcoux, E. J. J. Grabowski, *Org. Lett.* **2004**, *6*, 573; c) P. D. O'Shea, C.-Y. Chen, W. R. Chen, P. Dagneau, L. F. Frey, E. J. J. Grabowski, K. M. Marcantonio, R. A. Reamer, L. Tan, R. D. Tillyer, A. Roy, X. Wang, D. L. Zhao, *J. Org. Chem.* **2005**, *70*, 3021. For a nucleofugality scale and solvolysis rate constants of benzhydryl derivatives, see: d) B. Denegri, A. Streiter, S. Juric, A. R. Ofial, H. Mayr, *Chem. Eur. J.* **2005**, *11*, 1648; e) B. Denegri, A. R. Ofial, S. Juric, A. Streiter, O. Kronja, H. Mayr, *Chem. Eur. J.* **2005**, *11*, 1657.

33 a) M. Kitamura, T. Ohkuma, S. Inoue, N. Sayo, H. Kumobayashi, S. Akutagawa, T. Ohta, H. Takaya, R. Noyori, *J. Am. Chem. Soc.* **1988**, *110*, 629; b) E. J. Corey, C. J. Helal, *Tetrahedron Lett.* **1995**, *36*, 9153; c) E. J. Corey, C. J. Helal, *Tetrahedron Lett.* **1996**, *37*, 4837; d) E. J. Corey, C. J. Helal, *Tetrahedron Lett.* **1996**, *37*, 5675; e) E. J. Corey, C. J. Helal, *Angew. Chem.* **1998**, *110*, 2092; *Angew. Chem. Int. Ed.* **1998**, *37*, 1986; f) A. Marinetti, J.-P. Genêt, S. Jus, D. Blanc, V. Ratovelomanana-Vidal, *Chem. Eur. J.* **1999**, *5*, 1160; g) T. Ohkuma, M. Koizumi, H. Ikehira, T. Yokozawa, R. Noyori, *Org. Lett.* **2000**, *2*, 659; h) R. Noyori, T. Ohkuma, *Angew. Chem.* **2001**, *113*, 40; *Angew. Chem. Int. Ed.* **2001**, *40*, 40; i) C.-Y. Chen, R. A. Reamer, J. R. Chilenski, C. J. McWilliams, *Org. Lett.* **2003**, *5*, 5039; j) C.-y. Chen, R. A. Reamer, J. R. Chilenski, C. J. McWilliams, *Org. Lett.* **2003**, *5*, 5039; h) C.-y. Chen, R. A. Reamer, A. Roy, J. R. Chilenski, *Tetrahedron Lett.* **2005**, *46*, 5593.

34 a) K. Soai, Y. Kawase, A. Oshio, *J. Chem. Soc., Perkin Trans. I* **1991**, 1613; b) P. I. Dosa, J. C. Ruble, G. C. Fu, *J. Org. Chem.* **1997**, *62*, 444; c) W.-S. Huang, Q.-S. Hu, L. Pu, *J. Org. Chem.* **1999**, *64*, 7940; d) W.-S. Huang, L. Pu, *Tetrahedron Lett.* **2000**, *41*, 145; e) diastereoselectively: J. Hübscher, R. Barner, *Helv. Chim. Acta* **1990**, *73*, 1068; f) for a review, see: F. Schmidt, R. T. Stemmler, J. Rudolph, C. Bolm, *Chem. Soc. Rev.* **2006**, *35*, 454.

35 For related enantioselective arylations of ketones, which were developed *after* our initial work, see: a) O. Prieto, D. J. Ramón, M. Yus, *Tetrahedron: Asymmetry* **2003**, *14*, 1955; b) C. García, P. J. Walsh, *Org. Lett.* **2003**, *5*, 3641; reviews: c) D. J. Ramón, M. Yus, *Angew. Chem.* **2004**, *116*, 286; *Angew. Chem. Int. Ed.* **2004**, *43*, 284; d) C Nanya de Parrodi, P. J. Walsh, *Synlett* **2004**, 2417.

36 In our studies, diphenylzinc from Strem was used. Reagents obtained from other suppliers appeared to be less reliable.

37 For the application of this phenyl transfer protocol by others using different catalysts, see: a) G. Zhao, X.-G. Li, X.-R. Wang, *Tetrahedron: Asymmetry* **2001**, *12*, 399; b) D.-H. Ko, K. H. Kim, D.-C. Ha, *Org. Lett.* **2002**, *4*, 3759; c) M. G. Pizzuti, S. Superchi, *Tetrahedron: Asymmetry* **2005**, *16*, 2263; d) Y.-C. Qin, L. Pu, *Angew. Chem.* **2006**, *118*, 279; *Angew. Chem. Int. Ed.* **2006**, *45*, 273.

38 C. Bolm, N. Hermanns, J. P. Hildebrand, K. Muñiz, *Angew. Chem.* **2000**, *112*, 3607; *Angew. Chem. Int. Ed.* **2000**, *39*, 3465.

39 For the application of this phenyl transfer protocol by others using a different catalyst, see: M. Fontes, X. Verdaguer, L. Solà, M. A. Pericàs, A. Riera, *J. Org. Chem.* **2004**, *69*, 2532.

40 a) Use of zinc species obtained by mixing diphenylzinc with dialkylzinc had been described previously, but the ee_{max} was only 70%. J. Blacker, in *Proceedings of the 3rd International Conference on the Scale Up of Chemical Processes* (Ed.: T. Laird) Scientific Update, Mayfield, East Sussex, Great Britain, **1998**, p. 74. For other applications of mixed organozinc reagents, see: b) M. Srebnik, *Tetrahedron Lett.* **1991**, *32*, 2449; c) W. Oppolzer, R. N. Radinov, *Helv. Chim. Acta* **1992**, *75*, 170; d) W. Oppolzer, R. N. Radinov, *J. Am. Chem. Soc.* **1993**, *115*, 1593; e) S. Berger, F.

Langer, C. Lutz, P. Knochel, T. A. Mobley, C. K. Reddy, *Angew. Chem.* **1997**, *109*, 1603; *Angew. Chem. Int. Ed. Engl.* **1997**, *36*, 1496; f) C. Lutz, P. Knochel, *J. Org. Chem.* **1997**, *62*, 7895; g) C. Lutz, P. Knochel, *Synthesis* **1999**, 312; h) S. Dahmen, S. Bräse, *Org. Lett.* **2001**, *3*, 4119.

41 The different maxima in Fig. 2.1.2.1 can be explained by the experimental conditions, under which every spectrum consisted of 32 scans which were collected over approximately 12 s. For fast reactions the conversion during this period is averaged in the initial intensity. For details, see: N. Hermanns, Dissertation, RWTH Aachen University, **2002**. Similar results to those depicted in Fig. 2.1.2.1 have been obtained by Pericas and coworkers [39].

42 Silanols have been applied extensively in other areas. Reviews: a) K. Hirabayashi, A. Mori, *J. Synth. Org. Chem. Japan* **2000**, *58*, 926; b) V. Chandrasekhar, R. Boomishankar, S. Nagendran, *Chem. Rev.* **2004**, *104*, 5847. Polymers: c) P. D. Lickiss, *Adv. Inorg. Chem.* **1995**, *42*, 147. Cross-coupling: d) K. Hirabayashi, A. Mori, J. Kawashima, M. Suguro, Y. Nishihara, T. Hiyama, *J. Org. Chem.* **2000**, *65*, 5342; e) S. Chang, S. H. Yang, P. H. Lee, *Tetrahedron Lett.* **2001**, *42*, 4883; f) S. E. Denmark, R. F. Sweis, *Acc. Chem. Res.* **2002**, *35*, 835; g) S. E. Denmark, J. D. Baird, *Org. Lett.* **2004**, *6*, 3649; h) S. E. Denmark, M. H. Ober, *Adv. Synth. Catal.* **2004**, *346*, 1703. *Ortho*-directing group: i) M. C. Sieburth, L. Fensterbank, *J. Org. Chem.* **1993**, *58*, 6314. Bioactives: j) R. Tacke, K. Mahner, C. Strohmann, B. Forth, E. Mutschler, T. Friebe, G. Lambrecht, *J. Organomet. Chem.* **1991**, *417*, 339; k) R. Tacke, B. Forth, M. Waelbroeck, J. Gross, E. Mutscher, G. Lambrecht, *J. Organomet. Chem.* **1995**, *505*, 73; l) R. Tacke, T. Heinrich, R. Bertermann, C. Burschka, A. Hamacher, M. U. Kassack, *Organometallics* **2004**, *23*, 4468; m) J. O. Daiss, M. Penka, C. Burschka, R. Tacke, *Organometallics* **2004**, *23*, 4987; n) R. Tacke, T. Schmid, M. Hofmann, T. Tolasch, W. Franke, *Organometallics* **2003**, *22*, 370; o) M. wa Mutahi, T. Nittoli, L. Guo, S. McN. Sieburth, *J. Am. Chem. Soc.* **2002**, *124*, 7363; p) J. Kim, S. McN. Sieburth, *J. Org. Chem.* **2004**, *69*, 3008.

43 Steric bulk at the R group of silanols **13** appeared to increase the *ee* of the aryl transfer products. Unfortunately, however, all attempts to prepare silanol **13** with two *t*-butyl groups at the silicon atom failed.

44 a) C. Bolm, N. Hermanns, A. Claßen, K. Muñiz, *Bioorg. Med. Chem. Lett.* **2002**, *12*, 1795; for a subsequent investigation, see: b) D. Castellnou, M. Fontes, C. Jimeno, D. Font, L. Solà, X. Verdaguer, M. A. Pericàs, *Tetrahedron* **2005**, *61*, 12 111.

45 For general overviews on polymer-supported catalysis, see: a) *Chiral Catalyst Immobilization and Recycling* (Eds.: D. E. DeVos, I. F. J. Vankelecom, P. A. Jacobs), Wiley-VCH, Weinheim, **2000**; b) B. Clapham, T. S. Reger, K. D. Janda, *Tetrahedron* **2001**, *57*, 4637; c) S. Bräse, S. Dahmen, *Synthesis* **2001**, 1431.

46 For reviews of soluble polymer-supported catalysts, see: a) H. P. Wentworth, Jr., K. D. Janda, *Chem. Commun.* **1999**, 1917; b) L. Pu, *Chem. Eur. J.* **1999**, *5*, 2227; c) D. E. Bergbreiter, *Catal. Today* **1998**, *42*, 389; d) C. Bolm, A. Gerlach, *Eur. J. Org. Chem.* **1998**, 21.

47 For the use of other supported ferrocenes in asymmetric catalysis, see: a) C. Köllner, B. Pugin, A. Togni, *J. Am. Chem. Soc.* **1998**, *120*, 10 274; b) H.-U. Blaser, F. Spindler, in *Comprehensive Asymmetric Catalysis* ; Vol. 3 (Eds.: E. N. Jacobsen, A. Pfaltz, H. Yamamoto), Springer, Berlin, **1999**, p. 1427; c) B. Pugin, *Chimia* **2001**, *55*, 719; d) B. Gotov, S. Toma, D. J. Macquarrie, *New J. Chem.* **2000**, *24*, 597; e) B. F. G. Johnson, S. A. Raynor, D. S. Shephard, T. Mashmeyer, J. M. Thomas, G. Sankar, S. Bromley, R. Oldroyd, L. Gladden, M. D. Mantle, *Chem. Commun.* **1999**, 1167.

48 a) J. Rudolph, F. Schmidt, C. Bolm, *Adv. Synth. Catal.* **2004**, *346*, 867; b) see also: S. Dahmen, M. Lormann, *Org. Lett.* **2005**, *7*, 4597.

49 a) C. Bolm, J. Rudolph, *J. Am. Chem. Soc.* **2002**, *124*, 14 850; b) J. Rudolph, C.

Bolm, in *Catalysts for Fine Chemical Synthesis*, Vol. 3 (Eds.: S. M. Roberts, J. Xiao, J. Whittall, T. Pickett), Wiley-VCH, **2004**, 161; c) J. Rudolph, F. Schmidt, C. Bolm, *Synthesis* **2005**, 840.

50 J. Rudolph, N. Hermanns, C. Bolm, *J. Org. Chem.* **2004**, *69*, 3997.

51 a) J.-X. Ji, J. Wu, T. T.-L. Au-Yeung, C.-W. Yip, R. K. Haynes, A. S. C. Chan, *J. Org. Chem.* **2005**, *70*, 1093; b) K. Ito, Y. Tomita, T. Katsuki, *Tetrahedron Lett.* **2005**, *46*, 6083; c) A. L. Braga, D. S. Lüdtke, F. Vargas, M. W. Paixão, *Chem. Commun.* **2005**, 2512; d) A. L. Braga, D. S. Lüdtke, P. H. Schneider, F. Vargas, A. Schneider, L. A. Wessjohnn, M. W. Paixão, *Tetrahedron Lett.* **2005**, *46*, 7827; e) X. Wu, X. Liu, G. Zhao, *Tetrahedron: Asymmetry* **2005**, *16*, 2299; f) P.-Y. Wu, H.-L. Wu, B.-J. Uang, *J. Org. Chem.* **2006**, *71*, 833.

52 J. Rudolph, M. Lormann, C. Bolm, S. Dahmen, *Adv. Synth. Catal.* **2005**, *347*, 1361.

53 For general reviews of screening methods, see: a) M. T. Reetz, *Angew. Chem.* **2001**, *113*, 292; *Angew. Chem. Int. Ed.* **2001**, *40*, 284; b) M. T. Reetz, *Angew. Chem.* **2002**, *114*, 1391; *Angew. Chem. Int. Ed.* **2002**, *41*, 1335; d) J. G. de Vries, A. H. M. de Vries, *Eur. J. Org. Chem.* **2003**, 799.

54 For additive effects on reactions with organometallic reagents, see: a) H_2O: S. Ribe, P. Wipf, *Chem. Commun.* **2001**, 299; b) MeOH: P. I. Dosa, J. C. Ruble, G. C. Fu, *J. Org. Chem.* **1997**, *62*, 444. c) For a review of additive effects in catalyses: E. M. Vogl, H. Gröger, M. Shibasaki, *Angew. Chem.* **1999**, *111*, 1672; *Angew. Chem. Int. Ed.* **1999**, *38*, 1570. d) For a review of halide effects, see: K. Fagnou, M. Lautens, *Angew. Chem.* **2002**, *114*, 26; *Angew. Chem. Int. Ed.* **2002**, *41*, 26.

55 For a recent report on the application of this phenyl transfer protocol using a proline-derived catalyst with dendritic polyether wedges, see: X. y. Liu, C. y. Wu, Z. Chai, Y. y. Wu, G. Zhao, S. z. Zhu, *J. Org. Chem.* **2005**, *70*, 7432.

56 T. Focken, J. Rudolph, C. Bolm, *Synthesis* **2005**, 429.

57 a) For a review of Rh-catalyzed reactions of boronic acids and related organometallic reagents, see: K. Fagnou, M. Lautens, *Chem. Rev.* **2003**, *103*, 169. For reviews of Rh-catalyzed conjugate additions of arylboronic acids to α,β-unsaturated carbonyl compounds: b) T. Hayashi, *Synlett* **2001**, 879; c) T. Hayashi, K. Yamasaki, *Chem. Rev.* **2003**, *103*, 2761.

58 a) M. Sakai, M. Ueda, N. Miyaura, *Angew. Chem.* **1998**, *110*, 3475; *Angew. Chem. Int. Ed.* **1998**, *37*, 3279; b) M. Ueda, N. Miyaura, *J. Org. Chem.* **2000**, *65*, 4450.

59 For attempts to use chiral dinitrogen ligands in these reactions, see: C. Moreau, C. Hague, A. S. Weller, C. G. Frost, *Tetrahedron Lett.* **2001**, *42*, 6957.

60 A. Fürstner, H. Krause, *Adv. Synth. Catal.* **2001**, *343*, 343.

61 For reviews of NHCs and NHC–metal complexes, see: a) D. J. Cardin, B. Cetinkaya, M. F. Lappert, *Chem. Rev.* **1972**, *72*, 545; b) W. A. Herrmann, C. Köcher, *Angew. Chem.* **1997**, *109*, 2256; *Angew. Chem. Int. Ed. Engl.* **1997**, *36*, 2163; c) D. Bourissou, O. Guerret, F. P. Gabbaï, G. Bertrand, *Chem. Rev.* **2000**, *100*, 39; d) W. A. Herrmann, *Angew. Chem.* **2002**, *114*, 1342; *Angew. Chem. Int. Ed.* **2002**, *41*, 1290.

62 For reviews of the use of chiral NHCs in catalysis, see: a) M. C. Perry, K. Burgess, *Tetrahedron: Asymmetry* **2003**, *14*, 951; b) D. Enders, T. Balensiefer, *Acc. Chem. Res.* **2004**, *37*, 534; c) V. César, S. Bellemin-Laponnaz, L. H. Gade, *Chem. Soc. Rev.* **2004**, *33*, 619. Selected examples: d) hydrosilylation: W.-L. Duan, M. Shi, G. B. Rong, *Chem. Commun.* **2003**, 2916; e) L. H. Gade, V. César, S. Bellemin-Laponnaz, *Angew. Chem.* **2004**, *116*, 1036; *Angew. Chem. Int. Ed.* **2004**, *43*, 1014; f) olefin metathesis: J. J. Van Veldhuizen, D. G. Gillingham, S. B. Garber, O. Kataoka, A. H. Hoveyda, *J. Am. Chem. Soc.* **2003**, *125*, 12 502; g) conjugate additions: A. Alexakis, C. L. Winn, F. Guillen, J. Pytkowicz, S. Roland, P. Mangeney, *Adv. Synth. Catal.* **2003**, *345*, 345; h) α-arylations of amides: S. Lee, J. F. Hartwig, *J. Org. Chem.* **2001**, *66*, 3402; i) hydrogenations: X. Cui, K.

Burgess, *J. Am. Chem. Soc.* **2003**, *125*, 1421; j) M. C. Perry, K. Burgess, *Tetrahedron: Asymmetry* **2005**, *14*, 951; k) enone arylations: Y. Ma, C. Song, C. Ma, Z. Sun, Q. Chai, M. B. Andrus, *Angew. Chem.* **2003**, *115*, 6051; *Angew. Chem. Int. Ed.* **2003**, *42*, 5871.

63 For a review of [2.2]paracyclophanes in asymmetric catalysis, see: S. Gibson, J. C. Knight, *Org. Biomol. Chem.* **2003**, *1*, 1256.

64 a) J. Rudolph, T. Rasmussen, C. Bolm, P.-O. Norrby, *Angew. Chem.* **2003**, *115*, 3110; *Angew. Chem. Int. Ed.* **2003**, *42*, 3002; b) J. Rudolph, C. Bolm, P.-O. Norrby, *J. Am. Chem. Soc.* **2005**, *117*, 1548.

65 a) M. J. Bishop, R. W. McNutt, *Bioorg. Med. Chem. Lett.* **1995**, *5*, 1311; b) C. M. Spencer, D. Foulds, D. H. Peters, *Drugs* **1993**, *46*, 1055; c) S. Sakurai, N. Ogawa, T. Suzuki, K. Kato, T. Ohashi, S. Yasuda, H. Kato, Y. Ito, *Chem. Pharm. Bull.* **1996**, *44*, 765.

66 a) For a catalysis that gave rise to diarylmethylamines in very high enantioselectivities, but required the use of 5 equiv of the stannane in order to obtain the products in high yields, see: T. Hayashi, M. Ishigedani, *J. Am. Chem. Soc.* **2000**, *122*, 976. Such rhodium catalyses were later optimized, and now the use of chiral dienes as ligands allows the application of boroxines as aryl sources. For representative examples, see: b) N. Tokunaga, Y. Otomaru, K. Okamoto, K. Ueyama, R. Shintani, T. Hayashi, *J. Am. Chem. Soc.* **2004**, *126*, 13 584; c) Y. Otomaru, N. Tokunaga, R. Shintani, T. Hayashi, *Org. Lett.* **2005**, *7*, 307.

67 a) C. J. Opalka, T. E. D'Ambra, J. J. Faccone, G. Bodson, E. Cossement, *Synthesis* **1995**, 766; b) E. Cossement, G. Motte, G. Bodson, J. Gobert, UK Patent Appl. 2 225 321, **1990**; *Chem. Abstr.* **1990**, *113*, 191 396; c) M. Gillard, C. van der Perren, N. Moguilevsky, R. Massingham, P. Chatelain, *Mol. Pharmacol.* **2002**, *61*, 391; d) for a large-scale preparative HPLC (on chiral stationary phase) approach, see: D. A. Pflum, H. S. Wilkinson, G. J. Tanoury, D. W. Kessler, H. B. Kraus, C. H. Senanayake, S. A. Wald, *Org. Process Res. Dev.* **2001**, *5*, 110.

68 For diastereoselective syntheses of cetirizine, see: a) E. J. Corey, C. J. Helal, *Tetrahedron Lett.* **1996**, *37*, 4837; b) D. A. Pflum, D. Krishnamurthy, Z. Han, S. A. Wald, C. H. Senanayake, *Tetrahedron Lett.* **2002**, *43*, 923; c) N. Plobeck, D. Powell, *Tetrahedron: Asymmetry* **2002**, *13*, 303.

69 N. Hermanns, S. Dahmen, C. Bolm, S. Bräse, *Angew. Chem.* **2002**, *114*, 3844; *Angew. Chem. Int. Ed.* **2002**, *41*, 3692.

70 Earlier, [2.2]paracyclophane-based N,O-ligands had been employed in the dialkylzinc and the alkenylzinc addition to aldehydes: a) S. Dahmen, S. Bräse, *Chem. Commun.* **2002**, 26; b) S. Höfener, F. Lauterwasser, S. Bräse, *Adv. Synth. Catal.* **2004**, *346*, 755; c) review: S. Bräse, S. Dahmen, S. Höfener, F. Lauterwasser, M. Kreis, R. E. Ziegert, *Synlett* **2004**, 2647; d) for the original synthetic approaches toward these [2.2]paracyclophanes, see: V. Rozenberg, T. Danilova, E. Sergeeva, E. Vorontsov, Z. Starikova, K. Lysenko, Y. Belokon, *Eur. J. Org. Chem.* **2000**, 3295.

2.1.3
Substituted [2.2]Paracyclophane Derivatives as Efficient Ligands for Asymmetric 1,2- and 1,4-Addition Reactions
Stefan Bräse

2.1.3.1 [2.2]Paracyclophanes as Chiral Ligands

The synthesis of enantiomerically pure compounds is one of the major challenges in organic synthesis. On the basic principle of asymmetric catalysis using metal

2.1.3 Substituted [2.2]Paracyclophane Derivatives as Efficient Ligands

complexes, organic chemists are now able to conduct various transformations in a stereo-controlled, enantioselective, and atom-economic manner by using these complexes.

The element of planar chirality plays a pivotal role in many modern ligand systems. The particularly huge success of ferrocenyl ligands has not been matched by any other chiral backbone to date. Metallocene and metal-arene-based ligand backbones exhibit the common feature that they become planar chiral only upon addition of (at least) two substituents on one ring fragment. [2.2]Paracyclophanes, however, need only one substituent (Fig. 2.1.3.1) to be chiral.

Since the initial reports of Reich and Cram [1], the field of [2.2]paracyclophane chemistry has developed considerably, and the chemical behavior of [2.2]paracyclophanes is well understood nowadays. Furthermore, based on our experience, these compounds are coined "rocks": that is, they are usually extremely stable. This implies, however, that they are sometimes very unreactive toward desired reactions. In general, they are able to offer reasonable resistance against synthetic efforts and are therefore an interesting challenge for dedicated chemists.

When we first ventured into the field of [2.2]paracyclophane ligand synthesis, successful applications of such ligands were relatively rare [2]. The most prominent example was clearly the PHANEPHOS ligand developed by Rossen and Pye [3], who have found several successful applications in asymmetric hydrogenation reactions. A comprehensive survey of [2.2]paracyclophane-based ligands can be found in recent reviews [4, 5].

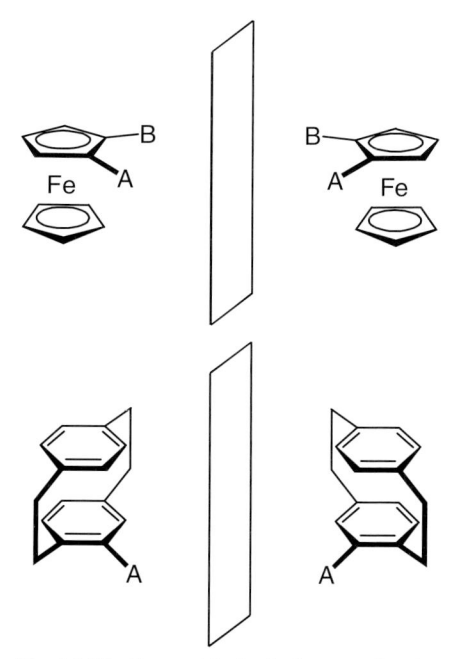

Fig. 2.1.3.1 Planar chirality in ferrocenyl and paracyclophanyl systems.

On one hand, the PHANEPHOS ligand exhibits the well-established C_2-symmetry, being present in various flourishing systems (BINOL, salen ligand, box system) [6]. On the other hand, in certain cases, a high degree of enantioselectivity is demonstrated in other successful ligand systems through other forms of symmetry.

The use of planar-chiral [7] and central-chiral ligands based on paracyclophane systems was still a relatively unexplored frontier, with notable exceptions in the reactions examined by the Rozenberg group [8] and the Berkessel group [9].

Central intermediates in our strategy were the known *ortho*-acylated hydroxy[2.2]paracyclophanes **1** (R^1 = H), **2** (R^1 = alkyl), and **3** (R^1 = aryl) [8, 10] (Fig. 2.1.3.2). These, in turn, can be condensed with amines to give imines **4** or ketimines **5** and **6**, or reduced to give amino alcohols **7–9**, respectively. The ligand structure is therefore vastly variable. Steric factors, such as flexibility of backbone and side-chains, as well as electronic factors (for example sp^2 versus sp^3 configuration of the *N*-donors) can be easily modulated. The introduction of central chirality via chiral amine side-chains is also possible. The interaction of planar and central chirality, usually referred to as chiral cooperativity [11–13], can thus be studied in a ligand system which has both planar and central chiral elements.

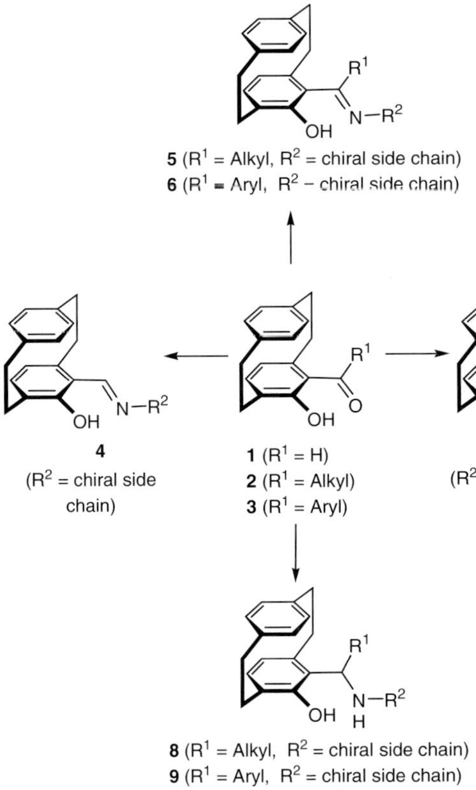

Fig. 2.1.3.2 The family of [2.2]paracyclophane-based ligands.

2.1.3.2 Synthesis of [2.2]Paracyclophane Ligands

2.1.3.2.1 Preparation of FHPC-, AHPC-, and BHPC-Based Imines

There were three important intermediates in our strategy. First was the 5-formyl-4-hydroxy[2.2]paracyclophane (FHPC, **1**), which can be regarded as a planar-chiral analogue of salicyl aldehyde [14] and was first prepared by Rozenberg, Belokon et al. [8]. The other two intermediates were 5-acetyl-4-hydroxy [2.2]paracyclophane (AHPC, **2**) and 5-benzoyl-4-hydroxy[2.2]paracyclophane (BHPC, **3**), which were synthesized by *ortho*-selective Friedel–Crafts acylation of 4-hydroxy[2.2]paracyclophane [10]. The resolution of racemic FHPC (**1**) [8] and AHPC (**2**) [10] was conducted via their imines **4a** or ketimines **5a** with (*S*)- or (*R*)-phenylethylamine, respectively (Scheme 2.1.3.1). Starting from enantiomerically pure FHPC (**1**) and AHPC (**2**), over 20 other imines were prepared by condensation with chiral amines [15] (Scheme 2.1.3.1), which resulted in good to excellent yields. Our initial plan was to study the influence of the size of the sidechain on the catalytic performance of the ligand.

2.1.3.2 Structural Information on AHPC-Based Imines

Although X-ray structures of noncomplexed ligands do not provide direct information about the structure of the catalytically active metal complex, they provide a starting point for the understanding and tuning of the catalysts. The X-ray

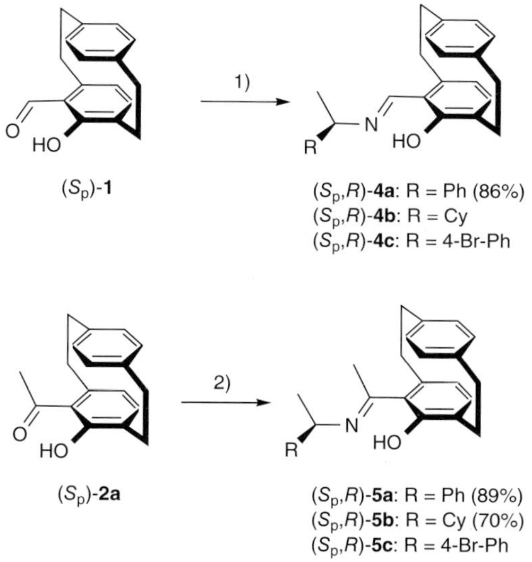

Scheme 2.1.3.1 Synthesis of planar-chiral and central chiral imines **4a–4c** and **5a–5c**. 1) Primary amine (2–5 equiv), toluene/EtOH, MS 4Å, Δ, 3 h. 2) Primary amine (2–5 equiv.), cat. Bu$_2$SnCl$_2$, toluene, MS 4Å, Δ, 40 h [15].

structures of the imines (R_p,S)-**6a** and (S_p,S)-**6a** derived from BHPC (**3**) are already given in the literature [10].

The structure of (R_p,S)-**6b** is shown in Fig. 2.1.3.3). Several other X-ray structures of related compounds were solved during our ongoing paracyclophane studies [15]. All of them showed certain similarities with respect to the hydroxyimine substructure. They showed a hydrogen bond between the imine nitrogen and the phenolic oxygen. This creates a six-membered ring that is nearly planar. The only distortion occurs because of the interaction between the methyl group (C18 for (R_p,S)-**5a**) or the phenyl group (for (R_p,S)-**6a** and (R_p,S)-**6b** respectively) and the [2.2]paracyclophane backbone.

2.1.3.3 Asymmetric 1,2-Addition Reactions to Aryl Aldehydes

2.1.3.3.1 Initial Considerations

The 1,2-addition of diethylzinc to aldehydes is a powerful method for C–C bond formation. As there is a wide variety of possible transition states, the reaction is very sensitive to changes in the ligand structure. For this reason the diethylzinc addition in Scheme 2.1.3.2 is a suitable test reaction for developing and establish-

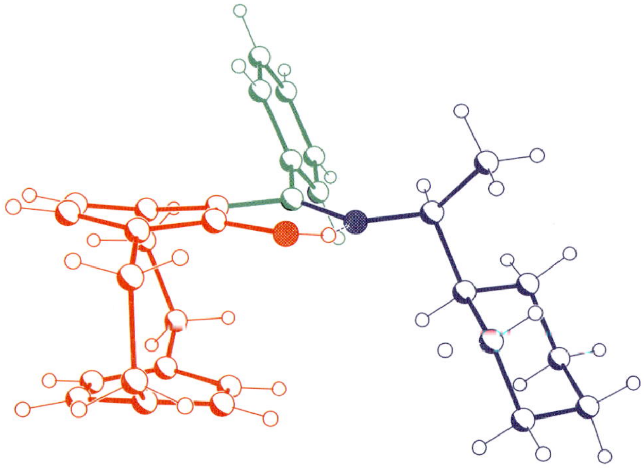

Fig. 2.1.3.3 X-ray structure of (R_p,S)-**6b**.

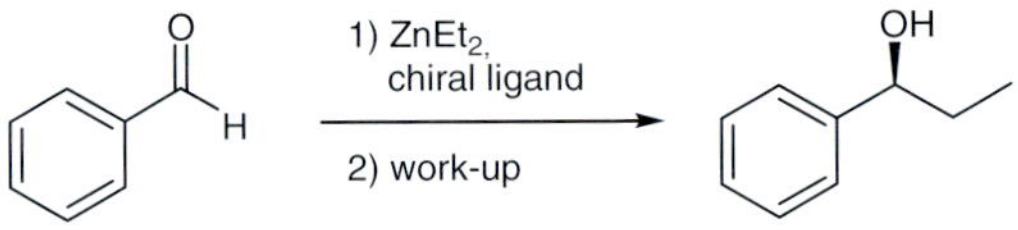

Scheme 2.1.3.2 Diethylzinc addition to benzaldehyde (**10a**). Conditions: benzaldehyde (0.5 mmol), ligand (5%), toluene (1 mL), diethylzinc (1.0 mL, 1.0 M in hexane, 2 equiv), 0 °C, 16 h under argon.

ing of a new class of ligands. This chapter gives the catalysis results of this test reaction.

The well-established planar- and central-chiral ferrocenyl ligands have shown high activity and selectivity in the reaction. However, to the best of our knowledge, these ligands have no central-chiral equivalent with similar bond length and bond angles. In comparison, the [2.2]paracyclophane structure creates central-chiral structures with similar properties, which offer the great opportunity to study its influence on different kinds of chiral elements.

Figure 2.1.3.4 shows the first generation of planar- and central-chiral imine ligand **4a** and ketimine ligands **5a** and **6a**. At the beginning, the FHPC-based imine ligands **4a** were used. The ligands produced only low to moderate enantiomeric excesses. The configuration of the desired product is controlled by the ligand backbone, not by the configuration of the central chirality. It is important to note that both enantiomers of the product can be produced by using the opposite diastereomer of the ligand.

Comparing the planar- and central-chiral ligand (S_p,S)-**4a** and a central-chiral equivalent, the configuration in the product is changed by the ligand backbone while keeping the same central chirality in the ligand side-chain. For the ligand **4a**, the chiral stereogenic centers in the phenylethyl residue and the planar-chiral paracyclophane moiety showed a positive cooperative effect (matched case,

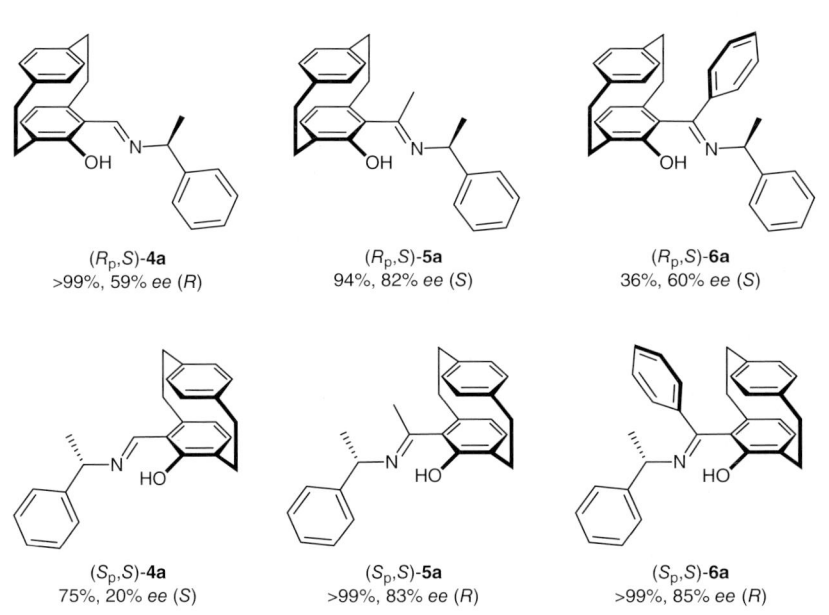

(R_p,S)-**4a**
>99%, 59% ee (R)

(R_p,S)-**5a**
94%, 82% ee (S)

(R_p,S)-**6a**
36%, 60% ee (S)

(S_p,S)-**4a**
75%, 20% ee (S)

(S_p,S)-**5a**
>99%, 83% ee (R)

(S_p,S)-**6a**
>99%, 85% ee (R)

Fig. 2.1.3.4 Central- and planar-chiral imine and ketimine ligands. Yields and selectivities were obtained using the conditions described in Scheme 2.1.3.2 [17].

55% ee for (R_p,S)-**4a** on one occasion only. For the diastereomer we observed a negative cooperative effect (mismatched case 20% ee for (S_p,S)-**4a**). The dominant chiral moiety is the planar chirality.

The combination of the planar and central chirality gave much better results than the central chirality alone, but the results from ligand **4a** were unsatisfactory. Substitution in the α-position of the side-chain of the paracyclophane ligands caused a remarkable change of selectivity and hence this ligand position proved to be very important for the catalytic results. The introduction of only a small substituent, such as a methyl group in ligand **5a**, generated a large increase in the selectivity. In this instance, it seemed that the stereogenic center of the side-chain had no influence at all.

Both the selectivity and the activity of the diastereomeric ligands, (R_p,S)-**5a** and (S_p,S)-**5a**, were nearly identical and both showed positive chiral cooperativity (there was no observation of a mismatched pair). Through introduction of a phenyl residue in the α-position of the side-chain as shown in ligand pair **6a**, there was again a small matched/mismatched effect, which was not as strong as in the unsubstituted case in ligand pair **4a**. Overall, we could derive a simple rule for the ligands:

> The paracyclophane backbone determines the configuration of the product.

The best results observed so far were with the central- and planar-chiral ketimine ligands, all of which were synthesized with the same chiral amine, (S)- or (R)-phenylethylamine, to result in the same central-chiral side-chain. The second generation of these ligands was synthesized with the various chiral amines depicted in Fig. 2.1.3.5 based on the AHPC **2** and BHPC **3**.

The first example synthesized was based on the AHPC structure with (S)-cyclohexylethylamine in the side-chain. The use of the more sterically hindered (S)-cyclohexylethylamine in the side-chain increased the enantiomeric excess from 83% ee (with (S_p,S)-**5a** to 90% ee with (S_p,S)-**5b** (complete conversion). The other diastereomer, (R_p,S)-**5b**, resulted in 94% yield and 87% ee. The ligand pair **6b**, which is based on BHPC with (S)-t-butylethylamine in the side-chain, resulted in a matched/mismatched pair. The (R_p,S) diastereomer gave a 98% yield and an excellent 89% ee, but the (S_p,S) diastereomer produced only a 25% yield with a moderate 42% ee. The diastereomeric pair **6c**, also based on BHPC **3** with (S)-naphthalen-1-ylethylamine in the side-chain, resulted similarly in a matched/mismatched pair with high activity, but only moderate enantiomeric excess.

In summary, the configuration of the desired product is controlled by the planar-chiral imine and ketimine ligand backbone. The selectivity of the reaction depends on both the chiral center and the communication of the side-chain with the ligand backbone. We tuned the side-chain to increase the enantioselectivity up to 90% ee. In the case of the amino alcohol ligands, chiral cooperativity is also observed. However, the influence of the planar chirality is much lower, whereas central chirality is dominant in this instance. In most cases the enantioselectivity is lower than for the ketimines.

(R_p,S)-**5b**
94%, 87% ee (S)

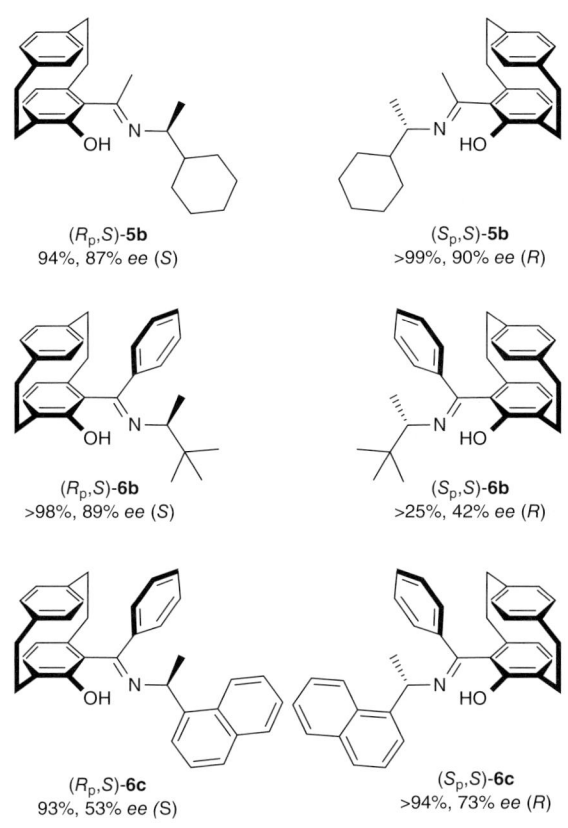

(S_p,S)-**5b**
>99%, 90% ee (R)

(R_p,S)-**6b**
>98%, 89% ee (S)

(S_p,S)-**6b**
>25%, 42% ee (R)

(R_p,S)-**6c**
93%, 53% ee (S)

(S_p,S)-**6c**
>94%, 73% ee (R)

Fig. 2.1.3.5 Second generation of central- and planar-chiral ligands **5b**, **6b**, and **6c**. Yields and selectivities were obtained under the conditions described in Scheme 2.1.3.5 [15].

In addition, we investigated a nonlinear-like effect (NLE), activity, temperature dependence, and kinetics of hydroxy[2.2]paracyclophane ketimine ligands with the 1,2-addition reaction of diethylzinc to cyclohexylcarbaldehyde. A linear correlation between the enantiomeric excess of AHPC ketimine ligands bearing a phenylethyl side group and the product was observed with 0.5 mol% of catalyst loading. When the catalyst loading of (S_P,S)/(R_P,R)-**4a** was increased to 4 mol%, a precipitate of the inactive heterochiral species was formed and resulted in a positive nonlinear like effect (Fig.2.1.3.6), while a linear behavior is observed with **5b** (Fig. 2.1.3.7). The enantiomeric ratio was found to have linear temperature dependence.[16]

2.1.3.3.2 Asymmetric Addition Reactions to Aromatic Aldehydes: Scope of Substrates

Through investigations of the reaction in Scheme 2.1.3.3, the substrate spectrum was unraveled [17]. Various aromatic aldehydes based on the ketimines **5a**, **6a**,

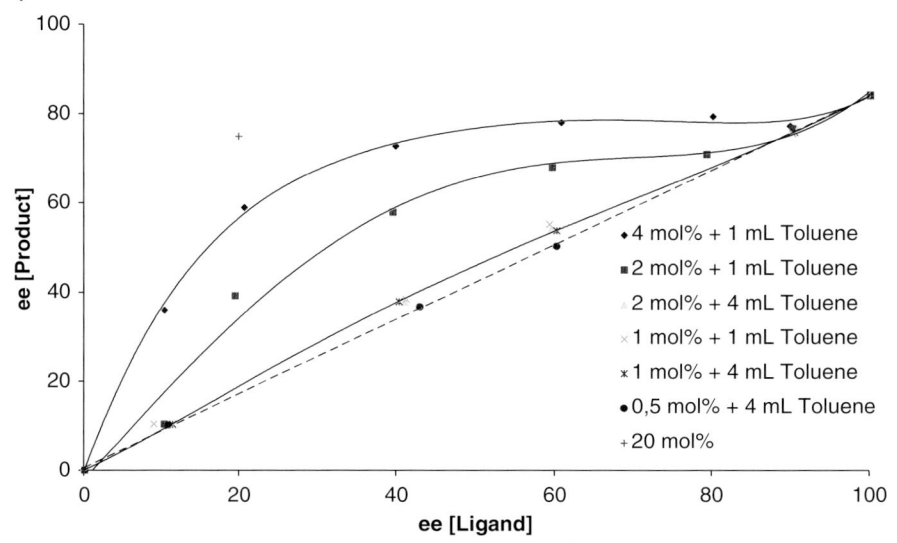

Fig. 2.1.3.6 Results of the studies of the nonlinear effect of the ligands $(S_P,S)/(R_P,R)$-**4a**. Conditions: 1) $C_6H_{11}CHO$ (0.5 mmol), ligand (0.5–4 mol%), Et_2Zn (1 mL; 1 M in hexanes), toluene (1–4 mL), 0 °C, 12 h; 2) Ac_2O, 24 h, rt.

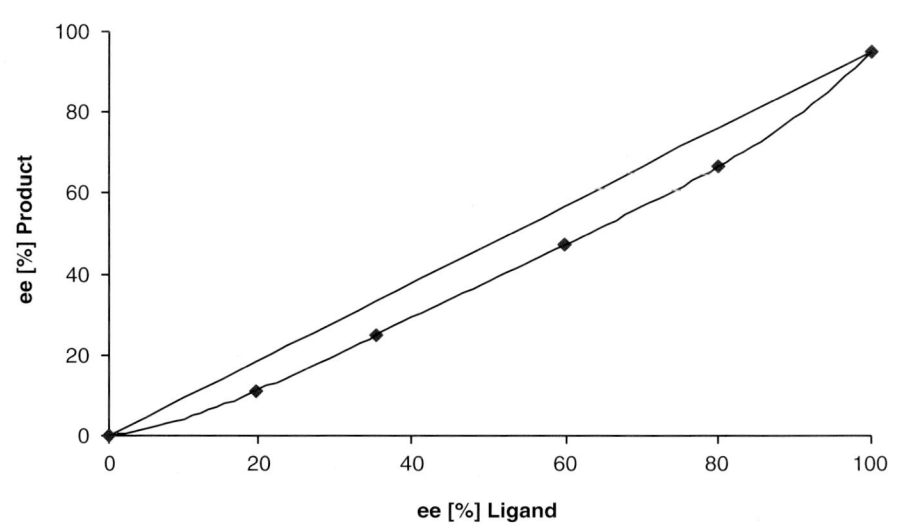

Fig. 2.1.3.7 Results for the nonlinear effect studies for the ligands $(S_P,S)/(R_P,R)$-**5b**. Conditions: 1) $C_6H_{11}CHO$ (0.5 mmol), ligand (4 mol%), Et_2Zn (1 mL, 1 M in hexanes), toluene (1 mL), 0 °C, 12 h; 2) Ac_2O, 24 h, rt.

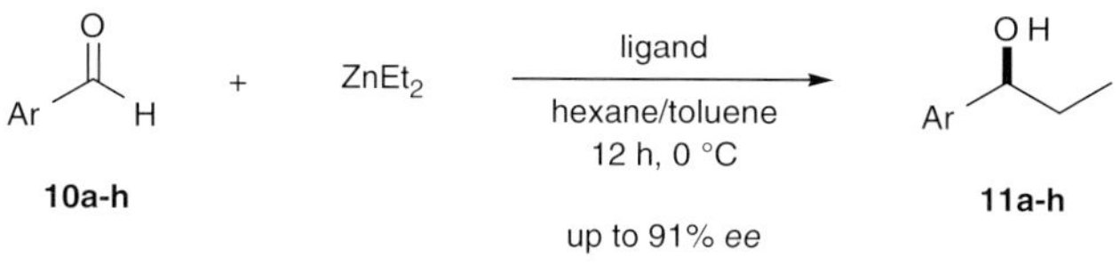

Scheme 2.1.3.3 The diethylzinc addition to aryl aldehydes.

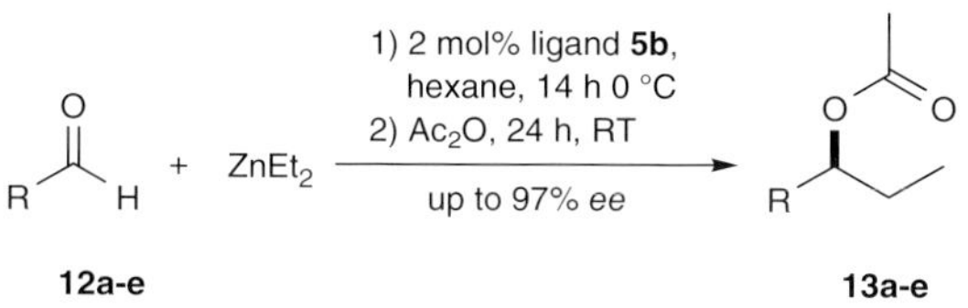

Scheme 2.1.3.4 Diethylzinc addition to aliphatic aldehydes.

and **5b** were used, such as substituted benzaldehydes and naphthaldehydes. For these substrates, the selectivities were in the 81–85% ee range, except for the *ortho*-substituted 2-methoxybenzaldehyde, in which a loss in selectivity was observed (55% ee). The mismatched diastereomer, (R_p,S)-**6a**, provided a reduction in yield and selectivity for benzaldehyde (36% conversion, 60% ee, S), with respect to (S_p,S)-**6a** (100% conversion, 85%, R).

2.1.3.4 Asymmetric Addition Reactions to Aliphatic Aldehydes

After successful application in the diethylzinc addition to aromatic aldehydes, we applied our ligands to aliphatic substrates. These substrates still pose a challenge for most types of ligands.

While cyclohexanecarbaldehyde and pivaldehyde, being α-branched aldehydes, are excellent substrates for most ligands leading to nearly perfect enantioselection, unbranched aldehydes still remain cumbersome substrates.

Taking this into consideration, we employed a further improved generation of ketimine ligands (R_p,S)-**5b** and (S_p,S)-**5b**, in which the cyclohexyl ring is more sterically demanding than the phenyl ring present in the ligands **5a** and **6a** (Scheme 2.1.3.4).

The application of diastereomeric ligands in the diethylzinc addition to aliphatic substrates demonstrated that two ligands, (S_p,S)-**5b** and (R_p,S)-**5b**, always yield appropriate esters **13** with opposite configurations.

A comparison of these results with those obtained from ligand (S_p,S)-**5a** demonstrates that the ligands (R_p,S)-**5b** and (S_p,S)-**5b** can raise the enantiomeric excesses by about 10% ee [18]. However, use of pivaldehyde as the substrate already showed excellent ee values (98%) when the first-generation ligand systems (S_p,S)-**5a** and (S_p,S)-**6a**, respectively, were employed [17].

In conclusion, one can say that although the effectiveness of our ligands is only moderately good for diethylzinc additions to aromatic aldehydes, it is, however, excellent in applications involving aliphatic aldehydes. The ee values for unbranched aliphatic aldehydes such as decanal, despite being diminished, are still within the mid-80% *ee* range.

2.1.3.5 Addition of Alkenylzinc Reagents to Aldehydes

The application of alkenylzinc reagents in asymmetric catalysis has been limited to very few examples, and its methodology is far from being a standard procedure. The addition of alkenylzinc to carbonyl compounds leads to the synthetically useful chiral allyl alcohols, which are the key intermediates in various reactions. Generally, alkenylzinc reagents are not temperature-stable, and therefore they are prepared in situ using transmetalation protocols. Oppolzer and Radinov reported the preparation of mixed alkylzinc–alkenylzinc reagents for the asymmetric addition to aldehydes by reaction of terminal alkynes with dicyclohexylborane, followed by a boron–zinc exchange. By using Noyori's DAIB ligand (3-*exo*-(dimethylamino)isoborneol) [19], good enantioselectivities were achieved for certain aromatic and aliphatic aldehydes [20, 21].

The [2.2]paracyclophane ligands above were thus used in asymmetric alkenylzinc addition to aldehydes [22]. The protocol used relies on the method initially developed by Oppolzer et al. [20] The four paracyclophane ligands **5a** and **6a** were tested under these optimized conditions with 1-octyne as the alkenylzinc precursor and benzaldehyde (**10a**) at −10 °C. At 2% catalyst loading, good enantioselectivities up to 86% were obtained. The diastereomers (S_p,S)-**5a** and (R_p,S)-**5a** gave essentially the same selectivities, while producing opposite enantiomers. Despite the use of excess diethylzinc, no 1-phenylpropanol (**11a**), the product of the diethylzinc addition to benzaldehyde (**10a**), was observed, consistently with the literature [20]. The conditions for the best ligand, (R_p,S)-**5a**, were further optimized by varying the reaction temperature and catalyst loading.

Electron-withdrawing substituents on aromatic aldehydes were well tolerated and led to increased enantioselectivity of 97% *ee* for *p*-chlorobenzaldehyde (**10b**). Aliphatic and especially unbranched aliphatic aldehydes are among the most problematic substrates for nearly all ligand systems. As far as we know, there is no report of a highly enantioselective alkenylzinc addition to these kinds of aldehydes. However, they are excellent substrates for the paracyclophane ligands, giving virtually complete enantioselectivity for cyclohexylcarbaldehyde and pivaldehyde. Bulky alkynes were examined next, and at first glance, they seemed to limit the wide applicability of the paracyclophane ligands. Only 75% *ee* was obtained with the internal alkyne 3-hexyne, and the sterically demanding *t*-butylethyne led to the desired product in a disappointing 64% *ee* (Scheme 2.1.3.5).

At this point, the other ligands were re-examined with *t*-butylethyne, but no improvement could be made. Clearly, the greater steric bulk of the zinc species resulting from *t*-butylethyne substantially diminished the stereoselection of the catalyst. In a final attempt to decrease the steric demand of the zinc species, dimethylzinc (3 equiv) was employed as the transmetalation reagent, which should have resulted in formation of the slightly less demanding zinc species (Table 2.1.3.1).

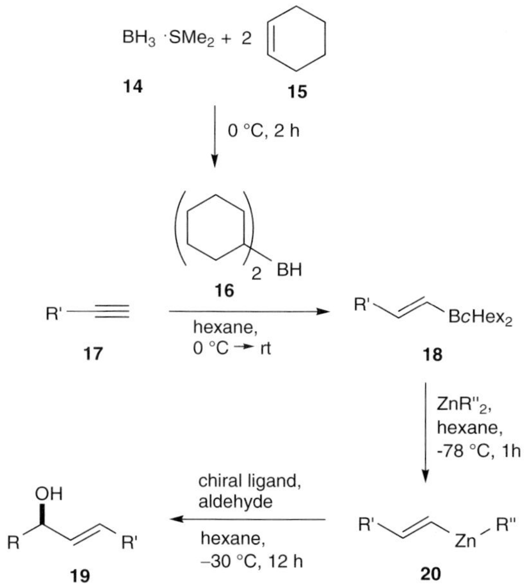

Scheme 2.1.3.5 Asymmetric addition of alkenylzinc reagents to aldehydes [18].

Table 2.1.3.1 Asymmetric addition of alkenylzinc reagents to aldehydes [22].

Alkyne	R in aldehyde RCHO	Zinc reagent	Yield [%]	ee [%]
3-Hexyne	phenyl (10a)	Et$_2$Zn	86	75 (S)[a]
3-Hexyne	phenyl (10a)	Me$_2$Zn	84	88 (S)[b]
tButylethyne	4-Cl-C$_6$H$_4$ (10b)	Et$_2$Zn	78	64 (S)[a]
t-Butylethyne	4-Cl-C$_6$H$_4$ (10b)	Me$_2$Zn	88	89 (S)[a]
1-Octyne	phenyl (10a)	Et$_2$Zn	71	86 (S)[a]
1-Octyne	phenyl (10a)	Et$_2$Zn	48	76 (S)[b]
1-Octyne	phenyl (10a)	Et$_2$Zn	71	86 (S)
1-Octyne	4-Cl-C$_6$H$_4$ (10b)	Et$_2$Zn	88	97 (S)
1-Octyne	4-MeO-C$_6$H$_4$ (10c)	Et$_2$Zn	62	91 (S)
1-Octyne	cyclohexyl (12b)	Et$_2$Zn	80	>98
1-Octyne	t-butyl (12d)	Et$_2$Zn	89	>98
1-Octyne	t-butyl (12d)	Et$_2$Zn	89	94[c]
1-Octyne	hexanal	Et$_2$Zn	55	72[c]

a Ligand (R_p, S)-5a was used. b Ligand (R_p, S)-6a was used. c Ligand (R_p, S)-5c was used.

Gratifyingly, this produced a breakthrough for bulky alkynes. For *t*-butylethyne 89% *ee* was achieved with the ligand (R_p,S)-5a, and 88% *ee* could be obtained for the symmetrical internal alkyne 3-hexyne.

The second generation of N,O-[2.2]paracyclophane ketimine ligands (R_p,S)-5b, 6b were investigated for their ability to catalyze the 1,2-addition of alkenylzinc reagents to aliphatic aldehydes with a special focus on functionalized substrates. For aliphatic aldehydes, which have always been a challenge in this field, remarkably high enantiomeric excesses could be determined (50–95% *ee*).

2.1.3.6 Asymmetric Conjugate Addition Reactions

A special ability in the class of [2.2]paracyclophaneketimine ligands is to catalyze the 1,4-addition of zinc reagents to α,β-unsaturated carbonyl compounds **21** [23] in the absence of copper ions [24]. While the reaction of α,β-unsaturated aldehydes such as **21** with zinc reagents in the presence of a small amount of the catalyst (4 mol%) gave excellent enantiomeric excesses (up to 97% *ee*) with moderate regioselectivity (Scheme 2.1.3.6), α,β-unsaturated ketones yielded β-alkylated ketones in good enantiomeric excesses (82–87% *ee*) and good yields.

2.1.3.7 Asymmetric Addition Reactions to Imines

The synthesis of α-branched amines caught our attention, as these compounds exhibit particular biological activity. In several of our ongoing projects involving the synthesis of biologically active compounds, we required the asymmetric synthesis of α-branched chiral amines. α-Branched amines can be prepared by various routes, all performed in an asymmetric fashion. Currently, enzymatic and chemical separation of racemic α-branched amines and also diastereoselective methods still play a major role on an industrial scale [25]. However, due to poor separation by the latter methods and for economic reasons, catalytic approaches will be favored.

Recently, various methods have been applied to synthesize α-branched amines. In general, imines serve as valuable starting materials since organyl groups or hydrogen can be delivered enantiospecifically at the C=N double bond.

The catalytic asymmetric preparation of α-chiral amines, by addition of organometallic reagents to C=N bonds, is one of the most important reactions in homogeneous catalysis [26]. However, the catalytic asymmetric addition of simple alkylmetals has been achieved only in recent years.

When we entered the field, Tomioka et al. had just introduced the dialkylzinc addition to *N*-sulfonylimines in the presence of chiral amidophosphine–copper(II) complexes, producing high levels of enantioselectivity (up to 94% *ee*) [27]. At the

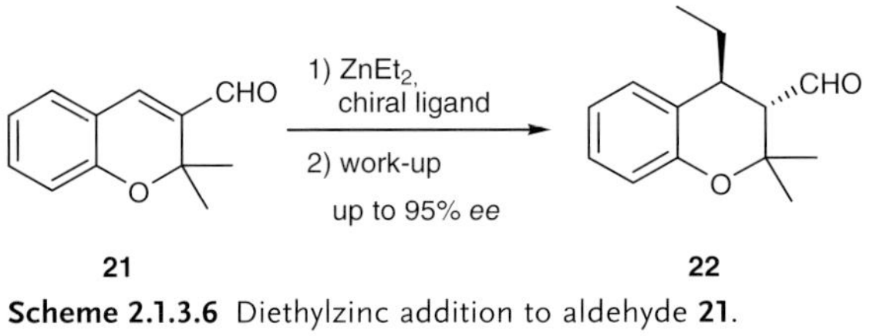

Scheme 2.1.3.6 Diethylzinc addition to aldehyde **21**. Conditions: aldehyde **21** (0.5 mmol), ligand (5%), toluene (1 mL), diethylzinc (1.0 mL, 1.0 M in hexane, 2 equiv), 0 °C, 16 h under argon.

same time, Hoveyda, Snapper et al. reported a zirconium-catalyzed variant using peptidic Schiff-base ligands, which were optimized in a combinatorial fashion (up to 97% ee was obtained for certain N-arylimines) [28]. The lack of a simple method, employing only a catalytic amount of an N,O-ligand and no additional central metal (other than zinc itself), is in sharp contrast to the asymmetric alkylation of aldehydes with organozinc reagents. This apparent deficit is not to be ascribed to selectivity problems in the addition reaction, but rather to the unreactivity of many imine substrates or precursors toward alkylzinc reagents. Additionally, reactive imine derivatives or their addition products tend to coordinate with the catalytically active zinc complexes and therefore prevent formation of the catalytic cycle.

At the outset of our study [29], a novel source for imines was used. We examined the reactivity of N-acyl-α-(p-tolylsulfonyl)benzylamines **23**, which can be applied successfully as N-acylimine precursors and are readily available in a one-pot-synthesis from benzaldehyde, amides, and p-tolylsulfinic acid. [30, 31] The reaction proceeds through deprotonation of the amide **23**, which leads to the elimination of the sulfinate to form the N-acylimine **24**. After several experiments, we found that the reaction with the formyl derivative **23a** proceeded cleanly to give the N-(1-phenylpropyl)formamide **25a** in 61–95% ee in the presence of 1 mol% of ligand (Scheme 2.1.3.7). Functionalized substrates are usually well tolerated in dialkylzinc additions. Electron-rich and electron-poor substrates gave comparably high enantiomeric excesses of 90–95%. *Ortho* and *para* substituents do not influence the selectivity of the catalysis and even hindered imines were recognized at a very high level of enantioselectivity. For *meta*-substituted substrates, however, ligand (R_p,S)-**5** gave superior results. A scaleup to 5 mmol of substrate gave identical results to those obtained on a 0.5 mmol scale.

An extension to aliphatic imines gave a certain level of success. The cyclohexyl-derived precursor **26** was successfully transformed into the corresponding amine **27** in 72% ee (Scheme 2.1.3.8).

Enantiomerically pure diarylmethylamines are important intermediates in the synthesis of biologically active compounds such as **28** or **29** (Fig. 2.1.3.8). Among several drug candidates, Cetirizine hydrochloride (**28**) stands out as a commercially important nonsedative antihistamine agent. Binding studies indicate that

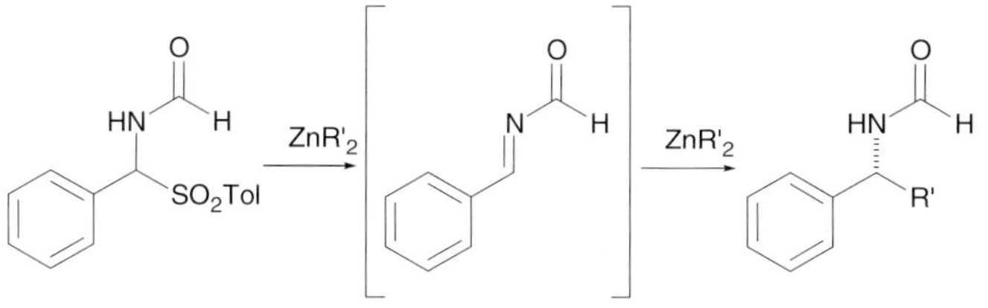

Scheme 2.1.3.7 1,2-Addition to imines.

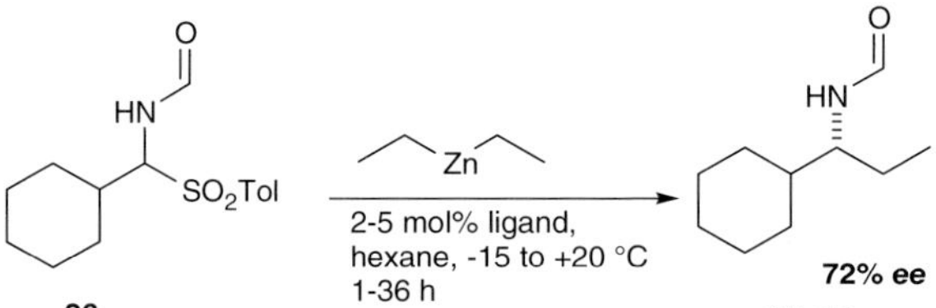

Scheme 2.1.3.8 Diethylzinc addition to aliphatic imines.

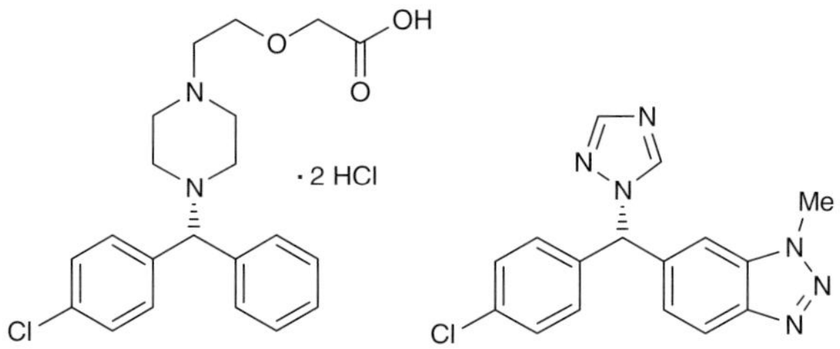

Cetirizine Hydrochloride (**28**)   (+)-Voronzole (**29**)

Fig. 2.1.3.8 Cetirizine hydrochloride **28**, a potent δ-opioid receptor agonist, and (+)-Voronzole **29**, an aromatase inhibitor.

the R enantiomer displays a better pharmacological profile than the racemate [32]. Despite the importance of enantiopure diarylmethylamines, synthetic access routes and especially asymmetric (catalytic) variants are rather limited [26]. There are several synthetic routes, e.g., to enantiopure Cetirizine employing either resolution techniques, stoichiometric amounts of chromium complexes, or diastereoselective approaches via chiral auxiliaries [33]. However, in our understanding, there is only a single report, by Hayashi et al. [34], on the asymmetric catalytic addition of an organometallic arylation agent to an imine derivative.

Although the phenyl transfer to imines was viable, the transfer to aldehydes was difficult, due to the much higher reactivity of diphenylzinc compared to dialkylzinc and the concomitant rapid uncatalyzed background reaction [35]. The catalytic procedure for the enantioselective addition of organozinc to masked N-formylimines employing catalytic amounts of [2.2]paracyclophane-based ketimines was nevertheless investigated with arylzinc reagents. In a fruitful cooperation of the Sonderforschungsbereich 380 with the Bolm group in Aachen, we were able to develop the first highly enantioselective phenylzinc addition to imines, giving rise to optically active diarylmethylamines in very high enantiomeric excesses.

At the outset of this study, a range of different N,O-ligands (Fig. 2.1.3.2) were examined in the phenylation of N-[p-tolylmethyl(toluene-4-sulfonyl)]formamide (**23b**, Table 2.1.3.2). We started out by employing the reaction conditions

Table 2.1.3.2 Substrate spectrum for the phenyl transfer to imines.[a]

Entry	R	Product	(R_p, S)-6 [mol%]	Yield[b] [%]	ee [%][c]
1	4-Me-C_6H_4	30b	10	99 (85)	97 (+)
2	4-Me-C_6H_4	30b	5	99	94 (+)
3	4-Cl-C_6H_4	30cb	10	99 (82)	94 (+)-(R)
4	4-Cl-C_6H_4	30c	5	99	81 (+)-(R)
5	4-Cl-C_6H_4	30c	1	98	69 (+)-(R)
6	4-MeO-C_6H_4	30d	10	99 (75)	97 (+)
7	3-Me-C_6H_4	30e	10	98	89 (+)
8	2,6-Cl_2-C_6H_4	30f	10	99 (89)	95 (+)
9	4-tBu-C_6H_4	30g	10	98 (81)	96 (+)
10	4-COOMe-C_6H_4	30h	10	99 (80)	95 (−)

a Reactions were carried out in toluene at −20 °C for 12 h with 2 equiv $ZnPh_2$, 2 equiv $ZnEt_2$, and 0.25 mmol of imine precursor **23b–23h**. b Determined by ^{1}H NMR. Yields in parentheses refer to yields after column chromatography. c Determined by HPLC using a chiral stationary phase.

developed for the enantioselective phenyl transfer to aldehydes, using a mixed zinc reagent formed in situ from diphenylzinc and diethylzinc. This reagent selectively transfers only the phenyl moiety to the substrate, yielding N-formylamide **30b** in a very high yield without formation of the corresponding ethylation product.

The ligand screening showed that ferrocenes and cyrhetrenes, which are the best ligands for the enantioselective phenyl transfer to aldehydes (up to 99% ee), gave only moderate enantioselectivities in the addition to imine **24a**. Catalysis with a diastereomeric ferrocene did not lead to any improvement. In contrast, the use of [2.2]paracyclophane-based imines **4–6** gave rise to N-formyldiarylmethylamines with good to excellent enantioselectivities (up to 97% ee). Interestingly, (S_p)-**5** bearing only the element of planar chirality also gave rise to the product amine in very high enantioselectivity. In our ongoing study of [2.2]-paracyclophane-based N,O-ligands, this is the first example of a ligand in which the combination of both planar and central chirality is *not* required to achieve high enantioselectivity.

In order to demonstrate the potentially wide application of this method, a wider range of substrates was applied in the title reaction (Table 2.1.3.2). The results reveal that aromatic imine precursors with different electronic properties as well as variable substitution patterns are equally well tolerated. The substrates can be electron-rich or electron-poor, and even sterically hindered imines with a double *ortho* substitution gave excellent results (95% *ee*, entry 8). Only *meta*-substituted starting material gave a product with a slightly lower *ee* value (89% *ee*, entry 7). Interestingly, the same effect was observed in the diethylzinc addition to imines using (R_p, S)-4 and (R_p, S)-6 [29].

The deprotection of *N*-formylamines **30** to the free amines can be achieved easily by acidic methanolysis. For the *N*-formylamide **30b**, it was shown that the deprotection proceeds quantitatively and without racemization.

These highly enantioselective Lewis-acid/Lewis-base-catalyzed dialkylzinc and phenylzinc addition reactions to imines give rise to arylalkylamides and diarylmethylamides in excellent yields and enantioselectivities. Due to the simplicity of the process and the good availability of the imine precursors **23** from the corresponding aldehydes, wide applicability of the reported catalytic reaction can be expected.

2.1.3.8 Asymmetric Addition Reactions on Solid Supports

We could now show that the asymmetric 1,2-addition using dialkylzinc reagents and chiral N,O-ligands with a paracyclophane backbone is also applicable on solid supports [36]. Therefore, we immobilized *ortho*-carboxybenzaldehyde **32** on Merrifield resin **31** under standard reaction conditions.

We then observed reactions of *ortho*-carboxybenzaldehyde resin **33** with different organometallic reagents under various conditions. Zinc reagents reacted selectively at a higher temperature in the presence of an amino alcohol such as *N,N*-dimethylaminoethanol to give phthalides of good purities and in moderate overall yields. The introduction of the chiral paracyclophane ligand (S_p, S)-**5a** into the 1,2-dialkylzinc addition reaction resulted in the phthalides **34** with moderate to good enantioselectivities (Scheme 2.1.3.9).

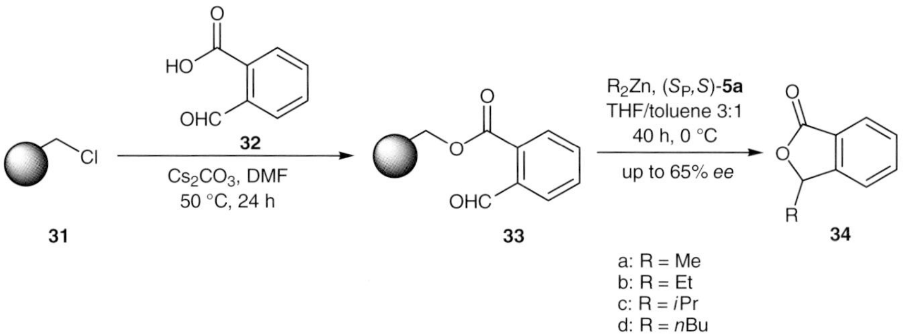

Scheme 2.1.3.9 Immobilization of carboxybenzaldehydes **32** on Merrifield resin (**31**) and synthesis of benzobutyrolactones **34** with dialkylzinc addition using a cyclative-cleavage approach [36].

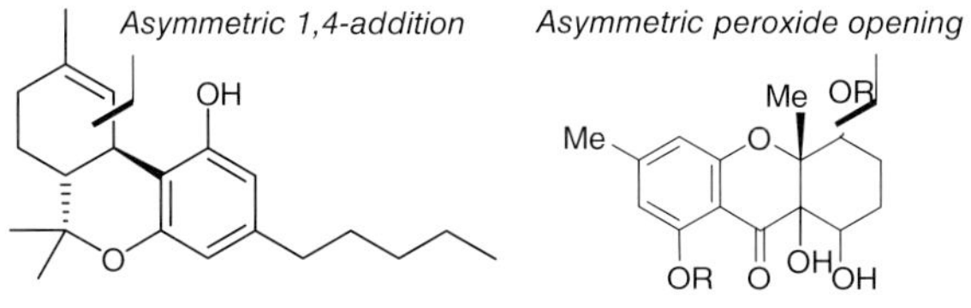

Fig. 2.1.3.9 Targets **35** and **36** for natural product syntheses.

2.1.3.8.1 Applications

Currently, we are using these strategies for the synthesis of natural products such as tetrahydrocannabinol (**35**) [37] and diversonol (**36**) (Fig. 2.3.1.9) [38, 39].

2.1.3.9 Conclusions and Future Perspective

Over the last five years, we have designed, synthesized, and applied new ligands for asymmetric 1,2- and 1,4-addition reactions. Suitable ligands were found for the addition of alkyl-, aryl-, and alkenylzinc reagents to α,β-unsaturated aldehydes and ketones, α-branched and unbranched aliphatic aldehydes, and imines. Although some substrates such as ketones and other carbonyl compounds have remained a challenge, we believe that this system provides an excellent entry into various classes of chiral intermediates. Application of these synthesized complex molecules is the current pursuit in our laboratories.

Acknowledgements

We acknowledge the fruitful cooperation and many stimulating discussions in the field of [2.2]paracyclophane chemistry with the groups of Professor Hopf, Professor Rozenberg (T. Danilova), and Professor Bolm (N. Hermanns, J. Rudolph). We thank the DFG (SFB 380) and the Fonds der Chemischen Industrie for financial support, and Dr. Klaus Ditrich for donation of chiral amines (Chipros).

References

1 H. J. Reich, D. J. Cram, *J. Am. Chem. Soc.* **1969**, *91*, 3527.
2 S. Bräse, S. Dahmen, S. Höfener, F. Lauterwasser, M. Kreis, R. E. Ziegert, *Synlett* **2004**, 2647.
3 P. J. Pye, K. Rossen, R. A. Reamer, N. N. Tsou, R. P. Volante, P. J. Reider, *J. Am. Chem. Soc.* **1997**, *119*, 6207.
4 a) S. E. Gibson, J. D. Knight, *Org. Biomol. Chem.* **2003**, *1*, 1256; b) A. de Meijere, B. König, *Synlett* **1997**, 1221; c) R. Gleiter, H. Hopf, *Modern Cyclophane Chemistry*, Wiley-VCH, Weinheim, **2004**.
5 S. Bräse, Planar chiral ligands based on [2.2]paracyclophanes, in *Asymmetric Synthesis – The Essentials* (Eds.: M. Christmann, S. Bräse), Wiley-VCH, Weinheim, **2006**.
6 S. Bräse, F. Lauterwasser, R. E. Ziegert, *Adv. Synth. Catal.* **2003**, *345*, 869.

7 For other contributions from our group, see: a) M. Kreis, C. J. Friedmann, S. Bräse, *Chem. Eur. J.* **2005**, *11*, 7387–7394. b) M. Kreis, M. Nieger, S. Bräse, *J. Organomet. Chem.* **2006**, *691*, 2171–2181.
8 V. Rozenberg, V. Kharitonov, D. Antonov, E. Sergeeva, A. Aleshkin, N. Ikonnikov, S. Orlova, Y. Belokon, *Angew. Chem.* **1994**, *106*, 106; *Angew. Chem. Int. Ed. Engl.* **1994**, *33*, 91.
9 A. H. Vetter, A. Berkessel, *Tetrahedron Lett.* **1998**, *39*, 1741.
10 V. Rozenberg, T. Danilova, E. Sergeeva, E. Vorontsov, Z. Starikova, K. Lysenko, Y. Belokon, *Eur. J. Org. Chem.* **2000**, 3295.
11 S. D. Pastor, A. Togni, *J. Am. Chem. Soc.* **1989**, *111*, 2333.
12 C. Bolm, K. Muniz-Fernández, A. Seger, G. Raabe, K. Günther, *J. Org. Chem.* **1998**, *63*, 7860.
13 For studies on chiral cooperativity in axial- and central-chiral ligands, see: a) G. J. H. Buisman, L. A. van der Veen, A. Klootwijk, W. G. W. de Lange, P. C. J. Kamer, P. W. N. M. van Leeuwen, D. Vogt, *Organometallics* **1997**, *16*, 2929; b) S. Cserépi-Szűcs, I. Tóth, L. Párkányi, J. Bakos, *Tetrahedron: Asymmetry* **1998**, *9*, 3135.
14 H. Hopf, D. G. Barrett, *Liebigs Ann.* **1995**, 449.
15 F. Lauterwasser, M. Nieger, H. Mansikkamäki, K. Nättinen, S. Bräse, *Chem. Eur. J.* **2005**, *11*, 4509–4525.
16 F. Lauterwasser, S. Vanderheiden, S. Bräse, *Adv. Synth. Catal.* **2006**, *348*, 443–448.
17 S. Dahmen, S. Bräse, *Chem. Commun.* **2002**, 26.
18 S. Höfener, F. Lauterwasser, S. Bräse, *Adv. Synth. Catal.* **2004**, *346*, 755.
19 M. Kitamura, S. Suga, K. Kawai, R. Noyori, *J. Am. Chem. Soc.* **1986**, *108*, 6071.
20 W. Oppolzer, R. N. Radinov, *Helv. Chim. Acta* **1992**, *75*, 10.
21 The addition of vinylzinc has also been studied: a) J. L. von dem Bussche-Hünnefeld, D. Seebach, *Tetrahedron* **1992**, *48*, 5719; b) K. Soai, K. Takahashi, *J. Chem. Soc., Perkin Trans. 1* **1994**, 1257.
22 S. Dahmen, S. Bräse, *Org. Lett.* **2001**, *3*, 4119.
23 S. Bräse, F. Lauterwasser, R. E. Ziegert, *Adv. Synth. Catal.* **2003**, *345*, 869.
24 S. Bräse, S. Höfener, *Angew. Chem.* **2005**, *117*, 8091; *Angew. Chem. Int. Ed.* **2005**, *44*, 7879.
25 M. Breuer, K. Ditrich, T. Habicher, B. Hauer, M. Keßeler, R. Stürmer, *Angew. Chem.* **2004**, *116*, 806.
26 For reviews on auxiliary controlled additions to C=N, see: a) D. Enders, U. Reinhold, *Tetrahedron: Asymmetry* **1997**, *8*, 1895; b) R. Bloch, *Chem. Rev.* **1998**, *98*, 1407; c) G. Alvaro, D. Savoia, *Synlett* **2002**, 651. For reviews on corresponding catalytic processes, see: d) S. Kobayashi, H. Ishitani, *Chem. Rev.* **1999**, *99*, 1069.
27 H. Fujihara, K. Nagai, K. Tomioka, *J. Am. Chem. Soc.* **2000**, *122*, 12 055.
28 J. R. Porter, J. F. Traverse, A. H. Hoveyda, M. L. Snapper, *J. Am. Chem. Soc.* **2001**, *123*, 984.
29 S. Dahmen, S. Bräse, *J. Am. Chem. Soc.* **2002**, *124*, 5940.
30 S. Schunk, D. Enders, *Org. Lett.* **2001**, *3*, 3177–3180.
31 J. Sisko, M. Mellinger, P. W. Sheldrake, N. H. Baine, *Tetrahedron Lett.* **1996**, *37*, 8113.
32 M. Gillard, C. van der Perren, N. Moguilevsky, R. Massingham, P. Chatelain, *Mol. Pharmacol.* **2002**, *61*, 391.
33 a) D. A. Pflum, D. Krishnamurthy, Z. Han, S. A Wald, C. H. Senanayake, *Tetrahedron Lett.* **2002**, *43*, 923; b) N. Plobeck, D. Powell, *Tetrahedron: Asymmetry* **2002**, *13*, 303.
34 T. Hayashi, M. Ishigedani, *J. Am. Chem. Soc.* **2000**, *122*, 976.
35 C. Bolm, N. Hermanns, J. P. Hildebrand, K. Muñiz, *Angew. Chem.* **2000**, *112*, 3607; *Angew. Chem. Int. Ed.* **2000**, *39*, 3465.
36 K. Knepper, R. E. Ziegert, S. Bräse, *Tetrahedron* **2004**, *60*, 8591.
37 B. Lesch, J. Toräng, M. Nieger, S. Bräse, *Synthesis* **2005**, 1888–1900.
38 C. F. Nising, U. K. Ohnemüller, S. Bräse, *Angew. Chem.* **2006**, *118*, 313; *Angew. Chem. Int. Ed.* **2006**, *45*, 307.
39 R. E. Ziegert, S. Bräse, *Synlett* **2006**, 2119–2123.

2.1.4
Palladium-Catalyzed Allylic Alkylation of Sulfur and Oxygen Nucleophiles – Asymmetric Synthesis, Kinetic Resolution and Dynamic Kinetic Resolution
Hans-Joachim Gais

2.1.4.1 Introduction

The separation of the enantiomers from a racemate (resolution) by various techniques is an important method for the attainment of enantio-enriched compounds on both a laboratory and an industrial scale [1]. Kinetic resolution methods employing chiral catalysts are especially attractive. While catalytic kinetic resolution has for a long time been the domain of enzymes [1k], much progress has been made since the mid-1990s toward the development of transition-metal and small-molecule catalysts. Kinetic resolution can give access either to both enantiomers of a racemate or to one enantiomer of a racemate and a further but structurally different chiral compound, depending on the reaction involved, in a maximum yield of 50% for each compound. Dynamic kinetic resolution is a very attractive method for the complete conversion of a racemate either to one enantiomer of the racemate or to a structurally different chiral target compound. It involves a kinetic resolution in combination with a racemization of the more slowly reacting enantiomer of the racemate. Currently both kinetic and dynamic kinetic resolution techniques employing chiral catalysts are being extensively investigated. Particularly interesting in this context is the Pd(0)-catalyzed allylic alkylation of nucleophiles with racemic allylic esters [2].

According to a simplified mechanistic scheme for Pd-catalyzed allylic alkylation involving a C_2-symmetric bisphosphane and a symmetrical allylic ester (for unsymmetrical substrates, see Section 2.1.4.5), both enantiomers of the substrate are converted by the chiral Pd(0) catalyst to the same π-allyl–Pd(II) complex, the reaction of which with the nucleophile affords the allylic substitution product and the catalyst (Scheme 2.1.4.1) [2, 3]. While the second step, the enantioselective alkylation, had been intensively studied at the beginning of our investigations, not much was known about the likelihood of a kinetic resolution in the first step and its selectivity. Kinetic resolution in Pd-catalyzed allylic alkylation was described for the first time by Hayashi et al. in 1986 [4]. Reaction of an unsymmetrical allylic ester with a carbon nucleophile in the presence of a Pd catalyst containing a chiral ferrocenylphosphane was found to proceed with medium selective kinetic resolution. In 1998 we observed high selectivities in both kinetic resolution and enantioselective alkylation in the reaction of racemic 1,3-diphenylallyl acetate with lithium *t*-butylsulfinate by employing an asymmetric phosphinooxazoline ligand [2e], and determined the stereochemical course of both the resolution and alkylation (Scheme 2.1.4.2) [5].

At the same time and in the years to follow, several other groups reported the observation of high selectivities in the Pd-catalyzed resolution of racemic substrates [6]. The kinetic resolution depicted in Scheme 2.1.4.2 gives access to both the enantio-enriched allylic acetate and sulfone. Because of the many applications chiral allylic alcohols and allylic sulphur derivatives have found in the synthesis

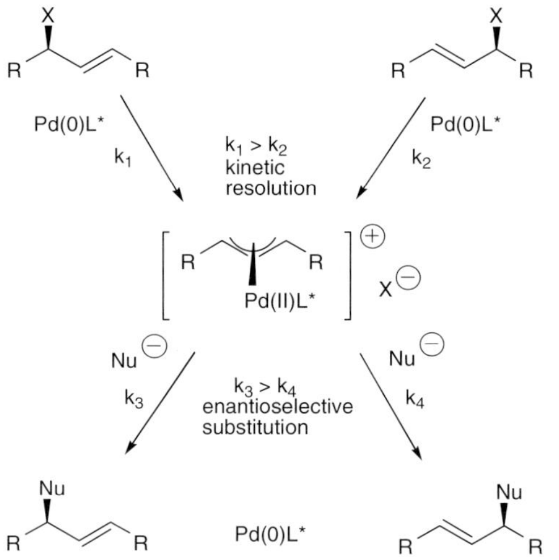

Scheme 2.1.4.1 Mechanistic scheme for the Pd(0)-catalyzed reaction of symmetrical racemic allylic substrates (X = leaving group) with nucleophiles (Nu) in the presence of a chiral C_2-symmetric ligand L*.

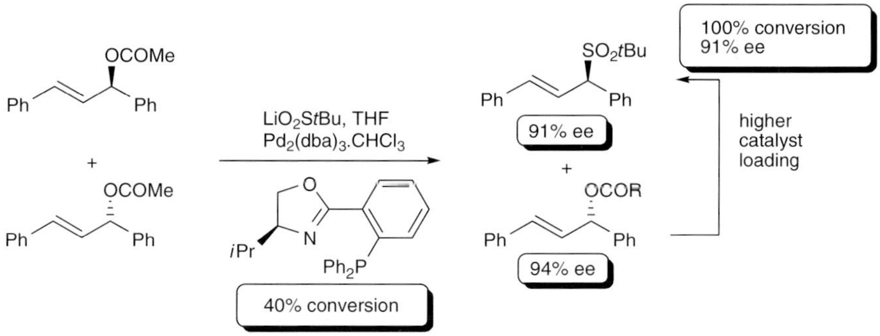

Scheme 2.1.4.2 Pd-catalyzed enantioselective allylic alkylation of a sulfinate ion and kinetic resolution of a racemic allylic ester.

of biologically active compounds, they are of high synthetic value. It was therefore of interest to explore the potential of the Pd-catalyzed allylic alkylation for both the asymmetric synthesis of allylic sulfones and sulfides, and the kinetic and dynamic kinetic resolution of allylic alcohols by using S- and O-nucleophiles [5, 7, 8].

2.1.4.2 Asymmetric Synthesis of Allylic Sulfones and Allylic Sulfides and Kinetic Resolution of Allylic Esters

2.1.4.2.1 Kinetic Resolution

The C_2-symmetrical bisphosphane **BPA** [9] (Scheme 2.1.4.3) was selected as a ligand for the Pd atom because it provided high enantioselectivities in the

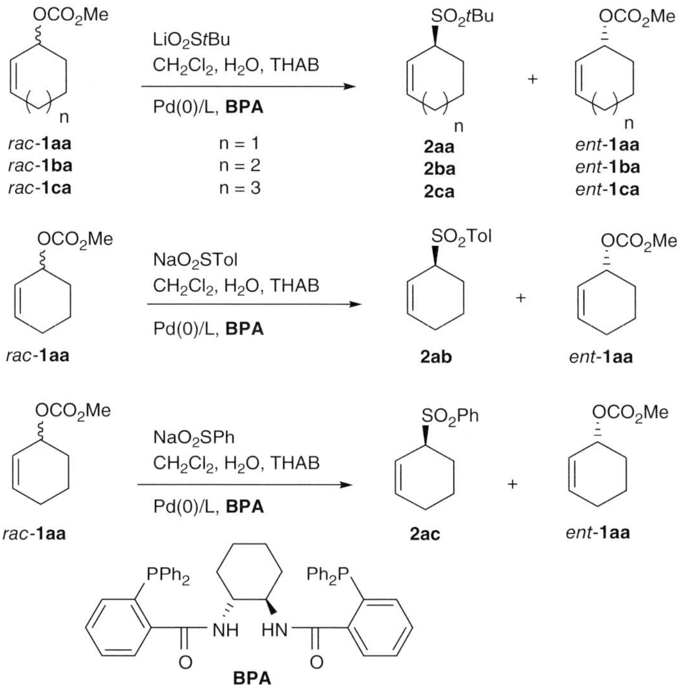

Scheme 2.1.4.3 Pd-catalyzed kinetic resolution of cyclic allylic allylic carbonates with lithium sulfinates.

Table 2.1.4.1 Pd(0)/BPA-catalyzed kinetic resolution of cyclic carbonates with lithium t-butylsulfinate.

Substrate	Conv. [%]	Carbonate	Yield [%]	ee [%]	Sulfone	Yield [%]	ee [%]
rac-**1aa**	54	ent-**1aa**	34	≥99	**2aa**	49	98
rac-**1ba**	53	ent-**1ba**	33	94	**2ba**	46	95
rac-**1ca**	58	ent-**1ca**	34	≥99	**2ca**	48	96

substitution of a range of symmetrical cyclic and acyclic substrates with various nucleophiles [2d,f–i].

Both the kinetic resolution and the enantioselective substitution of the cyclic carbonates rac-**1aa–1ca** in the reaction with lithium t-butylsulfinate in the presence of $Pd_2(dba)_3 \cdot CHCl_3$ (dba = dibenzylideneacetone) (Pd(0)/L) (1.5 mol%), **BPA** (4.5 mol%) and Hex_4NBr (THAB) gave the carbonates ent-**1aa–1ca** and sulfones **2aa**, **2ba**, and **2ca**, respectively, with high enantioselectivities in good yields (Table 2.1.4.1). After approximately 50% conversion of the carbonate, the reaction with the sulfinate salt came to a practically complete halt. In agreement with previous results, the faster reacting enantiomers of carbonates rac-**1aa–1ca** and the preferentially formed sulfones **2aa**, **2ba**, and **2ca** have the same absolute configuration [5, 10].

In addition to lithium *t*-butylsulfinate, sodium tolylsulfinate and sodium phenylsulfinate have been used as nucleophiles in the kinetic resolution of carbonate *rac*-**1aa**. Table 2.1.4.2 shows that the kinetic resolution and enantioselective substitution of carbonate *rac*-**1aa** by using the tolylsulfinate anion occurred with high selectivities and gave the highly enantio-enriched carbonate *ent*-**1aa** and sulfone **2ab** in good yields. Similar results were recorded by using sodium phenylsulfinate [10].

Next the kinetic resolution of the acyclic carbonate *rac*-**3aa** by using lithium *t*-butylsulfinate in the presence of Pd(0)/L (1.5 mol%) and **BPA** (4.5 mol%) was studied (Scheme 2.1.4.4). Both kinetic resolution and enantioselective substitution occurred in this case with high selectivities and gave the carbonate *ent*-**3aa** and sulfone **4aa** in good yields (Table 2.1.4.3).

A study of the Pd-catalyzed resolution of allylic substrates with thiols was undertaken to see whether they could function as nucleophiles [11]. The cyclic carbonates *rac*-**1aa** and *rac*-**1ba** and the acyclic carbonates *rac*-**3aa** and *rac*-**3ba** were selected as substrates and 2-pyrimidinethiol as nucleophile (Schemes 2.1.4.5 and 2.1.4.6). The reactions of carbonates *rac*-**1aa** and *rac*-**1ba** with 2-pyrimidinethiol

Table 2.1.4.2 Pd(0)/BPA-catalyzed kinetic resolution of the cyclic carbonate *rac*-**1aa** with arylsulfinates.

Sulfinate	Convn [%]	Carbonate	Yield [%]	ee [%]	Sulfone	Yield [%]	ee [%]
NaO$_2$STol	68	*ent*-**1aa**	24	≥99	**2ab**	60	≥99
NaO$_2$SPh	62	*ent*-**1aa**	27	≥99	**2ac**	56	≥99

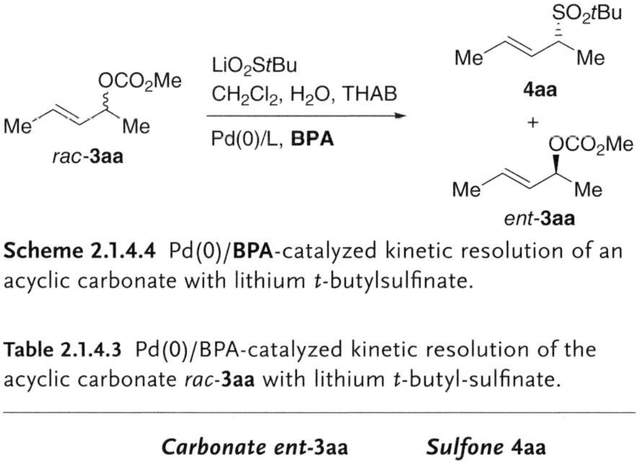

Scheme 2.1.4.4 Pd(0)/**BPA**-catalyzed kinetic resolution of an acyclic carbonate with lithium *t*-butylsulfinate.

Table 2.1.4.3 Pd(0)/BPA-catalyzed kinetic resolution of the acyclic carbonate *rac*-**3aa** with lithium *t*-butyl-sulfinate.

	Carbonate *ent*-**3aa**		Sulfone **4aa**	
Convn [%]	Yield [%]	ee [%]	Yield [%]	ee [%]
24	53	33	21	98
36	46	51	32	98
73	19	≥99	68	96

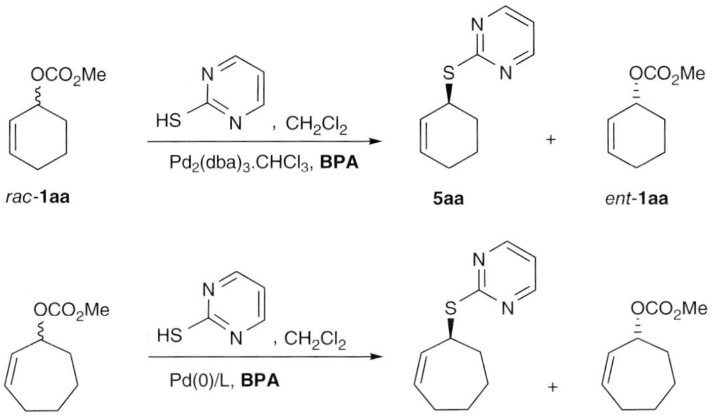

Scheme 2.1.4.5 Pd(0)/**BPA**-catalyzed kinetic resolution of cyclic carbonates with thiols.

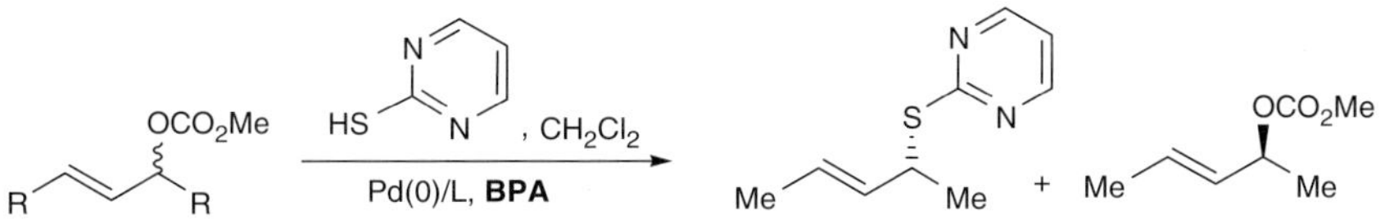

Scheme 2.1.4.6 Pd(0)/**BPA**-catalyzed kinetic resolution of acyclic carbonates with thiols.

Table 2.1.4.4 Pd(0)/**BPA**-catalyzed kinetic resolution of allylic carbonates with 2-pyrimidinethiol.

Entry	Substrate	Conv. [%]	Carbonate	Yield [%]	ee [%]	Sulfide	Yield [%]	ee [%]
1	rac-**1aa**	50	ent-**1aa**	41	≥99	**5aa**	46	84
2	rac-**1ba**	50	ent-**1ba**	39	97	**5ba**	38	84
3	rac-**3aa**	50	ent-**3aa**	36	≥99	**6aa**	36	93
4	rac-**3ba**	50	ent-**3ba**	28	≥99	**6ba**	44	92

in the presence of Pd(0)/L (2.5–5 mol%) and **BPA** (5.5–11 mol%) gave the carbonates ent-**1aa** and ent-**1ba**, respectively, and the sulfides **5aa** and **5ba**, respectively, with high enantioselectivities in good yields (Table 2.1.4.4, entries 1 and 2) [10].

The ee values of the sulfides **5aa** and **5ba** are significantly lower than those of the corresponding sulfones **2aa** and **2ba**. Kinetic resolution and substitution of the acyclic carbonates rac-**3aa** and rac-**3ba** with the thiol under the conditions used above proceeded with similar high selectivities and gave the carbonates ent-**3aa** and ent-**3ba**, respectively, and the sulfides **6aa** and **6ba**, respectively (entries 3 and 4). The reactions of the cyclic carbonates rac-**1aa** and rac-**1ba** went to 50%

conversion of the substrates in much shorter reaction times than those of the acyclic carbonates rac-**3aa** and rac-**3ba**.

2.1.4.2.2 Selectivity

All of the kinetic resolutions described above have been characterized in terms of yields and *ee* values of the recovered substrate and the product. In principle the efficiency of a kinetic resolution can also be described by the selectivity factor *S* [1u], the ratio of the rate constants for the reactions of the enantiomers of the substrate with the catalyst. For a Pd-catalyzed kinetic resolution of an allylic substrate obeying first-order kinetics in regard to the reaction of the substrate with the catalyst (unimolecularity) *S* can be calculated according to Eq. (1), which contains as variables the conversion (*c*) and the *ee* value of the substrate (ee_s).

$$S = \ln[(1-c)(1-ee_s)]/\ln[(1-c)(1+ee_s)] \quad (0 < c < 1, 0 < ee_s < 1) \tag{1}$$

Application of Eq. (1) requires, however, determination of *c* and *ee* of the substrate with high precision. Small errors in the measurement of *ee* and *c* can lead to major apparent changes of *S* with conversion, particularly in the case of high *S* values, a problem which is often underestimated in the measurement of *S* [6]. However, the overall error can be reduced by analysis of a series of *ee* versus *c* values [12].

Calculation of *S* for the kinetic resolution of the cyclohexenyl carbonate rac-**1aa** with 2-pyrimidinethiol and lithium *t*-butylsulfinate according to Eq. (1) gave large values for pairs of *ee* versus *c* (compare Tables 2.1.4.5 and 2.1.4.6). This is in accordance with the isolation of ent-**1aa** with a high *ee* at approximately 50% conversion in the preparative experiments (compare Tables 2.1.4.1 and 2.1.4.4). However, Tables 2.1.4.5 and 2.1.4.6 also reveal major and irregular changes of *S* with conversion. Since our measurements of *ee* and *c* had a precision of only ±0.5% and ±1.0%, respectively, we ascribe the change of *S* with conversion mainly to errors in the determination of both values.

2.1.4.2.3 Asymmetric Synthesis

It was then of interest to see whether an asymmetric synthesis of the cyclic sulfones **2aa**, **2ba**, **2ca**, and **2da** (Scheme 2.1.4.7) and of the acyclic sulfones **4aa** and

Table 2.1.4.5 Selectivity of the Pd(0)/BPA-catalyzed kinetic resolution of rac-**1aa** with lithium *t*-butyl-sulfinate.

Entry	t [min]	rac-1aa Conv. [%]	ent-1aa ee [%]	S
1	2	23	29	90
2	4	34	50	110
3	5	38	60	173
4	10	49	90	95
5	15	51	95	81
6	20	54	≥99	61

Table 2.1.4.6 Selectivity of the Pd(0)/BPA-catalyzed Kinetic resolution of rac-**1aa** with 2-pyrimidinethiol

Entry	t [min]	rac-1aa Convn [%]	ent-1aa ee [%]	S
1	10	29	40	141
2	15	36	51	34
3	20	40	64	95
4	25	46	77	46
5	30	48	89	164
6	35	52	97	76
7	40	53	≥99	80

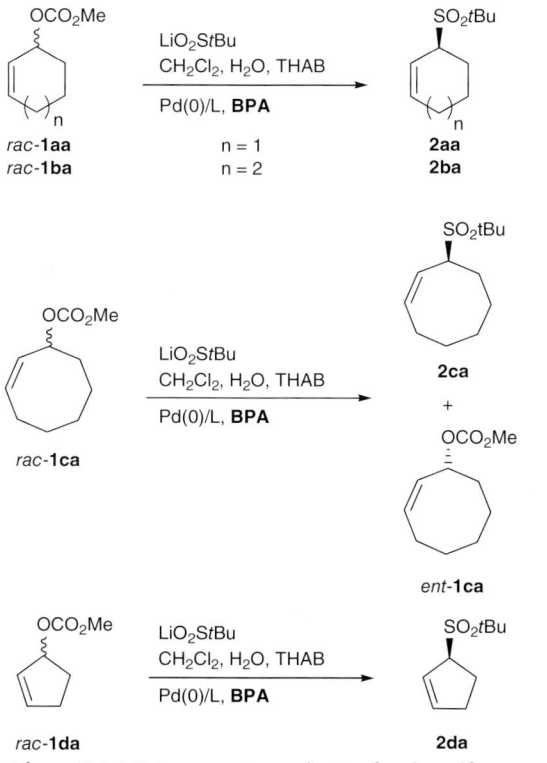

Scheme 2.1.4.7 Asymmetric synthesis of cyclic sulfones.

4ba (Scheme 2.1.4.8) could be achieved with complete transformation of the racemic substrates.

As shown by Table 2.1.4.7, the highly enantio-enriched cyclohexenyl and cycloheptenyl sulfones **2aa** and **2ba**, respectively, were obtained in high yields by using Pd(0)/L (1.5 mol%) and **BPA** (4.5 mol%) (entries 1 and 2). In the case of the eight-membered cyclic carbonate rac-**1ca** the selectivity of the kinetic resolution was so high that even after a prolonged reaction time the conversion of the allylic

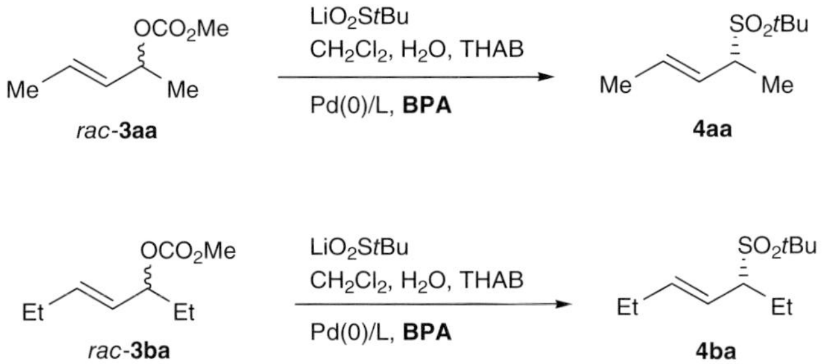

2.1 Chemical Methods

Scheme 2.1.4.8 Asymmetric synthesis of acyclic sulfones from carbonates.

rac-3aa (Me-CH=CH-CH(OCO$_2$Me)-Me) + LiO$_2$StBu, CH$_2$Cl$_2$, H$_2$O, THAB, Pd(0)/L, **BPA** → 4aa (Me-CH=CH-CH(SO$_2$tBu)-Me)

rac-3ba (Et-CH=CH-CH(OCO$_2$Me)-Et) + LiO$_2$StBu, CH$_2$Cl$_2$, H$_2$O, THAB, Pd(0)/L, **BPA** → 4ba (Et-CH=CH-CH(SO$_2$tBu)-Et)

Table 2.1.4.7 Pd(0)/BPA-catalyzed asymmetric synthesis of cylic allylic S-t-butylsulfones.

Entry	Carbonate	Sulfone	Yield [%]	ee [%]
1	rac-1aa	2aa	95	94
2	rac-1ba	2ba	89	93
3	rac-1ca	2ca	50	94
4	rac-1da	2da	76	89

Table 2.1.4.8 Pd(0)/BPA-catalyzed asymmetric synthesis of acyclic allylic S-t-butylsulfones.

Entry	Substrate	Conv. [%]	Sulfone	Yield [%]	ee [%]
1	rac-3aa	100	4aa	98	98
2	rac-3ba	100	4ba	97	97
4	rac-3ab	68	4aa	51	98
5	rac-3bb	53	4ba	43	96

substrate did not exceed 53% (entry 3). The reaction of the five-membered cyclic carbonate rac-1da with lithium t-butylsulfinate gave sulfone 2da with 89% ee in 76% yield (entry 4).

The acyclic carbonates rac-3aa and rac-3ba (Scheme 2.1.4.8) showed a higher reactivity than the cyclic carbonates rac-1aa–1da. After complete conversion of the substrates the highly enantio-enriched sulfones 4aa and 4ba were isolated in high yields (Table 2.1.4.8, entries 1 and 2).

When favorable results in the case of allylic sulfones had been obtained, asymmetric synthesis of the cyclic sulfides 5a, 5ba, 5ab and 5ca (Scheme 2.1.4.9) and of the acyclic sulfides 6aa, 6ab, 6ac, 6ba, and 6bb (Scheme 2.1.4.10) was investigated. As shown by Table 2.1.4.9, the pyrimidyl sulfide 5aa could be obtained in medium to good yield (entry 1). Interestingly, the corresponding pyridyl sulfide 5ab isolated from the reaction of rac-1aa with 2-pyridinethiol had a much lower ee value (entry 2). The reaction of the cycloheptenyl carbonate rac-1ba with 2-pyrimidinethiol afforded the sulfide 5ba in a similar yield to the cyclohexenyl derivative

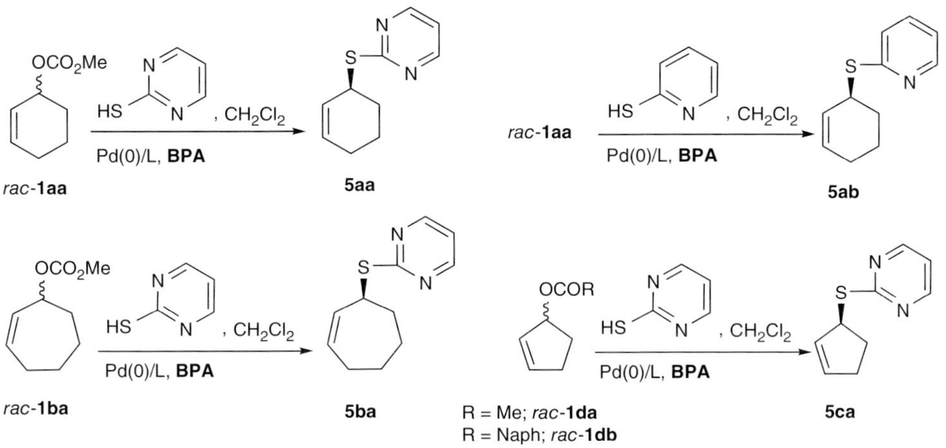

2.1.4 Palladium-Catalyzed Allylic Alkylation of Sulfur and Oxygen Nucleophiles

Scheme 2.1.4.9 Asymmetric synthesis of cyclic sulfides.

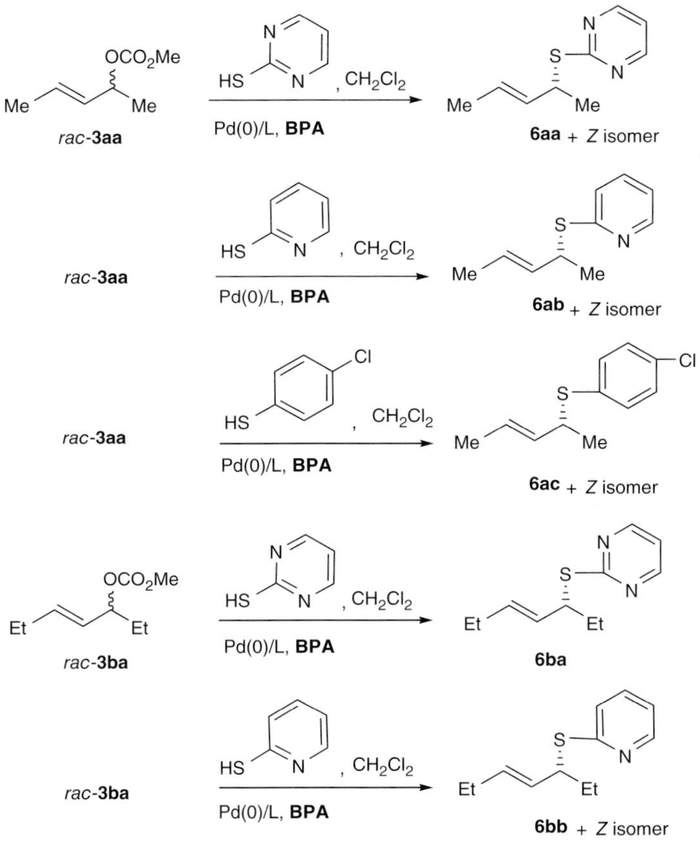

Scheme 2.1.4.10 Asymmetric synthesis of acyclic sulfides.

Table 2.1.4.9 Pd(0)/BPA-catalyzed asymmetric synthesis of cyclic and acyclic allylic sulfides.

Entry	Substrate	Thiol	Sulfide	Yield [%]	ee [%]
1	rac-1aa	2-pyrimidinethiol	5aa	63	84
2	rac-1aa	2-pyridinethiol	5ab	64	55
3	rac-1ba	2-pyrimidinethiol	5ba	61	84
4	rac-1da	2-pyrimidinethiol	5ca	80	34
5	rac-1db	2-pyrimidinethiol	5ca	96	36
6	rac-3aa	2-pyrimidinethiol	6aa	72	89
7	rac-3aa	2-pyridinethiol	6ab	87	68
8	rac-3aa	4-chlorothiophenol	6ac	73	90
9	rac-3ba	2-pyrimidinethiol	6ba	64	91
10	rac-3ba	2-pyridinethiol	6bb	24	50

Table 2.1.4.10 Synthesis of enantiopure cyclic and acyclic allylic alcohols.

Carbonate	Alcohol	Yield [%]	ee [%]
ent-1aa	9a	65	≥99
ent-1ba	9b	94	≥99
ent-1ca	9c	75	≥99
ent-4aa	10a	90	≥99
ent-4ba	10b	94	≥99

but with a lower *ee* value (entry 3). Surprising results were obtained in the reaction of the cyclopentenyl esters *rac-*1da and *rac-*1db with 2-pyrimidinethiol (entries 4 and 5). Not only were the reaction times much shorter than for the reaction of the cyclohexenyl analogues but also the *ee* values of the sulfide **5ca**, which was isolated in high yields, were much lower.

The reactions of the acyclic carbonates *rac-*3aa and *rac-*3ba with 2-pyrimidinethiol, 2-pyridinethiol, and 4-chlorothiophenol were carried out in a similar way to those of the cyclic carbonate (see Scheme 2.1.4.10). Table 2.1.4.9 shows that whereas pyrimidyl sulfides **6aa** and **6ba** were formed with high *ee* values (entries 6 and 9) the pyridyl sulfides **6ab** and **6bb** were obtained with low *ee* values (entries 7 and 10).

2.1.4.2.4 Synthesis of Enantiopure Allylic Alcohols

Hydrolysis of the cyclic carbonates *ent-*1aa–1ca and of the acyclic carbonates *ent-*3aa and *ent-*3ba gave the enantiopure alcohols **9a–9c**, **10a**, and **10b**, respectively, in medium to high yields (Scheme 2.1.4.11, Table 2.1.4.10) [13].

It should be noted that because of the high *S* values, the kinetic resolution gives access to allylic alcohols **9** and **10** with a high degree of enantio-enrichment not easily obtained by other methods.

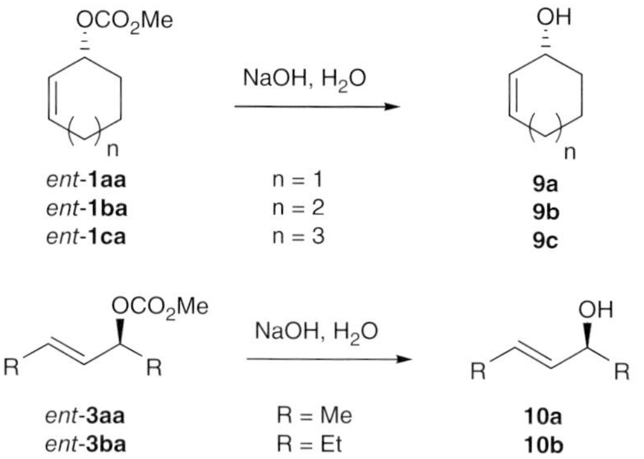
OCO2Me
NaOH, H2O
OH
n
ent-1aa n = 1 9a
ent-1ba n = 2 9b
ent-1ca n = 3 9c
OCO2Me
NaOH, H2O
OH
R
ent-3aa R = Me 10a
ent-3ba R = Et 10b

Scheme 2.1.4.11 Synthesis of enantiopure cyclic and acyclic allylic alcohols.

2.1.4.3 Asymmetric Rearrangment and Kinetic Resolution of Allylic Sulfinates

2.1.4.3.1 Introduction

Enantioselective Pd-catalyzed 1,3- and 3,3-rearrangements that interchange allylic heteroatoms, with the exception of those involving allylic imidates [14], have received only little attention [15]. We became interested in the enantioselective conversion of racemic allylic sulfinates into allylic sulfones (Scheme 2.1.4.12). Surprisingly, only a few studies of the enantioselective Pd-catalyzed rearrangement of the sulfinates of racemic allylic alcohols had been described at the beginning of our investigation. Hiroi et al. found that the tolylsulfinates of achiral acyclic allylic alcohols, in the presence of $Pd(PPh_3)_4$ and enantiopure 2,3-*O*-isopropylidene-2,3-dihydroxy-1,4-bis(diphenylphosphino)butane, suffered a facile rearrangement with formation of the corresponding optically active allylic sulfones together with the isomeric achiral allylic sulfones [16]. However, this ligand gave only a low enantioselectivity [17].

2.1.4.3.2 Synthesis of Racemic Allylic Sulfinates

The racemic acyclic allylic *t*-butylsulfinates *rac*-**11aa** and *rac*-**11ba** were prepared both as 1 : 1 mixtures of diastereomers from the racemic allylic alcohols *rac*-**10a** and *rac*-**10b**, respectively, and racemic 2-*t*-butylsulfinyl chloride [18], in high yields (Scheme 2.1.4.13).

Similarly the racemic cyclic allylic *t*-butyl- and tolylsulfinates *rac*-**12aa–12ca**, *rac*-**12ab**, and *rac*-**12bb**, respectively, were obtained as mixtures of two diastereomers in ratios of nearly 1 : 1 from the allylic alcohols *rac*-**9a**, *rac*-**9b**, and *rac*-**9c**, respectively, and racemic *t*-butylsulfinyl chloride and tolylsulfinyl chloride,

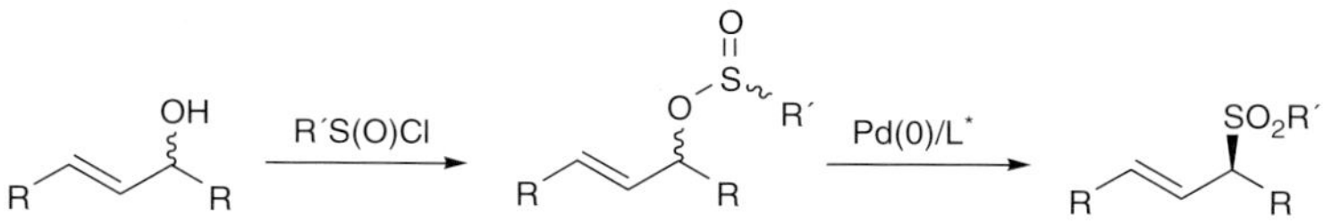
OH
R´S(O)Cl
O
S
R´
Pd(0)/L*
SO2R´
R

Scheme 2.1.4.12 Synthesis and Pd(0)-catalyzed rearrangement of allylic sulfinates to allylic sulfones.

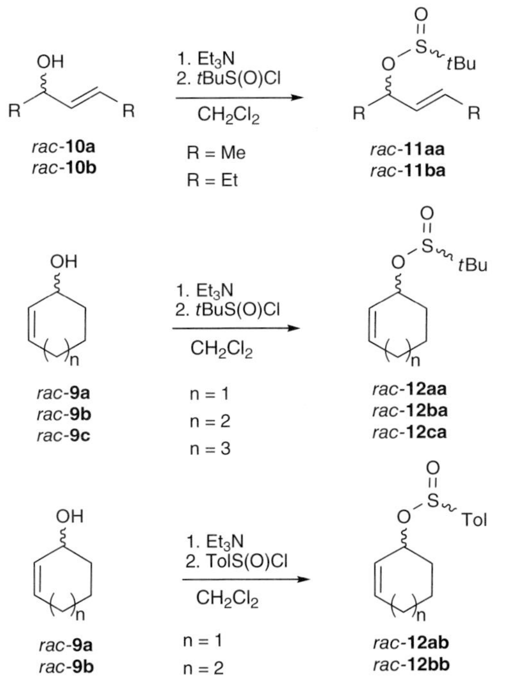

Scheme 2.1.4.13 Synthesis of racemic allylic sulfinates.

respectively, in high yields. The allylic sulfinates did not rearrange thermally at room temperature to the corresponding racemic sulfones.

2.1.4.3.3 Pd-Catalyzed Rearrangement

The bisphosphane **BPA** was selected as a ligand for the Pd atom because of the high enantioselectivities which had been recorded in the allylation of sulfinate ions with racemic allylic carbonates (vide supra). Besides the variations of the carbon skeleton, the substituent on the S atom of the racemic allylic sulfinates was varied in order to see whether both aryl and alkyl sulfinates are amenable to a highly selective rearrangement.

The Pd-catalyzed rearrangement of the acyclic S-t-butylsulfinates rac-**11aa** and rac-**11ba** proceeded quantitatively at room temperature and gave the allylic sulfones **4aa** and **4ba** with high enantioselectivities in high yields (Scheme 2.1.4.14 and Table 2.1.4.11, entries 1 and 2) [17].

Similarly effective was the Pd-catalyzed rearrangement of the S-t-butyl-substituted cyclohexenyl and cycloheptenyl sulfinates rac-**12aa** and rac-**12ba**, respectively, which gave the allylic sulfones **2aa** and **2ba**, respectively, with high enantioselectivities in high yields (entries 3 and 5). The rearrangement is not restricted to allylic S-t-butylsulfinates. Treatment of the racemic cyclic S-tolylsulfinates rac-**12ab** and rac-**12bb** with Pd(0)/L (2 mol%) and **BPA** (6 mol%) at room temperature afforded the allylic S-tolylsulfones **2ab** and **2bb**, respectively, with high enantioselectivities in high yields (entries 4 and 6).

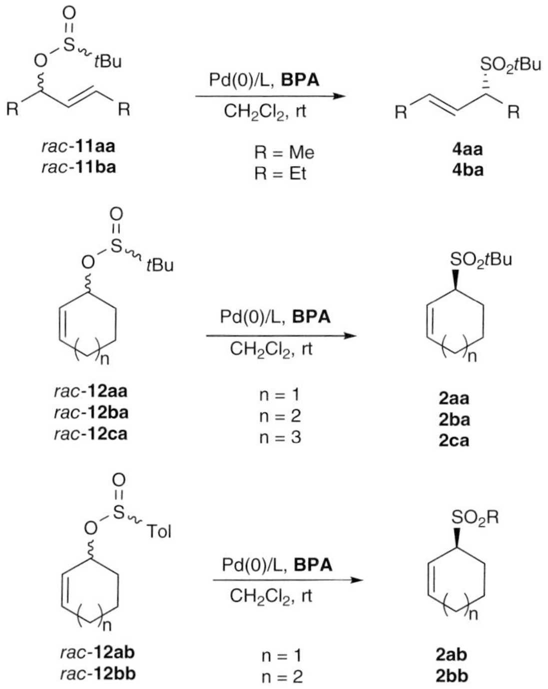

Scheme 2.1.4.14 Pd(0)/**BPA**-catalyzed enantioselective rearrangement of allylic sulfinates.

Table 2.1.4.11 Pd(0)/BPA-catalyzed enantioselective rearrangement of racemic allylic sulfinates.

Entry	Sulfinate	Pd/BPA [mol%]	Sulfone	Yield [%]	ee [%]
1	rac-11aa	4/6	4aa	86	93
2	rac-11ba	12/19	4ba	84	97
3	rac-12aa	4/6	2aa	92	95
4	rac-12aa	4/6	2ab	96	99
5	rac-12ba	4/6	2ba	82	98
6	rac-12bb	4/6	2bb	87	99
7	rac-12ca[a]	4/6	2ca	49	98
8	rac-12ca[a]	12/19	2ca	84	98

a Mixture of two diastereomers.

2.1.4.3.4 Kinetic Resolution

Interestingly, treatment of the racemic S-t-butyl substituted cyclooctenyl sulfinates rac-12ca and rac-12ca′ with Pd(0)/L (2 mol%) and **BPA** (6 mol%) gave the sulfone **2ca** with 98% ee in 49% yield and the sulfinates ent-12ca and ent-12ca′ with 92% ee and 84% ee, respectively, in 50% yield (Scheme 2.1.4.15) (Table 2.4.1.11, entry 7). The Pd-catalyzed conversion of rac-12ca and rac-12ca′ to **2ca**

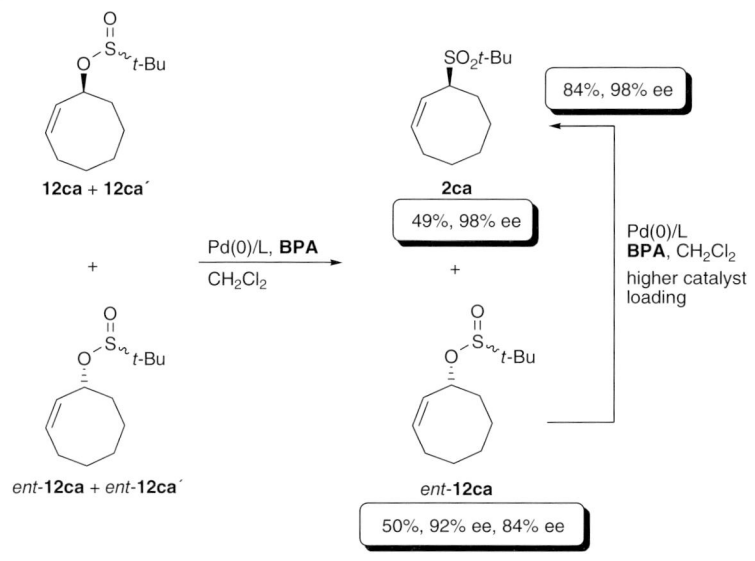

Scheme 2.1.4.15 Pd(0)/**BPA**-catalyzed kinetic resolution of the allylic sulfinate rac-**12ca**/rac-**12ca'** and asymmetric synthesis of sulfone **2ca**.

came to a practically complete halt at 50% conversion. We had observed previously that the Pd-catalyzed reaction of the racemic cyclooct-2-enyl carbonate with lithium t-butylsulfinate in the presence of **BPA** also proceeds with a highly selective kinetic resolution of the allylic carbonate, the R-configured enantiomer being the more slowly reacting one. However, in this case a complete conversion of the more slowly reacting enantiomer of the racemic allylic carbonate rac-**1ca** to the sulfone **2ca** was difficult to achieve and even an increase in the catalyst loading was only partially successful. In contrast, treatment of the mixture of rac-**12ca** and rac-**12ca'** with Pd(0)/L (6 mol%) and **BPA** (19 mol%) led to a quantitative rearrangement of both the fast- and the slow-reacting enantiomers of the sulfinate and gave sulfone **2ca** with 98% ee in 84% yield (entry 8) [17].

2.1.4.3.5 Mechanistic Considerations

A comparison between the Pd-catalyzed rearrangement of the racemic allylic sulfinates and substitution of the corresponding racemic allylic carbonates with lithium sulfinates in the presence of **BPA** (see Tables 2.1.4.1 and 2.1.4.2) revealed the same sense and a similar high degree of asymmetric induction in the formation of the corresponding cyclic and acyclic allylic sulfones. This indicates the following key steps of the rearrangement [16]. First, the Pd(0)/**BPA** reacts with the allylic sulfinates with formation of a π-allyl–Pd(II)/**BPA** complex and the sulfinate anion. Second, the former complex is substituted by the sulfinate ion at the S atom with formation of the allylic sulfone and the catalyst (Scheme 2.1.4.16).

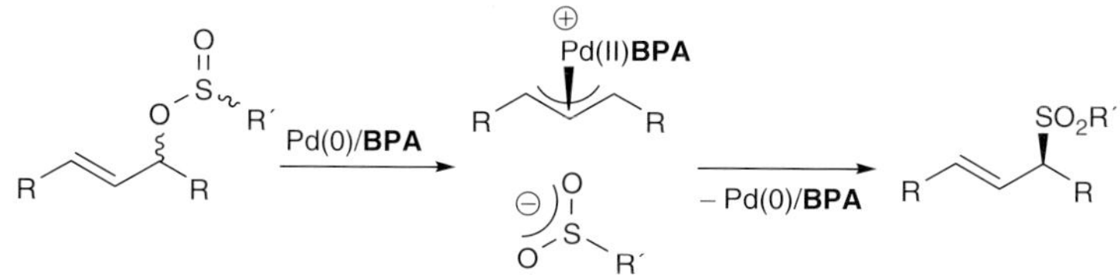

Scheme 2.1.4.16 Mechanistic scheme for the Pd(0)/**BPA**-catalyzed rearrangement of allylic sulfinates.

2.1.4.4 Asymmetric Rearrangment of Allylic Thiocarbamates

2.1.4.4.1 Introduction

Exploitation of the synthetic potential of chiral allylic sulfides and thioesters is hampered by the lack of general methods for their enantioselective synthesis. The facile Pd-catalyzed rearrangement of racemic O-allylic thiocarbamates in the presence of PPh$_3$ with formation of the corresponding racemic S-allylic derivatives [19] suggested that an enantioselective synthesis of S-allylic thiocarbamates from racemic allylic alcohols by using **BPA** as a ligand for the Pd atom would perhaps be efficient. The ready conversion of S-allylic thiocarbamates to the corresponding allylic thiols and the many methods available for their derivatization would thus provide an entry to the various chiral allylic sulfur compounds. Of the possible transformations, the rearrangement of N-monosubstituted O-allylic thiocarbamates to the corresponding S-allylic thiocarbamates seemed to be the most attractive since racemic N-monosubstituted O-allylic carbamates are easily accessible from the corresponding racemic allylic alcohols and isothiocyanates.

2.1.4.4.2 Synthesis of Racemic O-Allylic Thiocarbamates

The acyclic O-allylic thiocarbamates *rac*-**13aa–13af**, *rac*-**13ba**, *rac*-**13bd**, *rac*-**13bf**, and *rac*-**13cb** as well as the cyclic O-allylic thiocarbamates *rac*-**14aa–14ag** and *rac*-**14ba** were synthesized from the racemic allylic alcohols *rac*-**10a–10c**, *rac*-**9a**, and *rac*-**9b**, respectively, and the corresponding isothiocyanates in high yields (Scheme 2.1.4.17) [19].

While the cyclohexenyl and cycloheptenyl derivatives *rac*-**14aa–14ag** and *rac*-**14ba** were stable at room temperature, the cyclopentenyl O-allylic thiocarbamates already suffered a partial rearrangement at ambient temperatures [19, 20] while the acyclic O-allylic thiocarbamates showed an intermediate thermal stability.

2.1.4.4.3 Acyclic Carbamates

The racemic O-allylic thiocarbamates *rac*-**13aa–13af** bearing methyl groups on the allyl unit at room temperature in CH$_2$Cl$_2$ in the presence of Pd(0)/L (1.25 mol%) and **BPA** (3 mol%) suffered a quantitative and enantioselective rearrangement to the S-allylic thiocarbamates **15aa–15af** (Scheme 2.1.4.18 and Table 2.1.4.12) [21, 22].

Rearrangement of carbamate *rac*-**13aa**, which carries the least sterically demanding methyl group at the N atom, proceeded the most efficiently in terms of

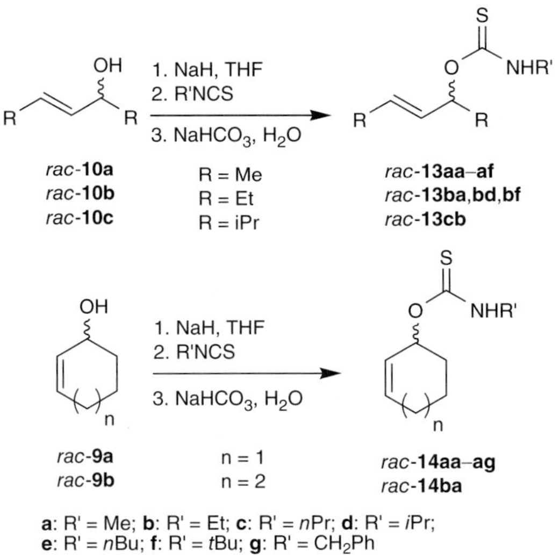

Scheme 2.1.4.17 Synthesis of racemic allylic thiocarbamates.

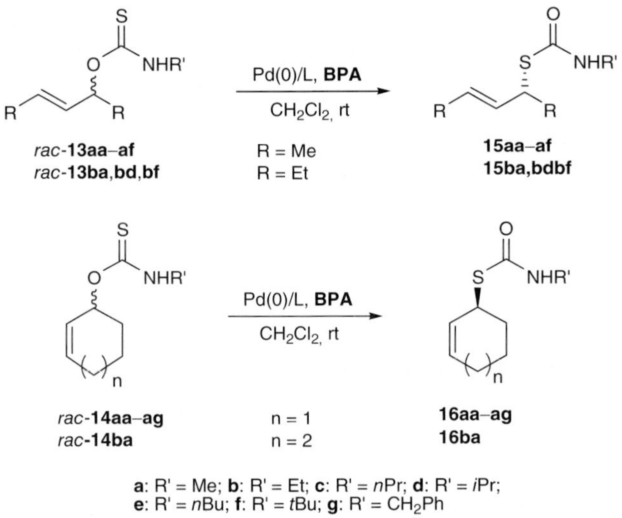

Scheme 2.1.4.18 Pd(0)/**BPA**-catalyzed rearrangement of allylic thiocarbamates.

Table 2.1.4.12 Pd(0)/BPA-catalyzed rearrangement of racemic acyclic O-allylic thiocarbamates.

Entry	Substrate	Pd[0]/BPA [mol%]	Conv. [%]	Product	Yield [%]	ee [%]
1	rac-13aa	2.5 : 3	100	15aa	92	91
2	rac-13ab	2.5 : 3	100	15ab	92	92
3	rac-13ac	2.5 : 3	100	15ac	89	90
4	rac-13ad	2.5 : 3	100	15ad	93	90
5	rac-13ae	2.5 : 3	100	15ae	89	85
6	rac-13af	2.5 : 3	100	15af	76	85
7	rac-13ba	7.5 : 9	100	15ba	92	91
8	rac-13bd	7.5 : 9	100	15bd	86	86
9	rac-13bf	7.5 : 9	100	15bf	91	64

enantioselectivity and reaction rate (entry 1). Rearrangement of *rac*-13ba and *rac*-13bd carrying ethyl groups at the allyl unit in the presence of Pd(0)/L (1.8 mol%) and **BPA** (9 mol%) in CH_2Cl_2 at room temperature also proceeded in a quantitative and enantioselective way to afford the S-allylic thiocarbamates 15ba and 15bd with *ee* values of 86% and 91%, respectively. Rearrangement of the N-methyl derivative *rac*-13ba, as in the case of *rac*-13aa, was not only the fastest but also occurred with the highest enantioselectivity (entry 7). The influence of the substituent at the N atom upon the rearrangement was most noticeable in the case of the N-*t*-butyl thiocarbamate *rac*-13af. While a complete conversion of *rac*-13af was observed after a reaction time of 15 h, the *ee* value of the S-allylic thiocarbamate 15bf was only 64% (entry 9). A control experiment with *rac*-13af under the same conditions but without the pre-catalyst revealed, however, a competing thermal rearrangement of the O-allylic thiocarbamate to *rac*-15bf.

2.1.4.4.4 Cyclic Carbamates

The Pd-catalyzed rearrangement of the racemic cyclic O-allylic thiocarbamates *rac*-14aa–14ag proceeded faster and with higher enantioselectivities than that of the acyclic O-allylic thiocarbamates. After reaction times of 0.5–16 h the S-allylic thiocarbamates 16aa–16ag were obtained with high enantioselectivities in high yields (Table 2.1.4.13) [21]. Except in the case of the thiocarbamate *rac*-14ae, which carries an *n*-butyl group on the N atom (entry 5), a complete rearrangement took place when only 0.62 mol% of Pd(0)/L and 1.5 mol% of **BPA** were used. Increasing the size of the substituent of the N atom of the cyclohexenyl derivatives led, as in the case of the acyclic O-allylic thiocarbamates, to a decrease in the reaction rate.

The Pd-catalyzed rearrangement of the seven-membered cyclic racemic N-methyl-substituted O-allylic thiocarbamate *rac*-14ba proceeded with high enantioselectivity and gave the corresponding S-allylic derivative 16ba in high yield (entry 7). The rate and the enantioselectivity of the rearrangement of *rac*-14ba were, however, somewhat lower than those of the corresponding 6-membered O-allylic carbamate *rac*-14aa (entry 1).

Table 2.1.4.13 Pd(0)/BPA-catalyzed rearrangement of racemic cyclic O-allylic thiocarbamates.

Entry	Substrate	Pd(0)/BPA [mol%]	Conv. [%]	Product	Yield [%]	ee [%]
1	*rac*-14aa	1.25 : 1.5	100	16aa	94	97
2	*rac*-14ab	1.25 : 1.5	100	16ab	96	95
3	*rac*-14ac	1.25 : 1.5	100	16ac	94	≥99
4	*rac*-14ad	1.25 : 1.5	100	16ad	92	92
5	*rac*-14ae	2.5 : 3	100	16ae	93	≥99
6	*rac*-14af	1.25 : 1.5	100	16af	92	97
7	*rac*-14ag	2.5 : 3	100	16ag	91	99
8	*rac*-14ba	1.25 : 1.5	100	16ba	94	92

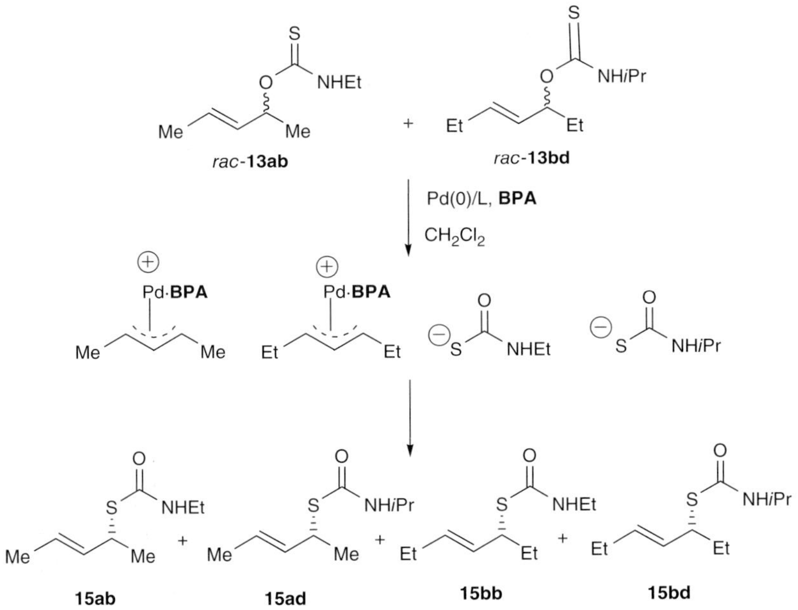

Scheme 2.1.4.19 Crossover experiment in the Pd(0)/**BPA**-catalyzed rearrangement of allylic thiocarbamates.

2.1.4.4.5 Mechanistic Considerations

The sense and degree of asymmetric induction of the Pd(0)-catalyzed rearrangement of the cyclic and acyclic O-allylic thiocarbamates in the presence of **BPA** are the same as, or similar to, those in the Pd-catalyzed substitutions of the corresponding cyclic and acyclic racemic allylic carbonates and acetates with sulfinates and thiols. It is therefore proposed that Pd(0)/**BPA** reacts with the racemic O-allylic thiocarbamate with formation of a π-allyl–Pd(II) complex, which contains as counter ion the corresponding thiocarbamate ion (Scheme 2.1.4.19) [23, 24]. Substitution of the π-allyl–Pd(II) complex by the thiocarbamate ion gives the S-allylic thiocarbamate and the Pd catalyst.

Verification for the ionization–substitution mechanism was sought by a crossover experiment [19]: a mixture of the racemic O-allylic thiocarbamates *rac*-**13ab** and *rac*-**13bd** (1 : 1) was treated with Pd(0)/L (2.5 mol%) and **BPA** (6 mol%) in CH$_2$Cl$_2$ at room temperature, which gave a mixture of the S-allylic thiocarbamates **15ab**, **15ad**, **15bb**, and **15bd** (30 : 20 : 18 : 30) in practically quantitative yield. Attainment of the mixture of carbamates can be rationalized by the formation of intermediates depicted in Scheme 2.1.4.19, and thus supports further the notion of a mechanism for the Pd-catalyzed rearrangement of the O-allylic thiocarbamates which includes as key steps an ionization followed by a substitution [21].

2.1.4.4.6 Synthesis of Allylic Sulfides

Finally, the application of S-allylic thiocarbamates to the synthesis of allylic sulfides and thiols was investigated (Scheme 2.1.4.20).

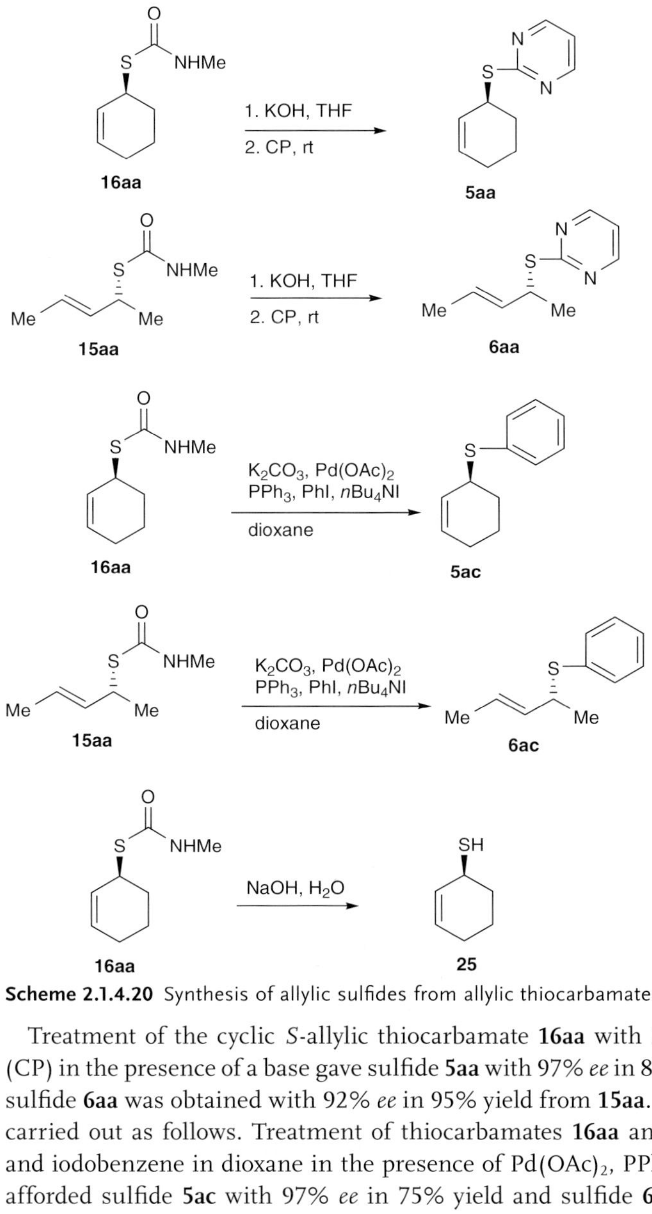

Scheme 2.1.4.20 Synthesis of allylic sulfides from allylic thiocarbamates.

Treatment of the cyclic S-allylic thiocarbamate **16aa** with 2-chloropyrimidine (CP) in the presence of a base gave sulfide **5aa** with 97% *ee* in 80% yield. Similarly, sulfide **6aa** was obtained with 92% *ee* in 95% yield from **15aa**. An S-arylation was carried out as follows. Treatment of thiocarbamates **16aa** and **15aa** with K_2CO_3 and iodobenzene in dioxane in the presence of $Pd(OAc)_2$, PPh_3, and nBu_4NI [19] afforded sulfide **5ac** with 97% *ee* in 75% yield and sulfide **6ac** (92% *ee* in 56% yield), respectively. As a last example, thiol **25** was prepared with 97% *ee* in 64% yield from thiocarbamate **16aa**.

2.1.4.5 Asymmetric Synthesis of Allylic Thioesters and Kinetic Resolution of Allylic Esters

2.1.4.5.1 Introduction
The Pd-catalyzed enantioselective allylic alkylation of thiols in the presence of **BPA** is limited to aryl and heteroaryl thiols [10]. However, allylic thioesters should

make further highly useful starting materials for the synthesis of allylic sulfides through saponification and alkylation and arylation at the S atom. The literature on the feasibility of a Pd-catalyzed enantioselective allylic alkylation of thiocarboxylate ions was, however, ambiguous [25] at the beginning of our investigation.

2.1.4.5.2 Asymmetric Synthesis of Allylic Thioesters

The Pd-catalyzed allylic alkylation of thiocarboxylate ions was carried out with potassium thioacetate (KSAc) and potassium thiobenzoate (KSBz) and the racemic cyclic and acyclic carbonates rac-3aa, rac-3ba, rac-1da, rac-1aa, rac-1ba, and rac-1ca, respectively (Scheme 2.1.4.21). The carbonates rac-3aa, rac-3ba, rac-1da, rac-1aa, and rac-1ba were treated with KSAc (1.4 equiv) or KSBz (2.0 equiv) in the presence of Pd(0)/L (2 mol%) and **BPA** (8 mol%) in CH_2Cl_2/H_2O. Under these conditions the acyclic carbonates rac-3aa and rac-3ba gave the thioesters **18aa**, **18ab** and **18ba**, respectively (Table 2.1.4.14, entries 1–3), with high enantioselectivities in high yields [26].

Treatment of the five-membered cyclic carbonate rac-1da with KSAc (entry 4) and KSBz (entry 5) furnished thioacetate **19da** with 73% ee and thiobenzoate **19db** with 59% ee, respectively, in only medium yields. A similar reaction of the six-membered cyclic carbonate rac-1aa with KSAc gave thioacetate **19aa** in an even lower yield of only 51% but with a higher ee value of 94% (entry 6). The somewhat low yields of **19da**, **19db**, and **19aa** are due to a competing Pd-catalyzed conversion of the allylic carbonates to the corresponding allylic alcohols involving a substitution by the O-nucleophile hydrogencarbonate formed in situ (vide infra, Section 2.1.4.5). However, the use of KSBz instead of KSAc in the reaction of the cyclohexenyl carbonate rac-1aa afforded thioester **19ab** with 89% ee in high yield (entry 7). The cycloheptenyl carbonate rac-1ba showed in the reaction with KSBz, in comparison to rac-1da and rac-1aa, a lower reactivity and gave thioester **30b** with 86% ee in good yield (entry 8).

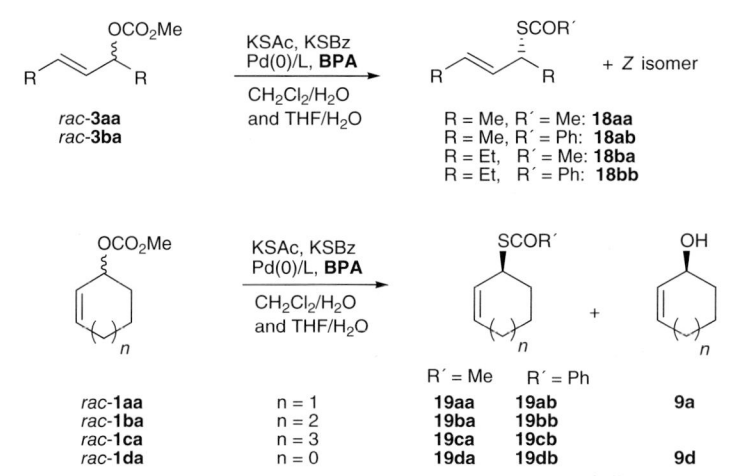

Scheme 2.1.4.21 Pd(0)/**BPA**-catalyzed asymmetric synthesis of allylic thioesters.

Table 2.1.4.14 Pd(0)/BPA-catalyzed asymmetric synthesis of allylic thioesters.

Entry	Substrate	Salt	Conv. [%]	Products	Yield [%]	ee [%]
1	rac-3aa	KSAc	≥99	18aa	76	84
2	rac-3aa	KSBz	92	18ab	85	87
3	rac-3ba	KSBz	≥99	18ba	97	90
4	rac-1da	KSAc	≥99	19da	62	73
				9d	22	57
5	rac-1da	KSBz	≥99	19db	68	59
				9d	20	55
6	rac-1aa	KSAc	≥99	19aa	51	94
				9a	30	82
7	rac-1aa	KSBz	≥99	19ab	92	89
				9a	5	
8	rac-1ba	KSBz	73	19bb	69	86
				9b		–

2.1.4.5.3 Kinetic Resolution of Allylic Esters

In the reaction of the eight-membered cyclic carbonate rac-1ca with KSAc in THF/H$_2$O (see Scheme 2.1.4.21) the conversion of the substrate, even at higher temperatures, did not exceed 53% and gave a mixture of thioacetate 19aa and carbonate ent-1ca in a ratio of 53 : 47. Formation of thioacetate 19aa with 84% ee and of carbonate ent-1ca with 72% ee in a ratio of approximately 1 : 1 (Table 2.1.4.15, entry 1) showed that an efficient kinetic resolution had occurred (see Scheme 2.1.4.21). Similar results were recorded in the reaction of carbonate rac-1ca with KSBz (entry 2). The results recorded after the termination of the reaction of the acyclic carbonate rac-3ba with KSBz in CH$_2$Cl$_2$/H$_2$O at 48% conversion also revealed the operation of kinetic resolution in this case (entry 3).

The kinetic resolution of the racemic allylic acetates rac-3ab, rac-1db, rac-1ab, and rac-1bb with thiocarboxylate ions and **BPA** were investigated in more detail (Scheme 2.1.4.22). The acetates were selected instead of the corresponding carbonates in order to avoid the competing formation of the corresponding allylic alcohols (vide supra). All reactions were carried out in CH$_2$Cl$_2$/H$_2$O (9 : 1) using 2 mol% of Pd(0)/L and 8 mol% of **BPA**. Termination of the reaction of the pentenyl acetate rac-3ab with KSAc at 35% conversion showed the operation of highly selective kinetic resolution (entry 4). However, 50% conversion of the substrate could be achieved neither at room nor at reflux temperature. This is in contrast to the reactivity of carbonate rac-3aa (cf. Table 2.1.4.14, entry 1) and perhaps reflects the lower reactivity of allylic acetates in Pd-catalyzed alkylation. This

Table 2.1.4.15 Pd(0)/BPA-catalyzed kinetic resolution of allylic carbonates and acetates with thiocarboxylate ions.

Entry	Substrate	Salt	Solvent	Conv. [%]	Products	Yield [%]	ee [%]
1	rac-**1ca**	KSAc	THF/H$_2$O	53	ent-**1ca** **19ca**	48 42	72 84
2	rac-**1ca**	KSBz	THF/H$_2$O	67	ent-**1ca** **19cb**	28 56	94 73
3	rac-**3ba**	KSBz	CH$_2$Cl$_2$/H$_2$O	48	**18aa** ent-**3ba**	44 47	93 57
4	rac-**3ab**	KSAc	CH$_2$Cl$_2$/H$_2$O	35	**18aa** ent-**3ab**	32 44	86 60
5	rac-**3ab**	KSBz	CH$_2$Cl$_2$/H$_2$O	56	**18ab** ent-**3ab**	53 22	83 99
6	rac-**1db**	KSAc	CH$_2$Cl$_2$/H$_2$O	62	**19da** ent-**1db**	52 40	92 62
7	rac-**1db**	KSAc	CH$_2$Cl$_2$/H$_2$O	≥99	**19da**	99	58
8	rac-**1ab**	KSAc	CH$_2$Cl$_2$/H$_2$O	51	**19aa** ent-**1ab**	48 43	97 ≥99
9	rac-**1bb**	KSAc	CH$_2$Cl$_2$/H$_2$O	51	**19ba** ent-**1bb**	50 48	98 ≥99

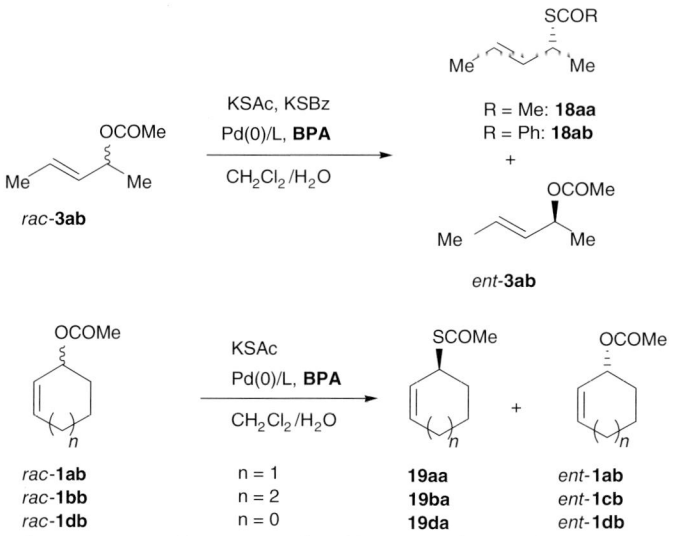

Scheme 2.1.4.22 Pd(0)/**BPA**-catalyzed kinetic resolution of allylic acetates with KSAc and KSBz.

problem could be overcome, however, by the use of the more reactive KSBz instead of KSAc. Here, at 56% conversion of *rac*-**3ab** the thiobenzoate **18ab** was isolated with 83% *ee* in 53% yield (Table 2.1.4.15, entry 5) and the acetate *ent*-**3ab** was recovered with 99% *ee* in 22% yield.

Surprising results were obtained in the case of the kinetic resolution of the cyclic acetates *rac*-**1db**, *rac*-**1ab**, and *rac*-**1bb**. The five-membered cyclic acetate exhibited by far the highest reactivity. Thioacetate **19aa** was obtained with 92% *ee* in 52% yield at 62% conversion of the substrate (entry 6). Strangely, however, the *ee* value of the recovered acetate *ent*-**1db** was only 62% (vide infra). Furthermore, the thioacetate **19da**, which was isolated in 99% yield, had an *ee* value of only 58% at full conversion of the substrate (entry 7) (vide infra).

The catalyst exhibited high enantiomer selectivity in the reaction of the six-membered cyclic acetate *rac*-**1ab** with KSAc on a 2.5 mmol scale. This led to the isolation of the thioacetate **19aa** with 97% *ee* in 48% yield and the acetate *ent*-**1ab** with ≥99% *ee* in 43% yield (entry 8). The reaction came to a practically complete halt after 51% conversion of the substrate. In order to determine the selectivity factor S, the kinetic resolution of *rac*-**1ab** was repeated and the *ee* values of the acetate and thioacetate were monitored over the whole course of the reaction (Fig. 2.1.4.1). Nonlinear regression of the selectivity factor S for ten pairs of *ee* and *c* values [6m, 27] gave $S = 72 \pm 19$. This value corresponds well with $S = 74 \pm 7$ for the Pd/**BPA**-catalyzed kinetic resolution of *rac*-**1aa** with lithium *t*-butylsulfinate in CH_2Cl_2/H_2O and with $S = 77 \pm 11$ for that of *rac*-**1aa** with 2-pyrimidine-thiol in CH_2Cl_2 [26].

A similar efficient kinetic resolution took place in the reaction of the cycloheptenyl acetate *rac*-**1bb** with KSAc. The reaction came to a practically complete halt at 51% conversion of the substrate. The acetate *ent*-**1bb** was obtained with ≥99% *ee* in 48% yield and the thioacetate **19ba** was isolated with 98% *ee* in 50% yield (entry 9, Table 2.1.4.15). Even an increase in the amount of Pd(0)/L from 2 to

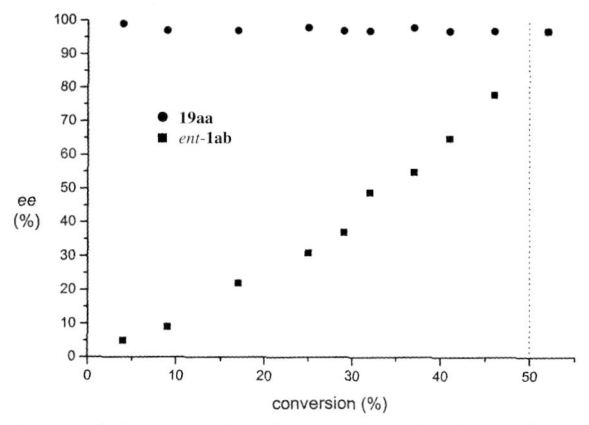

Fig. 2.1.4.1 Dependencies of the *ee* values of **19ab** and the mismatched acetate *ent*-**1ab** on the conversion of the substrate.

4 mol% and of **BPA** from 8 to 16 mol% did not lead to a further conversion of the remaining acetate *ent*-**1bb**.

2.1.4.5.4 Memory Effect and Dynamic Kinetic Resolution of the Five-Membered Cyclic Acetate

Reaction of the racemic substrate A comparison of the enantioselectivities of the kinetic resolution and substitution of the acetates *rac*-**1db**, *rac*-**1ab**, and *rac*-**1bb** with KSAc (see Table 2.1.4.15, entries 6–9, and Scheme 2.1.4.22) reveals interesting differences. While in all three cases the enantioselectivities of the allylic alkylation of the nucleophile with the matched acetates are apparently high, the selectivities of the kinetic resolution differ significantly. The selectivity of the kinetic resolution is high for the six- and seven-membered cyclic acetates but it appears to be low for the five-membered cyclic acetate. Furthermore, the enantioselectivity of the allylic alkylation of KSAc with the five-membered cyclic acetate *rac*-**1db** under complete turnover conditions is much lower than under incomplete turnover conditions, for example, at 62% conversion. A similar low enantioselectivity was observed in the case of alkylation of thiocarboxylate ions with the carbonate *rac*-**1da** under complete turnover conditions (see Table 2.1.4.14, entry 5). This stands in contrast to the reactivity of the six- and seven-membered carbonates *rac*-**1aa** and *rac*-**1ba**, respectively, which under high turnover conditions delivered the corresponding thioesters with high *ee* values (Table 2.1.4.14, entries 7 and 8). These differences in reactivity of *rac*-**1db**, *rac*-**1ab**, and *rac*-**1bb** point to the existence of a strong "memory effect" [28] in the case of *rac*-**1db**; that is, the two enantiomers of the substrate apparently react with different enantioselectivities. The alkylation with the matched acetate *ent*-**1db** proceeds with high selectivity, and that with the mismatched acetate *ent*-**1db** with low enantioselectivity, and the mismatched acetate **1db** racemizes during the course of the substitution reaction. In order to obtain verification of these assumptions, the *ee* values of **19da** and *ent*-**1db** were monitored over almost the whole course of the reaction of the racemic acetate *rac*-**1db** with KSAc: Figure 2.1.4.2 shows the dependencies

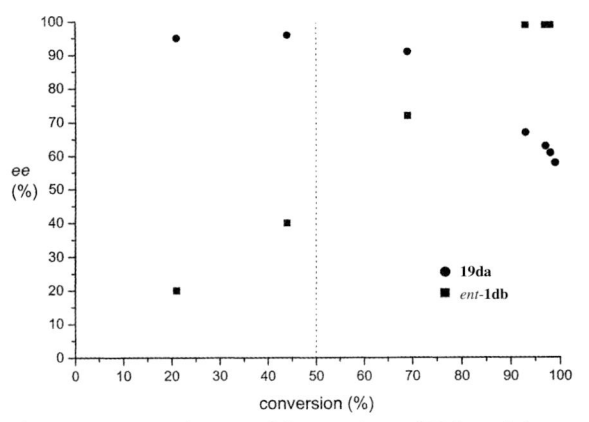

Fig. 2.1.4.2 Dependencies of the *ee* values of **19da** and the mismatched acetate *ent*-**1db** on the conversion of *rac*-**1db** in the reaction with KSAc.

of the *ee* values of **19db** and *ent*-**1db** on the conversion of the substrate. The *ee* of **19db** remained high (ca. 96%) until approximately 50% conversion. An increase in the conversion from 50% to ≥99% led to a strong decrease in the *ee* of the thioacetate. These results support the assumption of the operation of a strong "memory effec.t. A control experiment demonstrated that a partial racemization of **19db** under high turnover conditions did not occur [26].

Interestingly, the omission of the nucleophile KSAc (condition of no turnover) also did not lead to a significant racemization of **19da**.

Reactions of the enantiopure substrates In order to gain more information about the origin of the apparently low selectivity of the kinetic resolution of *rac*-**1db**, the stereochemical course of the reactions of both the matched acetate **1db** and the mismatched acetate *ent*-**1db** was investigated.

Preparative HPLC of the racemic cyclopentenyl naphthoate *rac*-**1dc** afforded the enantiopure α-naphthoates **1dc** and *ent*-**1dc**, each in 46% yield (Scheme 2.1.4.23). Hydrolysis of **1dc** and *ent*-**1dc** furnished the enantiopure alcohols **9d** and *ent*-**9d**, respectively, which were converted to the enantiopure acetates **1db** and *ent*-**1db**, respectively. Both acetates *ent*-**1db** and **1db** were submitted separately to a Pd-catalyzed reaction with KSAc (1.4 equiv) in the presence of Pd(0)/L (2 mol%) and **BPA** (4 mol%) in CH_2Cl_2/H_2O (9 : 1) (Scheme 2.1.4.24). The reaction of the matched acetate **1db** was faster than that of the mismatched acetate *ent*-**1db**.

The *ee* value of acetate **1db**, which was ≥98% at the beginning, decreased with increasing conversion (Fig. 2.1.4.3). At approximately 50% conversion only the racemic acetate *rac*-**1db** could be detected. From here on, the mismatched acetate

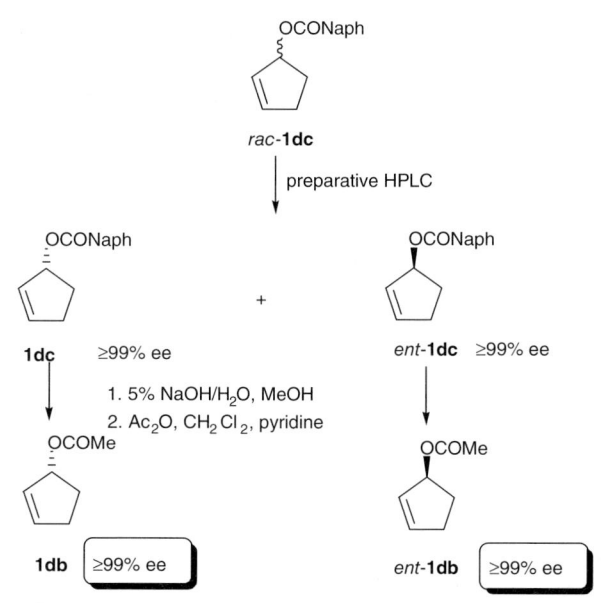

Scheme 2.1.4.23 Synthesis of the enantiopure cyclopentenyl acetates **1db** and *ent*-**1db**.

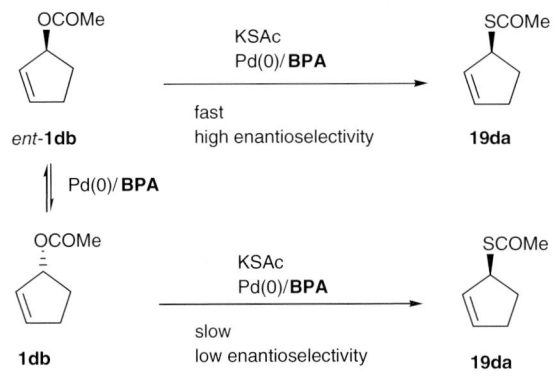

Scheme 2.1.4.24 Racemization and dynamic kinetic resolution of acetates *ent*-**1db** and **1db** in their Pd(0)/**BPA**-catalyzed reactions with KSAc.

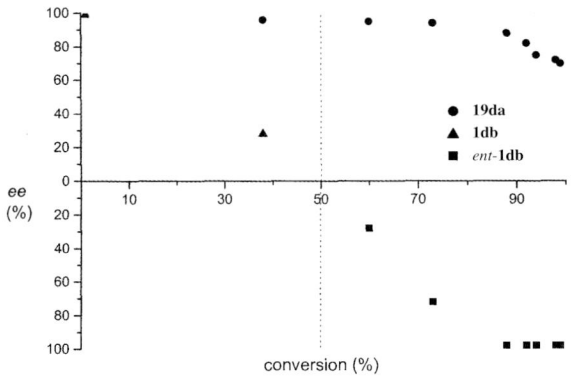

Fig. 2.1.4.3 Dependencies of the *ee* values of the thioacetate **19da** and the matched acetate **1db** on the conversion of **1db** in the reaction with KSAc.

ent-**1db** was in excess and its *ee* value increased with increasing conversion of the substrate to reach ≥99% at 88% conversion. The *ee* value of the thioacetate **19da** remained high up to approximately 75% conversion but decreased at higher conversion. These results show that at the beginning of the reaction of **1db** with KSAc a partial racemization of the matched acetate occurred. The matched acetate **1db** reacted preferentially and gave the thioacetate **19da** with high enantioselectivity up to 80% conversion. At the same time the concentration of the mismatched acetate *ent*-**1db** gradually increased. After the consumption of most of the matched acetate **1db**, the *ee* value of the thioacetate **19da** decreased, presumably because of the onset of a reaction of the mismatched acetate *ent*-**1db**, which proceeded with a lower enantioselectivity.

Figure 2.1.4.4 shows similar complex dependencies of the *ee* values of the components on the conversion of the substrate to those for the reaction of the enantiopure mismatched acetate *ent*-**1db** with KSAc. From the onset of the reaction the *ee* value of the mismatched acetate *ent*-**1db** decreased. The *ee* value of

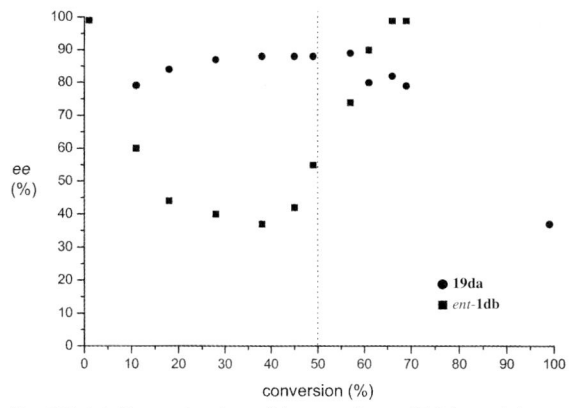

Fig. 2.1.4.4 Dependencies of the *ee* values of **19da** and the mismatched acetate *ent*-**1db** on the conversion of *ent*-**1db** in the reaction with KSAc.

ent-**1db** remained almost constant between approximately 18% and 45% conversion, and it increased again strongly at higher conversion. Surprisingly, the *ee* value of the thioacetate **19da** was much higher than expected, and it even increased with increasing conversion. Finally, the *ee* value of the thioacetate decreased at higher conversion where only the presence of *ent*-**1db** could be detected. These results suggest that at the beginning of the reaction a partial racemization of the mismatched acetate *ent*-**1db** occurred with formation of the matched acetate **1db** which reacted with high enantioselectivity. The observation of almost constant *ee* values of the acetate *ent*-**1db** and the thioacetate **19da** between 18% and 45% conversion of the substrate indicates strongly that at this point of the reaction a dynamic kinetic resolution [29] was established. It involves a racemization of the mismatched acetate and a preferential reaction of the matched acetate, both in the presence of the chiral catalyst. Surprisingly, however, the dynamic kinetic resolution apparently came to a halt at a conversion higher than 45%. From here on, the *ee* value of **1db** increased and, as a consequence of the lower enantioselectivity of the reaction of the mismatched acetate *ent*-**1db**, the *ee* value of the thioacetate **19da** decreased strongly. It was then of interest to see whether a Pd-catalyzed racemization of the cyclopentenyl acetate would also occur in the absence of the nucleophile, the condition of no turnover. Therefore, the enantiopure acetate **1db** was treated with Pd(0)/L (2 mol%) and **BPA** (8 mol%) in CH_2Cl_2/H_2O under the same conditions as used above but omitting KSAc. After a short reaction time the acetate **1db** had been completely racemized. The racemization of *ent*-**1db** and **1db** without nucleophile may proceed through a Pd(0)/**BPA**-catalyzed ionization like that shown in Scheme 2.1.4.1, followed by an addition of the acetate ion to the Pd atom and a subsequent reductive elimination. The racemization of the acetates in the presence of KSAc, the condition of complete turnover, is perhaps more complex, as indicated by the dependencies of the *ee* values of the substrates on their conversion (see Figs. 2.1.4.3 and 2.1.4.4).

2.1.4.5.5 Asymmetric Synthesis of Cyclopentenyl Thioacetate

An efficient asymmetric synthesis of the five-membered cyclic thioacetate **19da** from acetate *rac*-**1db** or carbonate *rac*-**1da** was precluded because of the operation of a strong "memory effect" and other effects. Therefore the Pd-catalyzed reaction of the naphthoate *rac*-**1dc** with KSAc in the presence of Pd(0)/L (2 mol%) and **BPA** (8 mol%) in CH_2Cl_2/H_2O was investigated. Indeed, reaction of the naphthoate *rac*-**1dc** gave thioacetate **19da** with 92% *ee* in high yield.

2.1.4.6 Kinetic and Dynamic Kinetic Resolution of Allylic Alcohols

2.1.4.6.1 Introduction

Chiral allylic alcohols are important intermediates in the synthesis of biologically active compounds [30]. Several methods are available for their synthesis in enantiopure form: for example, catalytic reduction and kinetic resolution [1k,q, 30b, 31–34]. Although catalytic resolution is frequently highly efficient, it suffers from the drawback of a maximum yield of 50%. Dynamic kinetic resolution is the method capable of overcoming this limitation. The lipase-catalyzed kinetic resolution of allylic acetates in combination with a Pd(0)-catalyzed racemization provides an efficient access to chiral allylic alcohols [31d,e], which so far has been applied only to nonfunctionalized allylic alcohols. An interesting alternative is the Pd(0)/**BPA**-catalyzed reaction of racemic allylic carbonates with carboxylates [34a,b]. This method, however, has been applied only to symmetrical cyclic substrates. During our investigation of the Pd(0)-catalyzed allylic alkylation of thiocarboxylate ions with racemic allylic carbonates in the presence of **BPA** (vide supra) and water we observed the formation of a mixture of the corresponding allylic alcohol and the thioester (see Scheme 2.1.4.21) [26]. Surprisingly, the alcohol was highly enriched enantiomerically and had the same absolute configuration as the thioester. Obviously, a competing Pd(0)-catalyzed allylic substitution with an O-nucleophile [35] had taken place, leading finally to the formation of the alcohol. This observation suggested the possibility of a Pd(0)-catalyzed asymmetric synthesis of allylic alcohols through treatment of the corresponding racemic carbonates with water in the presence of **BPA**.

2.1.4.6.2 Asymmetric Synthesis of Symmetrical Allylic Alcohols

Reaction of the cyclic carbonates *rac*-**1aa**, *rac*-**1ba**, and *rac*-**1da** with Pd(0)/L (2 mol%), **BPA** (3 mol%), and water in CH_2Cl_2 gave the allylic alcohols **9a**, **9b**, and **9d**, respectively, with high enantioselectivities in high yields (Scheme 2.1.4.25) [36].

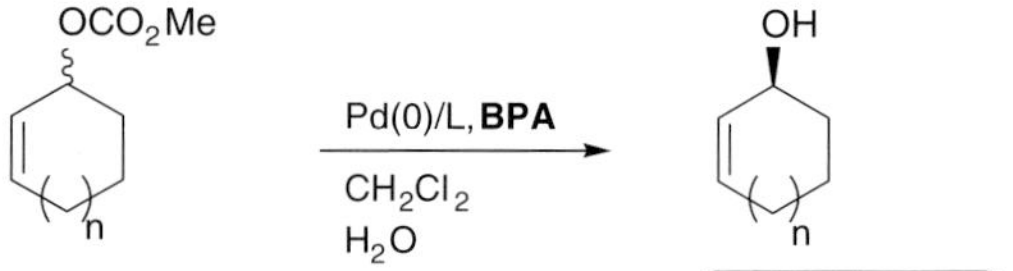

Scheme 2.1.4.25 Asymmetric synthesis of cyclic allylic alcohols.

Similar treatment of carbonate *rac*-**20** with Pd(0)/L (2 mol%), **BPA** (3 mol%), and water in CH_2Cl_2 afforded indenol **21** in 88% yield with 97% *ee* (Scheme 2.1.4.26) [37]. Indenol **21** and cyclopentenol **9d** are useful building blocks for the synthesis of an HIV-protease inhibitor and a chiral catalyst for asymmetric Diels–Alder reactions, respectively [38].

The reaction of the acyclic carbonates *rac*-**18aa** and *rac*-**18ba** with Pd(0)/L (2 mol%), **BPA** (3 mol%) and water in CH_2Cl_2 under similar conditions furnished the acyclic allylic alcohols **10a** and **10b**, respectively, in high yields with high *ee* values (Scheme 2.1.4.27). Interestingly, the use of $KHCO_3$ as an additive resulted in a higher reaction rate.

A mechanistic scheme for the Pd(0)-catalyzed reaction of allylic carbonates in the presence of **BPA** with water, leading to the corresponding allylic alcohols, is proposed in Scheme 2.1.4.28 [36]. Both enantiomers **1** and *ent*-**1** react with Pd(0)/**BPA** with formation of the π-allyl–Pd(II) complex **22** containing the methylcarbonate ion. Because of the symmetrical carbon skeleton of the substrates and the C_2 symmetry of **BPA**, only one π-allyl–Pd(II) complex **22** is formed. Water causes an irreversible hydrolysis of the methylcarbonate ion to the hydrogencarbonate

Scheme 2.1.4.26 Asymmetric synthesis of indenol.

Scheme 2.1.4.27 Asymmetric synthesis of acyclic allylic alcohols.

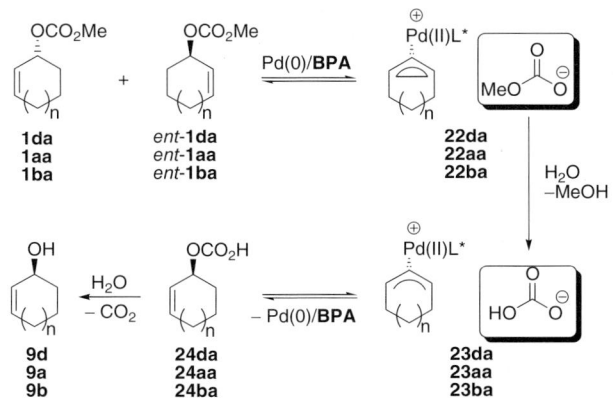

Scheme 2.1.4.28 Mechanistic scheme for the Pd(0)/**BPA**-catalyzed reaction of allylic carbonates with water.

ion, with formation of complex **23**. The hydrogencarbonate ion in turn reacts with the π-allyl–Pd/**BPA** complex as an O-nucleophile [36] to give the hydrogencarbonate **24**, which decomposes irreversibly with formation of alcohol **9** and CO_2 in the presence of water. This renders the transformation of allylic carbonate to allylic alcohol practically irreversible. Control experiments at different pH values, with different leaving groups, and with $MHCO_3$ as an external nucleophile showed that neither the hydroxide ion nor water but hydrogencarbonate acts as the O-nucleophile under the conditions used.

2.1.4.6.3 Asymmetric Synthesis of Unsymmetrical Allylic Alcohols

It was then of interest to see whether racemic unsymmetrical allylic carbonates are also capable of undergoing an efficient catalytic transformation to the corresponding allylic alcools with Pd(0)/**BPA** and water.

In this case both enantiomers **3** and *ent*-**3** react with Pd(0)/**BPA** with formation of the two diastereomeric π-allyl–Pd(II) complexes **25** and **26**, respectively (Scheme 2.1.4.29). Only if the following conditions exist can the racemic substrate be completely converted to the chiral alcohol with high efficiency: 1) the reactivity of the π-allyl–Pd(II) complexes **25** and **26** must be different; 2) a fast diastereomerization of **25** and **26** or racemization of **3** and/or *ent*-**3** must take place; 3) **BPA** must induce a high stereoselectivity; 4) the substituents of the allylic substrate have to provide for a high regioselectivity [39].

Treatment of the unsymmetrical racemic carbonate *rac*-**27a** with Pd(0)/**BPA**, $KHCO_3$ and water in CH_2Cl_2 gave alcohol **28a** in high yield and with high enantio-as well as regioselectivity (Scheme 2.1.4.30). Synthetically more relevant are the re-

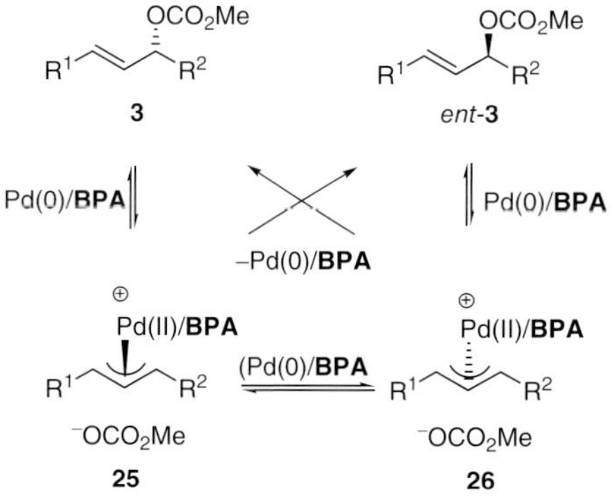

Scheme 2.1.4.29 Isomerization of diastereomeric π-allyl–Pd(II)/**BPA** complexes and racemization of the allylic carbonate.

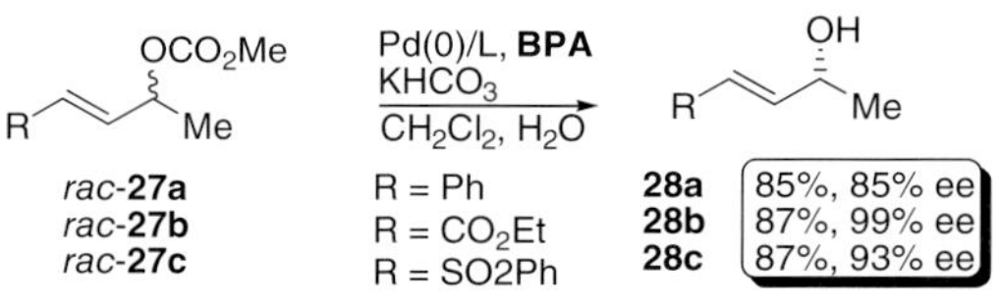

	Pd(0)/L, **BPA**		
rac-**27a**	R = Ph	**28a**	85%, 85% ee
rac-**27b**	R = CO_2Et	**28b**	87%, 99% ee
rac-**27c**	R = SO_2Ph	**28c**	87%, 93% ee

Scheme 2.1.4.30 Asymmetric synthesis of unsymmetrical allylic alcohols.

sults obtained with the functionalized racemic carbonates rac-**27b** and rac-**27c**. Similar treatment of these carbonates afforded alcohols **28b** and **28c** in high yields and with high enantio- as well as regioselectivities.

The high yields and enantioselectivities recorded in the Pd(0)/**BPA**-catalyzed reaction of rac-**27a–27c** with water and $KHCO_3$ show that not only a highly enantioselective alkylation but also an efficient dynamic kinetic resolution either via racemization of the more slowly reacting enantiomer of the substrate or isomerization of the diastereomeric π-allyl–Pd complexes (see Scheme 2.1.4.29) had occurred. Experiments with other racemic unsymmetrical allylic carbonates revealed a dependence of the enantioselectivity on the concentration of the Pd(0)/**BPA** catalyst. This seems to point to a Pd(0)/**BPA**-catalyzed isomerization of the π-allyl–Pd(II) complexes being responsible for the dynamic part of the kinetic resolution.

2.1.4.6.4 Asymmetric Synthesis of a Prostaglandin Building Block

Application of the Pd-catalyzed alkylation of hydrogencarbonate to the *meso*-biscarbonate **29** gave the allylic alcohol **30** in 87% yield with 96% *ee* (Scheme 2.1.4.31). Alcohol **30** has been converted via the silyl ether **31** and alcohol **32** to ketone **33**, the enantiomer of which is an important building block for the synthesis of prostaglandins [40]. Since both **BPA** and *ent*-**BPA** are readily available, access to *ent*-**33** is also provided.

2.1.4.6.5 Investigation of an Unsaturated Analogue of BPA

The bisphosphane **34** [41] (Scheme 2.1.4.32) is an unsaturated analogue of **6**, the potential of which for allylic substitution has not yet been studied. Treatment of the six-membered cyclic carbonate rac-**1aa** with Pd(0)/L (2 mol%), **BPA** (4 mol%), and $KHCO_3$ (1.4 equiv) in CH_2Cl_2/H_2O at room temperature furnished the alcohol **9a** with 93% *ee* in 90% yield. The use of the saturated ligand **BPA** in

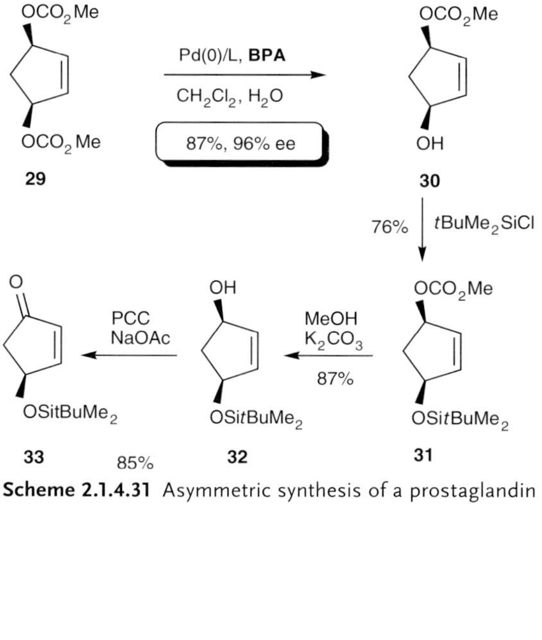

Scheme 2.1.4.31 Asymmetric synthesis of a prostaglandin building block.

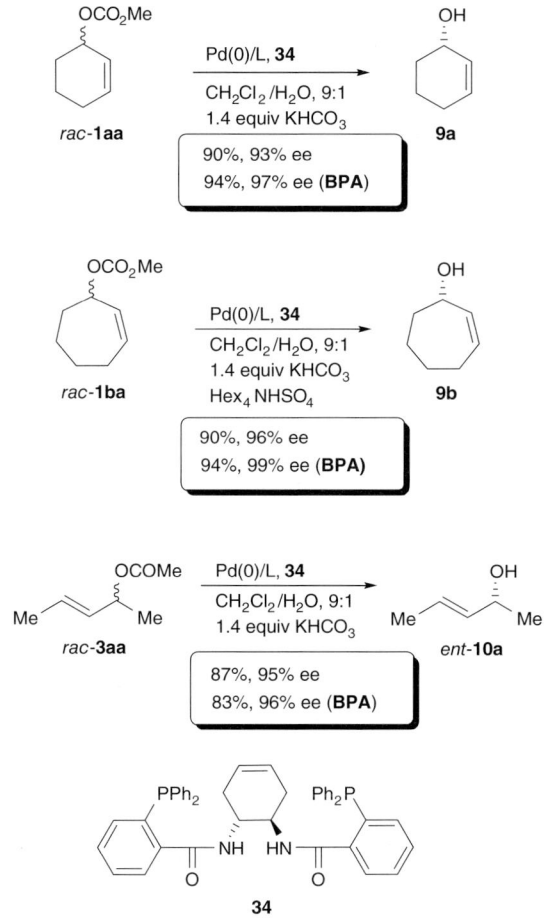

Scheme 2.1.4.32 Pd-catalyzed asymmetric synthesis of allylic alcohols in the presence of bisphosphane **34**.

this transformation had given **33** with 97% *ee*. Similarly, reaction of the seven-membered cyclic carbonate *rac*-**1ba** afforded the alcohol **9b** with 96% *ee* in 90% yield. The analogous reaction in the presence of **BPA** had furnished **35** with 99% *ee* in 94% yield. Finally, the acyclic carbonate *rac*-**3aa** was tested, which gave alcohol *ent*-**10a** with 95% *ee* in 87% yield, which in the case of **BPA** had been obtained with 96% *ee* in 83% yield.

2.1.4.7 Conclusions

The Pd-catalyzed allylic alkylation of sulfinate ions, thiols, and thiocarboxylate ions with racemic cyclic and acyclic allylic esters in the presence of bisphosphane **BPA** generally provides for an efficient asymmetric synthesis of allylic sulfones, sulfides, and thioesters. The Pd-catalyzed rearrangements of allylic sulfinates and allylic *O*-thiocarbamates, both of which proceed very efficiently in the presence of **BPA**, are attractive alternative ways to the asymmetric synthesis of allylic sulfones and allylic thioesters also starting from the corresponding racemic alcohols.

The Pd-catalyzed rearrangements most probably follow an ionization–substitution pathway with the intermediate formation of a π-allyl–Pd complex. The Pd-catalyzed allylic alkylation is generally accompanied by a highly selective kinetic and synthetically useful resolution of the racemic allylic acetates and carbonates. The kinetic resolution in the case of five-membered cyclic esters is accompanied by a dynamic kinetic resolution of the substrate featuring a Pd-catalyzed racemization of the substrate, the mechanism of which is not yet known. The Pd-catalyzed reaction of racemic symmetrical allylic carbonates with water in the presence of **BPA** leads to the generation of the O-nucleophile hydrogencarbonate and its highly enantioselective allylic alkylation to give finally the corresponding enantio-enriched allylic alcohols in high yields. Since the starting materials for the racemic allylic carbonates are the corresponding racemic allylic alcohols, the overall transformation may be regarded as deracemization of allylic alcohols. Unsymmetrical and functionalized racemic allylic carbonates are also amenable, occasionally with high efficiency, to Pd-catalyzed deracemization. In this case a dynamic kinetic resolution has to occur, the mechanism of which seems to involve a Pd-catalyzed isomerization of diastereomeric π-allyl–Pd(II) complexes.

Acknowledgements

This work was supported by the Deutsche Forschungsgemeinschaft (SFB 380 "Asymmetric Synthesis with Chemical and Biological Methods" and GK 440 "Methods in Asymmetric Synthesis"). The work described here would not have been possible without the dedicated efforts of my coworkers whose names are listed in the references and to whom I am grateful for their contributions.

References

1 a) *Chirality in Industry* (Eds.: A. N. Collins, G. N. Sheldrake, J. Crosby), Wiley, New York, **1992**; b) R. A. Sheldon, *Chirotechnology,* Marcel Dekker, New York, **1993**; c) R. Noyori, M. Tokunaga, M. Kitamura, *Bull. Chem. Soc. Jpn.* **1995**, *68*, 36–56; d) R. S. Ward, *Tetrahedron: Asymmetry* **1995**, *6*, 1475–1490; e) S. Caddick, K. Jenkins, *Chem. Soc. Rev.* **1996**, *25*, 447–456; f) U. T. Strauss, U. Felfer, K. Faber, *Tetrahedron: Asymmetry* **1999**, *10*, 107–117; g) G. R. Cook, *Curr. Org. Chem.* **2000**, *4*, 869–885; h) F. F. Huerta, A. B. Minidis, J.-E. Bäckvall, *Chem. Soc. Rev.* **2001**, *30*, 321–331; i) J. M. Keith, J. F. Larrow, E. N. Jacobsen, *Adv. Synth. Catal.* **2001**, *343*, 5–26; j) K. Faber, *Chem. Eur. J.* **2001**, *7*, 5005–5010; k) H.-J. Gais, F. Theil, in *Enzyme Catalysis in Organic Synthesis* (Eds.: K. Drauz, H. Waldmann), Wiley-VCH, Weinheim, **2002**, Vol. II, pp. 335–578; l) A. Bommarius, in *Enzyme Catalysis in Organic Synthesis* (Eds.: K. Drauz, H. Waldmann), Wiley-VCH, Weinheim, **2002**, Vol. II, pp. 741–761; m) T. Rein, T. M. Pedersen, *Synthesis* **2002**, *5*, 579–594; n) M.-J. Kim. Y. Ahn, J. Park, *Cur. Opin. Chem. Biol.* **2002**, *13*, 578–587; o) H. Pellissier, *Tetrahedron* **2003**, *59*, 8291–8327; p) D. E. J. E. Robinson, S. D. Bull, *Tetrahedron: Asymmetry,* **2003**, *14*, 1407–1446; q) O. Pàmies, J.-E. Bäckvall, *Chem. Rev.* **2003**, *103*, 3247–3261; r) N. J. Turner, *Cur. Opin. Chem. Biol.* **2004**, *8*, 114–119; s) O. Oamies, J.-E. Bäckvall, *Trends in Biotechnol.* **2004**, *22*, 130–135; t) M. Breuer, K. Ditrich, T. Habicher, B. Hauer, M. Keßeler, R. Stürmer, T. Zelinski, *Angew. Chem.* **2004**, *116*, 806–843; *Angew. Chem. Int. Ed.* **2004**, *43*, 788–824; u) E. Vedejs, M. Jure, *Angew. Chem.* **2005**, *117*, 4040–4069; *Angew. Chem. Int. Ed.* **2005**, *44*, 3974–4001;

v) M.-J. Kim, Y. Ahn, J. Park, *Bull. Kor. Chem. Soc.* **2005**, *26*, 515–522.

2 a) J. Tsuji, *Palladium Reagents and Catalysts*, Wiley, New York, **1997**; b) S. J. Sesay, J. M. J. Williams, *Adv. Asym. Synth.* **1998**, *3*, 235–271; c) A. Heumann, in *Transition Metals for Organic Synthesis* (Eds.: M. Beller, C. Bolm), Wiley-VCH, Weinheim, **1998**, Vol. 1, pp. 251–264; d) B. M. Trost, C. Lee, in *Catalytic Asymmetric Synthesis* (Ed.: I. Ojima), Wiley-VCH, Weinheim, **2000**, pp. 593–649; e) G. Helmchen, A. Pfaltz, *Acc. Chem. Res.* **2000**, *33*, 336–345; f) M. Moreno-Mañas, R. Pleixats, in *Handbook of Organopalladium Chemistry for Organic Synthesis* (Eds.: E. Negishi, A de Meijere), Wiley, New York, **2002**, Vol. 2, pp. 1707–1767; g) L. Acemoglu, J. M. J. Williams, in *Handbook of Organopalladium Chemistry* (Eds.: E. Negishi, A. de Meijere), Wiley, New York, **2002**, Vol. 2, pp. 1945–1979; h) B. M. Trost, M. L. Crawley, *Chem. Rev.* **2003**, *103*, 2921–2943; i) B. M. Trost, *J. Org. Chem.* **2004**, *69*, 5813–5837.

3 a) P. B. Mackenzie, J. Whelan, B. Bosnich, *J. Am. Chem. Soc.* **1985**, *107*, 2046–2054; b) K. L. Granberg, J.-E. Bäckvall, *J. Am. Chem. Soc.* **1992**, *114*, 6858–6863.

4 T. Hayashi, A. Yamamoto, I. Yoshihiko, *J. Chem. Soc. Chem. Commun.* **1986**, 1090–1092.

5 H.-J. Gais, H. Eichelmann, N. Spalthoff, F. Gerhards, M. Frank, G. Raabe, *Tetrahedron: Asymmetry* **1998**, *9*, 235–248.

6 a) M. Bourghida, M. Widhalm, *Tetrahedron: Asymmetry* **1998**, *9*, 1073–1083; b) H. Brunner, I. Deml, W. Dirnberger, K.-P. Ittner, W. Reißer, M. Zimmermann, *Eur. J. Inorg. Chem.* **1999**, 51–59; c) S. Ramdeehul, P. Dierkes, R. Aguado, P. C. J. Kamer, P. W. N. M. van Leeuwen, J. A. Osborn, *Angew. Chem.* **1998**, *110*, 3302–3304; *Angew. Chem. Int. Ed.* **1998**, *37*, 3118–3121; d) G. C. Lloyd-Jones, S. C. Stephen, *Chem. Commun.* **1998**, 2321–2322; e) B. M. Trost, E. J. Hembre, *Tetrahedron Lett.* **1999**, *40*, 219–222; f) B. M. Trost, F. D. Toste, *J. Am. Chem. Soc.* **1999**, *121*, 3543–3544; g) T. Nishimata, K. Yamaguchi, M. Mori, *Tetrahedron Lett.* **1999**, *40*, 5713–5716; h) M. T. Reetz, S. Sostmann, *J. Organomet. Chem.* **2000**, *603*, 105–109; i) J. M. Longmire, B. Wang, X. Zhang, *Tetrahedron Lett.* **2000**, *41*, 5435–5439; j) T. Okauchi, K. Fujita, T. Ohtagura, S. Ohshima, T. Minami, *Tetrahedron: Asymmetry* **2000**, *11*, 1397–1403; k) B. M. Trost, J. Dudash, Jr., E. J. Hembre, *Chem. Eur. J.* **2001**, *7*, 1619–1629; l) S. R. Gilbertson, P. Lan, *Org. Lett.* **2001**, *3*, 2237–2240; m) B. Dominguez, N. S. Hodnett, G. C. Lloyd-Jones, *Angew. Chem.* **1998**, *110*, 2092–2118; *Angew. Chem. Int. Ed.* **2001**, *40*, 4289–4291; n) J. W. Faller, N. Sarantopoulos, *Organometallics* **2004**, *23*, 2179–2185; o) J. W. Faller, J. C. Wilt, J. Parr, *Org. Lett.* **2004**, *6*, 1301–1304; p) S. Jansat, M. Gomez, K. Philippot, G. Muller, E. Guiu, C. Claver, S. Castillon, B. Chaudret, *J. Am. Chem. Soc.* **2004**, *126*, 1592–1593; q) K. Onitsuka, Y. Matsushima, S. Takahashi, *Organometallics* **2005**, *24*, 6472–6474.

7 a) H. Eichelmann, H.-J. Gais, *Tetrahedron: Asymmetry* **1995**, *3*, 643–646; b) M. Frank, H.-J. Gais, *Tetrahedron: Asymmetry* **1998**, *9*, 3353–3357; c) D. Vasen, A. Salzer, F. Gerhards, H.-J. Gais, N. H. Bieler, A. Togni, *Organometallics* **2000**, *19*, 539–546; d) K. N. Gavrilov, O. G. Bondarev, R. V. Lebedev, A. A. Shiryaev, S. E. Lyubimov, A. I. Polosukhin, G. V. Grintselev-Knyazev, K. A. Lyssenko, S. K. Moiseev, N. S. Ikonnikov, V. N. Kalinin, V. A. Davankov, A. V. Korostylev, H.-J. Gais, *Eur. J. Inorg. Chem.* **2002**, 1367–1376; e) O. G. Bondarev, K. N. Gavrilov, V. N. Tsarev, H.-J. Gais, *Russ. Chem. Bull., Int. Ed.*, **2002**, *51*, 1748–1750; f) K. N. Gavrilov, O. G. Bondarev, A. V. Korostylev, A. I. Polosukhin, V. N. Tsarev, N. E. Kadilnikov, S. E. Lyubimov, A. A. Shiryaev, S. V. Zheglov, H.-J. Gais, V. A. Davankov, *Chirality* **2003**, *15*, S97–S103; g) K. N. Gavrilov, V. N. Tsarev, A. A. Shiryaev, O. G. Bondarev, S. E. Lyubimov, E. B. Benetsky, A. A. Korlyukov, M. Y. Antipin, V. A. Davankov, H.-J. Gais, *Eur. J. Inorg. Chem.* **2004**, 629–634.

8 a) B. M. Trost, M. G. Organ, G. A. O´Doherty, *J. Am. Chem. Soc.* **1995**, *117*, 9662–9670; b) B. M. Trost, M. J. Krische, R. Radinov, G. Zanoni, *J. Am. Chem. Soc.* **1996**, *118*, 6297–6298; c) B. M. Trost, A. C. Krueger, R. C. Bunt, J. Zambrano, *J. Am. Chem. Soc.* **1996**, *118*, 6520–6521.
9 B. M. Trost, D. L. Van Vranken, C. Bingel, *J. Am. Chem. Soc.* **1992**, *114*, 9327–9343.
10 H.-J. Gais, T. Jagusch, N. Spalthoff, F. Gerhards, M. Frank, G. Raabe, *Chem. Eur. J.* **2003**, *9*, 4202–4221.
11 H.-J. Gais, N. Spalthoff, T. Jagusch, M. Frank, G. Raabe, *Tetrahedron Lett.* **2000**, *41*, 3809–3812.
12 J. M. Goodman, A.-K. Köhler, S. C. M. Alderton, *Tetrahedron Lett.* **1999**, *40*, 8715–8718.
13 M. J. Södergren, P. G. Andersson, *J. Am. Chem. Soc.* **1998**, *120*, 10760–10761.
14 a) L. E. Overman, C. E. Owen, M. M. Pavan, C. J. Richards, *Org. Lett.* **2003**, *5*, 1809–1812; b) Y. Uozumi, K. Kato, T. Hayashi, *Tetrahedron: Asymmetry* **1998**, *9*, 1065–1072; c) Y. Jiang, J. M. Longmire, X. Zhang, *Tetrahedron Lett.* **1999**, *40*, 1449–1453; d) P.-H. Leung, K.-H. Ng, Y. Li, A. J. P. White, J. D. Williams, *Chem. Commun.* **1999**, 2435–2436; e) J. Kang, K. H. Yew, T. H. Kim, D. H. Choi, *Tetrahedron Lett.* **2002**, *43*, 9509–9512.
15 a) P. Kočovský, I. Starý, in *Handbook of Organopalladium Chemistry for Organic Synthesis* (Eds.; E. Negishi, A. de Meijere), Wiley-Interscience: New York, **2002**; Vol. 2, pp. 2011–2025; b) H. Nakamura, Y. Yamamoto, in *Handbook of Organopalladium Chemistry for Organic Synthesis* (Eds.: E. Negishi, A. de Meijere), Wiley-Interscience, New York, **2002**; Vol. 2, pp. 2919–2934; c) U. Nubbemeyer, *Synthesis* **2003**, 961–1008; (d) M. Reggelin; C. Zur, *Synthesis* **2000**, 1–97.
16 a) K. Hiroi, K. Makino, *Chem. Lett.* **1986**, 617–620; b) K. Hiroi, K. Makino, *Chem. Pharm. Bull.* **1988**, *36*, 1744–1749.
17 T. Jagusch, H.-J. Gais, O. Bondarev, *J. Org. Chem.* **2004**, *69*, 2731–2736.
18 a) H. Prinzbach, T. Netscher, *Synthesis* **1987**, 683–688; (b) R. Hermann, J.-H. Youn, *Tetrahedron Lett.* **1986**, *27*, 1493–1494; (c) R. Hermann, J.-H. Youn, *Synthesis* **1987**, 72–73; d) K. B. Sharpless; H. Liu, A. V. Gontcharov, *Org. Lett.* **1999**, *1*, 783–786.
19 H. Harayama, T. Nagahama, T. Kozera, M. Kimura, K. Fugami, S.Tanaka; Y. Tamaru, *Bull. Chem. Soc. Jpn.* **1997**, *70*, 445–456.
20 F. Marr, R. Fröhlich, D. Hoppe, *Org. Lett.* **1999**, *1*, 2081–2083.
21 H.-J. Gais, A. Böhme, *J. Org. Chem.* **2002**, *67*, 1153–1161.
22 A. Böhme, H.-J. Gais, *Tetrahedron: Asymmetry* **1999**, *10*, 2511–2514.
23 P. R. Auburn; J. Whelan, B. Bosnich, *Organometallics* **1986**, *5*, 1533–1537.
24 Y. Tamaru, Z. Yoshida, Y. Yamada, K. Mukai, H. Yoshioka, *J. Org. Chem.* **1983**, *48*, 1293–1297.
25 a) D. Sinou, S. Divekar, M. Safi, M. Soufiaoui, *Sulfur Lett.* **1999**, *22*, 125–130; b) S. Divekar, M. Safi, M. Soufiaoui, D. Sinou, *Tetrahedron* **1999**, *55*, 4369–4376.
26 B. J. Lüssem, H.-J. Gais, *J. Org. Chem.* **2004**, *69*, 4041–4052.
27 D. G. Blackmond, N. S. Hodnett, G. C. Lloyd-Jones, *J. Am. Chem. Soc.* **2006**, *128*, 7450–7451.
28 a) J. C. Fiaud, J. L. Malleron, *Tetrahedron. Lett.* **1981**, *22*, 1399–1402; b) B. M. Trost, R. C. Bunt, *J. Am. Chem. Soc.* **1996**, *118*, 235–236; c) G. C. Lloyd-Jones, S. C. Stephen, *Chem. Eur. J.* **1998**, *4*, 2539–2549; d) C. P. Butts, J. Crosby, G. C. Lloyd-Jones, S. C. Stephen, *Chem. Commun.* **1999**, 1707–1708; e) I. J. S. Fairlamb, G. C. Lloyd-Jones, *Chem. Commun.* **2000**, 2447–2448; f) G. C. Lloyd-Jones, S. C. Stephen, M. Murray, C. P. Butts, Š. Vyskočil, P. P. Kočosvsky, *Chem. Eur. J.* **2000**, *6*, 4348–4357; g) J. M. Longmire, B. Wang, X. Zhang, *Tetrahedron Lett.* **2000**, *41*, 5435–5438; h) U. Kazmaier, F. L. Zumpe, *Angew. Chem. Int. Ed.* **2000**, *39*, 802–804; i) B. M. Trost, J.-P. Surivet, *J. Am. Chem. Soc.* **2000**, *122*, 6291–6292; j) I. J. S. Fairlamb, G. C. Lloyd-Jones, Š. Vyskočil, P. Kočosvsky, *Chem. Eur. J.* **2002**, *8*, 4443–4453; k) G. C. Lloyd-Jones; S. C. Stephen, I. J. S. Fairlamb, A. Martorell, B. Dominguez, P. M. Tomlin, M. Murray, J. M. Fernandez, J. C. Jeffery, T. Riis-Johannessen, T.

Guerziz, *Pure Appl. Chem.* **2004**, *76*, 589–601.
29 B. M. Trost, M. R. Machacek, H. C. Tsui, *J. Am. Chem. Soc.* **2005**, *127*, 7014–7024.
30 a) B. H. Lipshutz, S. Sengupta, *Org. React.* **1992**, *41*, 135–631; b) E. J. Corey, C. J. Helal, *Angew. Chem.* **1984**, *96*, 854–882; *Angew. Chem., Int. Ed.* **1998**, *37*, 1986–2012; c) A. B. Charette, A. Beauchemin, *Org. React.* **2001**, *58*, 1–415; d) R. M. Hanson, *Org. React.* **2002**, *60*, 1–156.
31 a) R. A. Johnson, K. B. Sharpless, in *Catalytic Asymmetric Synthesis* (Ed.: I. Ojima), Wiley-VCH, New York, **2000**; pp 231–285; b) M.-J. Kim, Y. Ahn, J. Park, *Cur. Opin. Biotechnol.* **2002**, *13*, 578–587; c) M. J. Williams, J. V. Allen, *Tetrahedron Lett.* **1996**, *37*, 1859–1862; d) Y. K. Choi, J. H. Suh, D. Lee, I. T. Lim, J. Y. Jung, M.-J. Kim, *J. Org. Chem.* **1999**, *64*, 8423–8424; e) D. Lee, E. A. Huh, M.-J. Kim, H. M. Jung, J. H. Koh, J. Park, *Org. Lett.* **2000**, *2*, 2377–2379.
32 a) R. Noyori, T. Ohkuma, *Angew. Chem.* **2001**, *113*, 40–75; *Angew. Chem., Int. Ed.* **2001**, *40*, 40–73; b) M. Itsuno, *Org. React.* **1998**, *52*, 395–576.
33 a) L. Pu, H.-B. Yu, *Chem. Rev.* **2001**, *101*, 757–824; b) H. Li, P. Walsh, *J. J. Am. Chem. Soc.* **2004**, *126*, 6538–6539.
34 a) B. M. Trost, M. G. Organ, *J. Am. Chem. Soc.* **1994**, *116*, 10 320–10 321; b) B. M. Trost, F. D. Toste, *J. Am. Chem. Soc.* **2003**, *125*, 3090–3100; c) A. Magnus, S. K. Bertilsson, P. G. Andersson, *Chem. Soc. Rev.* **2002**, *31*, 223–229; d) A. V. Malkov, D. Pernazza, M. Bell, M. Bella, A. Massa, F. Teplý, P. Mehani, P. Kočovský, *J. Org. Chem.* **2003**, *68*, 4727–4742; e) J. W. Faller, J. C. Wilt, *Tetrahedron Lett.* **2004**, *45*, 7613–7616; f) S. F. Kirsch, L. E. Overman, *J. Am. Chem. Soc.* **2005**, *127*, 2866–2867.
35 J. Muzart, *Tetrahedron* **2005**, *61*, 5955–6008.
36 B. J. Lüssem, H.-J. Gais, *J. Am. Chem. Soc.* **2003**, *125*, 6066–6067.
37 H.-J. Gais, O. Bondarev, R. Hetzer, *Tetrahedron Lett.* **2005**, *46*, 6279–6283.
38 a) T. Fukazawa, T. Hashimoto, *Tetrahedron: Asymmetry* **1993**, *4*, 2323–2326; b) S. K. Bertilsson, M. J. Soedergren, P. G. Andersson, *J. Org. Chem.* **2002**, *67*, 1567–1573; c) Q.-Y. Hu, P. D.Rege, E. J. Corey, *J. Am. Chem. Soc.* **2004**, *126*, 5984–5986; d) M. Takahashi, R. Koike, K. Ogasawara, *Chem. Pharm. Bull.* **1995**, *43*, 1585–1587.
39 a) A. Pfaltz, M. Lautens, in *Comprehensive Asymmetric Catalysis* (Eds.: E. N. Jacobson, A. Pfaltz, H. Yamamoto), Springer, Berlin, **1999**; Vol. II, pp. 833–884; b) H. Grennberg, V. Langer, J.-E. Bäckvall, *J. Chem. Soc., Chem. Commun.* **1991**, 1190–1192.
40 a) R. Noyori, M. Suziki, *Angew. Chem., Int. Ed. Engl.* **1984**, *23*, 847–876; b) A. G. Myers, M. Hammond, Y. Wu, *Tetrahedron Lett.* **1996**, *37*, 3083–3086.
41 W. Oppolzer, D. L. Kuo, M. W. Hutzinger, R. Léger, J.-O. Durand, C. Leslie, *Tetrahedron Lett.* **1997**, *38*, 6213–6216.

2.1.5
The QUINAPHOS Ligand Family and its Application in Asymmetric Catalysis
Giancarlo Franciò, Felice Faraone, and Walter Leitner

2.1.5.1 Introduction

The quest for new chiral ligands is of fundamental importance for progress in asymmetric catalysis [1]. The ultimate goal is the development of a chiral ligand that is readily available in both enantiomeric forms, which imparts high turnover numbers and enantioselectivity to the catalyst. Although such ligands are known

2.1.5 The QUINAPHOS Ligand Family and its Application in Asymmetric Catalysis

for several reactions, most of them often show a narrow scope. Only very few chiral ligands can be applied in different reactions and have a spectrum of substrates beyond the standard test molecules. Such "privileged" structures [2] are rare and almost impossible to predict. Therefore two different strategies have been developed to overcome this bottleneck.

The first one is based on a combinatorial approach and requires the preparation of a large number of ligands. Such ligands have to be synthesized in only a few steps and the procedure should be suitable for automation. This methodology is highly demanding not only for the synthetic effort but also because the evaluation of large amounts of catalytic data require high-throughput screening methods for testing and analysis. Therefore this approach has been pursued mostly by industrial companies, proving its efficiency for the selection of practical catalytic systems [3].

The second approach relies on the creation of a family of ligands using a modular design that makes it possible, with modest synthetic effort, to include a variety of substituents in a lead structure. The identification of a suitable lead structure is decisive for the success of this approach. Based on chemical intuition and on molecular understanding, the lead structure can be further developed and modified following a rational design. As a result, improved comprehension of structure–reactivity relationships can be gained and fine tuning of the ligand for specific substrates and/or reactions may be achieved.

After the first disclosure in 1996 by Feringa and coworkers [4], phosphoramidites have earned a prominent role in the ligand toolbox for homogeneously catalyzed reactions. It is noteworthy that phosphoramidites fulfill the requirements of both approaches described above and their potential has been explored extensively [5].

Some years ago we introduced a family of phosphoramidite ligands based on 1,2-dihydroquinoline as an easily accessible chiral backbone guaranteeing a certain degree of rigidity and a predefined geometry, in particular for bidentate ligands. In this chapter we will refer to this class of compounds as QUINAPHOS ligands, emphasizing the role of the shared backbone, whereas in the previous literature diverse trivial names have been assigned to the different members of the class (BINAPHOSQUIN, BIPHENPHOSQUIN, QUINAPHOS) [6, 7]. Important common features of the QUINAPHOS ligand family (Structure 2.1.5.1) can be summarized as follows.

- *Versatile coordination*: Depending on the group D situated in the 8-position of the quinoline ring and on the metal precursor, these ligands can act as monodentate, monodentate with an ancillary coordination site, or bidentate ligands. The position of the group at C8 relative to the phosphoramidite allows a facile chelation to the metal center, leading to the formation of a six-membered ring. In some cases, activation of the C8—D bond has been observed, resulting in a five-membered metallacycle.

- *Modular synthesis*: The group R located on the chiral C2 is introduced through an organometallic reagent *during* the synthesis. Thus, it can be chosen from a variety of reagents, allowing high structural freedom. The variation of the diol moiety is an additional obvious target for structural diversity, though this option has barely been explored till now.
- *Diastereomeric tuning*: The coexistence of two types of chiralities in the structure (one central and one axial) makes it possible to exploit the interplay of different stereogenic moieties. Interestingly, the matched and mismatched combination is strongly dependent on the type of reaction to which the diastereomers are applied.

The present chapter provides a summary of the current state of development for this ligand class covering the synthetic strategy and information on structure and coordination as well as typical applications.

2.1.5.2 Synthetic Strategy

QUINAPHOS ligands are usually synthesized in a one-pot-procedure from readily available 8-substituted quinolines [8] via nucleophilic addition of a lithium reagent [9] to the azomethinic double bond and direct quenching of the resulting 1,2-dihydroquinoline amide **1** with a phosphorochloridite derived from enantiomerically pure binaphthol (**l**) or from 3,3′-di-*t*-butyl-5,5′-dimethoxybiphenyl-2,2′-diol (**m**) [10] (Scheme 2.1.5.1, Method A). Alternatively, the anion **1** can be reacted with an excess (in order to avoid multiple substitution) of phosphorous trichloride to obtain the corresponding phosphorous dichloridite **2**, which can be isolated (Scheme 2.1.5.1, Method B). In a second step, **2** is converted into **4** by reaction with the desired diol in the presence of triethylamine.

R = alkyl or aryl
D = phosphine, halogen, ether group
⌒ = diol

Structure 2.1.5.1 QUINAPHOS structure: the arrows highlight the explored variation.

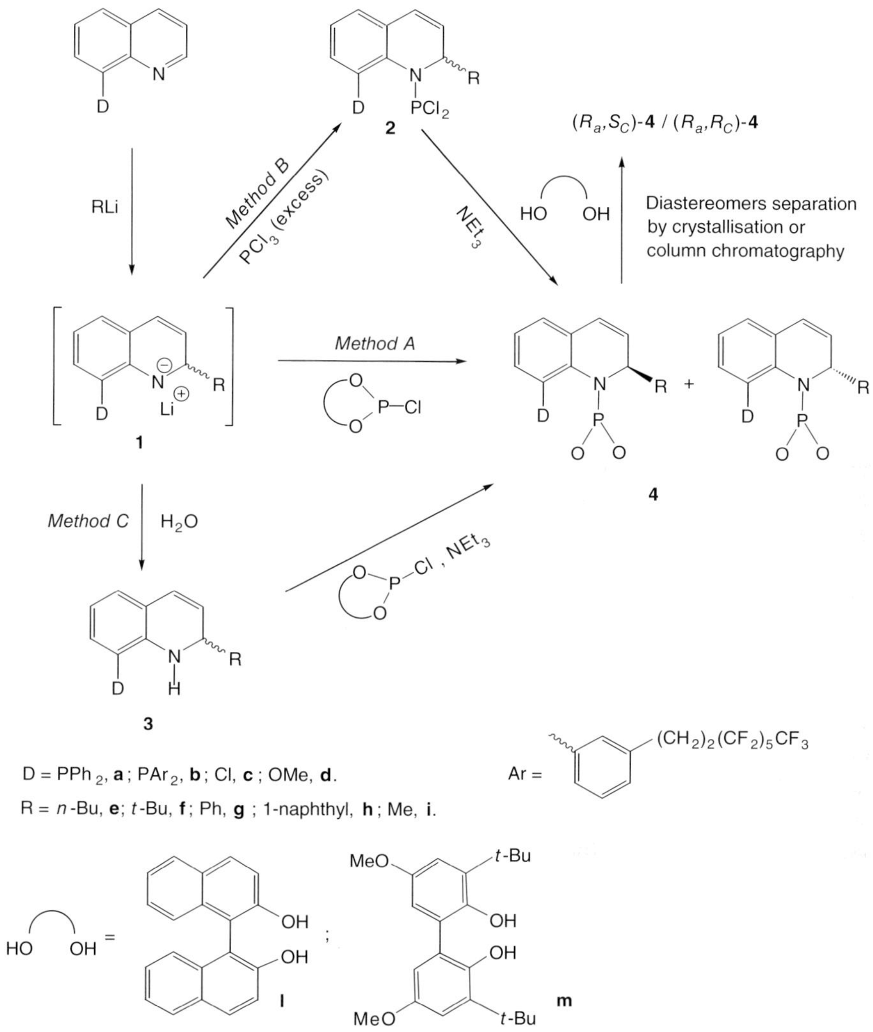

2.1.5 The QUINAPHOS Ligand Family and its Application in Asymmetric Catalysis | 253

Scheme 2.1.5.1 Synthetic procedures for the preparation of QUINAPHOS ligands.

To indicate the different QUINAPHOS derivatives the following notation will be used: (*chiral descriptors*)-**number, R, D, diol**; for example, (R_a,S_C)-**4cel** corresponds to (S)-2-butyl-8-chloro-1-(R)-3,5-dioxa-4-phosphacyclohepta[2,1-a;3,4-a']dinaphthalen-4-yl)-1,2-dihydroquinoline and (R_a,R_C)-**4ael** to (R)-2-butyl-8-diphenylphosphino-1-(R)-3,5-dioxa-4-phosphacyclohepta[2,1-a;3,4-a']dinaphthalen-4-yl)-1,2-dihydroquinoline, i.e., to the classical QUINAPHOS ligand itself.

(R_a,S_C)-**4cel**   (R_a,R_C)-**4ael**

The ligands containing **m** as the diol are obtained as racemic mixtures due to the tropoisomerism of this 6,6'-unsubstituted biphenyl moiety[11]. For instance, the compound **4cem** (named BIPHENPHOSQUIN in ref. [6]) shows a single resonance at δ = 141.1 in the ^{31}P{^{1}H} NMR spectrum and a single set of signals in the ^{1}H NMR spectrum at both 293 and 193 K. Such a behavior is commonly explained by the impossibility of slowing down the rapid tropoisomerism of the biphenyl on the NMR-timescale, even by lowering the temperature [12]. In some cases, however, the fixation of the conformation was observed at low temperature [13]. Alternatively, the central chirality at C2 can induce a preferential, energetically more stable conformation of the tropoisomeric moiety leading to the preponderant presence in solution of a single pair of diastereomers (which are in an enantiomeric relationship to each other) [14]. The latter argument was used to comment on the similar catalytic results obtained with chiral phosphoramidites bearing a tropo- and an atropoisomeric diol, respectively [15].

In contrast, the phosphoramidites comprising (R)-binaphthol or (S)-binaphthol moieties (**l**) are obtained as a diastereomeric mixture and can be separated. The separation of the diastereomers has been achieved for many derivatives exploiting the different solubility profiles of the two diastereomers. Alternatively, the separation could be accomplished by column chromatography. These methodologies need to be optimized for each diastereomeric pair, thus representing currently the synthetic bottleneck for the rapid generation of different QUINAPHOS derivatives.

Alternative synthetic approaches include enantioselective addition of the organometallic reagent to quinoline in the first step of the synthesis [16], the resolution of the racemic amines resulting from simple protonation of anions **1** (Scheme 2.1.5.1, Method C) by diastereomeric salts formation [17] or by enzymatic kinetic resolution [18], and the iridium-catalyzed enantioselective hydrogenation of 2-substituted quinolines [19]. All these methodologies would avoid the need for diastereomer separation later on, and give direct access to enantio-enriched QUINAPHOS derivatives bearing achiral or tropoisomeric diols. Current work in our laboratories is directed to the evaluation of these methods.

2.1.5.3 Stereochemistry and Coordination Properties

2.1.5.3.1 Free Ligands

The stereochemistry of QUINAPHOS ligands has been investigated through X-ray and NMR analysis. The absolute configuration at the stereocenter for

(R_a,R_C)-**4cel** was assigned through the structural determination by X-ray analysis of the platinum complex cis-[Pt{(R_a,R_C)-**4cel**}$_2$I$_2$] (**5**) (vide infra). Recently, crystals suitable for X-ray diffraction have been grown for the free ligand (R_a,S_C)-**4chl** (Fig. 2.1.5.1) [20]. The 1-naphthyl-substituent and the phosphoramidite moiety reside on opposite sides with respect to the mean plane of the heterocyclic backbone in order to reduce steric interaction. The torsion angle of the binaphthyl moiety responsible for the axial chirality measures -50.2° and is in the expected range.

In all QUINAPHOS ligands bearing (R)-binaphthol as the diol, the resonance in the ^{31}P NMR spectrum relative to the phosphoramidite group appears at higher field for the **A** with respect to the **B** diastereomer (Table 2.1.5.1). Also, the comparison of typical ^{1}H NMR-resonances of each diastereomeric pair of the QUINAPHOS family reveals clear trends. Starting from the NMR data of the diastereomers whose absolute configuration has been secured through X-ray analysis, an assignment of the stereochemistry of all other ligands can be achieved (Table 2.1.5.1).

The reliability of these assignments could be supported by means of a calculated conformational analysis followed by a detailed NMR investigation carried out for each diastereomer of **4cil** as examples [20, 21]. The conformation minima were located through molecular dynamics calculations (Monte Carlo method at PM3 level), followed by DFT geometry optimizations (Gaussian03, B3LYP/6-31G(d)) (Fig. 2.1.5.2). In accordance with the structures with lowest energy content predicted by the DFT calculations, the NOESY spectrum of the presumed (R_a,S_C)-**4cil** clearly proves the existence of an interaction between the protons of the methyl group and the *ortho* and *meta* protons of a naphthyl moiety, aside from

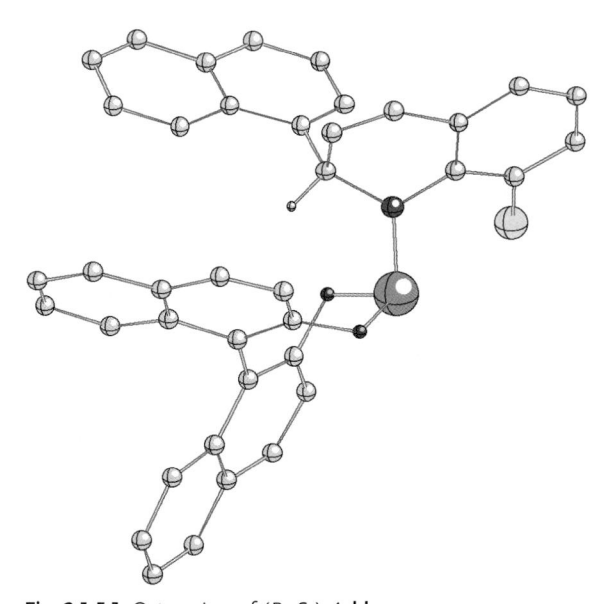

Fig. 2.1.5.1 Ortep view of (R_a,S_C)-**4chl**.

Table 2.1.5.1 Comparison of selected NMR data for various (R_a)-QUINAPHOS ligands.

Diastereomer **A**

Diastereomer **B**

Entry	Ligand	Diastereomer (config. at C2[a])	D	R	δ_pA	δ_pB	J_{pApB} [Hz]	$\delta H2$	$\delta H3$	$\delta H4$
1	(R_a)-4ael	A(R_c)	PPh$_2$	n-Bu	137.5	−17.8	192	4.13	5.85	6.52
2	(R_a)-4ael	B(S_c)	PPh$_2$	n-Bu	143.6	−16.4	131	3.99	5.65	6.28
3	(R_a)-4cel	A(R_c)[b]	Cl	n-Bu	138.1	–	–	4.09	5.74	6.40
4	(R_a)-4cel	B(S_c)	Cl	n-Bu	145.9	–	–	4.01	5.65	6.20
5	(R_a)-4cil	A(R_c)	Cl	Me	136.8	–	–	3.94	5.75	6.54
6	(R_a)-4cil	B(S_c)	Cl	Me	144.6	–	–	3.94	5.85	6.35
7	(R_a)-4agl	A(S_c)	PPh$_2$	Ph	132.8	−18.6	192	5.34	5.94	6.57
8	(R_a)-4agl	B(R_c)	PPh$_2$	Ph	141.2	−21.3	140	5.45	5.73	6.16
9	(R_a)-4chl	A(S_c)[b]	Cl	1-Naph	139.1	–	–	5.82	5.94	66.5
10	(R_a)-4chl	B(R_c)	Cl	1-Naph	143.8	–	–	6.08	6.16	6.50

a Due to the change in priority resulting from Cahn–Ingold–Prelog rules the stereochemical descriptors in entries 1, 3, 5, 7, 19 and 2, 4, 6, 8, 10 describe the same spatial arrangement A and B, respectively. b Configuration determined via X-ray crystal structure analysis.

that with the two vinylic protons, A and B, respectively. In contrast, the proton of the methyl group of the other diastereomer (R_a,R_C)-**4cil** does not show any NOE-interaction with the protons of the binaphthyl moiety, while an NOE is evident between the protons H3 and H-B. This result fits well with the calculated geometry in which the methyl group of (R_a,R_C)-**4cil** points away from the binaphthyl moiety whereas the vinylic proton H3 resides in its close proximity (Fig. 2.1.5.2).

The correctness of the stereochemical prediction based on the comparison of the NMR data of each diastereomeric pair could be assessed through this study. Furthermore, the observation that the methyl protons and the vinylic proton of the (R_a,S_C)- and of the (R_a,R_C)-diastereomer, respectively, show an NOE signal with the *ortho* and *meta* protons of *only one* binaphthyl moiety clearly indicates that the rotation around the P—N bond is hindered. In this structure the presence of a bulky chlorine substituent in the 8-position may account for the restricted rotation and a similar behavior would be expected for all 8-substituted QUINAPHOS ligands.

2.1.5.3.2 Complexes

The coordination chemistry toward transition metals of QUINAPHOS ligands with D = Cl has been investigated in detail [6]. Two coordination modes have

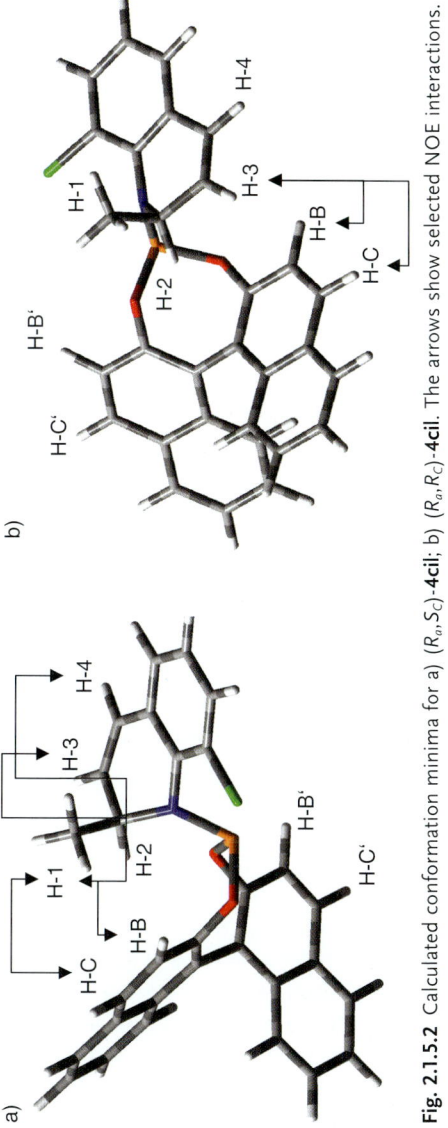

2.1.5 The QUINAPHOS Ligand Family and its Application in Asymmetric Catalysis | 257

Fig. 2.1.5.2 Calculated conformation minima for a) (R_a, S_C)-**4cil**; b) (R_a, R_C)-**4cil**. The arrows show selected NOE interactions.

been observed, depending on the diastereomer and on the metal precursor employed. In the first case, the ligand is bound to the metal exclusively through the phosphorus atom; in the second one an additional coordination occurs via the C8 carbon. This results from oxidative addition of the C–Cl bond to low-valent transition metals. Two complexes reflecting these different behaviors will be reported here as examples.

The monodentate coordination manner is evident in the platinum complex cis-[Pt{(R_a,R_C)-**4cel**}$_2$I$_2$] (**5**) obtained by reaction of cis-[Pt(cod)I$_2$] (cod = 1,5-cyclooctadiene) with 2 equiv of the ligand (R_a,R_C)-**4cel** (Fig. 2.1.5.3). The ligands are coordinated via phosphorus to the metal and the cis configuration of the metal precursor is also maintained in the resulting complex. The steric encumbrance of the two adjacent ligands causes a distortion from the ideal square-planar arrangement. The two phosphorous ligands are rotated away from each other with an angle of approximately 180° in order to minimize repulsive interaction. In comparison with the free ligand (R_a,S_C)-**4chl**, the nitrogen atoms of the coordinated (R_a,R_C)-**4cel** ligand exhibit a much more pronounced pyramidal arrangement (averaged angle sums 349°) with S-configuration.

The ability of QUINAPHOS ligands to oxidatively add to a metal in a low oxidation state forming cyclometalated complexes was evidenced in the reaction of **4cem** with [Rh(CO)$_2$Cl]$_2$. The resulting bimetallic complex **6** was fully characterized by NMR, IR and X-ray diffraction (Fig. 2.1.5.4). The cyclometalation results from the activation of the C–Cl bond and, consequently, the formation of a five-membered ring. The metals reside in octahedral coordination spheres comprising the chelated ligand, a terminal CO, and one apical and two bridging chlorines. The complex has an overall C_i symmetry. The two coordinated ligands have opposite central and axial chiralities (R_a,R_C and S_a,S_C, respectively). The formation of complex **6** proceeds most probably through a Rh(I) interme-diate in which the ligand is coordinated only via phosphorous, as in **5**. The subsequent intramolecular oxidative addition is favored by the proximity of the C–Cl bond to the metal imposed by the stereochemical arrangement. The conformation adopted by the tropisomeric biphenolic moieties with respect to the stereocenter seems to be mandatory for the accomplishment of the intramolecular addition process. In fact, a similar reaction outcome was observed between [Rh(CO)$_2$Cl]$_2$ and (R_a,R_C)-**4cel**, whereas (R_a,S_C)-**4cel** does not undergo oxidative addition to the Rhodium center, as demonstrated by IR spectroscopy. The strong impact of the stereochemistry on the coordination behavior may in many cases account for the dramatic difference in the catalytic outcomes of the two diastereoisomers (vide infra).

The reaction of (R_a,R_C)-**4ael** with [Rh(CO)$_2$Cl]$_2$ leads to the formation of a mixture of two compounds, one soluble, **7**, and one insoluble, **8**, in hexane. The ^{31}P NMR spectrum (C$_6$D$_6$) of **7** shows two doublets of doublets for the phosphoramiditic P and phosphinic P at δ = 143.3 (J_{RhP} = 257 Hz, $J_{P'P}$ = 68 Hz) and at δ = 20.1 ($J_{RhP'}$ = 124 Hz), respectively. In the IR spectrum, a single carbonyl streching band at 2036 cm^{-1} is present for **7**. Compound **8** gives rise to a similar

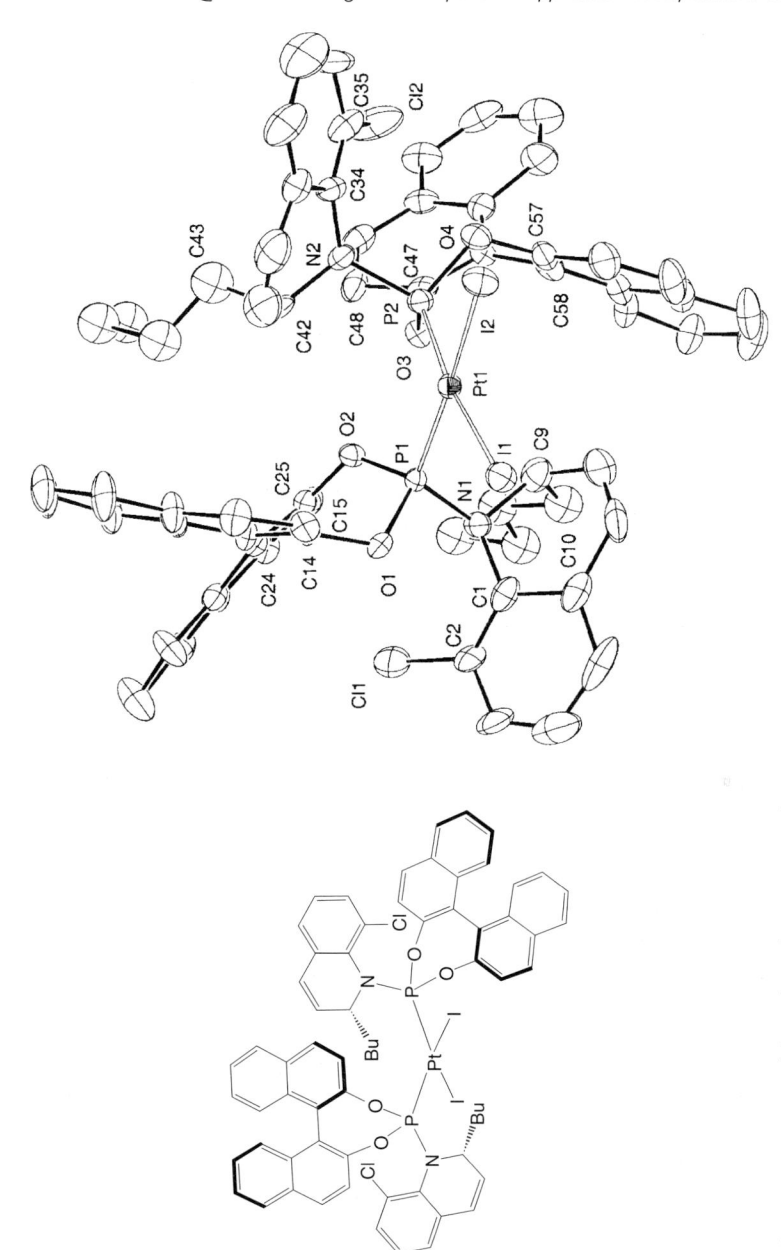

2.1.5 The QUINAPHOS Ligand Family and its Application in Asymmetric Catalysis | 259

Fig. 2.1.5.3 Structural formula and Ortep view of complex **5**.

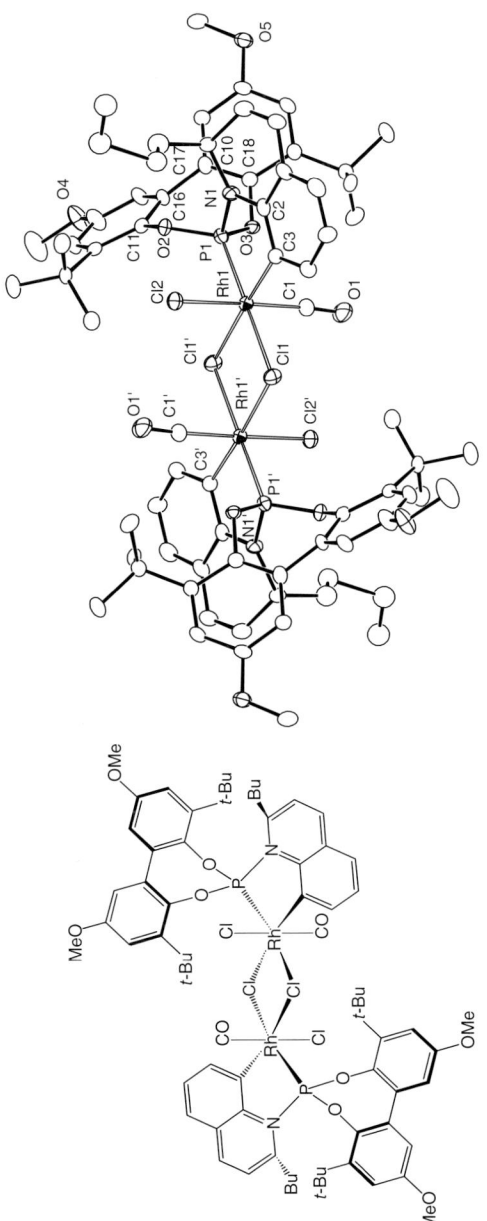

Fig. 2.1.5.4 Structural formula and Ortep view of complex **6**.

^{31}P NMR spectrum (C$_6$D$_6$) with two doublets of doublets for the phosphoramiditic P and phosphinic P at δ = 138.7 (J_{RhP} = 312 Hz, $J_{P'P}$ = 68 Hz) and at δ = 39.9 ($J_{RhP'}$ = 181 Hz), respectvely. The identical PP' coupling constant of 68 Hz in both complexes is a hint that the angle P-Rh-P' must be very similar. In contrast to 7, 8 presents two carbonyl bands in the IR spectrum at 2007 and 2078 cm^{-1}. Based on these spectroscopic data, 7 and 8 can be formulated as [Rh{(R_a,R_C)-4ael}(CO)Cl] and [Rh{(R_a,R_C)-4ael}(CO)$_2$Cl], respectively [22]. As a proof of this hypothesis, complex 8 was generated by bubbling CO into a dichloromethane solution of 7. This process is completely reversible. Bubbling argon through a solution of 8, complex 7 is regenerated. A trigonal-bipyramidal structure can be proposed for 8 with the ligand coordinated in an apical-axial fashion in which the more basic phosphine moiety is located in the equatorial position [23].

The catalyst precursors [{(R_a,R_C)-4ael}Rh(acac)]BF$_4$ (**9**, acac = acetylacetonate), and [{(R_a,R_C)-4ael}Rh(cod)]BF$_4$ (**10**) used in the hydroformylation and olefin hydrogenation, respectively, have usually been prepared in situ and characterized only by ^{31}P NMR spectroscopy. The characteristic coordination shifts with respect to the free ligand and the presence and magnitude of the RhP and PP' coupling confirmed for both complexes the ligand chelation to the rhodium center {[Rh{(R_a,R_C)-4ael}(acac)]: ^{31}P NMR (C$_6$D$_6$) δ = 145.9 (J_{RhP} = 303 Hz, $J_{P'P}$ = 99 Hz) and at δ = 47.5 ($J_{RhP'}$ = 172 Hz); [Rh{(R_a,R_C)-4ael}(cod)]BF$_4$: ^{31}P NMR (CD$_2$Cl$_2$) δ = 134.8 (J_{RhP} = 255 Hz, $J_{P'P}$ = 60 Hz) and at δ = 26.0 ($J_{RhP'}$ = 137 Hz)}.

2.1.5.4 Catalytic Applications

2.1.5.4.1 Rhodium-Catalyzed Asymmetric Hydroformylation of Styrene

The potential of QUINAPHOS ligands for asymmetric catalysis was assessed in rhodium-catalyzed enantioselective hydroformylation using styrene as a benchmark substrate (Table 2.1.5.2). The catalysts were prepared in situ from [(acac)Rh(CO)$_2$] and 4 equiv of the diastereomeric mixture or the single diaste-

Table 2.1.5.2 Rhodium catalyzed hydroformylation of styrene **11**.[a]

$$\text{11} \xrightarrow[\text{[Rh(acac)(CO)}_2\text{]/4ael}]{\text{H}_2/\text{CO}} \text{12 (PhCH(CHO)CH}_3\text{)} + \text{13 (PhCH}_2\text{CH}_2\text{CHO)}$$

Ligand	t [h]	Conv. [%]	TOF[b] [h⁻¹]	12/13	ee [%]
(R_a, R_c^*)-**4ael**	90	54.8	13.4	96.3 : 3.7	35.6 (S)
(R_a, R_c)-**4ael**	74	79.3	23.6	96.0 : 4.0	4.8 (S)
(R_a, S_c)-**4ael**	70	75	23.6	96.7 : 3.3	74.0 (S)

a Reaction conditions: $T = 40\,°C$; $p(CO/H_2 = 1:1) = 100$ bar; **4ael**/[(acac)Rh(CO)$_2$] = 4:1; styrene/Rh = 2200:1. b Average value over total reaction time.

reomers of (R_a,R_C^*)-**4ael**. Reactions were carried out in the neat substrate under otherwise typical hydroformylation conditions. The activities of all three catalysts were in the range observed with common chiral hydroformylation catalysts [24]. All the catalysts led to exclusive formation of the hydroformylation products with uniformly high regioselectivity for the desired branched aldehyde **12**.

The enantiomeric excess of **12** showed a dramatic dependence on the absolute configuration in the 2-position of the heterocyclic backbone of the ligand. The diastereomeric mixture (R_a,R_C^*)-**4ael** gave a moderate ee of 35.6% in favor of the S enantiomer, but an almost racemic product was obtained with the R_a,R_C diastereomer. The ligand (R_a,S_C)-**4ael**, however, gave a remarkable ee of 74% [24] for the S enantiomer under these nonoptimized reaction conditions. It is interesting to note that the ee observed with the diastereomeric mixture is almost exactly the mean of the two ee values for the single diastereomers. This indicates that the two diastereomers contribute in parallel and almost identical pathways to the overall reaction [25]; this is further supported by the observation that the separated ligands (R_a,S_C)-**4ael** and (R_a,R_C)-**4ael** resulted in almost identical reaction rates. The rate was significantly lower, however, when the two diastereomers were employed together in the form of the mixture (R_a,R_C^*)-**4ael**. This observation is best rationalized by partial formation of a fairly stable and catalytically nonproductive complex of type [{(R_a,R_C)-**4ael**}Rh{(R_a,S_C)-**4ael**}]⁺.

Taking into account the regioselectivity *and* the enantioselectivity, the overall selectivity of the hydroformylation process using (R_a,S_C)-**4ael** is 72% for (S)-**12**. When QUINAPHOS was first reported, only the phosphine/phosphite ligands BINAPHOS (83%) [26], 3-H$_2$F$_6$-BINAPHOS (89%) [27], and BIPHEMPHOS (85%) [28] provided higher overall selectivities [29]. These results can be rationalized in view of the structural similarity of 8-diphenyl-substituted-QUINAPHOS with BINAPHOS-type ligands, at least with regard to the binding sites.

QUINAPHOS (R_a,S_C)-**4ael**   (*R,S*)-BINAPHOS

2.1.5.4.2 Rhodium-Catalyzed Asymmetric Hydrogenation of Functionalized Alkenes

The rhodium-catalyzed hydrogenation of α,β-unsaturated carboxylic acid derivatives proceeded smoothly in CH_2Cl_2 solution at room temperature and elevated pressures, using either an in-situ catalyst formed from **4ael** and [Rh(cod)$_2$](BF$_4$) or the isolated complex [{(R_a,S_C)-**4ael**}Rh(cod)](BF$_4$) **10** as catalyst precursor (Table 2.1.5.3). The opposite configuration at the chiral carbon atom was required for high *ee* values in hydrogenation as compared to hydroformylation and (R_a,R_C)-**4ael** proved to be the more effective diastereomer (entries 1 and 2). The hydrogenation could also be accomplished in methanol, albeit with lower enantioselectivity (compare entries 3 and 1). Increasing the amount of ligand during in-situ preparation of the catalyst led to a marked increase in *ee* with dimethyl itaconate **14** as substrate (entry 4 versus 1). The preformed catalyst was required to achieve high enantioselectivities for the dehydroamino acid esters methyl 2-acetamidoacrylate **16a** (entry 5) and methyl (*Z*)-acetamidocinnamate **16b** (entry 6). Remarkably, the free acid (*Z*)-acetamidocinnamic acid **16c** could also be hydrogenated with high enantioselectivity (entry 7) [30].

For the hydrogenation of **14**, we noticed no change of pressure during the long reaction time used in the screening procedure, although the stoichiometric consumption of hydrogen should have caused a detectable pressure drop. Therefore, we quenched a hydrogenation experiment (**14/10** = 1000 : 1, $p(H_2)$ = 50 bar, room temperature) by venting after 5 min. Analysis of a small sample taken from the solution showed that the reaction was already complete and the *ee* of the product **8a** was determined as 98.2% (*R*). The reaction mixture remaining in the reactor was charged with a second, larger batch of substrate (**14/10** = 6000 : 1) and pressurized again with H_2 to 70 bar. Upon stirring, the pressure dropped immediately to reach a constant level of ca. 55 bar within less than 10 min. Standard workup and analysis revealed that hydrogenation was again quantitative and the saturated

Table 2.1.5.3 Asymmetric hydrogenation of α, β-unsaturated carboxylic acid derivatives.

MeOOC–C(=CH–COOMe) **14** → [Rh(cod)$_2$]BF$_4$/**4ael**, $p(H_2)$ = 30 bar, T = rt; t = 24 h → MeOOC–*CH–COOMe **15**

16a: R = Me; R' = H
16b: R = Me; R' = Ph
16c: R = H; R' = Ph
(R'–CH=C(NHCOMe)–COOR)

Entry	Ligand	Substrate	S/[Rh]/Lig	Solvent	Conv. [%]	ee [%]
1	(R_a, R_c)-**4ael**	14	1000:1:1.1	CH$_2$Cl$_2$	>99	96 (R)
2	(R_a, S_c)-**4ael**	14	1000:1:1.1	CH$_2$Cl$_2$	>99	63 (R)
3	(R_a, R_c)-**4ael**	14	1000:1:1.1	MeOH	>99	79 (R)
4	(R_a, R_c)-**4ael**	14	1000:1:2.2	CH$_2$Cl$_2$	>99	99 (S)
5	(R_a, R_c)-**4ael**	16a	1000:1:1.0[a]	CH$_2$Cl$_2$	>99	98 (S)
6	(R_a, R_c)-**4ael**	16b	500:1:1.0[a]	CH$_2$Cl$_2$	>99	97 (S)
7	(R_a, R_c)-**4ael**	16c	1000:1:1.1	CH$_2$Cl$_2$	>99	98 (S)

a [{(Ra, Rc)-**4ael**} Rh(cod)](BF$_4$) as the precatalyst.

ester **15** showed an *ee* of 99.4% (*R*). The total number of catalytic turnovers in the two consecutive runs was 7000, and an average turnover frequency of 36 000 h^{-1} could be estimated as the lower limit for the catalytic activity in the second run. Taking into account the somewhat lower *ee* of the product remaining in the mixture from the first run, this remarkably rapid hydrogenation proceeded with almost perfect selectivity for the *R* enantiomer (99.6% *ee*).

Following our initial report, a number of other highly unsymmetrical bidentate phosphorous ligands have been reported which provide good to excellent enantioselectivities in asymmetric hydrogenation [31]. BINAPHOS also was revealed to be a valuable ligand for this reaction [27b, 32], underlining again the analogies with QUINAPHOS. The simultaneous reports by Pringle [33], Reetz[34], and Feringa[5f] that rhodium complexes based on easily accessible monodentate phosphorous ligands [35] are efficient and highly enantioselective hydrogenation catalysts opened a new era [36]. It was demonstrated that hetero-combinations of monodentate ligands may outperform the homo-combinations [37]. For example, the addition of achiral monodentate phosphine ligand (typically triphenylphosphine) to a chiral monodentate phosphoramidite in a 1:2 ratio is highly beneficial for both the catalyst activity and the enantioselectivity in the rhodium-catalyzed hydrogenation of disubstituted acrylic acids [38]. It is interesting to note that the predominating catalytic species containing one chiral monodentate phosphoramidite and one achiral phosphine resembles very closely the structure (and the performance) of QUINAPHOS.

2.1.5.4.3 Ruthenium-Catalyzed Asymmetric Hydrogenation of Aromatic Ketones

The highly enantioselective Ru-catalyzed hydrogenation of unfunctionalized ketones represents a further application of QUINAPHOS **4ael** in asymmetric catalysis [39]. Selected catalytic runs are summarized in Table 2.1.5.4. As benchmark substrates, acetophenone **17a**, 4-fluoroacetophenone **17b**, and 4-methoxy-acetophenone **17c** have been chosen. The catalyst was prepared in situ from [RuCl$_2$(C$_6$H$_6$)]$_2$, **4acl** a diamine and sodium t-butoxide following a literature procedure [40]. In a first series of experiments (entries 1–8), ethylenediamine **19** was used as the co-catalyst. Using (R_a,R_C)-**4ael**, 1-phenylethanol **18a** was obtained with moderate conversion (45%) and an ee of 65% under standard reaction conditions (entry 1). By increasing the amount of base from 4 to 10 mol% almost quantitative conversion was achieved with an identical ee (entry 2). By using

Table 2.1.5.4 Asymmetric hydrogenation of acetophenones.[a]

$p(H_2)$ =30 bar; [Ru] / **4ael** / **19** or **20** / t-BuONa

17a: R = H
17b: R = F
17c: R = OMe

19 H$_2$N-CH$_2$CH$_2$-NH$_2$

(S,S)-**20** Ph-CH(NH$_2$)-CH(NH$_2$)-Ph

18a: R = H
18b: R = F
18c: R = OMe

Entry	Substrate	Ligand	Diamine	Conv. [%]	ee [%]
1[a]	17a	(R_a, R_c)-4ael	19	45	65 (S)
2	17a	(R_a, R_c)-4ael	19	91	65 (S)
3	17a	(R_a, S_c)-4ael	19	90	80 (R)
4[b]	17a	(R_a, S_c)-4ael	19	31	82 (R)
5	17b	(R_a, R_c)-4ael	19	32	69 (S)
6	17b	(R_a, S_c)-4ael	19	34	73 (R)
7	17c	(R_a, R_c)-4ael	19	65	67 (S)
8	17c	(R_a, S_c)-4ael	19	>99	86 (R)
9	17a	(R_a, R_c)-4ael	(S, S)-20	>99	6 (R)
10	17a	(S_a, R_c)-4ael	(S, S)-20	>99	36 (S)
11	17a	(S_a, S_c)-4ael	(S, S)-20	>99	75 (R)
12	17a	(R_a, S_c)-4ael	(S, S)-20	>99	94 (R)
13	17b	(R_a, S_c)-4ael	(S, S)-20	>99	94 (R)
14	17c	(R_a, S_c)-4ael	(S, S)-20	>99	94 (R)

a t-BuONa 4 mol%. b [Ru] = 0.02 mol%, t-BuONa = 4 mol%, t = 18 h, T = rt.

(R_a,S_C)-**4ael** as the ligand under otherwise identical conditions, the hydrogenation of **17a** was accomplished with similar conversion but with a significant higher *ee* of 80% (entry 3). Remarkably, an opposite absolute configuration of the alcohol was found (*R* versus *S*; entry 3 versus entry 2) indicating that the transfer of the chiral information from the ligand to the substrate is mainly controlled by the stereocenter in the dihydroquinoline backbone. The same level of enantioselectivity is retained also using an S/C (substrate/catalyst ratio) of 5000. After 18 h, 31% conversion was achieved, which corresponds to an average TOF of 86 h^{-1} (entry 4). An optimization of the reaction parameters toward higher catalyst activity was not pursued at this stage of the investigation.

The hydrogenation of substrate **17b** bearing an electron-withdrawing group proceeded more slowly with both diastereomer of QUINAPHOS leading to conversions between 32 and 34% after 16 h. With this substrate, (R_a,R_C)-**4ael** and (R_a,S_C)-**4ael** led to similar enantioselectivities (albeit with opposite signs) of 69% (*S*) and 73% (*R*), respectively (entries 5 and 6). The hydrogenation of 4-methoxy-substituted acetophenone **17c** carried out in the presence of (R_a,R_C)-**4ael** afforded **18c** with 65% conversion and an *ee* of 67% (*S*) (entry 7), whereas full conversion and an *ee* of 86% (*R*) resulted by using (R_a,S_C)-**4ael** (entry 8). For comparison, the hydrogenation of 1-acetonaphthone with the system (*S*)-BINAP/ethylenediamine affords the corresponding alcohol with an *ee* of only 57% (*R*) [41]. More recently, enantioselectivities in the same range as those obtained with (R_a, S_C)-**4ael**/**19** have been reported for a system comprising an achiral thioamine and the ligand (2*R*,2′*R*)-bis(diphenylphosphanyl)-(1*R*,1′*R*)-dicyclopentane {(*R*,*R*)-bicp} [40].

With the chiral diamine (*S*,*S*)-**20** as a co-catalyst full conversion was obtained in all cases, indicating that the amine has a pronounced influence on reactivity *and* selectivity (entries 9–14). The combination (R_a,R_C)-**4ael**/(*S*,*S*)-**20** afforded **18a** as an almost racemic mixture (entry 9). The value of 6% *ee* (*R*) obtained in this experiment reflects two opposite contributions. On one hand, the system chiral phosphorous ligand/achiral diamine (R_a,R_C) **4ael**/**19** led to **18a** with 65% *ee* (*S*) (entry 1). On the other hand, an *ee* value of 75% (*R*) in the hydrogenation of 1-acetonaphthone has been reported for the system achiral phosphine (PPh$_3$)/(*S*,*S*)-**20** [41] This indicates that two inductions are "canceled" in an almost additive way in the mixed system.

In order to gain a comprehensive picture of matching and mismatching effects, all possible stereoisomers of QUINAPHOS were tested in combination with (*S*,*S*)-**20**. The catalyst system based on (S_a,R_C)-**4ael**/(*S*,*S*)-**20** resulted in 36% *ee* of the *S* enantiomer (entry 10). This low enantioselectivity is again the result of combining the intrinsic *S*-selectivity of the phosphorous ligand with the *R*-selectivity of the diamine. Indeed, an increase in selectivity from 65% (see entry 1) to 75% *R* (entry 11) was observed upon matching the *R*-selective ligand (S_a,S_C)-**4ael** with (*S*,*S*)-**20**. Finally, the optimum match was identified for the pair (R_a,S_C)-**4ael**/(*S*,*S*)-**20**. This combination allowed the hydrogenation of **17a** with an enantioselectivity as high as 94% (*R*) (entry 12). Similarly, substrates **17b** and **17c** were converted quantitatively to the corresponding alcohols with an *ee* of 94% (entries 13 and 14).

With respect to the enantioselectivity, the results obtained with QUINAPHOS compares well with the best catalytic systems known [31] Significantly higher catalyst activity has been found using the Noyori system [42].

2.1.5.4.4 Copper-Catalyzed Enantioselective Conjugate Addition of Diethylzinc to Enones

The copper-catalyzed enantioselective Michael addition of organometallic reagents to enones was the first successful application of phosphoramidite chiral ligands in catalysis [4, 43]. Since this early report, substantial enhancement of the enantioselectivity and/or of the substrate scope has been achieved through an untiring effort to optimize the ligand structure [5a, 44].

The efficiency of a catalytic system formed in situ from $Cu(OTf)_2$ and ligand **4cel** was tested in the enantioselective conjugate addition of Et_2Zn to 2-cyclohex-en-1-one **21** as a standard substrate [45]. The two diastereomers led to strikingly different catalytic results. The reaction proceeds rapidly and is complete within 30 min when (S_a,S_C)-**4cel** is used as the ligand, although without any significant asymmetric induction (Table 2.1.5.5, entry 1). In contrast, in the presence of the (S_a,R_C)-**4cel** diastereomer under otherwise identical conditions the addition takes place slowly but selectively. After 5 h the addition product **22** is obtained almost quantitatively with an enantiomeric excess of 70% (*S*) (entry 2). By lowering the reaction temperature from −15 °C to −30 °C a decrease in the enantioselectivity was observed (entry 3) An increase of the ligand/metal ratio from 2 : 1 to 4 : 1 reduced the catalytic activity as well as the enantioselectivity (compare entries 2

Table 2.1.5.5 Cu-catalyzed enantioselective 1, 4-addition of diethylzinc to cyclohex-2-enone **21**.

Entry	Ligand	T [°C]	t [h]	L/Cu	Conv. [%]	ee [%]
1	(Sa, Sc)-**4cel**	−15	0.5	2 : 1	>99	0
2	(Sa, Rc)-**4cel**	−15	5	2 : 1	96	70 (S)
3	(Sa, Rc)-**4cel**	−30	24	2 : 1	90	54 (S)
4	(Sa, Rc)-**4cel**	−15	21	4 : 1	60	23 (S)

and 4). The striking difference in catalysis of the two diastereomers of **4ccl** may be related to a different coordination mode (see Section 2.1.5.3.2).

The results obtained using QUINAPHOS in this C–C bond-forming reaction are only modest in comparison with those achieved with other phosphoramidites. However, the testing of only one member of the QUINAPHOS family cannot be considered conclusive and further investigations including other QUINAPHOS derivatives are required to fully assess the potential of these ligands for this transformation.

2.1.5.4.5 Nickel-Catalyzed Asymmetric Hydrovinylation

Nickel-catalyzed hydrovinylation is a metal-mediated coupling reaction with a remarkable potential for enantioselective synthesis [46, 47]. It comprises the formal addition of hydrogen and a vinyl group to a prochiral olefin and gives access to high-value products in a very elegant and atom efficient way. Using ethene as the cheapest vinylic coupling partner, the transformation results effectively in a chain elongation of two carbon atoms simultaneously creating a stereogenic center in the allylic position (Scheme 2.1.5.1). In the search for new hydrovinylation catalysts that allow for systematic modification of a readily accessible ligand framework, we have turned our attention to chiral phosphoramidites. Following the successful structural motif of the Wilke ligand *all*-(R)-**25** [48, 49], the selection of the representative examples **4del**, **4cel**, and (R_a,S_C,S_C)-**26** was guided by the rather heuristic design principle that an efficient ligand system for asymmetric hydrovinylation should possess a P—N bond and contain more than one element of chirality, one of them preferably being an atropisomeric unit. In addition, the set of ligands should allow to assess the potential influence of additional donor groups. The most significant results obtained with catalysts formed from **4del**, **4cel**, or (R_a,S_C,S_C)-**26** and [Ni(allyl)Cl]$_2$ and using NaBARF (BARF = tetrakis-[3,5-bis(trifluormethyl)phenyl]borate) as activator [50] are summarized in Table 2.1.5.6 [51].

The (R_a,R_C) diastereomer of the ligand **4del** gave an active catalytic system for the hydrovinylation of styrene **11**, but the optical induction remained disappointingly low (entry 1). The C2 epimer (R_a,S_C)-**4del**, however, provided reasonable enantioselectivity at moderate conversion (entry 2). An even more dramatic difference in activity and selectivity was observed for the two diastereomers of the chloride-substituted **4cel**. The R_a,S_C configuration in **4cel** led to an extremely active catalyst, resulting preferentially in styrene dimerization and trimerization (entry 3). The high tendency toward oligomerization could not be suppressed even at −78 °C. In contrast, the other diastereomer (R_a,R_C)-**4cel** formed a less active catalyst, yielding **23** very selectively with a remarkable *ee* of 87% (*S*) (entry 4). It is interesting to note that the two ligand systems **4del** and **4cel** give rise to products with the opposite preferred stereochemistry and the opposite configuration at C2 is required to obtain high asymmetric induction. The results obtained with ligands **4del** and **4cel** further substantiate the cooperative effect of axial and central chirality and indicate also a strong influence of additional donor groups. Interestingly, the best result (87% *ee* at −30 °C) was, however, achieved

Table 2.1.5.6 Enantioselective hydrovinylation of styrene using NaBARF as activator.

Entry	Ligand	T [°C]	p(C$_2$H$_4$) [bar]	t [h]	11/Ni	Conv. [%]	Selectivity [%] 23	24	oligomers[a]	ee (23) [%]
1	(Ra, Rc)-**4del**	0	44	1	280	99.7	85.4	13.3	<1.5%	7.6 (R)
2	(Ra, Sc)-**4del**	−30	49	2	280	13.5	93.3	3.1	3.7	56.4 (R)
3	(Ra, Sc)-**4cel**	−32	12	2	300	>99	<1	<1	>99	–
4	(Ra, Rc)-**4cel**	−32	12	2	300	33.1	96.2	0.4	3.7	87.2 (S)
5	(Ra, Sc, Sc)-**26**	−70	~1	4	620	>99	84.9	4.3	8.1	94.8 (S)
6a	(Ra, Sc, Sc)-**26**	−65	~1	4	4600	89.2	>99	<1	<1	91.1 (S)
6b		−65→RT	~1	16	5490	82.7	96.4	<1	3.0	91.4 (S)

a Oligomerization products of **11** and secondary hydrovinylation products of **23** and **24** are summarized as "oligomers."

with ligand **4cel** containing the weaker donor group at C8. We therefore extended our study to Feringa's ligand system (R_a,S_C,S_C)-**26** [43] combining central and axial chirality but apparently lacking a strong additional donor group (vide infra).

Under particularly mild conditions, a Ni-catalyst based on (R_a,S_C,S_C)-**26** gave quantitative conversion of **11** with 84.9% selectivity for the desired product **23** and an excellent enantioselectivity of 94.8% (S) (entry 5). Moreover, the catalyst system proved extremely efficient and remarkably robust for the hydrovinylation of **11**. Almost 90% conversion and perfect chemoselectivity were achieved within 4 h at −65 °C even at a substrate/nickel ratio of 4600 : 1 (entry 6a). Further addition of substrate to this reaction mixture led again to almost complete conversion within 16 h (entry 6b) corresponding to a total turnover number of ca. 8340. Chemo- and enantioselectivity remained uniformly high under these conditions.

Density functional calculations revealed that the performance of the catalytically active system resulting from the Ni precursor and ligand (R_a,S_C,S_C)-**26** is largely affected by a hemilabile coordination mode of one phenyl ring of the phosphoramidite ligand allowing only for two chemically plausible orientations of styrene coordination in the active nickel hydride catalyst [52]. These two diastereomers differ in energy by 1.38 kcal mol^{-1} and are the starting points of two reaction paths for hydride transfer to styrene. One path is clearly preferred over the other, as can be deduced from the significantly lower activation energy (+17.57 versus +26.92 kcal mol^{-1}) corresponding to the enantiodiscriminating step of the reaction. The displacement of the part of the ligand coordinated in a hemilabile manner by ethylene initiates the subsequent reaction paths for C–C bond formation, yielding the *S*- and the *R*-configured reaction products, respectively. These pathways have activation energies of +6.59 and +5.90 kcal mol^{-1}, respectively. These results are in accordance with and help to rationalize the hydrovinylation results. Furthermore, the concept of hemilabile coordination has motivated other applications of ligand **26** [53].

At present, phosphoramidites represent the most promising ligand class for the hydrovinylation reaction, both for the excellent activities and enanioselectivities. The synthetic potential embraces not only styrene and styrene derivatives but also norbornene [54] and 1,3-dienes [55].

2.1.5.4.6 Nickel-Catalyzed Cycloisomerization of 1,6-Dienes

Chiral carbo- and heterocycles are widespread structural motifs in biologically active compounds. The cycloisomerization of 1,6-dienes (**A**) offers an elegant and atom-economic [56] approach to five- or six-membered carbo- or heterocycles [57]. Metal complexes based on Pd [58], Ni [59], Rh [60], Ru[61], and Ti[62] have been identified as promising lead structures for catalyst development. Some of the reported systems are highly chemo- and regioselective toward the formation of the individual five-membered ring compounds **B–D** (Scheme 2.1.5.2). Enantioselective cycloisomerization, however, has been assessed only sparsely so far, and remains a challenging task [46, 63].

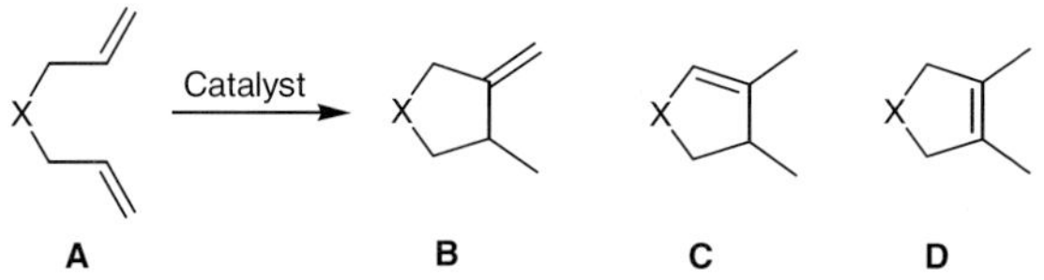

Scheme 2.1.5.2 Cycloisomerization of 1,6-dienes (X = CH$_2$, C(CO$_2$R)$_2$, O, N—R, etc.).

Based on the formal analogy between the intermolecular hydrovinylation and the intramolecular cycloisomerization process, we have chosen catalysts with proven potential for the first reaction type [48, 51] as the starting point of our study. The results are summarized in Table 2.1.5.7 [64]. Despite its excellent performance in the hydrovinylation of styrene [51], the [{Ni(allyl)Br}$_2$]/(R_a,S_C,S_C)-**26**/NaBARF system led to disappointingly low conversions and selectivities in the cycloisomerization of **27a** (entry 1). Similarly, the [{Ni(allyl)Cl}$_2$]/(R_a,R_C)-**4cel**/NaBARF system is not effective for the cycloisomerization of **27a** (entry 2) even though it is able to promote the hydrovinylation. The other diastereomer, (R_a,S_C)-**4cel**, however, which forms an active nickel catalyst for styrene oligomerization

Table 2.1.5.7 Cycloisomerization of diethyl diallylmalonate (**27a**) and N, N-diallyltosylamide (**27b**).[a]

27a X = C(CO$_2$Et)$_2$
27b X = N-Ts

28a X = C(CO$_2$Et)$_2$
28b X = N-Ts

Ligand = *all*-(R)-**25**; (R_a,S_C,S_C)-**26**; **4cel**; **4cgl**

Ni-precursor = **29** (Br bridge); **30** (Cl bridge) + NaY; **31**

Y = BARF, **a**; SbF$_6$, **b**; Al{OC(CF$_3$)$_3$}$_4$, **c**; Al{OC(CF$_3$)$_2$Ph}$_4$, **d**.

Entry	Precursor/ Activator	Ligand	Substrate S	S/[Ni]	t [h]	Conv. [%]	TOF [h^{-1}]	Select.[b]	ee [%]
1	29/NaBARF	(Ra, Sc, Sc)-26	27a	20	24	75	<1	65	39(−)
2	30/NaBARF	(Ra, Rc)-4cel	27a	20	17	3	≪1	50	–
3	29/NaBARF	(Ra, Sc)-4cel	27a	20	19	>99	~1	87	48(−)
4	31a/-	(Ra, Sc)-4cel	27a	20	0.5	94	37	97	46(−)
5	31a/-	(Ra, Rc)-4cgl	27a	20	0.5	90	36	93	40(−)
6	29/NaBARF	*all*-(R)-25	27a	20	17	81	~1	91	79(+)
7	31a/-	*all*-(R)-25	27a	200	1	72	144	91	72(+)
8	31c/-	*all*-(R)-25	27a	200	1	79	158	97	73(+)
9	31b/-	(Ra, Rc)-4cel	27b	20	0.5	89	36	88	40(−)
10	31b/-	(Ra, Sc)-4cgl	27b	20	0.5	42	17	95	50(−)
11	31b/-	*all*-(R)-25	27b	20	0.5	74	30	92	46(+)
12	31f/-	*all*-(R)-25	27b	90	1	>99	90	91	47(+)
13	31c/-	*all*-(R)-25	27b	200	1	11	22	99	54(+)
14	31a/-	*all*-(R)-25	27b	200	1	64	128	95	33(+)

a T = 20 °C, dichloromethane, P(Ligand)/Ni = 1:1. b Regioselectivity toward **28**.

but not for styrene hydrovinylation, gave promising results for the first time. In combination with [{Ni(allyl)Br}$_2$] and NaBARF, complete conversion has been achieved within 19 h (5 mol% Ni at room temperature) with a regioselectivity of 87% and an *ee* of 48% (entry 3).

In order to avoid the dehalogenation step during the formation of the in-situ catalyst we decided to use the cationic precursor [Ni(allyl)(cod)][BARF] [65, 66] instead of the [{Ni(allyl)X}$_2$]/NaBARF system. This led to a dramatic improvement of catalytic activity and almost complete conversion was reached within 30 min, corresponding to a lower limit for the turnover frequency of 37 h^{-1} (entry 4). Gratifyingly, the use of the cationic precursor also gave higher regioselectiviy with only marginal if any reduction of enantioselectivity (entries 3, 4) [67]. With this promising system in hand, we extended the structural variation of the ligand framework. Changing the *n*-Bu group in ligand **4cel** to a phenyl substituent in **4cgl** resulted again in an active system if the stereochemistry at the chiral carbon center in the 2-position of the quinoline backbone was maintained [68]. Slightly lower regio-, and enantioselectivities were observed, however, with the aromatic side-group (entries 4, 5).

The highest selectivity so far was obtained with Wilke's azaphospholene ligand. *All*-(*R*)-**25** afforded a very active and stereoselective catalyst system for the cycloisomerization of **27a**. With [{Ni(allyl)Br}$_2$] and NaBARF a conversion of 81%, a regioselectivity of 91%, and an *ee* value of 79% were achieved after 17 h (entry 6). With [Ni(allyl)(cod)][BARF] as precursor the activity was again strongly increased; this was accompanied by only a small decrease in enantioselectivity. At a catalyst loading of 0.5 mol% a conversion of 72% was reached after a 1 h reaction time, corresponding to an average turnover frequency (TOF$_{av}$) of 144 h^{-1} (entry 7).

As counter ion effects are often very pronounced in olefin dimerization [47], this aspect was assessed for the new cycloisomerization system. Catalyst activity was found to decrease somewhat in the order [Al{OC(CF$_3$)$_3$}$_4$]$^-$ > [BARF]$^-$ > [SbF$_6$]$^-$ ≳ [AsF$_6$]$^-$ ≫ [PF$_6$]$^-$, corroborating the common notion that weakly coordinating anions [69] are beneficial for the performance of the nickel catalysts. The highest activity so far for the cycloisomerization of **27a** was achieved using the new complex [Ni(η^3-allyl)(η^4-cod)][Al{OC(CF$_3$)$_3$}$_4$] (**31c**) in combination with *all*-(*R*)-**25**, resulting in a conversion of 79% and a (TOF$_{av}$) of 158 h^{-1} with almost perfect regioselectivity (entry 8).

In a second set of experiments, the cycloisomerization of *N,N*-diallyltosylamide (**27b**) was investigated as a new route to chiral five-membered N-heterocycles [70]. Thus, a brief ligand screening was carried out including (R_a,R_C)-**4cel**, (R_a,S_C)-**4cgl**, and *all*-(*R*)-**25** in combination with [Ni(η^3-allyl)(η^4-cod)][SbF$_6$] (**31b**) using a catalyst loading of 5 mol% and 30 min reaction time. The formation of the five-membered ring with an exocyclic double bond 3-methyl-4-methylene-1-tosylpyrrolidine (**28b**) was favored with all the catalyst systems used. In general, the reaction with the nitrogen-containing substrate **27b** tended to be slightly slower than with **27a**. In the presence of (R_a,R_C)-**4cel** high conversion (89%), good regioselectivity (88%) and an *ee* of 40% were obtained (entry 9). The use of the ligand

(R_a,S_C)-**4cgl** led to a more selective (regioselectivity 95%, *ee* 50%) but a less active (conversion 42%) catalyst (entry 10). A good compromise between activity (conversion 74%) and selectivity (regioselectivity 92%, *ee* 46%) could be found again by using *all*-(*R*)-**25** as the ligand (entry 11). Using [Ni(η^3-allyl)(η^4-cod)][Al{OC(CF$_3$)$_2$Ph}$_4$] (**31d**) as the precursor, full conversion was obtained after 1 h with a catalyst loading of 1.1 mol% (entry 12). Unexpectedly, a conversion of only 11% after 1 h was obtained with **31c** (entry 13). In sharp contrast, this precursor led to the highest turnover rate in the cycloisomerization of **27a** (entry 8). The most active catalyst for the cycloisomerization of **27b** was generated using the BARF-containing precursor **31a**. After 60 min, 64% of the substrate was consumed, corresponding to an average turnover frequency of 128 h^{-1} (entry 14).

The regio- and enantioselectivity of the cationic nickel catalyst seems not to be affected by the nature of the counter ion in the case of the cyclization of **27a** (entries 7, 8). In contrast, variation of the counter ion for substrate **27b** influenced not only the activity but also the catalyst selectivity. The enantioselectivity increased from 33% to 54% in the opposite order to that observed for the activity (entries 12–14). Intriguingly, the order of activity imparted from the different anions for the cyclization of **27a** differs significantly from that observed for **27b**.

These results represent the current state-of-the-art for this potentially synthetically useful transformation. The Wilke ligand forms the benchmark catalyst for the cycloisomerization of **27a**, whereas for **27b** the performance of QUINAPHOS-type ligands is on almost the same level. The obvious advantage of the latter ligands is the facile structure modification which is rather troublesome in the case of *all*-(*R*)-**25**. As evidenced with the catalyst systems based on *all*-(*R*)-**25**, the counter ion plays an important role in defining the catalyst performance. This aspect requires to be investigated for the QUINAPHOS ligands also.

2.1.5.5 Conclusions

The studies summarized in this review demonstrate the broad utility of the QUINAPHOS ligand family for a variety of reactions in combination with different metals. Both monodentate and bidentate ligands derived from 1,2-dihydroquinoline ensure high levels of enantioselectivity and/or activity, demonstrating the suitability of this chiral backbone.

The modular synthetic approach has allowed the preparation of many QUINAPHOS derivatives in both diastereomeric forms. Testing of the single diastereomers in catalysis has been revealed to be of crucial importance. Thus, a pronounced interplay between the axial and central chirality has been observed in all applications of QUINAPHOS in catalysis, whereby the choice of the diastereomer may affect not only the enantioselectivity but also the activity and the chemoselectivity of a catalyzed reaction.

It has to be noted, however, that in all reactions described till now, only a very limited selection of QUINAPHOS ligands have been explored so far. Therefore, future efforts need to be directed to optimize the synthetic methodology to generate a library of systematic variations of the lead structures. With these derivatives of the QUINAPHOS framework in hand, it should be possible to assess, for instance, the impact of the variation of the substituents in the 2-position.

Acknowledgements

We gratefully acknowledge the contributions of Carmela G. Arena, Christian Boeing, Simon Burk, GianPiero Calabrò, Christiane Diez-Holz, Markus Hölscher, and Thomas Pullmann in the course of the QUINAPHOS project (see the reference list for the individual contributions). We thank the Deutsche Forschungsgemeinschaft (SFB-380), the Fonds der Chemischen Industrie, the EU (LIGBANK, FP6-NMP3-CT-2003-505267) and MURST (Italy) for financial support.

References

1 *Comprehensive Asymmetric Catalysis* (E. N. Jacobsen, A. Pfaltz, H. Yamamoto, Eds.), Springer: Berlin, **2004**.
2 T. P. Yoon, E. N. Jacobsen, *Science* **2003**, *299*, 1691.
3 C. Gennari, U. Piarulli, *Chem. Rev.* **2003**, *103*, 3071.
4 A. H. M. de Vries, A. Meetsma, B. L. Feringa, *Angew. Chem. Int. Ed.* **1996**, *35*, 2374.
5 (a) B. L. Feringa, *Acc. Chem. Res.* **2000**, *33*, 346; (b) O. Huttenloch, J. Spieler, H. Waldmann, *Chem. Eur. J.* **2001**, *3*, 671; (c) B. Bartels, G. Helmchen, *Chem. Commun.* **1999**, 741; (d) C. A. Kiener, C. Shu, C. Incarvito, J. F. Hartwig, *J. Am. Chem. Soc.* **2003**, *125*, 14 272; (e) J. Jensen, B. Y. Svendsen, T. V. la Cour, H. L. Pedersen, M. Johannsen, *J. Am. Chem. Soc.* **2002**, *124*, 4558; (f) M. van den Berg, A. J. Minnaard, E. P. Schudde, J. van Esch, A. H. M. de Vries, J. G. de Vries, B. L. Feringa, *J. Am. Chem. Soc.* **2000**, *122*, 11 539; (g) Y. Fu, J. H. Xie, A. G. Hu, H. Zhou, L. X Wang, Q. L. Zhou, *Chem. Commun.* **2002**, 480; (h) A. G. Hu, Y. Fu, J. H. Xie, H. Zhou, L. X. Wang, Q. L. Zhou, *Angew. Chem. Int. Ed.* **2002**, *41*, 2348; (i) K. Tissot-Croset, D. Polet, A. Alexakis, *Angew. Chem. Int. Ed.* **2004**, *43*, 2426; (l) K. Biswas, O. Prieto, P. J. Goldsmith, S. Woodward, *Angew. Chem. Int. Ed.* **2005**, *44*, 2232; (m) A. Alexakis, V. Albrow, K. Biswas, M. d'Augustin, O. Prieto, S. Woodward, *Chem. Commun.* **2005**, 2843.
6 G. Franciò, C. G. Arena, F. Faraone, C. Graiff, M. Lanfranchi, A. Tiripicchio, *Eur. J. Inorg. Chem.* **1999**, 1219.
7 G. Franciò, F. Faraone, W. Leitner, *Angew. Chem. Int. Ed.* **2000**, *39*, 1428.
8 (a) R. D. Feltham, H. G. Metzger, *J. Organomet. Chem.* **1971**, *33*, 347; (b) P. Wehman, H. M. A. van Donge, A. Hagos, P. C. J. Kamer, P. W. N. M. van Leeuwen, *J. Organomet. Chem.* **1997**, *535*, 183.
9 C. E. Crawforth, O. Meth-Cohn, C. A. Russell, *J. Chem. Soc., Perkin I* **1972**, 2807.
10 S. D. Pastor, S. P. Shum, R. K. Rodebaugh, A. D. Debellis, F. H. Clarke, *Helv. Chim. Acta* **1993**, *76*, 900.
11 For phosphoramidite ligands bearing tropisomeric diols see, for example: (a) A. Alexakis, S. Rosset, J. Allamand, S. March, F. Guillen, C. Benhaim, *Synlett* **2001**, 1375; (b) C. Monti, C. Gennari, U. Piarulli, J. G. de Vries, A. H. M. de Vries, L. Lefort, *Chemistry* **2005**, *11*, 6701 and references therein.
12 G. J. H. Buisman, M. E. Martin, E. J. Vos, A. Klootwijk, P. C. J. Kamer, P. W.

N. M. van Leeuwen, *Tetrahedron: Asymmetry* **1995**, *6*, 719.
13 G. T. Whiteker, A. M. Harrison, A. G. Abatjoglou, *J. Chem. Soc., Chem. Commun.* **1995**, 1805.
14 M. J. Baker, K. N. Harrison, A. G. Orpen, P. G. Pringle, G. Shaw, *J. Chem. Soc., Chem. Commun.* **1991**, 803.
15 D. Polet, A. Alexakis, *Org. Lett.* **2005**, *7*, 1621.
16 (a) F. Amiot, L. Cointeaux, E. J. Silve, A. Alexakis, *Tetrahedron* **2004**, *60*, 8221; (b) L. Cointeaux, A. Alexakis, *Tetrahedron: Asymmetry* **2005**, *16*, 925.
17 J. Bálint, G. Egri, E. Fogassy, Z. Böcskei, K. Simon, A. Gajáry, A. Friesz, *Tetrahedron: Asymmetry* **1999**, *10*, 1079.
18 S. Hu, D. Tat, C. A. Martinez, D. R. Yazbeck, J. Tao, *Org. Lett.* **2005**, *7*, 4329.
19 (a) W. B. Wang, S. M. Lu, P. Y. Yang, X. W. Han, Y. G. Zhou, *J. Am. Chem. Soc.* **2003**, *125*, 10 536; (b) K. H. Lam, L. Xu, L. Feng, Q.-H. Fan, F. L. Lam, W.-h. Lo, A. S. C. Chan, *Adv. Synth. Catal.* **2005**, *347*, 1755; (c) L. Xu, K. H. Lam, J. Ji, J. Wu, Q.-H. Fan, W.-H. Lo, A. S. C. Chan, *Chem. Commun.* **2005**, 1390.
20 Christiane J. Diez-Holz, Diploma Thesis, University of Aachen (Germany), **2005**.
21 C. J. Diez-Holz, C. Boeing, J. Klankermayer, M. Hoelscher, G. Franciò, W. Leitner, in preparation.
22 G. Franciò, PhD Thesis, University of Messina (Italy), **2000**.
23 A similar arrangement was found for a rhodium complex based on the phosphine-phosphite BINAPHOS ligand: see ref. [26a].
24 F. Agbossou, J.-F. Carpentier, A. Mortreux, *Chem. Rev.* **1995**, *95*, 2485.
25 D. G. Blackmond, T. Rosner, T. Neugebauer, M. T. Reetz, *Angew. Chem. Int. Ed.* **1999**, *38*, 2196.
26 (a) K. Nozaki, N. Sakai, T. Nanno, T. Higashijima, S. Mano, T. Horiuchi, H. Takaya, *J. Am. Chem. Soc.* **1997**, *119*, 4413; (b) K. Nozaki, Y. Itoi, F. Shibahara, E. Shirakawa, T. Ohta, H. Takaya, T. Hiyama, *J. Am. Chem. Soc.* **1998**, *120*, 4051; (c) K. Nozaki, T. Matsuo, F. Shibahara, T. Hiyama, *Adv. Synth. Catal.* **2001**, *343*, 61.
27 (a) G. Franciò, W. Leitner, *Chem. Commun.* **1999**, 1663; (b) G. Franciò, K. Wittmann, W. Leitner, *J. Organomet. Chem.* **2001**, *621*, 130.
28 T. Higashizima, N. Sakai, K. Nozaki, H. Takaya, *Tetrahedron Lett.* **1994**, *35*, 2023.
29 A renewed interest in asymmetric hydroformylation driven by academic and industrial research laboratories as well recently enabled significant progress to be made in this synthetically extremely useful transformation. See: (a) A. T. Axtell, C. J. Cobley, J. Klosin, G. T. Whiteker, A. Zanotti-Gerosa, K. A. Abboud, *Angew. Chem. Int. Ed.* **2005**, *44*, 5834; (b) T. P. Clark, C. R. Landis, S. L. Freed, J. Klosin, K. A. Abboud, *J. Am. Chem. Soc.* **2005**, *127*, 5040; (c) C. J. Cobley, J. Klosin, C. Qin, G. T. Whiteker, *Org. Lett.* **2004**, *6*, 3277; (d) S. Breeden, D. J. Cole-Hamilton, D. F. Foster, G. J. Schwarz, M. Wills, *Angew. Chem. Int. Ed.* **2000**, *39*, 4106; (e) M. Dieguez, O. Pamies, C. Claver, *Chem. Commun.* **2005**, 1221; (f) M. Dieguez, O. Pamies, C. Claver, *Tetrahedron: Asymmetry* **2004**, *15*, 2113 and reference therein.
30 Simon Burk, Diploma Thesis, University of Aachen (Germany), **2004**.
31 W. J. Tang, X. M. Zhang, *Chem. Rev.* **2003**, *103*, 3029.
32 (a) K. Burgemeister, G. Franciò, H. Hugl, W. Leitner, *Chem. Commun.* **2005**, 6026; (b) Y. Yan, Y. Chi, X. Zhang, *Tetrahedron: Asymmetry* **2004**, *15*, 2173.
33 C. Claver, E. Fernandez, A. Gillon, K. Heslop, D. J. Hyett, A. Martorell, A. G. Orpen, P. G. Pringle, *Chem. Commun.* **2000**, 961.
34 M. T. Reetz, G. Mehler, *Angew. Chem. Int. Ed.* **2000**, *39*, 3889.
35 For the first asymmetric hydrogenation using monodentate P-ligands see: (a) W. S. Knowles, M. J. Sabacky, *Chem. Commun.* **1968**, 1445; (b) W. S. Knowles, M. J. Sabacky, B. D. Vineyard, *J. Chem. Soc., Chem. Commun.* **1972**, 10; (c) W. S. Knowles, *Angew. Chem. Int. Ed.* **2002**, *41*, 1998.
36 I. V. Komarov, A. Borner, *Angew. Chem. Int. Ed.* **2001**, *40*, 1197.

37 (a) M. T. Reetz, T. Sell, A. Meiswinkel, G. Mehler, *Angew. Chem. Int. Ed.* **2003**, *42*, 790; (b) D. Pena, A. J. Minnaard, J. A. F. Boogers, A. H. M. de Vries, J. G. de Vries, B. L. Feringa, *Org. Biomol. Chem.* **2003**, *1*, 1087.

38 R. Hoen, J. A. F. Boogers, H. Bernsmann, A. J. Minnaard, A. Meetsma, T. D. Tiemersma-Wegman, A. H. M. de Vries, J. G. de Vries, B. L. Feringa, *Angew. Chem. Int. Ed.* **2005**, *44*, 4209.

39 S. Burk, G. Franciò, W. Leitner, *Chem. Commun.* **2005**, 3460.

40 D. G. Genov, D. J. Ager, *Angew. Chem. Int. Ed.* **2004**, *43*, 2816.

41 T. Ohkuma, H. Ooka, S. Hashiguchi, T. Ikariya, R. Noyori, *J. Am. Chem. Soc.* **1995**, *117*, 2675.

42 R. Noyori, T. Ohkuma, *Angew. Chem. Int. Ed.* **2001**, *40*, 41.

43 B. L. Feringa, M. Pineschi, L. A. Arnold, R. Imbos, A. H. M. de Vries, *Angew. Chem. Int. Ed.* **1997**, *36*, 2620.

44 (a) A. Alexakis, C. Benhaim, *Eur. J. Org. Chem.* **2002**, 3221; (b) A. Alexakis, D. Polet, S. Rosset, S. March *J. Org. Chem.* **2004**, *69*, 5660.

45 C. G. Arena, G.; Calabrò, G. Franciò, F. Faraone, *Tetrahedron: Asymmetry* **2000**, *11*, 2387.

46 B. Bogdanović, *Adv. Organomet. Chem.* **1979**, *17*, 105.

47 For recent reviews see: (a) P. W. Jolly, G. Wilke, in *Applied Homogeneous Catalysis with Organometallic Compounds* (B. Cornils, W. A. Herrmann, Eds.), VCH, New York, **1996**; Vol. 2, p. 1024; (b) T. V. RajanBabu, *Chem. Rev.* **2003**, *103*, 2845.

48 G. Wilke, *Angew. Chem., Int. Ed. Engl.* **1988**, *27*, 185.

49 (a) A. Wegner, W. Leitner, *Chem. Commun.* **1999**, 1583; (b) A. Boesmann, G. Franciò, E. Janssen, M. Solinas, W. Leitner, P. Wasserscheid, *Angew. Chem. Int. Ed.* **2001**, *40*, 2697.

50 N. Nomura, J. Jin, H. Park, T. V. RajanBabu, *J. Am. Chem. Soc.* **1998**, *120*, 459.

51 G. Franciò, F. Faraone, W. Leitner, *J. Am. Chem. Soc.* **2002**, *124*, 736.

52 M. Hoelscher, G. Franciò, W. Leitner, *Organometallics* **2004**, *23*, 5606.

53 D. Huber, P. G. A. Kumar, P. S. Pregosin, A. Mezzetti, *Organometallics* **2005**, *24*, 5221.

54 (a) R. Kumareswaran, M. Nandi, T. V. Rajanbabu, *Org. Lett.* **2003**, *5*, 4345; (b) H. Park, R. Kumareswaran, T. V. RajanBabu, *Tetrahedron* **2005**, *61*, 6352.

55 A. Zhang, T. V. RajanBabu, *J. Am. Chem. Soc.* **2006**, *128*, 54.

56 B. M. Trost, *Angew. Chem. Int. Ed.* **1995**, *34*, 259.

57 (a) B. M. Trost, M. J. Krische, *Synlett* **1998**, 1; (b) G. C. Lloyd-Jones, *Org. Biomol. Chem.* **2003**, *1*, 215.

58 (a) P. Kisinga, L. A. Goj, R. A. Widenhoefer, *J. Org. Chem.* **2001**, *66*, 635; (b) K. L. Bray, I. J. S. Fairlamb, J.-P. Kaiser, G. C. Lloyd-Jones, P. A. Slatford, *Topics in Catalysis* **2002**, *19*, 49.

59 (a) A. Behr, U. Freudenberg, W. Keim, *J. Mol. Catal.* **1986**, *35*, 9; (b) B. Radetich, T. V. RajanBabu, *J. Am. Chem. Soc.* **1998**, *120*, 8007.

60 R. Grigg, J. F. Malone, T. R. B. Mitchell, A. Ramasubbu, R. M. Scott, *J. Chem. Soc., Perkin Trans. 1* **1984**, 1745.

61 (a) Y. Yamamoto, Y. Nakagai, N. Ohkoshi, K. Itoh, *J. Am. Chem. Soc.* **2001**, *123*, 6372; (b) I. Özdemir, E. Çetinkaya, B. Çetinkaya, M. Çiçek, D. Sémeril, C. Bruneau, P. H. Dixneuf, *Eur. J. Inorg. Chem.* **2004**, 418; (c) Y. Terada, M. Arisawa, A. Nishida, *Angew. Chem. Int. Ed.* **2004**, *43*, 4063.

62 S. Okamoto, T. Livinghouse, *J. Am. Chem. Soc.* **2000**, *122*, 1223.

63 (a) A. Heumann, M. Moukhliss, *Synlett* **1998**, 1211; (b) A. Heumann, M. Moukhliss, *Synlett* **1999**, 268.

64 C. Böing, G. Franciò, W. Leitner, *Chem. Commun.* **2005**, 1456.

65 M. Brookhart, E. M. Hauptmann, *PCT Int. Appl.*, WO98/22424, Priority November 18, **1996**.

66 The synthesis of the [Ni(allyl)(cod)][Y] complexes was accomplished by following the procedures in: (a) R. B. A. Pardy, I. Tkatschenko, *J. Chem. Soc., Chem. Commun.* **1981**, 49; (b) J. Ascenso,

A. R. Dias, P. T. Gomes, C. C. Romão, D. Neibecker, I. Tkatschenko, *Makromol. Chem.* **1989**, *190*, 2773.
67 [Ni(allyl)(cod)][BARF] as the catalyst without a P ligand led to 7% conversion and 22% regioselectivity within 24 h.
68 See footnote [a] in Table 2.1.5.1.
69 For a recent review on weakly coordinating anions see: I. Krossing, I. Raabe, *Angew. Chem. Int. Ed.* **2004**, *43*, 2066.
70 C. Boeing, G. Franciò, W. Leitner, *Adv. Synth. Catal.* **2005**, *347*, 1537.

2.1.6
Immobilization of Transition Metal Complexes and Their Application to Enantioselective Catalysis
Adrian Crosman, Carmen Schuster, Hans-Hermann Wagner, Melinda Batorfi, Jairo Cubillos, and Wolfgang Hölderich

2.1.6.1 Introduction

Asymmetric synthesis has received much attention because of the rapid growth of the pharmaceutical market in recent decades [1–3]. Homogeneous asymmetric catalysis has made tremendous progress during the same period and is well recognized in that field, a fact also expressed by dedication of the Nobel Prize for Chemistry in 2001 to Noyori, Sharpless, and Knowles [4]. However, the obstacles of homogeneous catalysis in catalyst separation and recycling often make its practical application difficult. Heterogeneous catalysis supplies the opportunity for easy separation and recycling of catalysts, easy purification of products, and, possibly, continuous or multiple processing of chiral compounds. Furthermore, in solid chiral catalysts, particularly chiral catalysts accommodated in the pores or cavities of porous materials, the effects of the surface area and the pores on the catalytic performance are quite different. Therefore, it is a strongly desired objective to immobilize homogeneous catalysts in porous materials, without losing its merits, and even enhancing the enantioselectivity.

Since the late 1960s, many approaches have been published by academic and industrial researchers to "heterogenize", "immobilize," or "anchor" homogeneous catalyst on solid supports [5–8]. Many excellent reviews have emerged in recent years which describe in detail the synthesis and use of polymer-supported catalysts [9–18] and catalysts on inorganic carriers [19–23], or both [24–27]. Covalent binding between the homogeneous catalysts and the supports is by far the most frequently used strategy. It can be effected either by copolymerization of functionalized ligands with a suitable monomer, or by grafting functionalized ligands or metal complexes with reactive groups onto a preformed support.

Another common and simple immobilization technique for catalytically active metal complexes is via ionic bonding, which is particularly useful for cationic rhodium or palladium catalysts. Various supports with ion-exchange capabilities can be used, including standard organic or inorganic ion-exchange resins, inorganic materials with polarized groups, and zeolites.

The advantage of immobilization by adsorption is the ease of preparation of the heterogenized catalyst, very often even without the need to previously functionalize the ligand. In this respect, immobilization by ionic bonding can be seen as a special case of heterogenization via adsorption. An innovative modular method was developed by Augustine, who used heteropolyacids as anchoring agents to attach various metal complexes on different supports [28].

When entrapment methods are being used for heterogenization, the size of the metal complex is more important than the specific adsorptive interaction. There are two different preparation strategies. The first is based on building up catalysts in well-defined cages of porous supports. This approach is also called the "ship in a bottle" method [29]. The other approach is to build up a polymer network around a preformed catalyst.

Some examples for the application of such immobilization strategies and the application of the immobilized complexes to asymmetric hydrogenation and epoxidation as well as to ring-opening reactions of epoxides will be presented in this chapter.

2.1.6.2 Immobilized Rh Diphosphino Complexes as Catalysts for Asymmetric Hydrogenation

The development of homogeneous asymmetric hydrogenation was initiated by Knowles and Horner in the late 1960s [30, 31], after the discovery of Wilkinson's homogeneous hydrogenation catalyst [RhCl(PPh$_3$)$_3$] [32]. By replacing triphenylphosphine of Wilkinson's catalyst with resolved chiral monophosphines, Knowles and Horner reported the earliest examples of enantioselective hydrogenation, albeit with poor enantioselectivity. Later, two breakthroughs were made in asymmetric hydrogenation by Kagan and Knowles, respectively. Kagan reported the first bisphosphine ligand, DIOP, for Rh-catalyzed asymmetric hydrogenation [33], while Knowles made his significant discovery of a C_2-symmetric chelating bisphosphine ligand, DIPAMP [34]. Due to its high catalytic efficiency in Rh-catalyzed asymmetric hydrogenation of dehydroamino acids, DIPAMP was quickly employed for the industrial production of L-DOPA.

Mesoporous molecular sieves have received much attention in the field of catalysis, especially for their use as supports. Ion-exchange, catalytic, and adsorptive properties of molecular sieve materials are based on the existence of acid sites which arise from the presence of accessible hydroxyl groups associated with tetrahedral aluminum in the silica–alumina framework. Our research was focused on the use of the M41S and SBA-15 type of materials as carriers, which are characterized by a well-defined pore structure and high surface area, offering new opportunities for the immobilization of large homogeneous catalyst species without any modification of their chemical structure [35–37]. MCM-48 has been investigated to a lesser extent even though it should be more applicable as a catalyst or adsorbent due to its three-dimensional pore architecture. The recently discovered pure silica phase, designated SBA-15, has long-range order, large monodispersed mesopores (up to 50 nm) and thicker walls (typically between 3 and 9 nm) which make it more thermally and hydrothermally stable than the M41S type of materials. Unfortunately, as the pure silica SBA-15 is synthesized in strong acid media (2M HCl solution), incorporation of framework aluminum into SBA-15 by direct synthesis seems to be impossible because most aluminum sources dissolve in strong acids. Previous studies have shown that aluminum can be effectively incorporated into siliceous SBA-15 via various post-synthesis procedures, e.g., by grafting aluminum onto SBA-15 wall surfaces with anhydrous $AlCl_3$ or aluminum isopropoxide in nonaqueous solutions, or sodium aluminate in aqueous solutions followed by calcination [38, 39].

Herein, we present a very straightforward technique for the immobilization of rhodium–diphosphine complexes. This heterogenization is based on an ionic interaction between the negatively charged Al-M41S or Al-SBA-15 framework and the cationic rhodium of the organometallic complex. Furthermore, activity of the obtained catalysts in the enantioselective hydrogenation of different prochiral olefins was investigated.

2.1.6.2.1 Preparation and Characterization of the Immobilized Rh–Diphosphine Complexes

Al-MCM-41 and Al-MCM-48 were synthesized according to slightly altered literature methods [40, 41]. SBA-15 was first prepared in the all-silica form, followed by impregnation with aluminum isopropoxide and calcination, making it possible to obtain the aluminum-containing SBA-15 [36, 39]. For all the molecular sieves used, the Si/Al ratio was set to 40 : 1. Before the immobilization of organometallic complexes on the solid support, the aluminum-containing molecular sieves were calcined overnight at 300 °C in order to remove all adsorbed water molecules. All the experiments involving diphosphines, rhodium–diphosphine complexes, and immobilized complexes were carried out under dry Ar using Schlenk techniques.

The complexes were prepared from $[CODRhCl]_2$ and 1.1 equiv of chiral diphosphine in dichloromethane as solvent. The chiral complex was added to a suspension of the support in dichloromethane. After being stirred for 24 h, the solid was filtered, washed with dichloromethane until the solvent showed no color, and afterward dried at room temperature for 16 h. In order to remove the excess of Rh complex not fixed to the solid carrier, the catalysts were extracted with methanol in a Soxhlet apparatus under reflux for 24 h (Scheme 2.1.6.1). Both ICP-AES analysis and FTIR spectra of the remaining solvent indicated no content of homogeneous complex. The resulting catalysts had a pale yellow color similar to that of the homogeneous complex.

Rhodium complex contents varied between 0.02 and 0.07 mmol (g catalyst)$^{-1}$, depending on the type of carrier and diphosphine used, whereas the theoretical content was 0.15 mmol g^{-1}. The color transfer during the immobilization in di-

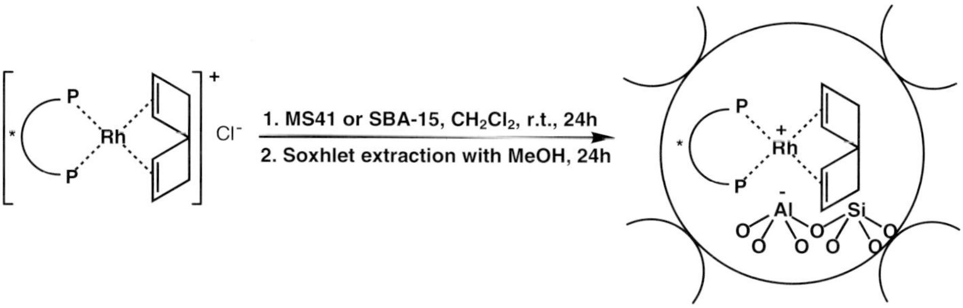

Scheme 2.1.6.1 Immobilization of rhodium–diphosphine complexes.

chloromethane from an orange/yellow solution to the white solid demonstrated a high degree of immobilization, but upon extraction with methanol, which in contrast to the nonpolar dichloromethane is adsorbed strongly on the carrier's surface, the organometallic complex desorbs from the silanol groups due to a competitive reaction with the polar alcohol.

A special case is the test of immobilization of rhodium–diphosphine complexes on all-silica materials. After the first immobilization step in dichloromethane, the solid material had a yellow color and a Rh content of 0.07 mmol g^{-1} was found. After the extraction with methanol, the entire amount of organometallic complex was washed out and the final material had again the original white color. No rhodium was detected in ICP-AES analysis of this sample. However, in the case of aluminum-containing materials the orange color obtained after the immobilization of the rhodium complexes in dichloromethane is clearly maintained even after extraction in methanol.

N$_2$ sorption isotherms were measured for the pure carrier and for all the immobilized complexes, respectively. Sorption isotherms and textural characteristics for pure and loaded samples are compared in Fig. 2.1.6.1 and Table 2.1.6.1. In all cases these isotherms presented the characteristic form of mesoporous materials. It is clear that the N$_2$ adsorbed/desorbed volume was higher in the case of pure carrier than in the case of immobilized complexes.

The X-ray powder diffraction patterns of the parent materials showed the hexagonal structure characteristic for MCM-41 and SBA-15, and the cubic structure for MCM-48, respectively. All the patterns matched well with the reported patterns, confirming the successful synthesis of the mesoporous molecular sieves. The intensity of the reflection did not change essentially upon loading the carrier with the organometallic complexes, nor after a catalytic cycle, showing that the mesoporous structures were not affected by incorporation of the catalyst.

Quantitative loading of the organometallic complex was demonstrated with thermal gravimetric analysis. For all immobilized complexes, thermogravimetric and differential scanning calorimetric measurements showed a thermal stability up to 200 °C. In all cases, oxidation of the organic structure took place in two steps, corresponding to about 250 °C and 400 °C, respectively. The loss of weight

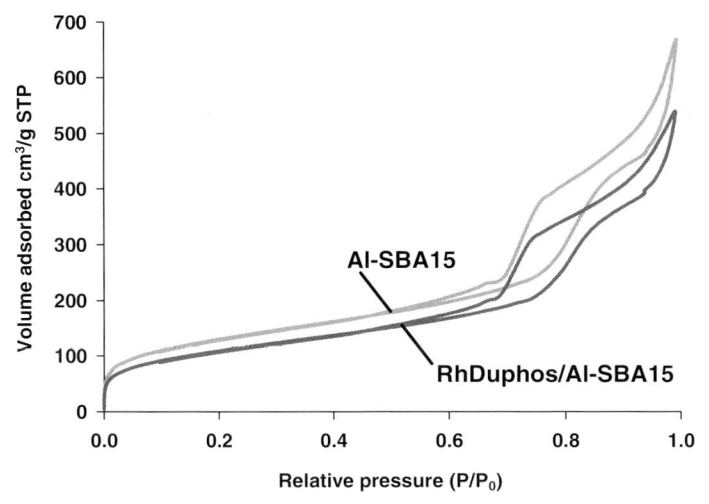

Fig. 2.1.6.1 N_2 sorption isotherms of Al-SBA-15 and CODRhDuphos/Al-SBA-15.

Table 2.1.6.1 Textural characteristics of parent and loaded materials

Sample	Surface area $[m^2 g^{-1}]$	Pore volume $[cm^3 g^{-1}]$	Pore diameter $[Å]$
Al-MCM-41	1190	0.97	34
RhDuphos/Al-MCM-41	1092	0.77	28
RhDiop/Al-MCM-41	1101	0.82	29
Al-MCM-48	1211	1.04	35
RhDuphos/Al-MCM-48	1082	0.81	30
RhDiop/Al-MCM-48	1097	0.87	31
Al-SBA-15	467	0.92	94
RhDuphos/Al-SBA-15	397	0.78	89
RhDiop/Al-SBA-15	401	0.82	87

of ca. 3 wt.% caused by oxidation of the complex was consistent with the content determined by chemical analysis. The oxidation of the pure carrier showed no characteristic peak.

^{31}P MAS NMR spectra of the solid homogeneous CODRhChiraphos complex and the same complex immobilized on Al-SBA-15 were measured in order to elucidate the nature of the immobilization (Fig. 2.1.6.2). The homogeneous complex gave a strong signal at 59 ppm and two weaker ones at 30 and 75 ppm due

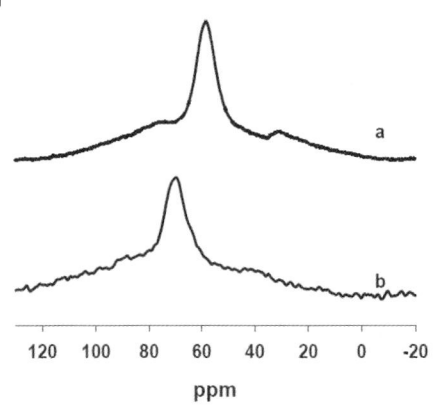

Fig. 2.1.6.2 ^{31}P MAS NMR spectra of a) homogeneous CODRhChiraphos and b) CODRhChiraphos immobilized on Al-SBA-1.5.

to trace amounts of oxidic impurities. The MAS NMR spectrum of the immobilized CODRhChiraphos complex showed only a single broad signal at 71 ppm. The absence of signals of the homogeneous complex and of the free phosphine ligand was an indication that the phosphine ligand was completely coordinated to the rhodium. The immobilization of this rhodium–Chiraphos complex led to a shift of the ^{31}P signal at 12 ppm to lower magnetic field than the homogeneous complex as result of the interaction of the guest complex with the surface of the Al-SBA-15 host.

The rhodium complexes used consisted of a chiral diphosphine and a cyclooctadiene ligand. There could be several forces involved in the bonding of the complex to aluminum-containing mesoporous materials. First, an electrostatic interaction of the cationic complexes with the anionic framework of the solid carrier could occur. A similar mechanism was reported for the immobilization of manganese complexes on Al-MCM-41 [42]. Furthermore, direct bridging of rhodium to surface oxygen of the mesoporous walls has also been observed and could occur after cleavage of the diene complex during the hydrogenation reaction [43]. However, in our case no evolution of cyclooctadiene during the immobilization reaction could be observed. FTIR spectra of the filtrate obtained after the immobilization did not contain bands attributable to free cyclooctadiene.

2.1.6.2.2 Enantioselective Hydrogenation over Immobilized Rhodium Diphosphine Complexes

Several diphosphine ligands have been applied and the corresponding complexes have been tested for the immobilization (Fig. 2.1.6.3). The activity of different free and immobilized complexes in the enantioselective hydrogenation of dimethyl itaconate and methyl α-acetamidoacrylate was investigated. In blank reactions over pure mesoporous materials no reaction took place. When rhodium supported on carriers was used as catalyst, no enantiomeric excess was observed,

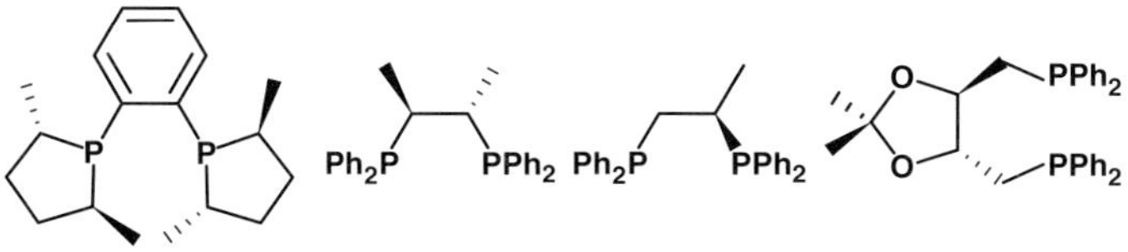

Fig. 2.1.6.3 Diphosphine ligands used.

S,S-Me-Duphos   S,S-Chiraphos   R-Prophos   R,R-Diop

Table 2.1.6.2 Hydrogenation over free and immobilized Rh complexes on Al-SBA-15 (3 bar H_2, 24 h reaction time).

Substrate	Complex	Support	Conversion [%]	TON	ee [%]
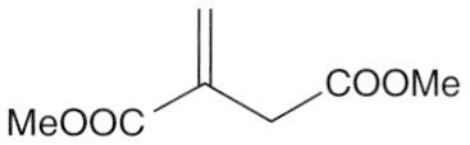 MeOOC-CH₂-C(=CH₂)-COOMe	RhDuphos	–	99	1000	94
		Al-SBA-15	99	1000	89
	RhChiraphos	–	99	1000	91
		Al-SBA-15	96	960	72
	RhProphos	–	99	1000	90
		Al-SBA-15	89	890	79
	RhDiop	–	99	1000	90
		Al-SBA-15	95	950	68
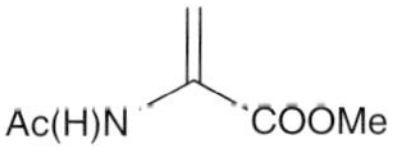 Ac(H)N-C(=CH₂)-COOMe	RhDuphos	–	99	1000	94
		Al-SBA-15	99	1000	93
	RhChiraphos	–	99	1000	89
		Al-SBA-15	92	920	77
	RhProphos	–	99	1000	90
		Al-SBA-15	94	940	78
	RhDiop	–	99	1000	86
		Al-SBA-15	95	950	62

although conversions up to 99% were found. Moreover, for all hydrogenation experiments the chemical selectivity was 100%. No side products were observed.

The catalytic results of free and immobilized rhodium complexes on Al-SBA-15 in the hydrogenation of dimethyl itaconate and methyl α-acetamidoacrylate are summarized in Table 2.1.6.2. For these reactions, the substrate/rhodium ratio was 1000:1. Using the (S,S)-Me-Duphos ligand, the best results in the hydrogenation of dimethyl itaconate were obtained with 94% *ee* in homogeneous media, and with 89% enantioselectivity for the immobilized complex at complete conversion. When the hydrogenation was carried out over RhDuphos immobilized on Al-MCM-41, an enantiomeric excess of 92% at a corresponding TON of 4000 was obtained (S/Rh = 4000:1), whereas the same complex immobilized on Al-MCM-48 gave 98% *ee* at a corresponding TON of 5600 (S/Rh = 6000:1).

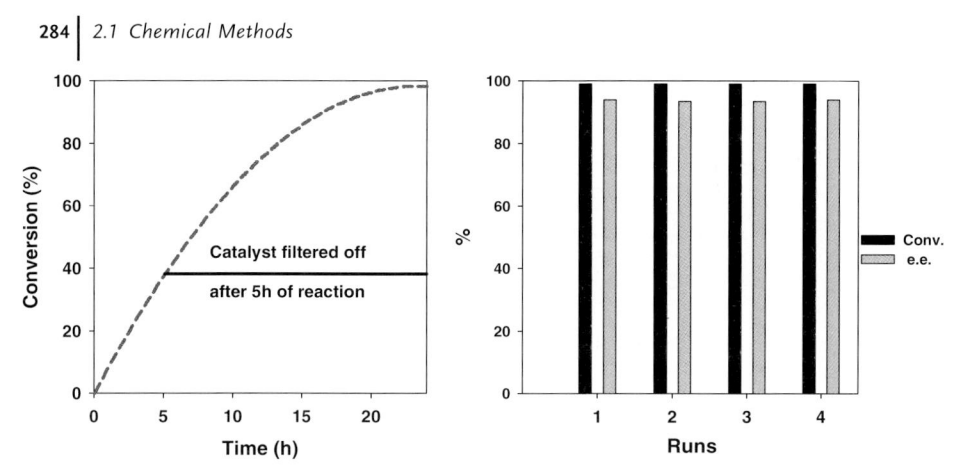

Fig. 2.1.6.4 Hot filtration test and catalyst recycling for hydrogenation of dimethyl itaconate over RhDuphos/Al-SBA-15.

For the hydrogenation of methyl α-acetamidoacrylate, the best results were again obtained with the (S,S)-Me-Duphos ligand. The complex showed complete conversion with 94% or 93% *ee* in homogeneous and heterogeneous media, respectively. When methyl α-acetamidoacrylate was used as a probe molecule for enantioselective hydrogenations over RhDuphos immobilized on M41S-type materials, 99% *ee* and a TON of 2000 were obtained over RhDuphos/Al-MCM-41 (S/Rh = 2000:1), whereas RhDuphos/Al-MCM-48 resulted in 97% *ee* and a TON of 2900 (S/Rh = 3000:1).

In order to prove that the reaction was catalyzed heterogeneously and to exclude the possibility of leaching and homogeneous catalysis, hot filtration tests were performed [44]. For the dimethyl itaconate hydrogenation, removal of the COD-RhDuphos immobilized on Al-SBA-15 effectively stopped the reaction after 5 h. After 24 h, the conversion of the filtered sample remained at 40%, whereas the original batch with catalyst was converted completely (Fig. 2.1.6.4).

The catalysts were easily recovered and reused without further treatment. The heterogenized RhDuphos on Al-SBA-15 was recycled four times. As the experiment showed, conversion as well as enantioselectivity of the catalyst remained high after several catalytic runs (Fig. 2.1.6.4).

To summarize, chiral heterogeneous catalysts were prepared from rhodium–diphosphine complexes and aluminum-containing mesoporous materials. The bonding occurred via an ionic interaction of the cationic complex with the host. These catalysts were suitable for asymmetric hydrogenation of functionalized olefins. The catalysts can be recycled easily by filtration or centrifugation with no significant loss of activity or enantioselectivity.

2.1.6.3 Heterogeneous Asymmetric Epoxidation of Olefins over Jacobsen's Catalyst Immobilized in Inorganic Porous Materials

The enantioselective epoxidation of olefins has received much attention since the corresponding chiral epoxides have become important building units for the

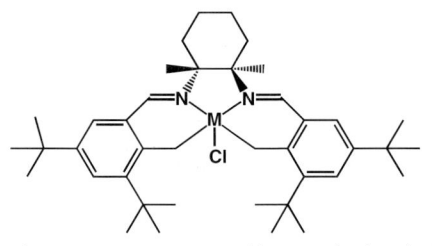

Fig. 2.1.6.5 (R,R)-(−)-N,N'-bis(3,5-di-t-butylsalicylidene)-1,2-cyclohexanediamine metal chloride (Jacobsen's catalyst).

synthesis of fine chemicals, drugs, agrochemicals, and food additives [45]. Chiral transition metal–salen complexes, first developed by Jacobsen and Katsuki, have been shown to be highly active and enantioselective catalysts for a number of conjugated unfunctionalized olefin epoxidation reactions [46]. Among them, the general catalyst depicted in Fig. 2.1.6.5 (well known as Jacobsen's catalyst) has shown the best results when using cis-disubstituted olefins as substrates [47].

The salen–manganese-catalyzed oxidation of olefins has long been investigated by various groups, Jacobsen and coworkers being only one of them [48]. Immobilization of the Jacobsen catalyst is still an attractive target. However, its outstanding activity, selectivity, and chiral induction are accompanied by several disadvantages, such as quick deactivation, difficult separation, and salt formation due to the use of sodium hypochloride as oxidant. Many attempts to immobilize the Jacobsen catalyst have been reported. The entrapment of a complex of this kind in zeolitic voids presents the problem that the average faujasite supercage is too small to host a complex occupying so much space. So far, only complexes of the less bulky bis-salicylidene-1,2-cyclohexanediamine have been occluded successfully in faujasites [49].

In the present work, the Jacobsen's catalyst was immobilized inside highly dealuminated zeolites X and Y, containing mesopores completely surrounded by micropores, and in Al-MCM-41 via ion exchange. Moreover, the complex was immobilized on modified silica MCM-41 via the metal center and through the salen ligand, respectively. cis-Ethyl cinnamate, (−)-α-pinene, styrene, and 1,2-dihydronaphtalene were used as test molecules for asymmetric epoxidation with NaOCl, m-CPBA (m-chloroperoxybenzoic acid), and dimethyldioxirane (DMD) generated in situ as the oxygen sources.

2.1.6.3.1 Preparation and Characterization of Immobilized Jacobsen's Catalysts

The zeolite dealumination mechanism is illustrated in Scheme 2.1.6.2. During treatment with silicon tetrachloride, a dealumination method first reported by Beyer et al. [50], the faujasite's framework aluminum was isomorphously replaced by silicon while maintaining the microporous structure. The reaction was self-

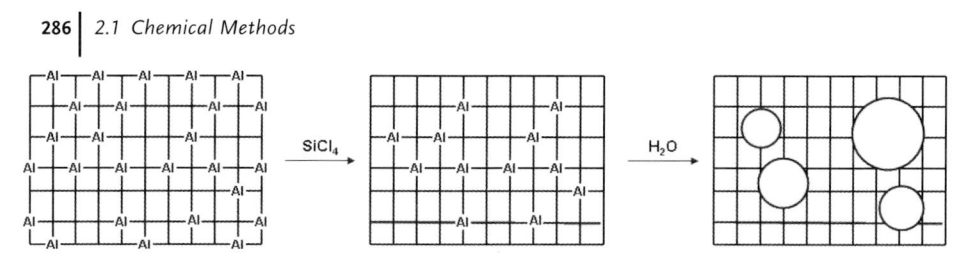

Scheme 2.1.6.2 Zeolite dealumination mechanism.

terminating due to the precipitation of NaAlCl$_4$ in the outer pores of the zeolite crystal. The temperature at which the reaction was conducted (250 °C) was relatively mild. Therefore it can be assumed that only the outer layers of the zeolite crystal were dealuminated by the treatment with SiCl$_4$. During the successive ion exchange, the chloro aluminum complexes were extracted and the zeolite was converted into the ammonium form. Afterward, the zeolite was dealuminated by means of steam. After this hydrothermal treatment, the amount of framework aluminum had considerably decreased to a high degree of dealumination with a SiO$_2$/Al$_2$O$_3$ ratio ranging from 125:1 to 190:1 for both zeolites X and Y. Using the described dealumination procedure, dealuminated faujasite zeolites were obtained regardless of the parent material. Final treatment of the host material with hydrochloric acid removed some of the extra framework aluminum in order to generate mesopores and thereby room for guest molecules. The overall procedure is described elsewhere in detail [50, 51].

The zeolitic host was ion-exchanged to introduce the desired transition metal cation. Synthesis of the salen ligands in the mesopores was conducted at room temperature under an inert atmosphere. To the ion-exchanged zeolitic material, the optically pure diamine in dichloromethane was added, in slight excess (1.2 equiv.) relative to the metal content. The appropriate amount of aldehyde was added to the slurry. The final material was Soxhlet-extracted with dichloromethane and toluene, respectively, until the solvent remained colorless. This procedure is described elsewhere in detail [51].

The immobilization by ionic bonding on Al-MCM-41 was carried out similarly to the immobilization of rhodium–diphosphine complexes as described above. This catalyst was named MCM1HC. For immobilization via the metal center and covalent bonding of salen, the all-silica MCM-41 was modified with (3-aminopropyl)triethoxysilane (APTES) [52]. The catalyst obtained by the metal center immobilization was denoted MCM2HC [53], whereas the material obtained by covalent bonding of the salen ligand was named MCM3HC [54]. Detailed procedures are described extensively elsewhere [55].

In Fig. 2.1.6.6, the FTIR spectra of the Jacobsen ligand (a), the Jacobsen catalyst (b), and the immobilized manganese salen complex in the cages of dealuminated faujasite zeolite (c) are compared. While spectra a and b have been measured using the standard KBr technique, the spectrum c of the "ship in a bottle" catalyst has been recorded using a self-supported wafer. The bands at wavenumbers 1466 cm^{-1}, 1434 cm^{-1}, 1399 cm^{-1} and 1365 cm^{-1} in spectrum c can be assigned to the

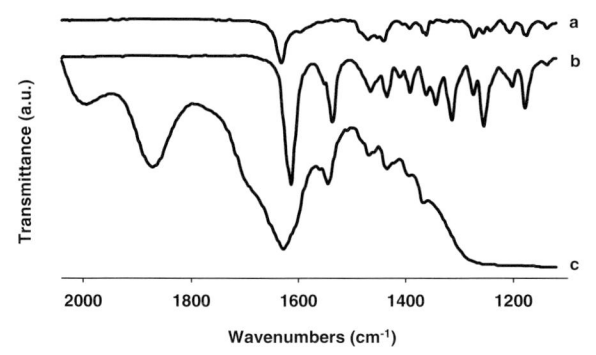

Fig. 2.1.6.6 FTIR spectra of a) the Jacobsen ligand, b) the Jacobsen catalyst, and c) the manganese–salen complex, in zeolite supercages.

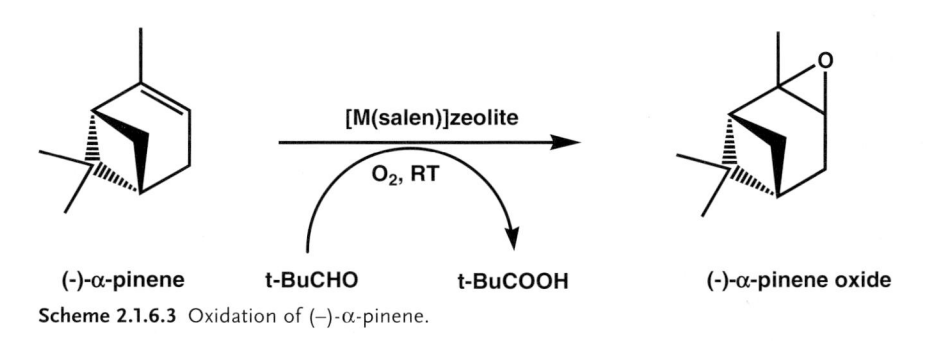

Scheme 2.1.6.3 Oxidation of (−)-α-pinene.

salen ligand because they also appear in spectra a and b. The band at 1542 cm^{-1} in spectrum c can only be assigned to the salen–manganese complex as it does appear in the spectrum of the Jacobsen catalyst (a) at 1535 cm^{-1}, too, whereas in the spectrum of the salen ligand (b) there is no band in this region between 1470 cm^{-1} and 1600 cm^{-1}. This can be considered as an indication of the presence of the Jacobsen catalyst in our "ship in a bottle" catalyst.

The immobilization of Jacobsen's catalyst on mesoporous materials was also investigated by FTIR analysis. Here, the characteristic signal from the structure of metal–salen complexes at 1540 cm^{-1} was found on material MCM1HC. On the other hand, the signal attributable to the metallosalen complex could not be observed clearly in the spectra of MCM2HC and MCM3HC because several signals originating from the aminopropyl groups appeared in the same area. However, UV/Vis analysis showed that spectra of the heterogenized catalysts were very similar to that of the free catalyst. This fact confirms the immobilization of Jacobsen's catalyst.

2.1.6.3.2 Epoxidation of Olefins over Immobilized Jacobsen Catalysts

For the catalysts prepared by entrapment in the zeolite supercages, only the epoxidation of (−)-α-pinene as depicted in Scheme 2.1.6.3 and of limonene was investigated. The oxidant applied in the reaction is somewhat similar to the one introduced by Mukaiyama et al. (O_2 and acetone). By using environmentally benign molecular oxygen at RT instead of NaOCl as oxidant at 0 °C the undesirable salt formation can be avoided. Therefore, this oxidation method is favored over the system used by Jacobsen and coworkers [14, 17], i.e., NaOCl.

Figure 2.1.6.7 shows that the salen complex of V (**V-S**) retained its catalytic properties in spite of the entrapment in the host material. In contrast to the manganese salen complex, which lost only some of its epoxide selectivity upon immobilization, the corresponding Co complexes (**Co-S**) showed an additional decrease in stereoselectivity as well. Strikingly, by replacing the cyclohexanediamine with diphenylethylenediamine in the salen structure, the immobilized Co complex (**Co-S1**), with 100% conversion, 96% selectivity and 91% *de*, achieved even better results in the epoxidation of (−)-α-pinene than its homogeneous counterpart. However, it is worth pointing out that among the cyclohexenediamine type of salen complexes neither the homogeneous nor the occluded Jacobsen complex catalyzed the epoxidation of (−)-α-pinene best.

Epoxidation of *cis*-ethyl cinnamate, styrene, and 1,2-dihydronaphtalene over MCM1HC, MCM2HC, and MCM3HC materials was carried out using different oxygen sources (e.g., NaOCl, *m*-CPBA, and DMD generated in situ). The catalyst stability was checked by recording the infrared spectra of the catalyst before and after the reaction. When NaOCl and *m*-CPBA were employed as oxygen sources,

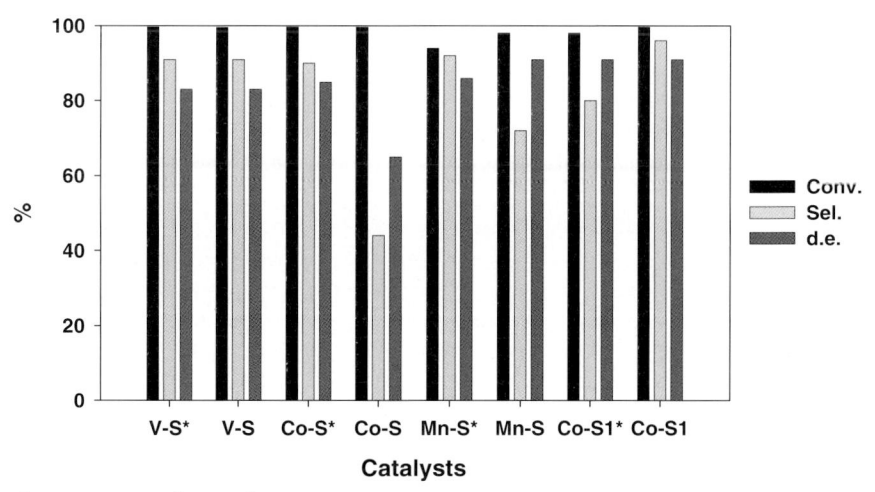

Fig. 2.1.6.7 Epoxidation of (−)-α-pinene over various homogeneous and immobilized transition metal–salen complexes. * = homogeneous complex (C_6H_6F, 4.6 mmol pivalic aldehyde, 1.85 mmol (−)-α-pinene, 25 mg catalyst, 30 bar O_2, RT).

the infrared spectra of the catalysts used showed that the immobilized complex degrades during the reaction. However, when DMD generated in situ was used as the oxygen source, the infrared spectra of the catalysts used matched well with spectra of the fresh catalysts. These experiments showed that it is possible to maintain the catalyst structure by using DMD generated in situ as the oxygen source.

For the substrates a different catalytic behavior was observed (Table 2.1.6.3). In the case of the most stable substrate (Z)-ethyl cinnamate the highest enantioselectivities have been found, while substrates which can be oxidized much easily, such as styrene and 1,2-dihydronaphthalene, showed lower enantioselectivities. Compared with the activity of the homogeneous catalysts, the heterogenized catalysts resulted only in a slight reduction of the *cis*-ethyl cinnamate conversion, whereas for the unfunctionalized olefins similar conversions were obtained. In all cases the high selectivities found for the homogeneous catalyst were also obtained for the heterogeneous systems. Additionally, the *cis/trans*-epoxide ratio for the *cis*-ethylcinnamate epoxidation remained almost unchanged. These results indicated that the side reactions catalyzed by the support did not influence the

Table 2.1.6.3 Catalytic activity in the epoxidation of three representative olefins.

Substrate	Catalyst	Conversion [%]	Selectivity [%]	ee [%]
COOEt	Homogeneous	38	99 (5.0)[a]	86[b] (72)[c]
	MCM1HC	32	97 (5.0)[a]	75[b] (50)[c]
	MCM2HC	25	98 (4.7)[a]	78[b] (60)[c]
	MCM3HC	30	95 (4.5)[a]	35[b] (30)[c]
styrene	Homogeneous	100	100	40
	MCM1HC	100	100	38
	MCM2HC	95	100	35
	MCM3HC	97	100	12
dihydronaphthalene	Homogeneous	94	89	37
	MCM1HC	95	88	32
	MCM2HC	96	90	34
	MCM3HC	95	90	8

Reaction conditions: 0.25 mmol substrate, 0.25 mmol Oxone®, 30 mg homogeneous catalyst or 300–500 mg heterogeneous catalyst, 0.075 mmol 4-PPNO, 4 mL CH$_3$COOH, pH 7–8 by use of aqueous NaHCO$_3$, RT, 25–50 min. **a** *cis/trans* ratio. **b** (2R, 3R) (*cis*-epoxide). **c** (2S, 3R) (*trans*-epoxide).

catalytic behavior of Jacobsen's catalyst immobilized into the mesopores of Al-MCM-41.

Enantioselectivities obtained over MCM1HC and MCM2HC were very close to the ones achieved in the homogeneous phase. However, the Mn content determined with ICP-AES showed leaching (between 10 and 30%), which indicates that the reaction takes place mainly in the homogeneous phase. This fact can be explained by the very weak interaction between the homogeneous catalyst and the support. In contrast, MCM3HC proved to be very stable against leaching. However, the enantioselectivities dropped tremendously. The reduced mobility of the immobilized complex inside the mesoporous framework of the MCM3HC sample might influence the catalytic performance negatively by inhibiting the appropriate geometric conformation needed for the chiral induction.

Finally, the recyclability of catalyst MCM3HC was investigated (Fig. 2.1.6.8). Both the conversion and enantiomeric excesses could be maintained for at least three cycles. This catalyst was stable against leaching, which proved that the use of covalent attachment of the salen ligand was a successful immobilization method. Also, the degradation of the salen complex was reduced by using DMD generated in situ as the oxidizing agent.

Different immobilization methods were applied for Jacobsen's catalyst. The entrapment of the organometallic complex in the supercages of the dealuminated zeolite was achieved without noticeable loss of activity and selectivity. The immobilized catalysts were reusable and did not leach. For the oxidation of (–)-α-pinene the system used only O_2 at RT instead of sodium hypochloride at 0 °C. There was a disadvantage in the use of pivalic aldehyde for oxygen transformation via the corresponding peracid. This results in the formation of pivalic acid, which has to be separated from the reaction mixture.

The heterogeneous catalysts as well as the homogeneous Jacobsen catalyst were found to be stable when DMD generated in situ was used as the oxygen source. By using covalent attachment on the salen ligand, the heterogenized catalyst MCM3HC was stable against leaching and could be reused four times without significant loss of activity or enantioselectivity.

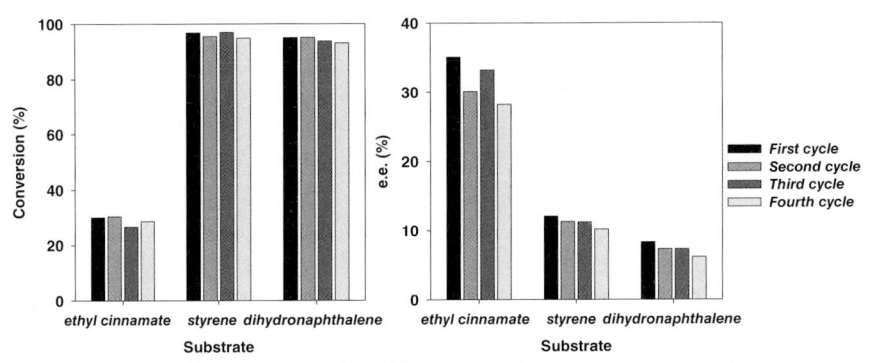

Fig. 2.1.6.8 Effect of the reuse of MCM3HC on conversion and enantioselectivity.

2.1.6.4 Novel Heterogenized Catalysts for Asymmetric Ring-Opening Reactions of Epoxides

A wide variety of highly selective asymmetric reactions catalyzed by chiral salen–metal complexes have been discovered since the mid-1990s. Salen-type ligands are among the synthetically most accessible molecular ligands which can be used for asymmetric catalysis and which can readily be tuned both sterically and electronically. One particular type of catalyst using such a chiral salen ligand is the so-called Jacobsen's catalyst. In combination with cobalt(II), cobalt(III), and chromium(III) ions as metal ion centers (R,R)-$(-)$-N,N'-bis(3,5-di-t-butylsalicylidene)-1,2-cyclohexanediamine (Fig. 2.1.6.5) [56, 59] was found to be a highly effective catalyst for ring-opening reactions of epoxides.

Epoxides are versatile building blocks for organic syntheses and are valuable intermediates for the stereocontrolled synthesis of complex organic compounds. The use of epoxides has expanded dramatically with the advent of practical asymmetric catalytic methods for their synthesis. Besides the enantioselective epoxidation of prochiral olefins, approaches for the use of epoxides in the synthesis of enantiomerically enriched compounds include the resolution of racemic epoxides.

Hydrolytic kinetic resolution (HKR) is an attractive strategy for the production of optically active epoxides, resulting in an economic and operationally simple preparation method. This attractive and elegant process uses only water as the other reagent; no solvents, and only low loadings of a recyclable catalyst, have to be added. The system results in the formation of highly valuable 1,2-diols combined with terminal epoxides in high yields with a high enantiomeric excess [56].

With this asymmetric ring opening of symmetrical epoxides, a technique has become available that begins from an easily prepared achiral starting material and leads to a functionalized product with two stereogenic centers which can be set simultaneously. Readily accessible synthetic catalysts (chiral metal–based salen complexes) have been found to catalyze asymmetric ring opening of *meso* and racemic terminal epoxides with a high degree of selectivity [59].

2.1.6.4.1 Synthesis and Characterization of the Heterogenized Catalysts

The large range of catalytic processes now involving soluble chiral metal–salen complexes makes the development of their heterogeneous counterparts very attractive. The immobilization of metal ions and metal complexes on inorganic matrices would open up more practical fine-chemical processes. Due to their mechanical and thermal stability, inorganic microporous and mesoporous materials can be used as supports for the heterogenization of such homogeneous complexes. Among such materials are the members of the M41S family type, such as MCM-41, MCM-48, MCM-50, and the various SBA-type materials. Recently, this family of mesoporous M41S materials has received much attention and their synthesis and characterization has been widely reported in the literature [37].

For this work, three types of mesoporous materials are the choice for the heterogenization of homogeneous catalysts: Al-MCM-41, Al-MCM-48, and

Al-SBA-15. For the heterogenization various methods are available: (1) covalent bond formation with the ligand; (2) adsorption or ion-pair formation; (3) encapsulation (see Scheme 2.1.6.4) and (4) entrapment [58].

One of the goals of such a heterogenization is to maintain the typical structure of the mesoporous materials. However, it has been found difficult to maintain the M41S type of structure during such heterogenization and to obtain an optimal metal loading. To determine the optimal immobilization parameters the solvents (ethanol, water, toluene, methanol, and dichloromethane), as well as the Co(II) and Cr(III) salts and their concentrations, were varied. For ion-exchange methods, alcohols and, for impregnation methods, toluene were found to be the optimal solvents, not resulting in destruction of the support structure. The metal contents were found to be 1.4 to 5.7 wt.%, respectively. The heterogeneous catalysts ((S,S)-Co(II)–Jacobsen complexes immobilized onto Al-MCM-41 by ion exchange and impregnation) were characterized using XRD, ICP-AES, TA, FTIR, UV/Vis, and physi- and chemisorption.

Figure 2.1.6.9 and Table 2.1.6.4 present the N_2-sorption isotherms and the results of the characterization of the Al-MCM-41 parent material and of the by ion exchange and impregnation onto Al-MCM-41 immobilized (S,S)-Co(II)-Jacobsen complexes. The N_2-sorption isotherms of the loaded materials clearly illustrate that (S,S)-Co(II)-Jacobsen complex is deposited in the inner surface of the pores

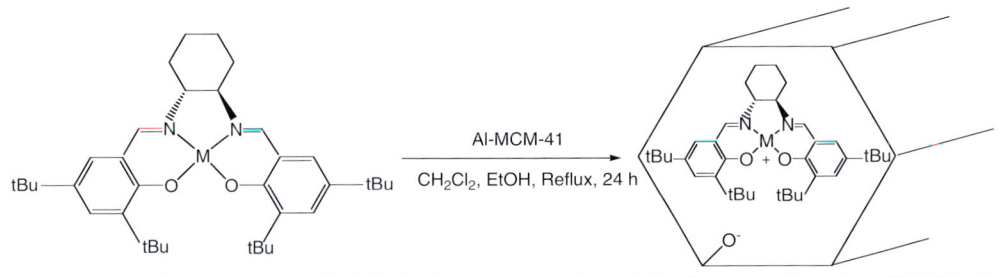

Scheme 2.1.6.4 Method for the impregnation of metal–Jacobsen complexes into Al-MCM-41.

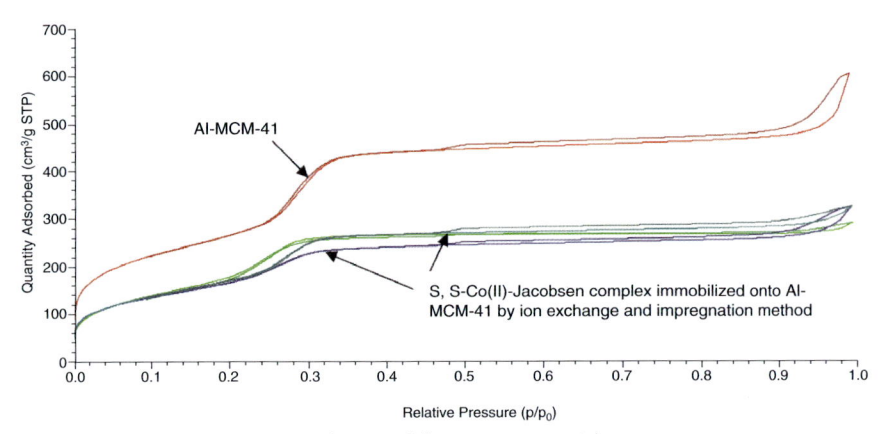

Fig. 2.1.6.9 Nitrogen sorption isotherms of the various materials.

Table 2.1.6.4 Nitrogen sorption data of the Co(II)-Jacobsen complex immobilized on Al-MCM-41 by ion exchange or impregnation.

Material	Surface area $[m^{2*}g^{-1}]$	Pore volume $[cm^{3*}g^{-1}]$	Pore diameter $[Å]$
Al-MCM-41	980	0.98	33
Immobilized ion-exchanged Co-Jacobsen complex	611	0.34	37
Immobilized impregnated Co-Jacobsen complex	415	0.24	27

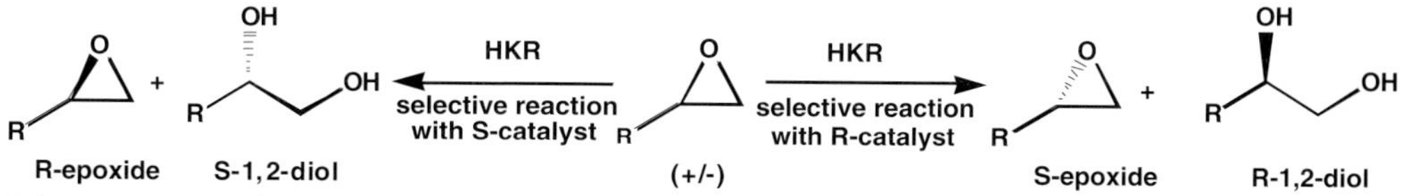

Scheme 2.1.6.5 Hydrolytic kinetic resolution (HKR) of terminal epoxides.

by both methods, causing a decrease of the mesoporous pore volume of heterogeneous catalysts.

2.1.6.4.2 Asymmetric Ring Opening of Epoxides over New Heterogenized Catalysts

To investigate the catalytic activity of the materials prepared, hydrolytic kinetic resolution of terminal epoxides such as styrene oxide, 1,2-epoxyhexane, and epichlorohydrin was carried out using only water, which acts as a nucleophilic agent according to Scheme 2.1.6.5 [56].

The results of the homogeneous series (see Table 2.1.6.5) were found to be comparable with those obtained by Jacobsen [56, 57]. For the heterogenized materials the best results in the asymmetric ring opening of styrene oxide (Table 2.1.6.5) and the asymmetric ring opening of *meso*-epoxides with a carboxylic acid (Table 2.1.6.6) were obtained over the (S,S)-Co(II)–Jacobsen complex immobilized on Al-MCM-41. The highest conversions, yields, and selectivities were obtained over the catalysts immobilized by impregnation (see Table 2.1.6.5). For the ring opening of styrene oxide the main products were found to be (S)-styrene oxide and (R)-styrene glycol.

The enantioselective ring opening of *meso*-epoxides such as cyclohexene oxide, which uses benzoic acid as the nucleophile, was also investigated (see Scheme 2.1.6.6) [57]. The main product was found to be the (R,R)-1,2-cyclohexanediol monobenzoate ester (see Table 2.1.6.6).

Table 2.1.6.5 HKR of styrene oxide.

	Conversion [%]	Yield [%]	Selectivity [%]	ee diol [%]	ee epoxide [%]
Literature (ref. [56])	–	95	–	99	99
Homogeneous (S, S)-Co(II)–Jacobsen complex	41	40	98	100	71
Immobilized ion-exchanged (S, S)-Co-(II)–Jacobsen complex	12	11	89	11	5
Immobilized impregnated (S, S)-Co(II)–Jacobsen complex	94	94	100	100	–

Table 2.1.6.6 HKR of cyclohexene oxide.

	Conversion [%]	Yield [%]	Selectivity [%]	ee ester [%]
Literature (ref. [57])	–	75	–	98
Homogeneous (S, S)-Co(II)–Jacobsen catalyst	51	51	100	90
Immobilized ion-exchanged (S, S)-Co(II)–Jacobsen complex	12	11	89	90
Immobilized impregnated (S, S)-Co(II)–Jacobsen complex	17	17	100	90

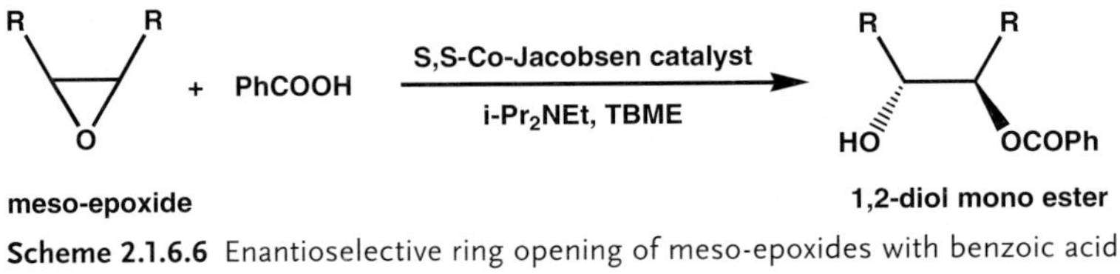

Scheme 2.1.6.6 Enantioselective ring opening of meso-epoxides with benzoic acid.

The (S,S)-Co(II)–Jacobsen complex was immobilized on Al-MCM-41 by both impregnation and ion exchange. For the ion-exchange and impregnation methods different solvents, several metal salts, and different concentrations were tested in order to avoid a structure collapse of the support material. The catalysts were indeed heterogenized inside the pores of Al-MCM-41, as could be shown by nitrogen sorption.

The catalytic activity of the (S,S)-Co(II)–Jacobsen complex immobilized by impregnation was found to be higher in the HKR of *meso* and terminal epoxides than the activity of the same complex immobilized by ion exchange.

Several different immobilization methods are currently under investigation in order to immobilize various types of Jacobsen complexes in the mesopores of Al-MCM-48, Al-MCM-41 and Al-SBA-15 types of support materials. The novel chiral heterogeneous catalysts obtained will be characterized and their activity in different test reactions will be investigated.

2.1.6.5 Conclusions

Different immobilization methods were employed, and the new heterogenized catalysts were applied in asymmetric hydrogenation and epoxidation of olefins, as well as in ring-opening reactions of epoxides. In some cases, the typical catalytic properties of homogeneous catalysts could be transferred to the heterogeneous systems. The immobilization of complexes on solid supports facilitated recovery and recycling of the catalysts.

The immobilization of chiral rhodium–diphosphine complexes on aluminum-containing mesoporous materials (e.g., Al-MCM-41, Al-MCM-48, Al-SBA-15) was demonstrated. The bonding occurred via an ionic interaction of the cationic complex with the host structure. A slight reduction of weak acidic sites of carrier materials was observed. These novel catalysts were suitable for asymmetric hydrogenation of functionalized olefins. The organometallic complexes remained stable within the mesopores of the carrier under reaction conditions. It was possible to achieve high yields and high enantiomeric excesses which were as good as in the case of the homogeneous counterparts.

The second example demonstrated immobilization via "ship in a bottle", ionic, metal center, and covalent bonding approaches of the metal–salen complexes. Zeolites X and Y were highly dealuminated by a succession of different dealumination methods, generating mesopores completely surrounded by micropores. This method made it possible to form cavities suitable to accommodate bulky metal complexes. The catalytic activity of transition metal complexes entrapped in these new materials (e.g, **Mn-S**, **V-S**, **Co-S**, **Co-S1**) was investigated in stereoselective epoxidation of (–)-α-pinene using O_2/pivalic aldehyde as the oxidant. The results obtained with the entrapped organometallic complex were comparable with those of the homogeneous complex.

The catalysts immobilized via ion exchange (MCM1HC), via the metal center (MCM2HC), or through the salen ligand (MCM3HC) were tested in epoxidation of (Z)-ethyl cinnamate, styrene, and 1,2-dehydronaphtalene using NaOCl, *m*-CPBA or dimethyldioxirane generated in situ as the oxygen source. Two of these materials (MCM1HC and MCM2HC) showed moderate leaching, probably due to degradation of the complex under severe oxidative reaction conditions. However, when DMD generated in situ was used as the oxidizing agent the degradation of salen ligand was reduced. By using covalent attachment on the salen ligand for immobilization, the heterogenized catalyst (MCM3HC) was stable against leaching and could be reused four times without significant loss of activity or enantioselectivity.

The third investigation track demonstrated the immobilization of metal–salen complexes in mesoporous materials and their use in the hydrolytic kinetic resolution of *meso* and terminal epoxides. The best results were obtained over cobalt–Jacobsen catalysts. The catalytic activity of the (S,S)-Co(II)–Jacobsen complex immobilized on Al-MCM-41 was comparable with that of the homogeneous counterpart. Several other immobilization methods are still under investigation.

Acknowledgement

The authors are very grateful for the financial support from the Sonderforschungsbereich SFB 380 of the Deutsche Forschungsgemeinschaft (DFG).

References

1 S. C. Stinson, *Chem. Eng. News* **2001**, *79*, 79.
2 A. M. Rouhi, *Chem. Eng. News* **2003**, *81*, 45.
3 A. M. Rouhi, *Chem. Eng. News* **2004**, *82*, 47.
4 Nobel lectures, *Angew. Chem. Int. Ed.* **2002**, *41*, 998.
5 F. J. Waller, *Chem. Ind.* **2003**, *89*, 1.
6 H. U. Blaser, B. Pugin, M. Studer, Enantioselective heterogeneous catalysis: academic and industrial challenges, in: D. E. de Vos, I. F. J. Vankelecom, P. A. Jacobs (Eds.), *Chiral Catalyst Immobilization and Recycling*, Wiley-VCH, Weinheim, **2000**, p. 1.
7 B. Pugin, H. U. Blaser, Catalyst immobilization: solid supports, in: E. N. Jacobsen, A. Pfaltz, H. Yamamoto (Eds.), *Comprehensive Asymmetric Catalysis I–III*, Vol. 3, Springer, Berlin, **1999**, p. 1367.
8 R. A. Sheldon, *Cur. Opin. Solid State Mater. Sci.* **1996**, *1*, 101.
9 S. Braese, F. Lauterwasser, R. E. Ziegert, *Adv. Synth. Catal.* **2003**, *345*, 869.
10 M. Benaglia, A. Puglisi, F. Cozzi, *Chem. Rev.* **2003**, *103*, 3401.
11 R. Arshady, *Microspheres Microcapsules Liposomes* **1999**, *1*, 197.
12 D. E. Bergbreiter in: D. E. de Vos, I. F. J. Vankelecom, P. A. Jacobs (Eds.), *Chiral Catalyst Immobilization and Recycling*, Wiley-VCH, Weinheim, **2 000**, p. 43.
13 R. Van Heerbeeck, P. C. J. Kamer, P. W. N. M. van Leeuwen, J. N. H. Reek, *Chem. Rev.* **2002**, *102*, 3717.
14 A. K. Kakkar, *Chem. Rev.* **2002**, *102*, 3717.
15 Q.-H. Fan, Y.-M. Li, A. S. C. Chan, *Chem. Rev.* **2002**, *102*, 3385.
16 D. E. Bergbreiter, *Chem. Rev.* **2002**, *102*, 3345.
17 C. A. McNamara, M. J. Dixon, M. Bradley, *Chem. Rev.* **2002**, *102*, 3275.
18 N. E. Leadbeater, M. Marco, *Chem. Rev.* **2002**, *102*, 3275.
19 IFJ. Vankelekom, P. A. Jacobs, Catalyst immobilization on inorganic supports, in: D. E. de Vos, I. F. J. Vankelecom, P. A. Jacobs (Eds.), *Chiral Catalyst Immobilization and Recycling*, Wiley-VCH, Weinheim, **2000**, p. 19.
20 C. E. Song, S.-G. Lee, *Chem. Rev.* **2002**, *102*, 3495.

21 D. E. de Vos, M. Dams, B. F. Sels, P. A. Jacobs, *Chem. Rev.* **2002**, *102*, 3615.
22 S. Anderson, H. Yang, S. K. Tanielyan, R. L. Augustine, *Chem. Ind.* **2001**, *82*, 557.
23 W. Keim, B. Dreissen-Holscher, Supported catalysts. Deposition of active component. Heterogenization of complexes and enzymes, in: G. Ertl, H. Knoezinger, J. Weitkamp (Eds.), *Preparation of Solid Catalysts*, Wiley-VCH, Weinheim, **1999**, p. 355.
24 D. E. de Vos, B. F. Sels, P. A. Jacobs, *Adv. Catal.* **2001**, *46*, 1.
25 W. F. Hoelderich, H. H. Wagner, M. H. Valkenberg, *Special Publication – Royal Society of Chemistry* **2001**, *266*, 76.
26 R. Haag, *Angew. Chem. Int. Ed.* **2002**, *41*, 520.
27 A. P. Wight, M. E. Davis, *Chem. Rev.* **2002**, *102*, 3589.
28 R. Augustine, S. Tanielyan, S. Anderson, H. Yang, *Chem. Commun.* **1999**, 1257.
29 D. E. de Vos, F. Thibault-Starzyk, P. P. Knops-Gerrits, R. F. Parton, J. A. Jacobs, *Macromol. Symp.* **1994**, *80*, 157.
30 W. S. Knowles, M. Sabacky, *Chem. Commun.* **1968**, 1445.
31 L. Horner, H. Buethe, *Angew. Chem. Int. Ed. Engl.* **1968**, *7*, 941.
32 J. A. Osborn, F. H. Jardine, J. F. Young, G. J. Wilkinson, *J. Chem. Soc.* **1968**, 1711.
33 H. Kagan, N. Langlois, T.-P. Dang, *J. Organomet. Chem.* **1975**, *90*, 353.
34 W. S. Knowles, M. Sabacky, B. D. Vineyard, *J. Chem. Soc.* **1972**, 10.
35 D. Zhao, J. Feng, Q Huo, N Melosh, G. H. Frederickson, B. F. Chmelka, G. D. Stucky, *Science* **1998**, *279*, 548.
36 D. Zhao, Q. Huo, J. Feng, B. F. Chmelka, G. D. Stucky, *J. Am. Chem. Soc.* **1998**, *120*, 6024.
37 J. S. Beck, J. C. Vartulli, W. J. Roth, M. E. Leonowicz, C. T. Kresge, K. D. Schmitt, C. T.-W. Chu, D. H. Olson, E. W. Sheppard, S. B. McCullen, J. B. Higgins, J. L. Schlenker, *J. Am. Chem. Soc.* **1992**, *114*, 10 834.
38 Z. Luan, M. Hartmann, D; Zhao, W. Zhou, L. Kevan, *Chem. Mater.* **1999**, *11*, 1621.
39 M. Cheng, Z. Wang, K. Sakurai, F. Kumata, T. Saito, T. Komatsu, T. Yashima, *Chem Lett.* **1999**, 131.
40 Z. Luan, C.-F. Cheng, W. Zhou, J. Klinowski, *J. Phys. Chem.* **1995**, *99*, 1018.
41 G. A. Eimer, L. B. Pierella, G. A. Monti, O. A. Anunziata, *Cat. Lett.* **2002**, *78*, 65.
42 S. S. Kim, W. Zhang, T. J. Pinnavaia, *Catal. Lett.* **1997**, *43*, 149.
43 A. Janssen, J. P. M. Niederer, W. F. Hölderich, *Catal. Lett.* **1997**, *48*, 165.
44 H. E. B. Lempers, R. A. Sheldon, *J. Catal.* **1998**, *175*, 62.
45 V. Schürig, F. Betschinger, *Chem. Rev.* **1992**, *92*, 873.
46 C. Limberg, *Angew. Chem. Int. Ed.* **2003**, *42*, 5932.
47 C. Baleizão, B. Gigante, M. J. Sabater, H. García, A. Corma, *J. Appl. Catal A* **2002**, *228*, 279.
48 E. N. Jacobsen, A. R. Muci, N. H. Lee, *Tetrahedron Lett.* **1991**, *32*, 5055.
49 W. Kahlen, H. H. Wagner, W. F. Hoelderich, *Catal. Lett.* **1998**, *54*, 58.
50 H. K. Beyer, I. M. Belenykaja, *Stud. Surf. Sci. Catal.* **1980**, *5*, 203.
51 C. Heinrichs, W. F. Hoelderich, *Catal. Lett.* **1999**, *58*, 75.
52 P. Piaggio, P. Morn, D. Murphy, D. Bethell, P. C. Pages, F. E. Hancock, C. Sly, O. J. Kerton, G. J. Hutchings, *J. Chem. Soc., Perkin. Trans.* **2000**, *2*, 2008.
53 G.-J. Kim, J.-H. Shin, *Tetrahedron Lett.* **1999**, *40*, 6827.
54 C. Schuster, E. Möllmann, A. Tompos, W. F. Hoelderich, *Catal. Lett.* **2001**, *74*, 69.
55 J. Cubillos, Ph. D. Thesis, University of Aachen, **2005**.
56 M. Tokunaga, J. F. Larrow, F. Kakiuchi, E. N. Jacobsen, *Science* **1997**, *277*, 936.
57 E. N. Jacobsen, F. Kakiuchi, J. F. Larrow, M. Tokunaga, *Tetrahedron Lett.* **1997**, *38*, 773.
58 G.-J. Kim, D.-W. Park, *Catal. Today* **2000**, *63*, 537.
59 J. M. Keith, J. F. Larrow, E. N. Jacobsen, *Adv. Synth. Catal.* **2001**, *343*, 5.

2.2
Biological Methods

2.2.1
Directed Evolution to Increase the Substrate Range of Benzoylformate Decarboxylase from *Pseudomonas putida*

Marion Wendorff, Thorsten Eggert, Martina Pohl, Carola Dresen, Michael Müller, and Karl-Erich Jaeger

2.2.1.1 Introduction

Benzoylformate decarboxylase (BFD; EC 4.1.1.7) belongs to the class of thiamine diphosphate (ThDP)-dependent enzymes. ThDP is the cofactor for a large number of enzymes, including pyruvate decarboxylase (PDC), benzaldehyde lyase (BAL), cyclohexane-1,2-dione hydrolase (CDH), acetohydroxyacid synthase (AHAS), and (1R,6R)-2-succinyl-6-hydroxy-2,4-cyclohexadiene-1-carboxylate synthase (SHCHC), which all catalyze the cleavage and formation of C–C bonds [1]. The underlying catalytic mechanism is summarized elsewhere [2] (see also Chapter 2.2.3).

The mandelate pathway, in which mandelate **1** is degraded to benzoic acid **5**, is known in *Pseudomonas* and *Acinetobacter* species enabling these microorganisms to grow on mandelate as a sole carbon source. BFD is the third enzyme in the mandelate pathway (Scheme 2.2.1.1) and catalyzes the nonoxidative decarboxylation of benzoylformate **3** to benzaldehyde **4** and carbon dioxide. Benzaldehyde **4** is then oxidized to benzoic acid, which is further metabolized in the β-ketoadipate pathway and the citric acid cycle. BFD encoded by the gene *mdlC* is located within the tricistronic operon *mdlCBA* together with the genes *mdlA* encoding mandelate racemase and *mdlB* encoding (S)-mandelate dehydrogenase [3].

BFD from *Pseudomonas putida* has been characterized in detail with respect to its biochemical properties [4, 5] and 3D structure [6, 7]. Like other enzymes of this class, BFD is a homotetramer with a subunit size of about 56 kDa. The four active sites are formed at the interfaces of two subunits. The structure was published in 2003 [7] and contains the competitive inhibitor (R)-mandelate bound to the active sites, allowing model-based predictions about the interactions between active site residues and the substrate.

As a side reaction, BFD catalyzes the carboligation of aldehydes to form chiral 2-hydroxy ketones (Scheme 2.2.1.2) [4, 8]. The physiological role of this additional enzymatic activity is still unknown. The carboligation of benzaldehyde **4**, benzaldehyde derivatives, and acetaldehyde **6** was studied in detail [4]. BFD accepts benzaldehyde **4** as a donor substrate and acetaldehyde **6** as the acceptor substrate. The ligation product 2-(S)-hydroxypropiophenone (2-HPP) **7** is formed with an enantiomeric excess (*ee*) of 82–94%, depending on the benzaldehyde concentration and on the reaction temperature [4, 5]. With regard to 2-HPP formation, the related ThDP-dependent enzyme benzaldehyde lyase (BAL) catalyzes the same carboligation reaction but with reverse stereoselectivity [9, 10]. By using these two

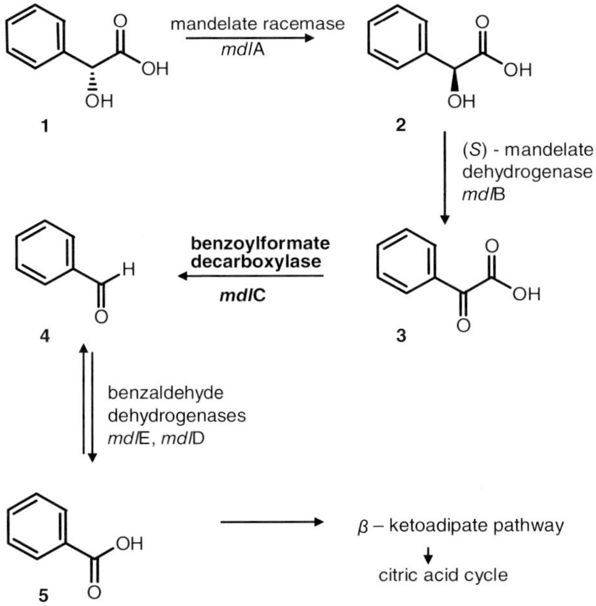

2.2.1 Directed Evolution to Increase the Substrate Range of Benzoylformate Decarboxylase

Scheme 2.2.1.1 Mandelate pathway: benzoylformate decarboxylase (BFD), encoded by the gene *mdlC*, catalyzes the conversion of benzoylformate **3** to benzaldehyde **4** and carbon dioxide.

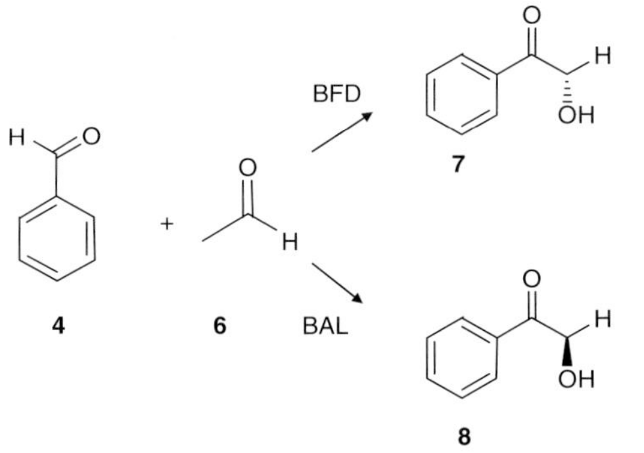

Scheme 2.2.1.2 BFD and BAL as enantiocomplementary catalysts: BFD and BAL act enantiocomplementarily in the formation of 2-HPP **7** and its derivatives, giving access to many 2-HPP analogues in either enantiomeric form.

enantiocomplementary enzymes, many 2-HPP analogues can be synthesized in both enantiomeric forms. The formation of (*R*)-benzoin (*R*)-**12** has also been reported as a ligation product of this reaction which is catalyzed by BFD, albeit with negligible activity [4].

The substrate range of BFD toward 2-substituted benzaldehydes could be increased by directed evolution, yielding a variant with a single amino acid

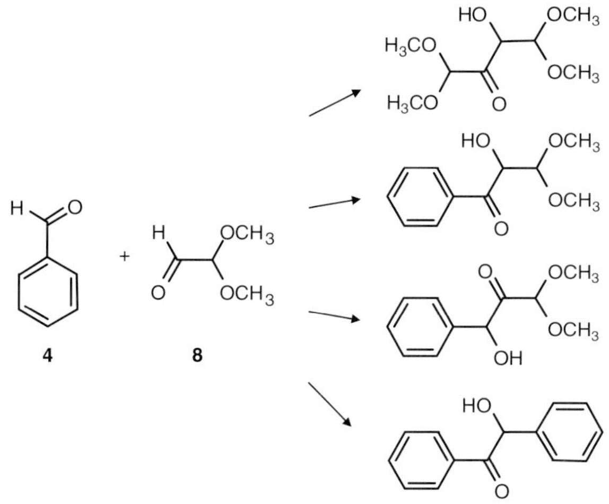

Scheme 2.2.1.3 Screening for novel carboligation activity: four enantiomeric products **9–12** are shown which can be formed starting from the substrates benzaldehyde **4** and dimethoxyacetaldehyde **8**.

exchange, L476Q. This variant also showed a higher stability in the presence of organic solvents [11]. By computer modeling studies amino acid L476 was identified as part of a surface-located loop close to the active site cavity. It was suggested that this residue may play a crucial role for substrate acceptance by exerting a gatekeeper-like function for the BFD active site [11]. At present, acetaldehyde is the only acceptor aldehyde known to be used by BFD variant L476Q.

Here we describe the results of an approach aimed to increase the substrate range of BFD with respect to more complex acceptor aldehydes. Firstly, a high-throughput screening assay had to be established making it possible to screen large libraries of BFD variants generated by directed evolution. Since we were interested in hydroxy ketones with additional functional side-chains, dimethoxyacetaldehyde **8** was used as an acceptor aldehyde which could lead to the carboligation products **9–12** shown in Scheme 2.2.1.3.

2.2.1.2 Materials and Methods

2.2.1.2.1 Reagents
Chemicals used in this work were purchased from Sigma-Aldrich (München, Germany). Restriction enzymes and T4 ligase were purchased from Eurogentec or Fermentas and used as recommended by the manufacturer. Oligonucleotides were obtained from Thermo Electron (Ulm, Germany).

2.2.1.2.2 Construction of Strains for Heterologous Expression of BFD and BAL
The gene encoding BFD variant L476Q was amplified by standard PCR using the upstream 33 bp primer PpBFD_*Nco*I, the downstream 34 bp primer PpBFD_*Hin*-

Table 2.2.1.1 Oligonucleotides used in this study.[a]

Oligonecleotide	Nucleotide sequence (5′ → 3′)	Modification
PpBFD_NcoI	ATAT**CCATGG**CTTCGGTACACGGCACCACATAC	NcoI
PpBFD_HindIII	ATAT**AAGCTT**CTTCACCGGGCTTACGGTGCTTAC	HindIII
MP_up_1.PCR	TTGTGGTGACCGT**CCATGG**CGATGATTAC	NcoI
MP_down_1.PCR	GGGCGCCGCCATCGCGACC	G → C
MP_up_2.PCR	TTGTGGTGACCGTCC	
MP_down_2.PCR	ATAT**CTCGAG**TGCGAAGGGGTCCATGC	XhoI
pKK233-2for	CACACAGGAAACAGA**CCATGG**	NcoI
pKK233-2rev	TCCGCCAAAACAGCC**AAGCTT**	HindIII

a Restriction sites are highlighted in bold letters, and base substitutions are underlined.

dIII (Table 2.2.1.1), and plasmid pKKBFDL476Q as the template. The unique NcoI and HindIII restriction sites located in the PCR primers were used for cloning the resulting 1632 bp PCR product into the corresponding restriction sites of the expression plasmid pET28a (Table 2.2.1.2) leading to an in-frame fusion of a C-terminal 6xHis-Tag. The resulting plasmid was named pETBFDL476Q (Table 2.2.1.2).

The gene *bznB* encoding BAL was mutated by site-directed mutagenesis in order to modify an internal NcoI restriction site by a silent base exchange. The resulting gene was also cloned into the expression vector pET28a using the restriction sites NcoI and XhoI, resulting in plasmid pETBAL_NcoI. Both plasmids were transformed into the expression host *E. coli* BL21 (DE3), resulting in the expression strains *E.coli* pETBFDL476Q and *E.coli* pETBAL (Table 2.2.1.2).

2.2.1.2.3 Polymerase Chain Reactions

Standard PCR Amplification of DNA fragments was performed in a 50 μL reaction mix containing 1 ng of plasmid DNA as the template, 25 pmol of each primer, 0.2 mM dNTPs, 2.5 U of *Pfu* polymerase (Stratagene, Heidelberg, Germany). The buffer was used as recommended by the manufacturer. Conditions for PCR were as follows: 1 × (5 min, 98 °C); 35 × (1 min, 95 °C; 1 min, 58 °C; 1.5 min, 72 °C); and 1 × (7 min, 72 °C). The PCR reaction was performed using a Mastercycler Gradient (Eppendorf, Hamburg, Germany).

Site-directed mutagenesis Site-directed mutagenesis was performed using megaprimer PCR in order to eliminate the internal NcoI restriction site located within the *bznB* gene encoding BAL. Two consecutive PCR reactions had to be carried out. In the first reaction, the primers MP_up_1.PCR and MP_down_1.PCR were used with MP_down_1.PCR carrying a single base mutation resulting in a silent base exchange. MP_up_1.PCR carried an artificial extension upstream of the start codon, producing a docking sequence for the upper primer MP_up_2.PCR in the second PCR reaction and thereby avoiding a template switch. The

Table 2.2.1.2 Bacterial strains and plasmids used in this study.

Strain or plasmids	Genotype or description	Reference
Strains		
E.coli BL21 (DE3)	F⁻ omp T hsdS$_B$(r$_B^-$ m$_B^-$) gal dcm (λclts857 ind1 Sam7 nin5 lacUV5-T7 gene 1)	Novagen, Madison, USA
E.coli BAL	E.coli BL21 (DE3) carrying the plasmid pETBAL_NcoI	this study
E.coli BFDL476Q	E.coli BL21 (DE3) carrying the plasmid pETBFDL476Q	this study
E.coli pET28a	E.coli BL21 (DE3) carrying the empty vector pET28a	this study
Plasmids		
pET28a	ColE1 P$_{T7lac}$ Kan^r	Novagen, Madison, USA
pKKBFDL476Q	mdlCL476Q encoding BFD variant L476Q cloned by NcoI/HindIII into plasmid pKK233-2	ref. [11]
pETBFDL476Q	mdlCL476Q cloned by NcoI/HindIII into pET28a, fusion of a C-terminal 6x-His-tag	this study
pETBAL	bznB from *Pseudomonas fluorescens* encoding BAL cloned by NdeI/XhoI into pET22b	M. Wendorff, unpublished results
pETBAL_NcoI	bznB without internal NcoI restriction site encoding BAL cloned by NcoI/XhoI to pET28a, fusion of a C-terminal H6 tag	this study

second PCR reaction was performed using the megaprimer produced in the first PCR reaction together with MP_up_2.PCR as the upper and MP_down_2.PCR as the lower primer. This PCR reaction yielded as the full length product the *bznB* gene without an internal *NcoI* restriction site. Plasmid pETBAL was used as the template for both PCRs, which were carried out under standard conditions. The annealing temperatures were 60 °C in the first and 55 °C in the second PCR reaction. All the primers are listed in Table 2.2.1.1.

2.2.1.2.4 Generation of a BFD Variant Library by Random Mutagenesis

Random mutagenesis of the gene encoding BFD variant L476Q was performed with the error-prone polymerase chain reaction (epPCR) [11, 14]. Oligonucleotides pKK233-2for and pKK233-2rev (Table 2.2.1.1) were used as primers. An error rate of two to four base substitutions per gene was achieved by using a reaction mix comprising: 5 pmol of each primer, 75 mM Tris/HCl buffer (pH 8.8), 20 mM (NH$_4$)$_2$SO$_4$, 1.5 mM MgCl$_2$, 0.5 mM MnCl$_2$, 0.2 mM dNTPs, 0.01% Tween 20, 1 ng template DNA (pKKBFDL476Q) (Table 2.2.1.2), and 10 U Goldstar *Taq*-polymerase (Eurogentec, Seraing, Belgium). The PCR reaction was performed as described above (annealing temperature 52 °C) using a Mastercycler Gradient (Eppendorf, Hamburg, Germany). A higher diversity of mutated BFD genes was achieved by mixing mutated genes obtained from three to five different epPCRs and subsequent ligation. The plasmids carrying the mutated DNA fragments were then transformed into the expression host *E.coli* BL21 (DE3) and the bacteria were plated on selective LB medium containing kanamycin (50 µg/ml).

2.2.1.2.5 High-Throughput Screening for Carboligation Activity with the Substrates Benzaldehyde and Dimethoxyacetaldehyde

The mutated BFD genes were cloned into the vector pET28a (Table 2.2.1.2) and the resulting plasmids were transformed into *E.coli* BL21 (DE3) (Table 2.2.1.2). The clones were transferred into 96 deep-well microtiter plates filled with 100 µL LB medium (10 g NaCl, 5 g yeast extract, 10 g trypton) supplemented with 50 µg mL^{-1} kanamycin. After overnight shaking (600 rpm) at 37 °C, 600 µL LB medium containing kanamycin (50 µg mL^{-1}) were added. After an additional 2.5 h of shaking (600 rpm) at 37 °C the expression was induced by adding isopropyl-β-D-thiogalactoside (IPTG) to a final concentration of 0.8 mM. The induced culture was grown for another 24 h at 37 °C. For storage, a masterplate was generated by transferring the clones on a fresh LB agarplate supplemented with kanamycin (50 µg mL^{-1}). The cells were separated from the culture supernatant by centrifugation at 4000 rpm for 20 min. The supernatant was removed and 300 µL of potassium phosphate buffer (50 mM KP$_i$, pH 7) containing cofactors (1 mM ThDP, 5 mM MgSO$_4$) was added. Cell lysis was achieved by adding 1 mg mL^{-1} lysozyme and incubating at 30 °C for 30 min on a shaker. A 100 µL aliquot of each crude extract was transferred into a fresh 96-well plate, and 100 µL of substrate solution (50 mM KP$_i$, pH 7, 40 mM benzaldehyde, 120 mM dimethoxyacetaldehyde) was added. After 24 h of incubation at 30 °C, the reaction mixture was loaded into a new microtiter plate containing 10 µL of 2,3,5-triphenyltetrazolium chloride (0.4% in EtOH). The color reaction was started by adding 30 µL 3 M NaOH [16].

2.2.1.2.6 Expression and Purification of BFD Variants

An overnight culture of *E. coli* pBFDL476Q (Table 2.2.1.2) containing pBFD55E4 was diluted 1 : 100 to inoculate 3 × 1 L TB medium (12 g tryptone, 24 g yeast extract, 17 mM KH$_2$PO$_4$, 72 mM K$_2$HPO4, 4 mL glycerin) supplemented with 50 µg mL^{-1} kanamycin in a 5 L Erlenmeyer flask. At an optical density of OD$_{580}$ = 1, IPTG was added to a final concentration of 1 mM. The culture was incubated for 24 h at 37 °C. The cells were harvested by centrifugation (Hettich Rotina 35R, 5000 rpm, 20 min), and potassium buffer (50 mM KP$_i$ pH 6, 2.5 mM MgSO$_4$, 0.5 mM ThDP) was added to prepare a 30% w/v solution. Cells were disrupted by ultrasonication at 140 W for 10 × 60 s and the insoluble components were separated by centrifugation (Sorvall RC-5B, SS34, 16 000 rpm, 30 min). The BFD variants were purified from the crude extracts using Ni-NTA (Qiagen, Hilden, Germany) and subsequent gel filtration chromatography (Amersham

Biosciences, Uppsala, Sweden). The purification was performed as described elsewhere [11]. Finally, the purified enzyme variants were lyophilized and stored at −20 °C.

2.2.1.2.7 Protein Analysis Methods

Protein determination Protein concentrations were determined according to Bradford [12] using BSA as a standard.

SDS-polyacrylamide gel electrophoresis Protein samples were analyzed under denaturing conditions in a discontinuous gel system as described by Laemmli [13] using a 5% stacking gel and a 12% separating gel in a vertical gel system (BioRad, Mini Protean II, CA, USA).

2.2.1.2.8 Enzyme Activity Assays

Decarboxylase activity The decarboxylation of benzoylformate was studied using a coupled enzyme test as described previously [4].

Carboligase activity toward benzaldehyde and dimethoxyacetaldehyde Two different assays were performed: (1) 0.6 mg of purified BFD variant L476Q or 0.5 mg of purified BFD variant 55E4 were incubated in 1.2 mL of 50 mM KP_i (pH 7.0), containing 0.1 mM ThDP, 2.5 mM $MgSO_4$ with 20% DMSO in the presence of 5 or 20 mM benzaldehyde at 30 °C and shaking at 300 rpm. After 19 h the reaction was stopped by extraction with 300 µL ethyl acetate. (2) 0.6 mg purified BFD variant L476Q or 0.5 mg BFD variant 55E4 were incubated in 1.2 mL of 50 mM KP_i (pH 7.0) containing 0.1 mM ThDP, 2.5 mM $MgSO_4$ with 20% DMSO in the presence of 5 or 20 mM benzaldehyde and 60 or 500 mM dimethoxyacetaldehyde at 30 °C and shaking at 300 rpm. After four days the reaction was stopped by extraction with 200 µL ethyl acetate.

The concentrations of benzaldehyde, the mixed product, and benzoin and the enantiomeric excesses (*ee* values) were determined by chiral-phase HPLC with a photodiode array detector. Chiral-phase HPLC was performed on a Chiracel OD-H (Daicel, Düsseldorf, Germany) using isohexane/isopropanol (90 : 10) as eluent, a flow rate of 0.5 mL min^{-1} and a column oven at 40 °C. The retention time for benzaldehyde was 10.2 min, for the mixed product DMA–HPP 15.1 and 16.4 min, for the (*S*)-benzoin 20.3 min, and for the (*R*)-benzoin 28.5 min.

2.2.1.3 Results and Discussion

2.2.1.3.1 Overexpression of BFD in *Escherichia coli*
In previous studies on benzoylformate decarboxylase (BFD) from *Pseudomonas putida*, the enzyme was expressed using the pKK233-3 vector system and *E. coli* SG13009 as the expression host. Here, expression was controlled by two plasmids,

namely the expression plasmid pKK233-2 and plasmid pRep4 [4]. In an attempt to avoid experimental problems occurring in directed evolution and high-throughput screening, we have constructed novel expression strains based on the pET system, thereby superseding the presence of a second plasmid which might cause problems during the isolation of the expression plasmid.

The gene encoding BFD variant L476Q was cloned into the *NcoI*/*Hind*III sites of pET28a leading to plasmid pETBFDL476Q, which was transferred into *E. coli* BL21(DE3) by transformation giving the parental clone for the random mutagenesis experiment. Benzaldehyde lyase (BAL) from *Pseudomonas fluorescens* is known to catalyze the carboligation of benzaldehyde and dimethoxyacetaldehyde; hence, a BAL-expressing *E. coli* BL21 (DE3) strain was constructed (Table 2.2.1.2) to serve as a positive control for the high-throughput screening assay. For BAL being cloned into pET28a, an internal *NcoI* restriction site had to be eliminated first by site-directed mutagenesis. Afterward the modified gene was cloned into pET28a by flanking *NcoI* and *XhoI* restriction sites.

2.2.1.3.2 Random Mutagenesis of BFD Variant L476Q

BFD variant L476Q was chosen as the parental enzyme for optimization by directed evolution. As compared to wild-type BFD, variant L467Q showed an increased substrate range toward 2-substituted benzaldehydes [14]. Error-prone PCR (epPCR) was used for the creation of a random mutant library. The entire gene encoding BFD variant L476Q was subjected to mutation, as previous studies had demonstrated that amino acids which affect given enzyme properties may not be located exclusively in the vicinity of the active site but occur scattered over the entire enzyme [15]. The reaction conditions for epPCR were adjusted by using increased $MgCl_2$ and $MnCl_2$ concentrations, resulting in one to three amino acid exchanges per BFD subunit.

2.2.1.3.3 Development of a High-Throughput Screening Assay for Carboligase Activity

Growth conditions in deep-well microtiter plates were optimized with respect to optimal expression of active enzymes (Fig. 2.2.1.1). The best results were obtained with an expression time of 20 h at 37 °C (Fig. 2.2.1.1, lanes 7–9). Subsequently, *E. coli* cells were enzymatically disrupted by lysozyme treatment, and the carboligase activity was monitored by a modified tetrazolium salt color assay [16]. This color assay is based on the reduction of the 2,3,5-triphenyltetrazolium chloride (TTC) **13** to the corresponding formazan **15**, which has an intense red color (Fig. 2.2.1.2A). Before screening of a BFD variant library, substrates and products were tested in the color assay. Neither substrate, benzaldehyde **4** nor dimethoxyacetaldehyde **8**, reduced TTC **13**; however, the product 2-hydroxy-3,3-dimethoxypropiophenone **10** already caused color formation at low concentrations of 2.5–10 mM (Fig. 2.2.1.2B). Benzoin **12** as the product also gave a color change at a similar concentration (data not shown).

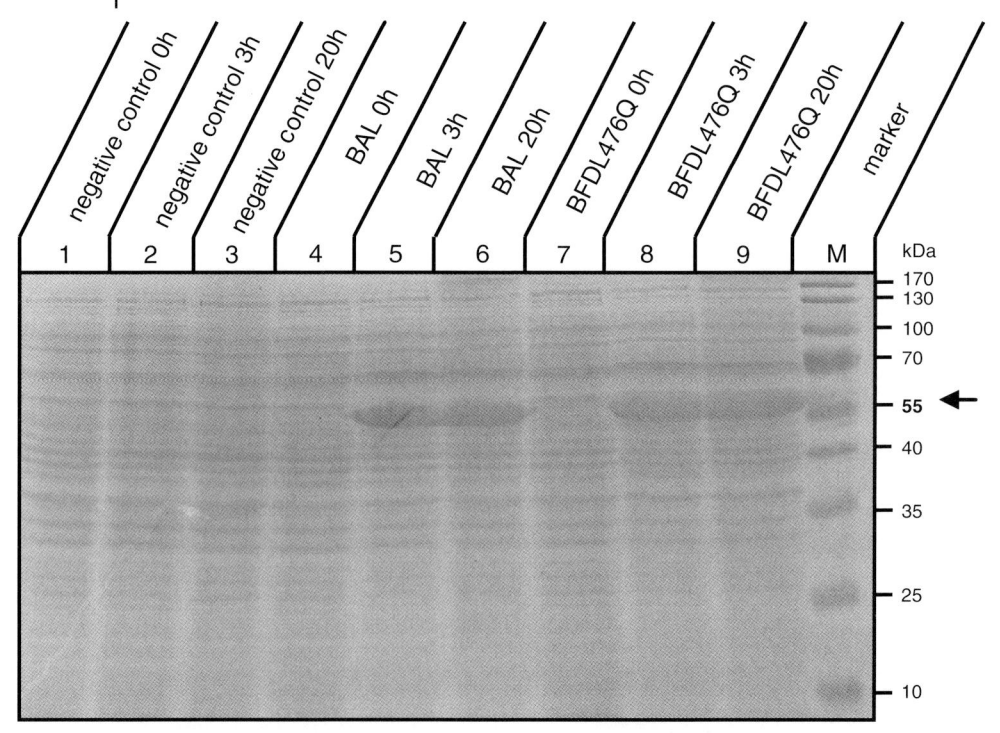

Fig. 2.2.1.1 SDS-PAGE analysis of cell extracts from *E. coli* overexpressing BAL and BFD. The expression cultures were induced (0 h) and grown for 3 and 20 h. Cell extracts (corresponding to $OD_{580} = 0.15$ of the bacterial culture) were harvested by centrifugation, and the proteins were separated on a 12% (w/v) polyacrylamide gel, and stained with Coomassie brilliant blue. The protein bands representing either BFD or BAL are marked with an arrow. Cell extracts from *E. coli* pET28, which served as a negative control, did not show a prominent protein band whereas the expression strains *E. coli* BAL and *E. coli* BFDL476Q showed a protein band of similar size after expression times of 3 h and 20 h.

Additionally, we tested cell extracts in the colorimetric assay: both negative control strains *E. coli* pET28a containing the empty vector and *E. coli* BFDL476Q (Table 2.2.1.2) did not show any enzyme activity in the TTC assay, whereas *E. coli* BAL (Table 2.2.1.2) led to the formation of an intense red color (data not shown).

2.2.1.3.4 Identification of a BFD Variant with an Optimized Acceptor Aldehyde Spectrum

The BFD mutant library generated by epPCR was expressed in microtiter plates and 8000 clones were subjected to the carboligation assay with benzaldehyde **4** and dimethoxyacetaldehyde **8** as the substrates. The reaction was incubated for 24 h at 30 °C as suggested from experiments using the positive control strain

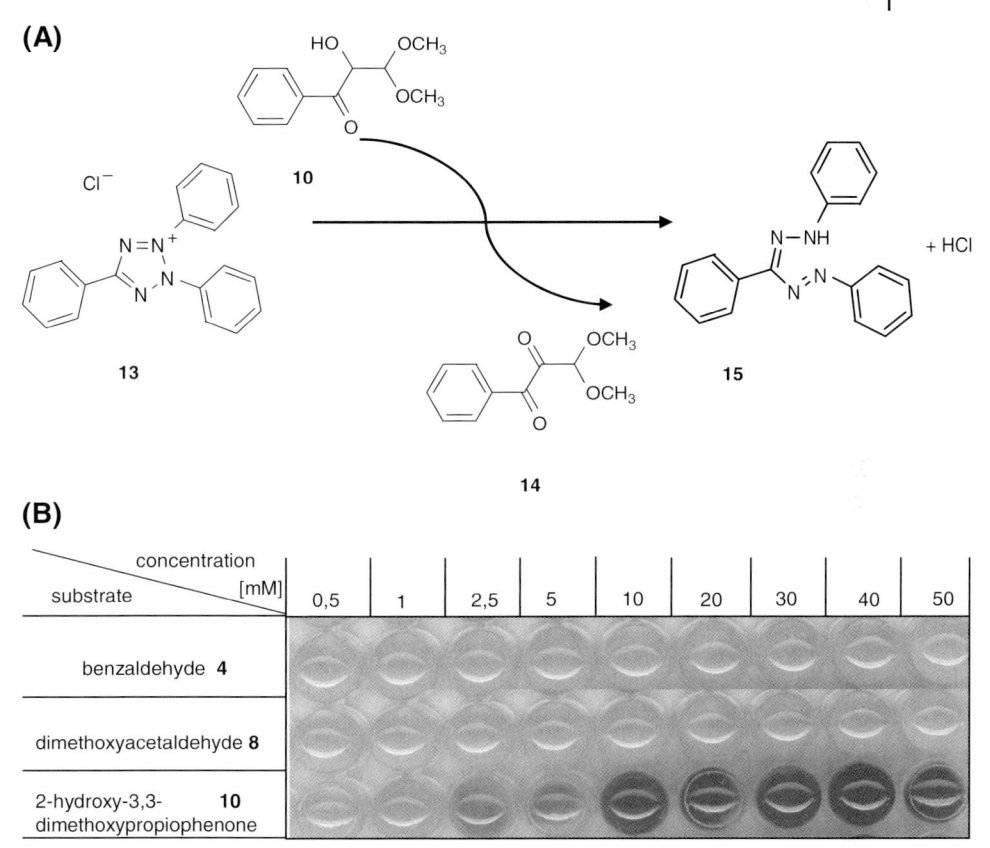

2.2.1 Directed Evolution to Increase the Substrate Range of Benzoylformate Decarboxylase

Fig. 2.2.1.2 Colorimetric assay for carboligase activity. A) DMA–HPP **10** reduces 2,3,5-triphenyltetrazolium **13** chloride to the respective formazan **15**, which has an intense red color. B) Formation of the red formazan **15** dye can be observed only in the presence of DMA–HPP **10**; neither substrate dimethoxyacetaldehyde **8** nor benzaldehyde **4** causes any change in color.

E.coli pETBAL (Table 2.2.1.2). Clone 55E4 showed activity with these substrates; therefore, the respective DNA insert was sequenced, revealing four base exchanges in addition to the CTG to CAG mutation which was present in the parental BFD variant L476Q. Two of the mutations resulted in amino acid exchanges: (1) the base triplet GCT at amino acid position 255 was mutated to GGT, now encoding glycine instead of alanine; and (ii) base triplet ATT at amino acid position 260 was mutated to ACT, exchanging isoleucine for threonine. In addition, two silent mutations were located at codons CTG (position 387) and GGT (position 491).

Both amino acid exchanges are located in the β-domain of the subunits close to the active site channel. The $Ala_{255}Gly$ exchange is found right at the beginning of an α-helix, whereas the $Ile_{260}Thr$ exchange is in the middle of the same α-helix. Both amino acids are located close to the substrate channel leading to the active site of BFD, so a direct effect on the active site functionality is therefore feasible. Modeling studies initiated in collaboration with Jürgen Pleiss's group (Institute of Technical Biochemistry, University of Stuttgart, Germany) will give more detailed insights into potential effects of these mutations.

2.2.1.3.5 Biochemical Characterization of the BFD Variants

Both BFD variants L476Q and 55E4 were overexpressed in *E. coli* BL21 (DE3) as His-tag fusion proteins and purified to electrophoretic homogeneity as detected by SDS-PAGE by metal chelate affinity chromatography using Ni-NTA (data not shown).

2.2.1.3.6 Decreased Benzoyl Formate Decarboxylation Activity of Variant 55E4

The influence of the mutations on the decarboxylation activity was investigated using the natural substrate benzoylformate **3** (Scheme 2.2.1.1). BFD variant L476Q showed a 1.4-fold higher decarboxylase activity (324 U mg^{-1}) toward benzoylformate **3** than variant 55E4 (233 U mg^{-1}).

2.2.1.3.7 Formation of 2-Hydroxy-3,3-dimethoxypropiophenone and Benzoin

A carboligation reaction with two different aldehydes as substrates may result in the formation of four different 2-hydroxy ketones (Scheme 2.2.1.3). Since the catalysis of benzoin **12** formation is known to be a weak side-activity of BFD variant L476Q, dimethoxyacetaldehyde was applied in excess (1 : 3 and 1 : 100) in order to suppress any possible benzoin formation. First investigations concerning the carboligase potential of variant 55E4 were performed in 1.2 mL batch reactions using the same substrate mixture as in the high-throughput screening assay (20 mM benzaldehyde, 60 mM dimethoxyacetaldehyde), and the reaction products were analyzed after four days by HPLC. The results showed that both DMA–HPP **10** and benzoin **12** were formed in the biotransformations and variant 55E4 showed higher enzyme activity. An estimate of the product yields revealed a 20-fold excess of DMA–HPP **10** obtained with variant 55E4 (Fig. 2.2.1.3).

Further studies, with the substrate ratio altered from 1 : 3 to 1 : 100, were performed in order to suppress the formation of benzoin **12** and increase the yield of DMA–HPP **10**. Although this strategy was successful, the results showed that even a 100-fold excess of dimethoxyacetaldehyde **8** (500 mM) relative to benzaldehyde **4** (5 mM) could not completely supress benzoin **12** formation catalyzed by variant 55E4. Furthermore, the formation of the mixed product DMA–HPP **10** was decreased by a factor of 2 relative to the 1 : 3 substrate mixture. Variant 55E4 showed a 55-fold higher productivity with respect to the formation of DMA–HPP **10** under these conditions as compared to variant L476Q. The overall carboligation

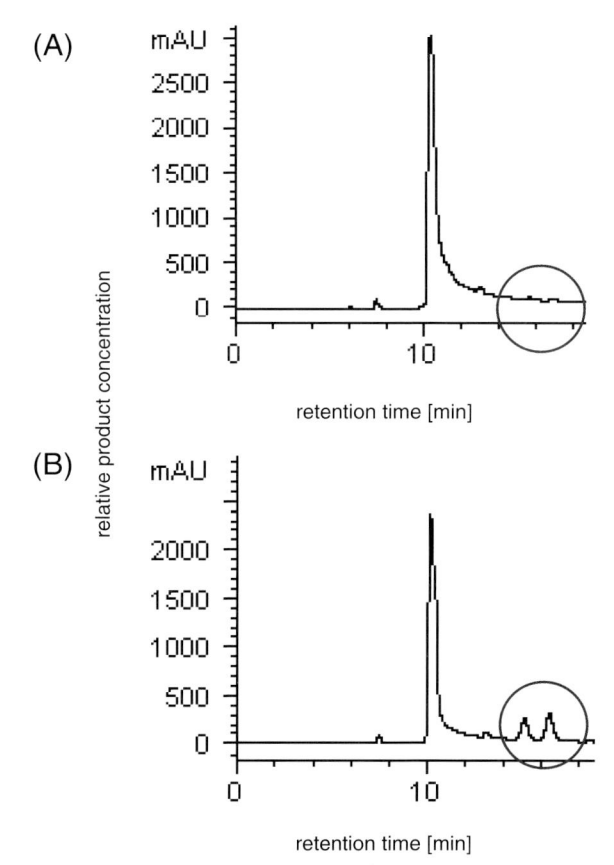

Fig. 2.2.1.3 HPLC diagram showing the formation of DMA–HPP catalyzed by variant BFD55E4. BFD variant enzymes were incubated with benzaldehyde **4** (20 mM) and dimethoxyacetaldehyde **8** (60 mM) for four days, and product formation was analyzed by chiral HPLC (see circles). A) BFD variant L476Q catalyzed the substrate conversion to DMA–HPP **10** only at a neglegible rate. B) BFD variant 55E4 catalyzed the formation of both enantiomers of DMA–HPP **10** with retention times of approx. 15.1 min for the (S)- and approx. 16.4 min for the (R)-enantiomer.

activity of BFD variant L476Q was hardly detectable. Also, the stability of both variants decreased in the presence of high aldehyde concentrations.

Benzoin formation was also investigated with benzaldehyde **4** as the substrate (at concentrations of 5 and 20 mM, respectively). Again, variant 55E4 showed a significantly higher benzoin-forming activity than variant L476Q (Table 2.2.1.3), which was even more pronounced at low benzaldehyde concentrations (5 mM), suggesting that the variant 55E4 has a higher affinity for benzaldehyde **4** than BFD variant L476Q. This observation coincides with the fact that benzoin forma-

Table 2.2.1.3 Synthesis of (R)-benzoin catalyzed by benzoylformate decarboxylase variants.

Substrates	BFD variant 55E4		BFD variant L476Q	
	Yield[a] [%]	ee[a] [%]	Yield[a] [%]	ee[a] [%]
5 mM benzaldehyde	48	99	3	>99
20 mM benzaldehyde	60	97	15	>99
20 mM benzaldehyde + 60 mM dimethoxyacetaldehyde	70	85	12	>99
5 mM benzaldehyde + 500 mM dimethoxyacetaldehyde	12	76	n.d.[b]	n.d.

a Benzaldehyde concentrations and the enantiomeric excesses (ee) of (R)-benzoin were determined by chiral-phase HPLC. Yields were calculated from amounts of benzaldehyde converted. b n.d.: the amount of benzoin formed was too low to calculate yield or ee.

tion is still catalyzed by variant 55E4 even in the presence of excess dimethoxyacetaldehyde **8** (see above).

2.2.1.3.8 Enantioselectivity of the Carboligation Reaction

Variant 55E4 proved to be more sensitive toward increasing aldehyde concentrations than variant L476Q in terms of the enantioselectivity of the (R)-**12** formation. The enantiomeric excess (ee) of the (R)-**12** products are summarized in Table 2.2.1.3. At the lowest benzaldehyde concentration (5 mM), variant 55E4 catalyzed the formation of enantiomerically pure (>99%) (R)-**12**. With increasing total aldehyde concentration, the ee decreased continuously to 76% at a total concentration of 505 mM (5 mM benzaldehyde and 500 mM acetaldehyde). Remarkably, increasing aldehyde concentrations had no effect on the enantioselectivity of the parental variant L476Q yielding enantiopure (R)-**12** independently of the reaction conditions. Surprisingly, variant 55E4 catalyzed the formation of both (R)- and (S)-DMA–HPP **10** in equal amounts as determined after four days of reaction time. As the enantioselectivity of (R)-**12** formation also decreased under these conditions (see above), it is at present difficult to explain the lack of enantioselectivity toward the product DMA–HPP **10**; it may represent an intrinsic property of variant 55E4 or it may result from destabilizing effects caused by the aldehyde which may finally lead to a distortion of the active site.

Effects of high concentrations of benzaldehyde **4** on the enantioselectivity of the carboligation reaction have been described previously [4, 5]. Siegert et al. described the increase of (R)-**7** for a BFD variant by increasing the concentration of benzaldehyde **4** from 1 to 10 mM, and Iding et al. were able to increase the enantioselectivity of wild-type BFD by keeping the concentration of benzaldehyde **4** at a low level. Surprisingly, acetaldehyde **6** showed no such effect. More recently, we have observed that the substrate dimethoxyacetaldehyde **8** affects the enantioselectivity of variant 55E4 in a similar way.

2.2.1.4 Conclusions

The new BFD variant 55E4 showed an increased enzyme activity and substrate range for acceptor aldehydes. Dimethoxyacetaldehyde **8** was accepted in carboligation reactions with benzaldehyde **4** as the donor aldehyde. The formation of DMA–HPP **10** was enhanced by a factor of 55 as compared to BFD variant L476Q, which was used as the parental enzyme for directed evolution. Furthermore, our studies clearly indicated that the enantioselectivity of the reaction was influenced by the total aldehyde concentration in the reaction and thus can be increased by alteration of the reaction conditions. At present, we are trying to correlate by molecular modeling the observed effects with the structural consequences of the amino acid exchanges that were randomly introduced into the BFD variant 55E4 by directed evolution.

References

1 M. Pohl, G. A. Sprenger, M. Müller, *Curr. Opin. Biotechnol.* **2004**, *15*, 335–342.
2 F. Jordan, *Nat. Prod. Rep.* **2003**, *20*, 184–201.
3 A. Y. Tsou, S. C. Ransom, J. A. Gerlt, D. D. Buechter, P. C. Babbitt, G. L. Kenyon, *Biochemistry* **1990**, *29*, 9856–9862.
4 H. Iding, T. Dünnwald, L. Greiner, A. Liese, M. Müller, P. Siegert, J. Grötzinger, A. S. Demir, M. Pohl, *Chem. Eur. J.* **2000**, *6*, 1483–1495.
5 P. Siegert, M. J. McLeish, M. Baumann, H. Iding, M. M. Kneen, G. L. Kenyon, M. Pohl, *Protein Eng. Des. Sel.* **2005**, *18*, 345–357.
6 M. S. Hasson, A. Muscate, M. J. McLeish, L. S. Polovnikova, J. A. Gerlt, G. L. Kenyon, G. A. Petsko, D. Ringe, *Biochemistry* **1998**, *37*, 9918–9930.
7 E. S. Polovnikova, M. J. McLeish, E. A. Sergienko, J. T. Burgner, N. L. Anderson, A. K. Bera, F. Jordan, G. L. Kenyon, M. S. Hasson, *Biochemistry* **2003**, *42*, 1820–1830.
8 R. Wilcocks, O. P. Ward, S. Collins, N. J. Dewdney, Y. Hong, E. Prosen, *Appl. Environ. Microbiol.* **1992**, *58*, 1699–1704.
9 A. S. Demir, M. Pohl, E. Janzen, M. Müller, *J. Chem. Soc. Perkin. Trans.* **2001**, *1*, 633–635.
10 A. S. Demir, O. Sesenoglu, E. Eren, B. Hosrik, M. Pohl, E. Janzen, D. Kolter, R. Feldmann, P. Dünkelmann, M. Müller, *Adv. Synth. Catal.* **2002**, *344*, 96–103.
11 B. Lingen, J. Grötzinger, D. Kolter, M. R. Kula, M. Pohl, *Protein Eng.* **2002**, *15*, 585–593.
12 M. M. Bradford, *Anal. Biochem.* **1976**, *72*, 248–254.
13 U. K. Laemmli, *Nature* **1970**, *227*, 680.
14 B. Lingen, D. Kolter-Jung, P. Dünkelmann, R. Feldmann, J. Grötzinger, M. Pohl, M. Müller, *ChemBioChem* **2003**, *4*, 721–726.
15 M. Bocola, N. Otte, K. E. Jaeger, M. T. Reetz, W. Thiel, *ChemBioChem* **2004**, *5*, 214–223.
16 M. Breuer, M. Pohl, B. Hauer, B. Lingen, *Anal. Bioanal. Chem.* **2002**, *374*, 1069–1073.

2.2.2
C–C-Bonding Microbial Enzymes: Thiamine Diphosphate-Dependent Enzymes and Class I Aldolases

Georg A. Sprenger, Melanie Schürmann, Martin Schürmann, Sandra Johnen, Gerda Sprenger, Hermann Sahm, Tomoyuki Inoue, and Ulrich Schörken

2.2.2.1 Introduction

Making and breaking of carbon–carbon (C–C) bonds is at the heart of asymmetric organic chemistry. However, especially in the field of carbohydrate chemistry with its highly functionalized polyhydroxy compounds, the tedious use of iterative protection and deprotection steps is often necessary. Here, due to their intrinsic regio- and stereospecificities, C–C-bonding enzymes could work as "able" tools in biocatalysis, provided that they are: avail*able*, in sustain*able*, afford*able*, and recogniz*able* supply. They also should be sufficiently st*able* and amena*ble* during handling (a necessity for cofactors, buffers, and so on). Enzymes from microbial sources could fulfill all these requirements. Enzymes are moreover mut*able*, so that variants can be gained and selected for improved performance (for example in substrate range and specificity, altered stereospecificity, stability in nonaqueous environments). Finally, however, these enzymes and the adjunct knowledge to handle and even improve them must be transfer*able* into the hands of synthetic bioorganic chemists to become really valu*able* for asymmetric syntheses.

From the viewpoint of basic science, it is of fundamental interest to learn how C–C-bonding enzymes handle their sugar substrates and how the stereospecific reactions are performed. Therefore, the analysis of structure–function relationships is of eminent importance. Data on three-dimensional (3D) structures, on reaction mechanisms, and on the deliberate changing of active sites by site-directed mutagenesis further contribute to our understanding of enzymes as tools in asymmetric syntheses. Project B21 was therefore intended not only to provide microbial C–C-bonding enzymes for asymmetric syntheses, but also to study these enzymes with respect to their structure–function relationships.

From the many enzymes that are known to make and break C–C bonds, we first chose the two transferases, transketolase (TKT) and transaldolase (TAL), both from the Gram-negative bacterium *Escherichia coli*. While project B21 evolved, we learned that this microorganism holds other and so far unknown enzymes which are of interest for asymmetric syntheses. One transketolase-like enzyme, 1-deoxy-D-xylulose 5-phosphate synthase (DXS), turned out to be the first enzyme of a novel biosynthetic pathway leading to isoprenoids in bacteria, algae, and plants. The other, fructose 6-phosphate aldolase (FSA) – while similar to transaldolase – allows the direct use of the inexpensive dihydroxyacetone in aldol condensations.

2.2.2.2 Thiamine Diphosphate (ThDP)-Dependent Enzymes

Thiamine diphosphate (ThDP) is an important cofactor in many enzymes which either transfer carbon units between molecules or decarboxylate organic acids.

Therefore, ThDP-dependent enzymes include the potential of both making and breaking of C–C bonds [1]. All enzymes have in common a ThDP-bound "active aldehyde" intermediate formed either by decarboxylation or by transfer from a suitable donor compound (e.g., from xylulose-5-phosphate by transketolase). Some of the decarboxylating enzymes also catalyze interesting side-reactions where two aldehydes are joined, resulting in so-called acyloin condensations [1, 2]. ThDP enzymes have been used for quite a while as catalysts in chemoenzymatic syntheses; for recent reviews see refs. [2] and [3].

In the framework of SFB380, two projects dealt extensively with acyloin-condensing ThDP-dependent enzymes such as pyruvate decarboxylase (PDC), benzoylformate decarboxylase (BFD), or benzaldehyde lyase (BAL) (see Chapters 2.2.3 and 2.2.7). Another ThDP-dependent decarboxylase, phosphonopyruvate decarboxylase (PPD) from *Streptomyces viridochromogenes*, became available only recently and was studied in project B21. We wanted to find out whether this PDC-related enzyme could be a valuable tool in the provision of acyloin condensations involving C–P bonds (see Section 2.2.2.2.3).

Other ThDP enzymes catalyze the formation of C–C bonds by transferring C_2 units from a donor compound to an acceptor compound (transferase activity). We have studied the *E. coli* transketolase with respect to structure–function relationships, as well as to possible applications in asymmetric syntheses (Section 2.2.2.2.1). During the late 1990s a transketolase-like enzyme, 1-deoxyxylulose 5-phosphate synthase, was discovered. Its structure and value in chemoenzymatic syntheses were also assessed in project B21 (Section 2.2.2.2.2).

2.2.2.2.1 Transketolase (TKT)

Transketolase (TKT, E.C. 2.2.1.1) is an enzyme from the pentose phosphate pathway of sugar metabolism. Its physiological role is the rearrangement of various sugar phosphates (chain lengths C_3 to C_7) to provide building blocks for cellular biosyntheses. The principal reactions of transketolase are depicted in Scheme 2.2.2.1. In general, a so-called "activated glycolaldehyde" (α, β-dihydroxyethyl group) is transferred from a ketose donor to an α-hydroxyaldehyde. Apart from the natural donor/acceptor pairs from the pentose phosphate pathway (xylulose 5-phosphate, fructose 6-phosphate, and sedoheptulose 7-phosphate), a wide variety of nonphosphorylated 2-hydroxyaldehydes is accepted at reasonable rates to yield ketoses with a 3*S*,4*R* configuration. The use of α-hydroxypyruvate as a donor substrate is of importance for synthetic purposes as this allows practically irreversible product formation with carbon dioxide leaving the assay (see Scheme 2.2.2.1). Thus, TKT, from yeast cells or from spinach leaves, had already attracted the interest of bioorganic chemists when SFB380 started in 1993 [4]. For example, yeast transketolase was used for one step in a chemoenzymatic synthesis of the beetle pheromone, α-*exo*-brevicomin [4a].

An impediment to a wider use of transketolases, however, was the lack of the catalyst itself, which had to be tediously purified from sources such as baker's yeast or spinach leaves [4]. With the advent of recombinant DNA technologies,

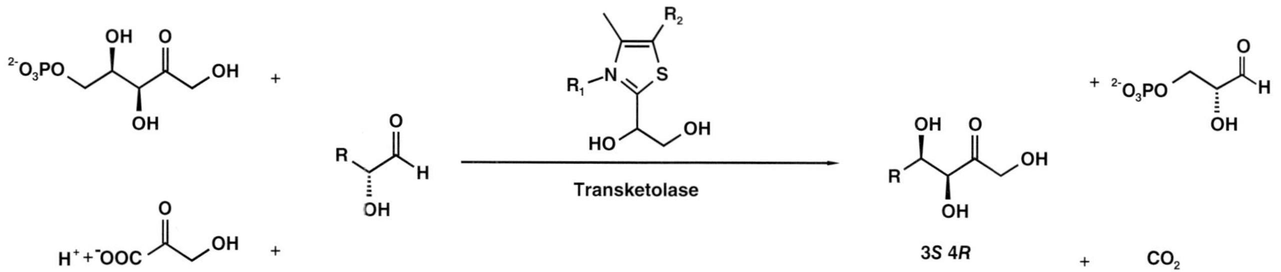

Scheme 2.2.2.1 Principal reactions of transketolase. Ketose donor substrates include xylulose 5-phosphate (upper left) or hydroxypyruvate (lower left). Acceptor substrates are α-hydroxyaldehydes. A C_2 unit ("activated glycolaldehyde") is transferred to the acceptor substrate via a ThDP-bound α,β-dihydroxyethyl group thereby forming a novel ketose of 3S,4R stereoconfiguration, and a shortened hydroxyaldehyde (glyceraldehyde 3-phosphate in the case of xylulose 5-P as donor). If hydroxypyruvate is used, carbon dioxide is evolved and leaves the reaction, thus shifting the equilibrium of the reaction to the products side. This leads to a virtually complete reaction.

cloning of transketolase genes supplied recombinant transketolases from *E. coli* or yeast, which could be purified with relatively simple protocols and resulted in sufficient amounts of biocatalyst for bioorganic syntheses [5, 6]. In particular transketolase A from recombinant *E. coli* is now available in sufficient amounts, either in homogeneous purity [6] or in enzyme-enriched crude extracts [7]. A wide range of reactions has been performed with transketolases from various origins, some of them on preparative scales [6, 8].

Aldehydes lacking an OH group at C2 are also transformed by transketolase, leading to a 3S configuration of the hydroxyl group in the deoxyketose product [7a, 9] albeit with a significantly lower rate than with the hydroxylated acceptors [6b, 10]. In contrast to the transketolases from spinach and yeast [9], no conversion of aromatic aldehydes, e.g., benzaldehyde or hydroxybenzaldehydes, could be detected with purified *E. coli* transketolase [6b].

For synthetic purposes the *E. coli* transketolase has a certain advantage over the enzymes from spinach and yeast, because the conversion of α-hydroxypyruvate with a rate of 60 U (mg of protein)$^{-1}$ [6b] is significantly higher than the rates of 2 U mg^{-1} and 9 U mg^{-1} reported for the spinach and yeast enzymes [9, 11].

The gram-scale preparation of rare sugars by *E. coli* transketolase was demonstrated successfully for (S)-erythrulose from glycolaldehyde and hydroxypyruvate in an enzyme membrane reactor which allowed the continuous production of (S)-erythrulose with high conversion and a space–time yield of 45 g L^{-1} d^{-1} was reached [12].

Xylulose 5-phosphate (D-*threo*-2-pentulose 5-phosphate) is the best donor compound for TKT [6]. As its cost is, however, prohibitively high for routine assays (formerly available batches from commercial suppliers were sold at ca. €1000 per 100 mg), a multi-enzymatic synthesis and subsequent chromatographic purification were developed which allowed the gram-scale synthesis of xylulose 5-phosphate with a 82% yield [13].

Studying structure–function relationships of transketolase was greatly enhanced by the X-ray determination of the 3D structure of yeast transketolase, which was the first structure of any ThDP enzyme to be solved [14]. While crystallization of both the apo- and holotransketolase from *E. coli* worked well [6b, 10], the solution of the 3D structure was not followed any further in the SFB project as another group had announced that the structure was under investigation [15]. Therefore the structure of the substrate-binding site of yeast transketolase [16] was taken to model the *E. coli* TKT A structure [6b, 10]. All amino acid residues of the yeast substrate-binding site were also present in the *E. coli* model. On this basis, site-directed mutagenesis was used to alter the amino acid residues. Mutant proteins were then purified and studied for their activity, substrate affinity, and reactivity [6b, 10]. Figure 2.2.2.1(a) depicts the model of the substrate binding site of *E. coli* transketolase A with the α,β-dihydroxyethyl group bound to the cofactor ThDP.

Analysis of the site-directed mutant proteins showed that the two arginine residues Arg359 and Arg528 are involved in the binding of the phosphate group of the acceptor compound. Asp477 (Fig. 2.2.2.1) is in close contact with the α-hy-

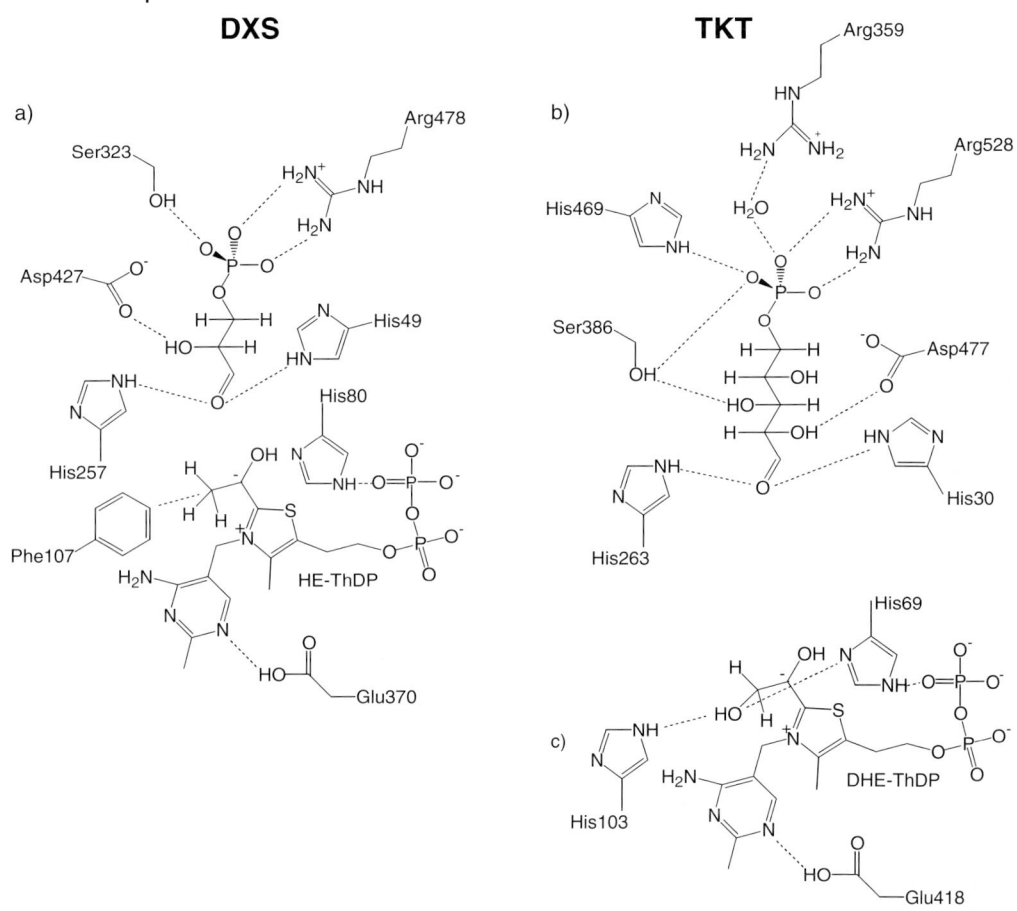

Fig. 2.2.2.1 Comparison of schematic substrate-binding site models for a) *E. coli* enzymes DXS [23a] and b) transketolase A [6b]. c) The cofactor ThDP with bound intermediates (hydroxyethyl: HE-ThDP; dihydroxyethyl: DHE-ThDP). Putative interactions between amino acid residues and bound acceptor substrates are indicated by dotted lines.

droxyl group of the acceptor compound. Interestingly, the alteration of this aspartyl residue to Glu477 (D477E) led to a sharp decrease in reactivity (K_{cat}/K_m) for all α-hydroxyaldehyde acceptors (less than 11% of wild-type K_{cat}/K_m), whereas α-deoxyaldehydes (e.g., acetaldehyde, propionaldehyde) gave drastic increases (up to 400% in V_{max}) compared to the wild-type values [6b]. This can be interpreted in terms of the way the Asp477 residue governs the stereospecificity of transketolase. This was corroborated by the analysis of the corresponding residue Asp477 in the yeast transketolase [17].

2.2.2.2.2 1-Deoxy-D-xylulose 5-Phosphate Synthase (DXS)

In the early stage of the SFB380 project, the enzyme 1-deoxy-D-xylulose 5-phosphate synthase had not been described. However, through labeling experiments with bacteria, the groups of Rohmer (in Strasbourg) and Hermann Sahm (in Jülich), and others, had already found that a novel, non-mevalonate biosynthetic pathway had to be assumed [18] which started with the formation of 1-deoxy-D-xylulose 5-phosphate (DXP) from a pyruvate-decarboxylating condensation reaction with D-glyceraldehyde 3-phosphate as acceptor (see Scheme 2.2.2.2). This C_2 transfer resembles the transketolase reaction, although the DXP synthase was still unknown. From DXP the novel biosynthetic pathway leads to the synthesis of isoprenoid compounds in many bacteria, in algae and plant chloroplasts, and in the malaria parasite, *Plasmodium falciparum*. This pathway has meanwhile been completely elucidated and is referred to as the non-mevalonate pathway, the DXP pathway, or the methylerythritol (MEP) pathway [19]. As no enzyme was known which performed this essential reaction as its main reaction, but transketolase reaction was rather similar, a genome mining approach with transketolase as a lead sequence resulted in the discovery of 1-deoxy-D-xylulose 5-phosphate synthase (DXS) [20a]. Independent work confirmed DXS [20b,c]. We detected that the novel enzyme of *E. coli* (1-deoxyxylulose 5-phosphate synthase, DXS) is ThDP-dependent and catalyzes the formation of 1-deoxy-D-xylulose from pyruvate and D-glyceraldehyde, or of 1-deoxy-D-xylulose 5-phosphate from pyruvate and D-glyceraldehyde-3-phosphate.

Although the purified enzyme is much less active than transketolase (3 U mg^{-1} versus 80–100 U mg^{-1}), it could be utilized as a tool in a one-pot multi-enzyme asymmetric synthesis yielding about 700 mg of the barium salt of DXP [21]. Our original paper on DXS has up to now been quoted in more than 200 articles. Purified DXS can be used further for the chemoenzymatic syntheses of various 1-deoxysugars and 1-deoxysugar phosphates, as shown by our group and others [22].

Attempts to crystallize DXS with the aim of determining the enzyme's 3D structure have failed so far. In the absence of a 3D structure, yeast transketolase structure was used as a model for DXS. Amino acid residues which are conserved

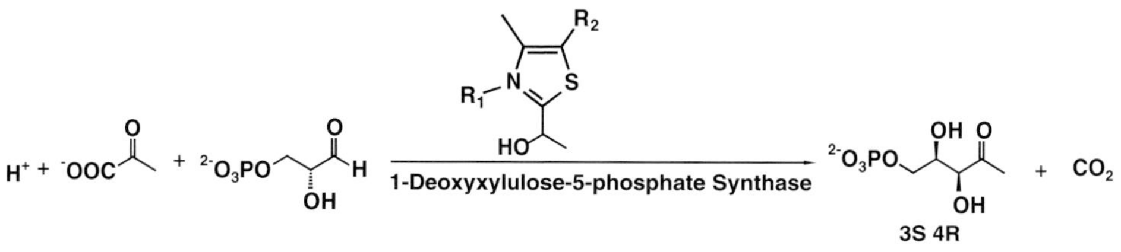

Scheme 2.2.2.2 DXP synthase transfers a C_2 group(hydroxyethyl) from pyruvate to acceptor compounds as D-glyceraldehyde 3-phosphate (GAP) or D-glyceraldehyde with an intermediate ThDP-bound stage. The reaction is drawn to the products side by decarboxylation, which is a virtually irreversible reaction.

between the two ThDP enzymes were used to build a DXS active site model (see Fig. 2.2.2.1a). The model was then the basis for site-directed mutageneses to study structure–function relationships at the substrate binding site of DXS [23a]. Boronat's group arrived at a similar model which was corroborated by the His-49Gln site-directed mutant, which was inactive [23b].

2.2.2.2.3 Phosphonopyruvate Decarboxylase (PPD) from *Streptomyces viridochromogenes*

C–P bonds are present in a range of natural compounds, for example, antibiotics. Through a reaction of PEP mutase, phosphoenolpyruvate is converted to phosphonopyruvate, which is the precursor of all natural phosphono compounds. The ThDP-dependent enzyme phosphonopyruvate decarboxylase (PPD) has been discovered in several Streptomycetes and other bacteria [24]; its reaction is depicted in Scheme 2.2.2.3.

PPD is similar in sequence to pyruvate decarboxylase (PDC). We wondered if PPD from *Str. viridochromogenes* could be used not only in the provision of phosphonoacetaldehyde from phosphonopyruvate, but also to perform acyloin condensations as PDC does. This would open the path to C–P bond-containing acyloins. The gene *ppd* from *Str. viridochromogenes* [24b] was subcloned and expressed in *E. coli* strains in the histidine-tagged version. PPD enzyme was prepared, characterized, and used successfully for the production of phosphonoacetaldehyde from chemically synthesized phosphonopyruvate (in collaboration with A. Cosp and M. Müller). However, so far we have no indications of an acyloin condensation activity of PPD [25].

PPD from another bacterium was characterized by the Dunaway-Mariano group [26]. Based on the PDC structure from *Zymomonas mobilis* and other ThDP enzymes [27], a hypothetical model for PPD was built. This model was then corroborated by site-directed mutagenesis of all the amino acid residues at the ThDP and phosphonopyruvate binding sites [25]. A hypothetical working model for the PPD active site is given in Fig. 2.2.2.2.

2.2.2.3 Class I Aldolases

A major group of C–C-bonding enzymes consists of aldolases which serve in various metabolic pathways, mainly in the direction of aldol cleavage. Aldol con-

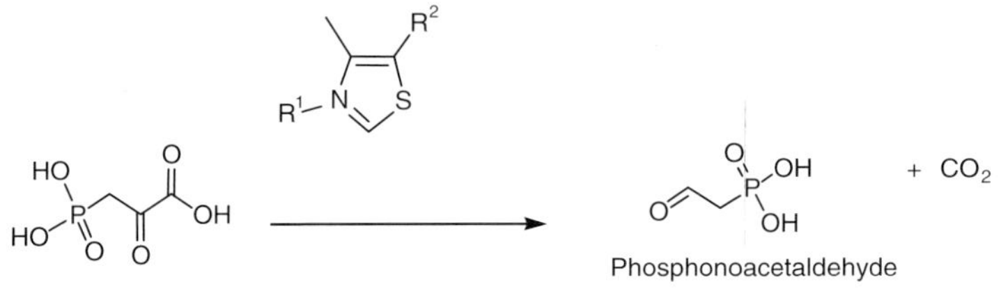

Scheme 2.2.2.3 Reaction of phosphonopyruvate decarboxylase (PPD).

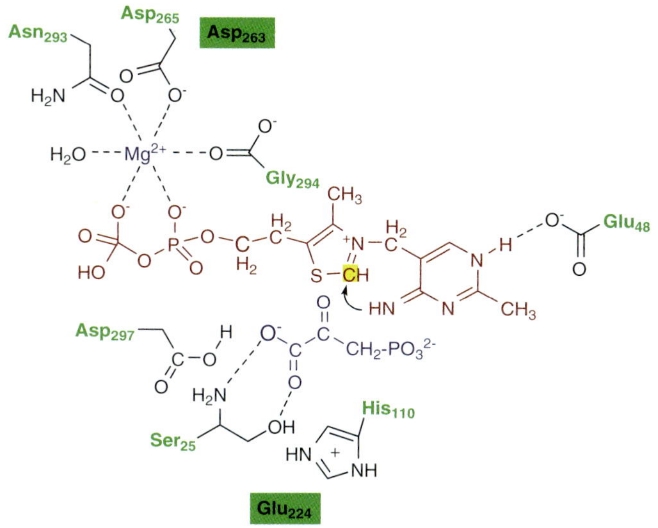

Fig. 2.2.2.2 Postulated model of the active site of phosphonopyruvate decarboxylase. PPD from *Str. viridochromogenes* Tü494 (based on the structure of PDC from *Zymomonas mobilis* [27] and data from analyis of site-directed mutants [25]). The model depicts the start of the catalytic reaction (deprotonation of the reactive C2 atom (yellow) on the thiazolium ring of ThDP (red). Putative interactions of essential amino acid residues (green) with the cofactors, ThDP and Mg^{2+} (blue) as well as the substrate molecule, phosphonopyruvate (blue), are depicted as broken lines. Residues Glu224 and Asp263 are conserved among phosphonopyruvate and sulfopyruvate decarboxylases and are shaded in green.

densation reactions of aldolases are favored thermodynamically and therefore of apparent interest for synthetic organic chemists. The classical fructose 1,6-bisphosphate aldolase (RAMA) and other DHAP aldolases were studied by Fessner's group (project B25) and were well established already as tools in asymmetric syntheses [28]. For transaldolase [TAL; E.C. 2.2.1.2], however, neither data for its use in biocatalysis nor structural data were available. Dihydroxyacetone phosphate (DHAP) is the donor substrate for DHAP aldolases [29]; however, it is rather expensive and unstable, so the search for aldolases which use the unphosphorylated donor, dihydroxyacetone, was of interest. When studying transaldolases from *E. coli*, we fortuitously discovered such an enzyme, fructose 6-phosphate aldolase (FSA). Aldolases are categorized in two classes – those which need a metal ion cofactor (Class II), and those which need no cofactor but have a reactive lysyl group in their active site [29]. The lysyl group forms a Schiff base during the aldol cleavage/aldolization process (Class I enzymes) [30]. Both TAL and FSA belong to the Class I aldolases.

2.2.2.3.1 Transaldolase (TAL)

Transaldolase (TAL, E.C. 2.2.1.2) is a transferase which, like TKT, acts in the pentose phosphate pathway. The reaction of transaldolase is the transfer of a dihydroxyacetone group (C_3) from a donor ketose (namely fructose 6-phosphate, sedoheptulose 7-phosphate) to a variety of α-hydroxyaldehydes [29b], thereby creating an aldol product that bears either one phosphate group or none, in contrast to DHAP aldolases. The resulting aldol product also has the 3S,4R stereoconfiguration. Transaldolase B (TAL B) from *E. coli* K-12 was cloned, and overexpressed in recombinant form; the enzyme was characterized biochemically [6b, 31]. Using fructose 6-phosphate (and to a lesser degree free fructose) as donor, various phosphorylated and unphosphorylated hydroxyaldehydes were used as acceptors [6b, 31]. In collaboration with Gunter Schneider's group (formerly at Uppsala University, now at the Karolinska Institute in Stockholm, Sweden), TAL B was crystallized and the 3D structure of the dimeric enzyme was determined by X-ray crystallography [32] (see Fig. 2.2.2.3a). At the center of the active site is a lysyl residue; its exchange for arginine abolishes enzyme activity. This lysyl residue therefore forms the Schiff base during a transaldolase reaction with ketose compounds [32]. After TAL B crystals were soaked with fructose 6-phosphate, the enzyme-substrate complex was trapped by reduction with borohydride. This made it possible, for the first time with a class I aldolase, to elucidate the structure of a Schiff base adduct [33]. Subsequent work was performed to analyze the functions of amino acid residues at the substrate binding site of TAL B. Various

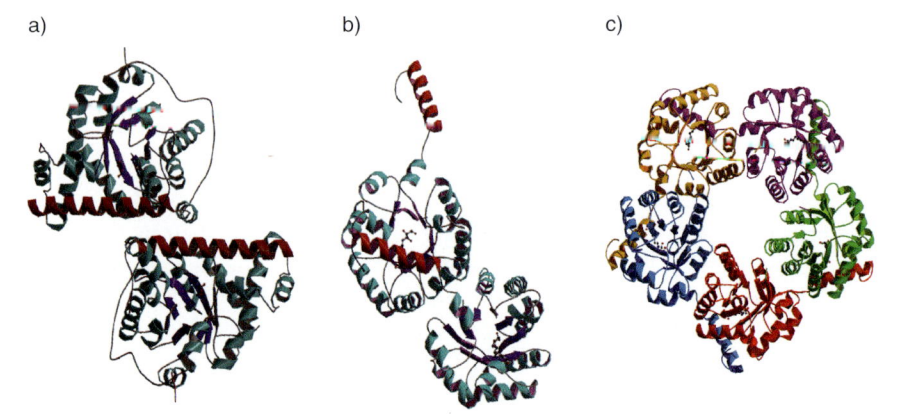

Fig. 2.2.2.3 Three-dimensional structures of transaldolase B (TAL B) and fructose 6-phosphate aldolase (FSA) from *E. coli*. a) TAL B forms a dimer structure; b), c) FSA is a decameric enzyme consisting of two pentamer rings (right) one on top of the other. b) Two adjacent subunits of FSA showing the C-terminal extension of one subunit which covers the active site of the neighboring subunit. The structure of the active site of FSA (bottom) and of a pentameric ring consisting of five subunits with an α/β barrel structure. The reactive lysyl residue Lys-85 is shown in the center [38]. Reproduced with the permission of the original publisher of ref. [38], Elsevier.

TAL B mutant proteins were characterized biochemically and their 3D structures were determined [34].

TAL B samples from *E. coli* as well as from yeast were assayed successfully with fluorogenic chemical probes that contained (3*S*,4*R*)-dihydroxy compounds [35]. This should open the search for novel stereospecificities for transaldolase mutant proteins.

2.2.2.3.2 Fructose 6-Phosphate Aldolase (FSA)

Through the genome sequencing efforts, a vast amount of gene information became available. In *E. coli*, two open reading frames (ORFs) with low sequence similarity to transaldolase B were discovered. The two ORFs were cloned and overexpressed. The recombinant proteins showed no transaldolase activity, however. Instead, a novel enzyme activity, fructose 6-phosphate aldolase, was discovered [36]. One of the two ORFs was analyzed further and the enzyme termed FSA. FSA uses dihydroxyacetone as donor and D-glyceraldehyde 3-phosphate is the best acceptor substrate tested so far. The aldol product has the 3*S*,4*R* stereo-configuration. A variety of other aldehydes are used as acceptors, among others formaldehyde and glycolaldehyde, but – interestingly – not D-glyceraldehyde [37]. The true enzyme function is still unknown as the wild-type enzyme is not expressed from its natural gene locus in *E. coli* cells [36]. Moreover, FSA uses hydroxyacetone (acetol) as a donor, thereby forming 1-deoxysugars or 1-deoxysugar phosphates [22a]. The 3D structure of FSA was solved in collaboration with Gunter Schneider's group [38]. The quaternary structure of FSA is quite different than that of transaldolase B and more similar to other transaldolases (e.g., *Bacillus subtilis* [36]). It is a decamer consisting of two pentamer rings (see Fig. 2.2.2.3c). The tight packing of FSA may explain its unusual thermostability (50% of activity retained after 16 h at 75 °C) [36, 37].

A schematic representation of the active sites of transaldolase B is given in Fig. 2.2.2.4, together with the corresponding amino acid residues in the other Class I aldolases, fructose 1,6-bisphosphate, and FSA. There is one amino acid residue in the active site of FSA which has no counterpart in TAL B, Tyr-131.

An exchange of Tyr-131 with either a phenylalanine or an alanine residue leads to an enzymatically inactive protein [39]. Exchange of Gln59 with Glu59 in FSA leads to no gross activity change [39]. Therefore, only two residues (Leu107 and Ala129 in the FSA sequence) are left which differ between TAL B and FSA.

Exchange of these FSA residues, either one or both, from Leu107 to Asn107 and Ala129 to Ser129 gave enzymatically improved FSA activities (see Table 2.2.2.1), but no transaldolase activity. Both single mutant variants showed improved enzyme activity and higher affinity for the substrates dihydroxyacetone (DHA) and fructose 6-phosphate, while affinity to the acceptor, glyceraldehyde-3-phosphate, was unaltered.

2.2.2.4 Summary and Outlook

Through recombinant DNA technology, we have made available microbial C–C-bonding enzymes which are interesting tools for asymmetric syntheses in bioca-

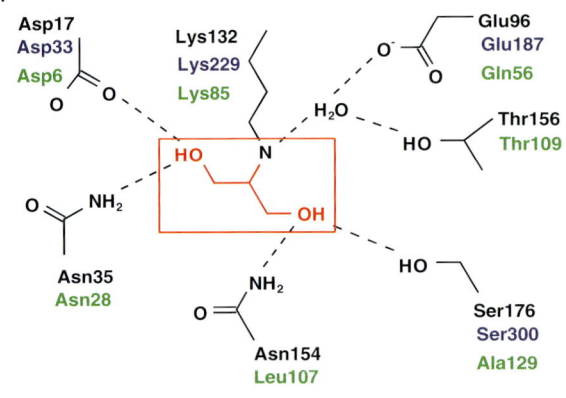

Fig. 2.2.2.4 Schematic representation of interactions of the reduced Schiff'base intermediate (red, boxed) with amino acid residues of the active site of transaldolase B. H-bridge bonds are shown by broken lines. The corresponding amino acid residues in the active sites are: top: TalB from *E. coli* (black); middle: Class I fructose 1,6-bisphosphate aldolase (blue); bottom: fructose 6-phosphate aldolase (green). There are no corresponding residues of FBP aldolase at several positions [37].

Table 2.2.2.1 K_M, V_{max}, and K_{cat}/K_M values of wild-type FSA compared with those of site-directed mutants.[a,b]

	Wild-type FSA	L107N	A129S	L107N/A129S
K_M DHA [mM]	35 (100)	10.3 (29)	6.3 (18)	7.8 (22)
V_{max} DHA [U mg^{-1}]	45 (100)	44 (98)	104 (231)	42 (94)
K_{cat}/K_M DHA [mM^{-1} s^{-1}]	5 (100)	17 (340)	63 (1260)	21 (420)
K_M GAP [mM]	0.8 (100)	0.8 (100)	0.9 (112)	0.9 (112)
V_{max} GAP [U mg^{-1}]	45 (100)	43 (96)	99 (220)	44 (98)
K_{cat}/K_M GAP [mM^{-1} s^{-1}]	216 (100)	206 (95)	422 (195)	187 (87)
K_M F6P [mM]	9 (100)	2.4 (27)	1.4 (16)	2.4 (27)
V_{max} F6P [U mg^{-1}]	7 (100)	11 (153)	22 (307)	8 (110)
K_{cat}/K_M F6P [mM^{-1} s^{-1}]	3 (100)	18 (600)	60 (2000)	13 (433)

a Values in brackets are percentages, with kinetic data of wild-type set as 100%.
b 1 U = 1 µmol of substrate converted per min [37].

talysis. The enzymes either need no addition of cofactors, or use the inexpensive cofactor ThDP (at micromolar concentrations). As far as we can judge, all enzymes are strictly regio- and stereospecific in their wild-type forms. From *Escherichia coli*, the well-known enzymes transketolase and transaldolase, as well as the newly described enzymes 1-deoxyxylulose 5 phosphate synthase (DXS; transketolase-like) and fructose 6-phosphate aldolase (FSA; transaldolase-like) were cloned, overexpressed, purified, and characterized in pure forms. They now enlarge the arsenal of C–C-bonding enzymatic tools for carbohydrate chemistry.

The supply of the recombinant enzymes is warranted and some enzymes can be obtained in large amounts (in the case of transketolase A more than a million units were gained from a large-scale fermentation) so that even minor, but nonetheless interesting, substrates can be converted.

We have prepared several chiral and high-priced compounds on a gram scale by transketolase, DXS, or FSA (partly in collaborations with projects B25 and C). Through site-directed mutagenesis, improved biocatalysts (FSA) were obtained, or the substrate range of C–C-bonding enzymes was altered (TKT, DXS). Site-directed mutagenesis also helped to elucidate structure–function relationships at the active sites of enzymes (transketolase, DXS, PPD), when no 3D structures were available.

In collaboration with Gunter Schneider's group at the Karolinska Institute in Stockholm, the 3D structures of transaldolase and of several of its mutant derivatives have been solved. For the first time, a Schiff base intermediate of an aldolase was analyzed crystallographically. The structure of FSA was solved too, and it was found that the enzyme forms a decameric structure out of two pentamer rings. FSA makes it possible to use dihydroxyacetone or hydroxyacetone as a donor compound for aldolization reactions; this opens up the field for novel carbohydrate compounds such as 1-deoxysugars which otherwise can be obtained by DXS through a different reaction.

It has been established that DXS catalyzes the first step in a novel biosynthetic pathway leading to isoprenoids in bacteria, algae, plant chloroplasts, and in the malaria parasite, *Plasmodium falciparum*. DXS is therefore a novel target for antibiotics, herbicides, or anti-malarials. Our work has contributed to an understanding of the novel biosynthetic pathway and could further open new perspectives on how to inhibit the pathway in pathogenic bacteria, protists, or weeds.

Further improvement of C–C-bonding enzymes or attempts to alter their stereospecificities (TAL, FSA) are subjects of ongoing research in our group at Stuttgart.

Acknowledgements

We thank the Deutsche Forschungsgemeinschaft for financial support in the scope of the SFB 380, project B21. We thank Mrs. Ursula Degner and Ms. Regine Halbach for excellent technical assistance. We are grateful, especially to Wolf-Dieter Fessner, Andreas Liese, Michael Müller, and Martina Pohl, for cooperation in the SFB 380. Special thanks go to Gunter Schneider's group at the Karolinska Institute in Stockholm for a fruitful cooperation on protein structures since 1995.

References

1 For a review on ThDP-enzymes see: A. Schellenberger, *Biochim. Biophys. Acta* **1998**, *1385*, 177–186; and other articles on ThDP enzymes in the same issue.

2 For some reviews of ThDP-dependent enzymes with regard to asymmetric synthesis, see: a) M. Pohl, *Adv. Biochem. Eng. Biotechnol.* **1997**, *58*, 16–43; b) G. A. Sprenger, M. Pohl, *J. Mol. Catal. B: Enzymatic* **1999**, *6*, 145–159; c) O. P. Ward, A. Singh, *Curr. Opin. Biotechnol.* **2000**, *11*, 520–526; d) M. Pohl, G. A. Sprenger, M. Müller, *Curr. Opin. Biotechnol.* **2004**, *15*, 335–342.

3 M. Müller, G. A. Sprenger, Thiamine-dependent enzymes as catalysts of C–C-bonding reactions: The role of "orphan" enzymes, in *Thiamine: Catalytic Mechanisms and Role in Normal and Disease States*, Marcel Dekker, New York, **2004**, pp. 77–91.

4 a) D. C. Myles, P. J. Andrulis III, G. M. Whitesides, *Tetrahedron Lett.* **1991**, *32*, 4835–4838; b) C. Demuynck, J. Bolte, L. Hecquet, V. Dalmas, *Tetrahedron Lett.* **1991**, *32*, 5085–5088. c) Y. Kobori, D. C. Myles, G. M. Whitesides, *J. Org. Chem.* **1992**, *57*, 5899–5907.

5 For mapping, cloning, and DNA sequence of *E. coli* transketolase A, see: a) K. M. Draths, J. W. Frost, *J. Am. Chem. Soc.* **1990**, *112*, 1657–1659. b) G. A. Sprenger, *J. Bacteriol.* **1992**, *174*, 1707–1708. c) G. A. Sprenger, *Biochim. Biophys. Acta* **1993**, *1216*, 307–310. For yeast TKL1 gene, see: d) M. Sundström, Y. Lindqvist, G. Schneider, U. Hellman, H. Ronne, *J. Biol. Chem.* **1993**, *268*, 24 346–24 352.

6 For purification and characterization of *E. coli* transketolase, see: a) G. A. Sprenger, U. Schörken, G. Sprenger, H. Sahm, *Eur. J. Biochem.* **1995**, *230*, 525–532; b) U. Schörken, Ph.D. Thesis, *Jül. Ber.* No. 3418, University of Düsseldorf, Germany, **1997**.

7 a) G. R. Hobbs, M. D. Lilly, N. J. Turner, J. M. Ward, A. J. Willets, J. M. Woodley, *J. Chem. Soc. Perkin Trans. I* **1993**, 165–166; b) K. G. Morris, E. B. Smith, N. J. Turner, M. D. Lilly, R. K. Mitra, J. M. Woodley, *Tetrahedron Asymm.* **1996**, *7*, 2185–2188; G. R. Hobbs, R. K. Mitra, R. P. Chauhan, J. M. Woodley, M. D. Lilly, *J. Biotechnol.* **1996**, *45*, 173–179.

8 For reviews on the synthetic use of TK, see: a) W.-D. Fessner, C. Walter, *Top. Curr. Chem.* **1996**, *1*, 97–194; a) S. Takayama, J. Glenn, G. J. McGarvey, C.-H. Wong, *Annu. Rev. Microbiol.* **1997**, *51*, 285–310. c) U. Schörken, G. A. Sprenger, *Biochim. Biophys. Acta* **1998**, *1385*, 229–243; d) N. J. Turner, *Curr. Opin. Biotechnol.* **2000**, *11*, 527–531.

9 C. Demuynck, J. Bolte, L. Hecquet, V. Dalmas, *Tetrahedron Lett.* **1991**, *32*, 5085–5088.

10 U. Schörken, H. Sahm, G. A. Sprenger, Substrate specificity, site-directed mutagenesis, modelling of the substrate channel, and preliminary X-ray crystallographic data of *E. coli* transketolase, in *Biochemistry and Physiology of Thiamin Diphosphate Enzymes, Proc. 4th Int.Meeting* (Eds.: H. Bisswanger, A. Schellenberger), A. & C. Intermann Wissenschaftlicher Verlag, Prien, **1996**, pp. 543–553.

11 J. Bolte, C. Demuynck, H. Samaki, *Tetrahedron Lett.* **1987**, *28*, 5525–5528.

12 a) J. Bongs, D. Hahn, U. Schörken, G. A. Sprenger, U. Kragl, C. Wandrey, *Biotechnol. Lett.* **1997**, *19*, 213–215; b) D. Vasic-Racki, J. Bongs, U. Schörken, G. A. Sprenger, A. Liese, *Bioproc. Eng.* **2003**, *25*, 285–290.

13 Starting from commercially available fructose 1,6-bisphosphate and hydroxypyruvate, three enzymes were in use: fructose 1,6-bisphosphate aldolase, triosephosphate isomerase, and transketolase; see: F. T. Zimmermann, A. Schneider, U. Schörken, G. A. Sprenger, W.-D. Fessner, *Tetrahedron Asymmetry* **1999**, *10*, 1643–1646.

14 Structural studies with yeast transketolase: a) M. Sundström, Y. Lindqvist, G. Schneider, *FEBS Lett.* **1992**, *313*, 229–251; b) Y. Lindqvist, G. Schneider, U. Ermler, M. Sundström,

EMBO J. **1992**, *11*, 2373–2379. TKL1 with bound ThDP analogue: c) U. Nilsson, Y. Lindqvist, R. Kluger, G. Schneider, *FEBS Lett.* **1993**, *326*, 145–148; d) G. Schneider, Y. Lindqvist, *Bioorg. Chem.* **1993**, *21*, 109–117; e) M. Nikkola, Y. Lindqvist, G. Schneider, *J. Mol. Biol.* **1994**, *238*, 387–404; f) C. Wikner, L. Meshalkina, U. Nilsson, M. Nikkola, Y. Lindqvist, M. Sundström, G. Schneider, *J. Biol. Chem.* **1994**, *269*, 32 144–32 150.

15 Crystallization of *E. coli* transketolase: a) J. Littlechild, N. Turner, *Acta Crystallogr. D* **1995**, *51*, 1074–1076; structure available as 1QGD, from Protein Data Bank (www.rcsb.org/pdb/).

16 Structure with erythrose 4-phosphate bound: a) U. Nilsson, L. Meshalkina, Y. Lindqvist, G. Schneider, *J. Biol. Chem.* **1997**, *272*, 1864–1869; b) C. Wikner, U. Nilsson, L. Meshalkina, C. Udekwu, Y. Lindqvist, G. Schneider, *Biochemistry* **1997**, *36*, 15 643–15 649. c) model of donor bound: G. Schneider, Y. Lindqvist, *Biochim. Biophys. Acta* **1998**, *1385*, 387–398.

17 a) U. Nilsson, L. Hecquet, T. Gefflaut, C. Guerard, G. Schneider, *FEBS Lett.* **1998**, *424*, 49–52; b) L. Hecquet, C. Demuynck, G. Schneider, J. Bolte, *J. Mol. Catal. B: Enzymatic* **2001**, *11*, 771–776.

18 a) M. Rohmer, M. Knani, P. Simonin, B. Sutter, H. Sahm, *Biochem. J.* **1993**, *295*, 517–524; b) S. T. J. Broers, Ph.D.Thesis No.10 978, ETHZurich, Switzerland, **1994**; c) M. Rohmer, M. Seemann, S. Horbach, S. Bringer-Meyer, H. Sahm, *J. Am. Chem. Soc.* **1996**, *118*, 2564–2566; d) H. K. Lichtenthaler, J. Schwender, A. Disch, M. Rohmer, *FEBS Lett.* **1997**, *400*, 271–274; e) D. Arigoni, S. Sagner, C. Latzel, W. Eisenreich, A. Bacher, M. H. Zenk, *Proc. Natl. Acad. Sci. USA* **1997**, *94*, 10 600–10 605.

19 For recent reviews see: a) M. Rodriguez-Concepcion, A. Boronat, *Plant Physiol.* **2002**, *130*, 1079–1089; b) M. Rohmer, *Pure Appl. Chem.* **2003**, *75*, 375–387; c) W. Eisenreich, A. Bacher, D. Arigoni, F. Rohdich, *Cell. Mol. Life Sci.* **2004**, *61*, 1401–1426. For malaria parasite: d) H. Jomaa, J. Wiesner, S. Sanderbrand, B. Altincicek, C. Weidemeyer, M. Hintz, I. Turbachova, M. Eberl, J. Zeidler, H. K. Lichtenthaler, D. Soldati, E. Beck, *Science* **1999**, *285*, 1573–1576.

20 a) G. A. Sprenger, U. Schörken, T. Wiegert, S. Grolle, A. A. de Graaf, S. V. Taylor, T. P. Begley, S. Bringer-Meyer, H. Sahm, *Proc. Natl. Acad. Sci. USA* **1997**, *94*, 12 857–12 862; b) L. M. Lois, N. Campos, S. R. Putra, K. Danielsen, M. Rohmer, A. Boronat, *Proc. Natl. Acad. Sci. USA* **1998**, *95*, 2105–2110; c) B. M. Lange, M. R. Wildung, D. McCaskill, R. Croteau, *Proc. Natl. Acad. Sci. USA* **1998**, *95*, 2100–2104.

21 Starting point was the aldolase-catalyzed cleavage of fructose 1,6-bisphosphate to yield dihydroxyacetone phosphate (DHAP) and D-glyceraldehyde 3-phosphate (Ga3P) in equal amounts. Ga3P was converted into DXP and CO_2 using pyruvate as the donor substrate and DXS as catalyst. Reversible interconversion of DHAP and Ga3P was achieved with triosephosphate isomerase. NMR analysis of the product confirmed that the D-*threo*-pentulose was formed with a high enantiomeric excess; see: a) S. V. Taylor, L. D. Vu, T. P. Begley, U. Schörken, G. A. Sprenger, S. Bringer-Meyer, H. Sahm, *J. Org. Chem.* **1998**, *63*, 2375–2377. An example of application of DXS to prepare labeled DXP for the elucidation of the vitamin B_6 pathway is given in: b) B. Laber, W. Maurer, S. Scharf, K. Stepusin, F. S. Schmidt, *FEBS Lett.* **1999**, *449*, 45–48.

22 Asymmetric synthesis of 1-deoxysugars and sugar phosphates using DXS and/or FSA enzymes: a) Me. Schürmann, Ma. Schürmann, G. A. Sprenger, *J. Mol. Catal. B: Enzymatic*, **2002**, *19–20*, 247–252. For use of DXS: b) J. Querol, C. Grosdemange-Billiard, M. Rohmer, A. Boronat, S. Imperial, *Tetrahedron Lett.* **2002**, *43*, 8265–8268.

23 a) Ma. Schürmann, Ph.D. Thesis, *Jül. Ber.* No. 3904, University of Düsseldorf, Germany, **2001**; b) J. Querol, M. Rodriguez-Concepcion, A. Boronat, S. Imperial, *Biochem. Biophys. Res. Commun.* **2001**, *289*, 155–160.

24 PPD: a) H. Nakashita, K. Watanabe, O. Hara, T. Hidaka, H. Seto, *J. Antibiot.* **1997**, *50*, 212–219; b) D. Schwartz, J. Recktenwald, S. Pelzer, W. Wohlleben, *FEMS Microbiol. Lett.* **1998**, *163*, 149–157; c) H. Nakashita, K. Kozuka, T. Hidaka, O. Hara, H. Seto, *Biochim. Biophys. Acta* **2000**, *1490*, 159–162.

25 S. Johnen, Ph.D. Thesis University of Düsseldorf, Germany, **2005**; *Schriften Forschungszentrum Jülich*, Vol. 17.

26 PPD enzyme from *Bacteroides fragilis*: G. Zhang, J. Dai, Z. Lu, D. Dunaway-Mariano, *J. Biol. Chem.* **2003**, *278*, 41 302–41 308.

27 PDC structure from *Z. mobilis*: a) D. Dobritzsch, S. König, G. Schneider, G. Lu, *J. Biol. Chem.* **1998**, *273*, 20 196–20 204. Site-directed mutants at the active site: b) J. M. Candy, R. G. Duggleby, *Biochem. J.* **1994**, *300*, 7–13; c) G. Schenk, F. J. Leeper, R. England, P. F. Nixon, R. G. Duggleby, *Eur. J. Biochem.* **1997**, *248*, 63–71; d) A. K. Chang, P. F. Nixon, R. G. Duggleby, *Biochem. J.* **1999**, *339*, 255–260. For BFD: e) E. S. Polovnikova, M. J. McLeish, E. A. Sergienko, J. T. Burgner, N. L. Anderson, A. K. Bera, F. Jordan, G. L. Kenyon, M. S. Hasson, *Biochemistry* **2003**, *42*, 1820–1830. ThDP-enzymes: f) K. Tittmann, R. Golbik, K. Uhlemann, L. Khailova, G. Schneider, M. Patel, F. Jordan, D. M. Chipman, R. G. Duggleby, G. Hübner, *Biochemistry* **2003**, *42*, 7885–7891.

28 For recent reviews on the use of DHAP aldolases, see: a) W. D. Fessner, C. Walter, *Bioorgan. Chem. Topics Curr. Chem.* **1997**, *184*, 97–194; b) T. D. Machajewski, C.-H. Wong, *Angew. Chem.* **2000**, *112*, 1406–1430; *Angew. Chem. Int. Ed.* **2000**, *39*, 1352–1374; c) M. G. Silvestri, G. DeSantis, M. Mitchell, C.-H. Wong, *Topics Stereochem.* **2003**, *23*, 267–342; d) J. Sukumaran, U. Hanefeld, *Chem. Soc. Rev.* **2005** *34*, 530–542.

29 a) W. J. Rutter, *Fed. Proc.* **1964**, *23*, 1248–1257; b) B. L. Horecker, O. Tsolas, C. Y. Lai, Aldolases, in: *The Enzymes*, Vol. VII, 3rd edn. (Ed.: P. D. Boyer), Academic Press, New York, **1972** pp. 213–258.

30 T. Gefflaut, C. Blonski, J. Perie, M. Willson, *Prog. Biophys. Mol. Biol.* **1995**, *63*, 301–340.

31 Cloning and biochemical characterization of TAL B: G. A. Sprenger, U. Schörken, G. Sprenger, H. Sahm, *J. Bacteriol.* **1995**, *177*, 5930–5936; see also ref. [6b].

32 a) J. Jia, Y. Lindqvist, G. Schneider, U. Schörken, H. Sahm, G. A. Sprenger, *Acta Crystallogr. Sect. D* **1996**, *52*, 192–193; b) J. Jia, W. Huang, U. Schörken, H. Sahm, G. A. Sprenger, Y. Lindqvist, G. Schneider, *Structure* **1996**, *4*, 715–724.

33 a) J. Jia, U. Schörken, Y. Lindqvist, G. A. Sprenger, G. Schneider, *Protein Sci.* **1997**, *6*, 119–124; b) G. Schneider, G. A. Sprenger, *Methods Enzymol.* **2002**, *354*, 197–201.

34 a) U. Schörken, J. Jia, H. Sahm, G. A. Sprenger, G. Schneider, *FEBS Lett.* **1998**, *441*, 247–250; b) U. Schörken, S. Thorell, M. Schürmann, J. Jia, G. A. Sprenger, G. Schneider, *Eur. J. Biochem.* **2001**, *268*, 2408–2415.

35 E. Gonzalez-Garcia, V. Helaine, G. Klein, M. Schürmann, G. A. Sprenger, W.-D. Fessner, J.-L. Reymond, *Chem. Eur. J.* **2003**, *9*, 893–899.

36 M. Schürmann, G. A. Sprenger, *J. Biol. Chem.* **2001**, *276*, 11 055–11 061.

37 Me. Schürmann, Ph.D. Thesis, *Jül. Ber.* No. 3903, University of Düsseldorf, Germany, **2001**.

38 S. Thorell, M. Schürmann, G. A. Sprenger, G. Schneider, *J. Mol. Biol.* **2002**, *319*, 161–171.

39 T. Inoue, PhD Thesis, Universität Stuttgart, 2006.

2.2.3
Enzymes for Carboligation – 2-Ketoacid Decarboxylases and Hydroxynitrile Lyases
Martina Pohl, Holger Breithaupt, Bettina Frölich, Petra Heim, Hans Iding, Bettina Juchem, Petra Siegert, and Maria-Regina Kula

2.2.3.1 Introduction
Enantioselective C–C bond formation is gaining more and more importance in bioorganic synthesis. This reaction is efficiently catalyzed by 2-ketoacid decarboxylases (E.C. 4.1.1.X) as well as by hydroxynitrile lyases (E.C. 4.1.2.X).

In order to increase the understanding of ThDP-dependent enzymes, the identification of amino acid side chains important for the catalysis of the carboligase reaction in pyruvate decarboxylase from *Zymomonas mobilis* (E.C. 4.1.1.1) and benzoylformate decarboxylase from *Pseudomonas putida* (E.C. 4.1.1.7) was a major task. Using site-directed mutagenesis and directed evolution, various enzyme variants were obtained, differing in substrate specificity and enantioselectivity.

The detailed characterization of hydroxynitrile lyases from *Sorghum bicolor* (E.C. 4.1.2.11) and *Linum usitassimum* (E.C. 4.1.2.37) has been hampered for a long time due to the lack of a recombinant expression system. Therefore our studies were focused on cloning of the coding genes, recombinant expression, and characterization of these enzymes.

2.2.3.2 2-Ketoacid Decarboxylases
The project encompassed the comparative characterization of pyruvate decarboxylase from *Z. mobilis* (PDC) and benzoylformate decarboxylase from *P. putida* (BFD) as well as their optimization for bioorganic synthesis. Both enzymes require thiamine diphosphate (ThDP) and magnesium ions as cofactors. Apart from the decarboxylation of 2-ketoacids, which is the main physiological reaction of these 2-ketoacid decarboxylases, both enzymes show a carboligase site reaction leading to chiral 2-hydroxy ketones (Scheme 2.2.3.1). A well-known example is

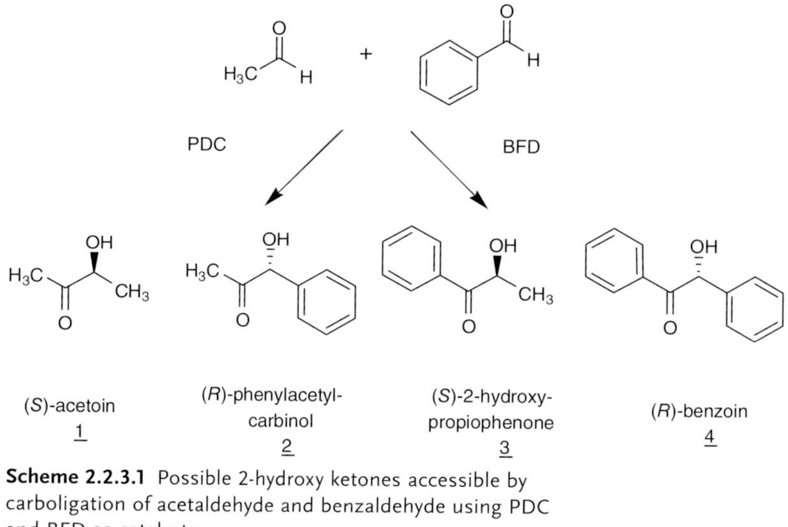

Scheme 2.2.3.1 Possible 2-hydroxy ketones accessible by carboligation of acetaldehyde and benzaldehyde using PDC and BFD as catalysts.

the formation of (R)-phenylacetylcarbinol **2**, the chiral pre-step toward (1R,2S)-ephedrine from benzaldehyde and acetaldehyde catalyzed by PDC [1]. Using BFD as a catalyst, further 2-hydroxy ketones are accessible, such as 2-hydroxypropiophenone **3**, benzoin **4**, and derivatives thereof (Scheme 2.2.3.1). Our studies were focused on the following points:

- comparative biochemical characterization of the wild-type enzymes
- identification of amino acid residues relevant to substrate specificity and enantioselectivity by site-directed mutagenesis
- optimization of the stability and substrate range of BFD by directed evolution.

2.2.3.2.1 Comparative Biochemical Characterization of Wild-Type PDC and BFD

Although the sequence similarity of both bacterial decarboxylases is low (<30%) their three-dimensional (3D) structures are highly similar, showing compact homotetramers [2, 3]. The main difference between the two structures is the length of the C-terminal helix, which is 40 amino acids longer in PDC (568 aa per subunit) than BFD (528 aa per subunit).

Decarboxylase reaction *Kinetic constants*: The optimum pH of the decarboxylase reaction was determined with the natural substrates of both enzymes, pyruvate (PDC) and benzoylformate (BFD). Both enzymes show a pH optimum at pH 6.0–6.5 for the decarboxylation reaction [4, 5] and investigation of the kinetic parameters gave hyperbolic $v/[S]$ plots. The kinetic constants are given in Table 2.2.3.1. The catalytic activity of both enzymes increases with the temperature up to about 60 °C. From these data activation energies of 34 kJ mol^{-1} (PDC) and 38 kJ mol^{-1} (BFD) were calculated using the Arrhenius equation [4, 6–8].

Substrate range: A broad variety of aliphatic and aromatic 2-keto acids have been tested as substrates, indicating that the substrate spectra of PDC and BFD are distinctly different. Whereas PDC exclusively decarboxylates linear aliphatic 2-

Table 2.2.3.1 Kinetic constants of PDC from *Z. mobilis* and BFD from *P. putida*, determined in potassium phosphate buffer (50 mM, pH 6.5) in the presence of ThDP (0.1 mM) and Mg^{2+} (5 mM).

Enzyme	K_M [mM]	V_{max} [U mg^{-1}]
PDC	1.1 ± 0.05 (pyruvate)	120 ± 20
BFD	0.8 ± 0.03 (benzoyl formate)	340 ± 20

2.2.3 Enzymes for Carboligation – 2-Ketoacid Decarboxylases and Hydroxynitrile Lyases

keto acids such as pyruvate, 2-ketobutyrate, and 2-ketopentanoate, BFD prefers benzoyl formate and shows some activity toward 2-ketobutyrate also. However, some branched-chain and aromatic substrates are also accepted by BFD with low activity. There is a clear correlation between activity and K_M values, indicating that differences in activity reflect different affinities to the active centers (Table 2.2.3.2) [9].

Table 2.2.3.2 Kinetic constants of PDC and BFD for various 2-keto acids R–CO–COO⁻ [8].

R	Specific activity [U mg^{-1}]/K_M [mM]			
	PDC	BFD	PDCI472A	BFDA460I
CH_3–	120/1.1	<0.05/≫50	50/7.8	1.4/≫50
C_2H_5–	79/4.7	1.6/>35	62.6/6.7	4.7/18.3
n-C_3H_7–	13/7.6	12/6.0	55/2.5	10.15/1.2
n-C_4H_9–	0.2/12.7	5.2/>35	32.5/0.2	18/1.0
$(CH_3)_2CH$–	0	2.6/n.d.	n.d.	n.d.
(C_2H_5)–$CH(CH_3)$–	0	1.6/n.d.	n.d.	n.d.
(C_2H_5)–$C(CH_3)_2$–	0	0	n.d.	n.d.
CH_3–$CH(CH_3)$–CH_2–	0.3/n.d.	2.6/n.d.	n.d.	n.d.
CH_3–$C(CH_3)_2$–CH_2–	0	0.2/n.d	n.d.	n.d.
(C_2H_5)–$CH(CH_3)$–CH_2–	0	2.5/n.d	n.d.	n.d.
Cyclohexyl–	0	<0.1/n.d	n.d.	n.d.
phenyl–CH_2–	0	0.1/n.d	n.d.	n.d.
phenyl–	0	360/0.8	1.7/1.8	15.5/0.2
phenyl–$(CH_2)_2$–	0	2.8/n.d	n.d.	n.d.
phenyl–$(CH_2)_3$–	0	3.8/n.d	n.d.	n.d.

a n.d. = not determined.

Carboligase reaction Carboligase activity of ThDP-dependent enzymes requires the formation of a so-called "activated aldehyde" at C2 of ThDP (enamine-carbanion), resulting in Umpolung of the carbonyl reactivity. This step is achieved by either decarboxylation of a 2-ketoacid or by direct addition of a donor aldehyde (Scheme 2.2.3.2). Subsequently, a second aldehyde molecule is attached nucleophilically, leading to C–C bond formation. Examples of the resulting products are given in Scheme 2.2.3.2. According to their substrate specificity, PDC accepts only acetaldehyde as a donor substrate, and BFD prefers aromatic aldehydes such as benzaldehyde and heteroaromatic aldehydes [8]. However, aliphatic and aromatic aldehydes are accepted as acceptor aldehydes by both enzymes. The carboligase activity of BFD has been studied intensively (see Chapters 2.2.7 and 3.1).

2.2.3.2.2 Identification of Amino Acid Residues Relevant to Substrate Specificity and Enantioselectivity

Comparison of the 3D structures of the two enzymes identified two residues in each, which are likely to influence the substrate specificity. Site-directed mutagenesis was used to exchange these residues in both BFD and PDC, resulting in six variants: PDC*I472A*, PDC*I476F*, PDC*I472A/I476F* and BFD*A460I*, BFD*F464I*, BFD*A460I/F464I*. The best results were obtained with two variants, PDC*I472A* and BFD*A460I*, both showing the desired cross-reactivity with benzoylformate and pyruvate, respectively. Additionally, both variants prefer the long-chain aliphatic substrates 2-ketopentanoic and 2-ketohexanoic acid (Table 2.2.3.2).

With respect to the carboligase activity, PDC*I472A* proved to be a real chimera between PDC and BFD, while BFD*A460I/F464I* provided the most interesting result with an almost complete reversal of the stereochemistry of its 2-hydroxy-propiophenone product [7, 9].

2.2.3.2.3 Optimization of the Substrate Range of BFD by Site-Directed Mutagenesis

Since wild-type BFD does not accept *ortho*-substituted benzaldehydes as donor substrates, the variant BFD*H281A* with an enlarged active site was prepared by site-directed mutagenesis of histidine-281 to alanine. This variant shows an activity higher by a factor of >140 with respect to the formation of (*R*)-benzoin. Furthermore, BFD*H281A* allows the selective formation of mixed substituted (*R*)-benzoins by employment of appropriate substrate combinations [10].

2.2.3.2.4 Optimization of Stability and Substrate Range of BFD by Directed Evolution

A major problem in the biotransformation of aromatic aldehydes is their low solubility in aqueous buffer. The same holds for some of the 2-hydroxy ketones

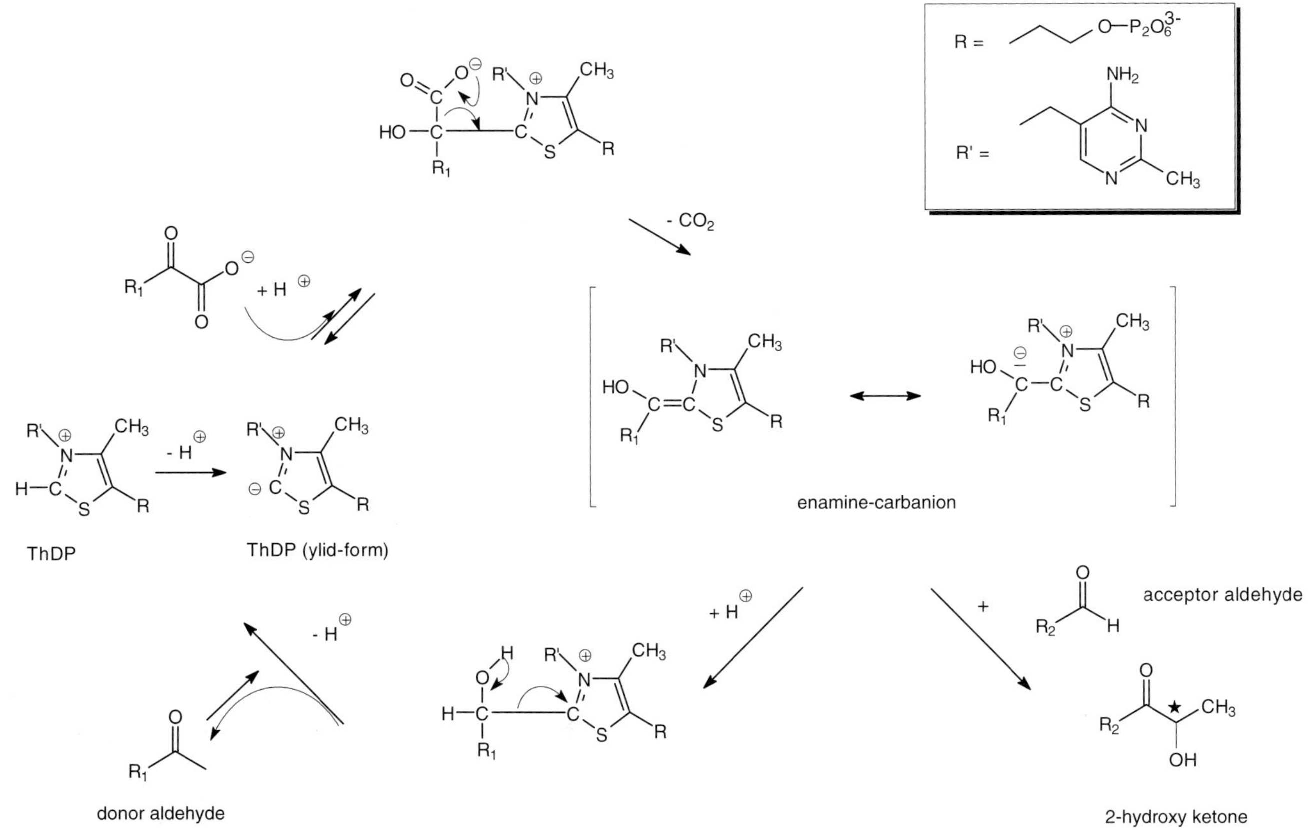

2.2.3 Enzymes for Carboligation – 2-Ketoacid Decarboxylases and Hydroxynitrile Lyases | 331

Scheme 2.2.3.2 Reaction mechanism of PDC and BFD. C2 of the cofactor ThDP is the real active site of both enzymes. The cofactor is regenerated during the reaction cycle. Decarboxylation of 2-keto acids and carboligation of aldehydes have a common reaction intermediate (enamine-carbanion = "active aldehyde").

formed. In general, low substrate and product solubility limit the application of enzymes in bioorganic syntheses. The solubility can be improved by addition of water-miscible organic solvents, such as DMSO and ethanol. Furthermore, carboligase activity of wild-type BFD with respect to the formation of 2-hydroxypropiophenone, at about 7 U mg^{-1}, is lower by a factor of about 50 than the decarboxylase activity of the enzyme (320–340 U mg^{-1}). In order to find more active BFD variants which are stable in the presence of water-miscible organic solvents, an enzyme library was generated by error-prone PCR (epPCR) of the BFD gene. The library was screened using benzaldehyde and acetaldehyde as substrates in the presence of 20 vol.% DMSO and 1.5 M ethanol, respectively. Variants with improved carboligase activity were detected using a colorimetric assay [11]. All improved BFD variants selected carry a single mutation at position Leu 476, which is close to the active site but does not directly interact with the active center. With respect to the increase of both benzaldehyde solubility and carboligase activity, DMSO is superior to ethanol. Compared to the wild-type enzyme the variant BFDL476Q shows a five-fold increased carboligase activity with respect to the formation of (S)-2-hydroxypropiophenone (32 U mg^{-1}) and catalyzes the carboligation with significantly higher enantioselectivity (97%) than wild-type BFD (85%) [12].

The same epPCR library was screened for altered substrate specificity. Since wild-type BFD does not accept *ortho*-substituted benzaldehydes as a donor substrate, *ortho*-methylbenzaldehyde and acetaldehyde were employed as substrates in the presence of 20 vol.% DMSO. Two variants were selected by screening the library, including again the variant BFD*L476Q*. Further investigation of the substrate range showed that this variant is able to accept a broad range of substituted benzaldehydes as donor aldehydes with acetaldehyde as acceptor. The resulting (S)-2-hydroxy ketones are produced enantioselectively [13].

2.2.3.3 Hydroxynitrile Lyases

Hydroxynitrile lyases (HNLs) have been found in plants and some insects. In plants their natural function is in the defence against herbivores by release of HCN from hydroxynitriles upon cell damage. The reverse reaction, the formation of (chiral) hydroxynitriles, is interesting for bioorganic chemistry, making HNLs important biocatalysts for technical applications [14–16].

HNLs comprise a heterogenous enzyme family, since hydroxynitrile lyase activity has evolved in different structural frames by convergent evolution [17, 18]. Thus, (S)-specific HNLs based on an α/β-hydrolase fold framework from *Manihot esculenta* (cassava) [19–21], *Hevea brasilensis* (rubber tree) [22–26], and *Sorghum bicolor* (millet) [27–33] have been described. (R)-specific HNLs based on the structural framework of oxidoreductases were isolated from *Linum usitatissimum* (flax) [30, 34–37] and Rosaceae (e.g., bitter almonds) [31, 38]. Despite their potential in biocatalysis only few HNLs (from cassava and rubber tree) are available by recombinant gene expression, which is a prerequisite for their technical application [20, 24]. Thus, cloning, recombinant expression, and

2.2.3.3.1 HNL from *Sorghum bicolor*

*Sb*HNL (Sordan 79 variety) is a heterotetrameric enzyme $(\alpha/\beta)_2$. Both α- (33 kDa) and β-subunits (23 kDa) are glycosylated and linked by two disulfide bridges [31, 33]. The α- and β-subunit are synthesized in one pre-protein, which is cleaved proteolytically after protein biosynthesis (Fig. 2.2.3.3).

Sequence analysis The N-terminal sequence of the small (23 kDa) subunit was easily accessible by Edman degradation, whereas the N-terminus of the big (33 kDa) subunit is blocked [28, 31, 32]. However, using a high protein concentration for Edman degradation allowed sequencing of a "background sequence" in the big subunit, showing high sequence similarity with carboxypeptidase II (CPII) from wheat [28, 32]. Later, the 3D structure of *Sb*HNL revealed the termini of the 33 kDa subunit [27] (Fig. 2.2.3.4).

Glycosylation Knowledge of possible glycosylation sites in *Sb*HNL is important for the selection of an appropriate recombinant host. Using conventional *N*-glycan detection kits, N-glycosylation of the small subunit was detected [33]. However, the electrophoretic mobility of both subunits was altered upon treatment of the denatured protein with N-glycosidase F, hinting that there were further glycosylation sites in the 33 kDa subunit also [32]. These data were later supported by the 3D structure of *Sb*HNL showing one glycosylation site in each subunits [27] (Fig. 2.2.3.4). The necessity of glycosylation for the enzyme activity was tested by enzymatic cleavage of *N*-glycans using N-glycosidase F. Since *N*-glycans were not

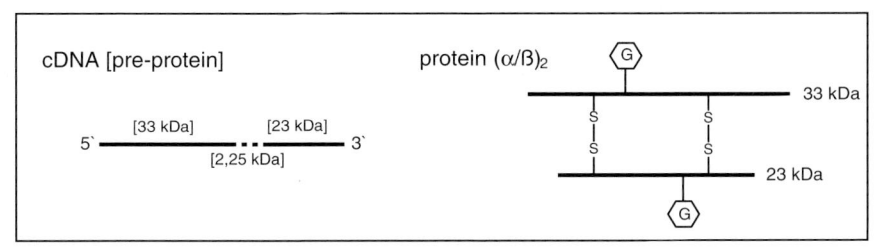

Fig. 2.2.3.3 *Sb*HNL belongs to the α/β-hydrolases. The heterotetramer is formed by two α/β-dimers, which are stabilized by two intermolecular disulfide bonds. Each subunit carries a glycosylation site. Both subunits are coded by one gene. Cleavage into α- and β-subunit is assumed to occur after folding of the pre-protein.

1						M	A	V	F	I	S	S	S	G	S	P	G	R	13
14	A	T	A	T	T	T	T	T	T	L	L	L	A	V	L	A	A		31
32	A	A	A	A	G	L	L	L	A	P	V	A	A	R	G	S	P	P	49
									N-Terminus of the 33 kDa subunit										
50	E	H	D	K	Q	L	Q	L	Q↓Q	Q	E	D	D	R	I	P	G		67
68	L	P	G	Q	P	N	G	V	A	F	G	M	Y	G	G	Y	V	T	85
86	I	D	D	N	N	G	R	A	L	Y	Y	W	F	Q	E	A	D	T	103
104	A	D	P	A	A	A	P	L	V	L	W	L	N	G	G	P	G	C	121
122	S	S	I	G	L	G	A	M	Q	E	L	G	P	F	R	V	H	T	139
								Sequence published in [28]											
140	N	G	E	S	L	L	L	↓N	E	Y	A	W	N	K	A	A	N	I	157
158	L	F	A	E	S	P	A	G	V	V	F	S	Y	S	**N**	T	S	S	175
176	D	L	S	M	G	D	D	K	M	A	Q	D	T	Y	T	F	L	V	193
194	K	W*	F	E	R	F	P	H	Y	N	Y	R	E	F	Y	I	A	G	211
212	E	**S**	G	H	F	I	P	Q	L	S	Q	V	V	Y	R	N	R	N	229
230	N	S	P	F	I	N	F	Q	G	L	L	V	S	S	G	L	T	N	247
248	D	H	E	D	M	I	G	M	F	E	L	W	W	H	H	G	L	I	265
266	S	D	E	T	R	D	S	G	L	K	V	C	P	G	T	S	F	M	283
284	H	P	T	P	E	C	T	E	V	W	N	K	A	L	A	E	Q	G	301
302	N	I	N	P	Y	T	I	Y	T	P	T	C	D	R	E	P	S	P	319
								C-Terminus of the 33 kDa subunit											
320	Y	Q	R	R	F	↓W	A	P	H	G	R	A	A	P	P	P	L	M	337
	↓N-Terminus of the 23 kDa subunit																		
338	↓L	P	P	Y	D	P	C	A	V	F	N	S	I	N	Y	L	N	L	355
356	P	E	V	Q	T	A	L	H	A	**N**	V	S	G	I	V	E	Y	P	373
374	W	T	V	C	S	N	T	I	F	D	Q	W	G	Q	A	A	D	D	391
392	L	L	P*	V	Y	R	E	L	I	Q	A	G	L	R	V	W	V	Y	409
410	S	G	**D**	T	D	S	V	V	P	V	S	S	T	R	R	S	L	A	427
428	A	L	E	L	P	V	K	T	S	W	Y	P	W	Y	W	A	P	T	445
446	E	R	E	V	G	G*	W	S	V	Q	Y	E	G	L	T	Y	V	S	463
464	P	S	G	A	G	**H**	L	V	P	V	H	R	P	A	Q	A	F	L	481
482	L	F	K	Q	F	L	K	G	E	P	M	P	A	E	E	K	N	D	499
500	I	L	L	P	S	E	K	A	P	F	Y	*							510

Fig. 2.2.3.4 Protein sequence of *Sb*HNL [28]. Residues of the potential catalytic triad (bold face) are indicated by an asterisk. The two glycosylation sites (bold face, underlined) and the N- and C-termini of the 33 kDa subunit have been deduced from the 3D structure [27].

accessible in the native state of *Sb*HNL, an unfolding–refolding process for *Sb*HNL was established using 6 M guanidinium hydrochloride (GdnHCl) as a denaturant [32]. The best renaturation results (40%) were obtained using 50 mM sodium phosphate buffer, pH 7.2, in the presence of 0.8 mg mL^{-1} bovine serum albumin (BSA). In order to prevent unfolding of N-glycosidase F, the enzyme was added during renaturation of GdnHCl-unfolded SbHNL. The efficiency of the cleavage was tested by SDS-PAGE. Following the regain of HNL activity we observed significant reactivation in the presence (30%) and absence (40%) of N-glycosidase F, proving that N-glycosylation is not essential for SbHNL activity, although an influence on the enzyme stability could not be excluded. However, the disulfide bonds are essential for the enzymatic activity, making a eukaryotic host still necessary for recombinant gene expression.

Identification of the gene and recombinant expression Identification of the SbHNL gene was hampered by secondary structure formation of the cDNA prepared for amplification of the gene by PCR [30]. Finally the complete coding sequence was solved in cooperation with H. Wajant [28]. The N-terminal sequence is probably a leader sequence targeting the enzyme into a cell organelle (Fig. 2.2.3.4). Due to the complex post-translational modifications of SbHNL, *Pichia pastoris* was chosen as a recombinant host. Trials to express the complete gene including the leader sequence were not successful. However, without the leader sequence a low SbHNL activity was detectable in only one clone. Further investigations revealed that the recombinant SbHNL was not processed proteolytically into two subunits. The best results were obtained with 5.2 U mg^{-1} in a 2 L pichia culture (72 h) with methanol induction. This correlates with the amount of enzyme available from 110 g of sorghum seedlings. Unfortunately, we did not succeed in further improving the recombinant expression system [30].

Carboxypeptidase side-activity Since the protein sequence and 3D structure of SbHNL are very similar to those of carboxypeptidase II (CPII) from wheat, we were interested in investigating both enzymes with respect to cross-activities [28, 39–42]. Further similarities have been found with respect to the structure of the substrates accepted by both enzymes: SbHNL prefers aromatic hydroxynitriles, such as 4-hydroxymandelonitrile [31], and CPII prefers C-terminal hydrophobic and basic amino acids in peptide substrates, such as lysine or arginine, as well as aromatic amino acids, e.g., phenylalanine and tyrosine [40]. Comparison of the preferred substrates of the two enzymes (Fig. 2.2.3.5) suggests that a common binding site for the aromatic ring might exist. Thus, CPII was tested for HNL activity and SbHNL was tested for carboxypeptidase activity. We were not able to detect any HNL activity in CPII. However, there is definitely a weak carboxypeptidase and esterase activity in SbHNL with a substrate preference for dipeptides such as Lys-Tyr and Arg-Tyr, the preferred substrates also of CPII (Table 2.2.3.3) [32].

Meanwhile the 3D structure of SbHNL was solved [27], showing the expected high similarity to the structure of CPII from wheat [41]. Both enzymes use the same active-site residues. However, in SbHNL a unique two-amino acid deletion

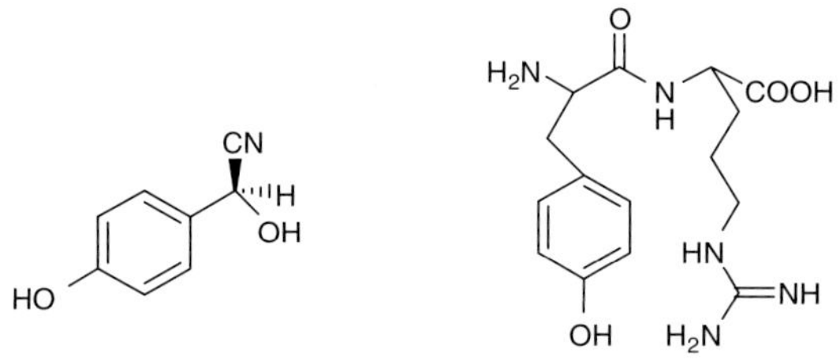

p-Hydroxymandelonitrile   Tyrosinyl-arginine

Fig. 2.2.3.5 Preferred substrates of *Sb*HNL. 4-Hydroxymandelonitrile and tyrosyl-arginine are the preferred substrates of the main hydroxynitrile lyase activity and the carboxypeptidase side-activity. Since both substrates resemble the aromatic ring of tyrosine, catalysis at the same active site can be assumed.

Table 2.2.3.3 Determination of the carboxypeptidase and esterase side-activities of *Sb*HNL in comparison with carboxypeptidase II from wheat.[a]

Substrate P_2-P_1-P_1'	***SbHNL***	***CPII***
Dipeptides		
Mal-Tyr-Arg[b]	++	+++
Tyr-Arg	++	+++
Tyr-Lys	++	+++
Phe-Arg	++	+++
Phe-Ala	–	+
Phe-Leu	–	+
Ala-Arg	–	++
Leu-Arg	–	+++
Amino acid esters		
Tyr-OMe	++	+++
Phe-OEt	++	++
Phe-OBut	–	–
Arg-OMe	–	–

[a] The cleavage of dipeptides was analyzed by thin layer chromatography on silica gel 60 F_{254} plates using a mixture of *n*-butane–acetic acid–water (70 : 30 : 20, by vol.) as the mobile phase. Detection was performed with 3% ninhydrin in isopropanol. A typical reaction mixture contained 100 μL of a 2 mM dipeptide in 50 mM sodium acetate buffer, pH 4.5, and 100 μL of enzyme solution (*Sb*HNL: 0.2 mg mL^{-1}, CPII: 0.02 mg mL^{-1}. Samples were incubated for 18 h at 37 °C. Reaction was stopped by heat inactivation. Transformations: +++: complete; ++: partial; +: weak; –: none. [b] Mal = maleyl.

is next to the putative active site Ser, removing thereby the putative oxyanion hole-forming Tyr residue. This deletion in *Sb*HNL defines a completely different *Sb*HNL active site architecture where the traditional view of a classic triad is not given any more [27].

2.2.3.3.2 HNL from *Linum usitatissimum*

*Lu*HNL is a nonglycosylated homodimer (84 kDa) which catalyzes the reversible cleavage of aliphatic (R)-cyanohydrins [34]. This HNL does not require complex protein modification after protein biosynthesis. Thus, expression in prokaryotic (*Escherichia coli*) and eukaryotic hosts (*Pichia pastoris*) is possible [35–37]. However, initial trials to express *Lu*HNL in *E. coli* were hampered by formation of inclusion bodies [36].

Optimization of recombinant LuHNL expression in E. coli Evaluation of different *E. coli* strains and optimization of the cell cultivation parameters resulted in expression of 75% soluble *Lu*HNL with and without N-terminal hexahistidine fusion. The fusion protein did not affect the catalytic activity negatively, but simplified protein purification significantly. The best results were obtained using the thioreductase-deficient *E. coli* strain AD494(DE3) at a growth and induction temperature of 30 °C and 0.1 M IPTG for induction of protein expression: 57 U of pure *Lu*HNL(His$_6$) are available from 4 g of *E. coli* cells. This correlates with the amount of enzyme accessible from 500 g of flax seedlings [30, 36].

Relationship of LuHNL to alcohol dehydrogenases The *Lu*HNL sequence is highly similar to long-chain Zn-dependent alcohol dehydrogenases (ADHs) of classes I and III. The similarity includes the binding sites of the structural and catalytically important Zn ions as well as the nucleotide binding region [36, 37]. However, *Lu*NHL showed no alcohol dehydrogenase activity [30]. A detailed investigation of the influences of Zn-ions and nucleotide cofactors such as NAD(P)H and NAD(P)$^+$ showed that the nucleotide cofactors are not specifically bound by the enzymes, whereas Zn^{2+} is definitely necessary to maintain the lyase activity of *Lu*HNL. Although the enzyme activity is not affected by chelators such as EDTA and 1,10-phenanthroline at ambient temperature [37], a rapid deactivation occurs at 50 °C [30]. Our studies also show that Zn^{2+} has a stabilizing effect on *Lu*HNL at elevated temperatures [30].

Homology model and putative active center Using the program SWISS Model a homology model of *Lu*HNL was created using the 3D structures of similar enzymes (human liver χχADH: 1TEHA.pdb; Leu141 mutant of human δδADH: 1D1TA.pdb) [30]. In order to identify the putative active center, the model was searched for a catalytic triad. D66, H85, and T111 were found to be in close vicinity to allow interaction. In ADHs D66 and H85 belong to the strictly conserved residues which bind the catalytically important Zn^{2+}. Two of these potential active site residues, H85 and T111, were exchanged by alanine using site-directed mutagenesis and both resulting variants were inactive, thus supporting our hypothesis [30].

Acknowledgements

The authors thank the Deutsche Forschungsgemeinschaft (DFG) for financial support in the SFB 380. We gratefully acknowledge our cooperation partners M. Müller, A. Liese, and C. Wandrey for fruitful discussions and H. Wajant for providing the SbHNL clone.

References

1 M. Pohl, Protein design on pyruvate decarboxylase (PDC) by site-directed mutagenesis. Application to mechanistical investigations, and tailoring PDC for the use in organic synthesis. *Adv. Biochem. Eng. Biotechnol.* **1997**, *58*, 15–43.
2 D. Dobritzsch, S. König, G. Schneider, G. Lu, High resolution crystal structure of pyruvate decarboxylase from *Zymomonas mobilis*. Implications for substrate activation in pyruvate decarboxylases. *J. Biol. Chem.* **1998**, *273*, 20 196–20 204.
3 M. S. Hasson, A. Muscate, M. J. McLeish, L. S. Polovnikova, J. A. Gerlt, G. L. Kenyon, G. A. Petsko, D. Ringe, The crystal structure of benzoylformate decarboxylase at 1.6 Å resolution, diversity of catalytic residues in thiamin diphosphate-dependent enzymes. *Biochemistry* **1998**, *37*, 9918–9930.
4 P. Iwan, G. Goetz, S. Schmitz, B. Hauer, M. Breuer, M. Pohl, Studies on the continuous production of (R)-(–)-phenylacetylcarbinol in an enzyme–membrane reactor. *J. Mol. Catal. B – Enzymatic* **2001**, *11*, 387–396.
5 H. Iding, P. Siegert, K. Mesch, M. Pohl, Application of alpha-keto acid decarboxylases in biotransformations. *Biochim. Biophys. Acta* **1998**, *1385*, 307–322.
6 G. Goetz, P. Iwan, B. Hauer, M. Breuer, M. Pohl, Continuous production of (R)-phenylacetylcarbinol in an enzyme–membrane reactor using a potent mutant of pyruvate decarboxylase from *Zymomonas mobilis*. *Biotechnol. Bioeng.* **2001**, *74*, 317–325.
7 M. Pohl, P. Siegert, K. Mesch, H. Bruhn, J. Grötzinger, Active site mutants of pyruvate decarboxylase from *Zymomonas mobilisi* – a site-directed mutagenesis study of L112, I472, I476, E473, and N482. *Eur. J. Biochem.* **1998**, *257*, 538–546.
8 H. Iding, T. Dünnwald, L. Greiner, A. Liese, M. Müller, P. Siegert, J. Grötzinger, A. S. Demir, M. Pohl, Benzoylformate decarboxylase from *Pseudomonas putida* as stable catalyst for the synthesis of chiral 2-hydroxy ketones. *Chem. Eur. J.* **2000**, *6*, 1483–1495.

9 P. Siegert, M. J. McLeish, M. Baumann, H. Iding, M. M. Kneen, G. L. Kenyon, M. Pohl, Exchanging the substrate specificities of pyruvate decarboxylase from *Zymomonas mobilis* and benzoylformate decarboxylase from *Pseudomonas putida*. *Protein Eng. Des. Sel.* **2005**, *18*, 345–357.

10 P. Dünkelmann, D. Kolter-Jung, A. Nitsche, A. S. Demir, P. Siegert, B. Lingen, M. Baumann, M. Pohl, M. Müller, Development of a donor–acceptor concept for enzymatic cross-coupling reactions of aldehydes, the first asymmetric cross-benzoin condensation. *J. Am. Chem. Soc.* **2002**, *124*, 12084–12085.

11 M. Breuer, M. Pohl, B. Hauer, B. Lingen, High-throughput assay of (*R*)-phenylacetylcarbinol synthesized by pyruvate decarboxylase. *Anal. Bioanal. Chem.* **2002**, *374*, 1069–1073.

12 B. Lingen, J. Grötzinger, D. Kolter, M. R. Kula, M. Pohl, Improving the carboligase activity of benzoylformate decarboxylase from *Pseudomonas putida* by a combination of directed evolution and site-directed mutagenesis. *Protein Eng.* **2002**, *15*, 585–593.

13 B. Lingen, D. Kolter-Jung, P. Dünkelmann, R. Feldmann, J. Grötzinger, M. Pohl, M. Müller, Alteration of the substrate specificity of benzoylformate decarboxylase from *Pseudomonas putida* by directed evolution. *ChemBioChem* **2003**, *4*, 721–726.

14 F. Effenberger, S. Forster, H. Wajant, Hydroxynitrile lyases in stereoselective catalysis. *Curr. Opin. Biotechnol.* **2000**, *11*, 532–539.

15 H. Griengl, H. Schwab, M. Fechter, The synthesis of chiral cyanohydrins by oxynitrilases. *Trends Biotechnol.* **2000**, *18*, 252–256.

16 D. V. Johnson, A. A. Zabelinskaja-Mackova, H. Griengl, Oxynitrilases for asymmetric C–C bond formation. *Curr. Opin. Chem. Biol.* **2000**, *4*, 103–109.

17 K. Gruber, G. Gartler, B. Krammer, H. Schwab, C. Kratky, Reaction mechanism of hydroxynitrile lyases of the χ/β-hydrolase superfamily. *J. Biol. Chem.* **2004**, *279*, 20 501–20 510.

18 H. Wajant, F. Effenberger, Hydroxynitrile lyases of higher plants. *Biol. Chem.* **1996**, *377*, 611–617.

19 J. Hughes, Z. Keresztessy, K. Brown, S. Suhandono, M. A. Hughes, Genomic organization and structure of χ-hydroxynitrile lyase in cassava (*Manihot esculenta* Crantz). *Arch. Biochem. Biophys.* **1998**, *356*, 107–116.

20 J. Hughes, J. H. Lakey, M. A. Hughes, Production and characterization of a plant χ-hydroxynitrile lyase in *Escherichia coli*. *Biotechnol. Bioeng.* **1997**, *53*, 332–338.

21 W. L. B. White, D. I. Arias-Garzon, J. M. McMahon, R. T. Sayre, Cyanogenesis in Cassava. *Plant Physiol.* **1998**, *116*, 1219–1225.

22 M. Hasslacher, M. Schall, M. Hayn, H. Griengl, S. D. Kohlwein, H. Schwab, (*S*)-hydroxynitrile lyase from *Hevea brasiliensis*. *Ann. N. Y. Acad. Sci.* **1996**, *799*, 707–712.

23 M. Hasslacher, M. Schall, M. Hayn, H. Griengl, S. D. Kohlwein, H. Schwab, Molecular cloning of the full-length cDNA of (*S*)-hydroxynitrile lyase from *Hevea brasiliensis*. Functional expression in *Escherichia coli* and *Saccharomyces cerevisiae* and identification of an active site residue. *J. Biol. Chem.* **1996**, *271*, 5884–5891.

24 M. Hasslacher, M. Schall, M. Hayn, R. Bona, K. Rumbold, J. Luckl, H. Griengl, S. D. Kohlwein, H. Schwab, High-level intracellular expression of hydroxynitrile lyase from the tropical rubber tree *Hevea brasiliensis* in microbial hosts. *Protein Expr. Purif.* **1997**, *11*, 61–71.

25 M. Hasslacher, C. Kratky, H. Griengl, H. Schwab, S. D. Kohlwein, Hydroxynitrile lyase from *Hevea brasiliensis*, molecular characterization and mechanism of enzyme catalysis. *Proteins* **1997**, *27*, 438–449.

26 U. G. Wagner, M. Hasslacher, H. Griengl, H. Schwab, C. Kratky, Mechanism of cyanogenesis, the crystal structure of hydroxynitrile lyase from *Hevea brasiliensis*. *Structure* **1996**, *4*, 811–822.

27 H. Lauble, B. Miehlich, S. Forster, H. Wajant, F. Effenberger, Crystal structure of hydroxynitrile lyase from *Sorghum bicolor* in complex with the inhibitor benzoic acid, a novel cyanogenic enzyme. *Biochemistry* **2002**, *41*, 12 043–12 050.

28 H. Wajant, K. W. Mundry, K. Pfizenmaier, Molecular cloning of hydroxynitrile lyase from *Sorghum bicolor* (L.). Homologies to serine carboxypeptidases. *Plant Mol. Biol.* **1994**, *26*, 735–746.

29 R. Woker, B. Champluvier, M. R. Kula, Purification of S-oxynitrilase from *Sorghum bicolor* by immobilized metal ion affinity chromatography on different carrier materials. *J. Chromatogr.* **1992**, *584*, 85–92.

30 P. Heim, Die Hydroxynitril-Lyasen aus *Linum usitatissimum* (Lein) und *Sorghum bicolor* (Hirse). Untersuchungen zur rekombinantion Expression und phylogenetischen Verwandtschaft [Doctoral Thesis], Heinrich-Heine University Düsseldorf, **2002**.

31 I. Jansen, R. Woker, M. R. Kula, Purification and protein characterization of hydroxynitrile lyase from Sorghum and Almond. *Biotechnol. Appl. Biochem.* **1992**, *15*, 90–99.

32 B. Juchem, Proteinchemische Charakterisierung der (S)-Oxynitrilase aus *Sorghum bicolor* (Hirse) [Diploma Thesis], FH Jülich, **1996**.

33 H. Wajant, K. W. Mundry, Hydroxynitrile lyase from *Sorghum bicolor*, a glycoprotein heterotetramer. *Plant Sci.* **1993**, *89*, 127–133.

34 J. Albrecht, I. Jansen, M. R. Kula, Improved purification of an (R)-oxynitrilase from *Linum usitatissimum* (flax) and investigation of the substrate range. *Biotechnol. Appl. Biochem.* **1993**, *17 (Pt. 2)*, 191–203.

35 H. Breithaupt, Klonierung und Expression der (R)-Hydroxynitrile Lyase aus *Linum usitatissimum* (Lein) [Doctoral Thesis], Heinrich-Heine University Düsseldorf, **1997**.

36 H. Breithaupt, M. Pohl, W. Bönigk, P. Heim, K. L. Schimz, M. R. Kula, Cloning and expression of (R)-hydroxynitrile lyase from *Linum usitatissimum* (flax). *J. Mol. Catal. B – Enzymatic* **1999**, *6*, 315–332.

37 K. Trummler, H. Wajant, Molecular cloning of acetone cyanohydrin lyase from flax (*Linum usitatissimum*). Definition of a novel class of hydroxynitrile lyases. *J. Biol. Chem.* **1997**, *272*, 4770–4774.

38 I. Dreveny, K. Gruber, A. Glieder, A. Thompson, C. Kratky, The hydroxynitrile lyase from almond, a lyase that looks like an oxidoreductase. *Structure* **2001**, *9*, 803–815.

39 T. L. Bullock, K. Breddam, S. J. Remington, Peptide aldehyde complexes with wheat serine carboxypeptidase II, implications for the catalytic mechanism and substrate specificity. *J. Mol. Biol.* **1996**, *255*, 714–725.

40 T. L. Bullock, B. Branchaud, S. J. Remington, Structure of the complex of L-benzylsuccinate with wheat serine carboxypeptidase II at 2.0 Å resolution. *Biochemistry* **1994**, *33*, 11 127–11 134.

41 D. I. Liao, K. Breddam, R. M. Sweet, T. Bullock, S. J. Remington, Refined atomic model of wheat serine carboxypeptidase II at 2.2 Å resolution. *Biochemistry* **1992**, *31*, 9796–9812.

42 D. L. Ollis, E. Cheah, M. Cygler, B. Dijkstra, F. Frolow, S. M. Franken, M. Harel, S. J. Remington, I. Silman, J. Schrag, et al., The alpha/beta hydrolase fold. *Protein. Eng.* **1992**, *5*, 197–211.

2.2.4
Preparative Syntheses of Chiral Alcohols using (R)-Specific Alcohol Dehydrogenases from *Lactobacillus* Strains
Andrea Weckbecker, Michael Müller, and Werner Hummel

2.2.4.1 Introduction

Alcohol dehydrogenases (ADHs) are becoming increasingly important for industrial applications and are of economic significance. Their products, enantiopure alcohols, are important building blocks for organic chemists. The most frequently used ADHs are the commercial available ones from yeast, horse liver, and *Thermoanaerobium brockii*. All these enzymes have disadvantages in technical applications, such as insufficient long-term stability and limited substrate acceptance. A recently described ADH from *Rhodococcus erythropolis* [1] cloned and overexpressed in *Escherichia coli* forms (S)-alcohols and is useful in technical applications [2]. From *Lactobacillus* strains new enzymes were described forming (R)-alcohols.

2.2.4.2 (R)-Specific Alcohol Dehydrogenase from *Lactobacillus kefir*

(R)-specific alcohol dehydrogenase (ADH) from *L. kefir* (LK-ADH) was found during a screening procedure for the enantioselective reduction of acetophenone [3]. The enzyme depends on NADP and needs the presence of Mg^{2+} ions for its activity. LK-ADH belongs to a group of dehydrogenases which accepts NADP as a cofactor, and catalyzes the reversible reduction of ketones to secondary alcohols. The substrate specificity of the enzyme is very broad. Syntheses of chiral aromatic, cyclic, polycyclic, and aliphatic alcohols proceed with high enantioselectivities. From the wild-type strain *L. kefir* this ADH was purified to homogeneity and its biochemical properties were determined [4]. Based upon the partially known amino acid sequence of LK-ADH the *adh* gene could be isolated [5]. The corresponding nucleotide sequence comprises 759 bp. The ORF encodes a protein of 252 amino acids which has a deduced molecular weight of 26 781 Da and a calculated isoelectric point of 5.09. A comparison of the deduced amino acid sequence of ADH with NCBI database entries demonstrates that the enzyme belongs to the family of short-chain alcohol dehydrogenases. This group of enzymes contains subunits which consist of approximately 250 amino acids, and a characteristic coenzyme binding site, GXXXGXG, in the N-terminal region. As shown by gel filtration, the molecular weight of the native enzyme was about 100 000 Da. This result suggests that LK-ADH appears to be a homotetramer. After cloning of the *adh* gene the recombinant protein was expressed in *E. coli* BL21(DE3). LK-ADH was purified by anion-exchange chromatography. The activity of the purified enzyme was determined to be 558 U mg^{-1} (Table 2.2.4.1).

Recombinant LK-ADH appears as a single protein band of about 28 kDa in SDS-PAGE (Fig. 2.2.4.1). The recombinant enzyme was characterized biochemically; it exhibits its maximum activity at 50 °C (Fig. 2.2.4.2). The optimum pH for reductions was determined as 7.0 (Fig. 2.2.4.3).

Table 2.2.4.1 Summary of purification of recombinant LK-ADH.

Purification step	Volume activity [U mL^{-1}]	Protein concentration [mg mL^{-1}]	Specific activity [U mg^{-1}]	Purification factor (-fold)
Crude extract	1449.9	13.0	111.5	1.0
Pool after anion-exchange chromatography	781.0	1.4	557.9	5.0

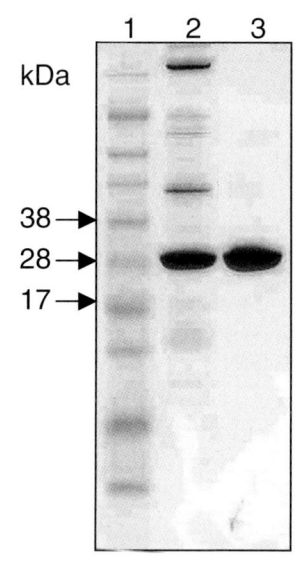

Fig. 2.2.4.1 SDS-PAGE analysis of recombinant LK-ADH purification. Lane 1: molecular weight standards (kDa); Lane 2: crude extract of recombinant E. coli BL21(DE3)/pADH; Lane 3: ADH after anion-exchange chromatography.

The kinetic parameters of recombinant LK-ADH were studied for various compounds. Data are shown in Table 2.2.4.2.

(R)-specific ADH from L. kefir was used for the reduction of various ketones to the corresponding secondary alcohols. Aliphatic, aromatic, and cyclic ketones as well as keto esters were accepted as substrates. The activities achieved with several substrates were compared with the activity obtained with the standard substrate of ADH, acetophenone (Fig. 2.2.4.4). As the figure shows, recombinant LK-ADH has a very broad substrate spectrum, including many types of ketones.

2.2.4.3 Comparison of (R)-Specific ADHs from L. kefir and L. brevis

Lactobacillus brevis also produces an (R)-specific ADH (LB-ADH) [4]. These two enzymes have similar properties. An alignment of the protein sequences of both short-chain dehydrogenases revealed an identity of 88% (Fig. 2.2.4.5). The nucleotide sequences of LK-ADH and LB-ADH show an identity of 78%.

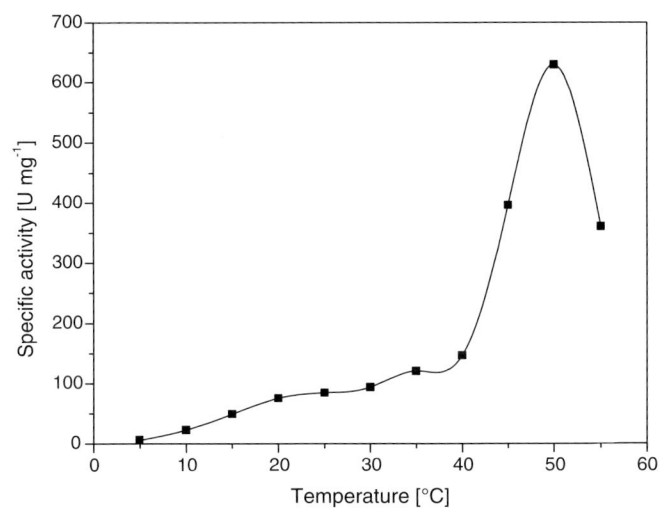

Fig. 2.2.4.2 Determination of the optimum temperature of recombinant (R)-specific ADH. The enzymatic activity was measured with acetophenone using the standard assay.

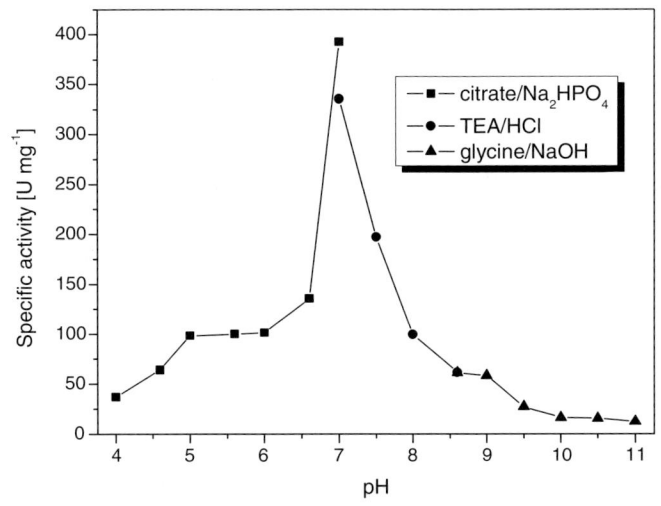

Fig. 2.2.4.3 Effect of pH on the activity of recombinant ADH. The enzymatic activity was measured using the following 250 mM buffers: citrate/Na_2HPO_4, pH 4.0–7.0; TEA/HCl, pH 7.0–8.6; and glycine/NaOH, pH 8.6–11.0.

Table 2.2.4.2 Kinetic characterization of recombinant LK-ADH for different substrates.[a]

Substrate	V_{max} ***[U mg^{-1}]***	K_M ***[mM]***
Acetophenone	281.4	1.90
(*R,S*)-Phenylethanol	23.4	3.09
NADPH	400.0	0.14
NADP	15.3	0.03
NADH	2.9	0.40
NAD	0.49	0.78

a All measurements were carried out under standard assay conditions with varying substrate concentrations.

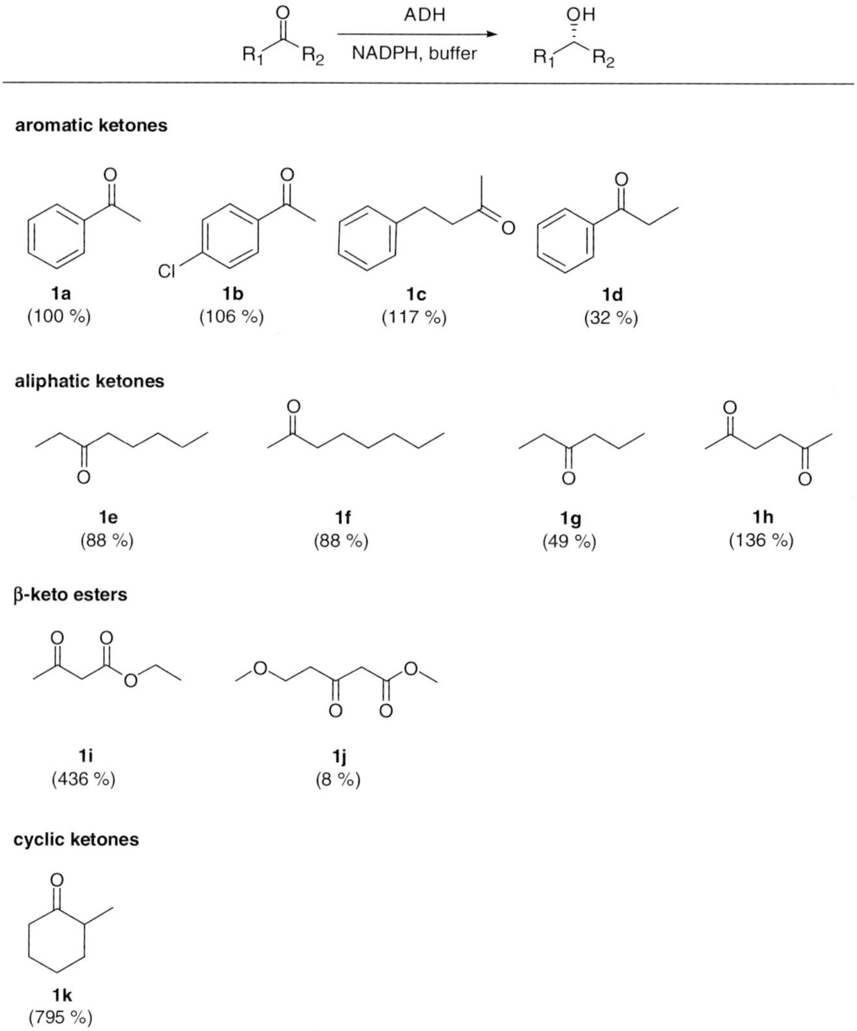
ADH
NADPH, buffer
aromatic ketones
1a (100 %)
1b (106 %)
1c (117 %)
1d (32 %)
aliphatic ketones
1e (88 %)
1f (88 %)
1g (49 %)
1h (136 %)
β-keto esters
1i (436 %)
1j (8 %)
cyclic ketones
1k (795 %)

Fig. 2.2.4.4 Determination of LK-ADH activity toward different substrate types.

2.2.4 Preparative Syntheses of Chiral Alcohols using (R)-Specific Alcohol Dehydrogenases | 345

The kinetic data of the two enzymes for important compounds such as NADPH are similar. A highly conserved region of short-chain dehydrogenases is the cofactor binding site, which is located at the N-termini. This characteristic sequence is shaded in gray in Fig. 2.2.4.5. A decisive advantage of ADH from *L. kefir* is its higher expression rate in pET expression systems [5]. Formate is a strong inhibitor of LB-ADH whereas it has no effect on LK-ADH [5]. A further advantageous feature of LK-ADH is the easier purification procedure for the recombinant enzyme. By using anion-exchange chromatography, pure enzyme is available in only one step, while the purification process of recombinant LB-ADH comprises four steps [4].

2.2.4.4 Preparative Applications of ADHs from *L. kefir* and *L. brevis*

LK-ADH and LB-ADH can both be used for enantioselective synthesis of chiral alcohols. Bradshaw et al. used LK-ADH for the synthesis of a number of chiral alcohols in good yields and high enantiomeric excesses (94–99%) [6]. For the regeneration of the expensive cofactor NADPH, 2-propanol was used according to Scheme 2.2.4.1.

```
LK-ADH    MTDRLKGKVAIVTGGTLGIGLAIADKFVEEGAKVVITGRHADVGEKAAKSIGGTDVIRFV  60
LB-ADH    MSNRLDGKVAIITGGTLGIGLAIATKFVEEGAKVMITGRHSDVGEKAAKSVGTPDQIQFF  60
          *::**.*****:*********** *********:*****:*********:* .* *:*.

LK-ADH    QHDASDEAGWTKLFDTTEEAFGPVTTVVNNAGIAVSKSVEDTTTEEWRKLLSVNLDGVFF  120
LB-ADH    QHDSSDEDGWTKLFDATEKAFGPVSTLVNNAGIAVNKSVEETTTAEWRKLLAVNLDGVFF  120
          ***:***.*******:**:*****:*:********.****:***:*****:********

LK-ADH    GTRLGIQRMKNKGLGASIINMSSIEGFVGDPTLGAYNASKGAVRIMSKSAALDCALKDYD  180
LB-ADH    GTRLGIQRMKNKGLGASIINMSSIEGFVGDPSLGAYNASKGAVRIMSKSAALDCALKDYD  180
          ******************************:*****************************

LK-ADH    VRVNTVHPGYIKTPLVDDLEGAEEMMSQRTKTPMGHIGEPNDIAWICVYLASDESKFATG  240
LB-ADH    VRVNTVHPGYIKTPLVDDLPGAEEAMSQRTKTPMGHIGEPNDIAYICVYLASNESKFATG  240
          ******************* ****.******************:******:*******

LK-ADH    AEFVVDGGYTAQ  252
LB-ADH    SEFVVDGGYTAQ  252
          :***********
```

Fig. 2.2.4.5 Sequence alignment of LK-ADH with LB-ADH. The typical coenzyme binding site, GXXXGXG, of short-chain ADHs is shaded in gray.

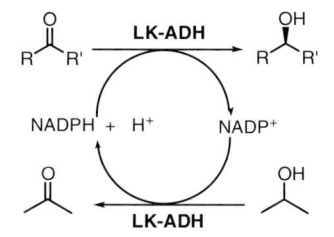

Scheme 2.2.4.1 Synthesis of chiral alcohols using LK-ADH. 2-Propanol was used as a co-substrate for cofactor regeneration.

2.2.4.4.1 Synthesis of (R,R)-Diols

A very important application of ADH from *L. kefir* is the diastereoselective reduction of diketones. The enzyme reduces both oxo functions highly diastereoselectively so that (2R,5R)-hexanediol could be produced starting from 2,5-hexanedione in quantitative yields with $ee > 99\%$ and $de > 99\%$ [7] (Scheme 2.2.4.2). For this process, resting whole cells of *L. kefir* were used.

2.2.4.4.2 Synthesis of Enantiopure 1-Phenylpropane-1,2-diols

LB-ADH was used for the synthesis of *vic*-diols. Starting from benzaldehyde and acetaldehyde, (1S,2S)-1-phenylpropane-1,2-diol ($de = 98\%$) and (1S,2R)-1-phenylpropane-1,2-diol ($de = 99\%$), respectively, could be produced in a stereoselective two-step enzymatic synthesis using benzaldehyde lyase (BAL) and accordingly benzoylformate decarboxylase (BFD) as well as LB-ADH [8] (Scheme 2.2.4.3).

2.2.4.4.3 Synthesis of Enantiopure Propargylic Alcohols

Chiral propargylic alcohols, which are important intermediates in the synthesis of a variety of natural products, including alkaloids, pheromones, antibiotics, and so forth, could be produced from alkynones using LB-ADH [9, 10]. Various alkynones, including aromatic, silyl, and *n*-alkyl alkynones, were accepted as substrates. The corresponding propargylic alcohols were obtained in good yields and excellent enantiomeric excesses. In this case 2-propanol was used as co-substrate for the regeneration of NADPH (Scheme 2.2.4.4).

2.2.4.4.4 Regioselective Reduction of *t*-Butyl 6-chloro-3,5-dioxohexanoate to the Corresponding Enantiopure (S)-5-Hydroxy Compound

t-Butyl (S)-6-chloro-5-hydroxy-3-oxohexanoate is an important building block in the synthesis of HMG-CoA reductase inhibitors. It can be produced from the corresponding 3,5-dioxocarboxylate using LB-ADH with $>99.5\%$ ee [11, 12]. In this process NADPH is regenerated by using 2-propanol as a co-substrate (see Chapter 2.2.7, Scheme 2.2.7.2).

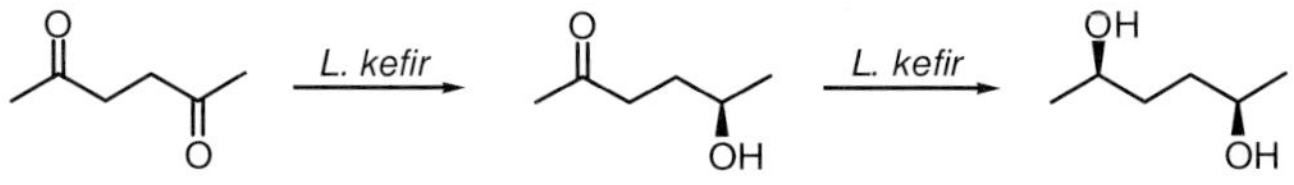

Scheme 2.2.4.2 Diastereoselective reduction of 2,5-hexanedione to (2R,5R)-hexanediol with whole cells of *L. kefir*.

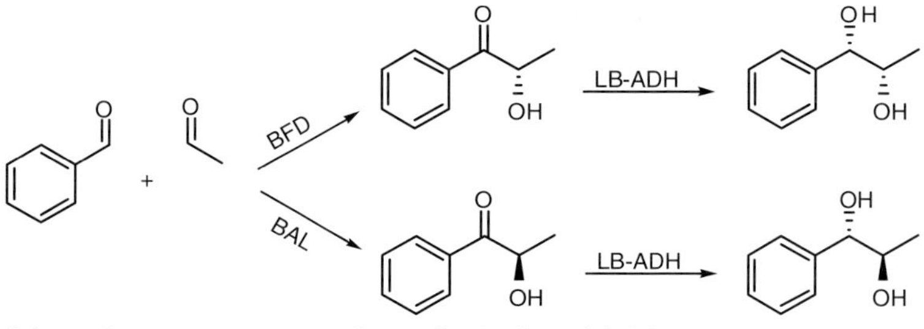

Scheme 2.2.4.3 Enzymatic synthesis of (1S,2S)- and (1S,2R)-1-phenylpropane-1,2-diol using BDD and BAL, respectively, as well as LB-ADH.

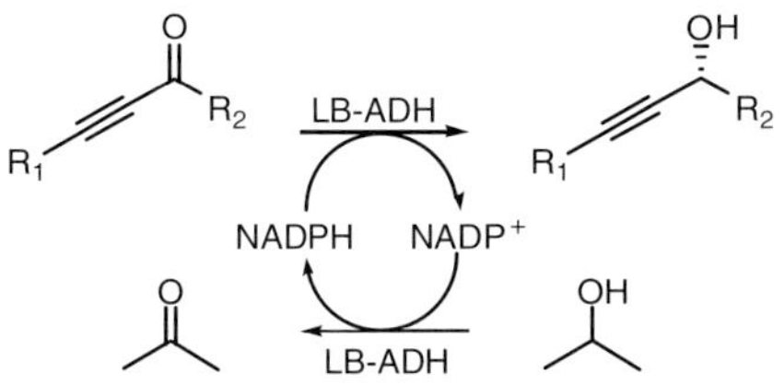

Scheme 2.2.4.4 Reduction of alkynones to chiral propargylic alcohols using LB-ADH. NADPH regeneration was performed using 2-propanol as co-substrate.

2.2.4.5 Coenzyme Regeneration and the Construction and Use of "Designer Cells"

The use of 2-propanol as a co-substrate to regenerate NAD(P)H has a few disadvantages. Both reactions are catalyzed by the same enzyme. In order to shift the balanced reaction, 2-propanol has to be available in excess. Due to the high concentration of 2-propanol, the ketone reduction is inhibited competitively, so it is difficult to achieve a complete conversion of the ketone. Another approach to cofactor regeneration, besides the substrate-coupled process, is the enzyme-coupled one. Dehydrogenases such as formate dehydrogenase (FDH) or glucose dehydrogenase (GDH) are often used for the regeneration of NAD(P)H [13, 14]. To avoid disadvantages arising through the use of isolated enzymes, such as their isolation and purification, we constructed "designer cells" which express both ADH from *L. kefir* and GDH from *Bacillus subtilis*. By using these coexpression systems, the advantages of whole-cell biotransformations such as the ease of use of biotransformations and cell internal cofactor regeneration can be exploited. For the construction of the whole-cell biocatalysts we ligated ADH as well as GDH in the expression vector pET-21a(+) [15]. The derived plasmid pAW-3 contains the *adh* gene behind the promoter and the first ribosomal binding site, and afterward the *gdh* gene behind a second ribosomal binding site, whereas pAW-4 contains these two both genes in the reverse order (Fig. 2.2.4.6). Both plasmids were transformed into *E. coli* BL21(DE3) cells.

After characterization of the systems, biotransformations were performed to produce chiral alcohols using 10 mM acetophenone, 15 mM 2,5-hexanedione, and 25 mM *t*-butyl 6-chloro-3,5-dioxohexanoate as substrates (Scheme 2.2.4.5).

The reaction mixtures contained 5 and 0.5 mg of cells, respectively, in a total volume of 1 mL. To half the batches 1 mM NADP was added to check whether the amount of intracellular cofactor was sufficient to perform efficient reactions. Starting from 10 mM acetophenone, this ketone compound was completely reduced after a short reaction time by *E. coli* BL21(DE3)/pAW-3 and by *E. coli* BL21(DE3)/pAW-4 cells (0.5 mg of each) (Fig. 2.2.4.7).

Figure 2.2.4.7 shows that after addition of NADP, reductions were more efficient. Conversion of acetophenone was complete after a reaction time of 20 min. ADH activity in *E. coli* BL21(DE3)/pAW-3 is threefold higher than in *E. coli* BL21(DE3)/pAW-4 cells. Enantioselective reductions of various ketones are more efficient using *E. coli* BL21(DE3)/pAW-3 than with pAW-4. All ketones were reduced completely in a stereoselective manner; alcohols were formed with >99% *ee* and *de*.

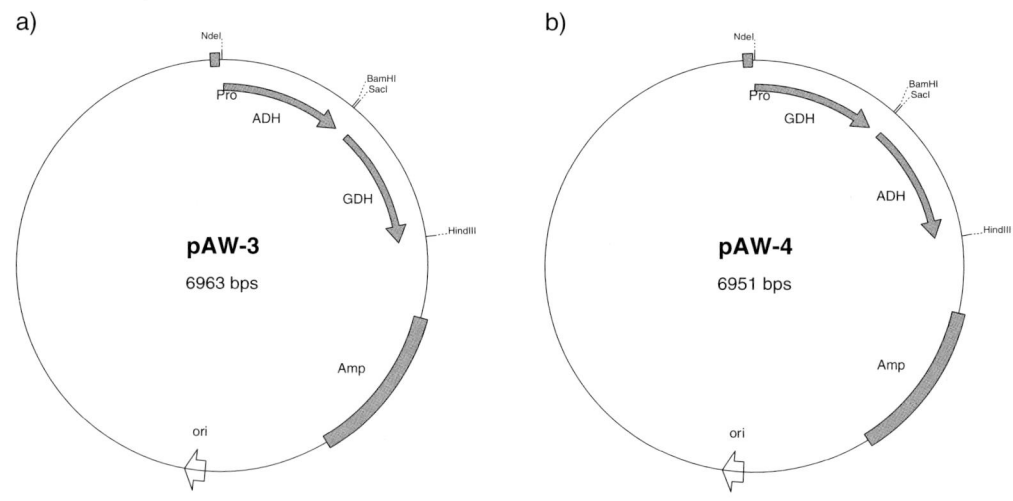

Fig. 2.2.4.6 Vector maps of a) pAW-3 and b) pAW-4. Both plasmids are based on pET-21a(+) and contain *adh* and *gdh* genes.

Scheme 2.2.4.5 Substrates of LK-ADH which were used in whole-cell biotransformations.

The specific cell activities for the reduction of 10 mM acetophenone, 15 mM 2,5-hexanedione, and 25 mM *t*-butyl 6-chloro-3,5-dioxohexanoate by *E. coli* BL21(DE3)/pAW-3 and *E. coli* BL21(DE3)/pAW-4 are listed in Table 2.2.4.3, which shows that the reduction of acetophenone is the most efficient one, and that excellent specific cell activities were reached particularly with *E. coli* BL21(DE3)/pAW-3 cells.

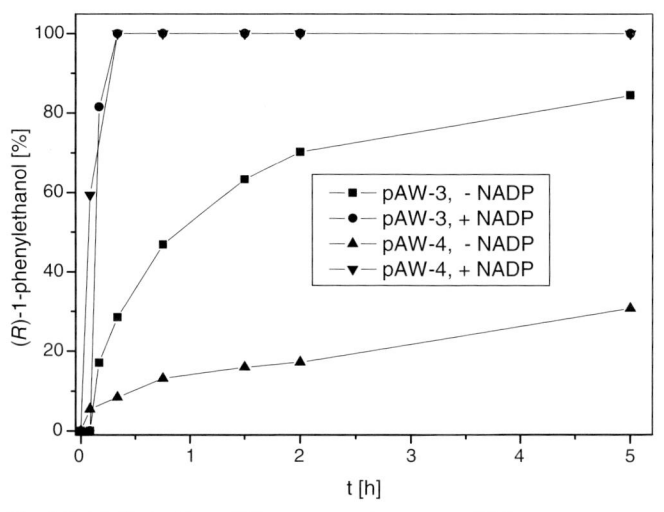

Fig. 2.2.4.7 Reduction of 10 mM acetophenone to (R)-1-phenylethanol using 0.5 mg E. coli BL21(DE3)/pAW-3 and E. coli BL21(DE3)/pAW-4 cells per 1 mL reaction mixture. NADP (1 mM) was added to half the batches.

Table 2.2.4.3 Specific cell activities achieved during conversions of 10 mM acetophenone, 15 mM 2,5-hexanedione, and 25 mM t-butyl 6-chloro-3,5-dioxohexanoate using E. coli BL21(DE3)/pAW-3 and E. coli BL21(DE3)/pAW-4 cells.

	Activity [$U\,mg^{-1}$ cell wet wt.]	
	E. coli BL21(DE3)/pAW-3	E. coli BL21(DE3)/pAW-4
Acetophenone	3047.16	1419.83
2,5-Hexanedione	185.80	49.88
t-Butyl 6-chloro-3,5-dioxohexanoate	345.60	221.20

2.2.4.6 Discussion

Alcohol dehydrogenases from *L. kefir* and *L. brevis* are enzymes with similar properties. Because of their wide substrate spectra and their high activities toward many carbonyl compounds their application potential is very broad. For example, they can be used in the synthesis of pharmaceuticals and natural products. Because LK-ADH has more advantageous properties than LB-ADH, such as an easier purification procedure and a higher expression rate in pET systems, we decided to use LK-ADH for the construction of whole-cell biocatalysts consisting of ADH as well as GDH for NADPH regeneration. Keeping in mind the high price of NADPH cofactor regeneration, it is necessary to perform efficient reaction processes. The application of whole-cell biocatalysts offers some advantages compared to isolated enzymes. The use of isolated enzymes requires the isolation and purification of the proteins. Furthermore, a second enzyme or a second substrate for cofactor regeneration has to be added to the reaction mixture. Because it is difficult to achieve quantitative conversion by application of the substrate-coupled approach, we designed catalysts that contain a second enzyme for cofactor regeneration. The application of these so-called designer cells is a novel approach to perform high-capacity biotransformations.

We designed two biocatalysts consisting of LK-ADH and GDH which differ in the order of the cloned genes. Plasmid pAW-3 contains *adh* directly behind the promoter, and *gdh* afterward, whereas the order of both genes is reversed in pAW-4. In *E. coli* BL21(DE3)/pAW-3 cells, ADH activity is threefold higher than in *E. coli* BL21(DE3)/pAW-4 cells. It is more efficient to use cells containing pAW-3 for conversions of different substrates. We examined the synthesis of chiral alcohols using acetophenone, 2,5-hexanedione, and *t*-butyl 6-chloro-3,5-dioxohexanoate as substrates. All the reactions proceeded completely stereoselectively with *ee* and *de* values of >99%. Acetophenone worked by far the best of these substrates but good yields were obtained with the other substrates, too.

Unlike the syntheses known hitherto, by applying these new biocatalysts it is possible to produce chiral alcohols with optical purities of >99% under moderate reaction conditions in an aqueous milieu.

References

1 K. Abokitse, W. Hummel, *Appl. Microbiol. Biotechnol.* **2003**, *62*, 380–386.
2 W. Hummel, K. Abokitse, K. Drauz, C. Rollmann, H. Gröger, *Adv. Synth. Catal.* **2003**, *345*, 153–159.
3 W. Hummel, *Appl. Microbiol. Biotechnol.* **1990**, 34, 15–19.
4 B. Riebel, Dissertation, Heinrich-Heine-Universität Düsseldorf, 1996.
5 A. Weckbecker, Dissertation, Heinrich-Heine-Universität Düsseldorf, 2005.
6 C. W. Bradshaw, W. Hummel, C.-H. Wong, *J. Org. Chem.* **1992**, *57*, 1532–1536.
7 J. Haberland, A. Kriegesmann, E. Wolfram, W. Hummel, A. Liese, *Appl. Microbiol. Biotechnol.* **2002**, *58*, 595–599.
8 D. Kihumbu, T. Stillger, W. Hummel, A. Liese, *Tetrahedron: Asymmetry* **2002**, *13*, 1069–1072.
9 T. Schubert, W. Hummel, M. Müller, *Angew. Chem.* **2002**, *114*, 656–659; *Angew. Chem. Int. Ed.* **2002**, *41*, 634–637.
10 T. Schubert, W. Hummel, M. R. Kula, M. Müller, *Eur. J. Org. Chem.* **2001**, 4181–4187.
11 M. Wolberg, W. Hummel, M. Müller, *Chemistry* **2001**, *7*, 4562–4571.
12 M. Wolberg, W. Hummel, C. Wandrey, M. Müller, *Angew. Chem.* **2000**, *112*, 4476–4478; *Angew. Chem. Int. Ed.* **2000**, *39*, 4306–4308.
13 W. Kruse, W. Hummel, U. Kragl, *Recl. Trav. Chim. Pays B* **1996**, *115*, 239–243.
14 M. Kataoka, K. Yamamoto, H. Kawabata, M. Wada, K. Kita, H. Yanase, S. Shimizu, *Appl. Microbiol. Biotechnol.* **1999**, *51*, 486–490.
15 A. Weckbecker, W. Hummel, in: *Microbial Enzymes and Biotransformations*, Vol. 17 (Ed.: J. L. Barredo), Humana Press Inc., Totowa, NJ, **2005**, pp. 241–253.

2.2.5
Biocatalytic C–C Bond Formation in Asymmetric Synthesis
Wolf-Dieter Fessner

2.2.5.1 Introduction

Asymmetric C–C bond formation is the most important and most challenging problem in synthetic organic chemistry. In Nature, such reactions are facilitated by lyases, which catalyze the addition of carbonucleophiles to C=O double bonds in a manner that is classified mechanistically as an aldol addition [1]. Most enzymes that have been investigated lately for synthetic applications include aldolases from carbohydrate, amino acid, or sialic acid metabolism [1, 2]. Because enzymes are active on unprotected substrates under very mild conditions and with high chemo-, regio-, and stereoselectivity, aldolases and related enzymes hold particularly high potential for the synthesis of polyfunctionalized products that are otherwise difficult to prepare and to handle by conventional chemical methods.

With respect to practical applications in asymmetric synthesis, the four stereochemically distinct dihydroxyacetone phosphate (DHAP, **1**)-dependent enzymes had been particularly appealing to us because these enzymes control the creation of two new asymmetric centers at the termini of a newly formed C–C bond, thus allowing an effective stereocombinatorial synthesis of stereoisomers (Scheme 2.2.5.1) [1, 3]. The individual aldolases are involved in the reversible cleavage of

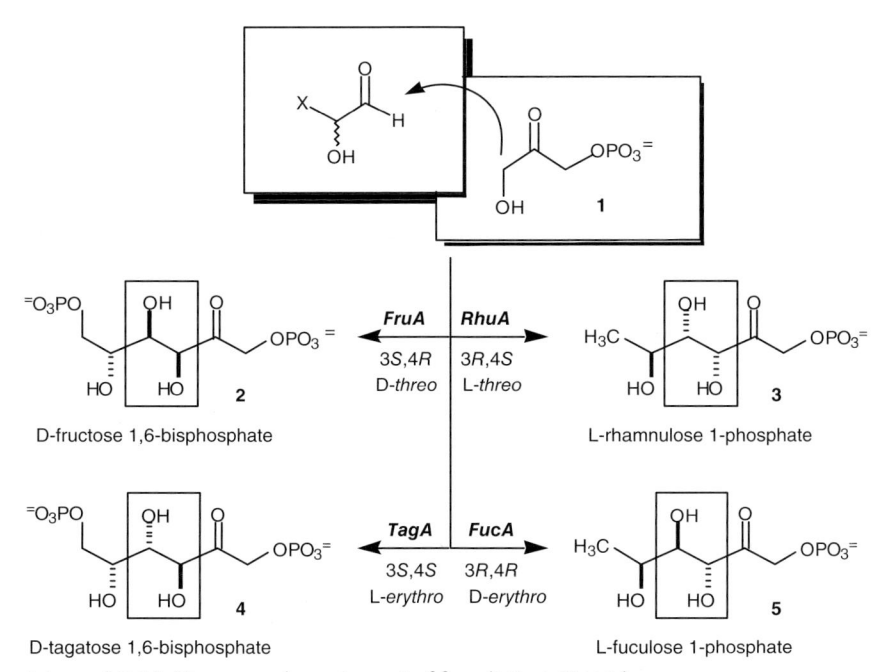

Scheme 2.2.5.1 Stereocomplementary set of four distinct DHAP lyases.

D-fructose 1,6-bisphosphate **2** (FruA; E.C. 4.1.2.13), D-tagatose 1,6-bisphosphate **4** (TagA; E.C. 4.1.2.40), L-fuculose 1-phosphate **5** (FucA, E.C. 4.1.2.17), and L-rhamnulose 1-phosphate **4** (RhuA, E.C. 4.1.2.19). From previous studies, we have DHAP aldolases with all four possible specificities readily available, we have demonstrated their broad substrate tolerance for variously substituted aldehydes, and we have investigated their stereoselectivity profile with non-natural substrates [3–6].

For preparative applications, it proved a limiting factor that the nucleophilic substrate DHAP is essential to the enzymes but is sensitive to decomposition and difficult to prepare by multi-step chemical routes [1]. Therefore we have developed several chemoenzymatic schemes for the in-situ generation of **1**, for example from glycerol based on a phosphorylation–dehydrogenation sequence including closed-loop cofactor regeneration [7], or by air oxidation using a glycerol phosphate oxidase (GPO) [8]. An alternative route starts from cheap sugars such as sucrose, which is processed by a cascade of reactions within an intricate "artificial metabolism" that involves up to seven enzymes [9].

2.2.5.2 Enzyme Mechanisms

2.2.5.2.1 Class II Aldolases

Aldolases have been classified into mechanistically distinct classes according to their mode of donor activation. Class I aldolases achieve stereospecific deprotonation via covalent imine/enamine formation at an active-site lysine residue, while Class II aldolases utilize a divalent transition metal cation for substrate coordination as an essential Lewis acid cofactor (usually Zn^{2+}) to facilitate deprotonation [1]. The metalloproteins show a very high stability in the presence of low Zn^{2+} concentrations, with half-lives in the range of months at room temperature, and the enzymes even tolerate the presence of large fractions of organic cosolvents ($\geq 30\%$) [6, 10]. Stereocontrol in general is highly effective with aldehydes carrying a 2- or 3-hydroxyl group, and many Class II aldolases offer a powerful kinetic preference for L-configured enantiomers of 2-hydroxyaldehydes **6** (Scheme 2.2.5.2) [11]. This latter feature facilitates racemate resolutions and allows the concurrent determination of three contiguous chiral centers in final products, which are obtained enantiopure and with high d.e. (≥ 95) even when starting from more readily accessible racemic material. For certain substrates, however, diastereoselectivity of Class II aldolases can be compromised in the control of the stereocenter at C4, which points to occasional inverse binding of the respective aldehyde carbonyl [1,

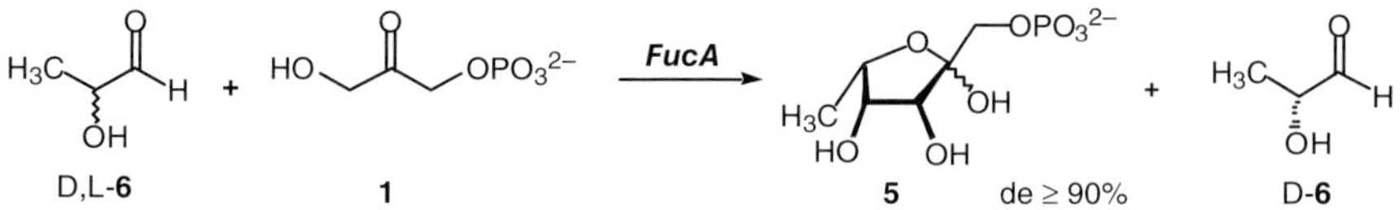

Scheme 2.2.5.2 Diastereoselective kinetic racemate resolution using the Class II aldolase FucA.

2.2.5 Biocatalytic C–C Bond Formation in Asymmetric Synthesis | 353

5, 6]. Certainly, a detailed understanding of the contributions of active-site residues in substrate recognition and the catalytic event was highly desirable.

For a distinction in the binding mode of substrate constitutional isomers, we first focused on the synthesis of the structurally related anhydroalditol derivatives **7** and **8** as potential inhibitors of FucA that lack the anomeric hydroxyl group of the natural substrate **5** and thus eliminate the possibility for ring opening and cleavage [12] (Scheme 2.2.5.3). Fucitol 1-phosphate **10** was included in the study as a potential mimic of the open-chain form. From kinetic data it became obvious that the aldolase binds preferentially a cyclic substrate, and selectively the more abundant β-anomer of the natural substrate that correlates with **7**.

Deprotonation of **1** at C3 yields an ene-diolate intermediate to which phosphoglycolohydroxamate (PGH) **9** bears a close structural resemblance (Scheme 2.2.5.3). In collaboration with J. V. Schloss, **9** was found to be a potent inhibitor not only of FucA, but indeed also of all currently accessible Class II aldolases with K_i in the nanomolar range [12]. Obviously, the hydroxamate mimics very effectively an advanced catalytic intermediate or transition state that is shared by these enzymes, and seems to be bound by all Zn^{2+}-dependent aldolases in a very similar fashion. Interestingly, when the active-site Zn^{2+} is replaced by Co^{2+} ions, catalytic activity is restored and actually becomes higher than with native zinc cofactor. However, the RhuA Co^{2+} complex catalyzes an oxygenase reaction that consumes **1** in a process similar to the photorespiration that is caused by the oxygen-consuming side reaction of D-ribulose 1,5-bisphosphate carboxylase [13], and therefore cannot be used for syntheses.

In concurrent efforts in collaboration with G.E. Schulz's group, the two Class II aldolases FucA and RhuA from *E. coli* have been crystallized; solution of their spatial structures confirmed a close similarity in their overall fold [14]. Both enzymes are homotetramers in which subunits are arranged in C_4 symmetry. The active site is assembled in deep clefts at the interface between adjacent subunits, and the catalytic zinc ion is tightly coordinated by three His residues. From X-ray

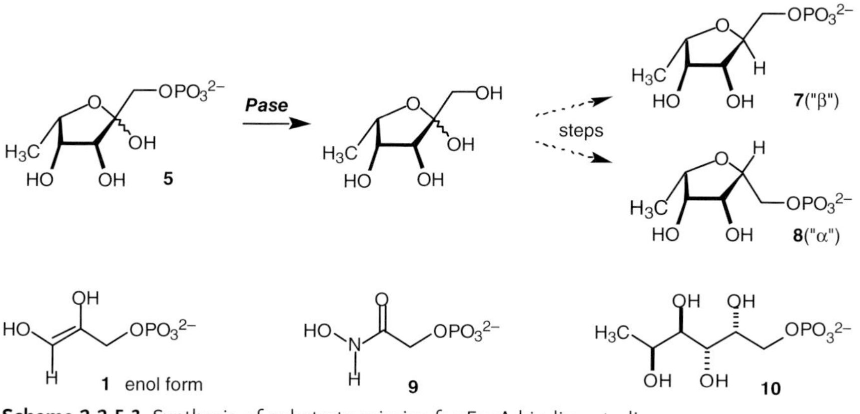

Scheme 2.2.5.3 Synthesis of substrate mimics for FucA binding studies.

structure determinations of FucA in the presence of the ligand **9**, the analysis revealed that the inhibitor forms a bidentate chelate in which both the ene-ol and hydroxamate oxygens are coordinated to Zn^{2+} (Fig. 2.2.5.1) [14, 15]. In line with the enzymological data [12], these results clearly indicate a role for the zinc ion in stabilization of the ene-diolate state. Bidentate binding of a DHAP ligand also helps to explain the natural preference of metal-dependent aldolases for an active-site Zn^{2+} atom because zinc readily adapts to trigonal-bipyramidal coordination and is able to change its coordination state with little energetic barrier, as compared with other metals.

In concert, structure determinations and enzymological studies for catalytic rates and product distributions with structurally varied aldehydes of native enzymes and numerous active-site mutants have allowed us to derive a conclusive blueprint for the catalytic cycle of FucA (Fig. 2.2.5.2). The proposed mechanism, which has general implications for other metal-dependent aldolases, is able to rationalize all key stereochemical issues successfully [15]. Independent work by other groups has recently provided further insight into related proteins with FruA and TagA specificity [16].

In the direction of aldol synthesis, DHAP enters FucA as the first substrate and coordinates to zinc by both hydroxyl and carbonyl oxygens in an extended con-

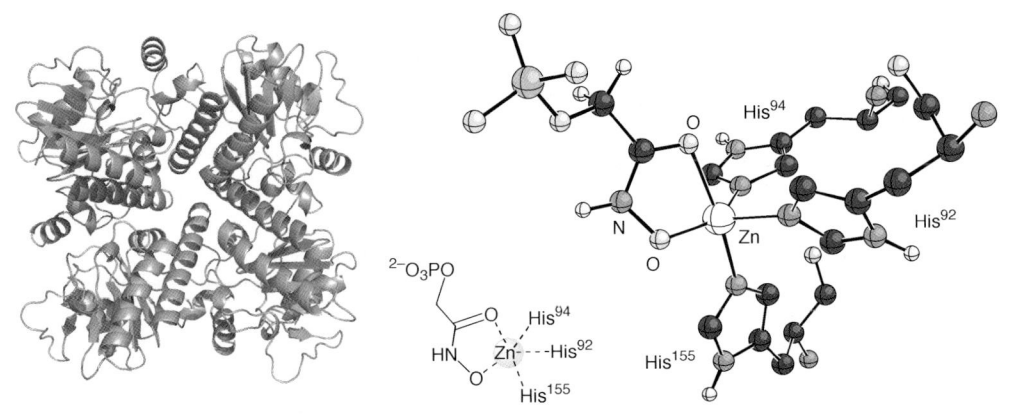

Fig. 2.2.5.1 Structure of FucA, as determined by X-ray analysis, and complexation of **9** (PGH) as an analogue to **1** (DHAP) by the catalytically active zinc ion.

Fig. 2.2.5.2 Proposed mechanism for the catalytic cycle of FucA, highlighting critical substrate–protein interactions.

formation. The phosphate group is held rigidly by a number of contacts to backbone amides and polar residues, most likely to suppress the facile ene-diolate fragmentation from an orthogonal conformation. Electrophilic polarization of the carbonyl bond serves to make the hydroxymethylene hydrogens more acidic and to facilitate abstraction of the *pro-R* proton by Glu73 acting as a general base. The nucleophilic *cis*-ene-diolate will then attack an incoming aldehyde carbonyl from its *Si* face. This is assisted by coordination of the aldehyde carbonyl by twofold hydrogen bonding to protonated Glu73 and Tyr113′, whereby the former can donate a proton to stabilize the developing charge. Hydrophobic binding of the aldehyde chain, together with steric constraints, is responsible for the asymmetric induction at C4, while shielding of the *Re* face of the DHAP ene-diolate by protein assures chirality at C3 of the product. The effect of some mutations on enantioselectivity and on the ratio of diastereomer formation with distinct aldehyde substrates, together with computational modeling studies for L-lactaldehyde binding, further point to a crucial role of hydrogen bonding of the 2-OH group to Tyr113′ in the kinetic discrimination of enantiomeric 2-hydroxyaldehydes [15].

2.2.5.2.2 Class I Fructose 1,6-Bisphosphate Aldolase

The synthetic potential of the Class I FruA from rabbit muscle (FruA$_{rab}$; for the systematic acronym nomenclature see ref. [1]) for preparative applications in the synthesis of various classes of chiral products is broadly documented [1, 2]. Commercial availability, broad substrate tolerance, and high D-*threo* diastereoselectivity make this enzyme highly attractive. However, major limitations as a practical catalyst remain in the relatively low stability of this mammalian enzyme under the usual reaction conditions with half-lives of only a few days in buffered solution, and in the narrow restrictions for application of cosolvents [4]. Stimulated by findings in M.-R. Kula's group for high pH and thermal stability of a related enzyme from *Staphylococcus carnosus* (FruA$_{sca}$) [17], we engaged in a collaborative effort for the construction of a new and highly efficient heterologous overexpression system for the FruA$_{sca}$ in *E. coli*. Thereby the enzyme, which in contrast to other known Class I DHAP aldolases is active as a monomer, can now be obtained by single-step purification in a state ready for synthetic applications [18]. A sequence comparison with other Class I FruA proteins suggests a common three-dimensional fold. Substrate screening studies confirmed a broad synthetic applicability similar to that of the rabbit muscle enzyme, and a high level of stereoselectivity for both newly created asymmetric centers; it also suggested a kinetic enantioselectivity for anionically charged 3-hydroxyaldehydes [18]. The extraordinary process stability of the enzyme was demonstrated by the preparation of bicyclic sugars from a glutaric dialdehyde derivative having strong protein cross-linking capacity (vide infra).

2.2.5.2.3 Sialic Acid Synthase

Some bacteria utilize phosphoenolpyruvate (PEP; **71**)-dependent lyases for the generation of sialic acids and related compounds (Scheme 2.2.5.4) [1]. By simultaneous release of inorganic phosphate from the preformed enolpyruvate nucleo-

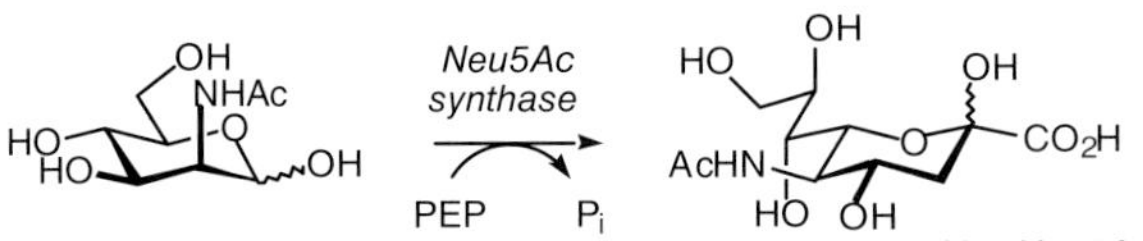

Scheme 2.2.5.4 Biosynthesis of N-acetylneuraminic acid (Neu5Ac, **12**) in bacteria.

phile during C–C bond formation, the additions are essentially irreversible, and therefore these lyases are often referred to as synthases. Owing to their prominent role as key elements in recognition processes during fertilization, embryogenesis, metastasis, inflammation, and host–pathogen adhesion, the preparation of sialic acids such as **12** and their derivatives is of high interest for pharmaceutical development [19]. However, no bacterial sialate synthase had hitherto been investigated for its synthetic potential [1, 2].

Thus, we had started an exploratory study with the N-acetylneuraminic acid synthase (Neu5Ac synthase or NeuS; E.C. 4.1.3.19), which was subcloned from *Neisseria meningitidis* [20] for recombinant production in *E. coli*. Preliminary data obtained with the highly expressed (ca. 40% of total protein; >1000-fold overexpression) and easily purified synthase indicate that it is Mn^{2+}-dependent and active in the pH range 7.0–9.0 as a dimer of 38 kDa subunits [21]. Most remarkably, the enzyme is highly tolerant to structural modifications of the natural substrate N-acetyl-D-mannosamine **11** and accepts many natural or modified sugars, particularly with respect to deoxygenation, substitution, truncation, or chain extension, as long as the aldehyde bears a correctly 3S-configured OH group. N-Acyl variations are well tolerated, including replacement by OH (D-mannose), which leads to the important natural sialic acid, 2-keto-3-deoxy-D-*glycero*-D-*galacto*-nononic acid (KDN).

2.2.5.2.4 Rhamnose Isomerase

While DHAP aldolases produce ketose derivatives, access to biologically important and structurally more diverse aldose isomers is achievable by use of enzymatic ketol isomerase interconversions. For this purpose, we had previously shown that L-rhamnose isomerase (RhaI; E.C. 5.3.1.14) and L-fucose isomerase (FucI; E.C. 5.3.1.3) from *E. coli* display a relaxed substrate tolerance. Both enzymes convert sugars and their derivatives with distinct stereopreference at C2 and common (3R)-OH configuration, but tolerate alterations in configuration or substitution pattern at subsequent positions of the chain (Scheme 2.2.5.5) [11, 12].

Knowledge of the three-dimensional structure of an enzyme is the basis for understanding its substrate specificity. Thus, the *E. coli* RhaI was crystallized and its X-ray crystal structure was solved in collaboration with I. Korndörfer [22]. The enzyme is a tight tetramer of four $(\beta/\alpha)_8$ barrels, with structural similarity to xylose isomerase (Fig. 2.2.5.3). The structures of complexes of rhamnose isomerase with the inhibitor L-rhamnitol and the natural substrate L-rhamnose were

Scheme 2.2.5.5 Stereodivergent isomerization of (3R)-configured aldolase products as an access route to aldoses.

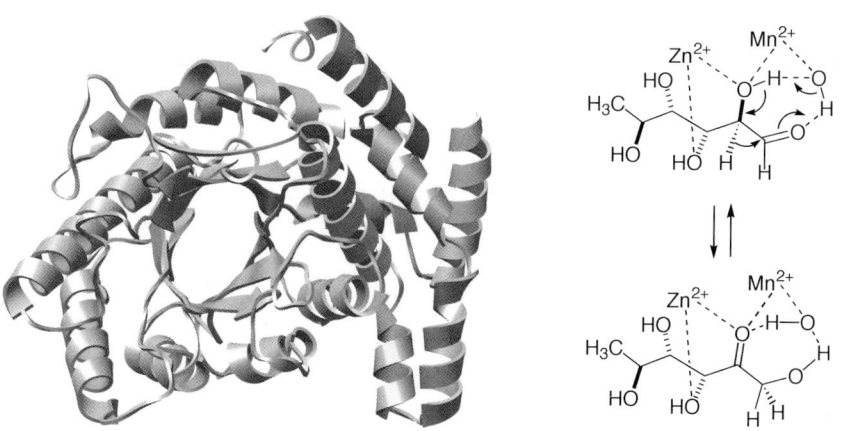

Fig. 2.2.5.3 Structure of rhamnose isomerase, and proposed hydride-shift mechanism for the ketol isomerization.

determined; they suggest that an array of hydrophobic residues, not present in xylose isomerase, is likely to be responsible for the recognition of L-rhamnose as a substrate. In the presence of substrate the enzyme also binds Mn^{2+} at a nearby "catalytic" site. The structural data available suggest that a metal-mediated hydride-shift mechanism, which is generally favored for xylose isomerase [23], is also feasible for rhamnose isomerase.

2.2.5.3 New Synthetic Strategies

2.2.5.3.1 Sugar Phosphonates

Phosphonic acids bio-isosteric with naturally occurring phosphate esters are attractive targets for synthesis because of their potential bioactivity, since replacement of the ester oxygen atom by a methylene group renders the corresponding phosphonates incapable of hydrolytic cleavage. Owing to their geometrical and polar similarity to the natural metabolites or effectors they can act as stable inhibitors or regulators of metabolic processes [24].

2.2 Biological Methods

Although DHAP aldolases are highly specific for **1** as the nucleophilic substrate, some enzymes had been found to be tolerant to small, mostly isosteric modifications at the phosphate ester linkage, such as to the phosphoramidate, phosphorothioate, or phosphonate (**14**) analogues. For this purpose, the synthetic entry to **14** by the glycerol phosphate oxidase (GPO) catalysis was particularly attractive (Scheme 2.2.5.6) [8]. To obviate problems encountered during product separation when starting from the racemic glycerol phosphate analogues, enantiomerically pure dihydroxybutylphosphonate (S)-**13** was prepared efficiently from (S)-malic acid **12** (five steps, ~50% overall yield, >99% ee). The latter could be oxidized to the corresponding ketone **14** by the GPO method in practically quantitative yield [25]. Since in principle the route via chiral intermediates is an unnecessary deviation, we further developed alternatively a new and highly effective chemical sequence that starts from inexpensive 2-butynediol **15** (five steps, 59% overall yield) [26].

Several DHAP aldolases having different stereospecificities were tested for their acceptance of this phosphonomethyl substrate mimic as the aldol donor. Individual enzymes belonging to both Class I and II types were found to catalyze the stereoselective addition of **14** to various aldehydes, providing bio-isosteric non-hydrolyzable analogues of sugar 1-phosphates in high yields (for example, **16/17**; Scheme 2.2.5.7) [25, 26].

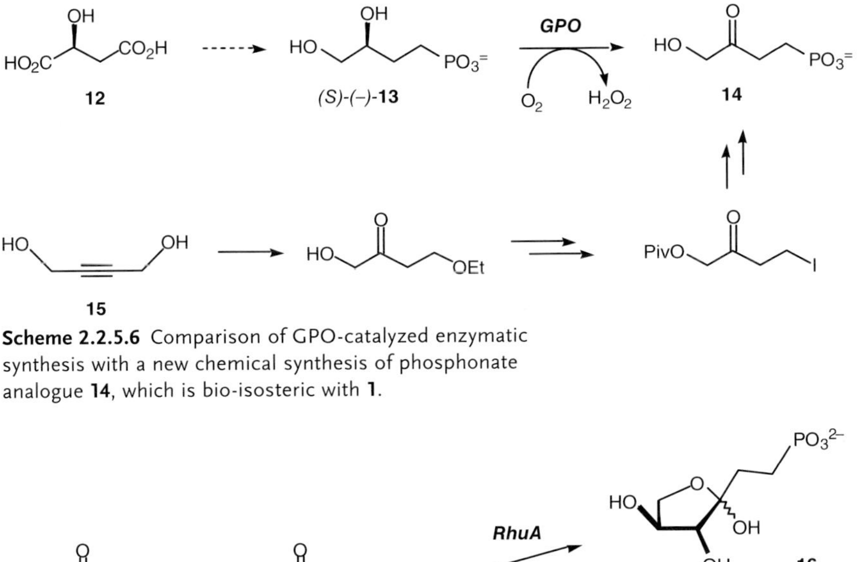

Scheme 2.2.5.6 Comparison of GPO-catalyzed enzymatic synthesis with a new chemical synthesis of phosphonate analogue **14**, which is bio-isosteric with **1**.

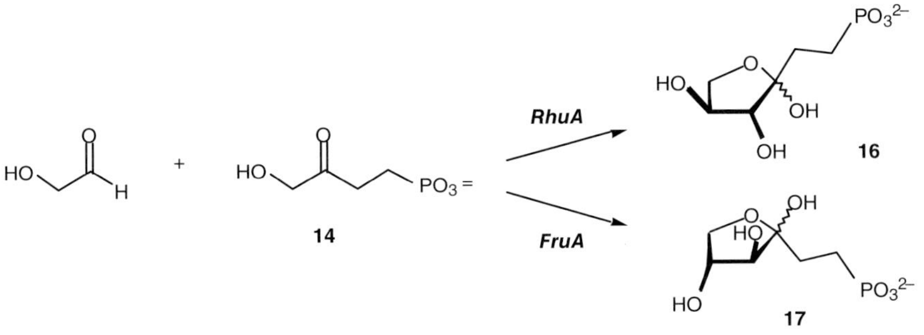

Scheme 2.2.5.7 Synthesis of sugar phosphonic acids as potential enzyme inhibitors.

2.2.5.3.2 Xylulose 5-Phosphate

The ready availability of the transketolase (TK; E.C. 2.2.1.1) from *E. coli* within the research collaboration in G. A. Sprenger's group suggested the joint development of an improved synthesis of D-xylulose 5-phosphate **19**, which was expensive but required routinely for activity measurements [27]. In vivo, transketolase catalyzes the stereospecific transfer of a hydroxyacetyl nucleophile between various sugar phosphates in the presence of a thiamine diphosphate cofactor and divalent cations, and the C_2 donor component **19** offers superior kinetic constants. For synthetic purposes, the enzyme is generally attractive for its high asymmetric induction at the newly formed chiral center and high kinetic enantioselectivity for 2-hydroxyaldehydes, as well as its broad substrate tolerance for aldehyde acceptors [28].

Based on the stereospecific transketolase-catalyzed ketol transfer from hydroxypyruvate (**20**) to D-glyceraldehyde 3-phosphate (**18**), we have thus developed a practical and efficient one-pot procedure for the preparation of the valuable ketosugar **19** on a gram scale in 82% overall yield [29]. Retro-aldolization of D-fructose 1,6-bisphosphate (**2**) in the presence of FruA with enzymatic equilibration of the C_3 fragments is used as a convenient in-situ source of the triose phosphate **18** (Scheme 2.2.5.8). Spontaneous release of CO_2 from the ketol donor **20** renders the overall synthetic reaction irreversible [29].

2.2.5.3.3 RhuA Stereoselectivity

Although generic 3-hydroxyaldehydes are usually converted by the aldolases stereospecifically [1, 4], an interesting, virtually complete reversal of stereoselectivity had been observed upon RhuA-catalyzed reaction of the conformationally restricted galactodialdose derivatives **21** and **23** (Scheme 2.2.5.9) [30]. While several

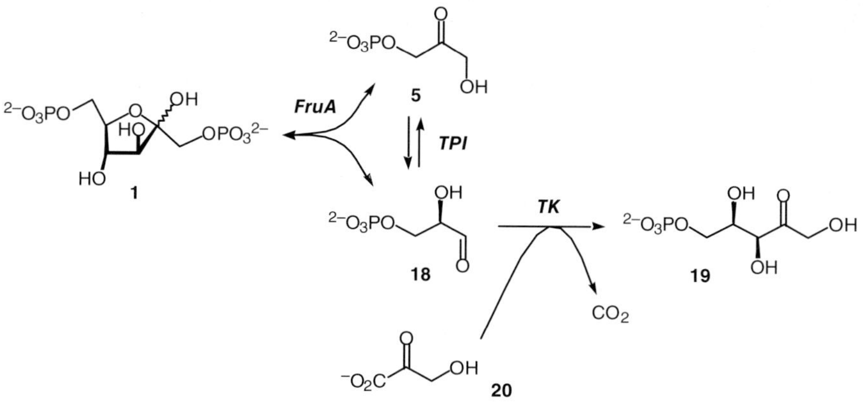

Scheme 2.2.5.8 Multi-enzymatic route for the stereoselective synthesis of D-xylulose 5-phosphate **19** from commercial D-fructose 1,6-bisphosphate (abbreviations: TPI, triose phosphate isomerase; TK, transketolase).

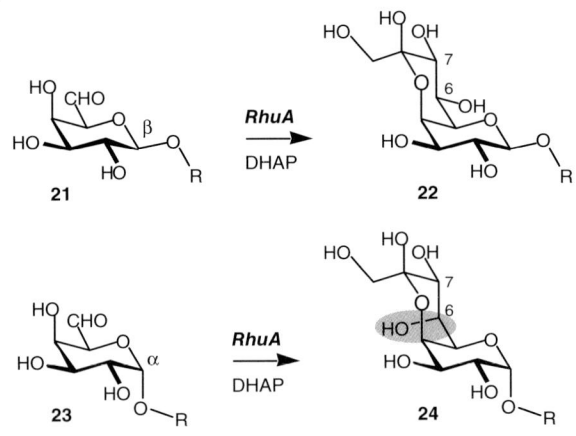

Scheme 2.2.5.9 Stereochemical bias in RhuA reactions with α-galactoside-derived C$_6$ aldehydes.

Scheme 2.2.5.10 RhuA stereoselectivity screening using structurally defined aldehydes based on carbohydrate scaffolds.

β-anomeric compounds led cleanly to the expected 6,7-*trans* adduct **22**, α-galactosides were practically completely converted to the corresponding 6,7-*cis* product **24**. This finding is particularly surprising, because the locus of structural change is relatively distant from the reaction center. As no conclusive pattern for substrate control of stereoselectivity was evident, a more systematic approach was desirable.

Toward a screening program for RhuA stereoselectivity, structurally more simplified dioxane derivatives **25–27** comprising enantiomeric and diastereomeric 3-hydroxyaldehyde geometries in a conformationally defined environment could be prepared easily from carbohydrate precursors (Scheme 2.2.5.10). First results from product analysis provide further evidence for occasionally biased fixation of the acceptor carbonyl group [31]. With X-ray structural data now available for the aldolase [14], an interpretation of the various results should be facilitated.

2.2.5.3.4 Aldolase Screening Assay

The stereoselectivity of aldolases is typically assayed in the direction of synthesis by NMR analysis of the product composition [4–6]. Although this option can be used to detect possible thermodynamic effects on aldolase product formation (due to the equilibrium nature of the catalytic reaction), the procedure is laborious and inconvenient, and inapplicable to rapid enzyme screening. As an example of a kinetic screen, we have therefore developed novel fluorogenic stereochemical probes for transaldolases and related enzymes. The fluorogenic assay (Scheme 2.2.5.11) is based on the retro-aldolization of diastereomeric substrates. The example shown, **29**, yields 3-*O*-coumarinylpropanal **31**, which subsequently undergoes a β-elimination to give the strongly fluorescent product umbelliferone **30** [32].

The reactive open-chain substrate **29** with the natural D-*threo* configuration was prepared along a chemoenzymatic route by making use of the common constitutional and stereochemical relationship which substrates of transaldolase share with those of transketolase. Thus, the *R*-configured 2-hydroxyaldehyde **28** was chain-extended under transketolase catalysis in the presence of **20** as ketol donor to yield the desired aldol. By this approach, several transaldolases could indeed be shown to display different levels of kinetic stereoselectivity.

2.2.5.3.5 Aldose Synthesis

In Nature, complex oligosaccharides serve to encode information for a broad spectrum of biological recognition processes. Of the sugars occurring most frequently in mammalian oligosaccharides, L-fucose (**34a**) is the most hydrophobic residue. Fucosylated oligosaccharides are involved in bacterial adherence or inflammation processes, and of importance for cancer diagnostics or immunotyping [33]. To facilitate investigations on the importance of hydrophobic contacts of

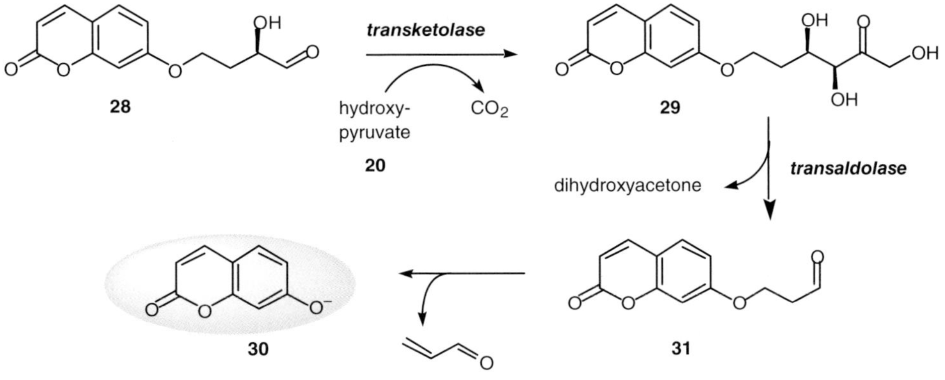

Scheme 2.2.5.11 Preparation of a fluorescent stereochemical probe useful for transaldolase screening.

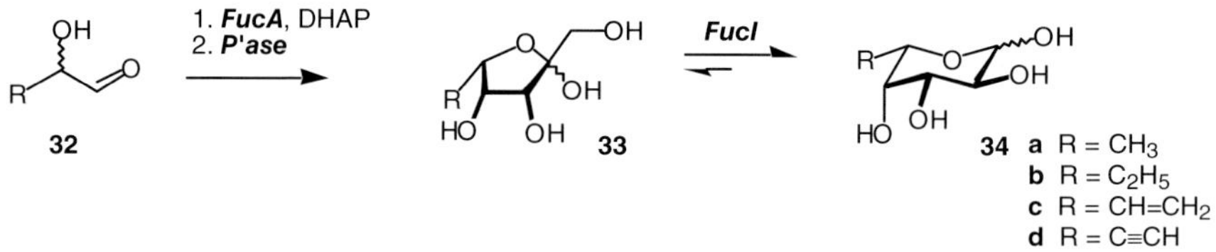

Scheme 2.2.5.12 Short enzymatic synthesis of L-fucose and hydrophobic analogues by aldolization–ketol isomerization, including kinetic resolution of racemic aldehyde precursors.

fucosylated structures upon receptor binding, we had set out to prepare L-fucose analogues modified at the nonpolar terminus. Our synthetic approach is based on a two-step enzymatic sequence (Scheme 2.2.5.12), which consists of a FucA-catalyzed preparation of ketose analogues **33** by asymmetric chain extension of appropriate aldehydes **32**, followed by a biocatalytic isomerization of analogues **33** using FucI to give the target aldose analogues. In particular, L-fucose derivatives that display extended linear (**34b**) and branched saturated, or different unsaturated (**34c,d**), aliphatic chains were of interest [34].

Indeed, from racemic 2-hydroxyaldehydes **32** only single diastereomers **33** having a natural relative and absolute configuration resulted. Thus, the kinetic enantioselectivity was not impaired by the steric bulk of the hydrophobic side-chains ($de \geq 95\%$), which allowed effective control over three adjacent stereogenic centers in a single step [11, 34]. All terminus-modified L-fuculose analogues **33** indeed proved to be FucI substrates. With increasing steric bulk and rigidity of the chain terminus, isomerization rates diminished considerably, which correlates with the FucI enzyme's rather low affinity even for its natural sugar substrate. Enantiomerically pure L-fucose analogues could be prepared in up to 30% overall efficiency by this short enzymatic route, which is completely stereoselective at each step in controlling the configuration of all four consecutive centers of chirality.

2.2.5.3.6 Tandem Chain Extension–Isomerization–Chain Extension

Transketolase catalyzes the reversible transfer of a hydroxyacetyl fragment from a ketose to an aldehyde. Because the ketose products formed by transketolase reactions are not acceptors for a consecutive transformation by the same enzyme, we have investigated the option to include a xylose(glucose) isomerase (XylI; E.C. 5.3.1.5), which has similar stereochemical specificity, for ketose to aldose equilibration (Scheme 2.2.5.13). Starting from racemic lactaldehyde **32a**, the transketolase forms 5-deoxy-D-xylulose **35a**, which indeed was accepted by the XylI in situ for diastereospecific conversion into 5-deoxy-D-xylose **36a**. The latter again proved to be a substrate of transketolase which completed a tandem operation to furnish 7-deoxy-sedoheptulose **37a** as the sole bisadduct in 24% overall yield and in enantio- and diastereomerically pure quality [35, 36]. All four stereocenters of the resulting product are completely controlled by the enzymes during this one-pot operation. The procedure profits from the limited tolerance of the isomerase

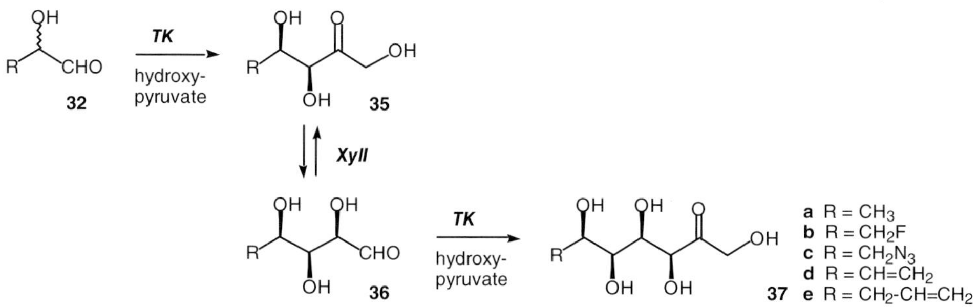

Scheme 2.2.5.13 Tandem conversions based on ketol transfer/ketol isomerization.

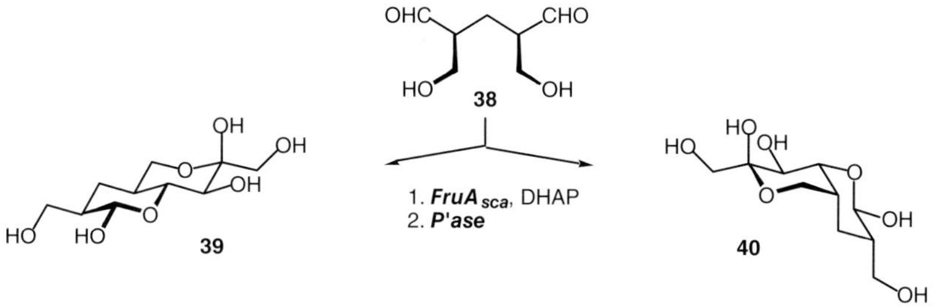

Scheme 2.2.5.14 Formation of bicyclic sugars by single extension of α,ω-dialdehydes upon attempted tandem addition to glutaric dialdehyde **38**.

for modifications at the sugar backbone, which excludes sugars of the size of heptuloses **37** and thus halts before the formation of mixed oligomers, but proved adaptable to substituted lactaldehyde derivatives **32b–32e**.

2.2.5.3.7 Tandem Bidirectional Chain Extensions

In principle, bifunctional aldehydes should be able to engage in twofold enzymatic aldol additions to both of their acceptor carbonyls in a fashion to be classified as a "tandem" reaction, that is, without the need for isolation of intermediates. Depending on the specificity of the enzyme used and on the functionalization in the starting material, the isomeric constitution as well as the absolute and relative stereochemistry should be deliberately addressable. Therefore, we engaged in a program to evaluate the scope and the limitations of such two-directional chain elongation processes for the construction of extended polyfunctional molecules [36].

From prior attempts at FruA-catalyzed DHAP additions to glyoxal or glutaric dialdehyde no product had been isolated, probably because the dialdehydes can cause cross-linking of the protein and thereby irreversibly destroy its enzymatic activity. On the other hand, hydroxylated aldehydes were assumed to form stable intramolecular hemiacetals in aqueous solution, which may mask the reactivity of free dialdehydes. Using the branched-chain glutaric dialdehyde **38** as a potential precursor to carbon-linked disaccharide mimetics (Scheme 2.2.5.14), we found that the monomeric FruA from *S. carnosus* (vide supra) proved sufficiently robust upon contact with this substrate and catalyzed a rapid conversion, whereas the rabbit muscle enzyme suffered immediate precipitation [18]. However, only monoadducts **39/40** resulted from this reaction, presumably because of the high stability of the bicyclic dipyranose structure which effectively precluded a second addition. Using the RhuA enzyme, which has an enantiocomplementary specificity, the corresponding antipodes *ent*-**39/40** could indeed be obtained [18].

Successful results could be realized when using linear, instead of branched-chain, 2-hydroxylated or 3-hydroxylated α,ω-dialdehydes. As a first example, the racemic *threo*-tartaric dialdehyde **41** was prepared by pinacol dimerization of acrolein, followed by ozonolysis. From the mixture of constitutional isomers of this precursor present in aqueous solution, as evident from NMR analysis, FruA-catalyzed DHAP addition proceeded smoothly to the desired bisadduct stage (Scheme 2.2.5.15) [36, 37]. As to be expected, transitory formation of a monoadduct (**42/43**) could be detected at low concentration, which indicates that the rate for the second addition step is of the same order of magnitude as that for the initial step. Surprisingly, the material isolated after enzymatic dephosphorylation proved to be only a single diastereomer having the C_2-symmetrical dipyranose constitution **44**. Although the Class I FruA used is unable to differentiate 2-hydroxyaldehydes with sufficient kinetic enantioselectivity [4], the corresponding bisadduct (**45**) derived from the opposite dialdehyde enantiomer could not be

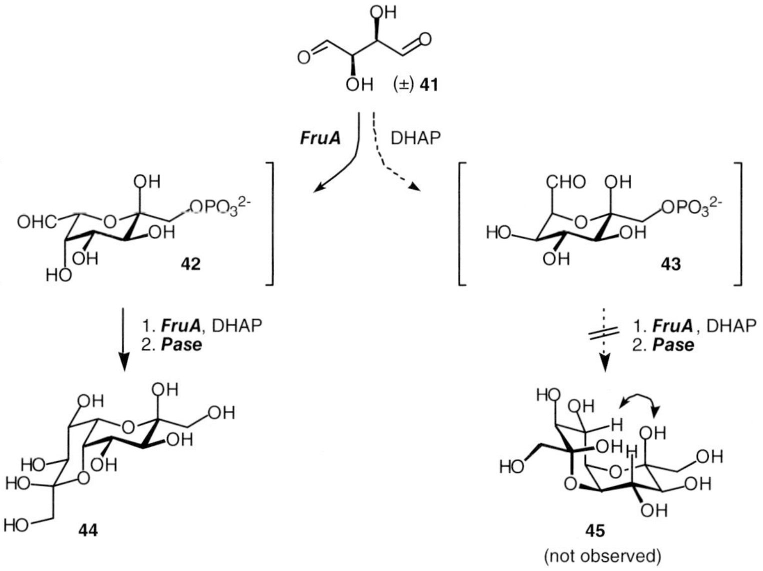

Scheme 2.2.5.15 Highly diastereoselective formation of a bicyclic decadiulose from racemic tartaric dialdehyde by bidirectional chain extension.

detected. However, inspection of molecular models showed that the elusive **45** would be highly destabilized in an annulated bipyranose form because of strong steric crowding effects.

For larger substrates, control of relative hydroxyl stereochemistry is facilitated when employing cycloolefinic precursors for mild ozonolytic generation of *threo*- or *erythro*-dialdehyde substrates. Following simple considerations, candidate substrates should make it possible to predict plausible product structure, where total chain length is determined by the precursor ring size plus twice the nucleophile length. Relative positioning of hydroxyl functions should determine the type of sugar ring: that is, 2-hydroxyl functions are expected to lead to furanoid rings, and 3-hydroxyl functions would lead to a pyranoid constitution.

Indeed, such predictions were verified by reaction exemplars using cyclopentene (**46**) and cyclohexene precursors (**48/50**) [38]. With racemic dialdehyde from cleavage of *trans*-3,5-dihydroxycyclopentene **46**, FruA-catalyzed twofold aldol addition of DHAP proceeded smoothly to furnish a single bisadduct (Scheme 2.2.5.16). The latter indubitably had a C_2-symmetrical bisfuranoid structure **47** interconnected by a methylene bridge in a *trans,trans* fashion. The high preference for only one of the substrate enantiomers can be rationalized by the considerable thermodynamic advantage of a bis-*trans*- over a bis-*cis*-connected linkage. Similarly, the homologous racemic dialdehyde from cleavage of *trans*-3,6-dihydroxycyclohexene **48** gave rise to a single all-*trans*-configured bisfuranose **49** – formally a C_2-symmetrical, 6,6'-linked dimer of D-fructose [38]. Again, the diastereomer stemming from the antipode was not detectable, plausibly again because of the implicated less favorable twofold *cis*-vicinal tether connection.

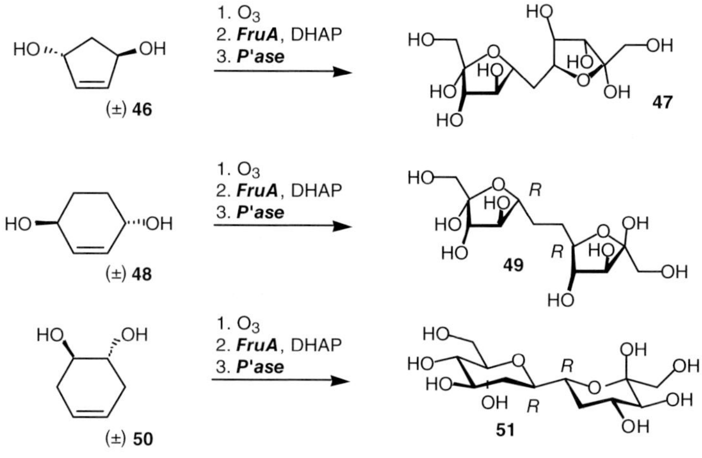

Scheme 2.2.5.16 Tandem aldolization of racemic dihydroxylated α,ω-dialdehyde precursors.

Transposition of the hydroxyl functions to a homoallylic location, as realized in *trans*-3,4-dihydroxycyclohexene **50**, produced a single bisadduct upon cleavage and enzymatic aldolization (Scheme 2.2.5.16). As was to be expected, it had a dipyranoid constitution **51** of C_2 symmetry, with sugar rings interconnected by a direct bis-equatorial linkage. Even with an excess of DHAP no bisadduct of the corresponding enantiomer could be obtained, probably because the latter would demand a strongly disadvantageous diaxial ring connection [38].

Although α,ω-dialdehydes are most conveniently produced from cyclic alkene precursors, alternatively 3-hydroxylated open-chain precursors **52** were made available via diastereoselective bidirectional chemical allylation of generic dialdehydes [36, 39]. By this route, the preparation is possible of diketoses (for example, **51/53/54**) interlinked by carbon chains of a variable tether length (Scheme 2.2.5.17). Interestingly, from racemic starting materials **52** – and even from *erythro/threo* mixtures – mostly single bisadduct diastereomers that are characterized by the thermodynamically most stable ring connectivity pattern were obtained [4, 38]. In addition to the expected kinetic stereoselectivity of the aldolase in generating new chiral centers [11], this observation certainly reflects the relative thermodynamic stability of an equatorial linkage in final bisadducts [38], but possibly also the relative prevalence of distinct equilibrating reaction intermediates from plausible constitutional isomerizations.

With *meso*-configured dialdehyde precursors, the enantiotopic nature of the termini must give rise to a configurational terminus differentiation upon twofold chain extension because the catalyst-controlled diastereoselective aldol additions will break the inherent σ symmetry. While the two enantiotopic termini cannot

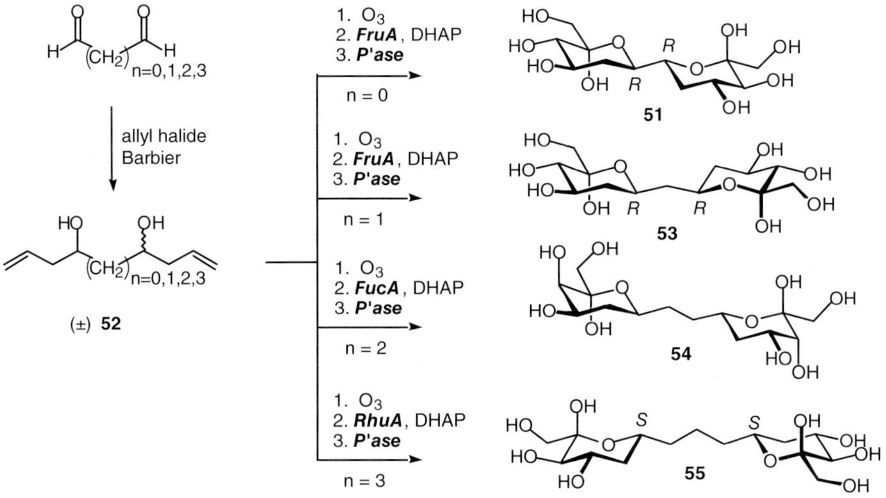

Scheme 2.2.5.17 Synthesis of disaccharide analogues with variable configuration and tether length from generic dialdehyde precursors by Barbier allylation and tandem α,ω-aldolization.

be distinguished well enough kinetically by the FruA enzyme [38], and thus analytical problems arise at the monoadduct stage, convergence of pathways is to be expected with progression to an identical, although nonsymmetrical, bisadduct. Indeed, the 2-hydroxylated and 3-hydroxylated *meso*-dialdehydes generated from **56** and **58** upon tandem aldolization with FruA catalysis delivered, as expected, the bisfuranoid and bispyranoid nonsymmetrical bisadducts **57** and **59** as the sole products, respectively (Scheme 2.2.5.18) [38]. Notably, spectroscopic data for **59** indicate that one of the pyranose subunits adopts a chair conformation that has to tolerate two axial oxygen substituents at C3/C4 in order to obviate an unfavorable equatorial to axial ring connection.

In both reactions with the *meso* substrates, no intermediary monoadduct could be detected. Consequently, a potential kinetic preference of the aldolase for either of the competing enantiotopic hydroxyaldehyde moieties within the starting substrates could not be investigated. No matter which of the enantiotopic aldehyde groups is attacked first, however, the second addition steps must be kinetically faster in each case, probably supported by the presence of an anionic charge in the intermediates, which should improve the substrate affinity to the enzyme.

Structural relationships suggest that the extended carbohydrate scaffolds obtainable by such tandem aldolizations may be regarded as metabolically stable mimetics of oligosaccharides, in particular of C-glycosides that are hydrolytically stable. The latter class of compounds shares an interest for potential therapeutic applications with the class of so-called "aza sugars" that have commanded attention in recent years as potent glycoprocessing inhibitors for the treatment of diabetes and other metabolic disorders, as well as for the blocking of viral or microbial infection and metastasis.

In this respect it is interesting to note that the tandem aldolization technique proved amenable also to the synthesis of a first C-glycosidic aza sugar (Scheme 2.2.5.19) [29, 36]. A rather simple dihydroxylated azido dialdehyde was generated from racemic azidocyclohexene **60** and subjected to FruA catalysis. The latter effected a smooth tandem addition to provide a diastereoisomerically pure bispyranoid azido C-disaccharide **61**, from which the pyrrolidine-type aza sugar **62** was

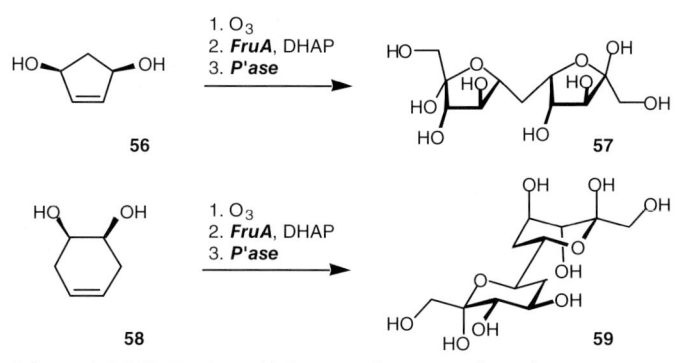

Scheme 2.2.5.18 Tandem aldolization of *meso*-configured precursors.

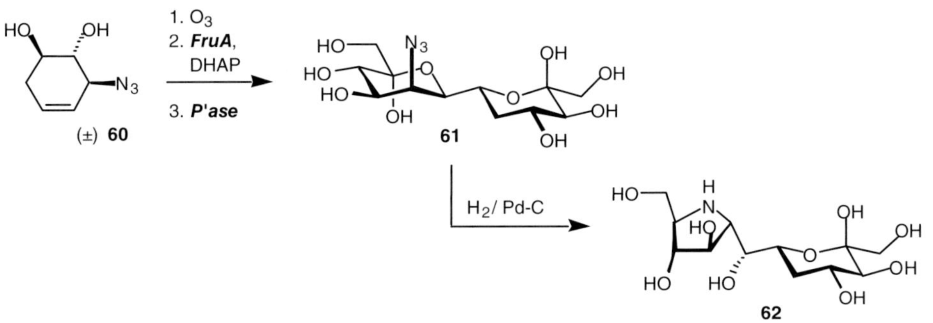

Scheme 2.2.5.19 Synthesis of an aza C-disaccharide by tandem aldolization and reductive amination.

produced highly stereoselectively upon standard reductive amination. Model considerations suggest a close resemblance of the protonated aza C-disaccharide **62** to transition states of saccharase or maltase. Indeed, **62** was found to inhibit the intestinal α-glucosidase with a K_i below 1 mM [29, 40].

Extension of the tandem aldolization concept to spirocyclic structures seems plausible, provided that the enzymes would still be able to act on substrates in which paired hydroxy and aldehyde functionalities are positioned as geminal substituents on an existing, but sterically demanding, cyclic support. From several conceivable probes, an easily accessible cyclohexane derivative was selected initially for its structural simplicity. The *trans*-configured dialdehyde generated from precursor **63** (Scheme 2.2.5.20) reacted under FruA catalysis to give a single enantio- and diastereomerically pure bisadduct **64**, still bearing an element of C_2 symmetry [36, 37]. From inspection of molecular models it is evident that the two spiropyranoses synthesized de novo are held at a distance that is quite similar to that in a 1,6-linked disaccharide, but with conformation and relative orientation of the sugar rings to one another rigidly confined by the cyclohexane core in a nonhydrolyzable manner.

An even more complex target with a central carbohydrate core structure could likewise be derived from the convex precursor **65**. The corresponding dialdehyde, in spite of its steric hindrance, was smoothly converted by FruA catalysis in the desired tandem fashion to give a single C_2-symmetrical bisadduct **66** (Scheme 2.2.5.20) [36]. A complementary reaction performed with the RhuA enzyme also yielded cleanly a diastereomeric bisadduct **67**. The combination of an enantiopure educt structure and opposite absolute configurations induced by the aldolases enforces a different relative spatial orientation of the sugar chairs with respect to the fused bicyclic core. While both tandem spiro frameworks **66/67** bear a certain resemblance to trisaccharides, each of them projects its polar functionalities rigidly, but in a topologically very distinct fashion, into three-dimensional space for potentially strongly discriminating molecular interactions.

Strikingly, this unique technique of enzymatic bidirectional chain synthesis leads in most cases to enantio- and diastereomerically pure higher-carbon diketoses of high structural complexity from rather simple α,ω-dialdehydes with good

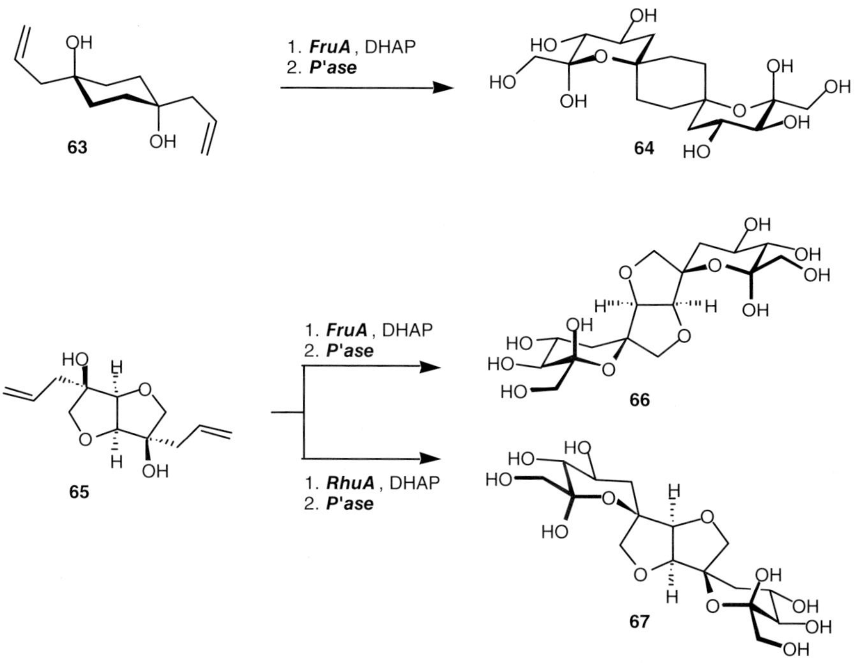

Scheme 2.2.5.20 Formation of rigid analogues of oligosaccharide structures by tandem aldolization of appropriately functionalized cyclic precursors.

overall yields. During the one-pot, tandem enzymatic operation a remarkable number of up to six asymmetric centers are determined: 2 × 2 stereocenters are newly created by the twofold C–C bond formation, and two more are selected from a racemic precursor due to kinetic and/or thermodynamic preference. No protecting groups need to be installed or removed, and the specific product structure and overall symmetry are determined by selection of the precursor constitution, relative configuration of hydroxyl substitutents, and the stereospecificity of the aldolase.

2.2.5.3.8 Non-Natural Sialoconjugates

Recent advances in glycobiology have revealed the significant role of complex sialylated oligosaccharide structures present on cell surfaces as key elements in recognition processes during fertilization, embryogenesis, metastasis, inflammation, and host–pathogen adhesion (for an example, see GM3 ganglioside **68**; Scheme 2.2.5.21) [19]. Investigations into the biological functions of glycoconjugates and the development of carbohydrate-based therapeutics are closely related to the accessibility of complex carbohydrate structures, with particular need also for non-natural modifications.

Owing to their prominent role in receptor interactions, the preparation of sialic acids and structural modifications is thus of great interest for pharmaceutical

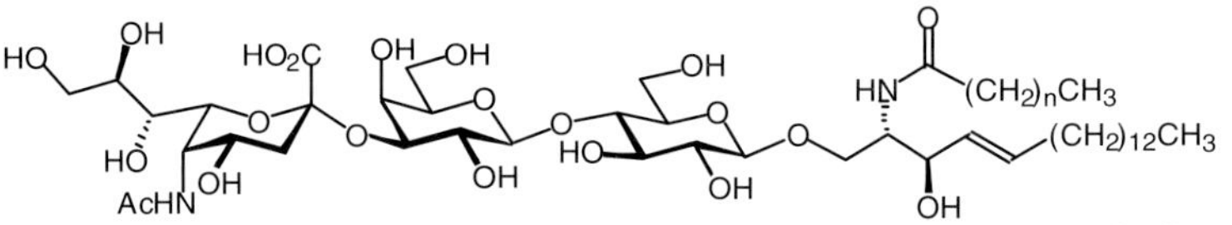

Scheme 2.2.5.21 Sialylated oligosaccharide substructure in GM$_3$ ganglioside.

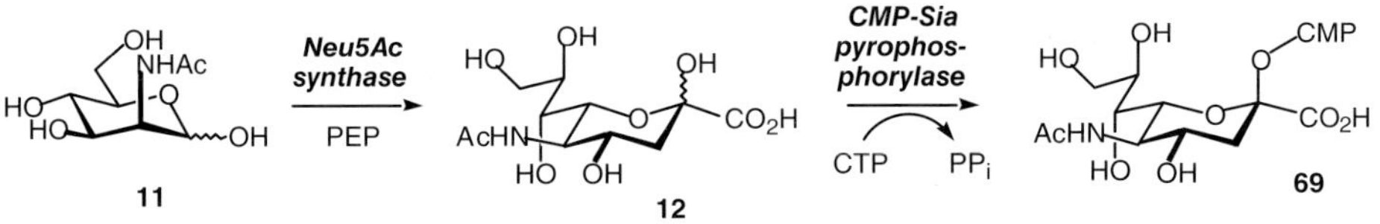

Scheme 2.2.5.22 Biosynthesis and activation of N-acetylneuraminic acid.

development. Chemical methods for both the synthesis of sialic acids and their incorporation into oligosaccharides are laborious and inefficient due to problems encountered in the control of product stereo- and regiochemistry [41]; this renders complementary enzymatic methods for oligosaccharide syntheses highly attractive [42]. To improve and broaden current methodology for sialoconjugate synthesis, we have started a program to investigate difficultly accessible enzymes from the sialoconjugate biosynthesis of pathogenic microorganisms for their preparative suitability (substrate tolerance, stereoselectivity, productivity, and so on). Particularly, enzymes from the colominic acid and serotype biosynthetic pathway of *Neisseria meningitidis* offered an attractive entry because complete genetic information for this organism is publicly accessible from whole-genome sequencing (vide supra, Section 2.2.5.2.3) [43].

Biosynthesis of oligosaccharides follows the general Leloir pathway. For sialic acids, a PEP-dependent lyase codes for N-acetylneuraminic acid synthesis, followed by a pyrophosphorylase-catalyzed CMP activation of the sugar and a final glycosyl transfer to a suitable acceptor saccharide (Scheme 2.2.5.22). In collaboration with M. Frosch's group the corresponding genes from *N. meningitidis* serogroup B were subcloned for overexpression in *E. coli*, and the corresponding recombinant enzymes could easily be purified in high yields [44].

Synthetic studies for sialic acid and its modifications have extensively used the catabolic enzyme N-acetylneuraminic acid aldolase (NeuA; E.C. 4.1.3.3), which catalyzes the reversible addition of pyruvate (**70**) to N-acetyl-D-mannosamine (ManNAc, **11**) to form the parent sialic acid N-acetylneuraminic acid (Neu5NAc, **12**; Scheme 2.2.5.23) [1, 2, 45]. In contrast, the N-acetylneuraminic acid synthase (NeuS; E.C. 4.1.3.19) has practically been ignored, although it holds considerable synthetic potential in that the enzyme utilizes phosphoenolpyruvate (PEP, **71**) as a preformed enol nucleophile from which release of inorganic phosphate during

Scheme 2.2.5.23 Alternative pathways for sialic acid synthesis using the catabolic aldolase (NeuA) or the anabolic synthase (NeuS) enzymes.

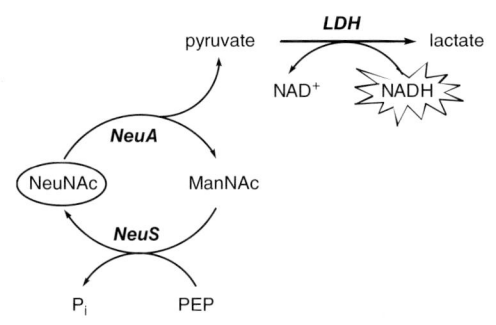

Scheme 2.2.5.24 Signal-amplifying assay for neuraminic acid quantification.

C–C bond formation renders the addition essentially irreversible. Our first preparative screening efforts have shown that the enzyme exhibits a broad substrate tolerance, particularly by accepting variously C5-modified Neu5Ac derivatives as substrates with complete stereocontrol [20, 21]. Thus, the enzyme seems to be practically equivalent to the commercial aldolase, except that reactions attain complete conversion without a need to drive an equilibrium by a large excess of substrate, which strongly simplifies product isolation.

The availability of both the catabolic aldolase and the uniquely synthetic anabolic synthase made it possible to assemble a novel continuous assay for the determination of the metabolite N-acetylneuraminic acid [46]. A combination of both enzymes, in the presence of an excess of PEP, will start a cycle in which the determinant sialic acid will undergo a steady conversion of cleavage and re-synthesis as a futile cycle (Scheme 2.2.5.24). With each progression, however, 1 equiv of pyruvate is liberated simultaneously, which causes time-dependent signal amplification. Pyruvate is quantified spectrophotometrically by a corresponding NADH consumption when the system is coupled to the standard pyruvate dehydrogenase assay. In essence, the velocity of signal decrease depends directly on the velocity of the sialate-recycling engine, which in turn is dependent on the sialate substrate concentration [46].

CMP-sialate synthetase [E.C. 2.7.7.43] catalyzes a CMP nucleotide transfer from CTP to the anomeric hydroxyl of **12**. The enzyme from *N. meningitidis*, in contrast to other known bacterial CMP-sialate synthetases, was found to exhibit a remarkably broad substrate tolerance for Neu5Ac derivatives **72**, in particular by accepting variously C5-acylated structures as substrates. This included neuraminic acid N-carbamoylated with typical protective groups of different lengths and bulkiness, an unsaturated acrylamide derivative, and the corresponding saturated moiety, as well as the deaminated KDN analogue (Scheme 2.2.5.25). Also, the latter structure could be varied by deoxygenation, epimerization at C5 or at the terminal chain, and shortening the chain length to an octulosonic acid [47]. The high expressivity of the recombinant production clone, the high catalytic efficiency of the enzyme, and its broad substrate tolerance make it the preferred catalyst for the enzymatic synthesis of CMP-Neu5Ac and of derivatives **73** modified in the sialic acid moiety.

Several CMP conjugates made available by this procedure could be transferred effectively to *N*-acetyl-lactosamine as an acceptor by using the α-2,6-sialyltransferase from rat liver (Fig. 2.2.5.4). In this manner, several trisaccharides **74–77** could be generated carrying natural as well as non-natural sialic acids in good yields [47], limited so far only by the high cost of the commercial glycosyltransferase.

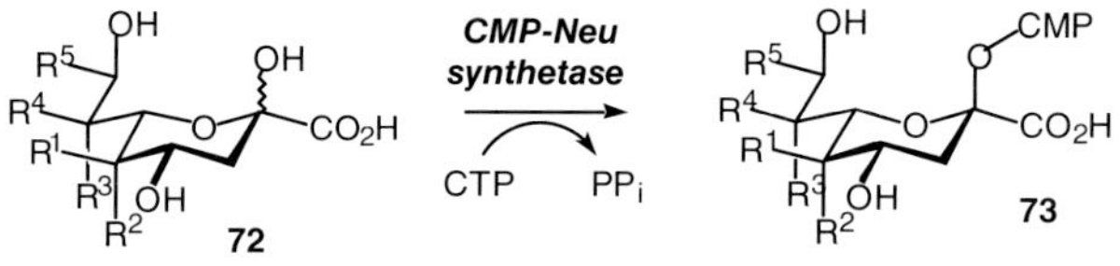

Scheme 2.2.5.25 Enzymatic activation of sialic acids to Leloir-type sugar nucleotides using the CMP-NeuAc synthetase of *Neisseria meningitidis*.

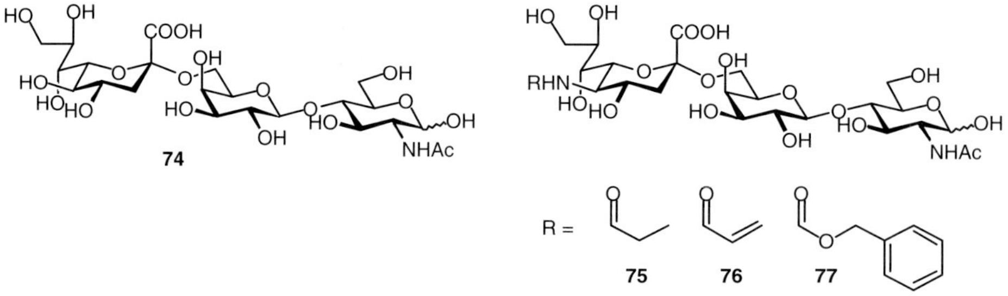

Fig. 2.2.5.4 Synthetic trisaccharides containing KDN and other Neu5Ac derivatives prepared by enzymatic glycosyl transfer.

2.2.5.4 Summary and Outlook

Bioprocesses mediated by enzymes or whole cells ("White Biotechnology") are attractive over the entire spectrum of organic synthesis, with tremendous potential to meet environmental challenges by reducing energy consumption, pollution by greenhouse gases, and the exhaustion of fossil fuels and raw materials according to the principles of "Green Chemistry". Clearly, in comparison with current chemical methods for asymmetric aldol reactions, the ecological advantage of biocatalytic procedures stems from the fact that they do not suffer from a need for harsh reaction conditions, corrosive reagents, costly protecting groups, high catalyst loading, consumption of chiral auxiliaries, or heavy-metal leakage to products.

Our group's experience in using C–C bond-forming enzymes for asymmetric synthesis has been extended profoundly during the research period at the RWTH. Several new enzymes either became available by scientific collaboration within the framework of the SFB380 or have been made available by recombinant DNA technology. These biocatalysts have been studied with respect to reaction mechanism, substrate breadth, stereoselectivity, and operation conditions for practical utilization. The examples of applications listed above demonstrate some of the many windows of synthetic opportunities that are opened for the convergent asymmetric synthesis of interesting polyfunctionalized products by aldolases and related C–C bond-forming biocatalysts. They also indicate useful interfaces to both chemical methods for asymmetric synthesis and to other types of enzymatic catalysis.

The examples further illustrate that the construction of new carbon–carbon bonds is feasible even with substrate analogues that are structurally very distant from the natural functions of these enzymes, which facilitates a rapid access to structures of high structural variability and of amazing complexity. Clearly, the most notable advantage thereby relies on the high asymmetric induction achievable with lyases for the preparation of products with high enantiomeric and diastereomeric purity. The exquisite reaction specificity under mild conditions not only eliminates the need for tedious protecting-group manipulations, but also opens up opportunities to develop highly integrated reaction schemes that comprise numerous chemical steps differing in nature ("artificial metabolisms"). Thus, contemporary tactical schemes of modern organic synthesis such as bidirectional chain synthesis [48] and tandem transformation reactions [49] can be transported readily to the arena of biocatalysis for asymmetric synthesis [36].

Acknowledgements

We gratefully acknowledge funding of this research by grants from the Deutsche Forschungsgemeinschaft (SFB380 B25), as well as through kind support with materials from Boehringer Mannheim GmbH (now Roche Diagnostics).

References

1 W.-D. Fessner, C. Walter, *Top. Curr. Chem.* **1996**, *184*, 97–194; W.-D. Fessner, *Curr. Opin. Chem. Biol.* **1998**, *2*, 85–97; T. D. Machajewski, C.-H. Wong, *Angew. Chem. Int. Ed.* **2000**, *39*, 1352–1374; W.-D. Fessner, V. Helaine, *Curr. Opin. Biotechnol.* **2001**, *12*, 574–586.

2 D. F. Henderson, E. J. Toone, in *Comprehensive Natural Product Chemistry*, Vol. 3 (Ed.: B. M. Pinto), Elsevier Science, Amsterdam, **1999**, pp. 367–440; E. J. Toone, E. S. Simon, M. D. Bednarski, G. M. Whitesides, *Tetrahedron* **1989**, *45*, 5365–5422; H. J. M. Gijsen, L. Qiao, W. Fitz, C.-H. Wong, *Chem. Rev.* **1996**, *96*, 443–473.

3 W.-D. Fessner, in *Microbial Reagents in Organic Synthesis*, Vol. 381 (Ed.: S. Servi), Kluwer Academic Publishers, Dordrecht, **1992**, pp. 43–55.

4 M. D. Bednarski, E. S. Simon, N. Bischofberger, W.-D. Fessner, M. J. Kim, W. Lees, T. Saito, H. Waldmann, G. M. Whitesides, *J. Am. Chem. Soc.* **1989**, *111*, 627–635.

5 W.-D. Fessner, O. Eyrisch, *Angew. Chem. Int Ed. Engl.* **1992**, *31*, 56–58; O. Eyrisch, G. Sinerius, W.-D. Fessner, *Carbohydrate Res.* **1993**, *238*, 287–306.

6 W.-D. Fessner, G. Sinerius, A. Schneider, M. Dreyer, G. E. Schulz, J. Badia, J. Aguilar, *Angew. Chem. Int. Ed. Engl.* **1991**, *30*, 555–558.

7 W.-D. Fessner, G. Sinerius, *Bioorg. Med. Chem.* **1994**, *2*, 639–645.

8 W.-D. Fessner, G. Sinerius, *Angew. Chem. Int. Ed. Engl.* **1994**, *33*, 209–212.

9 W.-D. Fessner, C. Walter, *Angew. Chem. Int. Ed. Engl.* **1992**, *31*, 614–616.

10 C. H. von der Osten, A. J. Sinskey, C. F. Barbas, R. L. Pederson, Y. F. Wang, C.-H. Wong, *J. Am. Chem. Soc.* **1989**, *111*, 3924–3927.

11 W.-D. Fessner, J. Badia, O. Eyrisch, A. Schneider, G. Sinerius, *Tetrahedron Lett.* **1992**, *33*, 5231–5234; W.-D. Fessner, A. Schneider, O. Eyrisch, G. Sinerius, J. Badia, *Tetrahedron: Asymmetry* **1993**, *4*, 1183–1192.

12 W.-D. Fessner, A. Schneider, H. Held, G. Sinerius, C. Walter, M. Hixon, J. V. Schloss, *Angew. Chem. Int. Ed. Engl.* **1996**, *35*, 2219–2221.

13 M. Hixon, G. Sinerius, A. Schneider, C. Walter, W.-D. Fessner, J. V. Schloss, *FEBS Lett.* **1996**, *392*, 281–284.

14 M. K. Dreyer, G. E. Schulz, *J. Mol. Biol.* **1996**, *259*, 458–466; M. Kroemer, G. E. Schulz, *Acta Crystallogr. Sect. D* **2002**, *58*, 824–832.

15 A. C. Joerger, C. Gosse, W.-D. Fessner, G. E. Schulz, *Biochemistry* **2000**, *39*, 6033–6041.

16 S. J. Cooper, G. A. Leonard, S. M. McSweeney, A. W. Thompson, J. H. Naismith, S. Qamar, A. Plater, A. Berry, W. N. Hunter, *Structure* **1996**, *4*, 1303–1315; D. R. Hall, C. S. Bond, G. A. Leonard, C. I. Watt, A. Berry, W. N. Hunter, *J. Biol. Chem.* **2002**, *277*, 22018–22024.

17 H. P. Brockamp, M. R. Kula, *Appl. Microbiol. Biotechnol.* **1990**, *34*, 287–291; F. Götz, S. Fischer, K.-H. Schleifer, *Eur. J. Biochem.* **1980**, *108*, 295–301.

18 M. T. Zannetti, C. Walter, M. Knorst, W.-D. Fessner, *Chem. Eur. J.* **1999**, *5*, 1882–1890.

19 A. Varki, *Glycobiology* **1993**, *3*, 97–130.

20 W.-D. Fessner, M. Knorst, DE 10034586, **2002**; *Chem. Abstr.* **2002**, *136*, 166160.

21 M. Knorst, Dissertation, RWTH Aachen, **1999**.

22 I. P. Korndörfer, W.-D. Fessner, B. W. Matthews, *J. Mol. Biol.* **2000**, *300*, 917–933.

23 M. Whitlow, A. J. Howard, B. C. Finzel, T. L. Poulos, E. Winborne, G. L. Gilliland, *Proteins Struct. Funct. Genet.* **1991**, *9*, 153–157.

24 R. Engel, The use of carbon–phosphorus analog compounds in the regulation of biological processes, in *Handbook of Organophosphorus Chemistry* (Ed.: R. Engel), Dekker, New York, **1992**, pp. 559–600.

25 H. L. Arth, G. Sinerius, W.-D. Fessner, *Liebigs Ann.* **1995**, 2037–2042.

26 H. L. Arth, W.-D. Fessner, *Carbohydrate Res.* **1997**, *305*, 313–321.

27 G. A. Sprenger, U. Schörken, G. Sprenger, H. Sahm, *Eur. J. Biochem.* **1995**, *230*, 525–532.

28 U. Schörken, G. A. Sprenger, *Biochim. Biophys. Acta.* **1998**, *1385*, 229–243; N. J. Turner, *Curr. Opin. Biotechnol.* **2000**, *11*, 527–531.
29 F. T. Zimmermann, A. Schneider, U. Schörken, G. A. Sprenger, W.-D. Fessner, *Tetrahedron: Asymmetry* **1999**, *10*, 1643–1646.
30 O. Eyrisch, Dissertation, University of Freiburg, **1994**; O. Eyrisch, M. Keller, W.-D. Fessner, *Tetrahedron Lett.* **1994**, *35*, 9013–9016.
31 K. Effertz, Diploma Thesis, RWTH Aachen, **1995**.
32 E. Gonzalez-Garcia, V. Helaine, G. Klein, M. Schuermann, G. A. Sprenger, W.-D. Fessner, J.-L. Reymond, *Chem. Eur. J.* **2003**, *9*, 893–899.
33 A. Kirschning, A. F.-W. Bechthold, J. Rohr, *Top. Curr. Chem.* **1997**, *188*, 1–84.
34 W.-D. Fessner, C. Gosse, G. Jaeschke, O. Eyrisch, *Eur. J. Org. Chem.* **2000**, 125–132.
35 F. Zimmermann, Dissertation, RWTH Aachen, **1999**.
36 M. Petersen, M. T. Zannetti, W.-D. Fessner, *Top. Curr. Chem.* **1997**, *186*, 87–117.
37 M. Petersen, Dissertation, RWTH Aachen, **1997**.
38 O. Eyrisch, W.-D. Fessner, *Angew. Chem. Int. Ed. Engl.* **1995**, *34*, 1639–1641.
39 C. Walter, Dissertation, RWTH Aachen, **1996**.
40 O. Lockhoff, personal communication.
41 K. Okamoto, T. Goto, *Tetrahedron* **1990**, *46*, 5835–5857; M. P. DeNinno, *Synthesis* **1991**, *8*, 583–593.
42 G. M. Watt, P. A. Lowden, S. L. Flitsch, *Curr. Opin. Struct. Biol.* **1997**, *7*, 652–660; R. Öhrlein, *Top. Curr. Chem.* **1998**, *200*, 227–254.
43 H. Tettelin, N. J. Saunders, J. Heidelberg, A. C. Jeffries, K. E. Nelson, J. A. Eisen, K. A. Ketchum, D. W. Hood, J. F. Peden, R. J. Dodson, W. C. Nelson, M. L. Gwinn, R. DeBoy, J. D. Peterson, E. K. Hickey, D. H. Haft, S. L. Salzberg, O. White, R. D. Fleischmann, B. A. Dougherty, T. Mason, A. Ciecko, D. S. Parksey, E. Blair, H. Cittone, E. B. Clark, M. D. Cotton, T. R. Utterback, H. Khouri, H. Qin, J. Vamathevan, J. Gill, V. Scarlato, V. Masignani, M. Pizza, G. Grandi, L. Sun, H. O. Smith, C. M. Fraser, E. R. Moxon, R. Rappuoli, J. C. Venter, *Science* **2000**, *287*, 1809–1815; J. Parkhill, M. Achtman, K. D. James, S. D. Bentley, C. Churcher, S. R. Klee, G. Morelli, D. Basham, D. Brown, T. Chillingworth, R. M. Davies, P. Davis, K. Devlin, T. Feltwell, N. Hamlin, S. Holroyd, K. Jagels, S. Leather, S. Moule, K. Mungall, M. A. Quail, M. A. Rajandream, K. M. Rutherford, M. Simmonds, J. Skelton, S. Whitehead, B. G. Spratt, B. G. Barrell, *Nature* **2000**, *404*, 502–506.
44 M. Petersen, W.-D. Fessner, M. Frosch, E. Lüneberg, *FEMS Microbiol. Lett.* **2000**, *184*, 161–164.
45 M. J. Kim, W. J. Hennen, H. M. Sweers, C.-H. Wong, *J. Am. Chem. Soc.* **1988**, *110*, 6481–6486.
46 W.-D. Fessner, M. Knorst, DE 10021019, **2001**; *Chem. Abstr.* **2001**, *135*, 354989.
47 M. Knorst, W.-D. Fessner, *Adv. Synth. Catal.* **2001**, *343*, 698–710.
48 C. S. Poss, S. L. Schreiber, *Acc. Chem. Res.* **1994**, *27*, 9–17; S. R. Magnuson, *Tetrahedron* **1995**, *51*, 2167–2213.
49 T.-L. Ho, *Tandem Organic Reactions*, Wiley-Interscience, New York, **1992**; L. F. Tietze, U. Beifuss, *Angew. Chem. Int. Ed. Engl.* **1993**, *32*, 131–163.
50 H. L. Arth, Dissertation, RWTH Aachen, **1997**.
51 C. Goße, Dissertation, RWTH Aachen, **1997**.

2.2.6
Exploring and Broadening the Biocatalytic Properties of Recombinant Sucrose Synthase 1 for the Synthesis of Sucrose Analogues
Lothar Elling

2.2.6.1 Introduction

The plant enzyme sucrose synthase (SuSy, E.C.2.4.1.13) catalyzes the synthesis and cleavage of sucrose 1 in vivo and in vitro (Scheme 2.2.6.1) and thus represents a unique case among the Leloir glycosyltransferases [1, 2]. As one of the key enzymes in plant carbohydrate metabolism, SuSy utilizes the conserved energy of the transport metabolite sucrose for the formation of nucleotide sugars as precursors for starch and cellulose biosynthesis [3, 4]. Two isoforms of SuSy are encoded by the genes *sus1* and *sus2* in all higher plants. In rice and in pea a third gene (*sus3*) is present [5]. The SuSy proteins consist of 808 amino acids (SuSy1) and 816 amino acids (SuSy2 and SuSy3), respectively. The isoenzymes SuSy1 and SuSy2 from different species share a homology of 80–95%, whereas within a single species, such as rice, homology lies between 90% (SuSy1/SuSy3) and 81% (SuSy1/SuSy2) [6]. SuSy isoenzymes from different plant species have a different acceptance for nucleoside diphosphates (NDP) and nucleotide-activated glucose (NDP-glucose) in the cleavage and synthesis reaction, respectively [7].

A characteristic feature of the SuSy isoforms is a conserved phosphorylated serine residue near the N-terminus [8–10]. In-vivo studies have demonstrated that phosphorylation and dephosphorylation direct the distribution of SuSy isoforms in the plant cell [10–12]. The soluble phosphorylated SuSy interacts with the actin cytoskeleton in the cytoplasm [13], and the dephosphorylated SuSy isoforms are targeted to the cell membrane to form complexes with other enzymes, e.g., glucan synthase, catalyzing cellulose biosynthesis from sucrose [4, 10, 14]. In this respect, recent studies on the dephosphorylated enzymes by cloning and expression of *sus* genes in *E. coli* have shown differences in some biochemical features when compared to the enzymes isolated from the corresponding plant material. Recom-

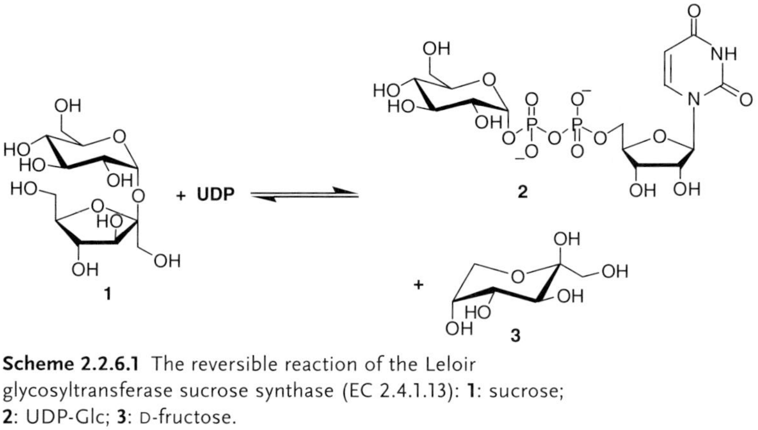

Scheme 2.2.6.1 The reversible reaction of the Leloir glycosyltransferase sucrose synthase (EC 2.4.1.13): **1**: sucrose; **2**: UDP-Glc; **3**: D-fructose.

binant nonphosphorylated SuSy isoforms from mung beans and maize leaves had lower affinities for sucrose in the cleavage reaction. After in-vitro phosphorylation the affinity for sucrose was increased, whereas the kinetics of the substrates in the synthesis reaction were not changed [8, 15, 16]. However; in-vitro phosphorylation of SuSy from soybean nodule expressed in E. coli altered the protein's hydrophobicity rather than its substrate affinity [12].

The prokaryotic SuSy2 isoform purified from Anabaena sp. strain PCC 7119 [17] and the recombinant enzyme expressed in E. coli [18] revealed unfavorable kinetics and characteristics compared to those of the higher plant enzymes, e.g., a low affinity for sucrose, donor-dependent affinities for D-fructose, and complete loss of cleavage activity in the presence of Mn^{2+}.

In conclusion, these studies underline that care has to be taken with reference to the expression system when aiming at a recombinant SuSy biocatalyst as a better and reliable enzyme source.

In our previous work we have demonstrated the wide substrate spectrum of SuSy purified from rice grains for the synthesis of nucleotide (deoxy)sugars and sucrose analogues [2, 19, 20]. In combination with UDP-Glc 4'-epimerase, a novel in-situ regeneration cycle for UDP-Gal as donor substrate for galactosyltransferases was established for the formation of the glycoconjugates Galβ1-4GlcNAc (LacNAc) and Galα1-3Galβ1-4GlcNAc (Galili-epitope) [21]. However, the SuSy preparation from rice grains suffered from the presence of low amounts of contaminating enzymes, e.g., invertase and UDP-Glc cleaving activities [20], decreasing the yields in the synthesis of activated sugars[22] and sucrose analogues [23], respectively.

The present overview summarizes our recent work on recombinant SuSy1 from potato and demonstrates that the biocatalytic properties are broadened by exploiting different expression systems as well as by site-directed mutagenesis of the conserved phosphorylation site.

2.2.6.2 Characteristics of Recombinant Sucrose Synthase 1 (SuSy1) Expressed in Saccharomyces cerevisiae

2.2.6.2.1 Expression and Purification of SuSy1 from Yeast

The sus1 gene from potato (Solanum tuberosum L.) was cloned into the plasmid pDR195. The PMA1 promoter of the expression vector allows the constitutive production of SuSy during cell growth, providing that the glucose feed is controlled and adapted to the logarithmic growth of the yeast cells [24]. With an optimized fermentation protocol a specific activity of 0.3 U mg^{-1} in the crude extract and a production rate of 402 U L^{-1} were obtained. A one-step purification by anion-exchange chromatography resulted in an enzyme preparation which is suitable for the synthesis of nucleotide sugars and sucrose analogues (Scheme 2.2.6.2). The nucleotide sugars ADP-Glc and dTDP-Glc were synthesized on a gram scale [25, 26]. dTDP-Glc is the precursor for dTDP deoxysugars and the utilization of SuSy1 opens up the possibility of in-situ regeneration of these complex sugars [26, 27].

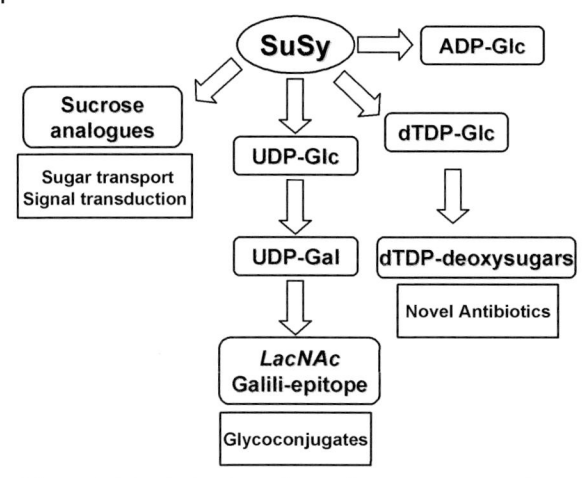

Scheme 2.2.6.2 Utilization of recombinant sucrose synthase 1 for carbohydrate engineering.

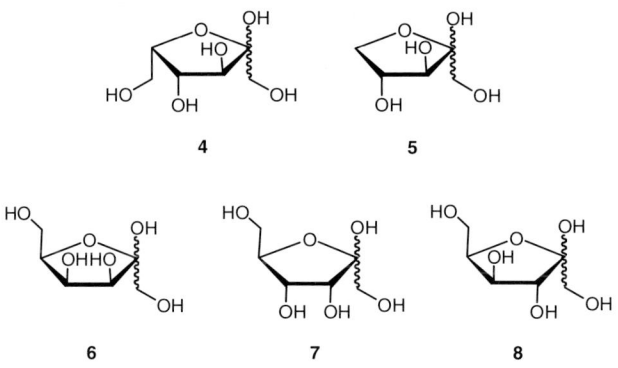

Fig. 2.2.6.1 Ketoses tested as acceptor substrates of recombinant sucrose synthase 1 expressed in yeast.

The inhibition of SuSy by the divalent cations Cu^{2+} (K_i 15 µM), Zn^{2+} (K_i 25 µM), and Ni^{2+} (K_i 37 µM) was exploited for further purification of SuSy1 by immobilized metal affinity chromatography (IMAC). A subsequent gel filtration yielded homogeneous SuSy1 suitable for crystallization experiments [24]. The protein chemical characterization revealed a homotetrameric organization of the 93 kDa subunit. Our kinetic data for the cleavage reaction and preliminary immunoblot analysis for phosphoserine suggested that SuSy1 may be phosphorylated in the yeast expression system [28].

2.2.6.2.2 The Substrate Spectrum of SuSy1 from Yeast

The substrate spectrum of SuSy1 from yeast is well documented for a variety of acceptors [24, 29, 30]. In the series of ketoses we concluded that SuSy1 favors the 3S,4R configuration because L-sorbose **4** and D-xylulose **5** are accepted and D-tagatose **6**, D-psicose **7**, and D-sorbose **8** are not substrates (Fig. 2.2.6.1 and

2.2.6 Exploring and Broadening the Biocatalytic Properties of Recombinant Sucrose Synthase 1

Table 2.2.6.1 Ketoses and aldoses tested as acceptor substrates of recombinant SuSy1 expressed in yeast.

Acceptor substrate	Relative activity[a] [%]
4	55
5	42
6	0
7	0
8	0
11	48
12	14
13	7
14	36
15	36
16	18
17	24
18	15
19	10
20	20
21	16
22	10

a The relative activity is referred to the conversion of 1 mM UDP-Glc with D-fructose after a 16 h incubation as described by Römer et al. [24, 29].

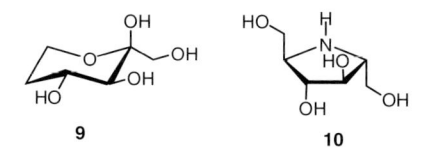

Fig. 2.2.6.2 Fructose analogues as inhibitors of sucrose synthase 1.

Table 2.2.6.1). The relatively strong inhibition of SuSy1 by 5-deoxy-D-fructose **9** (K_i 0.3 mM) (Fig. 2.2.6.2), an analogue of D-fructopyranose, led us to identify aldoses also as substrates (Fig. 2.2.6.3). We hypothesized that SuSy1 recognizes the $^1C_4$ chair conformation of D-fructose, enabling the acceptance of aldoses in their pyranose ring conformations also. Table 2.2.6.1 summarizes all the data obtained for the accepted ketoses and aldoses. The variety of the accepted UDP-activated nucleotide sugars, such as UDP-Glc, UDP-GlcNAc, and UDP-Gal, offers further variations for the synthesis of disaccharides [30]. In summary, it is possible to synthesize over 60 different sucrose analogues by combination of the identified donor and acceptor substrates. Figure 2.2.6.4 depicts the products synthesized so far. These sucrose analogues are not cleavable by invertase, which makes them valuable tools for sugar transport studies in plants [31].

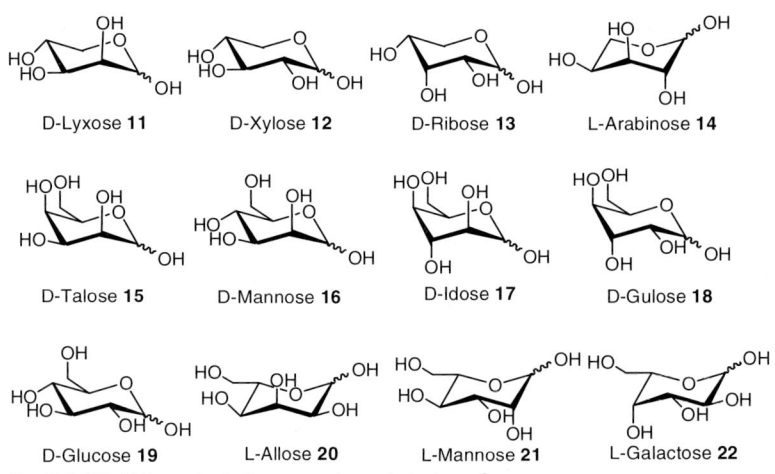

Fig. 2.2.6.3 Aldoses tested as acceptor substrates of recombinant sucrose synthase 1 expressed in yeast.

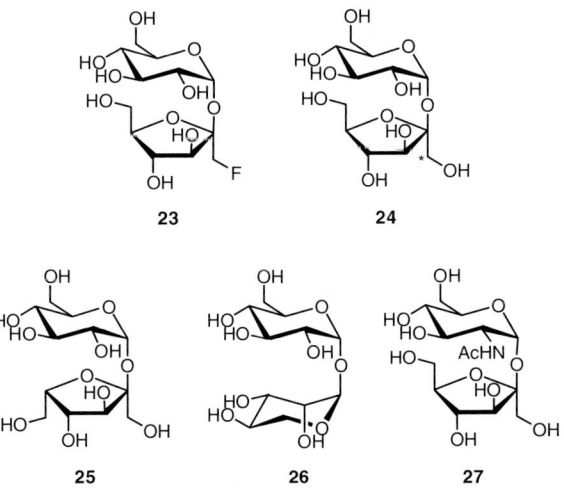

Fig. 2.2.6.4 Sucrose analogues synthesized with recombinant SuSy1 from yeast. **23**: 1'-deoxy-1'-fluoro-β-D-fructofuranosyl-α-D-glucopyranoside; **24**: [$^{13}C_1$]-β-D-fructofuranosyl-α-D-glucopyranoside; **25**: α-D-glucopyranosyl-α-L-sorbofuranoside; **26**: α-D-glucopyranosyl-α-D-lyxopyranoside; **27**: 2-acetamido-2-deoxy-α-D-glucopyranosyl-β-D-fructofuranoside.

2.2.6.3 Characteristics of Recombinant Sucrose Synthase 1 (SuSy1) Expressed in *Escherichia coli*

2.2.6.3.1 Expression and Purification of SuSy1 from *E. coli*

The aim of our further work was the production of the nonphosphorylated SuSy1 enzyme in *E. coli* in order to elucidate its protein chemical and biocatalytic characteristics (Sauerzapfe and Elling, unpublished results). The gene *sus1* from *Solanum tuberosum* was cloned in the expression vectors pET28a and pET16b under the control of the T7*lac* promoter. Expression of the vectors pET28a-*sus1*, pET28a-CHis$_6$*sus1*, and pET16b-*NHis$_6$sus1* in *E. coli* BL21DE3 yielded native SuSy1 and SuSy1 with C- or N-terminal His$_6$-tags (CHis$_6$SuSy1, NHis$_6$SuSy1), respectively. The specific cleavage activities of native SuSy1 (10 mU mg^{-1}), CHis$_6$SuSy1 (3 mU mg^{-1}) and NHis$_6$SuSy1 (0.6 mU mg^{-1}) expressed in *E. coli* were strikingly lower when compared to recombinant SuSy1 from yeast. Interestingly, in the synthesis reaction similar activities were obtained for native SuSy1 and CHis$_6$SuSy1 (both between 2 and 3 mU mg^{-1}), whereas NHis$_6$SuSy1 activity (0.6 mU mg^{-1}) was low. The purification by IMAC on a Ni^{2+}-NTA column also revealed striking differences. The main fraction of the His$_6$-tagged SuSy1 enzymes could already be eluted with 25 mM imidazole, which indicates that the His$_6$ tags are not accessible for binding to the immobilized Ni^{2+} cation. On the other hand, native SuSy1 from *E. coli* behaved like SuSy1 from yeast and could be eluted with 25 mM imidazole buffer. Activity assays of the eluted enzyme fractions revealed that for the His$_6$-tagged SuSy1 enzymes, only the synthesis activity was enriched (15- to 20-fold), whereas for native SuSy1 both the cleavage (160-fold) and synthesis (40-fold) activity were enriched. We concluded that recombinant native SuSy1 from *E. coli* is the enzyme that is suitable for further comparison of its biocatalytic characteristics with recombinant SuSy1 from yeast. For the tagged enzymes it can be concluded that the His$_6$ tags at least do not affect the synthesis reaction of SuSy1 and are deleterious to the cleavage activity.

Homogeneous purified native SuSy1 from *E. coli* was obtained in a similar quality by the protocol applied for the purification of SuSy1 from yeast [24]. However, analysis of both recombinant enzymes by gel filtration revealed that expression of *sus1* in *S. cerevisiae* and *E. coli* affects the oligomeric organization and properties of SuSy1. The main fraction of SuSy1 from yeast was eluted as a homotetramer, whereas SuSy1 from *E. coli* appears mainly as a trimeric protein. The major cleavage activities reside in these oligomeric forms. On the other hand we could demonstrate that the synthesis activity does not depend strictly on the oligomeric subunit organization because monomeric SuSy1 from *E. coli* showed approx. 90% of the activity of the trimeric enzyme.

In summary, we may conclude that nonphosphorylated SuSy1 produced in *E. coli* has strikingly different biochemical characteristics. It may further be concluded that phosphorylation could direct the subunit organization of SuSy1 and that our results may also give some insight into the regulation of the synthesis and cleavage reaction of SuSy1 in plants [10].

2.2.6.3.2 The Substrate Spectrum of SuSy1 from *E. coli*

The comparison of the kinetic data for recombinant SuSy1 from yeast and *E. coli* revealed no significant changes in the substrate affinities (Table 2.2.6.2) (Sauerzapfe and Elling, unpublished results). The influence of phosphorylation of SuSy on the enzyme's affinity for sucrose and UDP are discussed controversially in the literature. Nakai et al. found an increase in the substrates affinities [16]; however, Zhang et al. could not detect changes in the kinetic data for SuSy from soybean nodules [12]. With reference to our work, the expression in a eukaryotic or prokaryotic host influences the protein chemical characteristics of SuSy1. However, we cannot yet decide whether recombinant SuSy1 from yeast is phosphorylated.

The most significant indication that the biocatalytic properties also depend on the expression host comes from our studies on the fructose analogues DMDP **10** (2,5-dihydroxymethyl-(3*S*,4*R*)-dihydroxypyrrolidine) and 5-deoxy-D-fructose as inhibitors of SuSy1 (Fig. 2.2.6.2). In contrast to SuSy1 from yeast, the strongest inhibitor for recombinant SuSy1 from *E. coli* is **10** (IC_{50} 1.5 mM), suggesting that the enzyme prefers the D-fructofuranose conformation.

The ketoses D-tagatose **6** (43% relative activity) and D-ribulose **28** (24% relative activity) were identified as new acceptor substrates; they are not accepted by recombinant SuSy1 from yeast. However; the acceptance for D-xylulose **5** is lost. The most significant changes were observed for the aldoses tested: L-arabinose **14**, D-xylose **12**, and D-lyxose **11** are better substrates than D-fructose, with relative activities of 490%, 300%, and 151%, respectively. In the hexose series L-glucose **29** and L-rhamnose **30** were identified as new acceptor substrates, whereas the acceptance for D-and L-mannose, **16** and **21**, was improved (Fig. 2.2.6.5) (Sauerzapfe and Elling, unpublished results).

In summary, our results demonstrate that the biocatalytic properties of SuSy1 are broadened and improved by expression in different hosts. So far, four new acceptor substrates have been identified and the acceptance of three aldoses has been improved.

Table 2.2.6.2 Comparison of the affinity constants of recombinant SuSy1 expressed in yeast and *E. coli*.

Substrate	K_m [mM]	
	SuSy1, yeast	Susy1, *E. coli*
Cleavage reaction		
UDP	0.3	0.2
Sucrose	60.0	80.3
Synthesis reaction		
UDP-Glc	0.6	1.4
D-Fructose	5.0	5.5

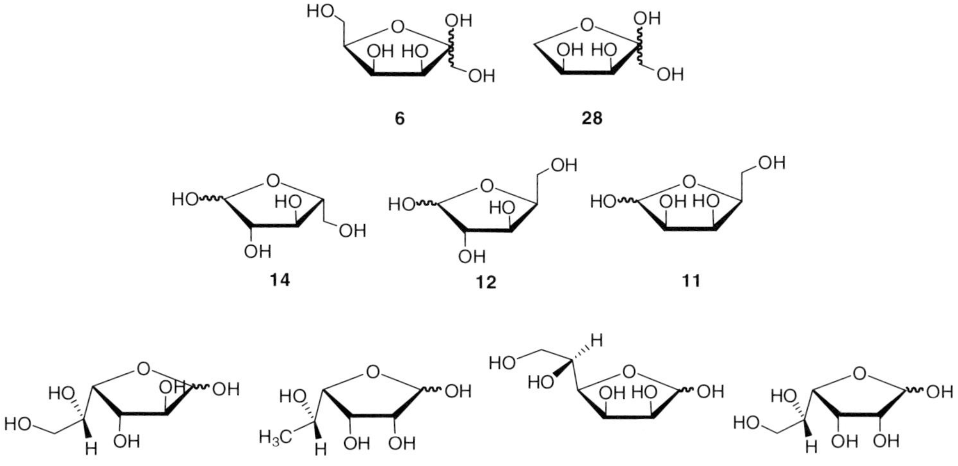

Fig. 2.2.6.5 Identified acceptor substrates of recombinant sucrose synthase 1 expressed in E. coli. The aldoses are depicted in their furanose conformations based on the conclusion that SuSy1 from E. coli accepts D-fructofuranose preferentially.

2.2.6.4 Sucrose Synthase 1 Mutants Expressed in *S. cerevisiae* and *E. coli*

As described above, SuSy enzymes possess a conserved serine phosphorylation site, being serine11 (S11) for SuSy1 from potato [12]. Our ongoing work focuses on the question of whether site-directed mutagenesis of this conserved S11 has an influence on the biocatalytic properties (Engels, Sauerzapfe, and Elling, unpublished results). In our first experiments we followed the strategy of imitating the anionic character of serine phosphorylation by a SuSy1(S11D) mutant expressed in *E. coli*. On the other hand, the putative phosphorylation site was disrupted in the SuSy1(S11A) mutant expressed in yeast. The kinetic characterization revealed interesting results. The SuSy1(S11D) mutant from *E. coli* showed a 10-fold improvement of the affinities for UDP-Glc and D-fructose, respectively. In contrast, the SuSy1(S11A) mutant from yeast showed improved (10-fold) affinities for UDP and sucrose. The K_m values of the other substrates in the synthesis and cleavage reaction, respectively, were not significantly changed.

Preliminary investigations on the substrate spectrum of the S11A mutant of SuSy1 expressed in yeast reveal L-glucose **29** and L-rhamnose **30** as new substrates, as well as D-xylulose **5** with a relative activity of 112%. Most interestingly, both mutants accept D-psicose and L-fucose as new acceptor substrates.

In summary, these first results already demonstrate the impact of site-directed mutagenesis of Ser11 on SuSy1. Work is in progress in our group to elucidate the effect of further amino acids substituting S11 and to identify further novel or improved biocatalytic properties of these mutants of SuSy1 from potato.

2.2.6.5 Outlook

Our recent results introduce novel aspects of the biocatalysis utilizing this Leloir glycosyltransferase from plants, sucrose synthase (SuSy). The number of possible sucrose analogues has been further expanded by expression in different hosts and by site-directed mutagenesis of the conserved phosphorylation site S11. What is the future of SuSy1? The most important aspect is to elucidate the reaction mechanism and the protein structure of sucrose synthase. SuSy is a retaining glycosyltransferase and grouped in family 4, where the protein structure of a family member has not yet been solved [32]. With respect to the different biocatalytic properties observed, recombinant SuSy1 from *E. coli* and from yeast should be investigated, including also phosphorylation analysis of the latter. With these data in our hands we may rationalize our findings on new or improved acceptor substrates. Meanwhile, further data on new mutants can be collected by saturation mutagenesis of the S11 site or by generating them via directed evolution strategies. Finally, the synthesis and structural analysis of novel sucrose analogues should complement the studies on the protein structure – together they could be used for the prediction of the regio- and enantioselectivity of the SuSy reaction using ketoses, aldoses, and probably also non-sugar substrates.

Acknowledgements

I thank my coworkers Dr. Ulrike Römer, Dipl.-Biol. Birgit Sauerzapfe, and Dipl.-Biol. Leonie Engels for their excellent contributions and ongoing fascinating results. Financial support by the DFG (SFB380, Teilprojekt B26: funding period 2000–2003, and project El 135/2-1) and by the Fonds der Chemischen Industrie is gratefully acknowledged. I thank also the cooperation partners within the SFB380, Prof. Dr. W. Fessner (TU Darmstadt), Prof. Dr. G. Sprenger (University of Stuttgart), Prof. Dr. A. Liese (TU Hamburg), and Prof Dr. M. Müller (University of Freiburg), for providing us with fructose analogues. Special thanks go to Prof. Dr. W. B. Frommer (Stanford University), Prof. Dr. T. Roitsch (Würzburg University), and Dr. W. Köckenberger (Nottingham University), for their stimulating discussions on applications of sucrose analogues in plant physiology.

References

1. P. Geigenberger, M. Stitt, *Planta* **1993**, *189*, 329.
2. L. Elling, in *Adv. Biochem. Eng./Biotechnol.*, Vol. 58 (Ed.: T. Scheper), Springer-Verlag, Berlin, **1997**, p. 89.
3. J. Pozueta Romero, P. Perata, T. Akazawa, *Crit. Rev. Plant Sci.* **1999**, *18*, 489; A. Sturm, G. Q. Tang, *Trends Plant Sci.* **1999**, *4*, 401; R. Zrenner, M. Salanoubat, L. Willmitzer, U. Sonnewald, *Plant J.* **1995**, *7*, 97.
4. Y. Amor, C. H. Haigler, S. Johnson, M. Wainscott, D. P. Delmer, *Proc. Natl. Acad. Sci. USA* **1995**, *92*, 9353.
5. A. Y. Wang, M. H. Kao, W. H. Yang, Y. Sayion, L. F. Liu, P. D. Lee, J. C. Su, *Plant Cell Physiol.* **1999**, *40*, 800; D. H. P. Barratt, L. Barber, N. J. Kruger, A. M.

Smith, T. L. Wang, C. Martin, *Plant Physiol.* **2001**, *127*, 655–664.
6 J.-W. Huang, J.-T. Chen, W.-P. Yu, L.-F. Shyur, A.-Y. Wang, H.-Y. Sung, P.-D. Lee, J.-C. Su, *Biosci. Biotechnol. Biochem.* **1996**, *60*, 233.
7 G. Avigad, in *Encyclopedia of Plant Physiology*, New Series, *Carbohydrates*, Vol. 13A (Eds.: F. A. Loewus, W. Tanner), Springer Verlag, Berlin, **1982**, p. 217; P.-Y. Lim, P. Perata, J. Pozueta-Romero, T. Akazawa, J. Yamaguchi, *Biosci. Biotechnol. Biochem.* **1992**, *56*, 695; D. Y. Huang, A. Y. Wang, *Biochem. Mol. Biol. Int.* **1998**, *46*, 107; C. S. Echt, P. S. Chourey, *Plant Physiol.* **1985**, *79*, 530.
8 S. C. Huber, J. L. Huber, P.-C. Liao, D. A. Gage, R. W. McMichael, P. S. Chourey, L. C. Hannah, K. Koch, *Plant Physiol.* **1996**, *112*, 793.
9 X.-Q. Zhang, R. Chollet, *FEBS Lett.* **1997**, *410*, 126.
10 H. Winter, S. C. Huber, *Crit. Rev. Plant Sci.* **2000**, *19*, 31.
11 H. Winter, J. L. Huber, S. C. Huber, *FEBS Lett.* **1997**, *420*, 151; S. C. Hardin, H. Winter, S. C. Huber, *Plant Physiol.* **2004**, *134*, 1427.
12 X. Q. Zhang, A. A. Lund, G. Sarath, R. L. Cerny, D. M. Roberts, R. Chollet, *Arch. Biochem. Biophys.* **1999**, *371*, 70.
13 H. Winter, J. L. Huber, S. C. Huber, *FEBS Lett.* **1998**, *430*, 205.
14 D. G. Robinson, *Bot. Acta* **1996**, *109*, 261.
15 T. Nakai, N. Tonouchi, T. Tsuchida, H. Mori, F. Sakai, T. Hayashi, *Biosci. Biotechnol. Biochem.* **1997**, *61*, 1500.
16 T. Nakai, T. Konishi, X. Q. Zhang, R. Chollet, N. Tonouchi, T. Tsuchida, F. Yoshinaga, H. Mori, F. Sakai, T. Hayashi, *Plant Cell Physiol.* **1998**, *39*, 1337.
17 A. C. Porchia, L. Curatti, G. L. Salerno, *Planta* **1999**, *210*, 34.
18 L. Curatti, A. C. Porchia, L. Herrera-Estrella, G. L. Salerno, *Planta* **2000**, *211*, 729.
19 L. Elling, in *Bioorganic Chemistry – Highlights and New Aspects* (Eds.: U. Diederichsen, T. K. Lindhorst, B. Westermann, L. Wessjohann), Wiley-VCH, Weinheim, **1999**, p. 166; A. Zervosen, L. Elling, in *Methods in Biotechnology: Carbohydrate Biotechnology Protocols*, Vol. 10 (Ed.: C. Bucke), Humana Press, Totowa, NJ, **1999**, p. 235.
20 L. Elling, B. Güldenberg, M. Grothus, A. Zervosen, M. Peus, A. Helfer, A. Stein, H. Adrian, M.-R. Kula, *Biotechnol. Appl. Biochem.* **1995**, *21*, 29.
21 L. Elling, M. Grothus, M.-R. Kula, *Glycobiology* **1993**, *3*, 349; A. Zervosen, L. Elling, *J. Am. Chem. Soc.* **1996**, *118*, 1836; C. H. Hokke, A. Zervosen, L. Elling, D. H. Joziasse, D. H. van den Eijnden, *Glycoconjugate J.* **1996**, *13*, 687.
22 A. Zervosen, A. Stein, H. Adrian, L. Elling, *Tetrahedron* **1996**, *52*, 2395.
23 J. Peters, H.-P. Brockamp, T. Minuth, M. Grothus, A. Steigel, M.-R. Kula, L. Elling, *Tetrahedron: Asymmetry* **1993**, *4*, 1173; M. Grothus, A. Steigel, M.-R. Kula, L. Elling, *Carbohydr. Lett.* **1994**, *1*, 83.
24 U. Römer, H. Schrader, N. Günther, N. Nettelstroth, W. B. Frommer, L. Elling, *J. Biotechnol.* **2004**, *107*, 135.
25 A. Zervosen, U. Römer, L. Elling, *J. Mol. Catal. B: Enzymatic* **1998**, *5*, 25.
26 L. Elling, C. Rupprath, N. Günther, U. Römer, S. Verseck, P. Weingarten, G. Dräger, A. Kirschning, W. Piepersberg, *ChemBioChem* **2005**, *6*, 1423. DOI: 10.1002/cbic.200500037.
27 C. Rupprath, T. Schumacher, L. Elling, *Curr. Med. Chem.* **2005**, *12*, 1637–1675.
28 U. Römer, Dissertation Thesis, Heinrich-Heine-Universität Düsseldorf, **2003**.
29 U. Römer, N. Nettelstroth, W. Köckenberger, L. Elling, *Adv. Synth. Catal.* **2001**, *343*, 655.
30 U. Römer, C. Rupprath, L. Elling, *Adv. Synth. Catal.* **2003**, *345*, 684.
31 A. K. Sinha, M. G. Hofmann, U. Römer, W. Köckenberger, L. Elling, T. Roitsch, *Plant Physiol.* **2002**, *128*, 1480–1489.
32 P. M. Coutinho, B. Henrissat, in *Recent Advances in Carbohydrate Bioengineering* (Eds.: H. J. H.J. Gilbert, G. Davies, B. Henrissat, B. Svensson), The Royal Society of Chemistry, Cambridge, **1999**, p. 3; P. M. Coutinho, B. Henrissat, Carbohydrate Active Enzymes database: http://afmb.cnrs-mrs.fr/CAZY/ **1999**.

2.2.7
Flexible Asymmetric Redox Reactions and C–C Bond Formation by Bioorganic Synthetic Strategies

Michael Müller, Michael Wolberg, Silke Bode, Ralf Feldmann, Petra Geilenkirchen, Thomas Schubert, Lydia Walter, Werner Hummel, Thomas Dünnwald, Ayhan S. Demir, Doris Kolter-Jung, Adam Nitsche, Pascal Dünkelmann, Annabel Cosp, Martina Pohl, Bettina Lingen, and Maria-Regina Kula

2.2.7.1 Introduction

The metabolic diversity generated through evolution by Nature is enormously broad. Up to now, about 200 000 different secondary metabolites have been described structurally [1]. However, the theoretically possible number of potential bioactive low-molecular compounds is incredibly large in comparison to the number of compounds actually realized by Nature. What chemists and biochemists have realized so far is that a simple combinatorial approach cannot be used to mimic the diversity observed with natural products. From an evolutionary point of view, Nature must have developed efficient methods to biosynthesize a broad diversity of metabolites by employing biocatalysts displaying broad diversity with regard to activity and specificity. In this chapter attention is drawn to some concepts based on investigation of different types of biotransformations, focusing on aspects of bioinspired diversity.

2.2.7.2 Diversity-Oriented Access to 1,3-Diols Through Regio- and Enantioselective Reduction of 3,5-Dioxocarboxylates

The biosynthesis of many hydroxylated natural products proceeds through regio- and enantioselective modification of polyketides, which are assembled through chain elongation via acetate or propionate units [2]. The enzymes responsible for the chain elongation and subsequent reduction, elimination, aromatization, and further modifications are classified as polyketide synthases [3]. These multifunctional enzymes have been used for whole-cell biotransformation toward "unnatural" metabolites that are within the scope of combinatorial biosynthesis [4].

We envisaged the biocatalytic regio- and stereoselective modification of polyketide-like compounds. This mimics the biosynthetic strategy of starting from a single substrate which after modification(s) affords a variety of different products. The regio- and enantioselective reduction of 3,5-dioxo esters **A** represents such a straightforward and flexible approach. All four stereoisomers of a dihydroxy ester **C** can be derived via hydroxy keto esters **B** from the same precursor **A** by this strategy (Scheme 2.2.7.1).

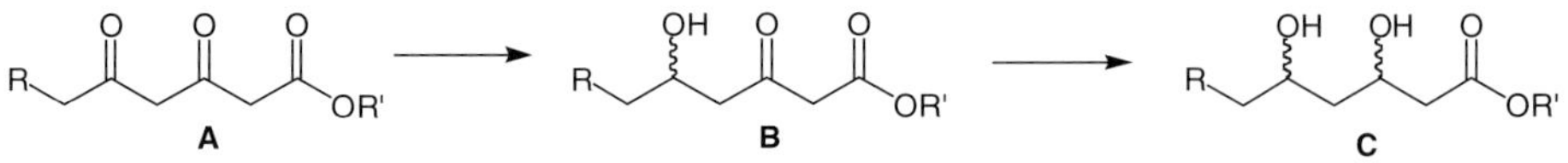

Scheme 2.2.7.1 The expected regio- and enantioselective reduction of 3,5-dioxo esters **A**.

Only a few publications dealing with this subject can be found in the literature. Hydrogenation of diketo esters **A** with chirally modified ruthenium catalysts resulted in mixtures of *syn-* and *anti-*dihydroxy esters **C** with varying enantiomeric excesses [5]. A notable exception to this is represented by the recent work of Carpentier et al., who succeeded in controlling the reduction of methyl 3,5-dioxohexanoate at the initial step, namely the reduction of the β-keto group. The enantiomeric excess achieved was, nevertheless, limited to 78% at best [5a].

Highly enantioselective reduction of ethyl 6-benzyloxy-3,5-dioxohexanoate by ADH of *Acinetobacter calcoaceticus* has been reported (97 to >99% *ee*) [6]. Regioselectivity was not encountered, however, as was the case in the reduction of a variety of 3,5-dioxohexanoates **A** with baker's yeast [7]. The application of isolated enzymes in an anticipated regio- and enantioselective reduction of diketo esters **A** seemed most promising to us.

2.2.7.2.1 Regio- and Enantioselective Enzymatic Reduction

An NADP(H)-dependent ADH of *Lactobacillus brevis* (LBADH) was identified as a suitable catalyst accepting a broad range of diketo esters **A** as substrate [8]. This stable enzyme is easily available in the form of a crude cell extract (recLBADH) from a recombinant *E. coli* strain [9]. The reaction with diketo esters **1a–1c** was performed on a preparative scale, using substrate-coupled regeneration of NADPH (Scheme 2.2.7.2).

Comparison of analytical data of the products (*R*)-**2b** and (*R*)-**2c** with literature data (NMR, [α]) clearly proved the formation of *R*-configured [*S* in the case of **2a**, R = Cl] δ-hydroxy-β-keto isomers of high optical purity. Neither the β-hydroxy-δ-keto regioisomer nor β,δ-di-hydroxy esters could be detected by NMR and GC-MS analysis of the crude enzymatic products.

The enzyme recLBADH is the first catalyst that has been found to allow the highly regio- *and* enantioselective synthesis of δ-hydroxy-β-keto esters by reduction of the respective diketo esters. This enzymatic reaction is of enormous preparative value. The substrates are readily available by acylation of β-keto ester bisenolates and the reaction only requires a simple batch technique which is easy to scale up. Reduction of the chlorinated compound **1a** has been performed routinely on a 75 g scale in our laboratory (8 L fed batch), yielding (*S*)-**2a** in an isolated yield of 84% [10].

In order to synthesize the *R*-enantiomer of hydroxy keto ester **2a** we reinvestigated the baker's yeast reduction of 3,5-dioxoesters [7]. The chlorinated diketo

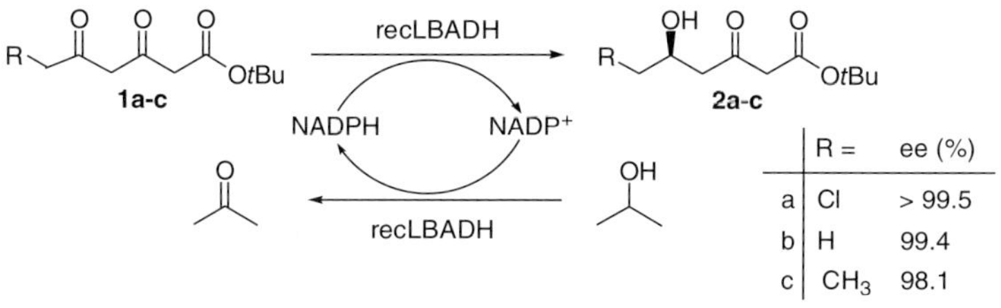

	R =	ee (%)
a	Cl	> 99.5
b	H	99.4
c	CH$_3$	98.1

Scheme 2.2.7.2 Reduction of diketo esters **1a–1c** by recLBADH (R = Cl, H, Me).

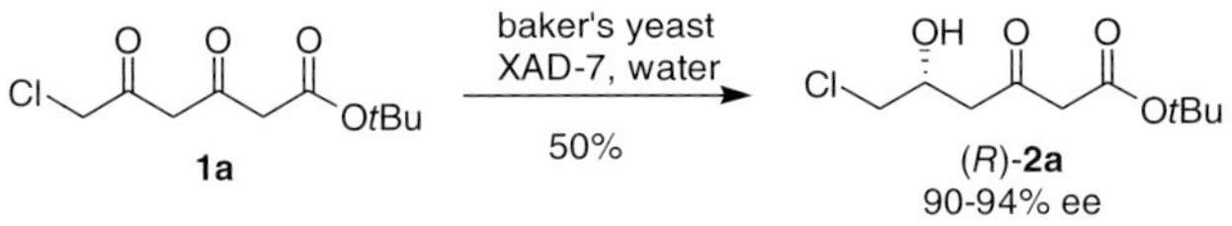

Scheme 2.2.7.3 Reduction of diketo ester **1a** by baker's yeast whole-cell transformation.

ester **1a** was reduced with high regioselectivity at C5 by baker's yeast, in contrast to its unchlorinated analogues **1b** and **1c**, which gave mixtures of regioisomers in this reaction [11]. The enantiomeric excess of the product (*R*)-**2a** was greatly enhanced by application of a biphasic system (hexane/water or Amberlite XAD-7/water). Additionally, it was found that the application of a dried baker's yeast and a high yeast/substrate ratio (10 g yeast per mmol **1a**) gave the best results with regard to enantioselectivity. A combination of these optimized parameters enhanced the enantiomeric excess of the product (*R*)-**2a** from 48% to 90–94% (Scheme 2.2.7.3). The application of the resin considerably facilitated the product isolation and the reaction could thus be carried out on a gram scale without difficulty [11]. Other substrates, such as the lipohilic heptanoate **1c** can be reduced selectively (recLBADH) using the two-phase resin method as well [12].

A set of *Saccharomyces cerevisiae* reductases was screened in collaboration with J. D. Stewart's group (University of Florida). It was demonstrated that diketo ester **1a** is accepted as substrate by at least three different NADP(H)-dependent reductases of this microorganism. Application of a cell-free system in preparative batches using enzyme-coupled coenzyme regeneration afforded (*R*)-**2a** with more than 99% enantiomeric excess [13].

In summary, it has been shown that the enzyme-catalyzed regio- and enantioselective reduction of 3,5-diketo esters can be performed advantageously on a preparative scale giving access to enantiopure (*R*)- and (*S*)-3,5-dihydroxyhexanoates.

2.2.7.2.2 Dynamic Kinetic Resolution

The enzyme-catalyzed regio- and enantioselective reduction of α- and/or γ-alkyl-substituted β,δ-diketo ester derivatives would enable the simultaneous introduction of up to four stereogenic centers into the molecule by two consecutive reduction steps through dynamic kinetic resolution with a theoretical maximum yield of 100%. Although the dynamic kinetic resolution of α-substituted β-keto esters by chemical [14] or biocatalytic [15] reduction has proven broad applicability in stereoselective synthesis, the corresponding dynamic kinetic resolution of 2-substituted 1,3-diketones is rarely found in the literature [16].

t-Butyl 4-methyl-3,5-dioxo-hexanoate (**3**) was prepared by acylation of the bisenolate of *t*-butyl 3-oxovalerate with commercially available Weinreb acetamide [12]. By enzymatic reduction of **3** with recLBADH in an aqueous buffer system, **4** was isolated in 66% yield (Scheme 2.2.7.4). NMR data of the major product *syn*-(4*S*,5*R*)-**4** clearly proved the regioselective single-site reduction of the keto group at C5. Additionally, no evidence for the reduction of the keto group at C3 could be found from GC-MS data of the crude product [17].

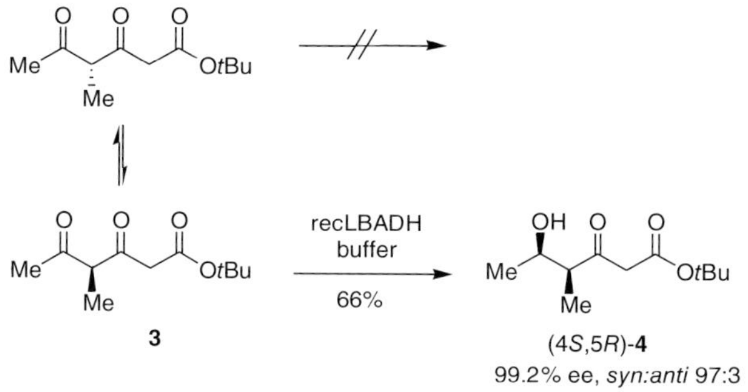

Scheme 2.2.7.4 Reduction of diketo esters **3** by recLBADH via dynamic kinetic resolution.

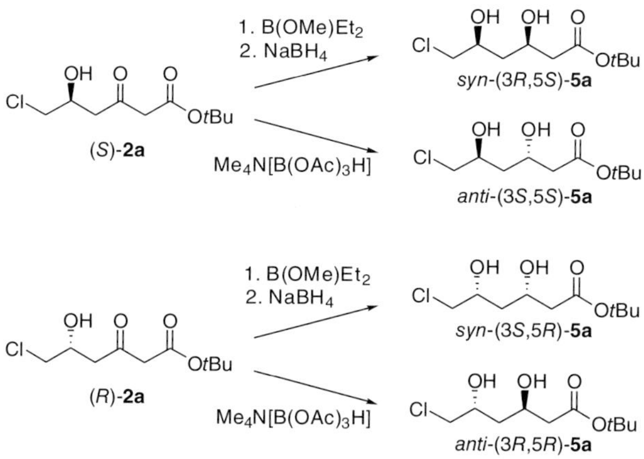

Scheme 2.2.7.5 Diastereoselective nonenzymatic reduction of hydroxy keto esters (S)-**2a** and (R)-**2a**.

This indeed verifies the dynamic kinetic resolution of *rac*-**3** through enzymatic reduction, representing the first example for the dynamic kinetic resolution of an open-chain 2-alkyl-substituted 1,3-diketone through reduction under neutral conditions.

2.2.7.2.3 Stereoselective Access to 1,3-Diols by Diastereoselective Reduction

Substituted 1,3-diols are valuable intermediates in the synthesis of drugs and natural products [18]. Starting from the regio- and enantioselective enzymatic reduction of diketo esters described above, various methods to obtain enantiomerically pure 3,5-dihydroxy esters were developed.

Diastereoselective reduction by chemical methods For the preparation of both enantiomers of dihydroxy ester *syn*-**5a** Prasad's [19] *syn*-selective borohydride reduction was applied, giving dihydroxy esters *syn*-**5a** in a diastereomeric ratio of $dr_{syn:anti} = 28:1$ to $45:1$. The enantiomers of dihydroxy ester *anti*-**5a** were synthesized according to Evans' method [20], which resulted in $dr_{anti:syn} = 14:1$ to $18:1$ (Scheme 2.2.7.5) [11].

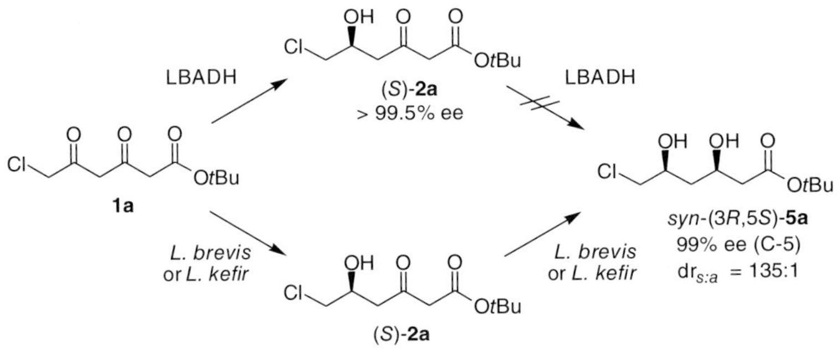

Scheme 2.2.7.6 Formation of dihydroxy ester syn-(3R,5S)-**5a** via hydroxy keto ester (S)-**2a** through whole-cell biotransformation of diketo ester **1a**.

Advantageously, the diastereomeric ratios could be raised to approximately 200 : 1 and more for all dihydroxy esters **5a** by a single crystallization step affording enantiopure diols **5a** on a gram scale [11]. Enantiopure syn-(3R,5S)-**5a** was prepared on a 200 g scale, enabling various applications in natural product total synthesis (see below) [21].

Lactobacillus brevis whole-cell biotransformation   When the reduction of diketo ester **1a** was performed with whole cells of Lactobacillus brevis or L. kefir, formation of the 3,5-dihydroxy ester (3R,5S)-**5a** was observed [10, 22]. This was surprising since it is known that the prevailing alcohol dehydrogenase in L. brevis is the one described as LBADH [23] and since, moreover, this enzyme does not reduce β-keto δ-hydroxy ester **2a** to the corresponding dihydroxy ester (Scheme 2.2.7.6). Under the conditions tested, further alcohol dehydrogenase activity is clearly present in L. brevis and L. kefir. Pfruender et al. optimized the production of L. kefir cells and used this biocatalyst for the one-pot synthesis of dihydroxy ester syn-(3R,5S)-**5a** using diketo ester **1a** as starting material [24].

2.2.7.2.4   Nucleophilic Substitution of Chlorine

In the previous sections it has been shown that all four stereoisomers of the chlorinated dihydroxy ester **5a** can be synthesized in an enantiopure form on a preparative scale. In order to extend the applicability of these compounds a nucleophilic substitution of chlorine was sought.

A two-step conversion of acetonide syn-(3R,5S)-**6a** to hydroxy compound syn-(3R,5S)-**12** is known from the patent literature [25]. This compound is an advanced intermediate in the synthesis of HMG-CoA reductase inhibitors [26]. Iodide syn-(3R,5S)-**13** has been utilized for this purpose, too.[27] We were able to substitute the chlorine of acetonide syn-(3R,5S)-**6a** by iodine in a single step under advanced halogen exchange conditions (52% yield; Scheme 2.2.7.7) [28]. However, conversion was incomplete (86%) and the remaining starting material could not be removed from the product syn-(3R,5S)-**13** [11].

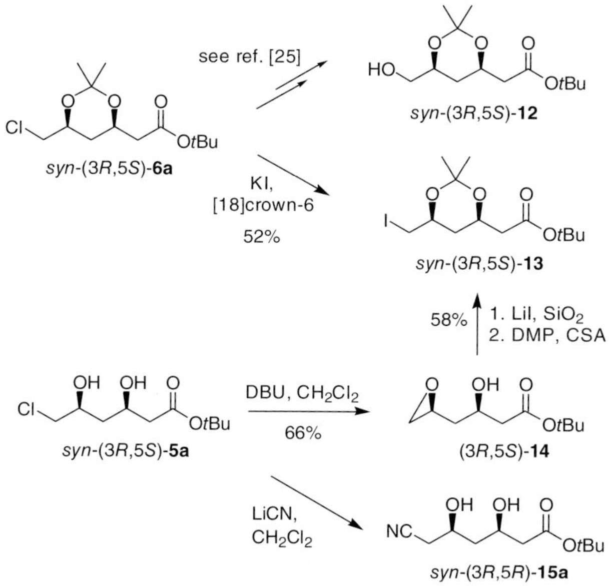

Scheme 2.2.7.7 Nucleophilic substitution of the chlorine of acetonide *syn*-(3R,5S)-**6a** or diol **5a**.

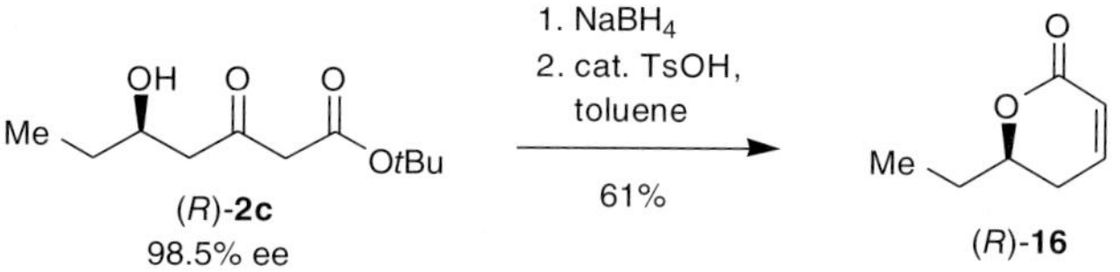

Scheme 2.2.7.8 Synthesis of lactone (R)-**16**, a natural fragrance compound occurring in the essential oil of labdanum resin [30].

Alternatively, epoxide (3R,5S)-**14** was opened regioselectively with lithium iodide on silica [29]. The crude product was immediately subjected to acetonide protection, which afforded the desired iodide *syn*-(3R,5S)-**13** with a 58% yield (44% from *syn*-(3R,5S)-**5a**; Scheme 2.2.7.7). Epoxide (3R,5S)-**14** was easily obtained from dihydroxy ester *syn*-(3R,5S)-**5a** by treatment with DBU (66% yield) [11]. Treatment of *syn*-(3R,5S)-**5a** with LiCN in CH$_2$Cl$_2$ gave nitrile (3R,5R)-**15** in almost quantitative yield [21].

2.2.7.2.5 Application in Natural Product Syntheses

Dihydroxy esters of the type **5a** can easily be lactonized with concomitant elimination of water to give the corresponding α,β-unsaturated lactones. Therefore, hydroxy keto ester (R)-**2c**, obtained by recLBADH-catalyzed reduction of **1c**, was reduced with NaBH$_4$ prior to acid-catalyzed lactonization for the synthesis of labdanum fragrance compound (R)-**16** (Scheme 2.2.7.8) [12].

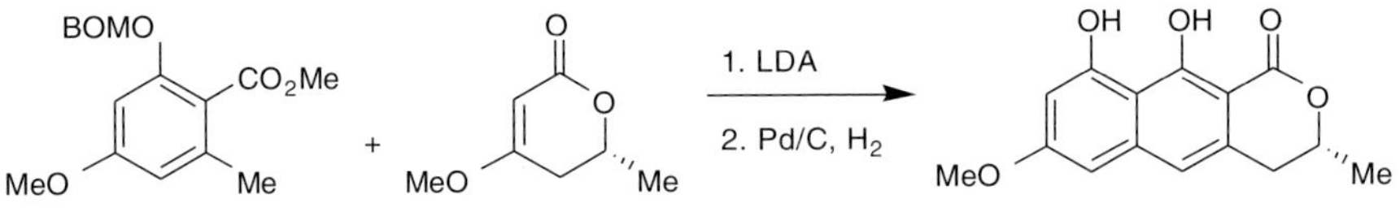

Scheme 2.2.7.9 Synthesis of (R)-semi-vioxanthin.

For the enantioselective synthesis of *semi*-vioxanthin the lactone **17** was required. To this end, hydroxy keto ester (R)-**2b** was lactonized to (R)-5,6-dihydro-4-hydroxy-6-methyl-2H-pyran-2-one by hydrolysis with trifluoroacetic acid [12, 31]. The synthesis was finished by O-methylation with dimethyl sulfate, affording lactone (R)-**17** [32]. The aromatic ring system of *semi*-vioxanthin was built up through a tandem Michael reaction of orsellinate **19** and pyranone (R)-**17** (27% yield). Selective cleavage of the benzyloxymethyl group was achieved by hydrogenolysis to give (R)-*semi*-vioxanthin in 44% yield (Scheme 2.2.7.9). (S)-*semi*-vioxanthin was synthesized in a similar way [32].

α,β-unsaturated δ-lactone moieties occur in many highly active substances [33]. To provide a useful building block for natural product syntheses an asymmetric synthesis of both enantiomers of δ-lactone equivalent **20** was developed in collaboration with Enders' group (University of Aachen), based on the regio- and enantioselective biocatalytic reduction of 3,5-dioxo esters [34]. Starting from **20** as the key intermediate, asymmetric synthesis of (S)-argentilactone, (S)-goniothalamin [34], (−)-callystatin, [35] and (+)-strictifolione [36] was realized. (Scheme 2.2.7.10).

2.2.7.2.6 Discussion and Outlook

The regio- and enantioselective reduction of 3,5-diketo esters with ADH of *L. brevis* is a reaction of broad applicability. A great variety of diketo esters is accepted by this readily available enzyme and the required reaction technique is remarkably simple. Due to the mild conditions the reaction can be carried out without special care even in the case of sensitive compounds such as **1a**. No serious limitations with regard to a further scale-up are apparent. Since cofactor regeneration can be perfomed in whole living cells it can be assumed that costs for catalysts and cofactor will decrease even more [37]. The hydroxy keto esters such as (S)-**2a** thus available are compounds of high functionalization, and we have developed several new chemical modifications and applications. Since baker's yeast reduction or application of the respective enzyme for the reduction of diketo ester **1a** enables formation of the enantiomeric products, a diversity-oriented approach toward all possible stereoisomers of the compounds depicted in Scheme 2.2.7.11 is conceivable. Remarkably, all transformations presented are built on one C–C bond formation step (synthesis of **1a**) using cheap starting materials and just 1 equiv of *n*-BuLi.

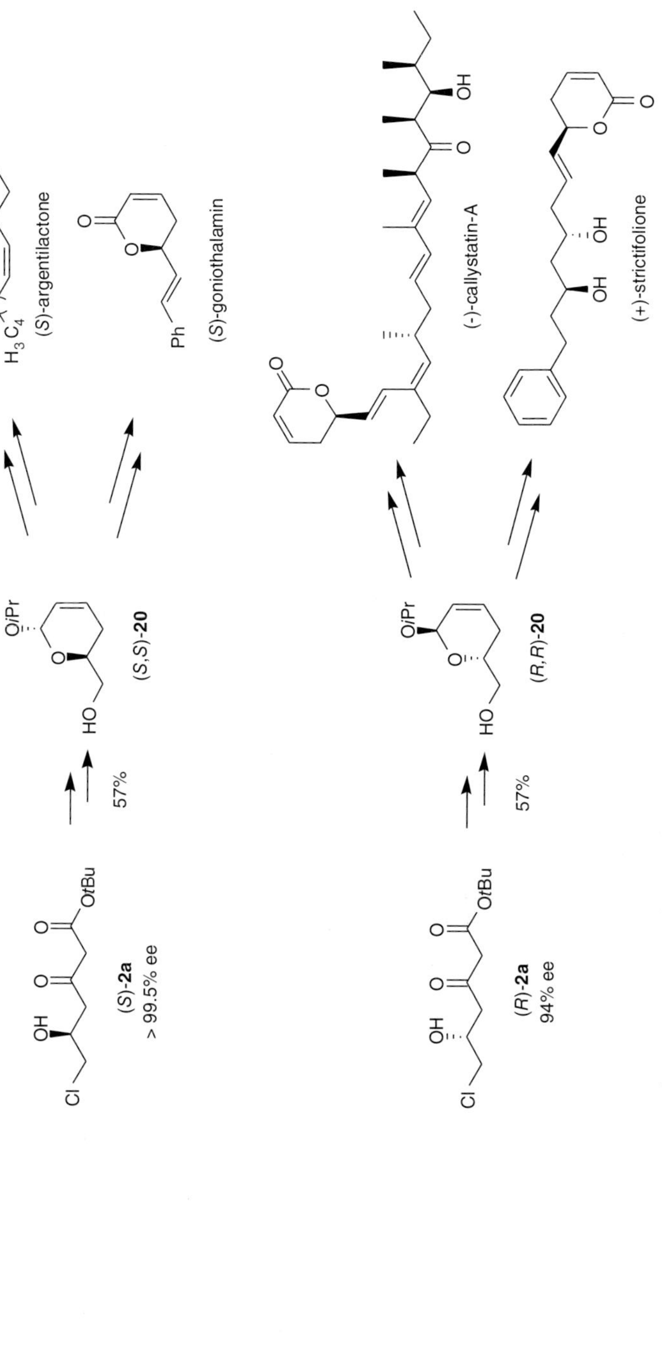

Scheme 2.2.7.10 Synthesis of δ-lactone equivalent **20** and application in the synthesis of (S)-argentilactone, (S)-goniothalamin, (−)-callystatin, and (+)-strictifolione [34–36].

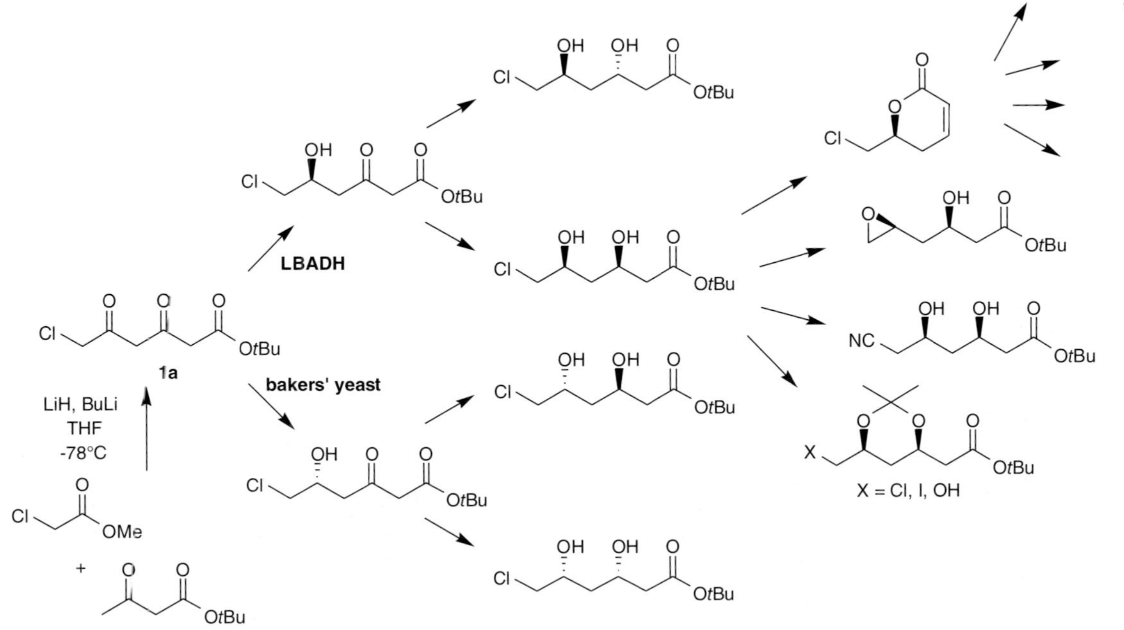

Scheme 2.2.7.11 Chiral building blocks evolved from diketo ester **1a** via regio- and enantioselective enzymatic reduction.

2.2.7.3 Chemo- and Enantioselective Reduction of Propargylic Ketones

Chiral, nonracemic propargylic alcohols are important synthetic intermediates in the synthesis of natural products and biologically active compounds. Since the C–C triple bond allows a flexible and diverse modification toward many other functional groups, enantiopure propargylic alcohols might be regarded as general chiral building blocks. Asymmetric reduction of α,β-acetylenic ketones is a straightforward approach to this class of compounds. A number of chemical reducing reagents have been developed which provide chiral propargylic alcohols in good yields. Nevertheless, almost all of these reagents afford only a small range of propargylic alcohols in high enantiomeric excess and many of them are limited to either hindered or unhindered alkynones. Furthermore, using hydrolytic enzymes such as lipases just a few of these alcohols could be obtained with an enantiomeric excess higher than 99% [38] and only a small number of α,β-acetylenic ketones were reduced by isolated oxidoreductases at all [39].

However, as alcohol dehydrogenases can react stereo- as well as chemoselectively under very mild conditions, they should provide good access to enantiopure propargylic alcohols. This strategy would make it possible to start from a single substrate **D** which after enzymatic reduction affords either enantiomer of propargylic alcohol **E** and after further modifications a variety of different enantiopure products in only two to three steps (Scheme 2.2.7.12).

2.2.7.3.1 Enantioselective Reduction of Aryl Alkynones

In a photometric assay NADP(H)-dependent LBADH (see above) [9] and NAD(H)-dependent *Candida parapsilosis* carbonyl reductase (CPCR) [40] were identified as suitable catalysts accepting a broad range of ynones as substrates. Both enzymes catalyze the reduction of various aryl alkynones **21** with high enantioselectivity and efficiency (Scheme 2.2.7.13) [41].

Since these two biocatalysts possess complementary stereoselectivity, they enable the synthesis of both enantiomers of the desired products. The applicability of enzymatic reduction of aryl alkynones on a preparative scale was optimized with regard to the amount of cofactor and enzyme, resulting in high total turnover numbers and almost quatitative conversion [41].

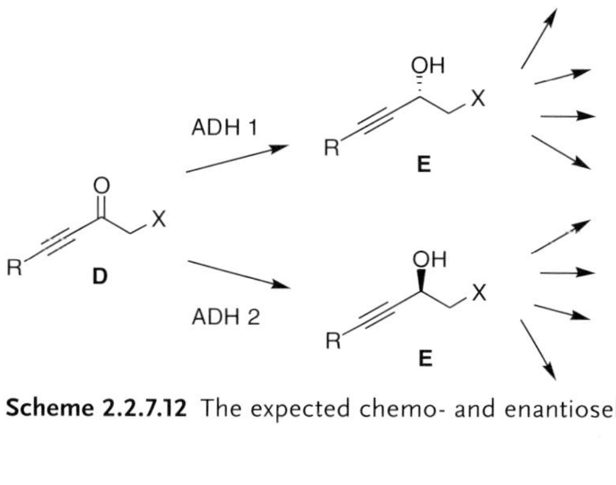

Scheme 2.2.7.12 The expected chemo- and enantioselective reduction of ynones **D**.

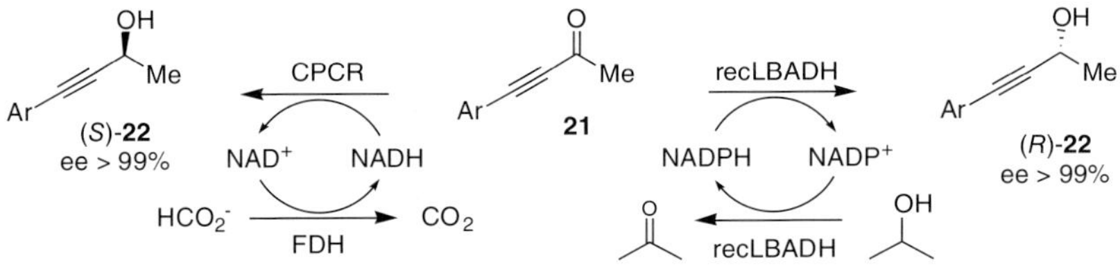

Scheme 2.2.7.13 Enzymatic reductions of aryl alkynones **21** using cofactor regeneration.

2.2.7.3.2 Synthesis of Enantiopure 3-Butyn-2-ol

The methyl and ethynyl residues of 3-butyn-2-one (**23a**) show similar steric demands, making it difficult for reducing reagents to distinguish between the two enantiofacial sides of the substrate. Thus, all methods for reducing **23a** fail to afford the enantiopure alcohol **24** [42].

Enzymatic reduction of **23a** with recLBADH and CPCR resulted in unsatisfactory results (60% and 49% *ee*) as well. The results mentioned above indicate that a bulky substituent at the alkyne moiety results in a higher selectivity of the reduction. Furthermore, Bradshaw et al. reported that *Lactobacillus kefir* ADH, an enzyme highly homologous to LBADH, affords (*R*)-4-trimethylsilyl-3-butyn-2-ol [(*R*)-**25**] with an *ee* of 94% in 25% yield [39b]. In our investigations ketone **23b** was reduced by recLBADH with almost quantitative conversion. The enantiomeric excess and absolute configuration of the product were determined by desilylation with borax converting alcohol (*R*)-**25** into enantiopure (*R*)-3-butyn-2-ol [(*R*)-**24**] (Scheme 2.2.7.14).

The synthesis of enantiomerically pure (*S*)-3-butyn-2-ol [(*S*)-**24**] was achieved via CPCR-catalyzed reduction by introducing a silyl group with an aromatic substituent into the substrate (compound **23c**) [41].

When recLBADH was tested with *n*-alkyl ethynyl ketones it was observed that the preferred stereochemistry of the resulting propargylic alcohol depends on the

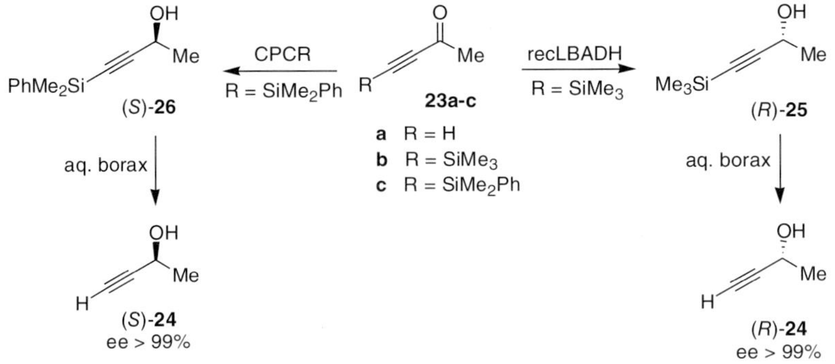

Scheme 2.2.7.14 Enzymatic reduction of silyl alkynones affording enantiopure 3-butyn-2-ol (**24**).

size of the alkyl unit. A similar observation was reported by Phillips et al. for SADH from *Thermoanaerobacter ethanolicus* [39e]. In the case of recLBADH a reversal of the enantioselectivity of the reduction already occurs when 1-pentyn-3-one is used as a substrate. Higher homologues such as 1-hexyn-3-one and 1-octyn-3-one are even reduced with complete enantioselectivity. Surprisingly, the enzymatic activity increases strongly for these substrates bearing longer alkyl chains (69% isolated yield). As a result, in the presence of phosphate buffer enantiopure propargylic alcohol could be easily obtained on a gram scale [41].

Thus, it was shown that substrate engineering can be used successfully to adjust the absolute configuration and the enantiomeric excess of the resulting propargylic alcohols by enzymatic reduction. Whereas the introduction of silyl-groups at C4 of the substrate 3-butyn-2-one (**23a**) resulted finally in enantiopure 3-butyn-2-ol (**24**), chain elongation at C1 afforded enantiopure higher homologues of opposite configuration.

2.2.7.3.3 Enzymatic Reduction of α-Halogenated Propargylic Ketones

Bradswaw et al. have published two examples of enzymatic activity observed with α-halogenated propargylic ketones **27c** and **27e** (UV assay) [39a,b]. We identified three ADHs accepting propargylic substrates with substituents the size of chloro- and bromomethyl groups. α-Halogenated propargylic ketones **27**, which were easily synthesized in one step [43], were reduced by horse liver alcohol dehydrogenase (HLADH), *Thermoanaerobium brockii* ADH (TBADH) and recLBADH (Table 2.2.7.1) [44].

All of these oxidoreductases reduce aromatic and aliphatic α-chloropropargylic ketones (**27a–27c**) with high activity. α-Bromopropargylic ketone **27d** is also accepted as a substrate; however, the enzymatic activity of TBADH and recLBADH decreases, probably for steric reasons. Due to the low solubility of substrate **27a** about 25% of a short-chained alcohol was added. The large excess of short-chained alcohol shifted the substrate/product equilibrium toward the desired propargylic alcohol **28a**, resulting in almost quantitative conversions with high total turnover numbers (TTN) of the cofactor.

Thus, millimolar quantities of substrate **27a** were converted using as little as 0.005 mol% cofactor and small amounts of enzyme (100 U recLBADH per g substrate; 200 U HLADH per g substrate, respectively). The analytically pure products (*S*)- and (*R*)-**28a** were obtained easily by extraction in an isolated yield of >95%. HPLC analysis revealed >99% conversion and, additionally, only a single enantiomer could be detected (*ee* > 99%).

Reduction of 1-(chloro or bromo)-3-butyn-2-one (**27e,f**) with recLBADH affords enantiopure *R*-alcohols **28e,f**, resulting in an interesting switch of the enantioselectivity of the enzymatic reduction. As the enantiomers (*S*)-**28e,f** can be obtained by recLBADH-catalyzed reduction of **27b–27d** and subsequent removal of the silyl-protecting group, this enzyme offers unique access to a pair of enantiomers via the same oxidoreductase. Due to the high volatility of the substrates (**27e,f**) these transformations were only performed on an analytical scale.

Table 2.2.7.1 Relative enzymatic activities of HLADH, TBADH and recLBADH.

R—≡—C(=O)—CH₂—X →[ADH] R—≡—CH(OH)*—CH₂—X
       27   28

Alcohol		Activity [%] (configuration)		
		HLADH	TBADH	recLBADH
28a	X = Cl R = Ph	32 (R)	53 (R)	70 (S)
28b	X = Cl R = TBS	24 (R)	51 (R)	37 (S)
28c	X = Cl R = TMS	15 (R)	74 (R)	28 (S)
28d	X = Br R = TMS	35 (R)	4 (R)	18 (S)
28e	X = Cl R = H	n.d.[a]	n.d.[a]	28 (R) [75]
28f	X = Br R = H	n.d.[a]	n.d.[a]	18 (R)
24	X = H R = Ph	6 (S)	47 (S)	142 (R)

a n.d. = not determined.

2.2.7.3.4 Modification of α-Halogenated Propargylic Alcohols

The high functionality of enantiomerically pure α-halogenated propargylic alcohols makes them ideal precursors for further modification; however, application of these chiral building blocks in target-oriented synthesis is scarcely found in the literature. Propargylic alcohols **28** can be converted easily with good yield to the corresponding epoxides **29** without racemization. The ring closure was achieved by treatment with DBU in EtOH/H_2O (Scheme 2.2.7.15). After workup by extraction no further purification was necessary since no by-products were detected. This is a general approach to enantiomerically pure terminal propargylic epoxides which offers multi-faceted applications in organic chemistry.

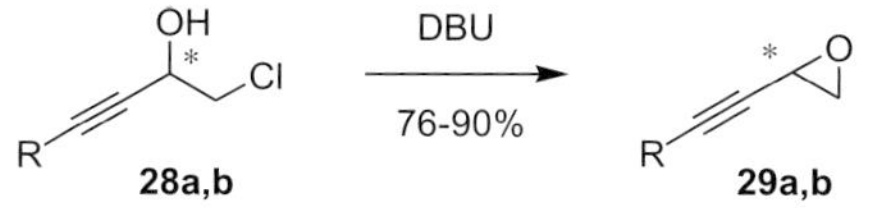

Scheme 2.2.7.15 Conversion of chlorohydrins **28** into epoxides **29**.

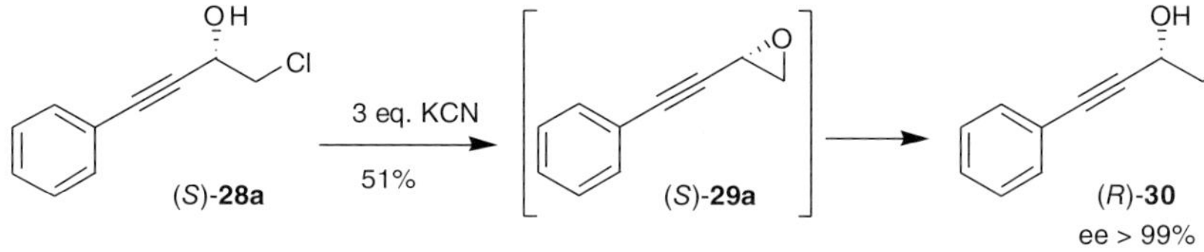

Scheme 2.2.7.16 Synthesis of nitrile (R)-**30** through intermediate epoxide (S)-**29a**.

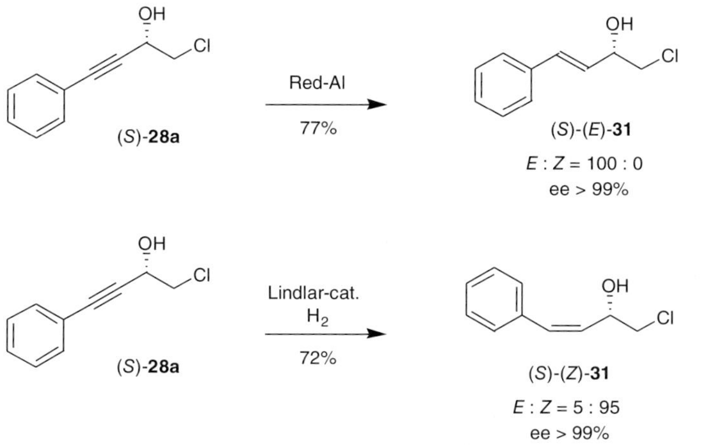

Scheme 2.2.7.17 Stereoselective synthesis of enantiopure E- and Z-olefins **31**.

Regioselective opening of the epoxides **29** was realized with various nucleophiles (CN^-, N_3^-, OH^-). Instead of a two-step process, direct conversion of chlorohydrin (S)-**28a** yielding cyanohydrin **30** was accomplished in 51% yield (Scheme 2.2.7.16) [45].

The C–C triple bond in propargylic alcohols allows access to different functional groups also at C3 and/or C4. As an example the chlorohydrin (S)-**28a** was transformed highly stereoselectively into the corresponding enantiopure E- and Z-olefins (S)-**31** in high yields (Scheme 2.2.7.17).

2.2.7.3.5 Olefinic Substrates

From the literature several examples of the nonenzymatic chemo- and enantioselective reduction of the carbonyl group of (E)-alkenones are known [46], but no example has been published for the similar reduction of a (Z)-methylalkenone such as (Z)-**32**. Furthermore, the enantioselective reduction of α-halogenated olefinic ketones **33** has not been described in the literature at all.

As depicted in Table 2.2.7.2, α,β-unsaturated alkenones possessing the Z configuration are only poorly accepted by the enzymes tested. In addition, with these substrates only low enantioselectivity was observed for the reduction step.

In the case of the (E)-configured olefinic ketones high conversion rates and enantiomeric excesses were achieved on the analytical and on the semi-preparative scale. With TBADH, which requires a slightly higher reaction temperature, decomposition of the halogenated substrates was observed. In the case of CPCR, which was used as a partially purified enzyme preparation from the parental strain, the formation of up to 50% of the fully saturated alcohols was observed (Scheme 2.2.7.18). Substrate (E)-**33** could be converted into enantiopure (S)- and (R)-(E)-**32**, respectively, by recLBADH- and HLADH/CPCR-catalyzed reaction.

The enzymatic reduction of E-configured alkenones represents a valuable alternative to the enzyme-catalyzed reduction of the corresponding alkynone with subsequent partial reduction of the C–C triple bond. Nevertheless, since in our results the Z-configured vinylic alcohols are not accessible in enantiopure form by enzyme-catalyzed reduction of the olefinic ketones, for this class of compounds the two-step process via the propargylic alcohols is a highly valuable new strategy.

Table 2.2.7.2 Coupled enzyme–substrate screening for olefinic ketones **32** and **33** with relative activities of recLBADH, HLADH, TBADH, and CPCR.

Substrate	Activity [%] (conversion)			
	recLBADH	HLADH	TBADH	CPCR
(E)-**32**	39 (90) [76]	very slow (5)	very slow (10)	12 (100) [77]
(Z)-**32**	0 (−)	0 (−)	very slow (5)	2 (10)
(E)-**33**	47 (80) ee > 99% (S)	5 (80) ee > 99% (R)	16 (30) ee > 99% (R)	9 (70) ee > 99% (R)
(Z)-**33**	31 (80) ee = 86% (S)	3 (20) ee = 6% (R)	5 (70) ee = 95% (R)	3 (60) ee = 86% (R)

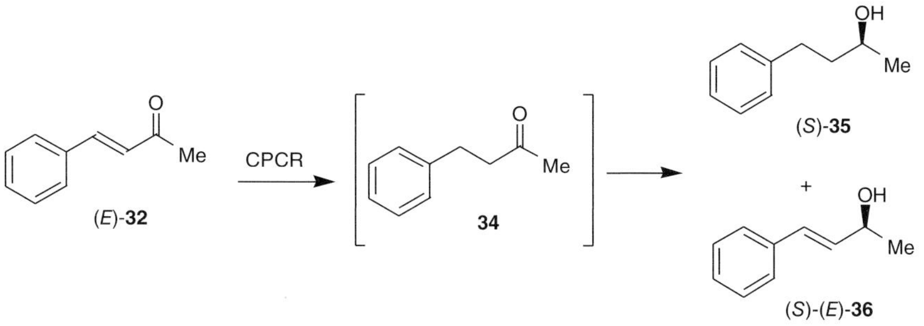

Scheme 2.2.7.18 Formation of (S)-**35**/**36** by CPCR-catalyzed reduction of (E)-**32**.

2.2.7.3.6 Discussion and Outlook

In conclusion, it has been shown that a broad variety of differently substituted acetylenic ketones can be reduced enantioselectively by the oxidoreductases recLBADH, CPCR, HLADH, and TBADH. Most propargylic alcohols were obtained with *ee* values higher than 99%, making this method superior to chemical reduction techniques. In the majority of cases, the alcohols could be obtained in either enantiomeric form, since *R*- and *S*-specific oxidoreductases were applied. The substrate spectrum includes aromatic ynones as well as a number of aliphatic derivatives. By varying the size of the substituents, the enantiomeric excess can be tuned, and even a reversal in enantioselectivity can be achieved. Last but not least, the enzymatic reductions were scaled up using low amounts of enzyme and cofactor, making this method highly attractive in ecological and economical terms. Thus, the use of α-halogenated substrates enables the synthesis of a new class of highly flexible chiral C4 building blocks in only two steps (Scheme 2.2.7.19).

From the results obtained in the substrate screening using α-chlorinated in comparison with the respective nonchlorinated ketones with different enzymes, two divergent conclusions can be drawn. In the case of the CPCR-catalyzed reductions an evident loss of activity is observed for α-halogenated substrates, whereas recLBADH and HLADH gain activity relative to the nonhalogenated substrates (see Tables 2.2.7.1 and 2.2.7.2). For the former, steric reasons can be assumed for the absence of electronic effects whereas the effect observed with the latter indicate a strong electronic influence resulting in an activation of the substrate [47].

2.2.7.4 Thiamine Diphosphate-Dependent Enzymes: Multi-purpose Catalysts in Asymmetric Synthesis

Various thiamine diphosphate (ThDP)-dependent α-keto acid decarboxylases have been described as catalyzing C–C bond formation and/or cleavage [48]. Extensive work has already been conducted on transketolase (TK) and pyruvate decarboxylase (PDC) from different sources [49]. Here attention should be drawn to some concepts based on the investigation of reactions catalyzed by the enzymes

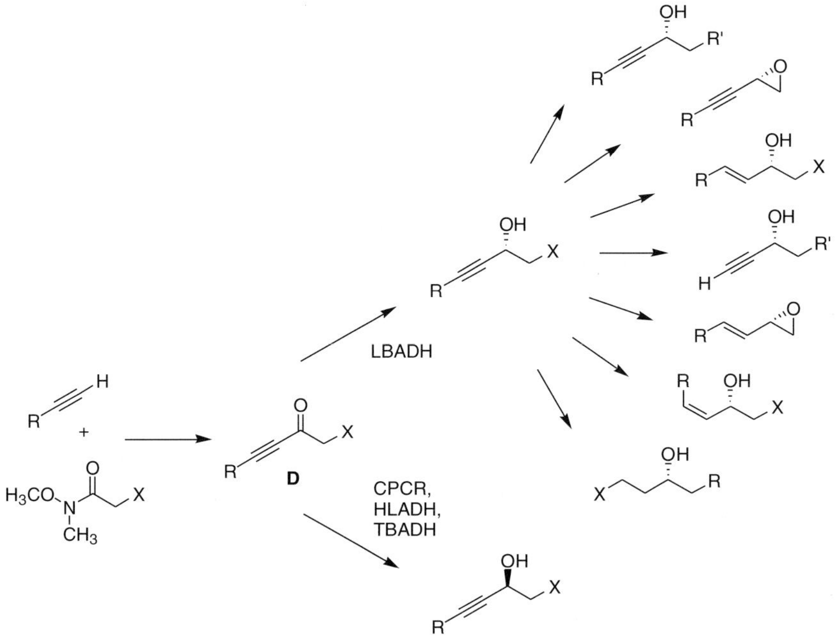

Scheme 2.2.7.19 Chiral building blocks evolved from enantioselective enzymatic reduction of propargylic ketones **D**.

benzoylformate decarboxylase (BFD) and benzaldehyde lyase (BAL), the genes of which were cloned and the proteins overexpressed in recombinant *E. coli* strains.

2.2.7.4.1 Formation of Chiral 2-Hydroxy Ketones Through BFD-Catalyzed Reactions

The potential of benzoylformate decarboxylase (BFD, E.C. 4.1.1.7) to catalyze C–C bond formation was first reported by Wilcocks at al. using crude extracts of *Pseudomonas putida* [50]. They observed the formation of (*S*)-2-hydroxy-1-phenylpropanone (*S*)-2-HPP when benzoyl formate was decarboxylated in the presence of acetaldehyde. Advantageously, aldehydes – without a previous decarboxylation step – can be used instead of the corresponding more expensive α-keto acids [51]. We could show that BFD is able to bind a broad range of different aromatic, heteroaromatic, and even cyclic aliphatic and conjugated olefinic aldehydes to ThDP before ligation to acetaldehyde or other aldehydes (Table 2.2.7.3) [52].

The best results with respect to the enantiomeric excess (*ee*) of the resulting 2-hydroxy ketones were obtained with *meta*-substituted benzaldehydes. When these substrates were used, the *ee* increased to more than 99%, indicating that the steric demand and electronic properties of the substituent play a decisive role in both conversion rate and *ee* (Table 2.2.7.3, entries 2–4). *Ortho*-substituted benzalde-

Table 2.2.7.3 Wild-type BFD-mediated carboligation on a preparative scale.

Entry	Ar	R	Yield [%]	ee [%]	Configuration
1	Ph	CH_3	90	92	S
2	3-$MeOC_6H_4$	CH_3	97	96	S
3	3-$iPrOC_6H_4$	CH_3	91	>99	S
4	3,5-di-$MeOC_6H_3$	CH_3	40	97	S
5	2-naphthyl	CH_3	32	88	S
6	Ph	Ph	70	>99	R
7	2-FC_6H_4	2-FC_6H_4	68	>99	R
8	4-MeC_6H_4	4-MeC_6H_4	69	>99	R

hyde derivatives, except 2-fluorobenzaldehyde, are only poorly accepted as donor substrates by the wild-type enzyme, probably because of steric hindrance. Nevertheless, in collaboration with Bettina Lingen and Martina Pohl's group, we were able to identify two BFD variants, L476Q and H353L-L461S-F512S, as potent catalysts accepting *ortho*-substituted benzaldehyde derivatives as donors [53]. Additionally, we demonstrated the first BFD-mediated stereoselective cross-coupling of two different aliphatic substrates, cyclohexane carbaldehyde and acetaldehyde [52]. An outstanding example of the value of these molecules in organic synthesis could be given by applying such α-hydroxy ketones in the synthesis of 3,6-dihydro-2*H*-pyran-3-ols [54].

In contrast to the large variety of aromatic, olefinic, and aliphatic aldehydes which can be used as donor substrates, wild-type BFD does not tolerate a modification of the methyl group of acetaldehyde in the case of aliphatic acceptor aldehydes. Apart from acetaldehyde, BFD shows activity with aromatic and heteroaromatic aldehydes as the acceptor substrate, forming enantiomerically pure (*R*)-benzoin and derivatives (Table 2.2.7.3, entries 6–8) [55].

Biotransformation of hydrophobic aldehydes is possible in the presence of water-miscible organic solvents. The best results with regard to an increased solubility of hydrophobic substrates together with the least loss of ligase activity of BFD were obtained by addition of DMSO [56]. In doing so, enantiomerically pure *(R)*-benzoin was obtained in 70% isolated yield [55].

Dialdehydes as substrates for BFD Being aware that *meta*-substituted aromatic aldehydes provide the highest *ee* values combined with good to excellent conversion rates, we subjected isophthalaldehyde 37 to the BFD-catalyzed coupling reaction (Scheme 2.2.7.20) [57].

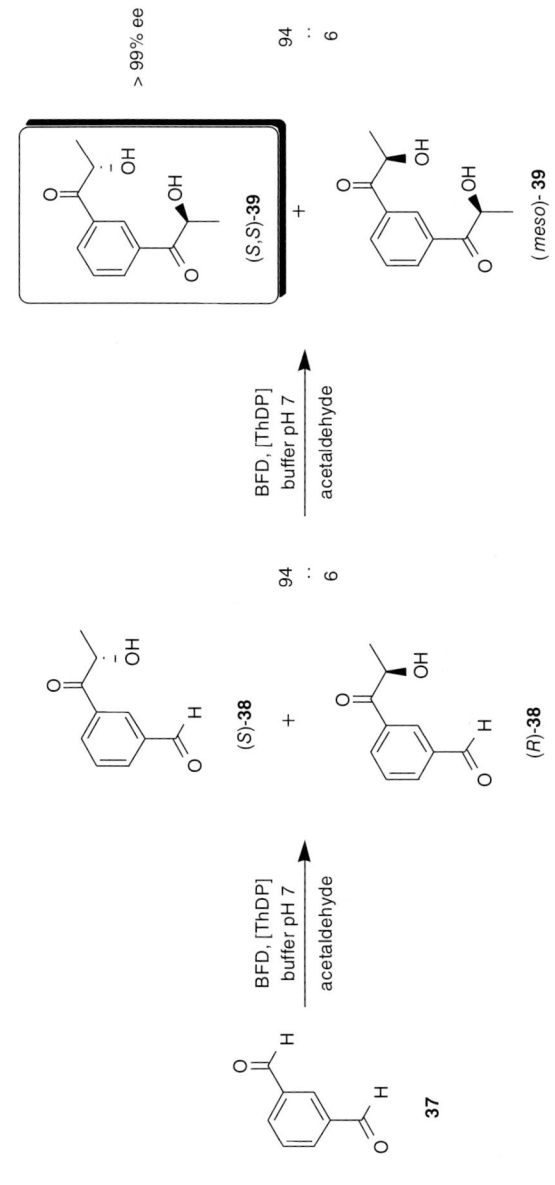

Scheme 2.2.7.20 BFD-mediated carboligation of isophthalaldehyde and acetaldehyde yielding (S)-**38** and (S,S)-**39**.

It is noteworthy that the *ee* of the monoadduct (*S*)-**38** increases to some extent with a progressive in-situ formation of the bisadduct **39**, which shows that BFD accepts both enantiomers of **38** as substrates. Therefore, in that case it is not practicable to use BFD for kinetic racemic resolution. Nevertheless, the second reaction step proceeds completely stereospecifically within the limits of detection. The monoadduct (*S*)-**38** is converted to (*S*,*S*)-**39** in enantiomerically pure form (>99% *ee*), whereas the minor enantiomer (*R*)-**38** leads to (*R*,*S*)-**39**, which is the diastereomeric *meso* form of the former (Scheme 2.2.7.20).

2.2.7.4.2 BAL as a Versatile Catalyst for C–C Bond Formation and Cleavage Reactions

Benzaldehyde lyase (BAL, E.C. 4.1.2.38) from *Pseudomonas fluorescens* Biovar I was first reported by Gonzáles and Vicuña [58]. They showed that this strain can grow on benzoin as a sole carbon and energy source, due to the ability of BAL to catalyze the cleavage of the acyloin linkage of benzoin. When racemic benzoin reacted with BAL [59] in potassium phosphate buffer, only a very small amount of benzaldehyde was formed. Addition of 20% DMSO as a cosolvent (or alternatively 15% poly(ethylene glycol), PEG 400) [60] resulted in enhanced formation of benzaldehyde [61]. Only (*R*)-benzoin is converted into benzaldehyde through BAL catalysis, although complete conversion of (*R*)-benzoin was not possible under the conditions tested. An equilibrium between cleavage and formation of (*R*)-benzoin apparently exists during this process. When just (*S*)-benzoin was applied in this reaction, no conversion could be detected at all [61].

From mechanistic considerations and assuming that cleavage and formation of (*R*)-benzoin are in equilibrium, BAL should also catalyze carboligation. Consequently, BAL-catalyzed acyloin condensation of benzaldehyde in an aqueous buffer/DMSO mixture resulted in almost quantitative formation of enantiomerically pure (*R*)-benzoin [Scheme 2.2.7.21, Eq. (1)]. The reaction was carried out on a preparative scale with different aromatic and heteroaromatic aldehydes [62]. From the viewpoint of the organic-preparative chemist, it is important to mention that crude cell extracts of the recombinant *E. coli* strain overexpressing the BAL gene are sufficient for catalysis, hence, purification of the enzyme is not necessary.

In contrast to wild-type BFD, BAL accepts aromatic aldehydes substituted in the *ortho* position as well. Only a few aromatic aldehydes, such as pyridine 3- and 4-carbaldehyde as well as sterically exceedingly demanding aldehydes, resulted either in very low yields or in no benzoin condensation at all [62]. Moreover, mono- and dimethoxyacetaldehyde are good substrates for BAL, leading to highly functionalized enantiopure hydroxypropiophenone derivatives (Scheme 2.2.7.22) [63].

Kinetic racemic resolution via BAL-catalyzed C–C bond cleavage For nonenzymatic benzoin condensations it is well established that benzoins can be used

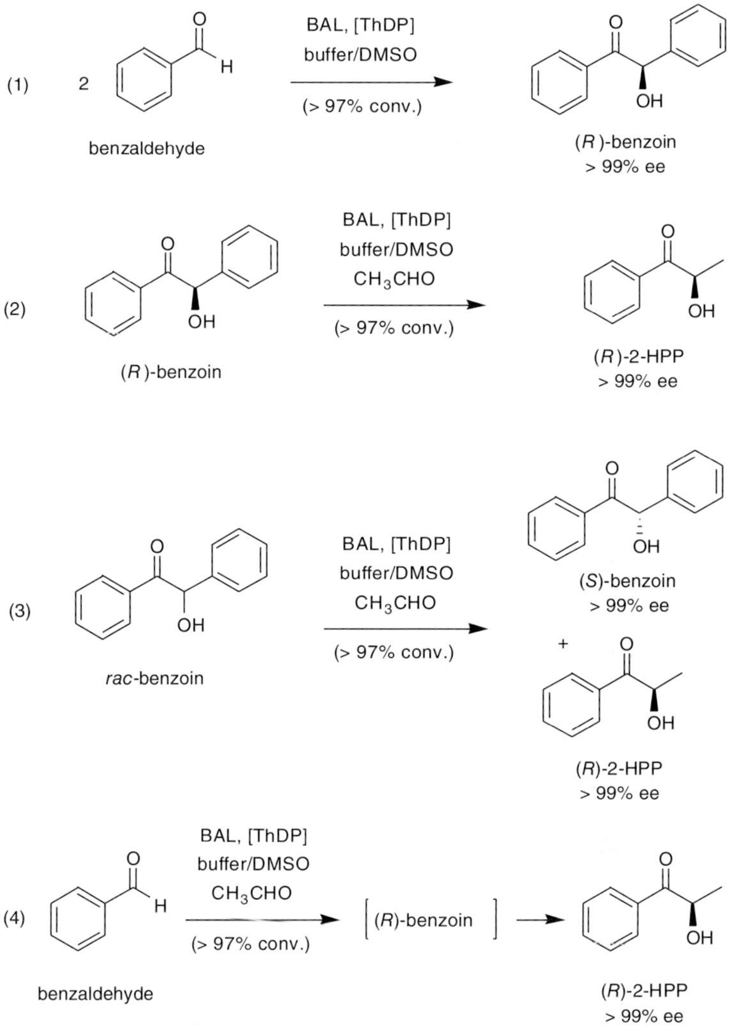

Scheme 2.2.7.21 Different types of reactions catalyzed by BAL.

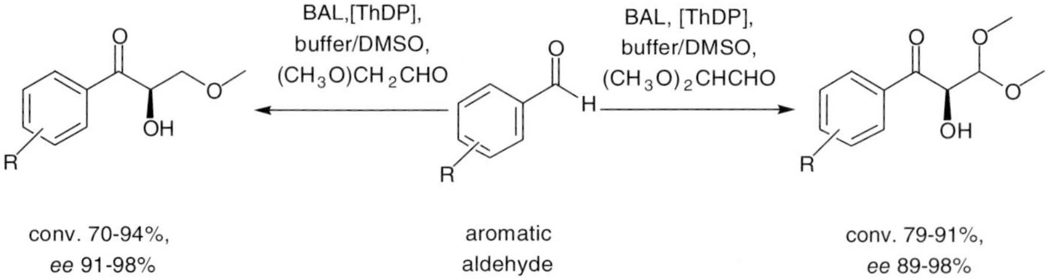

conv. 70-94%,
ee 91-98%

aromatic
aldehyde

conv. 79-91%,
ee 89-98%

Scheme 2.2.7.22 Application of monomethoxy- and dimethoxyacetaldehyde in BAL-catalyzed C–C bond formation reactions.

instead of aldehydes as substrates. When (R)-benzoin reacted with BAL in the presence of acetaldehyde [Scheme 2.2.7.21, Eq. (2)], quantitative formation of enantiopure (R)-2-HPP occurred [61]. The same reaction starting from (S)-benzoin failed. Repeating this reaction with racemic benzoin afforded enantiopure (R)-2-HPP and (S)-benzoin after separation of the products by column chromatography [Scheme 2.2.7.21, Eq. (3)]. As expected from these results, the BAL-catalyzed reaction of benzaldehyde and acetaldehyde also gave (R)-2-HPP in 95% yield [Scheme 2.2.7.21, Eq. (4)]. Different aromatic and heteroaromatic benzoin-like acyloins are accepted as substrates for the kinetic racemic resolution via C–C bond cleavage. The reactions work well in organic–aqueous medium, overcoming the solubility problem of lipophilic substrates and opening the way for large-scale preparations [62].

From the literature only a few examples of an enzyme-catalyzed kinetic racemic resolution via C–C bond-cleavage are known [64]. The broad substrate range of BAL in combination with the high extent of stereoselectivity observed for BAL-catalyzed resolutions impressively demonstrates the high potential of this strategy. Very recently, it has become possible to determine the three-dimensional structure of BAL via X-ray crystal structure analysis, which will enable a structure-based discussion of the observed stereocontrol [65].

2.2.7.4.3 Asymmetric Cross-Benzoin Condensation

Starting from the findings of the racemic cross-benzoin condensation [66], and assuming that aldehydes not accepted as donor substrates might still be suitable acceptor substrates, and vice versa, a mixed enzyme–substrate screening was performed in order to identify a biocatalytic system for the asymmetric cross-carboligation of aromatic aldehydes. For this purpose the reactions of 2-chloro- (**40a**), 2-methoxy- (**40b**) and 2-methylbenzaldehyde (**40c**), respectively, were studied with different enzymes in combination with benzaldehyde (Scheme 2.2.7.23) [67]. The three *ortho*-substituted benzaldehyde derivatives **40a–40c** were

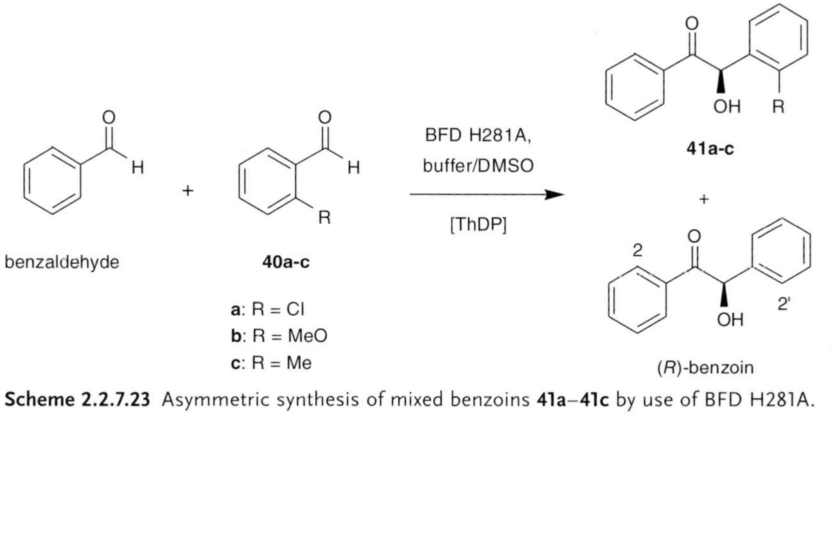

Scheme 2.2.7.23 Asymmetric synthesis of mixed benzoins **41a–41c** by use of BFD H281A.

chosen as putative selective acceptor substrates, particularly because of their inability to form symmetrical benzoins through the wild-type BFD-catalyzed reaction, meaning that these compounds are not accepted as donor substrates by this enzyme. The BFD-mutant BFD H281A [68] was identified as a potent catalyst, resulting in the formation of the mixed benzoins 2'-methoxybenzoin **41b** and 2'-methylbenzoin **41c**, accompanied by (R)-benzoin as the major product. In the case of the conversion of 2-chlorobenzaldehyde **40a** as acceptor substrate, the unsymmetrical benzoin (R)-**41a** (yield 74%, >99% ee) represents the major product [67].

Remarkably, the 2,2'-disubstituted benzoin or the mixed benzoin substituted in the 2 position was not generated in any of these reactions, revealing that the *ortho*-substituted benzaldehydes **40a–40c** react selectively as acceptors, as expected.

Subsequently, the concept was extended to selective donor molecules. With 2-chlorobenzaldehyde as the selective acceptor a vast variety of unsymmetrical benzoins was accessible, among which **41d–41f** were obtained selectively, proving that 4-(trifluoromethyl)benzaldehyde, 4-bromobenzaldehyde, and 3-cyanobenzaldehyde were selective donor substrates for BFD H281A under these reaction conditions [67, 69].

The selective donor–acceptor concept can be transferred to other ThDP-dependent enzymes. For example, enantiopure mixed benzoins were obtained when 2-chlorobenzaldehyde reacted with a variety of selective donor aldehydes in the presence of BAL [67]. By performing various cross-benzoin condensation reactions with this enzyme, not only new selective donors but also additional aldehydes reacting selectively as acceptors, such as 2-iodobenzaldehyde or 2,6-difluorobenzaldehyde, could be identified. Again all the mixed benzoins generated exhibited an *R*-configuration and were obtained with high to excellent enantiomeric excesses [69].

The selectivity is caused not only by the electronic properties of the substrates which are dependent on the nature of their substituents, as is the case in the chemical cross-benzoin condensation. Rather, steric demands of the aldehyde substituents and interactions of these with the active site of the biocatalyst, which (obviously) is different for each enzyme used, are also of significance.

2.2.7.4.4 Discussion and Outlook

From the results mentioned above, it can be proposed that a detailed investigation of different ThDP-dependent enzymes using substrate and protein engineering, with the focus on new transformations, will open new perspectives in catalytic asymmetric synthesis *and* in the biotechnology industry. As shown, it can be very rewarding to elucidate the potential of these biocatalysts, and additionally, to elaborate new chemical transformations such as kinetic racemic resolution via C–C bond cleavage or asymmetric cross-benzoin condensation [70]. At present, we are working toward an intermolecular asymmetric Stetter reaction [71].

Until now only TK [49a], PDC [49b], BFD, BAL [72], and mutants thereof have been investigated systematically with regard to preparative transformations. Numerous other ThDP-dependent enzymes are capable of catalyzing different asymmetric reactions as well [73].

2.2.7.5 Summary

As an outcome of the investigation of the biocatalytic strategies described in this chapter several new aspects with broad applicability in synthetic organic chemistry have been established:
- regio- and enantioselective reduction of 3,5-dioxocarboxylates
- dynamic kinetic resolution of 2-substituted 1,3-diketones
- diastereomer-differentiating hydrolysis of 1,3-diol acetonides
- halogenated propargylic alcohols as versatile C4 building blocks
- kinetic racemic resolution via C–C bond cleavage or C–C bond formation
- dialdehydes as substrates for umpolung reactions in stereospecific synthesis
- the selective donor–acceptor concept for asymmetric cross-coupling reactions.

In addition, new impacts with regard to biocatalysis have been determined:
- use of highly reactive substrates in enzyme-catalyzed reactions
- large-scale biotransformations using NAD(P)H-dependent enzymes
- identification of new enzymatic activities
- adjustment of the stereochemistry of products by use of directing groups
- combined substrate and protein engineering for new biotransformations
- application of thiamine diphosphate-dependent enzymes for diverse reactions
- the selective donor–acceptor concept for enzyme-catalyzed reactions.

Thus, the intrinsic property of diversity found in natural product biosynthesis has been used successfully as a model in the development of biocatalyc strategies.[74] This might enable future directions toward understanding the complex network of natural products biosynthesis.

Acknowledgement

We thank the Deutsche Forschungsgemeinschaft for financial support in the SFB 380.

References

1 See: *Dictionary of Natural Products*, CD-ROM, Chapman & Hall.
2 a) B. S. Moore, C. Hertweck, *Nat. Prod. Rep.* **2002**, *19*, 70–99; b) J. Staunton, K. J. Weissman, *Nat. Prod. Rep.* **2001**, *18*, 380–416.
3 a) *Chem. Rev.* **1997**, *97(7)*; b) B. S. Moore, J. N. Hopke, *ChemBioChem* **2001**, *2*, 35–38 and references cited therein.
4 a) P. F. Leadlay, *Curr. Opinion Chem. Biol.* **1997**, *1*, 162–168; b) R. McDaniel, S. Ebert-Khosla, D. A. Hopwood, C. Khosla, *Nature* **1995**, *375*, 549–554 and references cited therein.
5 a) V. Blandin, J.-F. Carpentier, A. Mortreux, *Eur. J. Org. Chem.* **1999**, 3421–3427; b) L. Shao, H. Kawano, M. Saburi, Y. Uchida, *Tetrahedron* **1993**, *49*, 1997–2010; c) L. Shao, T. Seki, H. Kawano, M. Saburi, *Tetrahedron Lett.* **1991**, *32*, 7699–7702; d) N. Sayo, T. Saito, H. Kumobayashi, S. Akutagawa, R. Noyori, H. Takaya (Takasago International Corp.), EP-A 297 752, **1989** [*Chem. Abstr.* **1989**, *111*, 114745n].
6 a) R. N. Patel, A. Banerjee, C. G. McNamee, D. Brzozowski, R. L. Hanson, L. J. Szarka, *Enzyme Microb. Technol.* **1993**, *15*, 1014–1021; b) R. N. Patel, C. G. McNamee, A. Banerjee, L. J. Szarka (Squibb & Sons, Inc.), EP-A 569 998, **1993** [*Chem. Abstr.* **1994**, *120*, 52826q].
7 M. Uko, H. Azuma, T. Sakai, S. Tsuboi (Mitsubishi Kasei Corp.), JP-Kokai 03-48641, **1991** [*Chem Abstr.* **1991**, *115*, 28713b].
8 M. Wolberg, W. Hummel, C. Wandrey, M. Müller, *Angew. Chem.* **2000**, *112*, 4476–4478; *Angew. Chem. Int. Ed.* **2000**, *39*, 4306–4308.
9 B. Riebel, Dissertation, University of Düsseldorf, **1996**. A 10 L fermentation of *E. coli* strain recADH-HB101+ yields approximately 600 000 U of recLBADH. LBADH was used in the form of a crude cell extract (recLBADH) from a recombinant *E. coli* strain.
10 M. Wolberg, Dissertation, University of Oldenburg, **2001**.
11 M. Wolberg, W. Hummel, M. Müller, *Chem. Eur. J.* **2001**, *7*, 4562–4571.
12 M. Wolberg, A. Ji, W. Hummel, M. Müller, *Synthesis* **2001**, 937–942.
13 M. Wolberg, I. A. Kaluzna, M. Müller, J. D. Stewart, *Tetrahedron: Asymmetry* **2004**, *15*, 2825–2828.
14 R. Noyori, M. Tokunaga, M. Kitamura, *Bull. Chem. Soc.* **1995**, *68*, 36–55 and references cited therein.
15 H. Stecher, K. Faber, *Synthesis* **1997**, 1–16 and references cited therein.
16 a) S. Tsuboi, E. Nishiyama, H. Furutani, M. Utaka, A. Takeda, *J. Org. Chem.* **1987**, *52*, 1359–1362; b) T. Fujisawa, B. I. Mobele, M. Shimizu, *Tetrahedron Lett.* **1992**, *33*, 5567–5570; c) *cf.* T. Zelinski, A. Liese, C. Wandrey, M.-R. Kula, *Tetrahedron: Asymmetry* **1999**, *10*, 1681–1687; d) see: R. Hayakawa, M. Shimizu, *Synlett* **1999**, 1298–1300 (α-substituted β-ketoaldehyde).
17 A. Ji, M. Wolberg, W. Hummel, C. Wandrey, M. Müller, *Chem. Commun.* **2001**, 57–58.
18 S. Bode, M. Wolberg, M. Müller, *Synthesis* **2006**, 557–588.
19 K.-M. Chen, G. E. Hardtmann, K. Prasad, O. Repič, M. J. Shapiro, *Tetrahedron Lett.* **1987**, *28*, 155–158.
20 D. A. Evans, K. T. Chapman, E. M. Carreira, *J. Am. Chem. Soc.* **1988**, *110*, 3560–3578.

21 M. Wolberg, S. Bode, R. Feldmann, M. F. Villela, M. Müller, in preparation.
22 M. Müller, M. Wolberg, W. Hummel, C. Wandrey, DE 199 37 825.8, **1999**.
23 W. Hummel, *Adv. Biochem. Eng./Biotechnol.* **1997**, *58*, 145–184.
24 H. Pfruender, M. Amidjojo, F. Hang, D. Weuster-Botz, *Appl. Microbiol. Biotechnol.* **2005**, *67*, 619–622.
25 J. K. Thottathill, Y. Pendri, W. S. Li, D. R. Kronenthal (Squibb & Sons, Inc.), US 5 278 313, **1994** [*Chem. Abstr.* **1994**, *120*, 217 700j]
26 M. Müller, *Angew. Chem.* **2005**, *117*, 366–369; *Angew. Chem. Int. Ed.* **2005**, *44*, 362–365.
27 H. Jendralla, E. Granzer, B. v. Kerekjarto, R. Krause, U. Schacht, E. Baader, W. Bartmann, G. Beck, A. Bergmann, K. Kesseler, G. Wess, L.-J. Chen, S. Granata, J. Herchen, H. Kleine, H. Schüssler, K. Wagner, *J. Med. Chem.* **1991**, *34*, 2962–2983.
28 a) T. I. Richardson, S. D. Rychnovsky, *J. Org. Chem.* **1996**, *61*, 4219–4231; b) S. D. Rychnovsky, G. Griesgraber, *J. Org. Chem.* **1992**, *57*, 1559–1563.
29 H. Kotsuki, T. Shimanouchi, *Tetrahedron Lett.* **1996**, *37*, 1845–1848.
30 G. Ohloff, *Scent and Fragrances*; Springer, Berlin, **1994**; pp. 182–183.
31 H. Tabuchi, T. Hamamoto, S. Miki, T. Tejima, A. Ichihara, *J. Org. Chem.* **1994**, *59*, 4749–4759.
32 D. Drochner, M. Müller, *Eur. J. Org. Chem.* **2001**, 211–215.
33 a) M. T. Davies-Coleman, D. E. A. Rivett, *Fortschr. Chem. Org. Naturst.* **1989**, *55*, 1–35; b) L. A. Collett, M. T. Davies-Coleman, D. E. A. Rivett, *Fortschr. Chem. Org. Naturst.* **1998**, *75*, 181–209.
34 A. Job, M. Wolberg, M. Müller, D. Enders, *Synlett* **2001**, 1796–1798.
35 a) J. L. Vicario, A. Job, M. Wolberg, M. Müller, D. Enders, *Org. Lett.* **2002**, *4*, 1023–1026; b) D. Enders, J. L. Vicario, A. Job, M. Wolberg, M. Müller, *Chem. Eur. J.* **2002**, *8*, 4272–4284.
36 D. Enders, A. Lenzen, M. Müller, *Synthesis* **2004**, 1486–1488.
37 M. Ernst, B. Kaup, M. Müller, S. Bringer-Meyer, H. Sahm, *Appl. Microbiol. Biotechnol.* **2005**, *66*, 629–634.

38 a) C. Waldinger, M. Schneider, M. Botta, F. Corelli, V. Summa, *Tetrahedron: Asymmetry* **1996**, *7*, 1485–1488; b) K. Nakamura, K. Takenaka, A. Ohno, *Tetrahedron: Asymmetry* **1998**, *9*, 4429–4439; c) S.-X. Xie, J. Ogawa, S. Shimizu, *Biotechnol. Lett.* **1998**, *20*, 935–938; d) M. Kurihara, K. Ishii, Y. Kasahara, N. Miyata, *Tetrahedron Lett.* **1999**, *40*, 3183–3184; e) K. Burgess, L. D. Jennings, *J. Am. Chem. Soc.* **1991**, *113*, 6129–6139; f) J. A. Marshall, H. R. Chobanian, M. M. Yanik, *Org. Lett.* **2001**, *3*, 3369–3372 and references cited therein.
39 a) C. W. Bradshaw, H. Fu, G.-J. Shen, C.-H. Wong, *J. Org. Chem.* **1992**, *57*, 1526–1532; b) C. W. Bradshaw, W. Hummel, C.-H. Wong, *J. Org. Chem.* **1992**, *57*, 1532–1536; c) C. Zheng, V. T. Pham, R. S. Phillips, *Catalysis Today* **1994**, *22*, 607–620; d) J.-M. Fang, C.-H. Lin, C. W. Bradshaw, C.-H. Wong, *J. Chem. Soc., Perkin Trans. 1* **1995**, 967–978; e) C. Heiss, R. S. Phillips, *J. Chem. Soc., Perkin Trans. 1* **2000**, 2821–2825.
40 J. Peters, T. Minuth, M.-R. Kula, *Enz. Microb. Technol.* **1993**, *15*, 950–958.
41 T. Schubert, W. Hummel, M.-R. Kula, M. Müller, *Eur. J. Org. Chem.* **2001**, 4181–4187.
42 The best result with regard to enantioselective reduction of **23a** has been obtained with TBADH [*ee* = 86%; (*S*)]: M. De Amici, C. De Micheli, G. Carrea, S. Spezia, *J. Org. Chem.* **1989**, *54*, 2646–2650.
43 Ketones were prepared according to: a) L. Birkofer, A. Ritter, H. Uhlenbrauck, *Chem. Ber.* **1963**, *96*, 3280–3288; b) S. Nahm, S. M. Weinreb, *Tetrahedron Lett.* **1981**, *22*, 3815–3818.
44 T. Schubert, W. Hummel, M. Müller, *Angew. Chem.* **2002**, *114*, 656–659; *Angew. Chem. Int. Ed.* **2002**, *41*, 634–637.
45 T. Schubert, W. Hummel, M. Müller, DE 101 05 866.7, **2001**.
46 a) H. C. Brown, M. Srebnik, P. V. Ramachandran, *J. Org. Chem.* **1989**, *54*, 1577–1583; b) J. Bach, R. Berenguer, J. Garcia, J. Lopez, J. Manzanal, J. Vilarrasa, *Tetrahedron* **1998**, *54*, 14 947–14 962; c) A. J. Souers, J. A. Ellmann, J.

Org. Chem. **2000**, *65*, 1222–1224; d) S. Terashima, N. Tanno, K. Kago, *J. Chem. Soc., Chem. Commun.* **1980**, 1026–1027; e) T. Sato, Y. Gotoh, Y. Wakabayashi, *Tetrahedron Lett.* **1983**, *24*, 4123–4126; f) H. C. Brown, W. S. Park, T. Byung, *J. Org. Chem.* **1987**, *52*, 5406–5412; g) T. Ohkuma, H. Ooka, T. Ikariya, R. Noyori, *J. Am. Chem. Soc.* **1995**, *117*, 10 417– 10 418; h) T. Ohkuma, M. Koizumi, H. Doucet, T. Pham, M. Kozawa, K. Murata, E. Katayama, T. Yokozawa, T. Ikariya, R. Noyori, *J. Am. Chem. Soc.* **1998**, *120*, 13 529–13 530; i) T. Ohkuma, M. Koizumi, M. Yoshida, R. Noyori, *Org. Lett.* **2000**, *2*, 1749–1752.

47 a) J. Peters, Dissertation, University of Düsseldorf, **1993**; b) T. Zelinski, Dissertation, University of Düsseldorf, **1995**; c) N. Itoh, N. Mizuguchi, M. Mabuchi, *J. Mol. Catal. B: Enzym.* **1999**, *6*, 41–50.

48 For some reviews of ThDP-dependent enzymes with regard to asymmetric synthesis see: a) G. A. Sprenger, M. Pohl, *J. Mol. Catal. B: Enzym.* **1999**, *6*, 145–159; b) U. Schörken, G. A. Sprenger, *Biochim. Biophys. Acta* **1998**, *1385*, 229–243; c) O. P. Ward, A. Singh, *Curr. Opinion Biotechnol.* **2000**, *11*, 520–526.

49 a) For a review on TK see: N. J. Turner, *Curr. Opinion Biotechnol.* **2000**, *11*, 527–531; b) for a review on PDC see: M. Pohl, *Adv. Biochem. Eng. Biotechnol.* **1997**, *58*, 16–43.

50 a) R. Wilcocks, O. P. Ward, S. Collins, N. J. Dewdney, Y. Hong, E. Prosen, *Appl. Environ. Microbiol.* **1992**, *58*, 1699–1704; b) R. Wilcocks, O. P. Ward, *Biotechnol. Bioeng.* **1992**, *39*, 1058–1063.

51 H. Iding, Dissertation, University of Düsseldorf, **1998**.

52 H. Iding, T. Dünnwald, L. Greiner, A. Liese, M. Müller, P. Siegert, J. Grötzinger, A. S. Demir, M. Pohl, *Chem. Eur. J.* **2000**, *6*, 1483–1495.

53 B. Lingen, D. Kolter-Jung, P. Dünkelmann, R. Feldmann, J. Grötzinger, M. Pohl, M. Müller, *ChemBioChem* **2003**, *4*, 721–726.

54 H. Wildemann, P. Dünkelmann, M. Müller, B. Schmidt, *J. Org. Chem.* **2003**, *68*, 799–804.

55 A. S. Demir, T. Dünnwald, H. Iding, M. Pohl, M. Müller, *Tetrahedron: Asymmetry* **1999**, *10*, 4769–4774.

56 T. Dünnwald, A. S. Demir, P. Siegert, M. Pohl, M. Müller, *Eur. J. Org. Chem.* **2000**, 2161–2170.

57 T. Dünnwald, M. Müller, *J. Org. Chem.* **2000**, *65*, 8608–8612.

58 a) B. Gonzáles, R. Vicuña, *J. Bacteriol.* **1989**, 2401–2405; b) P. Hinrichsen, I. Gómez, R. Vicuña, *Gene* **1994**, *144*, 137–138. We thank Dr. Vicuña for kindly providing us with the BAL gene.

59 The enzymes used in our studies were expressed and purified from a recombinant *E. coli* strain. For easier purification a hexahistidine tag was cloned to the C-terminus of the enzyme; E. Janzen, M. Müller, D. Kolter-Jung, M. M. Kneen, M. J. McLeish, M. Pohl, *Bioorg. Chem.* **2006**, *34*, 345–361.

60 E. Janzen, Dissertation, University of Düsseldorf, **2002**.

61 A. S. Demir, M. Pohl, E. Janzen, M. Müller, *J. Chem. Soc., Perkin Trans. 1* **2001**, 633–635.

62 A. S. Demir, Ö. Şeşenoglu, E. Eren, B. Hosrik, M. Pohl, E. Janzen, D. Kolter, R. Feldmann, P. Dünkelmann, M. Müller, *Adv. Synth. Catal.* **2002**, *344*, 96–103.

63 A. S. Demir, Ö. Şeşenoglu, P. Dünkelmann, M. Müller, *Org Lett.* **2003**, *5*, 2047–2050.

64 a) R. H. Dainty, *Biochem. J.* **1970**, *117*, 585–592; a) D. H. G. Crout, D. L. Rathbone, *J. Chem. Soc., Chem. Commun.* **1988**, 98–99; b) R. B. Herbert, B. Wilkinson, G. J. Ellames, E. K. Kunec, *J. Chem. Soc., Chem. Commun.* **1993**, 205–206; c) S. M. Roberts, P. W. H. Wan, *J. Mol. Catal. B: Enzym.* **1998**, *4*, 111–136; d) J. Q. Liu, M. Odani, T. Dairi, N. Itoh, S. Shimizu, H. Yamada, *Appl. Microbiol. Biotechnol.* **1999**, *51*, 586–591.

65 T. G. Mosbacher, M. Müller, G. E. Schulz, *FEBS Journal* **2005**, *272*, 6067–6076.

66 a) E. Fischer, *Annal. Chem.* **1882**, *211*, 214–232; b) J. S. Buck, W. S. Ide, in *Organic Reactions*, Vol. 4 (Eds.: R. Adams, W. E. Bachmann, H. A. Blatt, L. F. Fieser, J. R. Johnson, H. R. Snyder), Wiley, New York, **1949**, pp. 269–304, and references cited therein.

67 P. Dünkelmann, D. Kolter-Jung, A. Nitsche, A. S. Demir, P. Siegert, B. Lingen, M. Baumann, M. Pohl, M. Müller, *J. Am. Chem. Soc.* **2002**, *124*, 12 084–12 085.

68 E. S. Polovnikova, M. J. McLeish, E. A. Sergienko, J. T. Burgner, N. L. Anderson, A. K. Bera, F. Jordan, G. L. Kenyon, M. S. Hasson, *Biochemistry*, **2003**, *42*, 1820–1830. We thank Dr. McLeish for kindly providing us the BFD H281A gene.

69 P. Dünkelmann, *Dissertation*, University of Bonn, **2004**.

70 M. Pohl, B. Lingen, M. Müller, *Chem. Eur. J.* **2002**, *8*, 5288–5295.

71 A. Cosp, D. Kolter-Jung, R. Feldmann, C. Dresen, M. Müller, unpublished results.

72 On the basis of our work, other groups have broadened the possible applications of BAL in catalysis: a) D. Kihumbu, T. Stillger, W. Hummel, A. Liese, *Tetrahedron: Asymmetry*, **2002**, *13*, 1069–1072; b) M. Sanches-Gonzalez, J. P. N. Rosazza, *Adv. Synth. Catal.* **2003**, *345*, 819–824; c) A. S. Demir, P. Ayhan, A. C. Igdir, A. N. Duygu, *Tetrahedron* **2004**, *60*, 6509–6512; d) H. Trauthwein, P. D. de Maria, M. Pohl, D. Gocke, H. Gröger, H. Trauthwein, T. Stillger, M. Müller, submitted.

73 a) M. Müller, G. A. Sprenger, Thiamine-dependent enzymes as catalysts of C–C-bonding reactions: The role of "orphan" enzymes, in *Thiamine: Catalytic Mechanisms and Role in Normal and Disease States*, Marcel Dekker, New York, **2004**, pp. 77–91; b) M. Pohl, G. A. Sprenger, M. Müller, *Curr. Opin. Biotechnol.* **2004**, *15*, 335–342.

74 M. Müller, *Curr. Opin. Biotechnol.* **2004**, *15*, 591–599.

75 T. Schubert, Dissertation, University of Bonn, **2002**.

76 A. Liese, D. Hahn, personal communication.

77 *Cf.* T. Zelinski, A.Liese, C. Wandrey, M.-R. Kula, *Tetrahedron: Asymmetry*, **1999**, *10*, 1681–1687.

3
Reaction Technology in Asymmetric Synthesis

3.1
Reaction Engineering in Asymmetric Synthesis
Stephan Lütz, Udo Kragl, Andreas Liese, and Christian Wandrey

3.1.1
Introduction

The production of enantiopure compounds is of increasing importance to the chemical and biotechnological industries. Bioorganic transformations are predestined to meet this demand due to their inherent regio- and stereoselective nature. Indeed, growing amounts of enantiopure chemicals for pharmaceutical purposes are being produced biocatalytically today, in contrast to the production of racemic bulk commodities in the past. In this sense, biosynthesis needs to be understood as "chemistry by nature". The biological principles optimized over thousands of years experience a renaissance when applied to technical asymmetric catalysis. Nevertheless, one key tool or prerequisite for their application is the appropriate technology, namely reaction engineering. One of the most interesting targets of reaction engineering is the field of asymmetric catalysis, with either chemical or biological catalysts. In the chemical industry several asymmetric catalytic processes have been established, some of which are summarized in Table 3.1.1 [1].

The main task in technical application of asymmetric catalysis is to maximize catalytic efficiency, which can be expressed as the *ttn* (total turnover number, moles of product produced per moles of catalyst consumed) or biocatalyst consumption (grams of product per gram biocatalyst consumed, referring either to wet cell weight (wcw) or alternatively to cell dry weight (cdw)) [2]. One method of reducing the amount of catalyst consumed is to decouple the residence times of reactants and catalysts by means of retention or recycling of the precious catalyst. This leads to an increased exploitation of the catalyst in the synthesis reaction.

Several methods can be employed to achieve the decoupling, namely immobilization (in a solid matrix or a second phase) or membrane filtration (Fig. 3.1.1) [2, 3].

Asymmetric Synthesis with Chemical and Biological Methods. Edited by Dieter Enders and Karl-Erich Jaeger
Copyright © 2007 WILEY-VCH Verlag GmbH & Co. KGaA, Weinheim
ISBN 978-3-527-31473-7

Table 3.1.1 Asymmetric chemical and biotechnological processes in industry (modified and adapted from ref. [1]).

Year	Process
1880	L-lactic acid formation by fermentation of glucose or lactose with *Lactobacilli*
1912	enantioselective hydrocyanation
1930	L-ephedrine intermediate through asymmetric C–C bond formation with yeast
1930	oxidation of sorbitol to L-sorbose in vitamin C production by *Gluconobacter oxydans*
1932	support of metallic catalysts in asymmetric (de)hydrogenations on quartz
1939	modification of a *heterogeneous* Pt hydrogenation catalyst by a chinchona alkaloid
1950	L-amino acids through fermentation by *Brevi-*, *Corynebacterium* sp. fermentation
1966	homogeneous asymmetric catalysis (of cyclopropanation of styrene) by a soluble chiral metal complex
1968	hydrogenation of olefins with modified Wilkinson catalyst
1970	L-aspartic acid by ammonia addition to fumaric acid by aspartase (from *E. coli*)
1971	synthesis of L-dopa catalyzed by Rh-DIPAMP complex
1980	asymmetric epoxidation of allylic alcohols with Ti^{IV} tartrate complex
1980	L-dopa from ammonia, pyruvate, and catechol by bacterial enzymes
1980	L-malic acid by addition of water to fumaric acid by fumarase from *Brevibacterium* sp.
1981	reduction of ketones catalyzed by oxazaborolidines
1984	ZnEt$_2$ addition to aldehydes catalyzed by β-amino alcohols

Table 3.1.1 *Continued*

Year	Process
1988	asymmetric dihydroxylation via ligand-accelerated catalysis with chinchona alkaloid
1991	epoxidation of unfunctionalized olefins catalyzed by Mn^{III} salen complex
1993	L-menthol synthesis with (S)-BINAP
2000	(1S)-hydroxy(3-phenoxyphenyl)acetonitrile by hydrocyanation with oxynitrilase

immobilization   two-phase system   membrane reactor

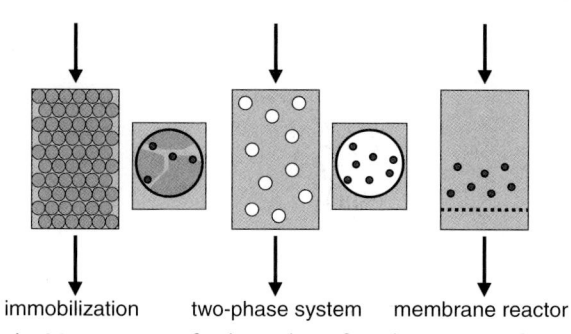

Fig. 3.1.1 Strategies for decoupling of residence times of catalysts and reactants.

Since membrane filtration methods such as ultrafiltration or nanofiltration can discriminate according to the size of a given molecule, they can easily be used to retain biocatalysts (which are macromolecules). For chemical catalysts additional procedures have to be applied mostly. Immobilization on a solid support is used

3.1.2
Membrane Reactors with Chemical Catalysts

To use membrane filtration for residence time decoupling the molecular weight of the chemical catalyst has to be increased, for example, by binding the catalyst to a homogeneously soluble polymer [4]. This allows for separation of reactants and catalysts by size. Due to their similarity to biological catalysts, the term "chemzyme" (*chem*ical en*zyme*) [5, 6] has been coined for these polymer-enlarged but still homogeneously soluble chemical catalysts (Fig. 3.1.2) [7].

In analogy to the enzyme membrane reactors (EMRs) [8], a chemzyme membrane reactor (CMR) is used to retain a polymer-enlarged chemical catalyst of this kind. Tremendous progress could be made in the recycling of polymer-enlarged catalysts (Fig. 3.1.3) by employing different types of catalysts for both the enantioselective C–C bond formation and redox reactions.

Using the addition of diethylzinc to benzaldehyde catalyzed by an α,α-diphenyl-L-prolinol bound to a methacrylate polymer **1** ($M = 96\,000$ g mol^{-1}) (Fig. 3.1.3, left) the *ttn* for the catalyst could be increased 10-fold (from 50 to 500) in comparison with the free catalyst via membrane retention [9]. The product was obtained in 80% enantiomeric excess (*ee*).

A modified oxazaborolidine **2** catalyzing the enantioselective reduction of acetophenone or tetralone with borane proved to give *ttn* values in the same order of magnitude [10, 11]. Using a special hydroxyproline-based polymer-enlarged oxazaborolidine **3**, a *ttn* of 1400 for the reduction of tetralone was achieved (Fig. 3.1.3, **3**) [5, 12].

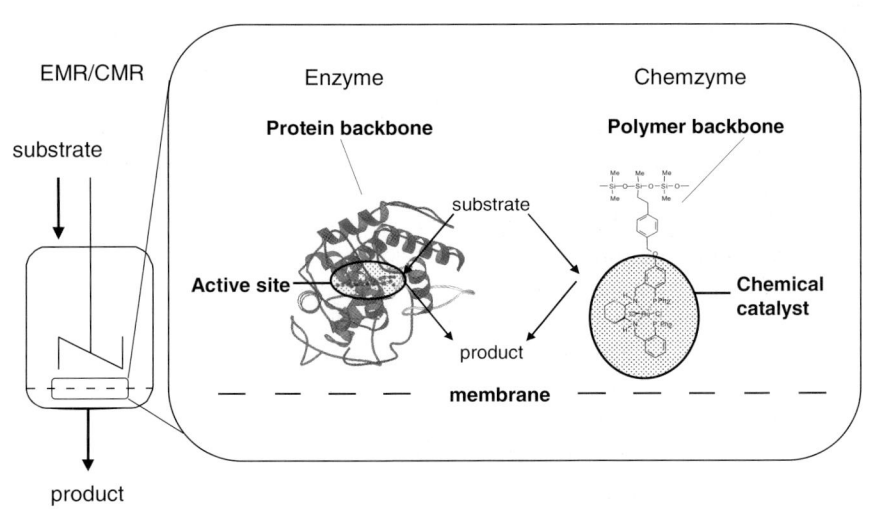

Fig. 3.1.2 Retention of enzymes and chemzymes in membrane reactors.

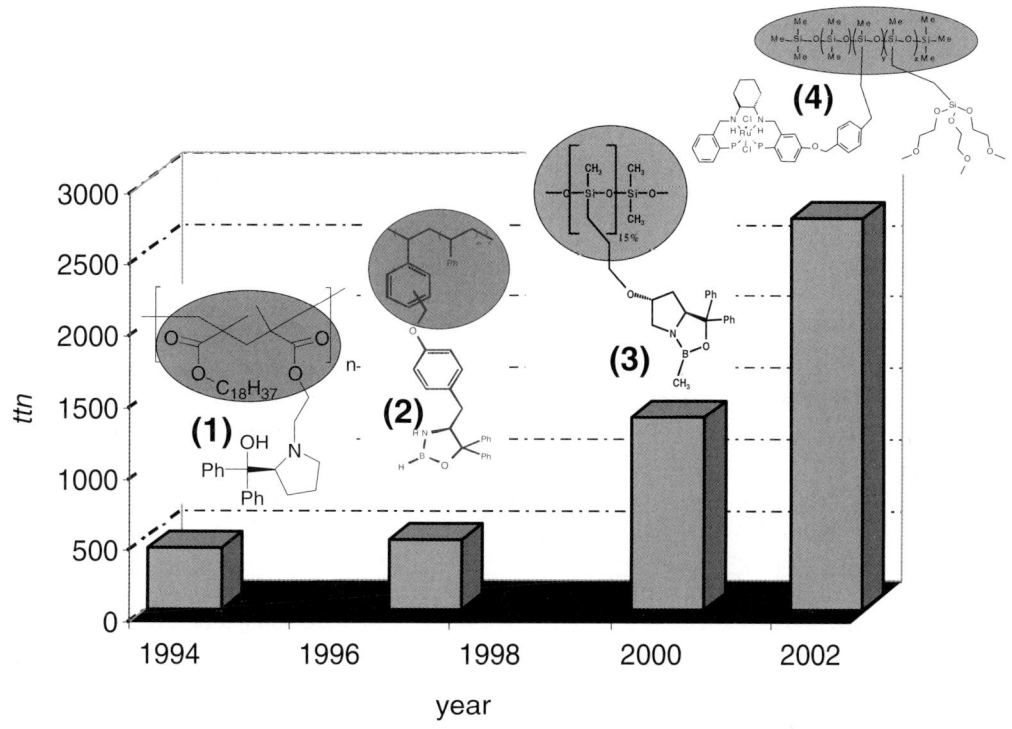

Fig. 3.1.3 Total turnover numbers (*ttn*) for different types of chemzymes: **1**, α,α-diphenyl-L-prolinol chemzyme; **2**, oxazaborolidine chemzyme; **3**, hydroxyproline-based oxazaborolidine chemzyme; **4**, Gao–Noyori chemzyme.

The highest *ttn* published to our knowledge so far for chemzymes (in the sense of polymer-enlarged chemical catalysts) is found in the transfer hydrogenation process catalyzed by Gao-Noyori's catalyst bound to a siloxane polymer (Fig. 3.1.3, 4) [13, 14]. In this transfer hydrogenation acetophenone is reduced to (S)-phenylethanol using isopropanol as hydrogen donor. The product is produced in a CMR with 91% *ee* at a space–time yield of 578 g L^{-1} d^{-1}; the *ttn* for the catalyst is 2633.

Besides the increase in the *ttn*, use of the membrane filtration as an integrated process offers additional benefits.

- The workup of the solution leaving the membrane reactor is simplified as only the solvent has to be removed. No tedious separation steps to remove the catalyst are needed.
- Due to the behavior of the membrane reactor as a continuously operated stirred tank reactor (CSTR) [2], it can be used effectively to suppress side-reactions, for example the noncatalyzed reduction yielding the racemate in the oxazaborolidine reaction [11].

By choosing the right combination of catalyst, polymer, and membrane reactor, tremendous progress has been made in continuous homogeneous catalysis.

3.1.3
Membrane Reactors with Biological Catalysts

3.1.3.1 Membrane Reactors with Whole Cells

Porous membranes for the retention of microorganisms are also available, which allow the establishment of continuously operated whole-cell biotransformations. These are of special interest in the case of redox reactions since the expensive cofactors are regenerated internally in the cell [16]. A particularly interesting reaction is the reduction of diketones to the corresponding optically active diols (for example, the reduction of 2,5-hexanedione to (2R,5R)-hexanediol). These diols are interesting building blocks for the synthesis of phospholane ligands for chemical catalysts [17]. One microorganism capable of catalyzing this enantio- and diastereoselective reduction is *Lactobacillus kefiri* [18]. Resting cells of *L. kefiri* were used in a continuously operated membrane reactor with integrated product separation and purification (see Fig. 3.1.4) [19]. As redox-equivalent glucose was add-

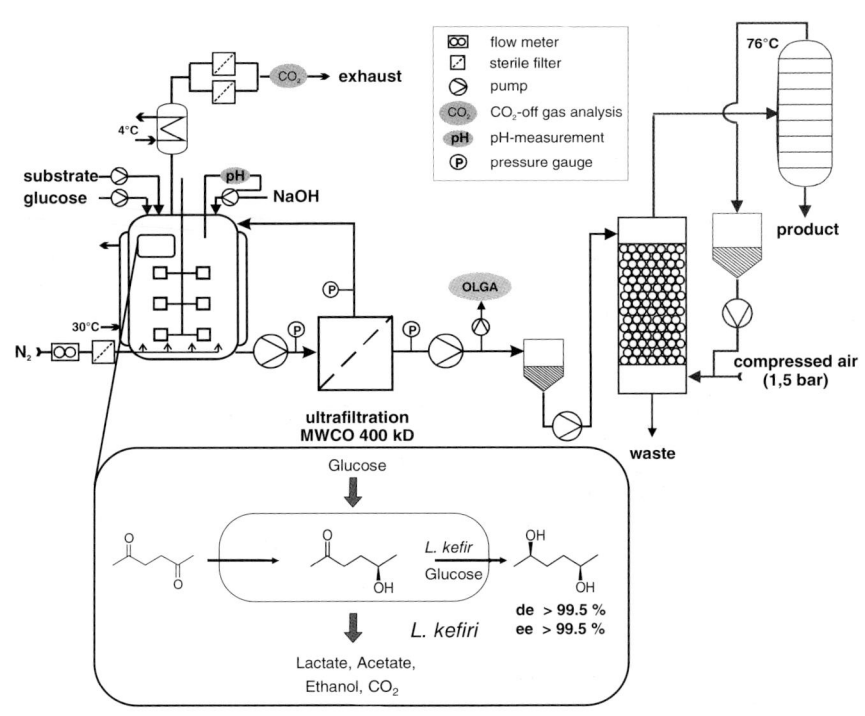

Fig. 3.1.4 Whole-cell biotransformation with a membrane reactor.

ed, this was converted by the cell metabolism to lactate, acetate, ethanol, and carbon dioxide. The oxidation of glucose regenerates the intracellular cofactor pools for the reduction of the diketone.

In this reactor the product could be synthesized with a space–time yield of 64 g L^{-1} d^{-1} with an excellent enantiomeric and diastereomeric excess (*ee* and *de* > 99%). The biocatalyst consumption could be decreased 30-fold to 15 $g_{product}\, g_{wcw}^{-1}$ by using the membrane reactor as compared with a batch reactor. The corresponding (2*S*,5*S*)-hexanediol can also be obtained via biocatalysis [20].

3.1.3.2 Membrane Reactors with Isolated Enzymes

Relatively complex compounds with two stereogenic centers, such as enantiopure diols, can also be synthesized using biocatalysis in reaction sequences starting from readily available building blocks. This can be demonstrated by combining an enzymatic C–C bond formation and a redox reaction in a cascade of two membrane reactors (see Fig. 3.1.5) [1, 21].

In a first reactor, where benzoylformate decarboxylase (BFD) is retained, benzaldehyde and acetaldehyde are coupled to yield (*S*)-hydroxy-1-phenylpropanone. This hydroxy ketone is then reduced to the corresponding diol in a second reactor by an alcohol dehydrogenase (ADH). Regeneration of the necessary cofactor is achieved by formate dehydrogenase (FDH) or by other methods. To avoid additional consumption of redox equivalents by unselective reduction of residual starting material from the first reactor, the volatile aldehydes are removed via an inline stripping module between the two membrane reactors. In this setup the diol was produced with high optical purity (*ee*, *de* > 90%) at an overall space–time yield of 32 g L^{-1} d^{-1}.

For this type of C–C bond formation both stereoisomers of the hydroxyphenylpropanone can be obtained using either the BFD mentioned above or the benzaldehyde lyase (BAL). Both of these enzymes are dependent on thiamine diphosphate (ThDP) as cofactor [22]. For the enantioselective reduction of the intermediate also, both stereoisomers can be obtained by using two different ADH enzymes. Thus all four possible stereoisomers of the diol can be obtained in high optical purity (see Scheme. 3.1.1) [23].

The products could be obtained in isolated overall yields of >80% in all cases.

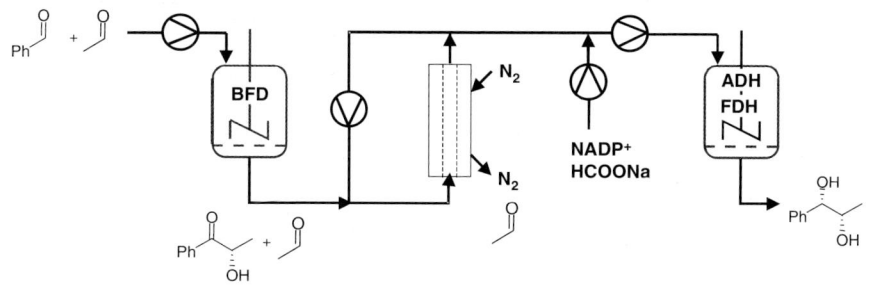

Fig. 3.1.5 Cascade of membrane reactors with isolated enzymes (BFD: benzoylformate decarboxylase; ADH: alcohol dehydrogenase; FDH: formate dehydrogenase).

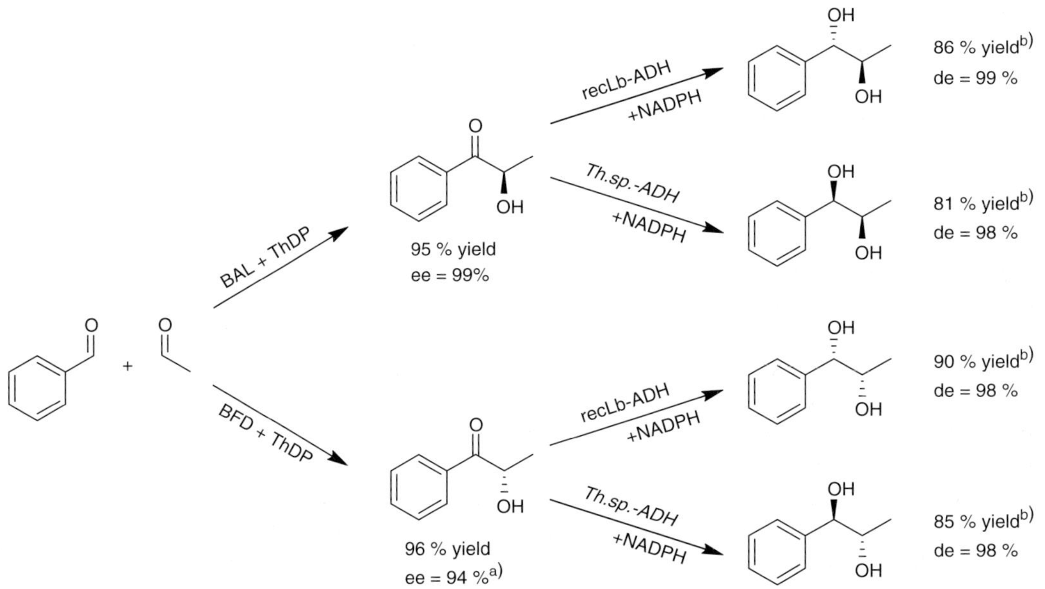

Scheme 3.1.1 Enantio- and diastereoselective enzymatic synthesis of 1-phenylpropane-1, 2-diol stereoisomers in a reaction sequence. recLbADH, recombinant ADH from *Lactobacillus brevis*; *Th.sp.*-ADH, ADH from *Thermoanaerobium species*. [a] After crystallization ee > 99%. [b] Yields of diols are given as overall yields for both reaction steps.

3.1.4
Two-Phase Systems

Solubility of interesting substrates for biocatalysis in water is often a problem. Therefore organic solvents which are not miscible with water are often used in two-phase systems to serve as substrate reservoir and for *in-situ* product extraction. Aqueous–organic two-phase systems combine the benefit of enhancing the substrate solubility with the immobilization of the biocatalyst in the aqueous phase. In these systems the biocatalyst might get deactivated at the phase boundary, depending on its characteristics and those of the solvent. The log P (P = partition coefficient of the biocatalyst in the standard *n*-octanol/water two-phase system) is often used as a criterion for the choice of an organic solvent which is compatible with the biocatalyst [24, 25]. Usually solvents with log $P > 4$ are considered biocompatible. In recent research it could be shown that not only the polarity but also the chemical structure and functional groups should be considered for the choice of the solvent. It is of fundamental importance to differentiate between a real deactivation and instantaneous inhibition introduced by the organic solvent, as was demonstrated for two different enzyme classes of biocatalysts [26].

Aliphatic ethers with branched side chains such as MTBE (methyl *t*-butyl ether), especially, deactivate enzymes only to a very small degree in incubation experiments; for example, the BAL mentioned above has a half-life $t_{1/2}$ of up to 500 h in aqueous–organic two-phase systems (see Fig. 3.1.6) [21]. This may not hold true for a special enzyme/solvent combination under process conditions. When incubated at higher temperatures or even in the presence of the substrate benzaldehyde the deactivation of the enzyme is much higher (see Table 3.1.2) [27].

There is a destabilizing effect on the BAL from the temperature, the organic solvent, and the substrate. These influences are summarized in Fig. 3.1.7.

When the enzyme is incubated at 4 °C with aqueous buffer, a very small deactivation constant is found. In the presence of a second phase of MTBE, deactivation is 30% higher (4 °C). The highest increase in enzyme deactivation is due to the temperature; the deactivation constants are 22-fold (1.3×17) or 255-fold (15×17) higher when incubating the enzyme at 20 °C in pure buffer or in a two-phase system respectively. In the presence of the substrate benzaldehyde almost the entire enzyme is deactivated within 1 h. This deactivation in a two-phase system is to some extent dependent on the size of the phase boundary and can

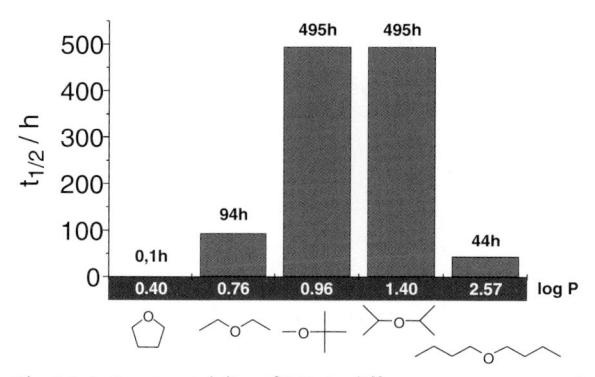

Fig. 3.1.6 Storage stability of BAL in different aqueous–organic two-phase systems (4 °C).

Table 3.1.2 Stability of benzaldehyde lyase under process conditions.

System	K_{des} [h^{-1}]	
	at 20 °C	at 35 °C
Buffer/MTBE (90 mL 50 mM TEA, pH 8)/25 mL MTBE	0.197	0.372
Buffer/MTBE and substrate (as above, +20 mM benzaldehyde in organic phase)	0.887	1.578

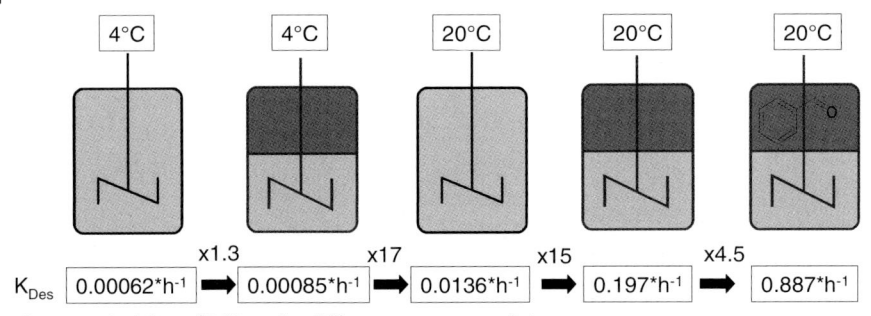

Fig. 3.1.7 Stability of BAL under different process conditions. 50 mM TEA buffer (pH 8), 0.5 mM ThDP, 0.5 mM MgCl$_2$.

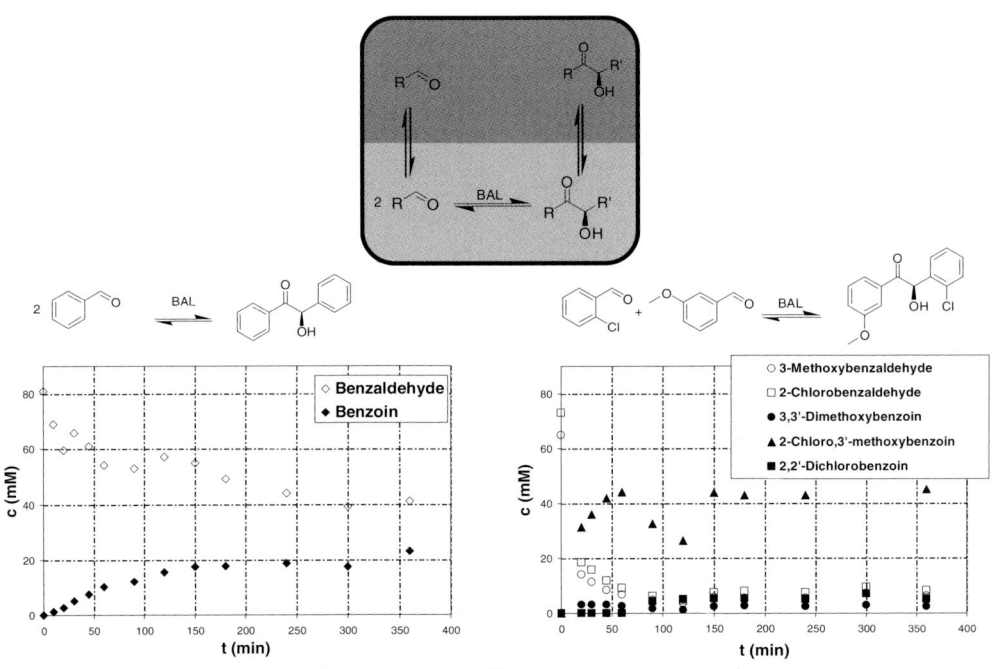

K_{Des}: 0.00062*h^{-1} →×1.3→ 0.00085*h^{-1} →×17→ 0.0136*h^{-1} →×15→ 0.197*h^{-1} →×4.5→ 0.887*h^{-1}

Fig. 3.1.8 Enantioselective synthesis of benzoins in a two-phase system. KP$_i$ buffer (25 mL, 50 mM, pH 8), ThDP (0.5 mM, MgCl$_2$ (0.5 mM), MTBE (15 mL), substrate (40 mM with respect to the whole reaction volume of 40 mL: 40 mM benzaldehyde or 20 mM 2-chlorobenzaldehyde and 20 mM 3-methoxybenzaldehyde, respectively). Concentrations of reactants in the diagram are given for the organic phase only.

thus be partially avoided if a membrane-separated two-phase system is used. This nevertheless imposes mass transfer limitations.

Despite the deactivation, the BAL-catalyzed synthesis of benzoins and derivatives could be carried out in a two-phase system (see Fig. 3.1.8) [27] in a batch reactor.

Using only benzaldehyde as substrate (see Fig. 3.1.8a), (R)-benzoin can be obtained on a 200 mg scale from a 40 mL two-phase system. This corresponds to a volumetric productivity of 53 g L^{-1} d^{-1} for the organic phase, or 20 g L^{-1} d^{-1} with respect to the overall reaction volume. When using 2-chlorobenzaldehyde and 3-methoxybenzaldehyde as substrate in equimolar amounts, (R)-2-chloro-3'-methoxybenzoin is formed as the main product with approximately 80% selectivity (see Fig. 3.1.8b). The symmetric benzoins (2,2'-dichlorobenzoin, 3,3'-dimethoxybenzoin) are formed as side-products. The main product is obtained on a 150 mg scale, which corresponds to a volumetric productivity of 40 g L^{-1} d^{-1} with regard to the organic phase.

However, due to the deactivating influence, application in a fed-batch mode or a CSTR would be favorable, since in these reactor types the stationary concentration of the substrate is lower than in a batch reactor [2].

3.1.5 Conclusions

Reaction engineering helps in characterization and application of chemical and biological catalysts. Both types of catalyst can be retained in membrane reactors, resulting in a significant reduction of the product-specific catalyst consumption. The application of membrane reactors allows the use of non-immobilized biocatalysts with high volumetric productivities. Biocatalysts can also be immobilized in the aqueous phase of an aqueous–organic two-phase system. Here the choice of the enzyme–solvent combination and the process parameters are crucial for a successful application.

But there is still another point, not yet discussed but with considerable potential, which may also impact eventually on technical asymmetric catalysis. Even though biocatalysts are efficient, active, and selective, there still remains one big disadvantage: At present, there is not yet an appropriate enzyme known or available for every given chemical reaction. It is estimated that about 25 000 enzymes exist in Nature, and 90% of these have still to be discovered [28, 29]. New biocatalysts are made available nowadays not only from screening known organism but also via screening metagenomic libraries and directed evolution techniques [30].

Furthermore, approximately only 10% of all known/investigated enzymes are already available on larger scale. This represents a challenge and opens up a new "playground" for chemists. The challenge of the future will be to design chemzymes that can catalyze even those reactions which are not yet feasible with enzymes. And an even further projection might be the successful combination of chemzymes with enzymes in continuously operated reactors to synthesize complex products stereoselectively starting from bulk chemicals.

In summary, understanding biological principles and processes both in organic chemical and physical terms is a prerequisite and represents the first step toward the development of novel engineered systems. To successfully implement technical asymmetric catalysis on a preparative scale requires scientists who integrate the knowledge of the different fields involved, namely biocatalysis, organic chemistry, and biochemical and process engineering.

Acknowledgements

Thanks are due to all the cooperation partners within SFB380 throughout its four periods for the fruitful cooperation, and to all the coworkers who have contributed to these results. Some of their names can be found on the publications listed in the references.

Abbreviations

ADH	alcohol dehydrogenase
BAL	benzaldehyde lyase
BFD	benzoylformate decarboxylase
cdw	cell dry weight
CMR	chemzyme membrane reactor
CSTR	continuously operated stirred tank reactor
EMR	enzyme membrane reactor
FDH	formate dehydrogenase
K_{des}	deactivation constant (assuming first-order exponential decay)
KP_i	potassium phosphate
log P	logarithm of the partition coefficient of a given compound in the standard n-octanol/water two-phase system
MTBE	methyl t-butyl ether
$NAD(P)^+/NAD(P)H$	nicotinamide adenine dinucleotide (oxidized/reduced form), (phosphorylated)
$t_{1/2}$	half-life
TEA	triethanolamine
ThDP	thiamine diphosphate
ttn	total turnover number
wcw	wet cell weight

References

1 A. Liese, *Adv. Biochem. Eng. Biotechnol.* **2005**, *92*, 197–224.
2 N. Rao, S. Lütz, K. Seelbach, A. Liese, Basics of bioreaction engineering, in *Industrial Biotransformations* (Eds.: A. Liese, K. Seelbach, C. Wandrey), Wiley-VCH, Weinheim, **2006**, pp. 115–145.
3 U. Kragl, T. Dwars, *Trends Biotechnol.* **2002**, *20*, 45.
4 U. Kragl, C. Dreisbach, C. Wandrey, in *Applied Homogeneous Catalysis with Organometallic Compounds*, Vol. 2 (Eds.: B. Cornils, W. A. Herrmann), VCH, Weinheim, **1996**, p. 832.
5 S. Laue, L. Greiner, J. Wöltinger, A. Liese, *Adv. Synth. Catal.* **2001**, *343*, 711.
6 J. Wöltinger, A. S. Bommarius, K. Drautz, C. Wandrey, *Org. Proc. Res. Dev.* **2001**, *5*, 241.
7 Enzyme visualization from http://metallo.scripps.edu/PROMISE
8 S. Lütz, N. Rao, C. Wandrey, *Chem. Ing. Tech.* **2005**, *77*, 1669–1682.

9 U. Kragl, C. Dreisbach, *Angew. Chem.* **1996**, *108*, 684; *Angew. Chem. Int. Ed. Engl.* **1996**, *35*, 642.
10 M. Felder, G. Giffels, C. Wandrey, *Tetrahedron: Asymmetry* **1997**, *8*, 1975.
11 G. Giffels, J. Beliczey, M. Felder, U. Kragl, *Tetrahedron: Asymmetry* **1998**, *9*, 691.
12 J. Wöltinger, K. Drauz, A. S. Bommarius, *Appl. Catal. A* **2001**, *221*, 171.
13 S. Laue, L. Greiner, J. Wöltinger, A. Liese, *Adv. Synth. Catal.* **2001**, *343*, 711.
14 L. Greiner, S. Laue, A. Liese, C. Wandrey, *Chem. Eur. J.* **2006**, *12*, 1818.
15 A. Bruggink, R. Schoevaart, T. Kieboom, *Org. Proc. Res. Dev.* **2003**, *7*, 622.
16 C. Wandrey, *Chemical Record* **2004**, *4*, 254.
17 W. Braun, B. Calmuschi, J. Haberland, W. Hummel, A. Liese, T. Nickel, O. Stelzer, A. Salzer, *Eur. J. Inorg. Chem.* **2004**, 2235.
18 J. Haberland, A. Kriegesmann, E. Wolfram, W. Hummel, A. Liese, *Appl. Microbiol. Biotechnol.* **2002**, *58*, 595.
19 J. Haberland, W. Hummel, T. Dausmann, A. Liese, *Org. Proc. Res. Dev.* **2002**, *6*, 458.
20 S. Rissom, Ph.D. Thesis, University of Bonn, **1999**.
21 A. Liese, *Biological Principles Applied to Technical Asymmetric Catalysis*, Verlag Forschungszentrum Jülich, Jülich, **2003**.
22 M. Pohl, G. Sprenger, M. Müller, *Curr. Opin. Biotechnol.* **2004**, *15*, 335.
23 D. Kihumbu, T. Stillger, W. Hummel, A. Liese, *Tetrahedron: Asymmetry* **2002**, *13*, 1069.
24 G. Carrea, *Trends Biotechnol.* **1984**, *2*, 102.
25 C. Laane, S. Boeren, K. Vos, C. Veeger, *Biotechnol. Bioeng.* **1987**, *30*, 81.
26 M. Villela, T. Stillger, M. Müller, A. Liese, C. Wandrey, *Angew. Chem. Int. Ed.* **2003**, *42*, 2993.
27 S. Kühl, M. Pohl, M. Müller, C. Wandrey, S. Lütz, unpublished results.
28 R. D. Appel, A. Bairoch, D. F. Hochstrasser, *Trends Biochem. Sci.* **1994**, *19*, 258.
29 R. D. Appel, A. Bairoch, D. F. Hochstrasser, *Trends Biochem. Sci.* **1994**, *19*, 258.
30 T. Drepper, T. Eggert, W. Hummel, C. Leggewie, M. Pohl, F. Rosenau, K. E. Jaeger, *Chem. Ing. Technik* **2006**, *78*, 239.

3.2
Biocatalyzed Asymmetric Syntheses Using Gel-Stabilized Aqueous–Organic Two-Phase Systems
Marion B. Ansorge-Schumacher

Despite the outstanding abilities of isolated enzymes in terms of specificity and stereoselectivity for asymmetric syntheses [1], the overall number of technical processes using these catalysts for the production of chiral compounds is rather limited to date [2]. A major reason for this can probably be found in the poor solubility of many theoretically suitable reactants in aqueous solution [3] (the standard medium for biocatalyzed syntheses), as the resulting dilution prevents the reaction from gaining maximum velocity, decreases productivity, and complicates downstream processing. On the other hand, performing the reaction in more suitable hydrophobic solutions is restricted by the difficulties arising from the poor solubility of necessary cofactors in hydrophobic solutions, the problematic enzyme recovery after transformation, and particularly the rapid deactivation of many synthetically valuable biocatalysts by organic solvents and aqueous–organic interfaces [4]. Attempts to circumvent these difficulties include the use of cyclodextrins as alternative mediators of solubility in aqueous solutions [5], and stabilization of biocatalysts by cross-linking [6] or genetic engineering [7]. However, the application of all these approaches remains limited to particular biocatalysts or specific groups of enzymes due to the structure-defined individual interaction with additives such as cyclodextrins or crosslinkers and specific effects of amino acid exchanges.

A novel and more general method to enable biocatalyzed conversion and synthesis of hydrophobic compounds involves the use of gel-stabilized aqueous–organic two-phase systems [8]. Features, advantages, disadvantages, and perspectives of this method in asymmetric synthesis will be discussed in this chapter, illustrated for the stereoselective benzoin condensation and the reduction of ketones catalyzed by thiamine pyrophosphate (TPP)-dependent lyases and NAD(P)H-dependent alcohol dehydrogenases, respectively.

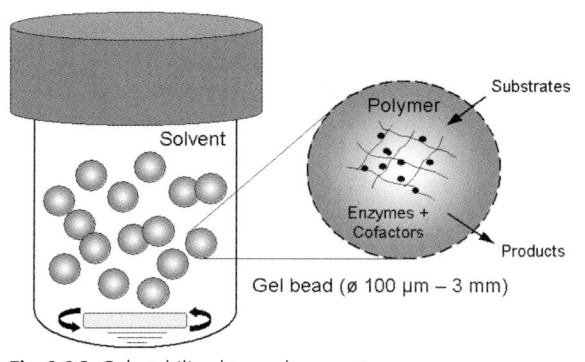

3.2.1
Gel-Stabilized Two-Phase Systems

Gel-stabilized two-phase systems consisting of small solidified aqueous compartments in a nonmiscible organic solvent (Fig. 3.2.1) combine the advantages of standard two-phase systems (that is, provision of favourable aqueous surroundings for the biocatalyst, a constant supply of an optimal concentration of hydrophobic substrates from a reservoir in the organic solvent, and beneficial effects on conversion and downstream processing by extraction of products from the aqueous phase [9]) with protection of enzymes against rapid deactivation at the interface between the aqueous and organic phases [10–12]. Consequently, they provide a general basis for extending the application range of delicate enzymes in organic synthesis. This is further supported by the concomitant immobilization of the biocatalyst in small particles, facilitating its reuse in subsequent reaction cycles or continuous operation in standard-type reactors such as batch, plug-flow, or fluidized-bed reactors. Thus, higher activities and productivities can

Fig. 3.2.1 Gel-stabilized two-phase system.

be achieved than in alternative membrane reactors [13] and application as a catalytic unit in a multi-step process becomes feasible.

Solidification of the aqueous phase in a two-phase system involves the addition of a gel-forming polymer before or after adapting the composition of the aqueous phase with regard to ionic strength, pH, cofactors, and so forth. Individual demands of the entrapped biocatalysts can thus be accounted for, almost regardless of the polymer employed. The type of polymer, however, has a strong influence on the operational stability of the two-phase system in terms of resistance against organic solvents, participating reactants, and mechanical forces [10, 14]. Among the many gel-forming polymers available [15], only a few combine all essential properties for an application in technical synthesis. While polymers from natural sources such as alginate, carrageenan, or gellan enable the entrapment of enzymes under very gentle conditions and consequently with a very high residual activity, they lack the mechanical strength required for long-term operation in standard-type reactors [10]. Synthetic polymers such as acrylate derivatives, on the other hand, reveal excellent resistance against mechanical stress [16], but often cause a considerable deactivation of enzymes before or during polymerization [16, 17]. Table 3.2.1 gives an idea of the decrease in activity of formate dehydrogenase from *Candida boidinii* [E.C. 1.2.1.2], an important enzyme catalyzing the regeneration of the cofactor $NADH+H^+$ in reduction processes [18], in the presence of components involved in the formation of polyacrylamide gels.

A material with promising properties has been found in poly(vinyl alcohol) (PVA), a synthetic polymer that forms elastic gels when a concentrated solution of the polymer (10% w/v; degree of polymerization 1400) and poly(ethylene glycol) 1000 (10% w/v) is dropped into chilled silicone oil (–20 °C) and slowly thawed to room temperature [19]. It therefore combines the mild gelation conditions of natural polymers with the advantageous operational properties of synthetic gels, and was applied successfully to the entrapment of benzaldehyde lyase from *Pseudomonas fluorescens* Biovar I (BAL; E.C. 4.1.2.38) and alcohol dehydrogenase from

Table 3.2.1 Half-life ($t_{1/2}$) of FDH in the presence of compounds involved in polyacrylamide formation.

Composition	$t_{1/2}$ [min]
Buffer	1346
0.1% APS	763
0.1% TEMED	890
0.1% APS, 0.1% TEMED	84
8% Acrylamide*	111
8% Acrylamide,[a] 0.1% APS	54
8% Acrylamide,[a] 0.1% TEMED	38

a Mixture of acrylamide and bisacrylamide in a ratio of 37.5:1.

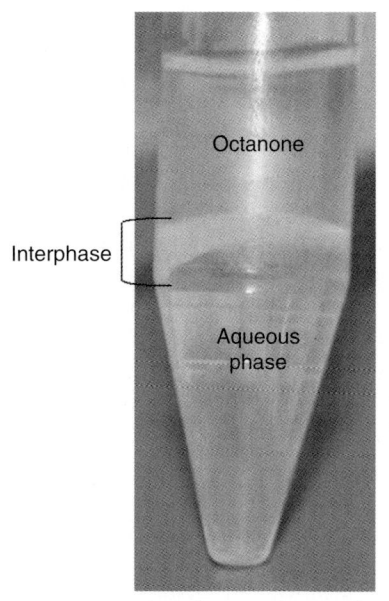

Fig. 3.2.2 Denatured BAL at the interface between a buffered aqueous solution and octanone after short mixing (10 s).

Scheme 3.2.1 BAL-catalyzed benzoin condensation.

*Thiaminepyrophosphate

Lactobacillus kefir (ADH; E.C. 1.1.1.1) for use in organic solvents [11, 12]. Both biocatalysts are characterized by a very low stability in pure organic solvents or standard aqueous–organic two-phase systems [20], though their broad substrate ranges include many hydrophobic compounds [21, 22]. Figure 3.2.2 illustrates the denaturation of native BAL at the interface between a buffered aqueous solution and octanone.

3.2.2
Benzoin Condensation with Entrapped Benzaldehyde Lyase

Native BAL is a versatile catalyst in the stereoselective carboligation of aryl aldehydes (Scheme 3.2.1), yielding *R*-configured benzoins with 99% *ee* [21].

3.2.2 Benzoin Condensation with Entrapped Benzaldehyde Lyase | 431

With the PVA-entrapped BAL suspended in hexane as a standard solvent, this reaction was performed with an at least threefold higher productivity, relative to the amount of catalyst applied, than in aqueous phases containing 20% DMSO [11]. The entrapped biocatalyst also exhibited a considerable stability toward the organic solvent, retaining full activity for one week during storage at 4 °C and for more than one month at −20 °C. The stereoselectivity of the BAL-catalyzed reaction was not affected by the immobilization, and conversion of a number of novel hydrophobic substrates (Table 3.2.2) was easily obtained. In batch syntheses, this conversion was particularly high (up to 95%) for substrates with a log P (P = partition coefficient in a two-phase system of octanol and water) below 1, and also good (25–50%) for substrates with a log P between 1 and 3. More hydrophobic substrates such as 4-isopropylbenzaldehyde, however, were hardly converted. At a hexane/PVA ratio of 4 : 1 (v/v), around 30% of the products were spontaneously extracted into the hexane phase and could easily be purified on a silica column [11].

Continuous operation of the entrapped BAL in a fluidized-bed reactor realized a maximum conversion of 3-furanyl carboxaldehyde (Table 3.2.2) of 42% and a

Table 3.2.2 Substrates 1 for the BAL-catalyzed benzoin condensation.

Substrate	log P_{ow}[a]	Substrate	log P_{ow}[a]
2-furanyl carboxaldehyde	0.34	3-ethoxybenzaldehyde	1.99
3-furanyl carboxaldehyde	0.34	4-ethoxybenzaldehyde	1.99
2-benzofuranyl carboxaldehyde	1.45	3-methylbenzaldehyde	2.27
3-thienyl carboxaldehyde	1.71	4-isoproylbenzaldehyde	3.02

a Calculated according to Ghose's and Crippen's fragmentation methodology [23].

space–time yield of $STY_{max} = 3.6$ g (L d)$^{-1}$. The initially low stability of the enzyme under these conditions ($k_{deact} = 0.8\%$ h^{-1}; $t_{1/2} = 52$ h), increased at lower inlet concentrations ($k_{deact} = 1.3\%$ h^{-1}, $t_{1/2} = 103$ h), achieving a steady state for 20 h. A particular advantage of the operation was found in the crystallization of the product in the organic solvent when the temperature at the outlet was set to −20 °C, which overcame the previously observed spontaneous racemization of compounds with a 3-furancarbonyl moiety [11, 24] and resulted in an enantiomeric excess (*ee*) of the product *(R)*-3,3′-furoin above 99%.

Further improvement of the synthetic performance of the reaction system was attempted by adapting the composition of the organic phase in such a way that partition of products into the organic phase was enhanced. In the synthesis of *(R)*-3,3′-furoin, product extraction was increased by 11% when hexane was mixed with 2-octanone in a ratio of 3 : 1, while replacing hexane with pure 2-octanone, 2-pentanone, 2-butanone, heptane, or decane yielded considerably lower product concentrations. No apparent rule was underlying these results.

3.2.3
Reduction of Ketones with Entrapped Alcohol Dehydrogenase

Native ADH catalyzes the asymmetric reduction of carbonyl compounds to the corresponding *R*-configured alcohols using NADPH+H$^+$ as cofactor (Scheme 3.2.2).

As previously observed for BAL, entrapment of ADH in PVA and its subsequent application in hexane as a standard organic solvent enabled the conversion of a number of interesting hydrophobic substrates (Table 3.2.3) by stabilizing the delicate biocatalyst against deactivating effects of the aqueous–organic interface.

As an added benefit, the immobilization resulted in an enhanced half-life of the biocatalyst at elevated temperatures and during storage [12], and thus in an overall improvement of technical applicability. With the entrapped ADH, *(R)*-phenylethanol was synthesized from acetophenone with a productivity of 5.0 g (L h)$^{-1}$ and *ee* above 98%.

Cofactor regeneration, an economically essential step in the synthetic use of ADH, was accomplished within the PVA matrix using isopropanol as cosubstrate for the ADH itself or for a second alcohol dehydrogenase from *Thermoanaerobium brockii* (E.C. 1.1.1.2). An overall turnover number of 10^2 was achieved, which is a promising magnitude for technical application. However, while the presence of the cosubstrate in the gel-stabilized two-phase system improved the solubility of substrates in the gel phase and consequently enhanced the productivity of the

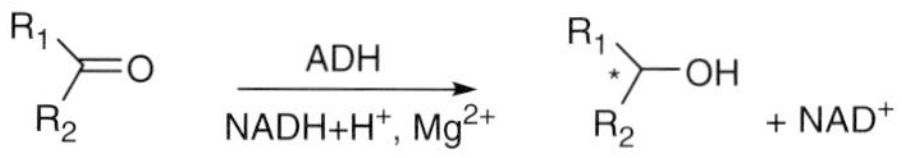

Scheme 3.2.2 ADH-catalyzed reduction of prochiral carbonyls.

Table 3.2.3 Substrates **3** of the ADH-catalyzed carbonyl reduction.

Substrate	log P_{ow}^a	Substrate	log P_{ow}^a
cyclopropyl methyl ketone	0.21	1-benzyl-4-piperidone	1.19
3-chloro-5-norbornen-2-one	0.41	5-acetyl-2-norbornene	1.45
5-oxohexanoic acid ethyl ester	0.86	acetophenone	1.67
5-hexen-2-one	1.02	1-octen-3-one	2.17
phenoxyacetone	1.04	3-octanone	2.50
3,4-methylenedioxyacetophenone	1.08	α-tetralone	2.61

a Calculated according to Ghose's and Crippen's fragmentation methodology [23].

reaction, cofactor regeneration also resulted in a rapid loss of catalytic performance, caused by the accumulation of the coproduct acetone in the PVA matrix. Thus, even more than in the BAL-catalyzed synthesis of 3,3′-furoin, the rapid and efficient extraction of products from the aqueous phase by the organic solvent plays a crucial role in the synthetic performance of the reaction system.

3.2.4
Conclusion

Biocatalyzed asymmetric conversion and synthesis of compounds with a low solubility or stability in aqueous solutions can be performed straightforwardly by using gel-stabilized two-phase systems as media. Consequently, these offer a tremendous opportunity to extend the scope of biocatalyzed reactions. For an efficient technical application, however, further knowledge about the complex influences on the performance of the systems and a subsequent directed design will be mandatory.

References

1. M. Breuer, K. Ditrich, T. Habicher, B. Hauer, M. Keeler, R. Stürmer, T. Zelinski, *Angew. Chem.* 2005, *4A–5A*, A318–A355; *Angew. Chem. Int. Ed.* **2004**, *43*, 788–824.
2. A. Liese, K. Seelbach, C. Wandrey, *Industrial Biotransformations*, Wiley-VCH, Weinheim, **2006**.
3. K. Faber, *Biotransformations in Organic Chemistry*, Springer, Berlin, **2004**.
4. A. S. Ghatorae, G. Bell, P. J. Halling, *Biotechnol. Bioeng.* 1994, *43*, 331–336; L. Kvittingen, *Tetrahedron* 1994, *50(28)*, 8253–8274.
5. T. Zelinski, M.-R. Kula, *Biocatal. Biotrans.* 1997, *15*, 57–74.
6. I. Migneault, C. Dartiguenave, M. J. Bertrand, K. C. Waldron, *BioTechniques* **2004**, *37*, 790–802.
7. F. H. Arnold, *Current Opin. Biotechnol.* **1993**, *4*, 450–455; J. C. Moore, F. H. Arnold, *Nature Biotechnol.* **1996**, *14*, 458–467.
8. M. B. Ansorge-Schumacher, B. Doumèche, D. Metrangolo, W. Hartmeier, *Minerva Biotec.* **2000**, *12*, 265–269.
9. E. Antonini, G. Carrea, P. Cremonesi, *Enzyme Microb. Technol.* 1981, *3*, 291–296; G. Carrea, *Trends Biotechnol.* **1984**, *2*, 102–106.
10. M. B. Ansorge-Schumacher, G. Ple, D. Metrangolo, W. Hartmeier, *Landbaufg. Völkenrode* **2002**, *SH241*, 99–102.
11. T. Hischer, D. Gocke, M. Fernandez, P. Hoyos, A. R. Alcantara, J. V. Sinisterra, W. Hartmeier, M. B. Ansorge-Schumacher, *Tetrahedron* **2005**, *61*, 7378–7383.
12. D. Metrangolo-Ruiz de Temino, W. Hartmeier, M. B. Ansorge-Schumacher, *Enzyme Microb. Technol.* **2005**, *36*, 3–9.
13. K. Hosono, S. Kajiwara, Y. Yamazaki, H. Maeda, *J. Biotechnol.* **1990**, *14*, 149–156; P. J. Cunnah, J. M. Woodley, *BHR Group Conf. Ser. Publ.* 1993, *3*, 81–92.
14. B. Doumeche, M. Küppers, S. Stapf, B. Blümich, W. Hartmeier, M. B. Ansorge-Schumacher, *J. Microencaps.* **2004**, *21*, 565–573.
15. R. G. Willaert, G. V. Baron, *Rev. Chem. Eng.* 1996, *12*, 1–205.
16. H. Nilsson, R. Mosbach, K. Mosbach, *Biochim. Biophys. Acta* 1972, *268*, 253–256; A. C. Johansson, K. Mosbach, *Biochim. Biophys. Acta* **1974**, *370*, 339–347.
17. K. Mosbach, R. Mosbach, *Acta Chem. Scand.* **1966**, *20*, 2807–2810; Y. Degani, T. Miron, *Biochim. Biophys. Acta* **1970**, *212*, 362–364.
18. H. K. Chenault, E. S. Simon, G. M. Whitesides, *Biotechnol. Gen. Eng. Rev.* **1988**, *6*, 221–270.
19. U. Prüße, K.-D. Vorlop, *Chem. Ing. Tech.* **1997**, *69*, 100.
20. W. Hummel, *Trends Biotechnol.* 1999, *17*, 487–492.
21. A. S. Demir, M. Pohl, E. Janzen, M. Müller, *J. Chem. Soc. Perkin Trans.* **2001**, *1*, 633–635; A. S. Demir, O. Sesenoglu, E. Eren, B. Hosrik, M. Pohl, E. Janzen, D. Kolter, R. Feldmann, P. Dünkelmann, M. Müller, *Adv. Synth. Catal.* **2002**, *344*, 96–103.
22. C. W. Bradshaw, W. Hummel, C.-H. Wong, *J. Am. Chem. Soc.* **1992**, *57*, 1532–1539.
23. A. K. Ghose, G. M. Crippen, *J. Chem. Inf. Comput. Sci.* **1987**, *27*, 21–35.
24. J. E. T. Corrie, *Tetrahedron* 1998, *54*, 5407–5416.

Index

a

Accufluor® 2
5-acetyl-4-hydroxy[2.2]paracyclophane (AHPC) 199
Acinetobacter 298
Acinetobacter calcoaceticus 387
Actinidia polygama 53
Actinidia polygama Miq. 51
activated glycolaldehyde 313
acyclic carbamates 229
acyclic imines 168
acyloin condensation 313, 405
1,4-addition 208
alcohol dehydrogenase 337, 341, 421, 432
aldol reaction 8
aldolases 318, 351
α,ω-aldolization 366
aldoses 380
alkenyloxiranes 78, 100
alkenylsulfones 20
alkenylzinc reagents 206
α-alkylation 18
alkylidene carbenes 97–98
allyl alcohols 206
allyl sulfoximine 75, 79–80, 84
allylic alcohols 216, 242
allylic alkylation 215
allylic substitution 150
allylic sulfides 216
allylic sulfinates 225
allylic sulfonation 128
allylic sulfones 216, 225
allylic thiocarbamates 229
allylic thioesters 233
allyltitanium complexes 80
Al-MCM-41 279
Al-MCM-48 279
Alzheimer's disease 102
ambidoselective cyclization 18
(R,R,R)-2-amino-3-methoxymethylazabicyclo [3.3.1]octane (RAMBO) 9
(S)-1-amino-2-methoxymethylpyrrolidine (SAMP) 40
α-amino acids 78, 86–87
β-amino acids 5, 78, 108
amino alcohols 11, 14
amino ester 14
β-aminoethylations 21
α-amino ketones 14
β-amino ketones 4
amino lactones 16
aminoalkylation 77, 85
α-aminoacylation 14
γ-aminobutyric acid (GABA) 18
α-aminomethylation 3
α-aminonitriles 14
β-aminosulfonates 5
β-aminosulfones 9–10
5-amino-4-oxo esters 14
aminosulfoximines 167
(3-aminopropyl)triethoxysilane (APTES) 286
(2S,12′R)-2-(12′-aminotridecyl)pyrrolidine 57
anabolic synthase (NeuS) 371
Ando 45
antihistamine agent 189–190, 209
(arene)chromiumtricarbonyl 131
argentilactone 392
Arndt–Eistert homologation 108
aryl transfer reactions 176, 178, 180–181, 184
arylation 156
asymmetric hydrogenation 145, 263, 278
asymmetric reductions 181
asymmetric ring-opening (ARO) 124
Attenol A and B 62

Index

aza-Michael additions 5
aza sugar 367

b

Baeyer–Villiger oxidation 18, 66
Baeyer–Villiger reaction 40
baker's yeast 45, 47, 387
Baker–Venkataraman rearrangement 41
Barbier allylation 366
Barton and McCombie's 62
Barton–McCombie 26, 69
batrachotoxins 53
benzaldehyde lyase (BAL) 298, 313, 402, 421
benzoin 299, 308, 310, 328, 405, 424, 425
benzoin condensation 430–431
benzoylformate decarboxylase (BFD) 298, 313, 402, 421
5-benzoyl-4-hydroxy[2.2]paracyclophane (BHPC) 199
bicyclic amino acids 105
bicyclic prolines 102
BINAP 115, 123–124, 131, 159–160, 266
BINAPHOS 262, 264
BINAPHOSQUIN 251
binaphthol 252, 254–255
BINOL 169–170, 198
BIPHEMPHOS 262
BIPHENPHOSQUIN 251
4-biphenyl-substituted 2,2-dimethyl-1,3-dioxan-5-amine 28
bipyridine 176
bis(allyl)titanium complexes 80–82, 85–87
α,α'-bisalkylation 26
bissulfoximine 158–159, 161, 163
BPA 216
Brevetoxin B 10
Brevibacterium sp. 416
Buchwald–Hartwig-type coupling reactions 159
3-Butyn-2-ol 396
γ-butyrolactams 18, 21

c

callistatin A 41, 392
Callyspongia truncata 41
camphorsulfonic acid 151
Candida boidinii 429
Candida parapsilosis 395
carbamoylation 109
carbenes 178
carbinoxamine 180

carboligase 330
carboligation 176, 305, 310, 327
carbonates 217
carboxypeptidase 335
catabolic aldolase (NeuA) 371
catalytic asymmetric synthesis 149
ceric ammonium nitrate (CAN) 40
cetirizine hydrochloride 209
C-glycosides 367
"chemzyme" (chemical enzyme) 418
chiral pool 115
chlorodiphenylphosphine–borane adduct 23
chlorophosphines 23
chromiumtricarbonyl 116
chromiumtricarbonyl arenes 177
chromone 39
Chrysopidae 51
cicaprost 75
citronellal 124
Class I aldolases 312
Class II aldolases 352
CODRhChiraphos 282
conjugate addition 267
conjugate hydroxylation 11
Corey–Seebach reaction 62
Corynebacterium sp. 416
Cosmopolites sordidus Germar 66
Cp–phosphaferrocene hybrid systems 139
Cp–phosphaferrocene ligand 143
cross-benzoin condensation 407
cross-coupling 160–162, 166, 168
Curtin–Hammett principle 91–92
cyanohydrin 150, 337
cyclic carbamates 231
cycloaddition 163, 163
(*R*,*R*)-1,2-cyclohexanediol monobenzoate ester 293
cycloisomerization 270
cyclopentanoid natural products 51
cyclopentenyl thioacetate 242
cyrhetrene 178, 184–185, 190, 211

d

DAIB ligand (3-*exo*-(dimethylamino)isoborneol) 206
Daniphos 115
dehydroiridodial 51
dehydroiridodiol 51
dendrobatid alkaloids 53
deoxygenation 26
5-deoxy-D-xylulose 362
1-deoxy-D-xylulose 5-phosphate (DXP) 317

1-deoxy-D-xylulose 5-phosphate synthase 317
deoxynojirimycin 58
deracemization 247
derivatives 131
designer cells 347
D-fructofuranose 383
D-fructose 365, 382
D-fructose 1,6-bisphosphate 351, 359
D-glucose 380
D-glyceraldehyde 3-phosphate 321, 359
1,2:5,6-di-*O*-isopropyliden-α-D-allofuranose 28
dialkylaminolactams 21
diaryl ketones 181
diarylmethylamines 189–190, 209
D-idose 380
Diels–Alder reactions 243
diethylzinc 183
differential scanning calorimetric 280
1,2-dihydronaphthalene 288, 289
dihydroxyacetone (DHA) 321
dihydroxyacetone phosphate (DHAP) 351
dihydroxybutylphosphonate 358
dimethoxyacetaldehyde 311
2,2-dimethyl-1,3-dioxane-5-one 26, 62, 65, 69, 71
2,3-dihydrofurans 78, 96
1,3-diketone 389
DiMPEG 187
1,3-diols 24, 386, 389
DIOP 130, 278
3,5-dioxo esters 386
DIPAMP 278
diphenylzinc 182–183
directed evolution 330
disaccharide mimetics 363
D-lyxose 380, 382
D-mannose 380
D-ribose 380
D-ribulose 382
D-tagatose 382
D-tagatose 1,6-bisphosphate 351
D-talose 380
DUPHOS 115, 131
D-xylose 380
D-xylulose 382
dynamic kinetic resolution 215, 241

e

ebelactone A 49
electrophilic fluorination 2
enantioselective hydrogenations 120
enantioselective protonation 18

entrapped benzaldehyde lyase 430
enzymatic reduction 45
enzyme membrane reactor 315
2-epi-deoxoprosopinine 58
epibatidine 53
Epilachna varivestis (Coccinellidae) 57
epoxidation 284
Escherichia coli 304, 312, 337, 381
Eschweiler–Clark conditions 158
ester enolate alkylation 8
ethyl cinnamate 288–289
Evans aldol 39, 43, 45
exo-brevicomin 313

f

ferrocene 131, 176–177
2-ferrocenyl oxazoline 177
ferrocenylphosphane 215
Finkelstein reaction 161
fluorination 2
fluoro ketones 2
(S)-folic acid 123
folic acid ester 122
formate dehydrogenase 421
5-formyl-4-hydroxy[2.2]paracyclophane (FHPC) 199
Friedel–Crafts acylation 199
fructose 1,6-bisphosphate 321
fructose 1,6-bisphosphate aldolase (RAMA) 319, 355
fructose 6-phosphate aldolase (FSA) 321

g

α-galactosides 360
galactosyltransferases 377
ganglioside 370
gel-stabilized two-phase systems 427, 428
geranylamine 124
Gluconobacter oxydans 416
glutamine synthetase inhibitors 149
glycerol phosphate oxidase 358
glycolaldehyde 321
glycosylation 333
glycosyltransferase 384
goniothalamin 392
greenhouse gases 373
Grignard-addition 66
Grubbs' olefin metathesis 155

h

half-sandwich complex 134
heavy-metal leakage 373
Heck coupling 45

heptuloses 363
herbicide 121
hetero-Diels–Alder reaction 160, 162–163, 165
heterogeneous catalysis 277
heterogenization 291–292
heterometallocenes 130
hetero-Michael additions 5
Hevea brasiliensis 332
(2R,5R)-hexanediol 346, 420–421
2,5-hexanedione 346, 420
histrionicotoxin 53
HIV-protease inhibitor 243
HMG-CoA reductase inhibitors 390
homoallyl alcohols 81, 83–84
homopropargylic alcohols 78
Horner–Wadsworth-Emmons olefination 41, 47
hydrocyanation 417
hydroformylation 18, 146, 261–262
hydrogenation 168
hydrolytic kinetic resolution (HKR) 291, 293
hydrovinylation 126, 268
hydroxamic acids 178
2-hydroxyaldehydes 359
δ-hydroxy-β-amino acids 109
2-hydroxy-3,3-dimethoxypropiophenone 308
hydroxy ketones 24, 308
δ-hydroxy-β-keto esters 387
2-hydroxypropiophenone 328
(S)-2-hydroxy-1-phenylpropanone 402
hydroxyalkylation 77, 80, 83–84, 88
hydroxynitrile lyases 327, 332
2-(S)-hydroxypropiophenone 298
hydroxypyruvate 359

i
imination 151
immobilization 277, 280
immobilization of rhodium–diphosphine complexes 279
indolizidines 53
invertase 379
isocarbacyclin 75, 77

j
Jacobsen's catalyst 284–285, 287, 290
Josiphos 115
Julia olefination approach 43

k
β-keto esters 24
2-keto-3-deoxy-D-*glycero*-D-*galacto*-nononic acid (KDN) 356

2-ketoacid decarboxylases 327
kinetic resolution 215, 225, 352, 389

l
labdanum resin 391
lactaldehyde 362
β-lactam antibiotics 7
β-lactams 108
δ-lactams 16
lactams 18
Lactobacillus brevis 342, 387, 390
Lactobacillus kefir 341, 396, 420, 430
Lactobacillus strains 341
lactone hydrazones 18
lactones 18
L-allose 380
L-arabinose 380
L-DOPA 278
Leloir glycosyltransferase 376, 384
L-fucose 361
L-fuculose 1-phosphate 351
L-galactose 380
L-glucose 382
lignans 96
Linum usitassimum 327
Linum usitatissimum 332–333, 337
lipase 242
lithium tri-*t*-butoxyaluminum hydride 15
L-mannose 380
L-rhamnose 356–357
L-rhamnulose 1-phosphate 351
L-selectride 15

m
mandelate pathway 299
Manihot esculenta (cassava) 332
Mannich bases 3
Mannich reactions 3
McMurry reaction 137
Me-Duphos 283
Meerwein–Eschenmoser rearrangement 18
membrane filtration 417
membrane reactor 417–418
memory effect 238
menthol 124
Merrifield resin 101, 212
mesoporous molecular sieves 278
metal catalysis 149
metal–salen complexes 285
methionine 149
(S)-2-methoxymethyl-1-trimethylsilylaminopyrrolidine (TMS-SAMP) 7
(S)-metolachlor 118, 169

Mexican bean beetle 57
Meyery reaction 31
Michael addition 51, 267
Michael-initiated ring closure (MIRC) 9
Michaelis–Arbuzov rearrangement 41
mono(allyl)titanium complex 82, 84–85, 87
mono(allyl)tris(diethylamino)titanium complexes 82
MPEG 185, 188
Mukaiyama-type aldol reaction 166
mutagenesis 301

n
N-acetyl-D-mannosamine 370
N-acetyl-lactosamine 372
N-acetylneuraminic acid 356, 370–371
N-acetylneuraminic acid synthase 356
N-acylimine 209
N-alkylation 159
nanofiltration 417
N-arylation 159–160, 165
Nazarov cyclizations 16
Neisseria meningitidis 356, 370, 372
neobenodine 180
neonepetalactone 53
neuraminic acid 371
NF reagents 2
N-fluorosulfonamide 2
N-formyldiarylmethylamines 211
N-formylnorephedrine 11
NFSI 2
N-heterocyclic carbene (NHC) 188
nitroalkenes 10
β-nitrophosphonic acids 13
γ-nitro-methyl sulfonates 29
N-methylsulfonamides 29
nonlinear-like effect (NLE) 203, 204
N-sulfonyl imino esters 85
N-sulfonylimines 208
N-tolylsulfonyl imino ester 85
nucleophilic alkenoylation 16
nucleosides 96
N-vinylation 162

o
3-O-coumarinylpropanal 361
O-mesitylenesulfonylhydroxylamine (MSH) 151
oligonucleotides 302
organozinc reagents 176
orphenadrine 180
ortho-metallations 157
oxabenzonorbornadiene 124
oxa-Michael additions 10
oxazaborolidines 416
oxazoline 179–180
oxazoline derivatives 146
oxazoline method 1
oxynitrilase 417

p
P,N ligands 133, 168
P,P ligands 135
palladium-catalyzed cross-coupling 159
paracyclophane 177, 198
[2.2]paracyclophane 188, 190, 196
[2.2]paracyclophaneketimine 208
Pauson–Khand cycloaddition 106
Pauson–Khand reaction 108
Pd-catalyzed rearrangement 225
PDE-IV inhibitors 180
pentose phosphate pathway 313
β-peptides 7, 108
peptidase A 150
Pfu polymerase 301
PHANEPHOS 197–198
phenyl transfer 184
phenylsulfonimidoyl group 75
1-phenylpropane-1,2-diol 346
pheromone 66
phosphaferrocenes 131
phospha-Michael additions 11
phosphane–borane adducts 11
phosphine oxides 168
2-phosphino alcohols 22
α-phosphino ketones 22
phosphino sulfoximine 168–169
phosphinylation 23
phosphoenolpyruvate 318, 355, 370
phospholane 131, 358
phosphonoacetaldehyde 318
phosphonopyruvate decarboxylase (PPD) 313, 318
phosphoramidate 358
phosphoramidite 251, 267–268
phosphorothioate 358
phosphorylation 376, 382, 383
p-hydroxymandelonitrile 336
Pichia pastoris 337
(−)-α-pinene 287, 295
Pinna attenuata 62
piperidine alkaloids 58
planar chirality 178, 197
Plasmodium falciparum 317
poly(vinyl alcohol) 429
polyether antibiotics 96
polyketides 386
polymerase chain reactions 301

polymer-supported catalysts 277
1,3-polyols 24
Prelactone B and V 69
prolines 102
propargylic alcohols 346, 401
propargylic ketones 395, 397
(+)-prosophylline 58
(+)-prosopinine 58
Prosopis alkaloids 58
prostaglandin 245
Pseudomonas 298
Pseudomonas fluorescens Biovar I 429
Pseudomonas putida 298, 304, 327, 402
pseudotripeptide 150
Pudovik reaction 13
pungent principle 51
pyridine 176
pyridinium dichromate (PDC) 41
2-pyrimidinethiol 224
pyruvate 317
pyruvate decarboxylase (PDC) 313, 318, 327, 401

q

QUINAPHOS 250
quinolines 252

r

random mutagenesis 302
random mutant library 305
reaction engineering 415
reaction technology 415
recombinant DNA technology 321
redox reactions 386
reductive amination 166
Rh diphosphino complexes 278
rhamnose isomerase 356
RhDuphos 284
rheniumtricarbonyl 178
Rhodococcus erythropolis 341
ring-opening reactions 291

s

Saccharomyces carnosus 364
Saccharomyces cerevisiae 381, 383
SADP 18
SAEP 18
salen 198
SAMP hydrazones 21
SAMP/RAMP hydrazone method 1, 38
Schiff-base ligands 209
SDS-polyacrylamide gel electrophoresis 304

Selectfluor® 2
serine phosphorylation site 383
Sharpless asymmetric dihydroxylation 62
Sharpless epoxidation 24
Shell Higher Olefin Process (SHOP) 22
ship in a bottle 286–287, 295
sialate synthetase 372
sialic acid 356, 369
sialic acid synthase 355
sialoconjugates 369
α-2,6-sialyltransferase 372
silica–alumina framework 278
silyl ketone 1
silyl trick 2
site-directed mutagenesis 301, 305, 330
Solanum tuberosum 377, 381
solid supports 212
Sonderforschungsbereich 380 1
Sordidin 66
Sorghum bicolor 327, 333
Staphylococcus carnosus 355
Stetter reaction 409
Stigmatella aurantiaca 38
Stigmatellin A, 1 38
Stigmatellin B 38
Still-Gennari 45
(+)-streptenol A 65
Streptenol A 64
Streptomyces griseus 69
Streptomyces viridochromogenes 313, 318
strictifolione 392
styrene 288–289
(R)-styrene glycol 293
(S)-styrene oxide 293
β-substituted β-phosphonomalonates 13
sucrose 376, 379
sucrose synthase 376
sugar phosphonates 357
sulfides 216
sulfilimines 153–154
sulfinamide 99–100
sulfonamides 26
sulfonates 26
sulfondiimides 153–154
sulfones 9, 31, 149
sulfonimidoyl group 75
sulfonimidoyl-substituted allyltitanium complexes 75
sulfoxides 151
sulfoximidoyl ferrocenes 177
sulfoximine 79–80, 87, 149, 176
supercages 287

Suzuki coupling 45, 126–127
Swern oxidation 15, 46, 51, 53, 57

t
t-butyl 6-chloro-3,5-dioxohexanoate 346
t-butylsulfinates 225
t-leucine 177
taddol 11 ,130
Takasago process 124
tetrahydrofurans 98
Thermoanaerobacter ethanolicus 397
Thermoanaerobium brockii 341, 397, 432
thermogravimetric 280
thiamine diphosphate (ThDP) 312, 401
thiamine diphosphate (ThDP)-dependent enzymes 298, 312
thiocarbamates 229
thioesters 234
tiglinaldehyde 41
Tishchenko reduction 24
tosylaziridine 18
total turnover number 415
transaldolase (TAL) 312
transketolase (TKT) 312–313, 359, 401
transmetallation 186
tricarbonyl complexes 143
triphenylborane 186
two-phase systems 422
tyrosinyl-arginine 336

u
Ugi's amine 118
ultrafiltration 417

umbelliferone 361
Umpolung 330

v
Vilsmeyer formylation 132
vinyl sulfoximine 75, 78
vioxanthin 392

w
Weinreb acetamide 388
White Biotechnology 373
whole-cell transformation 388
Wilke's azaphospholene ligand 272
Wilkinson's catalyst 278
Wittig olefination 57, 62
Wittig reaction 43, 45, 49, 71

x
Xyliphos 118, 121–122
xylulose 5-phosphate 315, 359

y
ylide 91
ynones 395

z
zeolite 286, 295
zeolite dealumination 285
Zimmerman–Traxler model 4
Zymomonas mobilis 318, 327

Name Index

a
Ahlbrecht   17
Ando   45
Ansorge-Schumacher   427
Augustine   278

b
Barrett   10
Batorfi   277
Belmont   54
Belokon   199
Berkessel   198
Bertagnolli   163
Bettray   1, 38
Beyer   285
Blechert   65–66
Bode   386
Bogdanovic   126
Bolm   149, 176, 210, 213
Bräse   190, 196
Brade   10
Bradshaw   345, 396, 397
Braga   188
Braun   115
Brückner   24
Buchwald   128
Buchwald–Hartwig   159
Budenberg   66

c
Cammidge   126
Carpentier   387
Chan   188
Chung   18
Claisen   10
Corey   41, 181
Cosp   318, 386
Craig   150

Cram   197
Crépy   126
Crimmins   43, 45
Crosman   277
Cubillos   277
Cui   152

d
Dahmen   188
Danilova   213
Davis   2, 28
Degner   323
Demir   386
Dias   45
Diaz   43
Ditrich   213
Domagk   26
Dresen   298
Dünkelmann   386
Dünnwald   386
Ducrot   66

e
Eggert   298
Elling   376, 381–383
Enders   1, 38, 45, 392
Engels   383, 384
Eppendorf   303
Eschweiler–Clark   158
Eurogentec   303
Eustache   62
Evans   24, 40, 43, 163, 389

f
Fang   17
Faraone   250
Feldmann   386

Feringa 251, 264
Fessner 323, 351, 384
Fleischhauer 150
Franciò 250
Frommer 384
Fu 181
Fürstner 188

g
Gais 150, 169, 215
Ganter 130
Geilenkirchen 386
Gescheidt 163
Ghosez 28
Gonzáles 405
Grubbs 155

h
Halbach 323
Harada 24
Harmata 150
Hayashi 210, 215
He 152
Heppert 116
Hermanns 213
Hiroi 225
Höfle 38–40
Hölderich 277
Hopf 213
Horner 278
Hoveyda 209
Huart 28
Hummel 341, 386
Husson 18

i
Inoue 312

j
Jacobsen 285, 293
Jaeger 298
Jagow et al. 39
Jøgensen 163
Johnen 312
Johnson 150

k
Kagan 278
Kalesse 43, 45
Kamimura 10
Katsuki 188, 285
Kitching 66
Knowles 277–278

Köckenberger 384
Kobayashi 18, 41, 43, 45
Koenig 13
Koga 18
Kolter-Jung 386
Komnenos 10
Korndörfer 356
Kragl 415
Kula 327, 355, 386

l
Lautens 43, 124
Leitner 250
Ley 66
Liese 323, 338, 384, 415
Lingen 386, 403
Lormann 188
Loydl 10
Lütz 415

m
Marshall 43, 45
Mathey 131
McMurry 137
Meyers 1, 18
Michael 10
Miyaura 188
Mock 149
Mori 66–67
Morita 13
Müller 45, 298, 318, 323, 338, 384, 386
Mukaiyama 166

n
Nakai 382
Naso 157
Nicolaou 10
Nitsche 386
Noltes 12
Norrby 188
Noyori 181, 277, 206

o
Oehlschlager 66
Oku 24
Oppolzer 206

p
Panek 43
Pauson–Khand 106, 108
Pfruender 390
Phillips 397
Pleiss 308

Pohl 298, 323, 327, 386, 403
Polniaszek 54
Pringle 264
Pu 181
Pudovik 13
Pye 197
Pyne 150

q
Quirion 18

r
Raabe 150
Radinov 206
Reetz 169, 264
Reggelin 77, 150
Reich 197
Richards 177
Römer 384
Rohmer 317
Roitsch 384
Rossen 197
Royer 18
Rozenberg 198–199, 213
Rudolph 213
Rychnovsky 24

s
Sahm 312, 317
Sakan 51, 53
Salzer 115
Sammakia 177
Sauerzapfe 381–384
Schloss 353
Schneider 320–321,
Schörken 312
Schürmann 312
Schubert 386
Schulz 353
Schuster 277
Schweiger 163
Sharpless 277
Siegert 310
Smith 43, 45
Snapper 209
Soai 181

Sprenger, Georg A. 312, 359
Sprenger, Gerda 312, 384
Steglich 3
Stevenson 18
Stewart 388
Stille 157
Stratagene 301
Suzuki 126, 157

t
Takasago 124
Togni 117
Tomioka 208
Tsujihara 150

u
Uemura 62, 177
Ugi 118

v
Van de Weghe 62
Vasella 10
Vemura 62
Vicuña 405
Vilsmeyer 132

w
Wagner 277
Wajant, H. 335, 338
Walter 386
Wandrey 338, 415
Wardrop 66
Weckbecker 341
Weinhold 150
Wendorff 298
Wilcocks 402
Wilke 126
Wolberg 386

y
Yamamoto 7

z
Zeeck 69
Zhang 382
Zhao 188

Related Titles

Christmann, M., Bräse, S. (Eds.)

Asymmetric Synthesis – The Essentials

2006
ISBN: 978-3-527-31399-0

Francotte, E., Lindner, W., Mannhold, R., Kubinyi, H., Folkers, G. (Eds.)

Chirality in Drug Research

2006
ISBN: 978-3-527-31076-0

Berkessel, A., Gröger, H.

Asymmetric Organocatalysis

From Biomimetic Concepts to Applications in Asymmetric Synthesis

2005
ISBN: 978-3-527-30517-9

Beller, M., Bolm, C. (Eds.)

Transition Metals for Organic Synthesis

Building Blocks and Fine Chemicals

2004
ISBN: 978-3-527-30613-8

Blaser, H. U., Schmidt, E. (Ed.)

Asymmetric Catalysis on Industrial Scale

Challenges, Approaches and Solutions

2004
ISBN: 978-3-527-30631-2

Nierhaus, K. H., Wilson, D. N. (Eds.)

Protein Synthesis and Ribosome Structure

Translating the Genome

2004
ISBN: 978-3-527-30638-1

Shibasaki, M., Yamamoto, Y. (Eds.)

Multimetallic Catalysts in Organic Synthesis

2004
ISBN: 978-3-527-30828-6

de Meijere, A., Diederich, F. (Eds.)

Metal-Catalyzed Cross-Coupling Reactions

2004
ISBN: 978-3-527-30518-6